Organic Reactions

Organic Reactions

VOLUME 30

JOHN WILEY & SONS, INC.

New York · Chichester · Brisbane · Toronto · Singapore

Published by John Wiley & Sons, Inc.

Library of Congress Catalog Card Number 42–20265

ISBN 0-471-89013-8

Printed in the United States of America

10 9 8 7 6 5 4 3 2 1

BORIS WEINSTEIN

March 31, 1930–July 29, 1983

Boris Weinstein, Co-Secretary to the Editorial Board of *Organic Reactions* since 1972 and Secretary-Treasurer of *Organic Reactions* since 1978, died after a brief illness on July 29, 1983, in Seattle, Washington at the age of 53.

Professor Weinstein was born in New Orleans, Louisiana on March 31, 1930, and received his B.S. degree from Louisiana State University in 1951, his M.S. degree from Purdue University in 1953, and his Ph.D. degree in organic chemistry from Ohio State University in 1959. He was a postdoctoral fellow at the University of California, Berkeley, 1959–1960 and a research chemist at Stanford Research Institute, 1960–1961.

From the period 1961–1967 he was Director of the Chemical Laboratory at Stanford University. He became Associate Professor of Chemistry at the University of Washington in 1967 and Professor in 1974, a position he held at the time of his death.

During his career, he was active in research related to the synthesis of biologically active peptides and proteins and in the chemistry of natural products in the field of neurochemistry. Professor Weinstein was an editor of *Chemistry and Biochemistry of Amino Acids, Peptides, and Proteins*. His substantial contributions to *Organic Reactions* can be fully appreciated only by the many editors and authors with whom he worked.

As a teacher of Boris Weinstein, I had the pleasure of observing his scientific and administrative abilities, and the honor of being his associate and friend. His contributions and friendship are missed by all of his colleagues in *Organic Reactions*.

WILLIAM G. DAUBEN

August 1983

PREFACE TO THE SERIES

In the course of nearly every program of research in organic chemistry the investigator finds it necessary to use several of the better-known synthetic reactions. To discover the optimum conditions for the application of even the most familiar one to a compound not previously subjected to the reaction often requires an extensive search of the literature; even then a series of experiments may be necessary. When the results of the investigation are published, the synthesis, which may have required months of work, is usually described without comment. The background of knowledge and experience gained in the literature search and experimentation is thus lost to those who subsequently have occasion to apply the general method. The student of preparative organic chemistry faces similar difficulties. The textbooks and laboratory manuals furnish numerous examples of the application of various syntheses, but only rarely do they convey an accurate conception of the scope and usefulness of the processes.

For many years American organic chemists have discussed these problems. The plan of compiling critical discussions of the more important reactions thus was evolved. The volumes of *Organic Reactions* are collections of chapters each devoted to a single reaction, or a definite phase of a reaction, of wide applicability. The authors have had experience with the processes surveyed. The subjects are presented from the preparative viewpoint, and particular attention is given to limitations, interfering influences, effects of structure, and the selection of experimental techniques. Each chapter includes several detailed procedures illustrating the significant modifications of the method. Most of these procedures have been found satisfactory by the author or one of the editors, but unlike those in *Organic Syntheses* they have not been subjected to careful testing in two or more laboratories.

Each chapter contains tables that include all the examples of the reaction under consideration that the author has been able to find. It is inevitable, however, that in the search of the literature some examples will be missed, especially when the reaction is used as one step in an extended synthesis. Nevertheless, the investigator will be able to use the tables and their accompanying bibliographies in place of most or all of the literature search so often required.

Because of the systematic arrangement of the material in the chapters and the entries in the tables, users of the books will be able to find information desired by reference to the table of contents of the appropriate chapter. In the interest of economy the entries in the indices have been kept to a minimum, and, in particular, the compounds listed in the tables are not repeated in the indices.

The success of this publication, which will appear periodically, depends upon the cooperation of organic chemists and their willingness to devote time and effort to the preparation of the chapters. They have manifested their interest already by the unanimous acceptance of invitations to contribute to the work. The editors will welcome their continued interest and their suggestions for improvements in *Organic Reactions*.

Chemists who are considering the preparation of a manuscript for submission to *Organic Reactions* are urged to write either secretary before they begin work.

CONTENTS

Organic Reactions

CHAPTER 1

PHOTOCYCLIZATION OF STILBENES AND RELATED MOLECULES

FRANK B. MALLORY AND CLELIA W. MALLORY

Bryn Mawr College
Bryn Mawr, Pennsylvania

CONTENTS

INTRODUCTION

On irradiation in solution with ultraviolet (UV) light, *cis*-stilbene (**1**) undergoes reversible photocyclization to give *trans*-4a,4b-dihydrophenanthrene (**2**), an intermediate that can be trapped oxidatively with hydrogen acceptors such as iodine, oxygen, or tetracyanoethylene to give phenanthrene (**3**) in high yield.

$$\mathbf{1} \underset{h\nu \text{ or dark}}{\overset{h\nu}{\rightleftharpoons}} [\mathbf{2}] \xrightarrow{\text{Oxidant}} \mathbf{3}$$

This type of photoreaction occurs with a wide range of substituted stilbenes and related molecules, including various polycyclic and heterocyclic analogs; in addition, certain systems with a single heteroatom (nitrogen, oxygen, or sulfur) in place of the central π bond undergo photocyclization, creating a new five-membered heterocyclic ring. In some of these photocyclizing systems, the transient dihydroaromatic intermediates analogous to **2** are trapped without added oxidants by the elimination of suitable leaving groups or by various hydrogen shifts.

Photocyclization is the preferred method for the synthesis of many different polynuclear aromatic systems.

The discovery and early development of stilbene photocyclization has been surveyed,[1] and several other general reviews have appeared.[2–6c]

MECHANISM

The mechanism of the photocyclization of stilbenes has been studied extensively.[7–37b] The photocyclization of stilbenes and the trapping of the resulting dihydrophenanthrenes are complicated transformations, and many mechanistic details are not yet firmly established. The current mechanistic picture has emerged largely from experimental studies of simple stilbenes and a few other systems; modifications may turn out to be appropriate for some of the stilbene analogs covered in the Tabular Survey at the end of this chapter.

Dihydrophenanthrenes from *cis*-Stilbenes

Photocyclization reactions, with one possible exception,[38a] proceed only by absorption of a photon by the *cis* isomer and not by the *trans* isomer of the stilbene derivative. However, the synthetically more accessible *trans* isomer is commonly used as the starting material since reversible *cis–trans* photoisomerization generates the mechanistically required *cis* isomer.

$$\text{Ar–CH=CH–Ar}' \ (trans) \underset{h\nu}{\overset{h\nu}{\rightleftharpoons}} \text{Ar–CH=CH–Ar}' \ (cis)$$

Typically, stilbene photocyclizations are carried out at concentrations of 10^{-2} *M* or less to minimize the competing photodimerization of the *trans* isomer to give a 1,2,3,4-tetraarylcyclobutane derivative.[38b] Photochemical [2 + 2] cycloaddition involving C-9 and C-10 of the product phenanthrene and the double bond of the starting stilbene, particularly the *trans* isomer, is another potentially troublesome side reaction that is minimized at low concentrations.[38c] Photocyclizations of stilbenes and related systems (with some exceptions as noted below) proceed only by way of molecules in their lowest excited singlet state and not their lowest excited triplet state. Excited triplet-state *cis*-stilbenes, which can be produced by energy transfer from a suitable excited sensitizer such as triplet-state benzophenone, generally fail to cyclize to dihydrophenanthrenes and instead function only as intermediates in sensitized *cis–trans* isomerization.

Photochemically produced dihydrophenanthrenes such as **2** are unstable because both photochemical and thermal ring-opening reactions regenerate the starting *cis*-stilbene derivative. In the rigorous absence of an oxidant (the last traces of dissolved oxygen are very difficult to remove), photostationary mixtures are obtained in which the ratio of the dihydrophenanthrene to the *cis*-stilbene depends on the wavelength distribution of the irradiating light and on the temperature. These dihydrophenanthrenes generally are not isolable, but under appropriate conditions many are sufficiently long-lived in solution to study their chemical reactions and spectral properties.[14,22,23,27,37b] The assignment of *trans* rather than *cis* stereochemistry for the two allylic hydrogens on the carbon atoms at the site of ring closure is based on orbital symmetry considerations and on experimental evidence in a few systems.[24, 39–41]

A current view of the photocyclization mechanism is illustrated for *cis*-stilbene (**1**) in Scheme 1. In this rough energy–reaction coordinate diagram, the important pathways are shown for both the photocyclization of **1** to dihydrophenanthrene **2** and the competing photoisomerization of **1** to *trans*-stilbene (**4**).

Photoexcitation of *cis*-stilbene (**1**) by absorption of a photon of UV light transforms the molecule from its ground singlet electronic state S_0 to its lowest excited singlet electronic state S_1. The resulting excited species **1*** is drawn partly with dotted lines in Scheme 1 to indicate that the distribution of π-electron density in the S_1 state differs from that in the S_0 state. By considering the nodal properties of the highest-energy bonding π molecular orbital of **1**, and also those of the lowest-energy antibonding π molecular orbital into which one electron from the former orbital is promoted in the excitation process $\mathbf{1} + h\nu \rightarrow \mathbf{1^*}$, predictions can be made about the distribution of π-electron density in **1***. For example, in comparison to **1**, **1*** is expected to have a smaller π-bond order between the two olefinic carbon atoms C-α and C-α' and a greater π-bond order between C-1 and C-α (and also between C-1′ and C-α').

The excited *cis*-stilbene **1*** is thought to have two principal fates: rotation around the bond between C-α and C-α' followed by an $S_1 \rightarrow S_0$ electronic

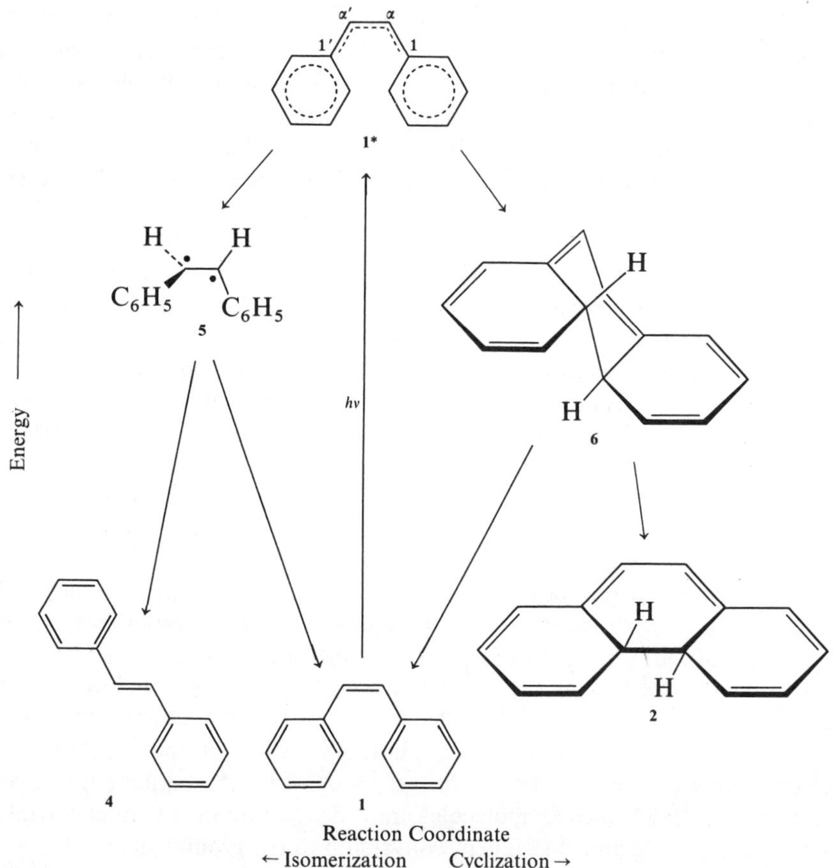

Scheme 1

transition to give the twisted ground-state stilbene species **5**, and ring closure followed by an $S_1 \rightarrow S_0$ electronic transition to give the skewed ground-state dihydrophenanthrene species **6**. Both **5** and **6** are at energy maxima on the ground-state potential-energy surface: **5** is considered identical with the transition state for the thermal *cis–trans* isomerization reaction **1** → **4**, and **6** is considered identical with the transition state for the thermal ring-opening reaction **2** → **1**. The dihedral angle between the bond from C-α to C-1 and the bond from C-α′ to C-1′ is assumed to be about 90° in the twisted stilbene **5**. The bond between the saturated carbon atoms in the skewed dihydrophenanthrene **6** is regarded as abnormally long and weak compared to the corresponding bond in the relaxed dihydrophenanthrene **2**. Species **5** relaxes rapidly to either **1** or **4**, and species **6** relaxes rapidly to either **1** or **2**.

The importance of $S_1 \rightarrow T_1$ intersystem crossing in excited *trans*-stilbene (**4***) is controversial.[42–45] For excited *cis*-stilbene (**1***), however, and also for its derivatives that lack a perturbing substituent such as a nitro group, we consider the $S_1 \rightarrow T_1$ process kinetically insignificant in competition with the chemical decay pathways of rotation (**1*** → **5**) and ring closure (**1*** → **6**).

The overall quantum yield for the formation of 4a,4b-dihydrophenanthrene (**2**) from *cis*-stilbene (**1**) is approximately 0.1, which means that each individual excited reactant molecule **1*** has only about a 10% probability of ending up as the ground-state product molecule **2** by the pathway **1*** → **6** → **2** and has about a 90% probability of returning to the ground state as *cis*-stilbene (**1**) or *trans*-stilbene (**4**) by the pathway **1*** → **5** → **1** + **4**. Photocyclization quantum yields depend on the molecular structure of the stilbene derivative, with values ranging from zero (for molecules that fail to react) to about 0.5 (for molecules such as 1,2-diphenylcyclopentene, in which the decay pathway analogous to **1*** → **5** is restricted structurally). Although the *quantum* yields (which measure how efficiently *photons* are used in forming products) are much less than unity for stilbene photocyclizations, the *chemical* yields (which measure how efficiently *reactant molecules* are used in forming products) can approach 100% for the isolated phenanthrene derivatives obtained by trapping the dihydrophenanthrene intermediates through oxidation, elimination, or hydrogen shifts.

With some possible exceptions,[46,47] stilbenes with nitro or amino substituents fail to undergo photocyclization. These substituents may interfere with photocyclization by drastically altering the nature of the lowest-energy excited singlet state of the stilbene. For example, if a substituent-derived ^{1}n,π* excited singlet state (a state lacking the requisite electron distribution for ring closure) happens to lie lower in energy than the stilbene-like $^1\pi,\pi^*$ excited singlet state in one of these stilbene derivatives, photocyclization of the $^1\pi,\pi^*$ excited molecules may be thwarted by a very fast $^1\pi,\pi^* \rightarrow {}^1n,\pi^*$ internal conversion. The resulting $^1n,\pi^*$ excited molecules may decay rapidly by singlet–triplet intersystem crossing and subsequent conversion to the ground state.

The problem of a $^1n,\pi^*$ lowest excited singlet state also precludes the photocyclization of azobenzene under ordinary conditions. However, the transformation of an azobenzene

derivative to the corresponding 9,10-diazaphenanthrene can be achieved by carrying out the irradiation in sulfuric acid solution.[48] Under these conditions the azobenzene is protonated, and the resulting cation (e.g., **7**) has a stilbene-like $^1\pi,\pi^*$ lowest excited singlet state.[49,50] The overall conversion of cation **7** to the protonated product **8** probably involves photocyclization of **7** to the dihydroaromatic intermediate **9**, isomerization of **9** to the partially aromatized intermediate **10** by a series of proton transfers to and from the sulfuric acid medium, and oxidation of **10** to **8**.[13]

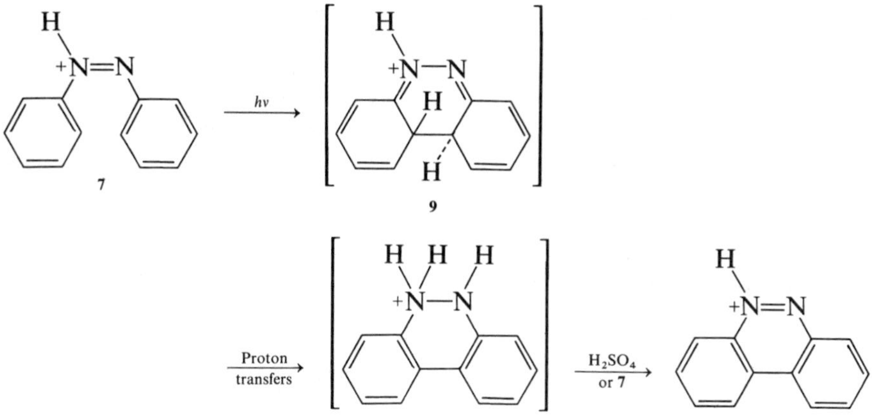

This oxidation can be accomplished by either sulfuric acid or protonated azobenzene **7**. In the first instance, the reduction product is sulfur dioxide; in the second, the ultimate reduction products are derived from a benzidine rearrangement of the protonated hydrazobenzene **11**. The *cis* cation **7** is a much more effective oxidant than the isomeric *trans* cation.[51]

$$\mathbf{10} + C_6H_5\overset{+}{N}H{=}NC_6H_5 \;(\mathbf{7}) \longrightarrow \mathbf{8} + C_6H_5\overset{+}{N}H_2NHC_6H_5 \;(\mathbf{11})$$

Certain stilbene derivatives fail to undergo oxidative photocyclization because of steric constraints imposed by additional rings in these molecules. These constraints apparently cause the relevant *ortho* carbon atoms to lie too far apart for the new bond to form in competition with decay of the excited stilbene. For example, irradiation of 1,2-diphenylcyclobutene causes fluorescence, photofragmentation to diphenylacetylene and ethylene, and photodimerization, but apparently not photocyclization to 9,10-cyclobutenophenanthrene.[52–54] In contrast, the less strained 1,2-diphenylcyclopentene (**12**) readily undergoes oxidative photocyclization to 9,10-cyclopentenophenanthrene (**13**).[22,55,56]

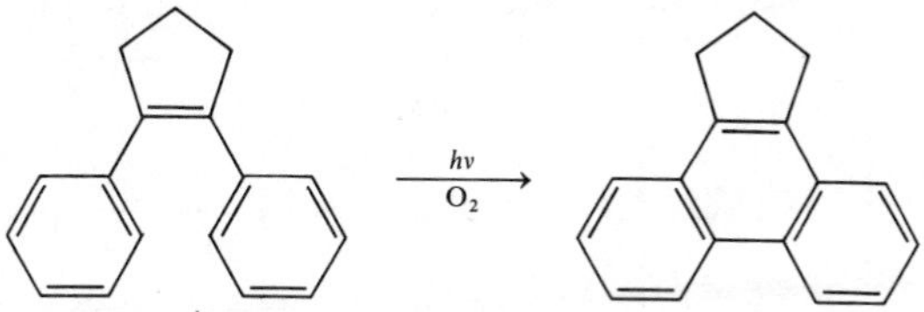

12 $\xrightarrow[O_2]{h\nu}$ **13** (82%)

Furthermore, substituted diphenylcyclobutenes such as **14** undergo oxidative photocyclization,[54] perhaps because the methyl groups exert a buttressing effect that forces the two phenyl groups into sufficiently close proximity to permit ring closure.

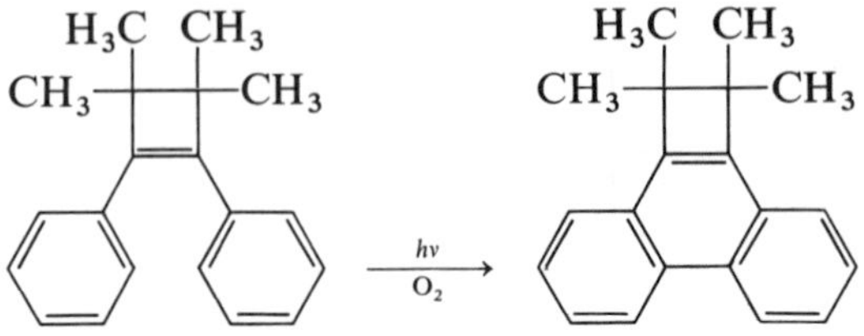

H_3C CH_3 / CH_3 CH_3 — $\xrightarrow[O_2]{h\nu}$ —

14

Oxidative photocyclization to the corresponding 1-azaphenanthrene derivative fails for the constrained indane **15** but succeeds for the more flexible six-membered analog **16**.[57] Similarly, oxidative photocyclization fails for 9-benzylidenefluorene (**17**)[58] but succeeds for triphenylethylene (**18**)[55,59,60]

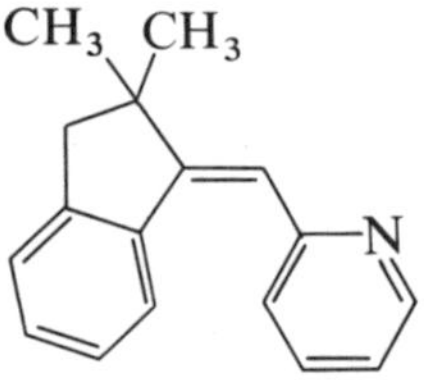

15 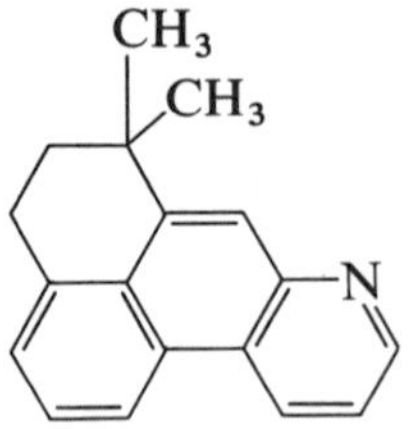 16

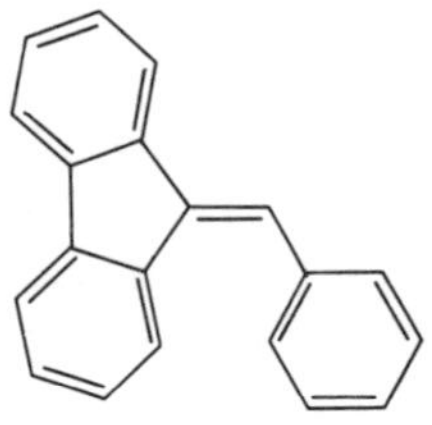

17 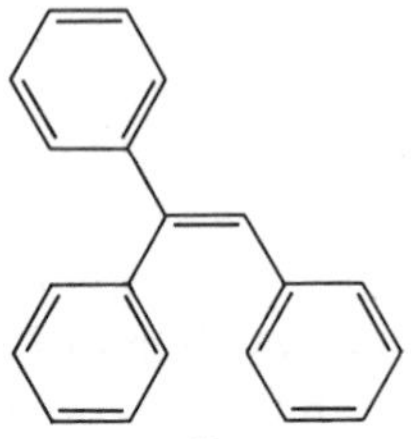 18

Phenanthrenes from Dihydrophenanthrenes by Oxidative Trapping

The conversion of *trans*-4a,4b-dihydrophenanthrene (**2**) to phenanthrene (**3**) in air-saturated solution proceeds by a radical chain mechanism involving two successive hydrogen abstractions with radical **19** as an intermediate.[10,27a,61,62]

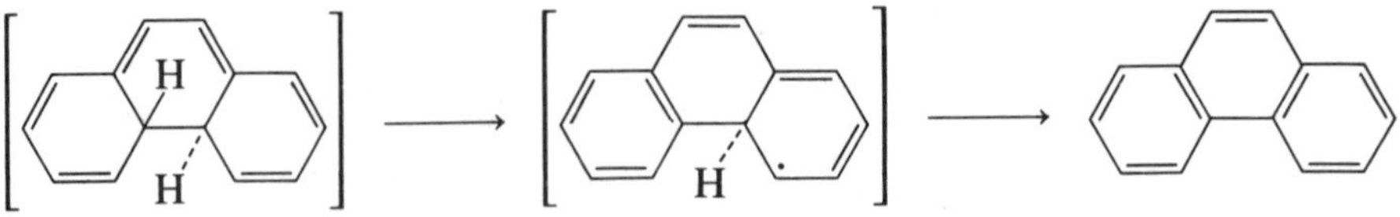

2 (PH_2) 19 ($PH\cdot$) 3 (P)

$$\text{Initiation:}\quad PH_2 + O_2 \longrightarrow PH\cdot + HO_2\cdot$$

$$\text{Propagation:}\quad PH_2 + HO_2\cdot \longrightarrow PH\cdot + H_2O_2$$

$$PH\cdot + O_2 \longrightarrow P + HO_2\cdot$$

$$PH_2 + O_2 \longrightarrow P + H_2O_2$$

Although oxygen acts cleanly as the oxidant in the photocyclizations of very dilute solutions of stilbenes (10^{-5}–10^{-4} *M*), its use with the more concentrated solutions (10^{-2} *M*) needed for practical preparative work often results in low yields of impure phenanthrenes. This problem may stem from unwanted reactions of the accumulated hydrogen peroxide.

Usually it is advantageous to carry out preparative-scale photocyclizations in an air-saturated solution with a small amount of added iodine. This generally shortens the irradiation time for complete conversion and gives a cleaner product in higher yield. Iodine provides another chain mechanism for trapping dihydrophenanthrenes.

$$\begin{array}{llcl} \text{Initiation:} & I_2 + h\nu & \longrightarrow & I\cdot + I\cdot \\ \text{Propagation:} & PH_2 + I\cdot & \longrightarrow & PH\cdot + HI \\ & PH\cdot + I_2 & \longrightarrow & P + HI + I\cdot \\ \hline & PH_2 + I_2 & \longrightarrow & P + 2HI \end{array}$$

The photochemical dissociation of molecular iodine may involve predominantly the visible component of the emission from the light source used in the irradiation. The hydrogen iodide from the propagation steps may undergo oxidation under the reaction conditions to regenerate iodine. Another conceivable initiation step involving iodine is $PH_2 + I_2 \rightarrow PH\cdot + HI + I\cdot$.

Hydrogen abstraction by iodine atoms is more exothermic than by oxygen molecules by 24 kcal/mol. [This is the difference in the bond dissociation energies of H—I (71 kcal/mol) and H—$O_2\cdot$ (47 kcal/mol).][63] This can account for the greater success of iodine as compared to oxygen in trapping certain dihydrophenanthrene derivatives for which the competing ring opening is especially fast. For example, iodine is considerably more effective than oxygen in the oxidative photocyclization of tetramethylstilbene **20** to the corresponding phenanthrene **21**.[64] Dihydro intermediate **22** should have an unusually large driving force for thermal ring opening to regenerate stilbene **20** because that process relieves the crowding between the C-4 and C-5 methyl groups in **22**.

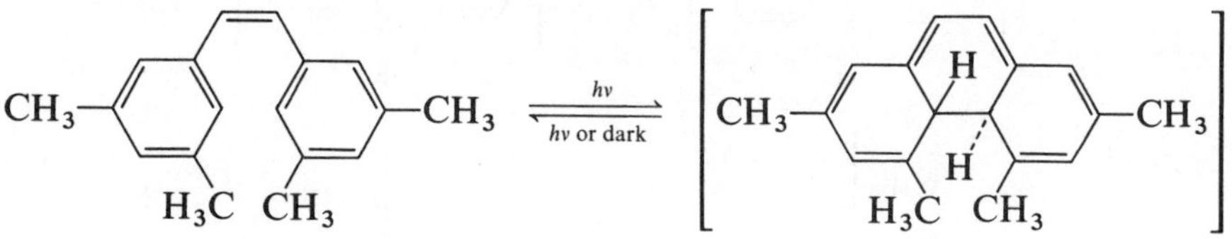

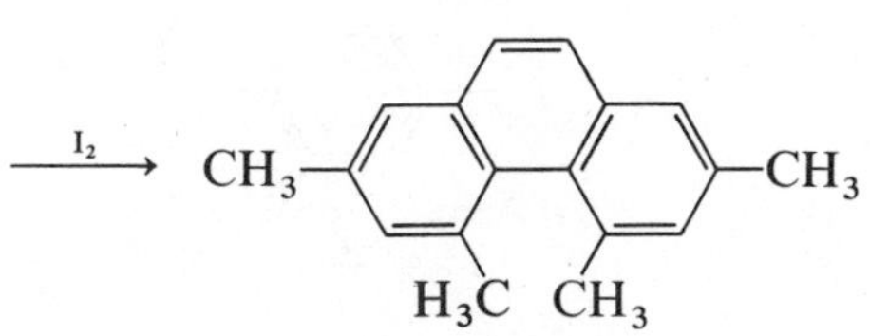

A recent investigation indicates that tetracyanoethylene may be particularly useful as a hydrogen acceptor in the conversions of 4a,4b-dihydrophenanthrenes to the corresponding phenanthrenes.[65]

Scattered reports claim the formation of phenanthrenes from the irradiation of stilbenes in solution under conditions designed to exclude oxidants. The dihydroaromatic intermediates in these reactions probably are being trapped oxidatively, however, either by residual dissolved oxygen or other adventitious oxidants or by hydrogen transfers to reactant or product molecules.

Phenanthrenes from Dihydrophenanthrenes by Nonoxidative Trapping

4a,4b-Dihydrophenanthrenes with suitably placed leaving groups can undergo elimination reactions leading to phenanthrenes in the absence of oxidants. This behavior is found, for example, in the photocyclizations of certain *ortho*-substituted stilbenes. As shown in Scheme 2, an *ortho*-substituted stilbene exists in solution as a mixture of conformers **23a** and **23b**; therefore, UV irradiation produces two isomeric dihydrophenanthrenes, **24a** and **24b**, respectively. For certain substituents X, dihydrophenanthrene **24b** can be trapped to give phenanthrene (**3**) by a highly exothermic elimination reaction;

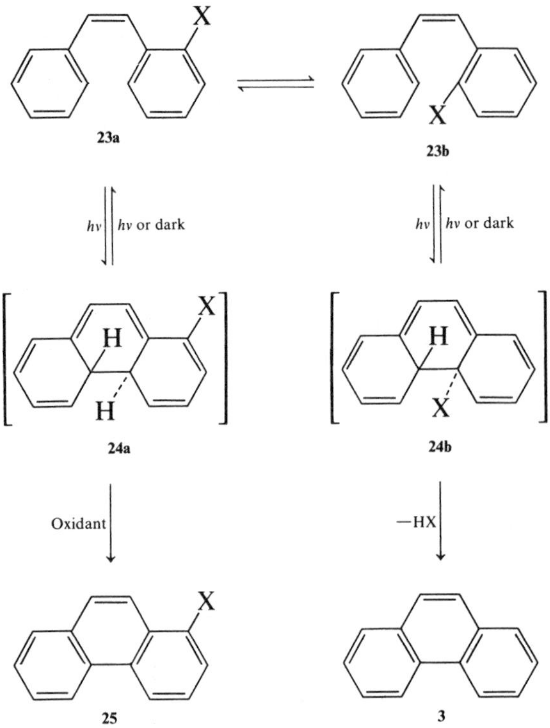

23a ⇌ **23b**

$h\nu$ / $h\nu$ or dark

24a **24b**

Oxidant | $-HX$

25 **3**

Scheme 2

the trapping of the isomeric dihydrophenanthrene **24a** requires an oxidant. If oxidants are excluded during irradiation in such systems, the reversibility of formation of the 1-substituted dihydrophenanthrene **24a** allows the overall photoreaction to proceed in high yield by way of eliminative trapping of the 4a-substituted dihydrophenanthrene **24b**. Thus stilbenes with *o*-methoxy substituents photocyclize under a nitrogen atmosphere with elimination of methanol.[28]

Stilbenes with *o*-bromo substituents also undergo eliminative photocyclization. Although this type of photoreaction has been demonstrated for only a few bromostilbenes (e.g., **26**),[25] it probably has wide scope. One mechanistic suggestion, illustrated for the conversion of **26** to **27**, is that irradiation of the bromostilbene under nitrogen leads reversibly to a mixture of dihydrophenanthrenes with the bromine atom at either C-1 or C-4a, and in the presence of a base such as potassium *tert*-butoxide the latter dihydrophenanthrenes are trapped by an E2 elimination of the elements of hydrogen bromide.[25]

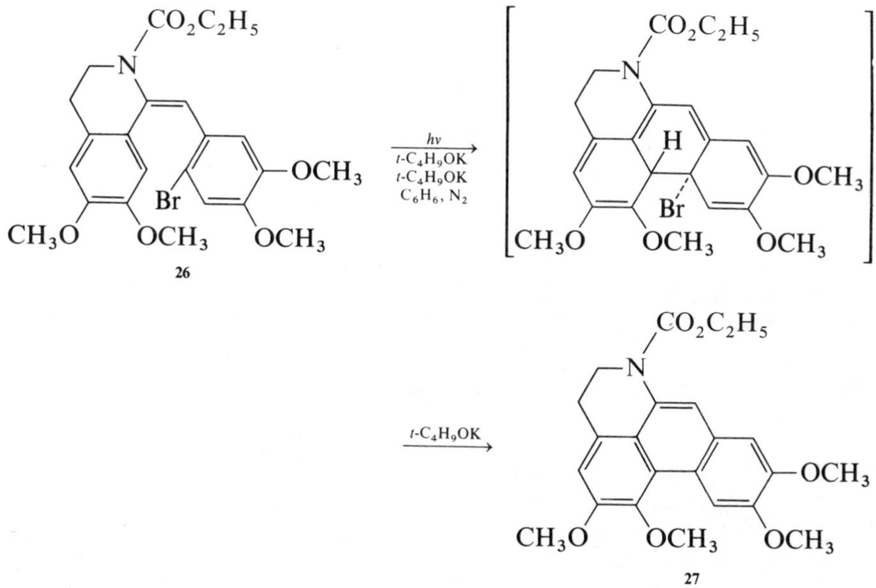

$CO_2C_2H_5$, N, Br, OCH_3, CH_3O, OCH_3, OCH_3

26

$\xrightarrow[t\text{-}C_4H_9OK,\ C_6H_6,\ N_2]{h\nu}$

[$CO_2C_2H_5$, N, H, Br, OCH_3, CH_3O, OCH_3, OCH_3]

$\xrightarrow{t\text{-}C_4H_9OK}$

$CO_2C_2H_5$, N, OCH_3, CH_3O, OCH_3, OCH_3

27
(59%)

An alternative mechanism has been suggested, involving as the first step photolysis of the carbon–bromine bond in the starting bromostilbene assisted by intramolecular radical complexation.[37a]

The overall transformation of stilbene **28** to phenanthrene **29** involves nonoxidative trapping of dihydrophenanthrene **30** by the elimination of tetrachlorocatechol. The sequential elimination steps **30** → **31** and **31** → **29** presumably follow ionic mechanisms (vinylogous E1 or E2) because the overall reaction takes place in isopropyl alcohol but not in cyclohexane.[66]

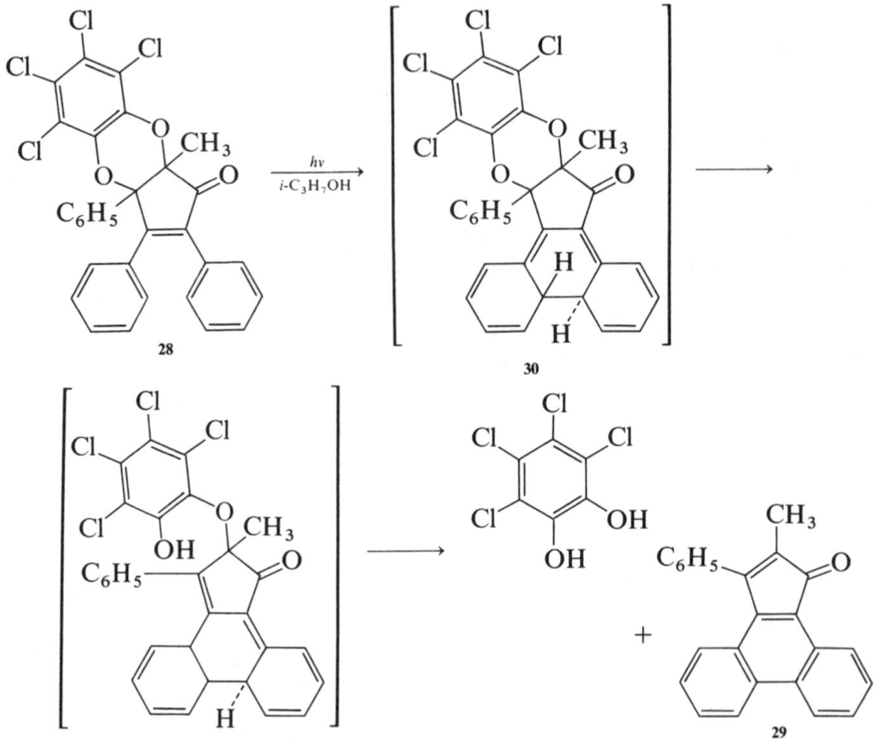

28 30 31 29 (49%)

Photocyclization of acetoxy lactone **32** to lactone **33** can be accounted for by a similar mechanism in which the dihydrophenanthrene **34** undergoes elimination and the partially aromatized intermediate **35** isomerizes to the isolated product **33** by an enolization–ketonization sequence of proton transfers.[67]

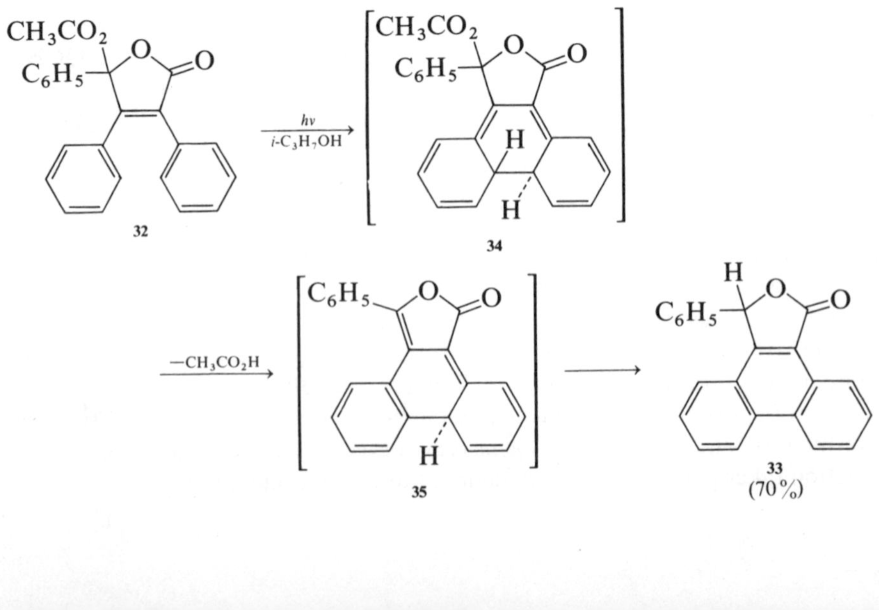

32 34 35 33 (70%)

Self-Trapping of Dihydro Intermediates by Hydrogen Shifts

In the absence of oxidants, certain dihydrophenanthrene derivatives undergo hydrogen shifts to give isomeric dihydrophenanthrenes that do not undergo rapid ring opening and thus can be isolated.

1,3 Shifts. With rare exceptions,[13,68a,b] self-trapping reactions in which 4a,4b-dihydrophenanthrenes are transformed to 9,10-dihydrophenanthrenes (reactions formally involving 1,3-hydrogen shifts) appear to require special structural features such as carbonyl or cyano substituents at C-9 and/or C-10. Specific examples include the conversion of dinitrile **36** to *trans*-9,10-dicyano-9,10-dihydrophenanthrene (**37**)[69] by way of intermediate **38**,[70] and the conversion of diphenylmaleic anhydride (**39**) to anhydride **40** by way of dihydro intermediate **41**.[69]

Unimolecular thermal mechanisms for hydrogen shifts of this type are ruled out by the observation of deuterium incorporation from deuterated solvents and by orbital symmetry considerations. Two mechanisms can be suggested for these hydrogen shifts: an ionic enolization–ketonization sequence involving proton transfers and a radical chain process involving hydrogen atom transfers; the former may dominate in hydroxylic solvents and the latter in other solvents.

For example, the photocyclization of keto acid **42** in a solvent mixture of pyridine and deuterium oxide presumably follows an ionic mechanism of the type depicted in Scheme 3 with proton and deuteron transfers to and from solvent molecules because the isolated product **43** contains deuterium atoms at both C-9 and C-10.[67]

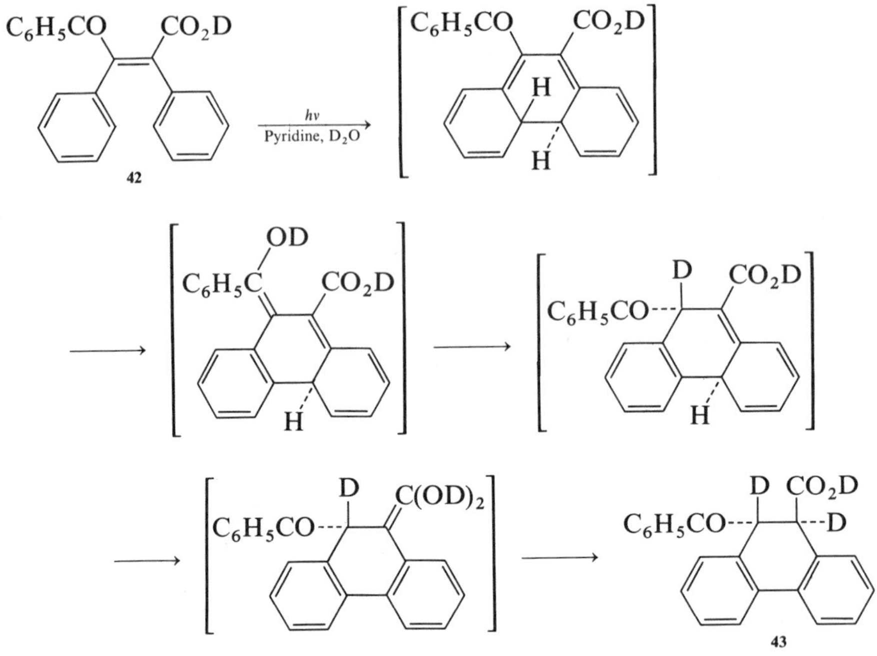

Scheme 3

An analogous ionic mechanism accounts for the acid catalysis observed for the transformation of 4a,4b-dihydrophenanthrene **38** to 9,10-dihydrophenanthrene **37**.[71a]

In contrast, in the photocyclization of dinitrile **36** in degassed benzene solution,[69] the isomerization of intermediate **38** to product **37** may proceed by a radical-chain mechanism involving steps of the type outlined in Scheme 4. In this scheme, RH stands for species such as **38** and **45**, and R· stands for species such as **44** and **46**; in solvents with easily abstractable hydrogens (e.g., chloroform or methanol), RH may also be a solvent molecule and R· a solvent-derived radical (e.g., trichloromethyl or hydroxymethyl). The production of dinitrile **37** with 40–44% deuterium content at C-9 and C-10 from the irradiation of stilbene **36** in deuteriochloroform is consistent with the notion that deuteriochloroform competes with dihydrophenanthrenes **38** and **45** as a donor molecule RH.

A proposed mechanism is shown in Scheme 5 for the photochemical conversion in methanol solution of methyl α-phenylcinnamate (**47**) to 9-carbomethoxy-9,10-dihydrophenanthrene (**48**).[32] In this mechanism, an enolization–ketonization sequence converts the first-formed 4a,4b-dihydrophenanthrene **49** to the half-aromatized intermediate **50**. The further transformation of **50** to the isolated product **48** is an unusual example of an overall 1,3-hydrogen shift because the assistance of a carbonyl or cyano substituent is lacking. Perhaps

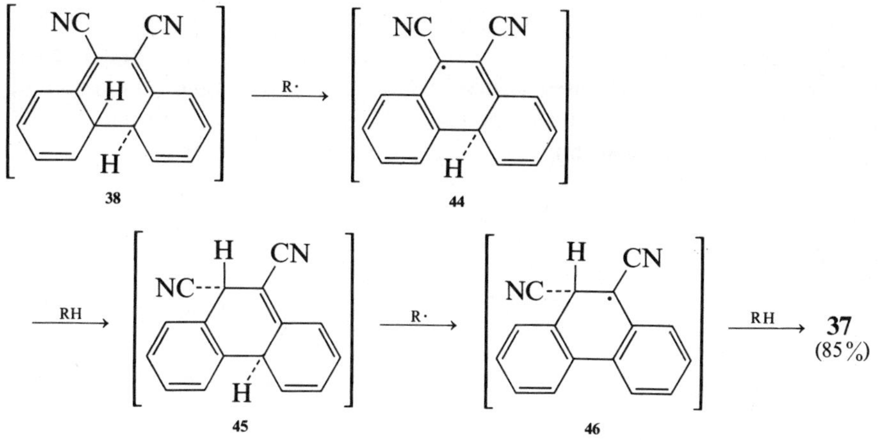

38 44

45 46 37 (85%)

Scheme 4

such assistance is not critical because the lifetime of intermediate **50** is not limited by a ring-opening reaction. In accordance with the proposed mechanism, the isolated product contains one deuterium atom at C-9 and none at C-10 when the photocyclization of ester **47** is carried out in CH_3OD.[32]

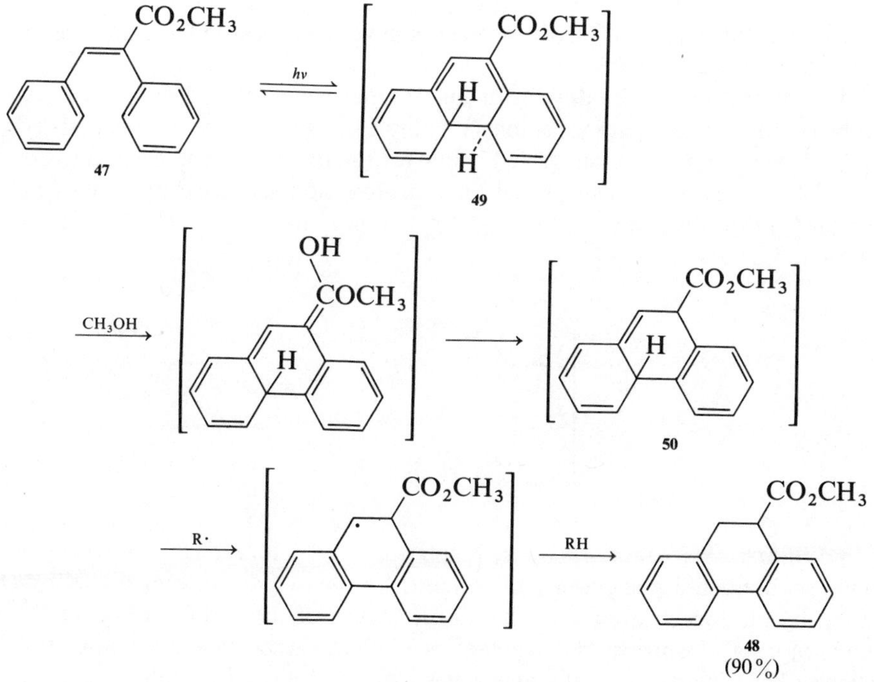

47 49 50 48 (90%)

Scheme 5

Photocyclization of fulgide **51** gives the crowded 1,8a-dihydronaphthalene **52** in which both the thermal ring opening to **51** and the 1,5-shift of hydrogen from C-8a to C-2 are inhibited sterically. On prolonged irradiation at 366 nm, intermediate **52** undergoes a photochemical 1,3-hydrogen shift (it can also be viewed as a 1,7-hydrogen shift) to give 1,4-dihydronaphthalene **53**.[71b]

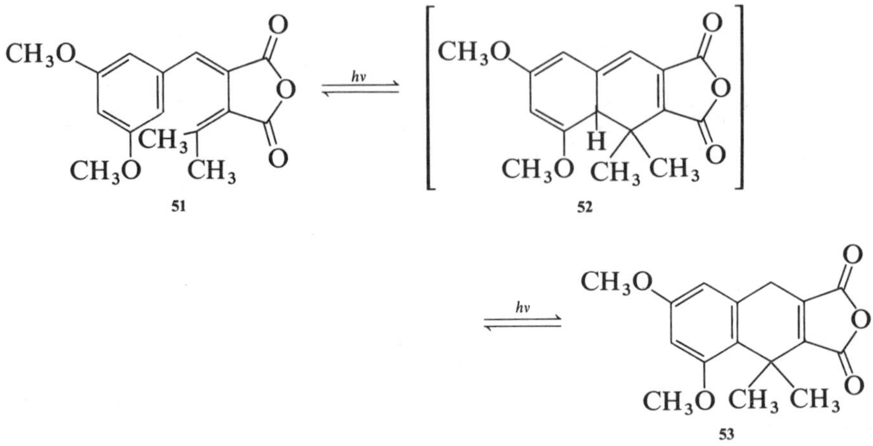

Similar photochemical 1,3-hydrogen shifts have been reported as minor pathways in two other systems.[68,70]

The formation of 1,4-dihydrophenanthrenes by photocyclization of *cis*-stilbenes in *n*-propylamine solution[72] appears to involve amine-assisted 1,3-hydrogen transfers as suggested below for the parent system. Each transfer may follow a two-step deprotonation–reprotonation sequence by way of an ion-pair intermediate consisting of *n*-propylammonium cation and a delocalized carbanion.

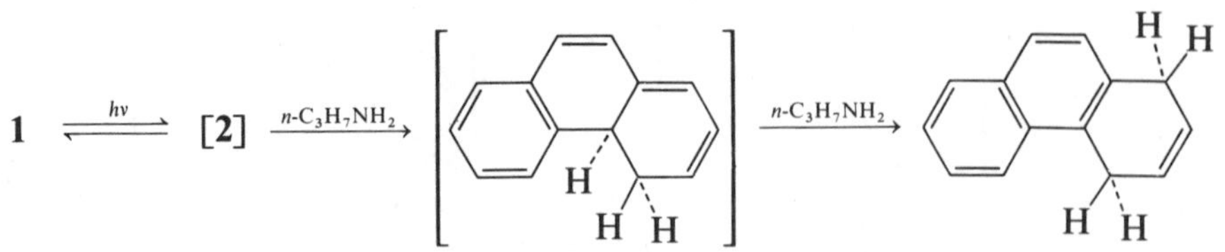

1,5 Shifts. Self-trapping by symmetry-allowed suprafacial 1,5-hydrogen shifts is observed in the absence of oxidants for the dihydro intermediates produced by photocyclizations in which at least one of the carbon atoms at the point of ring closure in the reactant is olefinic rather than aromatic. For example, the irradiation of vinylbiphenyl **54** in degassed cyclohexane gives 9,10-dihydro-9-phenylphenanthrene (**55**) by a fast 1,5-hydrogen shift in the initial intermediate **56**.[36,73]

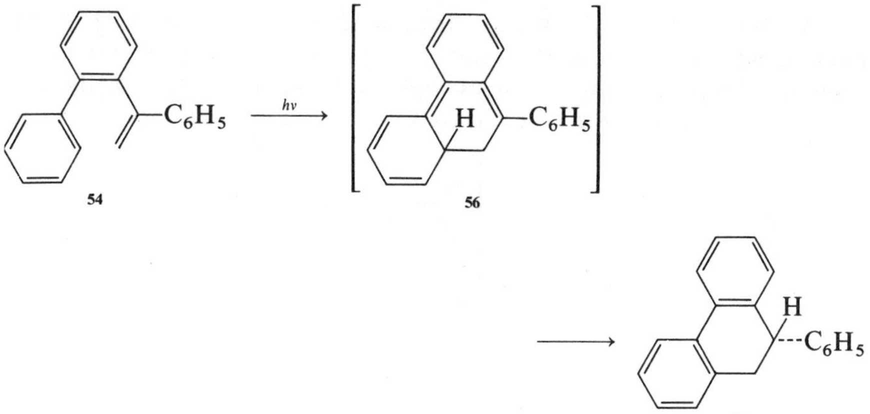

$-C_6H_5$ $\xrightarrow{h\nu}$ [H $-C_6H_5$]

54 56

$\longrightarrow$ H $-C_6H_5$

55
(85–98%)

Analogous 1,5-hydrogen shifts are not important pathways in photocyclizations in which ring closure links two aromatic rings in the reactant. For example, dihydrophenanthrene **57** is not detected as a product of the irradiation of *cis*-stilbene (**1**) under nonoxidizing conditions.

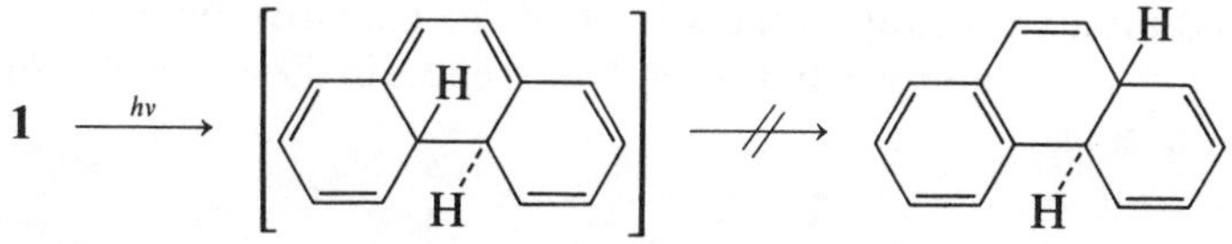

1 $\xrightarrow{h\nu}$ [H H] $\not\longrightarrow$ H H

2 57

This contrasting behavior has a straightforward thermochemical explanation. A 1,5-hydrogen shift will be observed only if it competes successfully with ring opening of the first-formed dihydroaromatic intermediate. Two major factors determine the driving forces for these competing reactions: the energetically unfavorable net exchange of one carbon–carbon σ bond for one carbon–carbon π bond in the ring opening (a factor not operative in the 1,5-hydrogen shift) and the energetically favorable creation of resonance-stabilized aromatic rings in both processes. Because of the first factor, 1,5-hydrogen shifts should have an intrinsic kinetic advantage over ring openings, but this advantage can be overcome in some systems. For example, the resonance stabilization factor for the ring opening **2** → **1** is so much more favorable (two benzene rings are generated) than that for the 1,5-hydrogen shift **2** → **57** (only one benzene ring would be generated) that the shift does not compete with the ring opening. In contrast, the 1,5-hydrogen shift **56** → **55** *is* observed because the resonance stabilization factor is comparable for both the hydrogen shift **56** → **55** and the competing ring opening **56** → **54** (two benzene rings are generated in each reaction).

1,4 Shifts. Aryl vinyl sulfides, ethers, and amines photocyclize to give zwitterionic intermediates that can undergo self-trapping by symmetry-allowed 1,4-hydrogen shifts. For example, the photocyclization of the deuterium-labeled sulfide **58** gives exclusively dihydrothiophene **59**.[74]

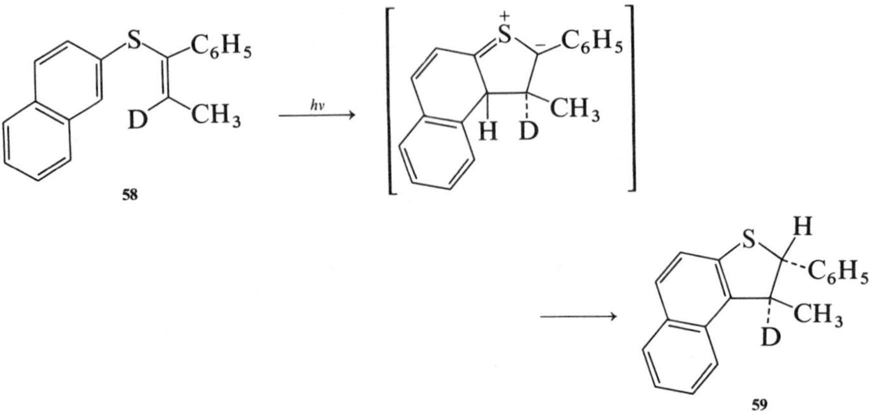

S C6H5 D CH3 hv S+ C6H5 H D CH3

58

S H C6H5 CH3 D

59

Oxidative Photocyclization of Arene Analogs of Stilbenes

The oxidative photocyclization of *o*-terphenyl to triphenylene[20] can be formulated mechanistically as the analog of the oxidative photocyclization of *cis*-stilbene to phenanthrene, but several differences in these two photoreactions are worth noting.

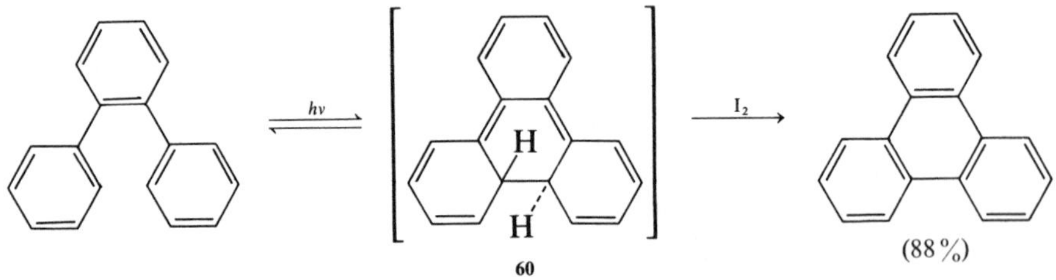

hv H H I_2

60

(88%)

For example, oxygen fails as an oxidant in the conversion of *o*-terphenyl to triphenylene; this conversion requires iodine, and high concentrations are especially effective. This suggests that the thermal ring opening of dihydrotriphenylene **60** is so fast (*three* aromatic rings are generated) that oxidative trapping of **60** competes with its ring opening only when the trapping agent is a very effective hydrogen abstractor. Similar trapping behavior is found for other intermediates with unusually exothermic ring openings.

Furthermore, the irradiation time required to convert a given amount of *o*-terphenyl is much shorter and the yield of triphenylene is higher in an aromatic solvent such as benzene or chlorobenzene rather than in a solvent such as cyclohexane.[20] This unusual solvent effect can be explained by considering the UV spectral properties of *o*-terphenyl, triphenylene, and benzene. Throughout most of the important wavelength region for the photoexcitation of *o*-terphenyl (230–300 nm), the absorptivity of triphenylene is roughly an order of magnitude greater than that of *o*-terphenyl. Therefore, for irradiations in cyclohexane, which is transparent in this wavelength region, the fraction of light captured by the *o*-terphenyl decreases rapidly as triphenylene is produced. Since the singlet energy of triphenylene (84 kcal/mol) is significantly less than that of *o*-terphenyl (90–100 kcal/mol), photoexcited triphenylene is incapable of transferring its excitation energy to *o*-terphenyl, and instead it is dissipated through radiationless decay of the excited triphenylene. This inner filtering effect of the triphenylene accounts for the marked retardation of the photoconversion of *o*-terphenyl in cyclohexane. If the irradiation is carried out in benzene solution with light in the region of 230–300 nm, most of the light actually is absorbed by the solvent. Because the singlet energy of benzene (108 kcal/mol) is greater than that of *o*-terphenyl (90–100 kcal/mol), singlet energy transfer is possible. In benzene solution the photocyclization of *o*-terphenyl to dihydro intermediate **60** is thus primarily a sensitized reaction involving excited benzene as the singlet sensitizer.[20]

$$\text{Benzene } (S_0) + h\nu \longrightarrow \text{Benzene } (S_1)$$

$$\text{Benzene } (S_1) + o\text{-Terphenyl } (S_0) \longrightarrow \text{Benzene } (S_0) + o\text{-Terphenyl } (S_1)$$

$$o\text{-Terphenyl } (S_1) \longrightarrow \text{Products}$$

Inner filtering by triphenylene does not retard the photocyclization of *o*-terphenyl in benzene as drastically as it does in cyclohexane because benzene, as a result of its high concentration, competes very successfully with the triphenylene for the capture of photons in the UV region.

The polycyclic hydrocarbon **61** fails to undergo oxidative photocyclization,[75] although dibenzo[*c,g*]phenanthrene (**62**) gives benzo[*ghi*]perylene (**63**).[76]

$h\nu$, I_2 (no reaction)

61

$h\nu$, I_2

62

63
(59%)

An explanation for this contrasting photochemical behavior of hydrocarbons **61** and **62** has been suggested on the basis of molecular orbital calculations.[75] The coefficients of the highest-occupied and lowest-unoccupied orbitals on the two interior carbon atoms that are relevant for bond formation are such that on excitation, the weak overlap interaction between these formally nonbonded carbons becomes more bonding for **62** but more antibonding for **61**. Thus it is energetically favorable in **62** but energetically unfavorable in **61** for the two carbon atoms of interest to move toward one another as required for successful photocyclization. A similar explanation may apply to the lack of oxidative photocyclization of 2,3-diphenylnaphthalene (**64**).[76]

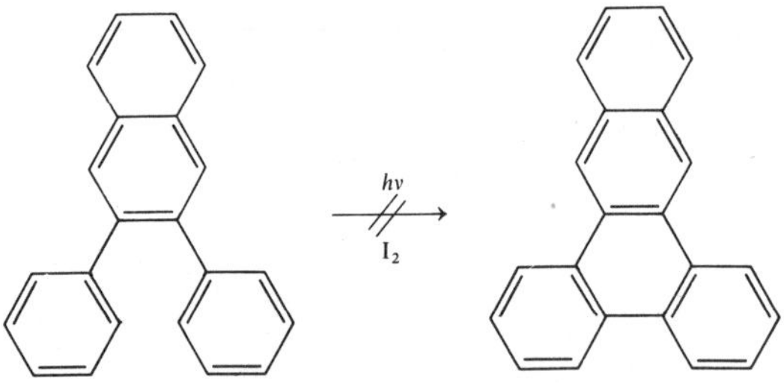

hv, I_2 (no reaction)

64

Photocyclization of Amides

The oxidative photocyclization of benzanilide (**65**) gives 6(5*H*)-phenanthridinone (**66**).[77]

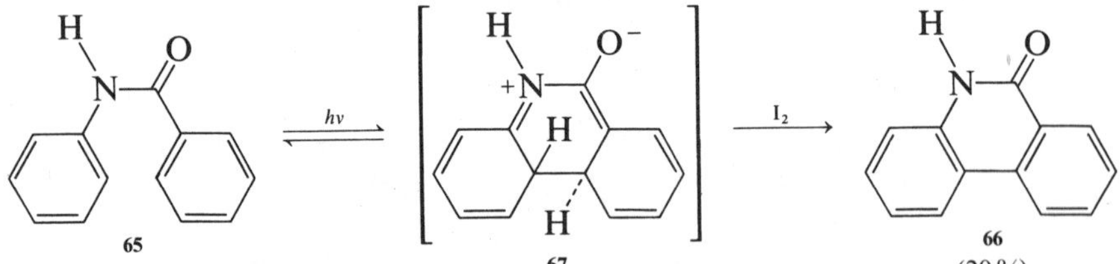

hv; I_2

65 **67** **66** (20%)

The partial double-bond character of the amide linkage

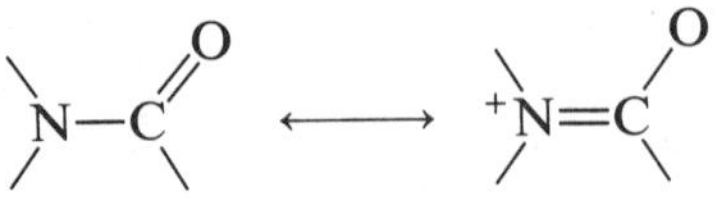

$$\text{N—C(=O)} \longleftrightarrow {}^{+}\text{N=C—O}^{-}$$

allows benzanilide to be viewed as a stilbene analog that can be photocyclized to zwitterionic intermediate **67**.

As with stilbene systems, *o*-methoxy substituents permit a nonoxidative photocyclization of benzanilides with the elimination of methanol (e.g., **68** → **69**).[78]

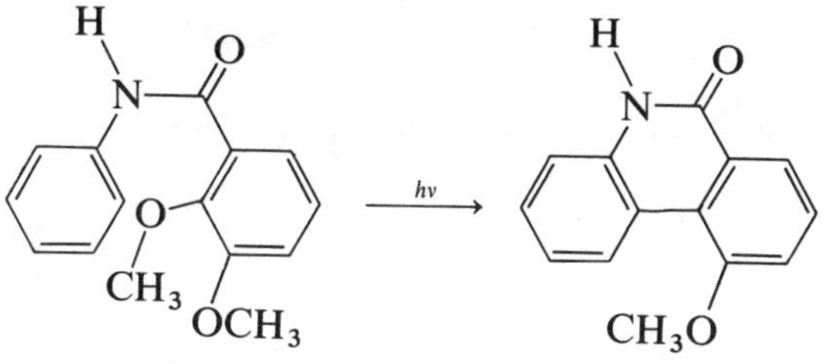

The photocyclization behavior of olefinic amides also parallels that of their hydrocarbon counterparts, as illustrated for enamide **70** in Scheme 6.

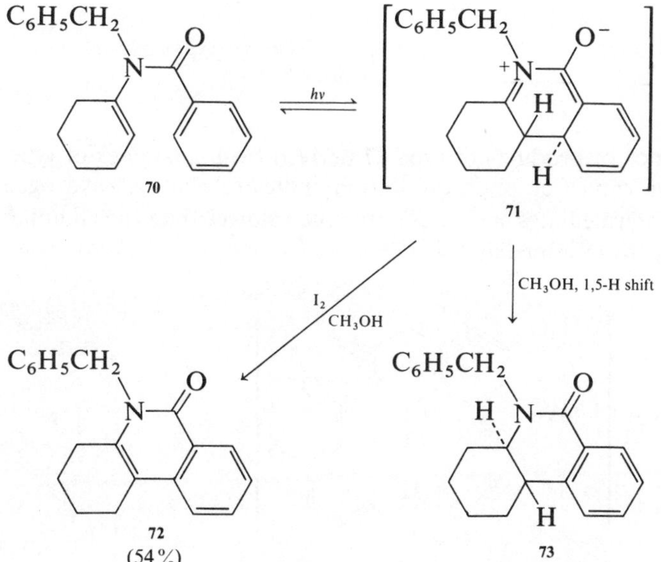

Scheme 6

Photocyclization of **70** leads to a dihydro intermediate **71** that gives the dehydrogenated lactam **72** in the presence of iodine but gives lactam **73** under nonoxidizing conditions.[79]

Enamides with appropriate *ortho* substituents undergo eliminative photocyclization as illustrated in Scheme 7 by the conversion of amides of type **74** to the cyclized product **75** in 50–85% yields for X = F, Cl, Br, CH_3CO_2, and CH_3S.[80]

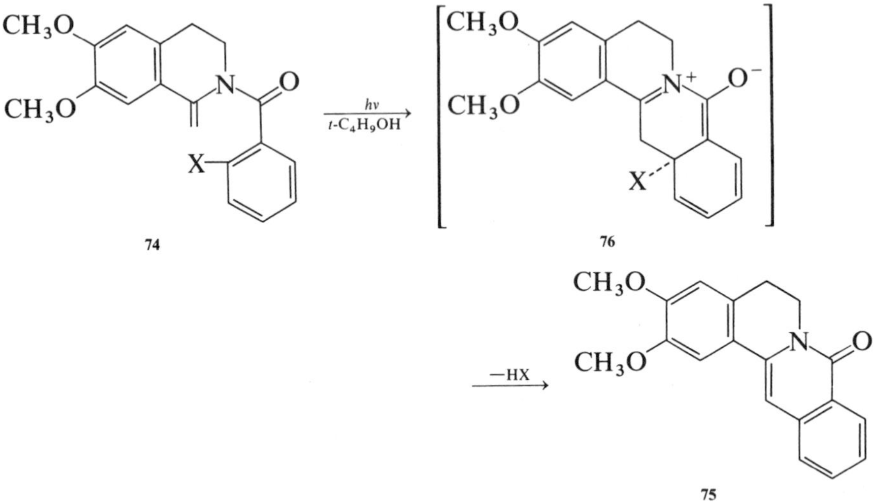

Scheme 7

The absence of products of type **77** derived from enamides of type **74** by the alternative mode of photocyclization indicates that 1,5-hydrogen shifts in dihydro intermediates such as **78** are much slower than the elimination of HX from dihydro intermediates like **76**.

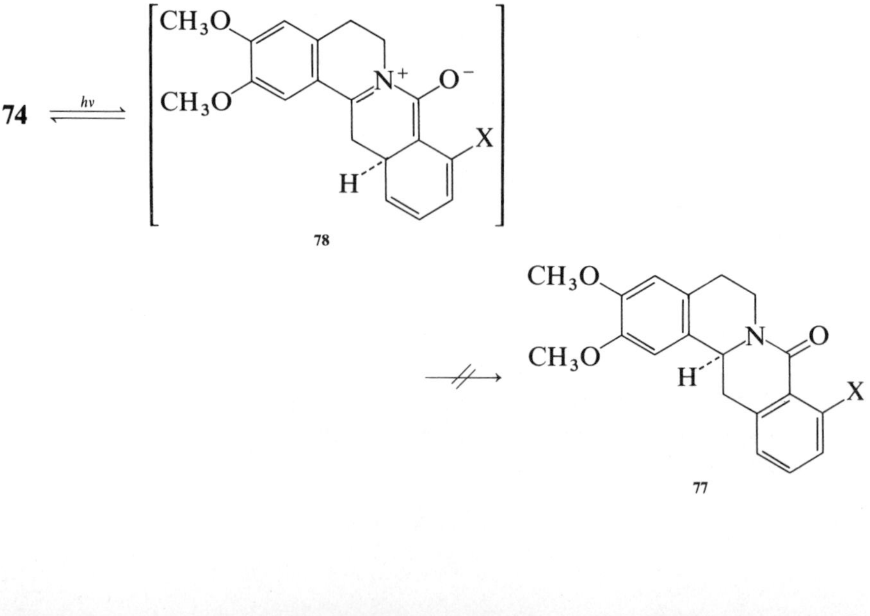

Such 1,5-hydrogen shifts become important pathways, however, if a good leaving group X is not present, as indicated by the photocyclization of enamide **79** to give product **80**.[81]

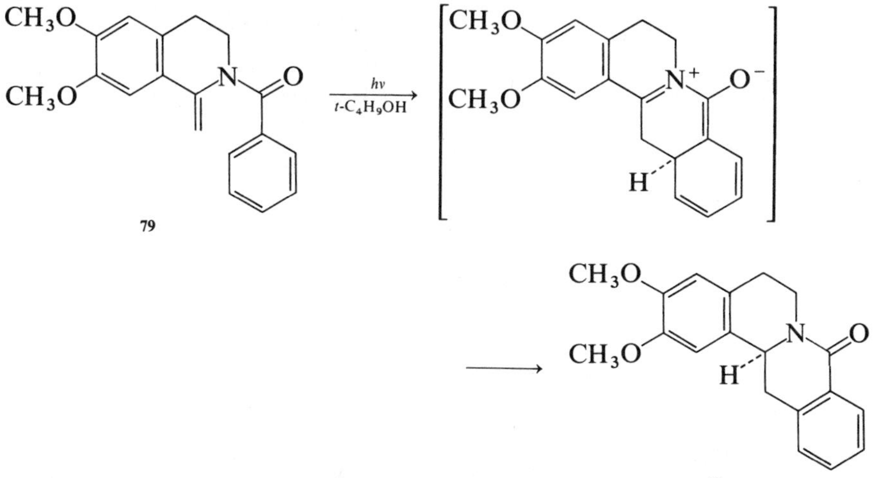

Triplet-State Photocyclizations

The photochemical conversion of diarylamines to carbazoles proceeds by an adiabatic triplet-state mechanism, as shown for *N*-methyldiphenylamine (**81**) in Scheme 8.[19,26,82a,b] The transient dihydrocarbazole **82** reacts with dissolved oxygen to give *N*-methylcarbazole (**83**) and hydrogen peroxide; under anaerobic conditions **82** undergoes disproportionation to give carbazole **83** and a tetrahydrocarbazole. There are contradictory reports about the photochemical formation of carbazole itself from diphenylamine.[83a] Thus the evolution of molecular hydrogen was reported in an early study of this conversion under anaerobic conditions.[83b] In a more recent study, however, the photocyclization of diphenylamine was found to be analogous in all respects to that of the *N*-methyl derivative **81** as illustrated in Scheme 8.[84]

Aryl vinyl amines also photocyclize by a triplet-state mechanism analogous to that depicted in Scheme 8.[85] The nonoxidative conversion of enamine **84** to hexahydrocarbazole **85** is an example.[19,86]

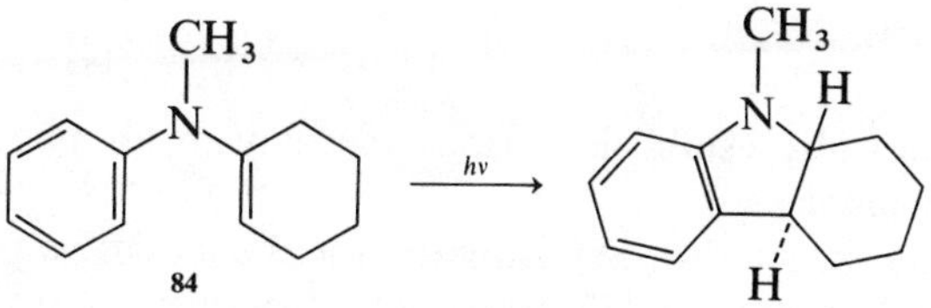

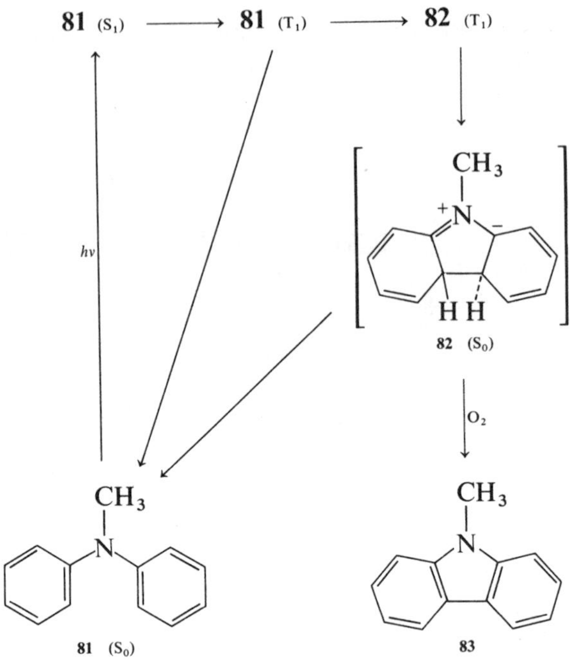

Scheme 8

The related photocyclizations of aryl vinyl ethers and aryl vinyl sulfides also appear to be triplet-state processes in view of the fact that the ring closures **86** → **87**[35] and **88** → **89**[34] can be accomplished by triplet sensitization. Flash photolytic studies also establish a triplet mechanism for photocyclization of some aryl vinyl sulfides.[87a]

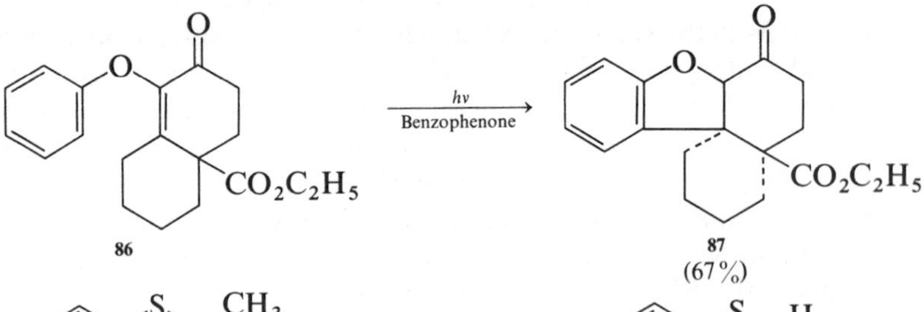

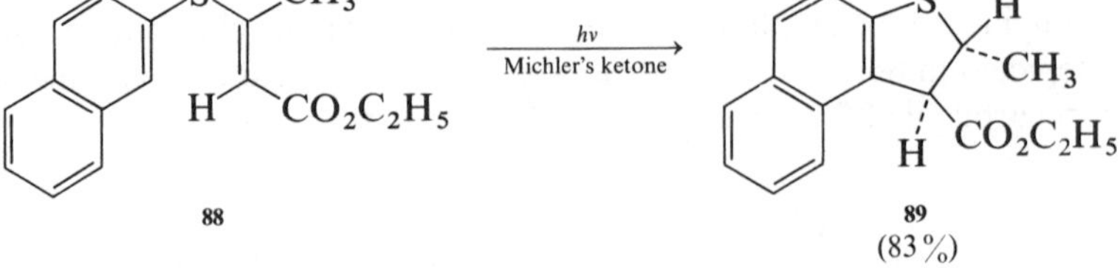

A triplet-state mechanism is postulated for the oxidative photocyclization of dibenzo[*c,g*]phenanthrene (**62**) to benzo[*ghi*]perylene (**63**).[87b]

Certain derivatives of 2-vinylbiphenyl undergo both triplet-state and singlet-state photocyclization.[33,36] As illustrated in Scheme 9 for the conversion

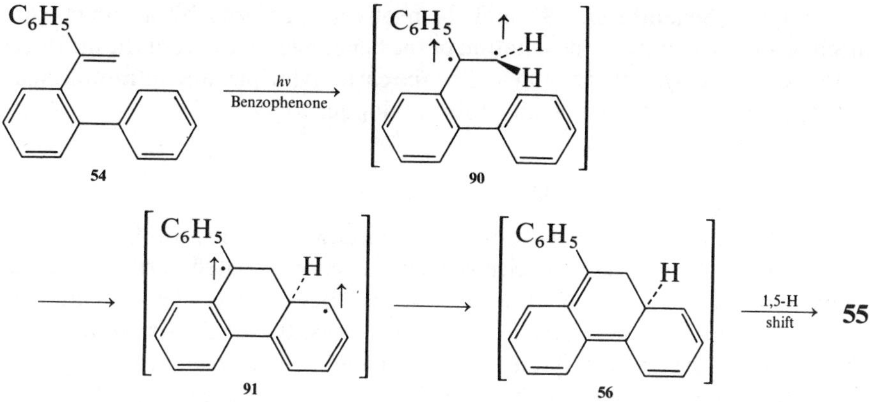

C_6H_5 hv Benzophenone C_6H_5 H H

54 90

C_6H_5 H C_6H_5 H 1,5-H shift 55

91 56

Scheme 9

of 2-(1-phenylvinyl)biphenyl (**54**) to 9,10-dihydro-9-phenylphenanthrene (**55**),[36a] the ring-closure step in the triplet-state reactions can be viewed as a radical-like attack by the terminal carbon atom of a twisted olefin such as **90** to give a cyclized triplet biradical like **91**; spin inversion and a 1,5-hydrogen shift complete the transformation. As illustrated in Scheme 10, the photocyclization of ketone **92** with long-wavelength light to give a mixture of isomeric

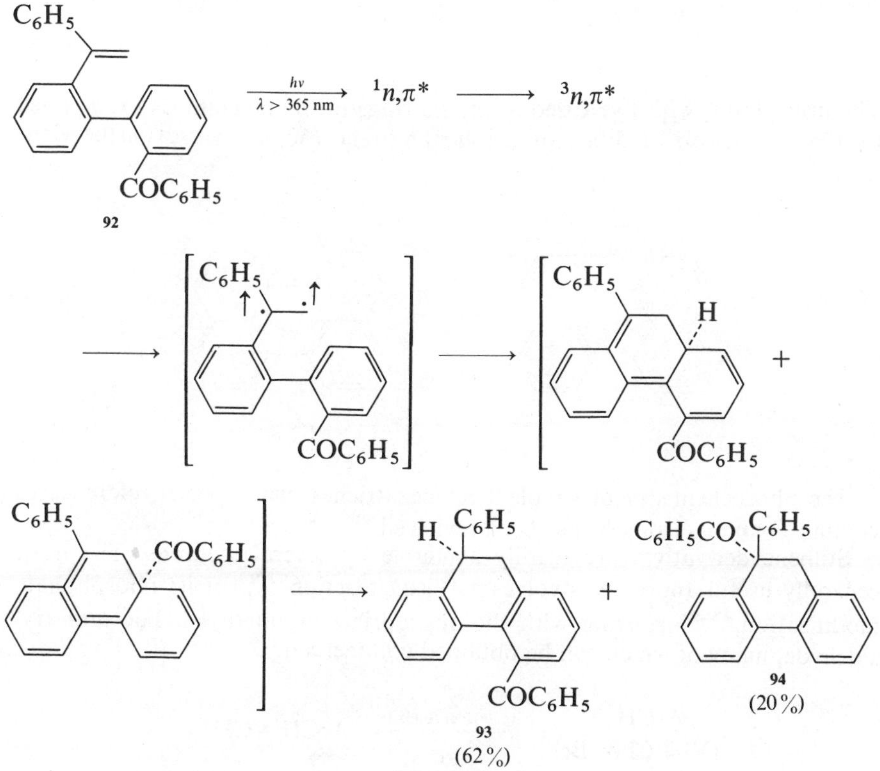

C_6H_5 hv $\lambda > 365$ nm $^1n,\pi^*$ $\longrightarrow$ $^3n,\pi^*$

COC_6H_5

92

C_6H_5 C_6H_5 H +

COC_6H_5 COC_6H_5

C_6H_5 COC_6H_5 H C_6H_5 + C_6H_5CO C_6H_5

COC_6H_5

93 (62%) 94 (20%)

Scheme 10

9,10-dihydrophenanthrenes **93** and **94** probably proceeds by a triplet-state mechanism involving photoexcitation of the benzophenone part of the molecule followed by intersystem crossing to give the carbonyl triplet and intramolecular energy transfer to give the olefin triplet analogous to **90**.[33]

SCOPE AND LIMITATIONS

The photocyclization of *cis*-stilbene (**1**) to *trans*-4a,4b-dihydrophenanthrene (**2**) is one of a large class of electrocyclic reactions related formally to the conversion of *cis*-1,3,5-hexatriene to 1,3-cyclohexadiene. This chapter is limited arbitrarily to those systems in which at least two of the six atoms in the photocyclizing hexatriene system are part of an aromatic or heteroaromatic ring. Thus the conversion of the phenylbutadiene derivative **95** to the naphthalene derivative **96**[88] exemplifies one limit among the photocyclizations under consideration because the reactant has only one aromatic ring fused to the fundamental hexatriene system.

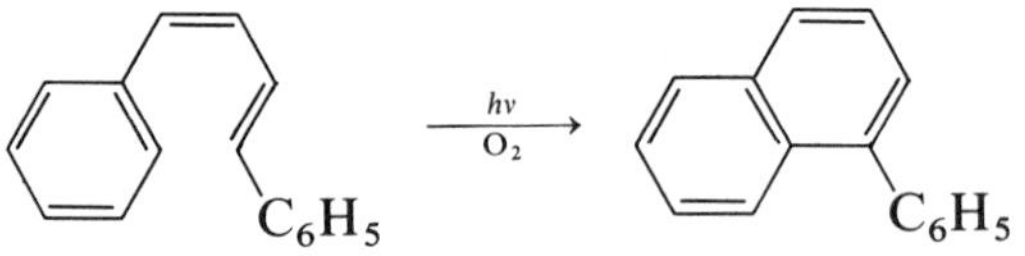

C_6H_5 $\xrightarrow[O_2]{h\nu}$ C_6H_5

95 **96** (50%)

The other limit, with five fused aromatic rings in the reactant, is exemplified by the conversion of dibenzo[*c*,*g*]phenanthrene (**62**) to benzo[*ghi*]perylene (**63**).[76,89]

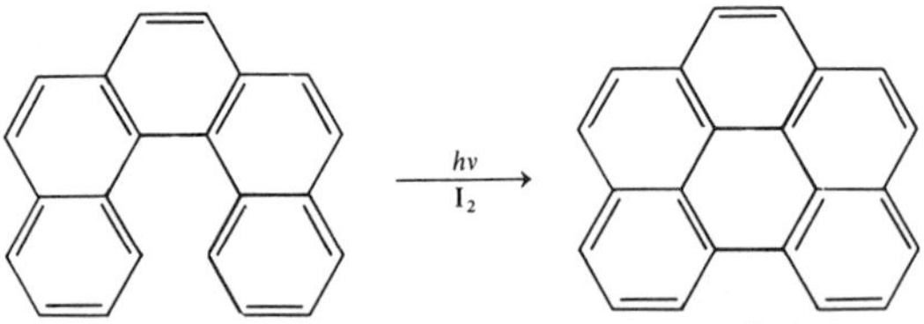

$\xrightarrow[I_2]{h\nu}$

62 **63** (59%)

The photochemistry of simple 1,3,5-hexatrienes, with special reference to vitamin D and its isomers, has been reviewed.[90]

Stilbene derivatives are readily available by several methods. The most generally useful approach involves a Wittig reaction[91,92] or its phosphonate modification,[93–95] starting with the appropriate arylmethyl halide and aryl aldehyde, many of which can be obtained commercially.

$$\underset{(X = Cl\text{ or }Br)}{ArCH_2X} \xrightarrow[\text{2. Base; 3. Ar'CHO}]{\text{1. }Ph_3P\text{ or }(EtO)_3P} ArCH{=}CHAr'$$

Other routes to ArCH=CHAr′ include Grignard reactions of $ArCH_2Cl$ and Ar′CHO, or ArBr and $Ar'CH_2CHO$, followed by dehydration; a Perkin condensation of $ArCH_2CO_2H$ and Ar′CHO followed by decarboxylation;[96,97] a Meerwein arylation reaction of $ArCH{=}CHCO_2H$ and $Ar'NH_2$;[98,99] and a Siegrist reaction of $ArCH_3$ and $Ar'CH{=}NC_6H_5$.[100–102] These methods often produce a mixture of the *cis* and *trans* isomers of the stilbene derivative. Although such a mixture can be photocyclized directly, it may be desirable to convert the mixture to the *trans* isomer for easier handling or for characterization. (*trans*-Stilbenes usually are crystalline; *cis*-stilbenes often are not.) One particularly convenient way to accomplish this thermodynamically favorable *cis* → *trans* isomerization is to irradiate a benzene solution containing the stilbene isomers and a trace of molecular iodine with *visible* light from an ordinary 100-W bulb.

$$\text{Ar-CH=CH-Ar}'\ (cis) \xrightarrow[I_2,\ C_6H_6]{h\nu\ (\text{visible})} \text{Ar-CH=CH-Ar}'\ (trans)$$

Although various reasons can be advanced to account for the reports that certain stilbene analogs fail to undergo photocyclization, an earlier cautious admonition[1] seems worth quoting: "a proper outlook in regard to failures would perhaps be that the proper conditions for the particular cyclization have not yet been found."

Phenanthrenes from Simple Stilbenes

Photocyclization is used widely to produce substituted phenanthrenes from the corresponding stilbenes (see Table I). Phenanthrenes substituted at specific positions frequently are difficult to synthesize by other methods. The reported yields of stilbene photocyclizations vary considerably, but values of 60–90% often can be achieved.

Many common substituents, including alkyl (R), aryl, CN, $CONH_2$, CHO, COR, CO_2H, CO_2R, CF_3, NHCOR, $NHCO_2R$, OH, OR, OCOR, F, Cl, and Br, survive photocyclization intact, but there are problems with a few substituents. With some exceptions,[46,47,103] the reaction fails for stilbenes with nitro or amino substituents; and iodo substituents are lost by photolysis of the carbon–iodine bond to give an aryl radical.

From *para*-Substituted Stilbenes. The photocyclizations of *para*-substituted stilbenes produce phenanthrenes substituted at the 3 (or 6) positions.

$$\text{4-X-stilbene} \xrightarrow[\text{Oxidant}]{h\nu} \text{3-X-phenanthrene}$$

Many examples are known, including systems with additional substituents at other positions.

Contrary to earlier reports that 4-acetylstilbene (**97**) failed to undergo oxidative photocyclization to give 3-acetylphenanthrene (**98**),[10,55] a recent reinvestigation has shown that this conversion can be achieved.[76]

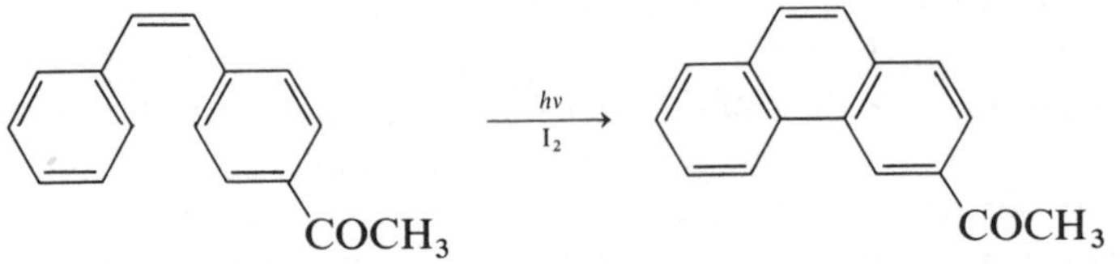

$\xrightarrow[I_2]{h\nu}$

$COCH_3$ **97** → $COCH_3$ **98** (67%)

Aldehyde derivatives of stilbene also undergo oxidative photocyclization, as illustrated by the conversions of **99** to **100**[104] and **101** to **102**.[105]

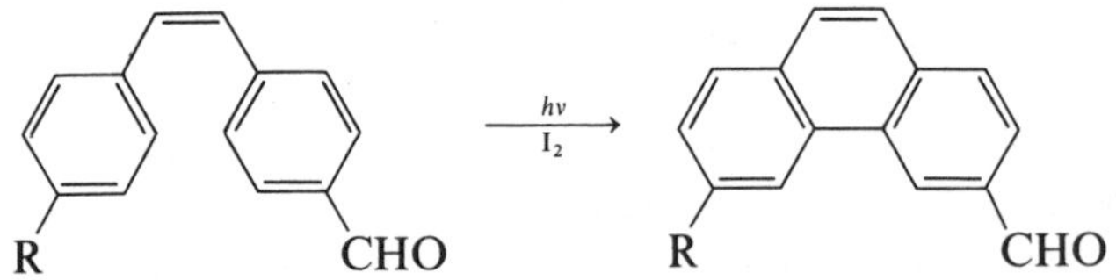

R, CHO $\xrightarrow[I_2]{h\nu}$ R, CHO

99 R = H
101 R = CHO

100 R = H (40%)
102 R = CHO (40%)

A novel set of polymethylene-bridged phenanthrenes of type **103** is obtained from stilbenes of type **104** by oxidative photocyclization.[106]

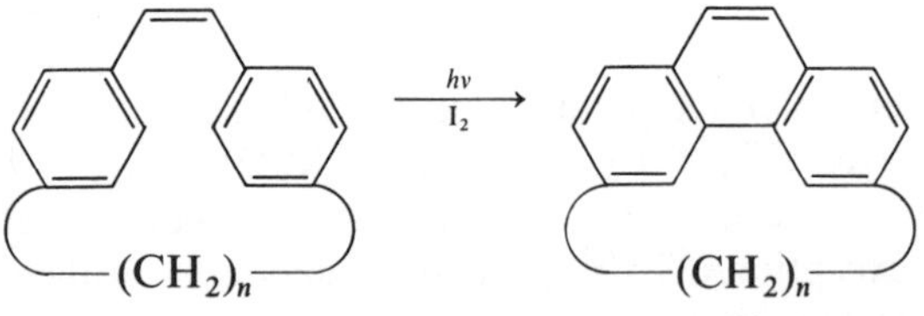

$(CH_2)_n$ **104** $\xrightarrow[I_2]{h\nu}$ $(CH_2)_n$ **103**

n = 7, 8, 9, 10

From *meta*-Substituted Stilbenes. *meta*-Substituted stilbenes follow two photocyclization pathways, one leading to 2-substituted phenanthrenes and the other to 4-substituted phenanthrenes as illustrated in Scheme 11.

In spite of the greater steric crowding associated with the latter pathway, these photocyclizations often give product mixtures consisting of almost as much of the 4-substituted phenanthrene as the 2-substituted phenanthrene for many substituents. For example, in the irradiation of stilbenes of type **105** with X = CH_3, CF_3, or Cl, the **107b**:**107a** product ratio is about 0.95;[21] even with X = C_6H_5 this ratio is 0.75.[107] These initially surprising results can be

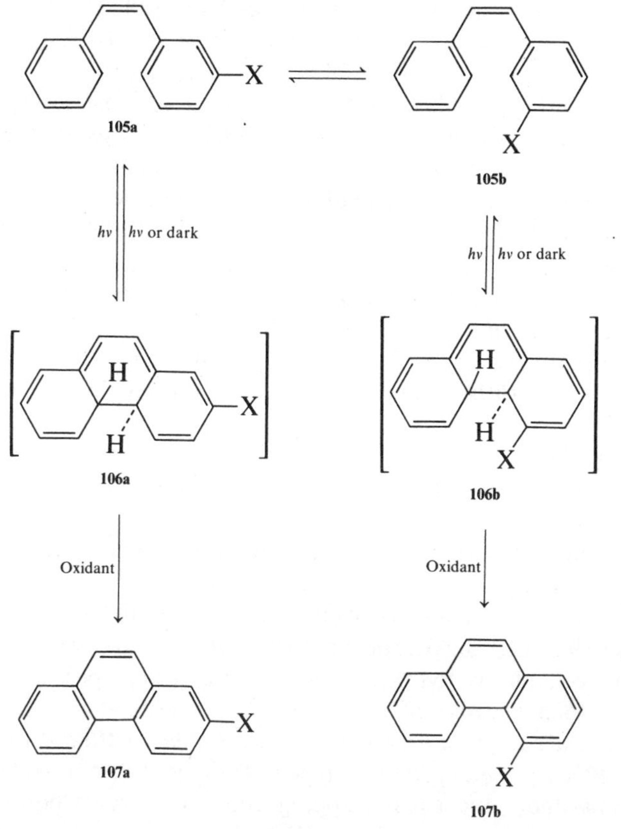

Scheme 11

accounted for by two postulates.[21] First, photoexcitation of a mixture of ground-state stilbene conformers such as **105a** and **105b** gives essentially equal amounts of the corresponding S_1 excited-state stilbenes **105a*** and **105b*** because **105a** and **105b** have essentially equal population at equilibrium. (Steric interactions between inside *meta* and *meta′* substituents in *cis*-stilbenes are minimal because of the twisting around the 1-α and 1′-α' bonds that is required to avoid the potentially more severe steric interactions between inside *ortho* and *ortho′* substituents.) Second, the conversions of the excited stilbenes **105a*** and **105b*** to the corresponding dihydrophenanthrenes **106a** and **106b** take place with nearly equal probability because the transition states for these two conversions each occur early along the reaction coordinate before the greater crowding of the substituent X in **106b** than that in **106a** becomes an important destabilizing factor.

Under some circumstances, four of which are considered in the four paragraphs that follow, the ordinarily slight predominance of the 2-substituted rather than the 4-substituted phenanthrene in the isolated product mixture can become more pronounced.

First, in the photocyclization of certain *meta*-substituted stilbenes, prolonged irradiation selectively destroys small amounts of the 4-substituted phenanthrene.[21]

Second, in certain highly crowded systems, in contrast to the moderately crowded systems considered above, *meta* substituents *do* interfere with the formation of the more hindered dihydrophenanthrene. For example, the photocyclization of *m,m'*-dimethylstilbene (**108**) (Scheme 12) in the presence of sufficient iodine to trap the intermediate dihydrophenanthrenes **109** with complete efficiency gives a mixture of dimethylphenanthrenes consisting of 28% of the 2,7 isomer **110a**, 54% of the 2,5 isomer **110b**, and 18% of the 4,5 isomer **110c**. Steric hindrance from methyl–methyl crowding in the photocyclization of stilbene conformer **108c** apparently limits the yield of the 4,5 isomer to only 18% rather than the statistically expected 25%. (The 54:28 ratio of the yields of the 2,5 and 2,7 isomers is close to the statistical value of 2:1, providing another example of the insignificance of the crowding effect of a single inside methyl substituent on the photocyclization of a conformer such as **108b** as compared to **108a**.)

A third factor that can lead to nonstatistical product ratios is inefficient trapping of one of the isomeric dihydrophenanthrene intermediates. In the *m,m'*-dimethylstilbene system, for example, the rate of the thermal ring opening of the highly strained dihydrophenanthrene **109c** to give *cis*-stilbene **108c** is greatly enhanced relative to the rates of the thermal ring openings **109a** → **108a** and **109b** → **108b** because of the relief of the destabilizing methyl–methyl crowding that accompanies the conversion of **109c** to **108c**. In fact, the ring opening of **109c** competes successfully with the oxidative removal of its 4a and 4b hydrogens under the usual trapping conditions. As a consequence, the amount of 4,5-dimethylphenanthrene (**110c**) in the product mixture varies with iodine concentration, ranging from none in the absence of iodine (with oxygen as the only oxidant) to 4% for $[I_2] = 5 \times 10^{-4}$ *M* and a maximum of 18% for $[I_2]$ greater than about 3×10^{-2} *M*.[21] Photocyclization of *m,m'*-dimethoxystilbene (**111**) with oxygen as the oxidant also gives only 2,7-dimethoxyphenanthrene (**112a**) and 2,5-dimethoxyphenanthrene (**112b**) and no 4,5-dimethoxyphenanthrene.[108] The fact that 4,5-disubstituted dihydrophenanthrenes are

CH_3O — **111** — OCH_3

$\xrightarrow[O_2]{h\nu}$ **112a** (45%) + **112b** (28%)

CH_3 CH_3 **108a** ⇌ CH_3 CH_3 **108b** ⇌ H_3C CH_3 **108c**

hv | *hv* or dark

[CH_3 H H CH_3] **109a** [H H CH_3 CH_3] **109b** [H H H_3C CH_3] **109c**

I_2

CH_3 CH_3 **110a** CH_3 CH_3 **110b** H_3C CH_3 **110c**

110a:**110b**:**110c** = 28:54:18

Scheme 12

difficult to trap oxidatively can be advantageous synthetically. If the 4,5-disubstituted phenanthrene is an unwanted side product, it can be avoided by using low concentrations of iodine (e.g., 5×10^{-4} *M*). This probably accounts for the observations that the oxidative photocyclizations of stilbenes **113**,[109,110] **114**,[111] and **115**[112] each give only one product as shown.

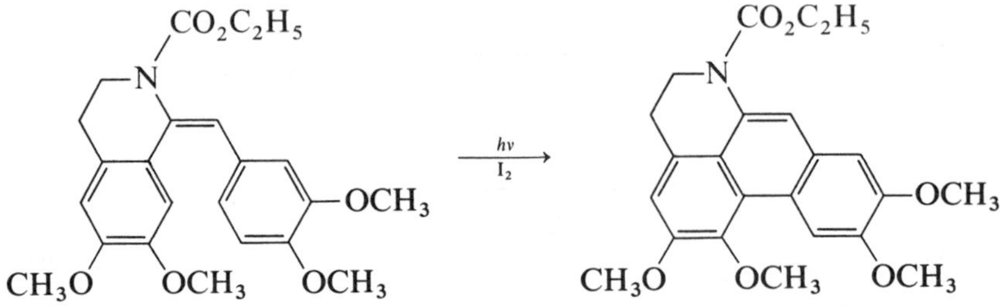

113 → ($h\nu$, I_2) (19–25%)

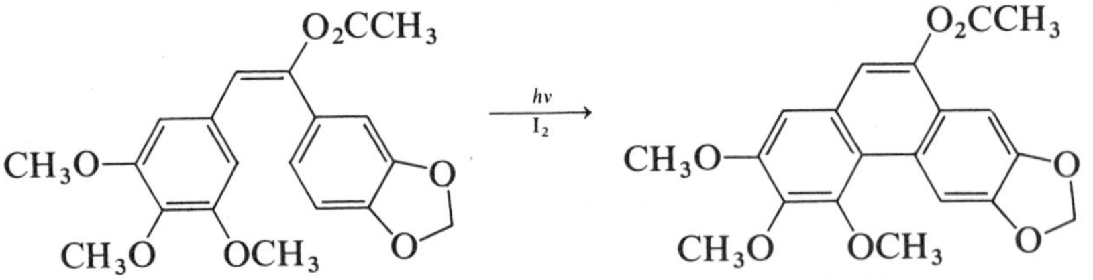

114 → ($h\nu$, I_2) (49%)

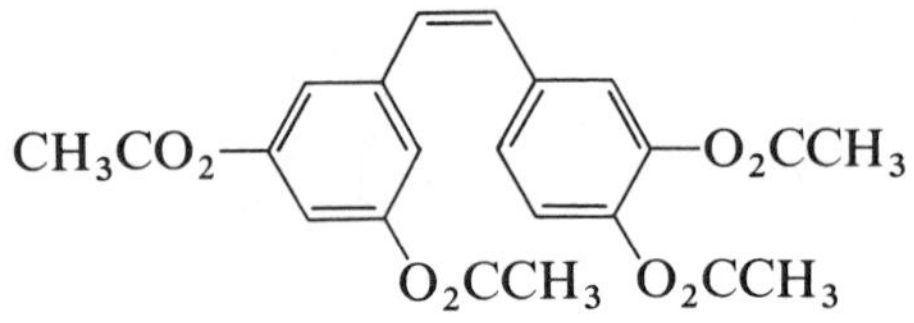

115 → ($h\nu$, I_2) (48%)

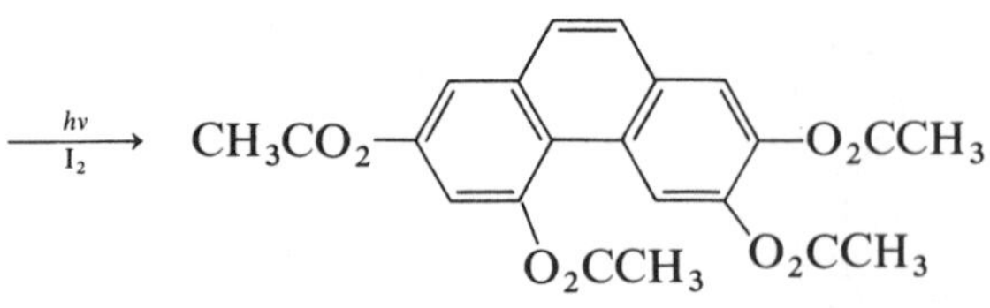

An exception is stilbene **116**, which unaccountably gives comparable amounts of both possible products **117** and **118**.[113,114]

Finally, the oxidative photocyclizations of stilbenes with *m*-cyano and *m*-carbethoxy substituents proceed with an exceptionally high degree of regioselectivity. For example, *m*-cyanostilbene gives 71% of 2-cyanophenanthrene and 19% of 4-cyanophenanthrene;[115] this result can be rationalized by molecular orbital calculations.[47] Similarly, nitrile **119** gives products **120** and **121** in a

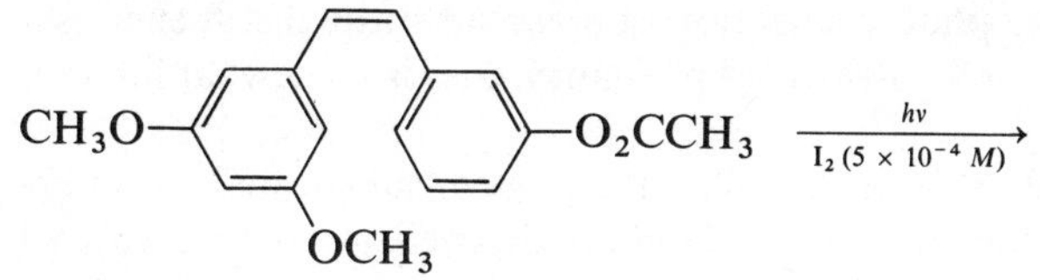

CH_3O OCH_3 O_2CCH_3 $\xrightarrow[I_2\,(5 \times 10^{-4}\,M)]{h\nu}$

116

CH_3O OCH_3 O_2CCH_3 + CH_3O CH_3O O_2CCH_3

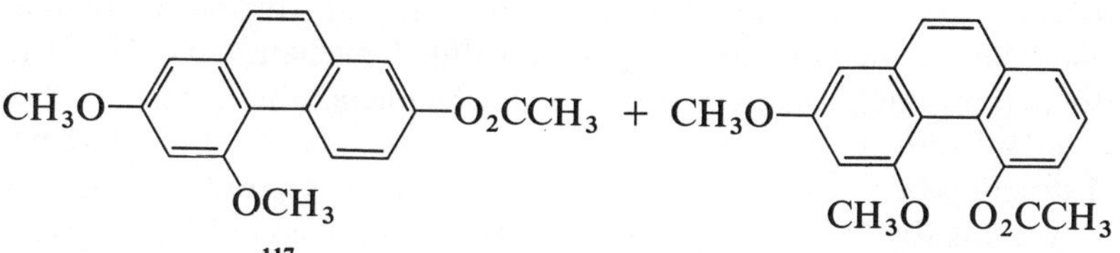

117 (28–35%) **118** (31–36%)

7:1 ratio,[116] and ester **122** gives phenanthrene **123** as the only isolated product.[117]

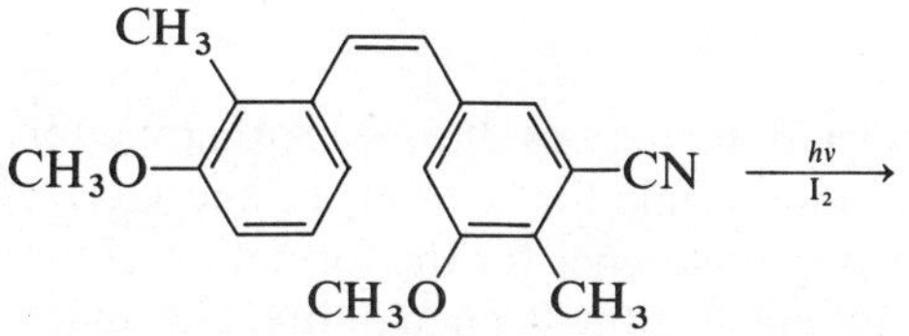

CH_3 CH_3O CN CH_3O CH_3 $\xrightarrow[I_2]{h\nu}$

119

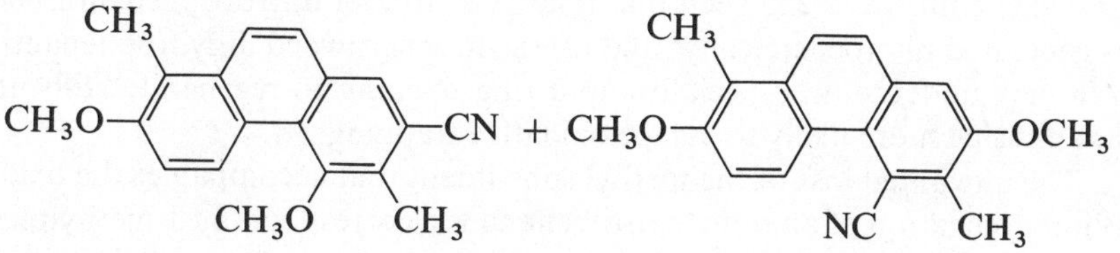

CH_3 CH_3O CN CH_3O CH_3 + CH_3 CH_3O OCH_3 NC CH_3

120 **121** (9%)

CH_3 $CO_2C_2H_5$ $\xrightarrow[I_2]{h\nu}$ CH_3 $CO_2C_2H_5$

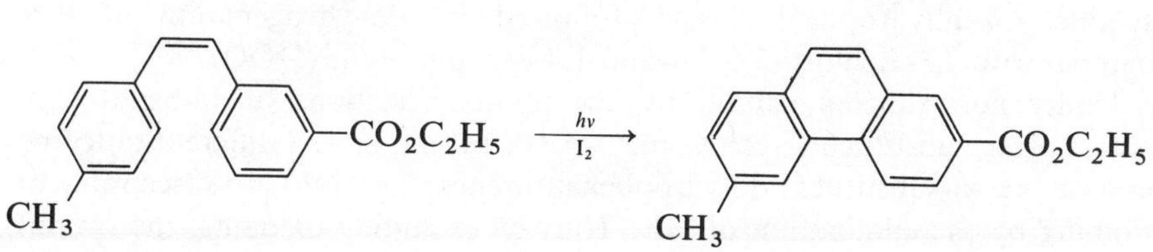

122 **123** (60%)

Because mixtures of isomeric phenanthrenes sometimes are tedious to separate (e.g., by column chromatography on alumina or silicic acid), the practicality of the oxidative photocyclization of *meta*-substituted stilbenes as a synthetic route to either 2- or 4-substituted phenanthrenes is somewhat limited.

From *ortho*-Substituted Stilbenes. Like their *meta*-substituted counterparts, *ortho*-substituted stilbenes can undergo photocyclization to give two 4a,4b-dihydrophenanthrenes as shown in Scheme 2. In the presence of an oxidant, oxidative trapping of a 1-substituted dihydrophenanthrene of type **24a** leads to the corresponding 1-substituted phenanthrene **25** as the major isolated product, usually in good yield. The phenanthrene lacking the original *ortho* substituent often is formed as a minor side product, presumably by abstraction of the 4b hydrogen from the 4a-substituted dihydrophenanthrene of type **24b** (by O_2, $HO_2\cdot$, I·, X·, or a solvent-derived radical) to give a radical of type **124** that subsequently undergoes unimolecular expulsion of X·. This

124 ⟶ **3** + X·

expulsion is estimated from bond dissociation energies[63] to be exothermic for X = CH_3, CH_3O, Cl, Br, and I but endothermic for X = H, F, and C_6H_5. Two factors could contribute to the strong preference for the formation of the 1-substituted phenanthrene **25** over the unsubstituted phenanthrene **3** under oxidizing conditions: (1) stilbene conformer **23a** may predominate over the more crowded conformer **23b** such that relatively little of dihydrophenanthrene **24b** is generated photochemically; and (2) the more crowded dihydrophenanthrene **24b** may undergo an especially rapid ring opening to regenerate stilbene **23b** and thus be more likely to escape oxidative trapping.

The unwanted loss of the methyl substituent that accompanies the oxidative photocyclization of an *o*-methylstilbene to the corresponding 1-methylphenanthrene is minimized with an improved procedure involving two stages: UV irradiation of the *o*-methylstilbene in *n*-propylamine under nitrogen to give an isolable 1,4-dihydrophenanthrene, followed by dehydrogenation of that material with 2,3-dichloro-5,6-dicyano-1,4-benzoquinone (DDQ).[72b]

Under nonoxidizing conditions, the photocyclizations of stilbenes with certain *ortho* substituents such as methoxy or bromo have a different outcome because the 4a-substituted dihydrophenanthrenes (e.g., **24b**) are susceptible to trapping by the elimination of HX. Thus by excluding oxidants, the overall photocyclization reaction can be forced to proceed in high yield along the pathway involving loss of HX.

Ortho substituents can impose regioselectivity on photocyclizations of *meta*-substituted stilbenes. For example, the nonoxidative photocyclization of an *o*-methoxystilbene such as **125** would give exclusively the 2-substituted phenanthrene **107a**, and the nonoxidative photocyclization of the isomeric

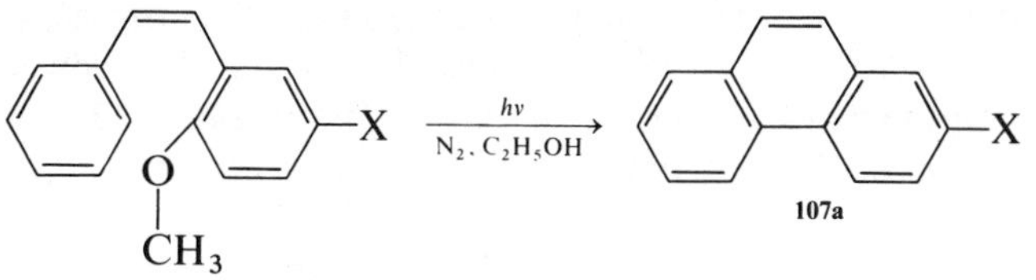

X
hv
N_2, C_2H_5OH
X
O
CH_3
125
107a

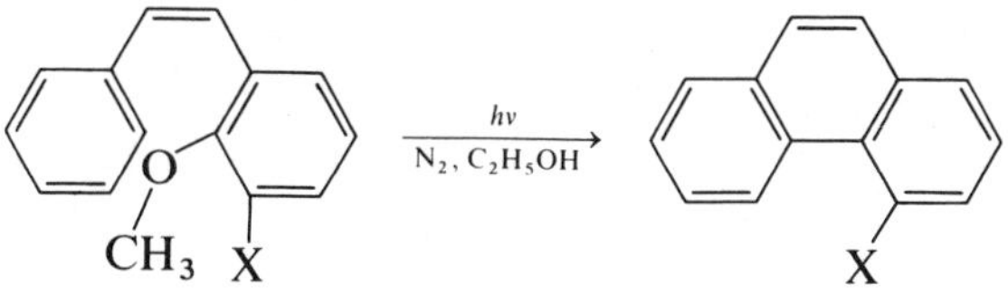

hv
N_2, C_2H_5OH
O
CH_3 X
X
126
107b

stilbene **126** would give exclusively the 4-substituted phenanthrene **107b**. This approach is useful for the syntheses of aporphine alkaloids as illustrated by the conversion of bromostilbene **127** to the neolitsine derivative **128**.[25]

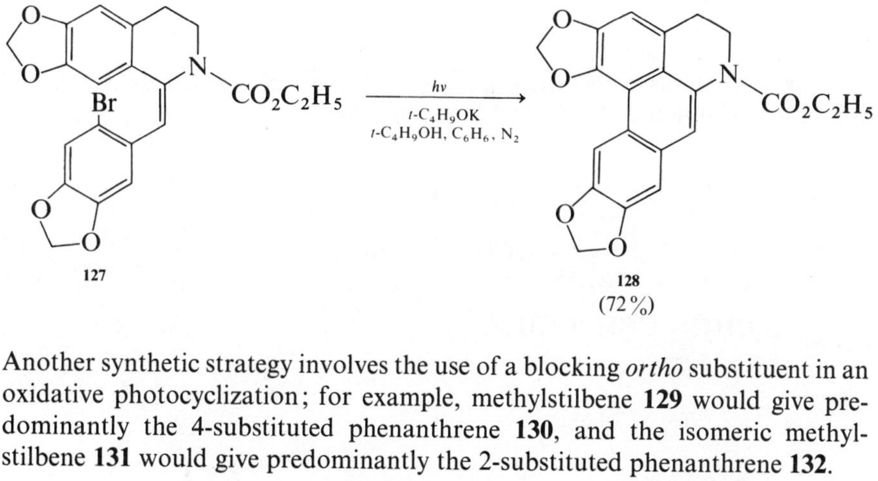

O
O
N
$CO_2C_2H_5$
Br
O
O
hv
t-C_4H_9OK
t-C_4H_9OH, C_6H_6, N_2
127
128
(72%)

Another synthetic strategy involves the use of a blocking *ortho* substituent in an oxidative photocyclization; for example, methylstilbene **129** would give predominantly the 4-substituted phenanthrene **130**, and the isomeric methylstilbene **131** would give predominantly the 2-substituted phenanthrene **132**.

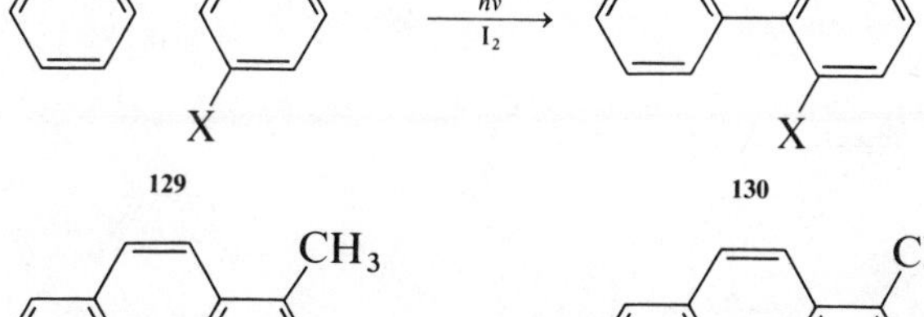

CH_3
hv
I_2
CH_3
X
X
129
130

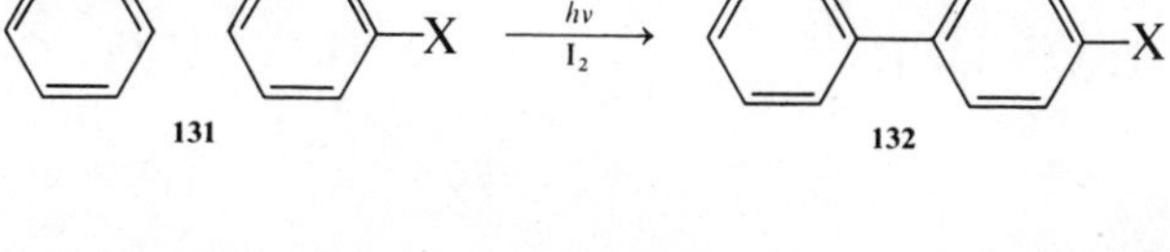

CH_3
X
hv
I_2
CH_3
X
131
132

The oxidative photocyclization of stilbene **133** to phenanthrene **134** proceeds in very low yield; this may stem from unusually severe steric crowding between the methoxy groups at C-4 and C-5 in dihydro intermediate **135** because of the buttressing effect of the adjacent methyl groups at C-3 and C-6.[118]

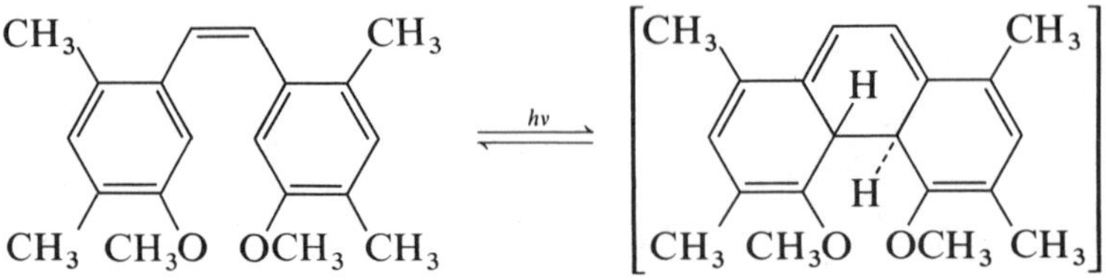

CH_3 CH_3 CH_3 CH_3O OCH_3 CH_3
133

$\xrightarrow{h\nu}$

[CH_3 CH_3 H H CH_3 CH_3O OCH_3 CH_3]
135

$\xrightarrow{I_2}$

CH_3 CH_3 CH_3 CH_3O OCH_3 CH_3
134
(4%)

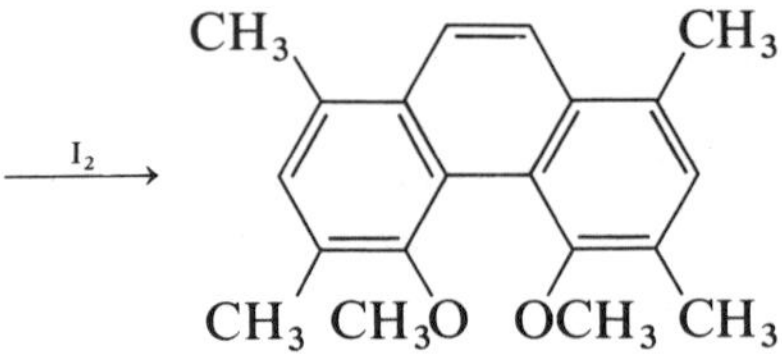

Arguing against this explanation, however, are the successful oxidative photocyclizations of **136** to **137**[119] and **138** to **139**.[120]

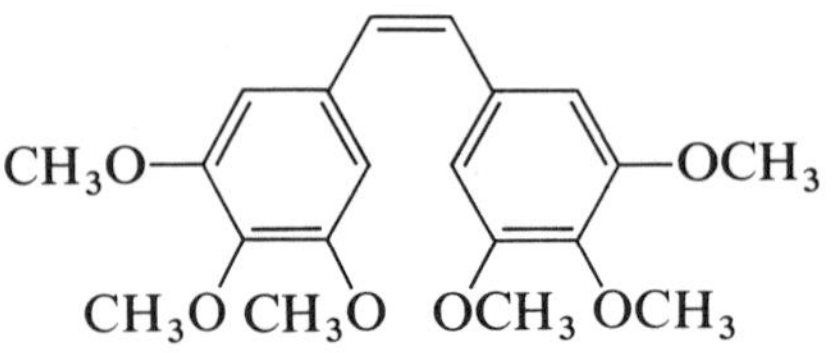

CH_3O OCH_3 CH_3O CH_3O OCH_3 OCH_3
136

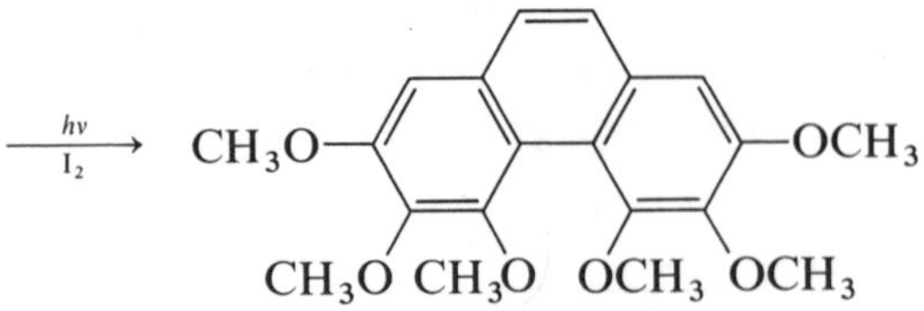

$\xrightarrow[I_2]{h\nu}$

CH_3O OCH_3 CH_3O CH_3O OCH_3 OCH_3
137
(82%)

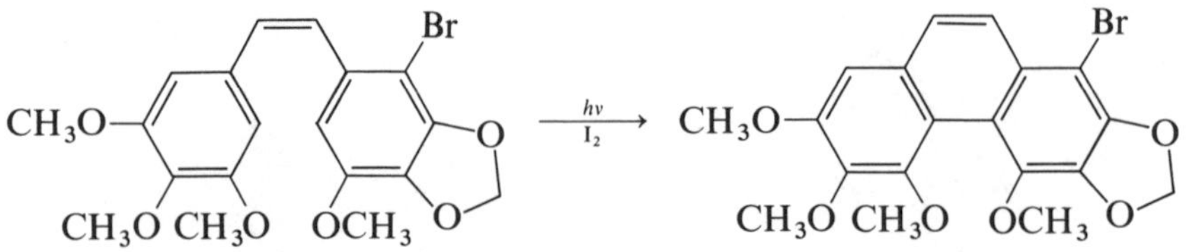

Br CH_3O O CH_3O CH_3O OCH_3 O
138

$\xrightarrow[I_2]{h\nu}$

Br CH_3O O CH_3O CH_3O OCH_3 O
139
(84%)

Stilbene **140**, with no bromine atom as a blocking substituent, gives a 1:1 mixture of phenanthrenes **141** and **142**.[120]

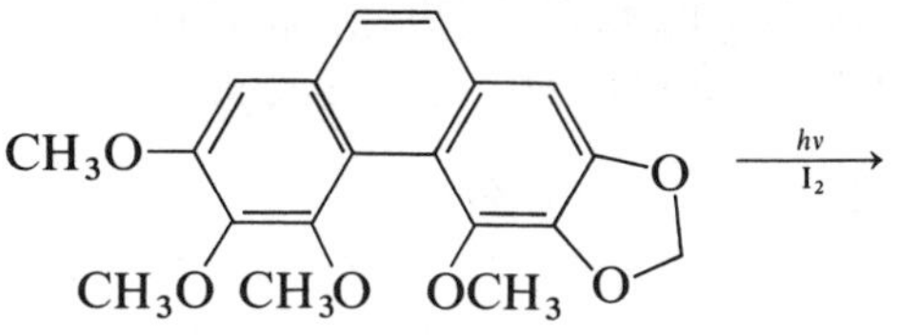

CH_3O CH_3O CH_3O OCH_3 O O

140

$\xrightarrow[I_2]{h\nu}$

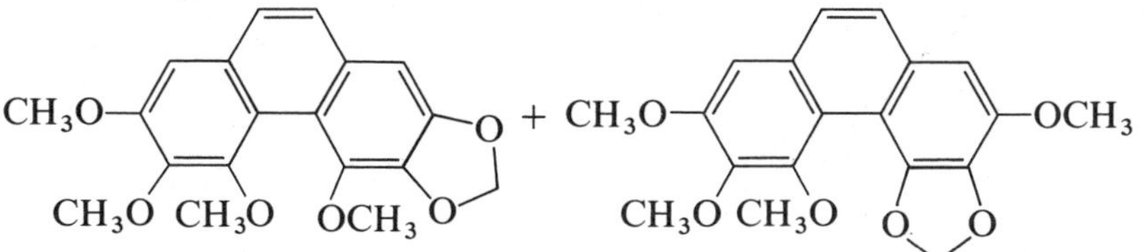

141 (40%) + **142** (40%)

From *o*-Iodostilbenes. The photochemistry of stilbenes with *o*-iodo substituents is complicated by another possible mechanistic pathway leading to phenanthrenes in addition to that proceeding by way of dihydrophenanthrene intermediates. This other pathway involves photolysis of the carbon–iodine bond followed by an intramolecular free-radical arylation as illustrated in Scheme 13 for the parent *o*-iodostilbene system **143**. Sometimes this alternative

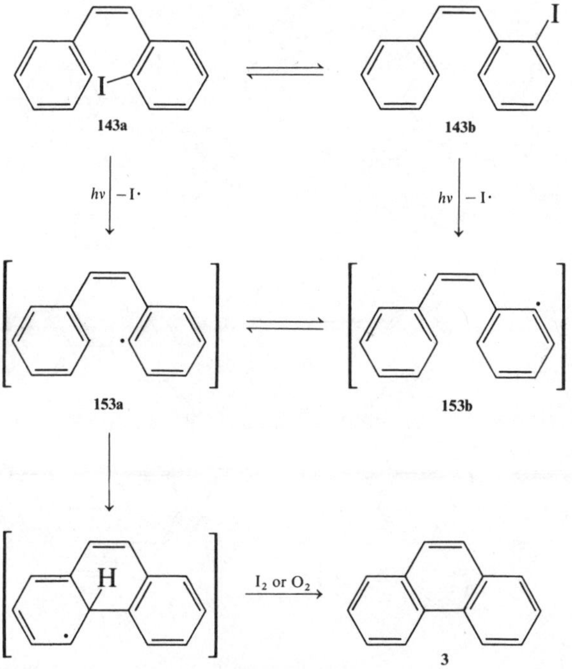

143a ⇌ **143b**

$h\nu$ | $-I\cdot$

153a ⇌ **153b**

I_2 or O_2

3

Scheme 13

route to phenanthrenes is advantageous synthetically. For example, nitro-substituted phenanthrenes are not produced from the corresponding stilbenes by the electrocyclic pathway, but UV irradiation of iodostilbene **144** gives phenanthrene **145**, the methyl ester of the natural product aristolochic acid I.[121]

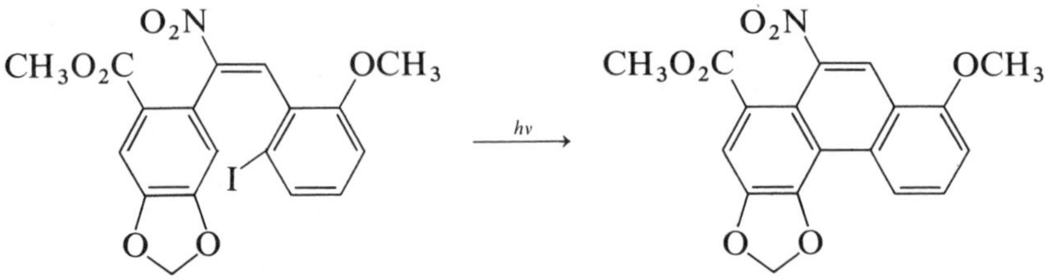

144 → (hν) **145** (54%)

Similarly, systems **146**,[122] **147**,[123] and **148**[124] are obtained by irradiation of the *o*-iodo precursors **149**, **150**, and **151**, respectively, but are not available by oxidative photocyclization of the analogs of **149–151** lacking the iodo substituents.

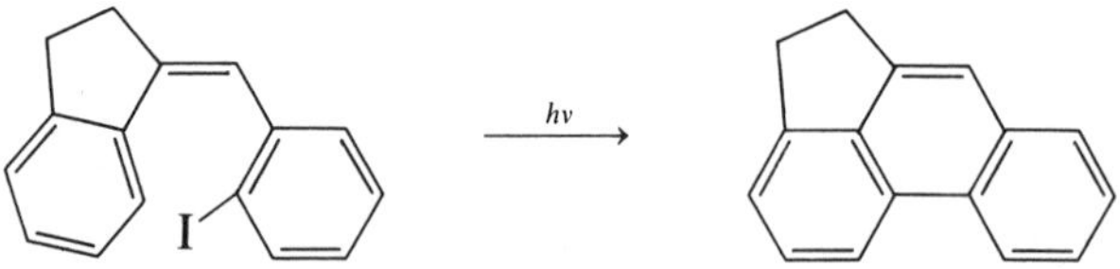

149 → (hν) **146** (25%)

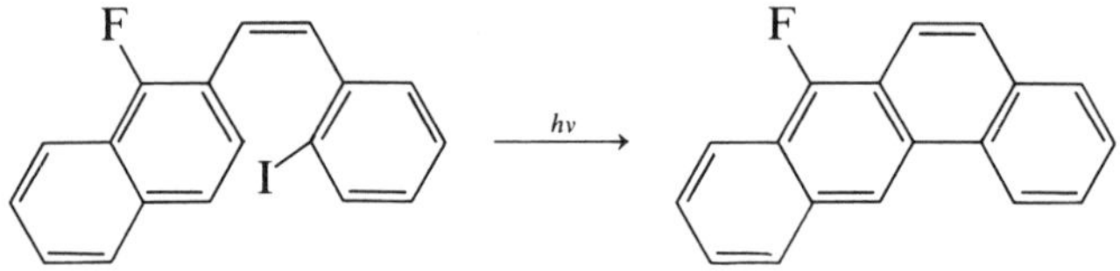

150 → (hν) **147** (52%)

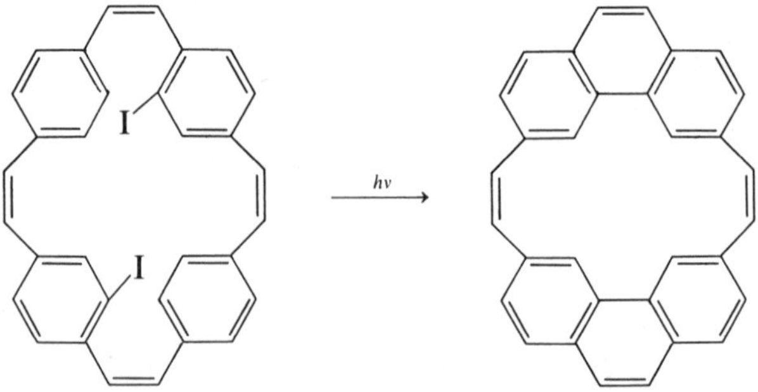

151 → (hν) **148** (15%)

Several pathways are conceivable for the formation of phenanthrenes with loss of the iodo substituent from the irradiation of *o*-iodostilbenes; often it is not clear which pathway predominates. For example, the ring closure of *o*-iodostilbene (**143a**) might occur by a free-radical mechanism as illustrated in Scheme 13[125] or by an electrocyclic mechanism with subsequent elimination of HI from intermediate **152**.

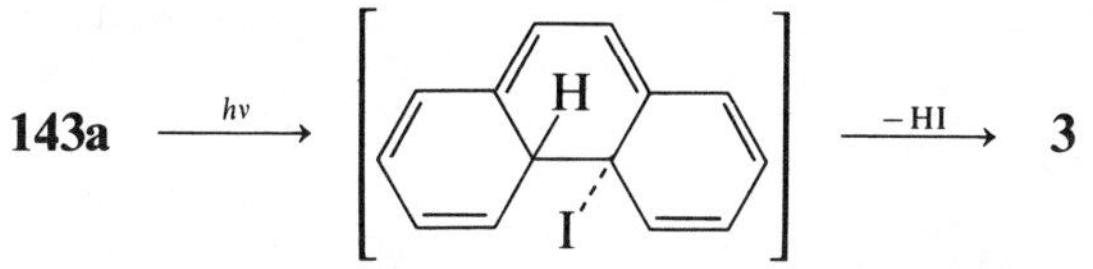

143a $\xrightarrow{h\nu}$ [**152**] $\xrightarrow{-HI}$ **3**

Furthermore, in a solvent with easily abstractable hydrogen atoms (e.g., cyclohexane), the formation of phenanthrene (**3**) might follow one or more of the following pathways in which the *o*-iodo substituent plays a trivial role: (1) photolysis of **143b** to give radical **153b** (Scheme 13), followed by hydrogen transfer from the solvent and oxidative photocyclization of the resulting stilbene, perhaps with the iodine generated *in situ* serving as the oxidant; (2) a sequence analogous to pathway 1, starting with photolysis of the *trans* isomer of the iodostilbene; and (3) oxidative photocyclization of stilbene **143b** to give 1-iodophenanthrene (**154**), with subsequent photolysis of the carbon–iodine bond and hydrogen transfer from the solvent to give phenanthrene (**3**).

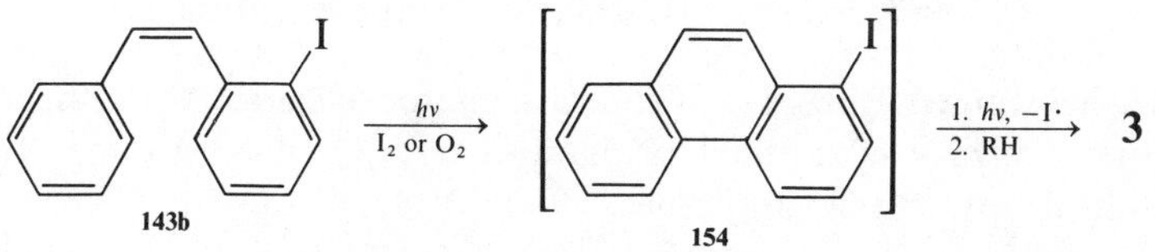

143b $\xrightarrow[I_2 \text{ or } O_2]{h\nu}$ [**154**] $\xrightarrow[2.\ RH]{1.\ h\nu,\ -I\cdot}$ **3**

At least one of the latter three types of pathway *must* be involved, for example, in the photocyclization of *m*-acetoxystilbene **155** in cyclohexane because the major product is 7-acetoxyphenanthrene **156a** rather than the 5-acetoxy isomer **156b**.[114]

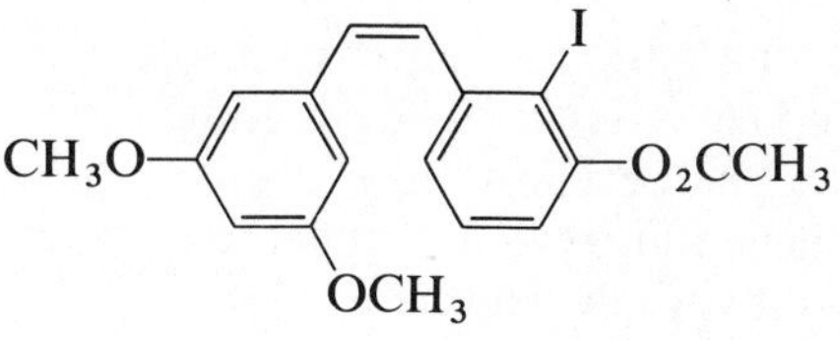

155 $\xrightarrow{h\nu}$ **156a** (57%) + **156b** (14%)

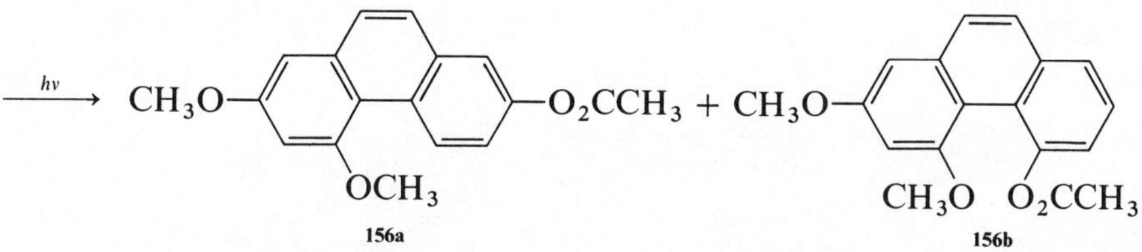

The ambiguous role of an *o*-iodo substituent is illustrated by the photocyclization of iodostilbene derivative **157** in cyclohexane because comparable results are obtained from the oxidative photocyclization of stilbene **158**.[126]

F, I, $h\nu$

157 (91%)

F, $h\nu$, I_2

158 (82%)

It is seldom worth the extra effort to introduce an *o*-iodo substituent in an attempt to facilitate or regiochemically control a stilbene photocyclization. However, the *o*-iodo approach has special value if the stilbene lacking the iodo substituent fails to undergo normal oxidative photocyclization.

From α-Substituted Stilbenes. Oxidative photocyclizations of stilbenes with α substituents give 9-substituted phenanthrenes; α,α′-disubstituted stilbenes give 9,10-disubstituted phenanthrenes.

X, $h\nu$, Oxidant, X

Stilbenes with certain electron-withdrawing α (or α,α′) substituents can be converted photochemically to 9,10-dihydrophenanthrenes by irradiating under nonoxidizing conditions. For example, the photocyclization of α,α′-dicyanostilbene (**36**), which gives 9,10-dicyanophenanthrene (**159**) under oxidizing conditions, gives *trans*-9,10-dicyano-9,10-dihydrophenanthrene (**37**) in 85% yield in degassed benzene solution.[69]

NC CN, $h\nu$, I_2, air, $CHCl_3$, NC CN

36 **159** (60%)

For unknown reasons, some oxidative photocyclizations of α,α′-dialkylstilbenes give higher yields when cupric chloride is present. For example, the irradiation of 4,4′-dimethoxy-α,α′-dimethylstilbene (**160**) in ethanol containing iodine and cupric chloride gives 3,6-dimethoxy-9,10-dimethylphenanthrene (**161**) in 47% yield, whereas in cyclohexane containing iodine alone the yield of **161** is only 11%.[60]

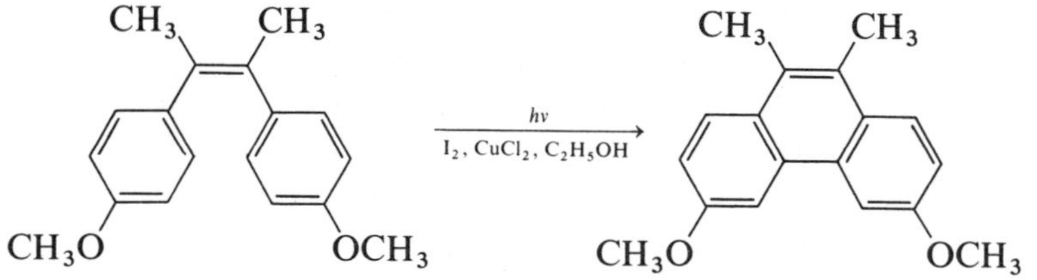

$$\xrightarrow[I_2,\ CuCl_2,\ C_2H_5OH]{h\nu}$$

160 **161** (47%)

In contrast, cupric chloride has no significant effect on the yields of the oxidative photocyclization products from 4,4′-dimethoxy-α-methylstilbene, 4-methoxystilbene, or tetraphenylethylene.[60]

The initially surprising failure of α-fluorostilbene to undergo oxidative photocyclization to 9-fluorophenanthrene can be rationalized by molecular orbital calculations.[126] Both α-chlorostilbene and α-bromostilbene undergo photoelimination in diethyl ether to give diphenylacetylene as the major product.[127] α,α′-Dichlorostilbene (**162**) undergoes oxidative photocyclization to give 9,10-dichlorophenanthrene (**163**).[128] In contrast, α,α′-dibromostilbene (**164**) undergoes photoreduction by hydrogen abstraction from the solvent to

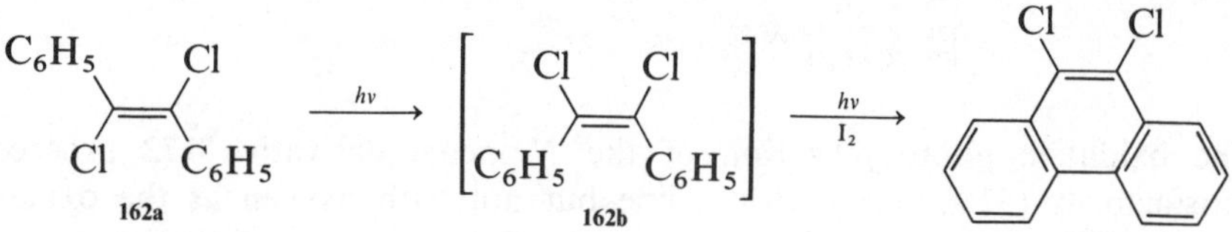

$$\xrightarrow{h\nu} \quad \xrightarrow[I_2]{h\nu}$$

162a **162b** **163** (12%)

give bibenzyl derivative **165**; presumably the rate of the $S_1 \rightarrow T_1$ intersystem crossing for stilbene **164** is so enhanced by the heavy-atom effect of the two bromine atoms that S_1 photocyclization is not competitive.[128]

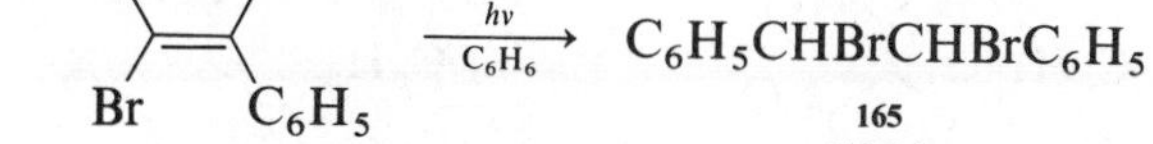

$$\xrightarrow[C_6H_6]{h\nu} C_6H_5CHBrCHBrC_6H_5$$

164 **165** (48%)

α,α′-Diiodostilbene undergoes photolysis to give diphenylacetylene and iodine.[76]

The vinylene carbonate **166** undergoes oxidative photocyclization with iodine in cyclohexane to give phenanthrene **167** in 77% yield;[129] this transformation takes place in only 13% yield with oxygen in ethanol.[130] The oxidative photocyclization of the dimethoxy derivative **168** to product **169** is best carried out (50% yield) with iodine in refluxing cyclohexane.[131]

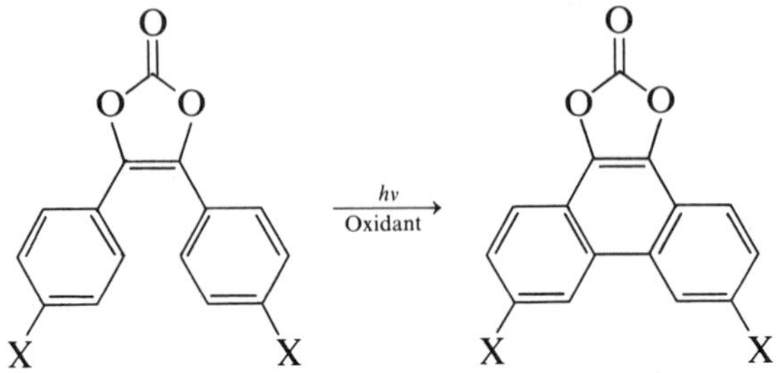

166 X = H
168 X = CH_3O

$\xrightarrow[\text{Oxidant}]{h\nu}$

167 X = H
169 X = CH_3O

Oxidative photocyclization of oxazolone **170** proceeds in disappointing yield (8–39%);[132,133] an 80% yield is obtained for the protected derivative **171** with its easily removable *N*-benzyl group.[133]

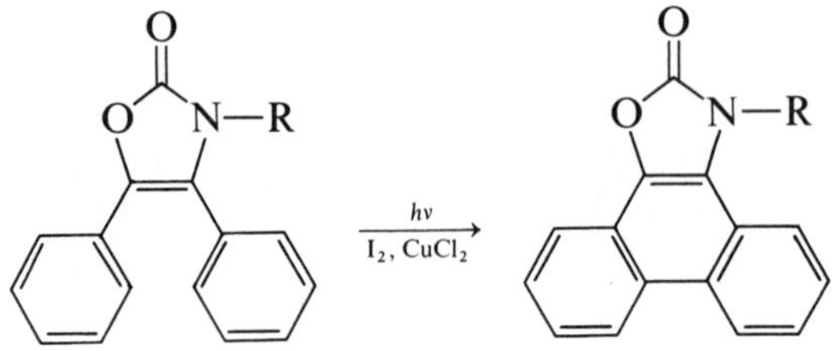

170 R = H
171 R = $CH_2C_6H_5$
172 R = C_6H_5

$\xrightarrow[I_2,\ CuCl_2]{h\nu}$

The oxidative photocyclization of the *N*-phenyl derivative **172** proceeds satisfactorily (52% yield) with iodine but not with oxygen as the oxidant because **172** serves as a photosensitizer for the generation of singlet oxygen, which in turn transforms **172** into a variety of products by way of *N*,*N*-dibenzoylaniline.[134]

Ketostilbene **173** undergoes oxidative photocyclization to give ketophenanthrene **174**.[135]

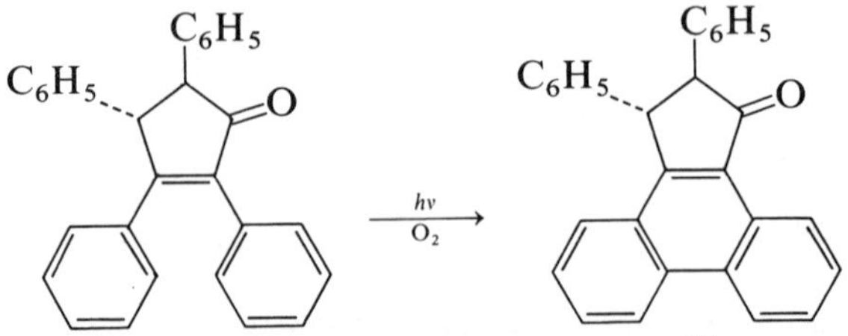

173

$\xrightarrow[O_2]{h\nu}$

174
(54–55%)

This same product is obtained from the nonoxidative photocyclization of tetraphenylcyclopentadienone (**175**) in isopropyl alcohol; the formation of **174** from the expected dihydro intermediate **176** can be accounted for by two successive solvent-mediated enolization–ketonization sequences.[136]

C_6H_5 C_6H_5 O **175** $\xrightleftharpoons{h\nu}$ [C_6H_5 C_6H_5 O H H **176**] $\xrightarrow{i\text{-}C_3H_7OH}$ **174** (45%)

Oxidative photocyclization of tetraphenylethylene (**177**) gives 9,10-diphenylphenanthrene (**178**);[60,76] further photochemical transformation of the *o*-terphenyl analog **178** to the corresponding phenanthrophenanthrene fails.[55,69]

177 $\xrightarrow[I_2]{h\nu}$ **178** (88–90%) $\xrightarrow[I_2]{h\nu}$ (no reaction)

The phenyl groups in 9,10-diphenylphenanthrene (**178**) are twisted out of conjugation with the phenanthrene ring system, as indicated by the close resemblance of the UV absorption spectrum of **178** to that of phenanthrene itself. This indicates that the excitation (i.e., the perturbed distribution of electron density that gives the S_1 excited state a chemical reactivity different from that of the S_0 ground state) in the diphenyl derivative **178** may be confined almost entirely to the phenanthrene system and excluded from the phenyl substituents where it would be required for ring closure. A similar explanation may account for the lack of photocyclization of phenanthropyranone **179**, itself produced by oxidative photocyclization of tetraphenylpyranone **180**;[137] that is, the excitation in **179*** may reside principally in the phenanthrene part of the molecule and not significantly in the stilbene part.

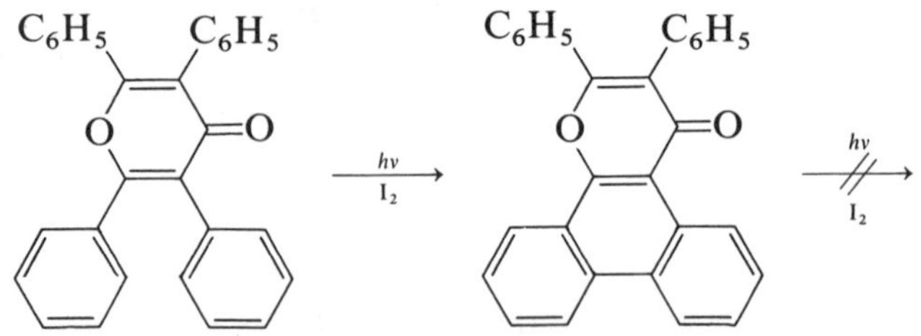

180 → (hν, I_2) → 179 (50%) → (hν, I_2, no reaction)

From Dianthrones and Other Doubly Bridged Tetraphenylethylenes. In contrast to the behavior of tetraphenylethylene itself (**177**), certain bridged tetraphenylethylenes of type **181** are converted photochemically by way of phenanthrenes of type **182** to doubly cyclized products of type **183**.

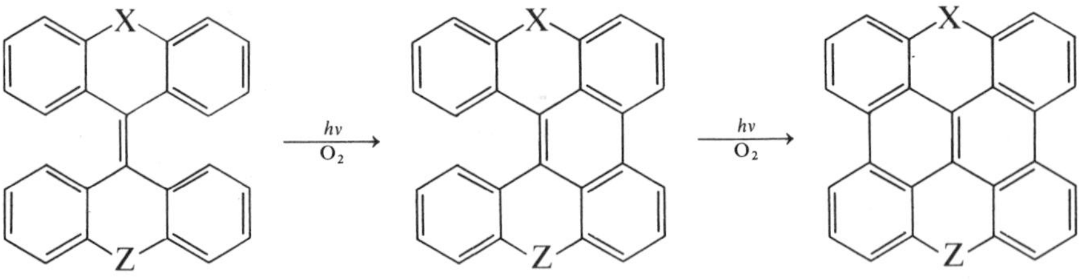

181 → (hν, O_2) → 182 → (hν, O_2) → 183

Examples of bridging groups X and Z include methylene,[138] substituted methylene,[138,139] carbonyl,[140,141a] oxygen,[141b,142,143] and sulfur[141c,142] (Table IV). Presumably the second photocyclization in this sequence, **182** → **183**, takes place because groups X and Z severely limit the twisting of the benzene rings relative to the phenanthrene system in **182**, with the result that the excitation in the S_1 excited state of **182** spreads significantly onto the benzene rings.

Dianthrones, compounds in which both X and Z are carbonyl groups, are the most thoroughly studied compounds of type **181**.[141a, 144–149]

The irradiation of dianthrone (**184**) in polar solvents gives rise to a dihydro intermediate **185** that rapidly undergoes isomerization to the fully aromatic diol **186** by two successive enolizations.[141a]

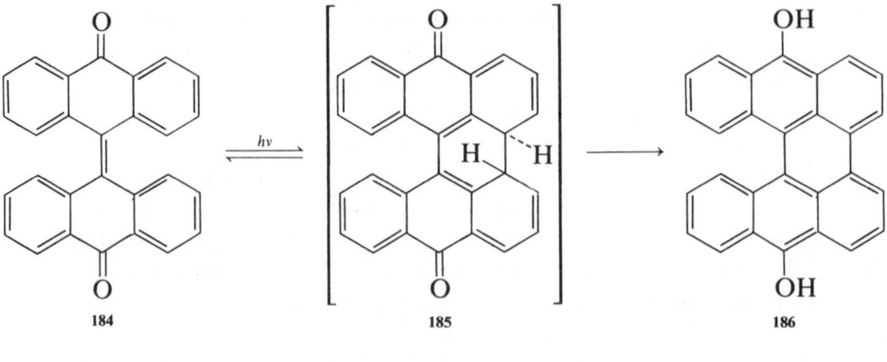

184 ⇌ (hν) [185] → 186

In the absence of dissolved oxygen, this diol **186** is oxidized to helianthrone (**187**) by a hydrogen-transfer reaction in which a second molecule of dianthrone (**184**) is reduced to dianthranol (**188**).[141a]

186 + **184** ⟶ [structure **187**] + [structure **188**]

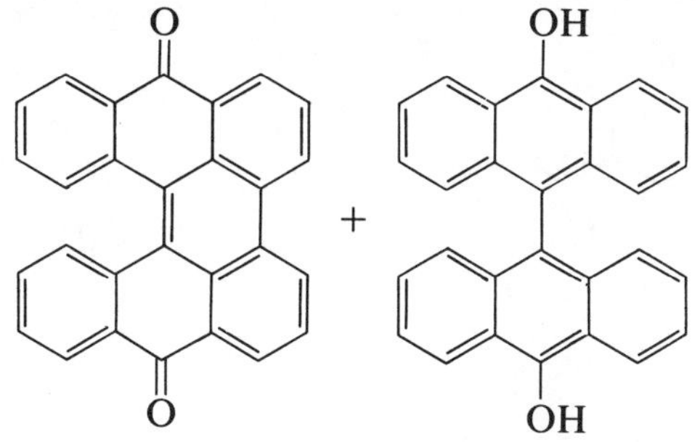

187 188

On further irradiation, helianthrone (**187**) undergoes an analogous photocyclization with dianthrone as oxidant to give mesonaphthodianthrone (**189**) and another equivalent of dianthranol (**188**).

$$\mathbf{187} \xrightarrow[\mathbf{184}]{h\nu} \mathbf{189} + \mathbf{188}$$

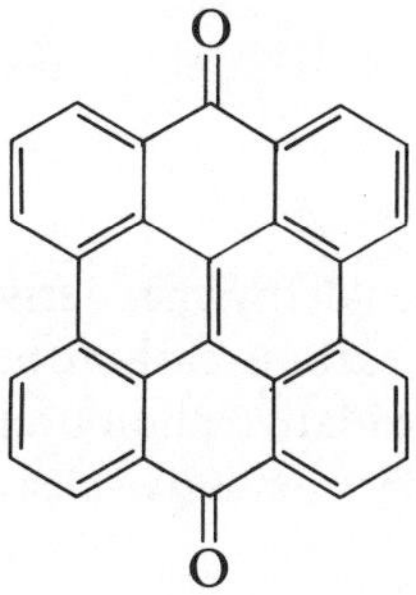

189

Thus the maximum yield of the doubly cyclized product **189** under these conditions is 33%

$$3\ \mathbf{184} \xrightarrow{h\nu} \mathbf{189} + 2\ \mathbf{188}$$

When dianthrone (**184**) is irradiated in air-saturated solutions, nearly quantitative yields of **189** can be obtained; under these conditions, the oxidation of intermediates such as **186** is accomplished by dissolved oxygen rather than by dianthrone.

Diol **190** undergoes oxidative photocyclization with iodine as oxidant to give hydrocarbon **191**.[139] The removal of the two hydroxy groups, the final step in the overall transformation, involves reduction of diol **192** by the hydrogen iodide that is generated in the preceding two oxidative photocyclization steps.

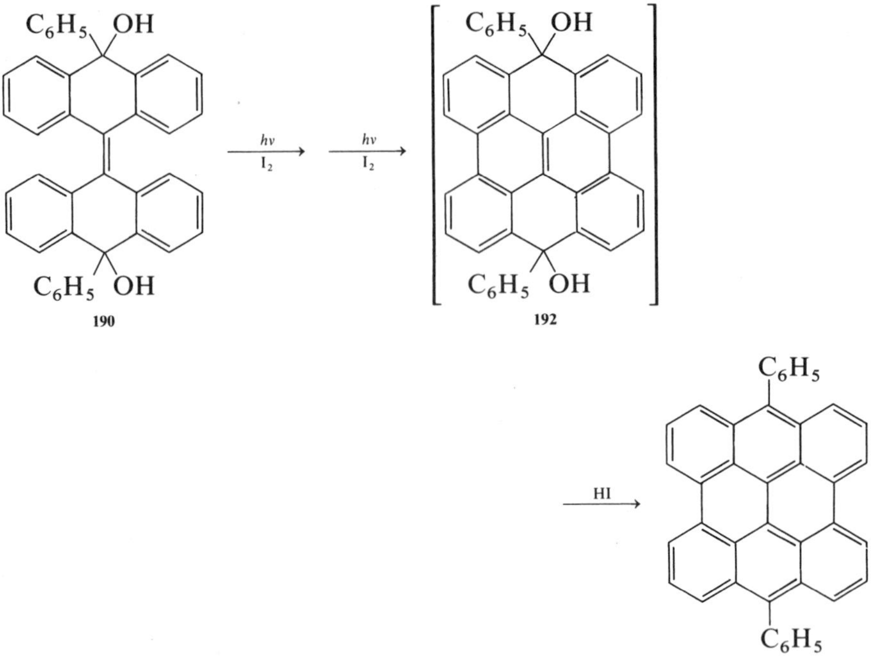

C_6H_5 OH / C_6H_5 OH

190 $\xrightarrow[I_2]{h\nu}$ $\xrightarrow[I_2]{h\nu}$ [192] $\xrightarrow{HI}$ 191 (20%)

Other bridged tetraphenylethylene derivatives in which the bridging is accomplished by a direct carbon–carbon bond instead of an X or Z group generally fail to undergo oxidative photocyclization.[150–152] For example, the fluorene derivative **193** suffers oxidative cleavage.[150]

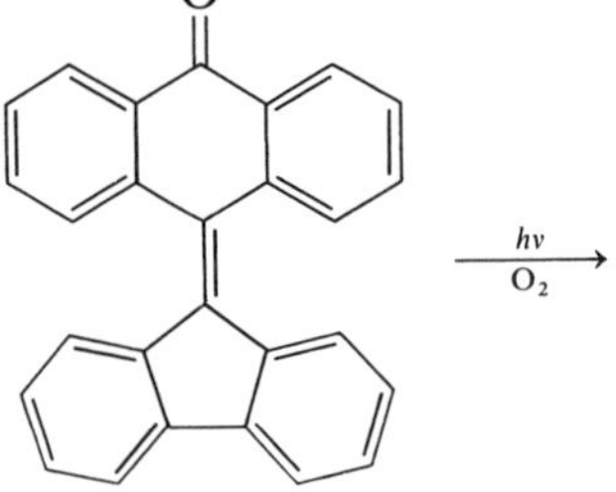

193 $\xrightarrow[O_2]{h\nu}$ 9,10-Anthraquinone (74%) + Fluorenone (77%)

Miscellaneous. The formation of certain chromophoric systems in beech lignin may involve oxidative photocyclization of various stilbene-like moieties.[153a]

The antiestrogenic triphenylethylene derivative **194** (tamoxifen) is used clinically for the treatment of advanced breast cancer. The oxidative photocyclization of **194** to the fluorescent phenanthrene derivative **195** is the basis of a sensitive analytical method for determining the concentration of this drug in a patient's serum.[153b]

C_2H_5 C_6H_5 $OCH_2CH_2N(CH_3)_2$ **194** $\xrightarrow[\text{Air}]{h\nu}$ C_2H_5 C_6H_5 $OCH_2CH_2N(CH_3)_2$ **195**

Photocyclization of stilbene **196** takes place in a multilayer assembly.[154a]

$(CH_2)_{15}CO_2H$ **196** $\xrightarrow[\text{Air}]{h\nu}$ $(CH_2)_{15}CO_2H$

Phenanthrene is formed by UV irradiation of *cis*-stilbene enclathrated in crystalline tri-*o*-thymotide.[154b]

The irradiation of 1,1,1-trichloro-2,2-bis(4-chlorophenyl)ethane (DDT) with 300-nm light gives 3,6,9,10-tetrachlorophenanthrene as a minor product, presumably by way of α,α',4,4′-tetrachlorostilbene.[155a]

4,4′-Dihydroxystilbenes such as diethylstilbestrol (DES) undergo photochemical conversion to isolable diketones through self-trapping of the first-formed enolic dihydrophenanthrenes by ketonization.[39,155b]

C_2H_5 C_2H_5 HO OH $\underset{}{\overset{h\nu}{\rightleftharpoons}}$ [C_2H_5 C_2H_5 H H HO OH] $\longrightarrow$ C_2H_5 C_2H_5 H H O O

Various Carbocyclic Systems from Stilbene Analogs

From Polynuclear Diaryl Olefins. 1,2-Diarylethylenes in which one or both aryl groups are polynuclear undergo oxidative photocyclization to give fused aromatic systems larger than phenanthrene (see Table II), making readily available various molecules that would otherwise be extremely tedious to prepare. For example, 1-styrylnaphthalene (**197**) gives chrysene (**198**),[55] 1,2-di-1-naphthylethylene (**199**) gives picene (**200**),[59,156a,156b] 1-styrylphenanthrene also gives picene (**200**),[157] and 9-styrylphenanthrene (**201**) gives benzo[*g*]chrysene (**202**).[158]

197 $\xrightarrow[I_2]{h\nu}$ **198** (77%)

199 $\xrightarrow[I_2]{h\nu}$ **200** (91%)

201 $\xrightarrow[I_2]{h\nu}$ **202** (50%)

When photocyclization can occur at either of two different ring carbon atoms in a polynuclear system, the regiochemistry is controlled by both electronic and steric factors.[159] In sterically unhindered systems, the product derived from the dihydroaromatic intermediate with the greater aromatic resonance stabilization is preferred. This regioselectivity has also been accounted for by molecular orbital calculations of the differences in excited-state free-valence indices[2, 159–161] and Mulliken electronic overlap populations[75,162] at the two sets of carbon atoms where closure occurs. This preference for the more stable dihydro intermediate can be overriden when the electronically favored pathway is severely hindered sterically and when a sterically unhindered alternative pathway exists that is not prohibitively unfavorable electronically.

The simplest polynuclear system with dual photocyclization pathways is 2-styrylnaphthalene (**203**). As illustrated in Scheme 14, photocyclization proceeds either through conformer **203a** to give dihydrobenzo[*c*]phenanthrene **204**, which has the resonance stabilization of a benzene ring, or through

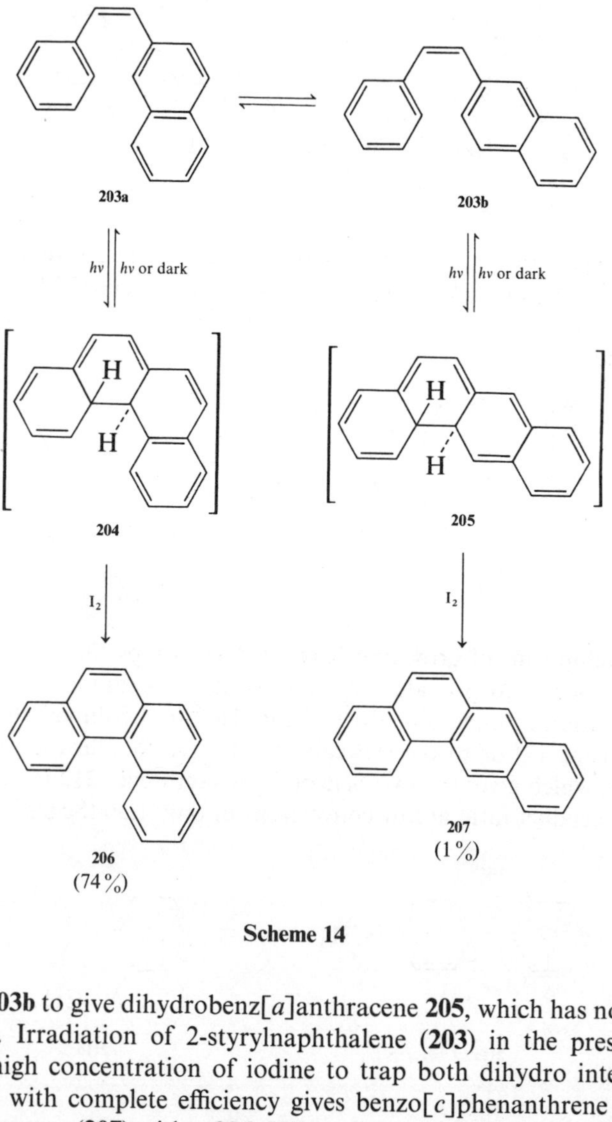

Scheme 14

conformer **203b** to give dihydrobenz[*a*]anthracene **205**, which has no aromatic stabilization. Irradiation of 2-styrylnaphthalene (**203**) in the presence of a sufficiently high concentration of iodine to trap both dihydro intermediates **204** and **205** with complete efficiency gives benzo[*c*]phenanthrene (**206**) and benz[*a*]anthracene (**207**) with a **206**:**207** product ratio of 98.5:1.5.[76,163a]

At lower iodine concentrations, the product ratio becomes even more skewed in favor of benzo[*c*]phenanthrene (**206**),[76,163a] presumably because ring opening of dihydro intermediate **205** to **203b**, which is clearly much more exothermic and hence much faster than the ring opening of the isomeric dihydro intermediate **204** to **203a**, competes with oxidative trapping of intermediate **205** to give product **207**. With oxygen as the trapping agent, the oxidative photocyclization of 2-styrylnaphthalene (**203**) gives benzo[*c*]phenanthrene (**206**) as the only detected product,[76,163a] and the attempted oxidative photocyclization of the conformationally restricted analog **208** fails.[163b]

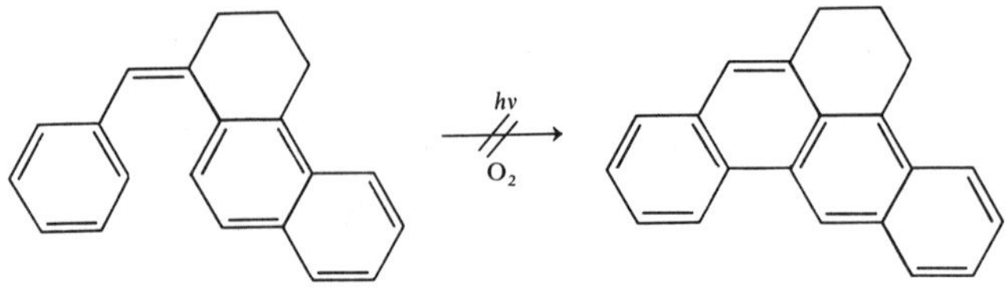

$\xrightarrow[O_2]{h\nu}$ (no reaction)

208

A high degree of regioselectivity also is found in the conversion of the unsymmetrical dinaphthylethylene **209** to benzo[*c*]chrysene (**210**);[163a] similarly, 2-styrylphenanthrene gives benzo[*c*]chrysene (**210**) as the only isolated product in 81% yield.[164]

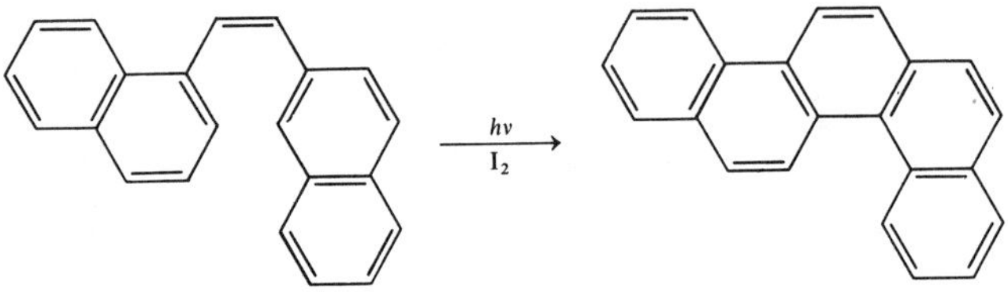

$\xrightarrow[I_2]{h\nu}$

209 **210** (76%)

The development of crowding between CH groups that have a cyclic 1,6 relationship seems to provide only modest steric hindrance to the oxidative photocyclization of *m,m'*-dimethylstilbene (**108**) to 4,5-dimethylphenanthrene (**110c**) (Scheme 12) or to the sterically similar photocyclization of the benzo analog **211**, which gives the two benzo[*c*]phenanthrenes **212a** and **212b** with a **212a**:**212b** product ratio at low conversions of only 1.56 (Scheme 15).[21]

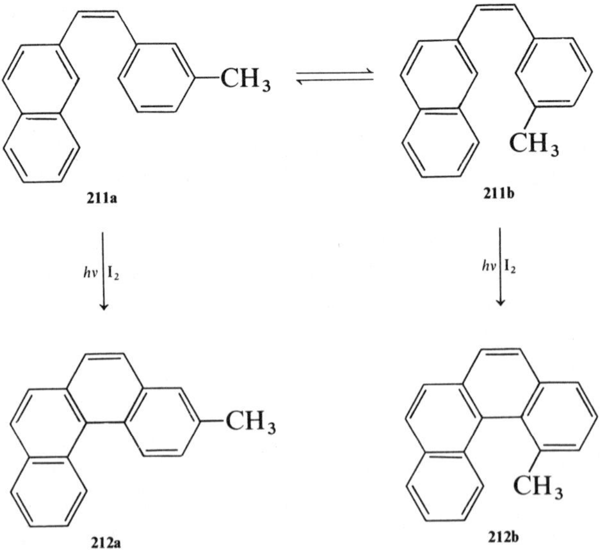

CH_3 ⇌ CH_3

211a **211b**

$h\nu$ | I_2 ↓ $h\nu$ | I_2 ↓

CH_3 CH_3

212a **212b**

Scheme 15

This type of crowding does appear to be important, however, in dinaphthylethylene **213** (Scheme 16). Irradiation of **213** gives 38% of dibenzo[*b,g*]phenanthrene (**214**), the product derived from the electronically disfavored intermediate **215**, along with 25% of dibenzo[*c,g*]phenanthrene (**62**) and 37% of benzo[*ghi*]perylene (**63**). The latter products are derived from the electronically favored intermediate **216**.[76,163a]

213a ⇌ **213b**

213a $\xrightarrow{h\nu}$ **216** ($h\nu$ or dark, reverse)

213b $\xrightarrow{h\nu}$ **215** ($h\nu$ or dark, reverse)

216 $\xrightarrow{I_2}$ **62**

215 $\xrightarrow{I_2}$ **214**

62 $\xrightarrow{h\nu,\ I_2}$ Benzo[*ghi*]perylene (**63**)

$$\frac{\mathbf{62} + \mathbf{63}}{\mathbf{214}} = \frac{62}{38}$$

Scheme 16

Pentaphene (**217**), the product requiring photocyclization of dinaphthylethylene **213** to the dihydro intermediate **218** that is doubly unfavorable electronically, is not obtained.[76]

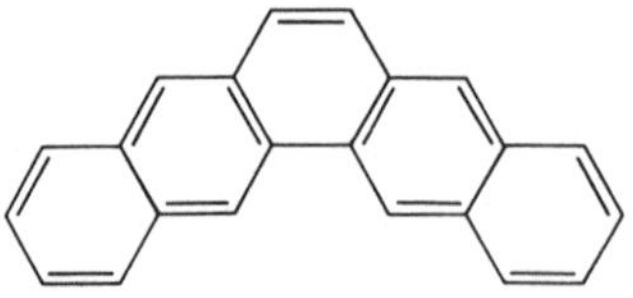

217

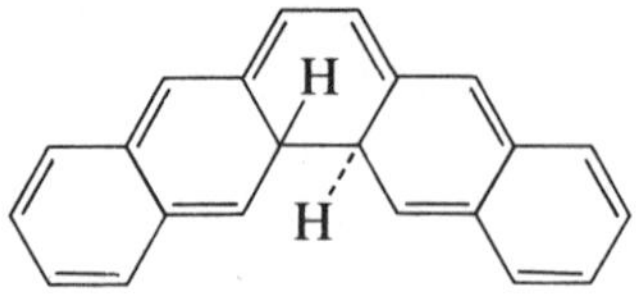

218

Evidently the photocyclization of dinaphthylethylene **213a** to intermediate **216** is sufficiently impeded by the nonbonded steric crowding that develops between the 8 and 8′ positions during that ring closure that the electronically less favored closure of **213b** to **215** becomes competitive.

In the dual-pathway photocyclization of dinaphthylethylene **213**, the product ratio depends on the irradiation conditions. For example, flash spectroscopy reveals that the relative extent to which the two observable pathways are followed is governed by the wavelength distribution of the exciting light and the temperature, with shorter wavelengths and lower temperatures favoring the dihydro[*b,g*]phenanthrene intermediate **215** over the dihydro[*c,g*]phenanthrene intermediate **216**.[27a] Furthermore, the yield of the oxidatively trapped product **214** depends on the nature and the concentration of the oxidant[76] because the ring opening of the higher-energy intermediate **215** is vastly faster than that of the lower-energy intermediate **216**.[27a] In fact, photocyclization of dinaphthylethylene **213** with oxygen as the oxidant (rather than iodine) is reported to give only benzo[*ghi*]perylene (**63**).[165]

Similarly, steric problems in the photocyclization of diarylethylene **219** apparently hinder the electronically favored pathway to [6]helicene (**220**) and allow the alternative pathway to hydrocarbon **221** to compete as illustrated in Scheme 17.[159]

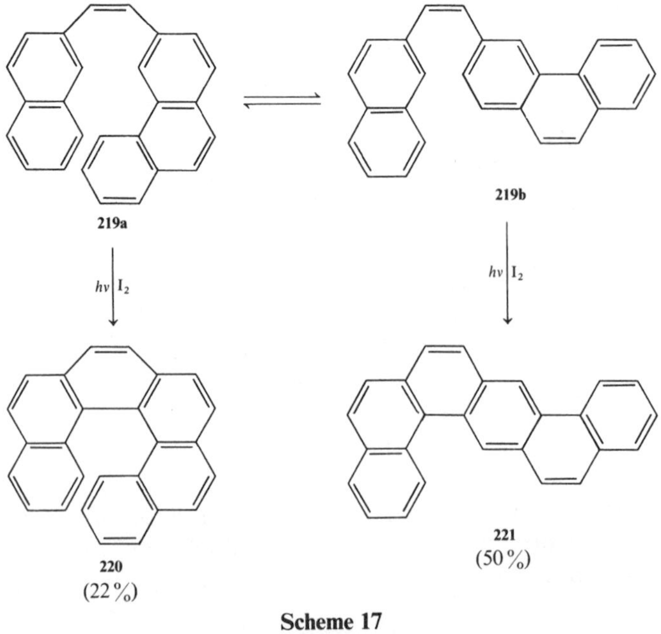

219a

219b

$h\nu$ | I_2

$h\nu$ | I_2

220
(22%)

221
(50%)

Scheme 17

A particularly striking example of a sterically controlled photocyclization is the conversion of the unsymmetrical diphenanthrylethylene **222** to polycyclic hydrocarbon **223** as the only isolated product;[159] the electronically favored pathway leading to helicene **224** apparently does not compete at all (Scheme 18).

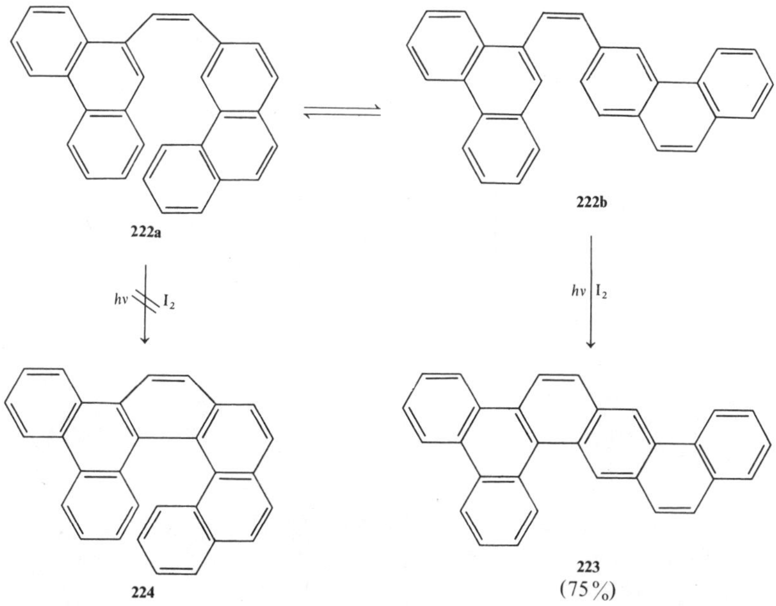

222a ⇌ **222b**

222a —$h\nu$, I_2 (crossed out)→ **224**

222b —$h\nu$, I_2→ **223** (75%)

Scheme 18

In spite of steric hindrance, the oxidative photocyclization of the di-3-phenanthrylethylene **225** proceeds exclusively by way of intermediate **226** to give [7]helicene (**227**) (Scheme 19).[159,166] This apparent anomaly can be explained by noting that the alternative closure of **225b** to give intermediate **228** *also* involves a sterically hindering cyclic 1,6 relationship between the CH groups at the 1 and 5′ positions. A similar explanation applies to the photocyclization of stilbene derivative **229**, which follows the electronically preferred pathway to give [8]helicene (**230**) as the only isolated product.[167,168]

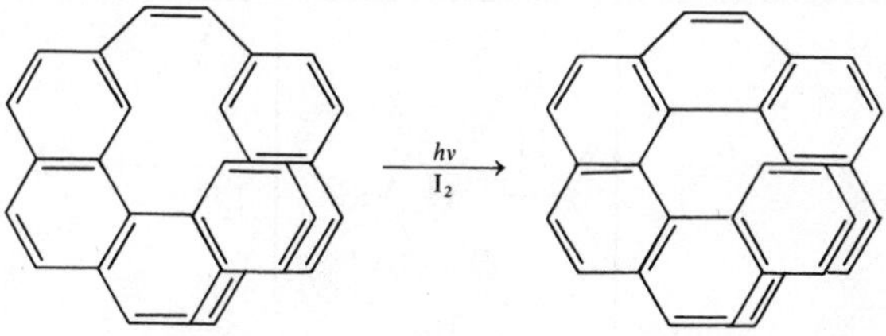

229 —$h\nu$, I_2→ **230** (80–85%)

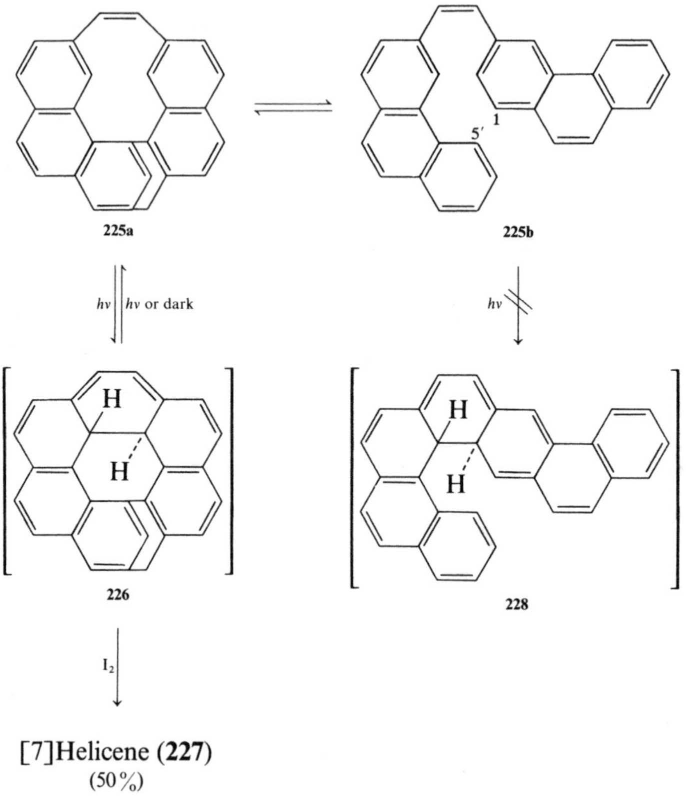

225a ⇌ 225b

hν ⇅ hν or dark

226 — 228

I_2

[7]Helicene (**227**)
(50%)

Scheme 19

Oxidative photocyclization of 3-styrylphenanthrene (**231**) gives benzo[*ghi*]perylene (**63**).[169] This product obviously arises from two successive oxidative photocyclizations, the first converting 3-styrylphenanthrene (**231**) to dibenzo[*c,g*]phenanthrene (**62**) and the second converting **62** to benzo[*ghi*]perylene (**63**). The first photocyclization apparently proceeds exclusively by the electronically preferred pathway, **231** → **232**, even though a cyclic 1,6 relationship develops between the 3′ and 5 positions during this mode of ring closure and a sterically unhindered alternative pathway is available involving closure at C-2 rather than C-4 on the phenanthryl group in **231**.

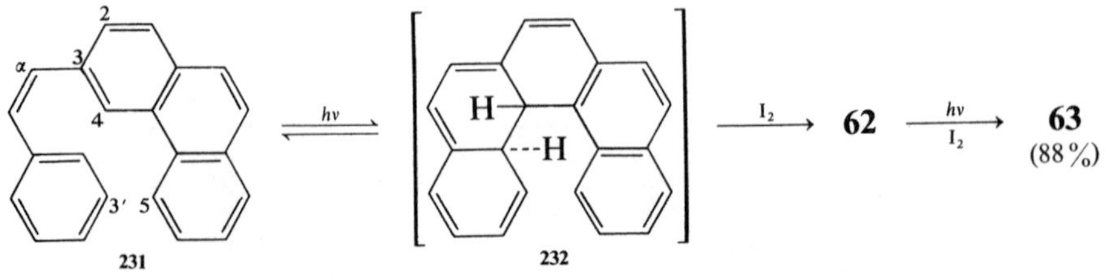

231 $\xrightleftharpoons{h\nu}$ [**232**] $\xrightarrow{I_2}$ **62** $\xrightarrow[I_2]{h\nu}$ **63** (88%)

Perhaps in the excited-state transition state for cyclization of **231** to **232** the phenanthryl group is tilted by rotation around the bond between C-α and C-3 so that the crowding between the 3′ and 5 carbon atoms is smaller than the crowding between the 8′ and 8 carbon atoms that destabilizes the excited-state transition state for cyclization of dinaphthylethylene **213a** to **216** (Scheme 16). The plausibility of this postulated difference in steric crowding in the closure of 3-styrylphenanthrene (**231**) compared to dinaphthylethylene **213a** can be supported by molecular models. The oxidative photocyclization of diphenanthrylethylene **233** is analogous to that of 3-styrylphenanthrene (**231**), giving exclusively products derived from initial photocyclization at C-4 rather than C-2 of the 3-phenanthryl group.[159]

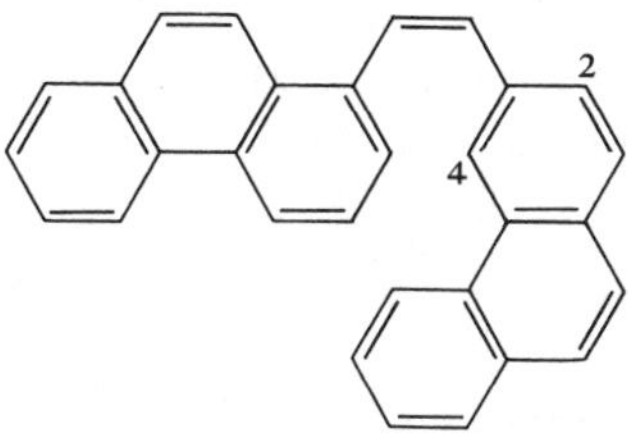

233

The nature of the presumed tilt of the 3-phenanthryl group in the photocyclizations of **231** and **233** is such that an analogous tilt of the 2-benzo[*c*]phenanthryl group in the photocyclization of **234** would place the terminal ring of that benzo[*c*]phenanthryl group even further removed from internal crowding than the terminal ring of the phenanthryl group of **231** or **233**. Thus the closure of **234** at C-1 of the benzo[*c*]phenanthryl group should be no more hindered sterically than the closure of **231** or **233** at C-4 of the phenanthryl group. This accounts for the observation that oxidative photocyclization of **234** gives [6]helicene (**220**) as the only isolated product.[68a,159,168,170]

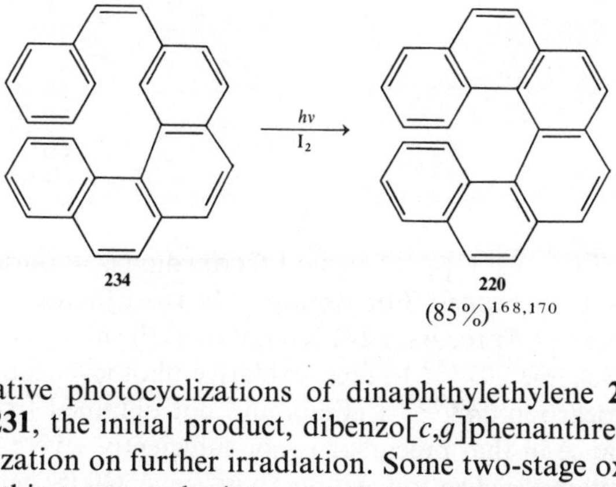

234 $\xrightarrow[I_2]{h\nu}$ **220** (85%)[168,170]

In the oxidative photocyclizations of dinaphthylethylene **213** and styrylphenanthrene **231**, the initial product, dibenzo[*c*,*g*]phenanthrene (**62**), undergoes photocyclization on further irradiation. Some two-stage oxidative photocyclizations of this type can be interrupted and the initially formed product isolated in good yield.[76,171] This requires careful monitoring [e.g., by gas–liquid chromatography (GLC)] to determine when to stop irradiating. It may also be advantageous to use Pyrex-filtered light, because UV light with a wavelength of less than about 300 nm is often more strongly absorbed by the initial product (e.g., **62**) than by the starting stilbene derivative (e.g., **213** or **231**).

Bis-olefin **235** exhibits anomalous photocyclization behavior with no apparent explanation. The ring closure of **235** would be expected to occur primarily at C-2 rather than C-4 of the photocyclizing phenanthryl group by analogy with systems **219** (Scheme 17) and **222** (Scheme 18). Similarly, the photocyclization of the mono-olefin **236** would be expected to occur preferentially at C-2 rather than C-4 of the phenanthryl group. Experimentally, however, the oxidative photocyclization of bis-olefin **235** gives predominantly a mixture of racemic and meso **237**, the products from C-4 closure at each stage; the two isomeric products from C-2 closure at one or both stages each are obtained in only 5% yield.[172] Closure at C-2 is blocked by bromo substituents in the oxidative photocyclization of **238** to **239**.[173]

235 X = H
238 X = Br

$\xrightarrow[I_2]{h\nu}$

236 X = H

$\xrightarrow[I_2]{h\nu}$

237 X = H (70%)
239 X = Br (89%)

Recognition of the two types of cyclic 1,6 crowding is important in planning synthetic routes to helicenes. For example, benzo[6]helicene **240** is readily available from the styryl precursor **241** but not from the more sterically hindered precursor **242** (Scheme 20).[159] In fact, oxidative photocyclization of **242** gives polycyclic hydrocarbon **243**,[159] a compound not obtained from **241** because the steric hindrance in that precursor is not sufficiently severe to preclude the electronically favored cyclization leading to helicene **240** (Scheme 20).

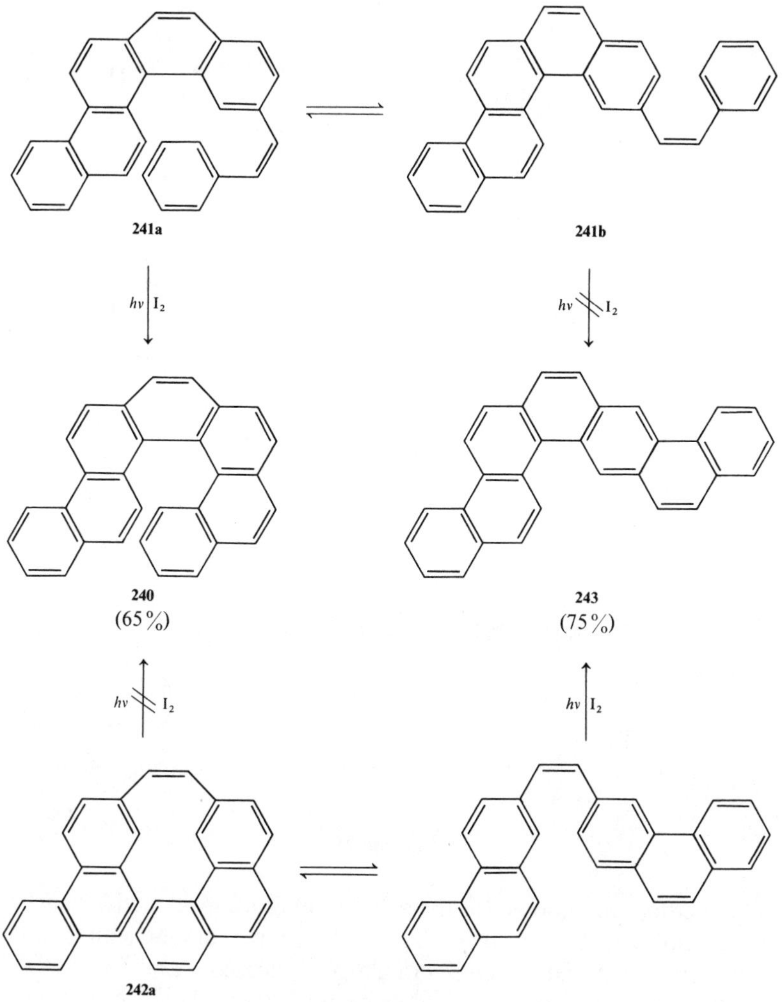

241a **241b**

$h\nu$ I_2

240 (65%) **243** (75%)

$h\nu$ I_2

242a

Scheme 20

In general, photocyclization allows the elaboration of a benzylic methyl group into a new fused aromatic ring by a sequence of three reactions: benzylic bromination with *N*-bromosuccinimide (NBS); a Wittig reaction with an appropriate aromatic aldehyde to give a stilbene derivative; and oxidative photocyclization. The multiple use of this sequence to construct the interesting double helicene **244** is illustrated in Scheme 21.[174]

The extensive use of oxidative photocyclization in the syntheses of various helicenes has been reviewed.[175a] An interesting recent example involves the synthesis of a [7]helicene analog with terminal five-membered rings that serves as a precursor to a novel helical ferrocene.[175b]

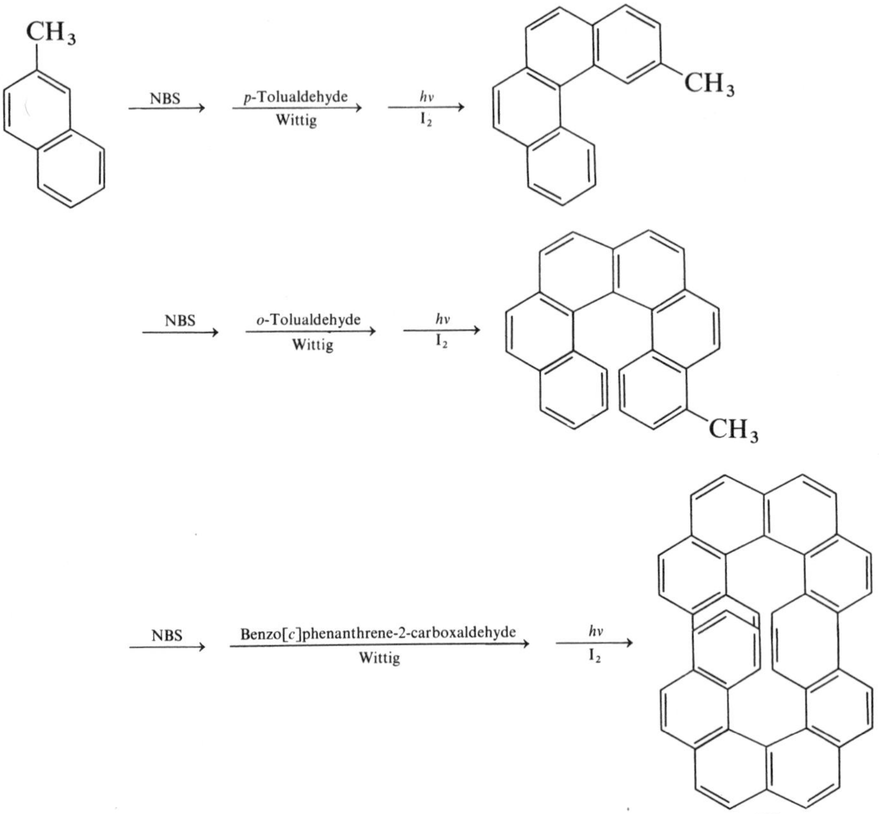

Scheme 21

Three methods are known for the generation of optically active helicenes by oxidative photocyclization: use of stilbene-like precursors with chiral substituents;[176,177] irradiation with circularly polarized light;[40,168,170,178,179] and irradiation in chiral solvents,[179b] chiral liquid crystals,[180a] or mechanically twisted nematic liquid crystals.[180b]

From Distyrylbenzenes and Other Bis-olefins. *m*-Distyrylbenzene (**245**) undergoes oxidative photocyclization normally, giving as the major product benzo[*c*]chrysene (**210**) by way of 2-styrylphenanthrene as an intermediate.[169,181,182]

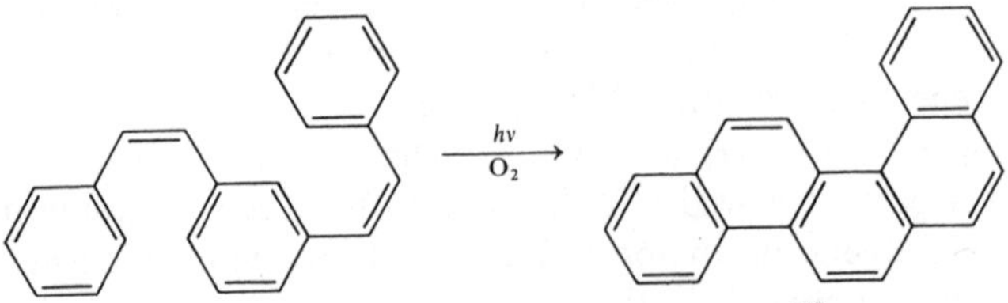

In contrast, *p*-distyrylbenzene (**246**) fails to undergo *direct* oxidative photocyclization to 3-styrylphenanthrene (**231**) but instead gives cyclobutane **247** by photodimerization.[181] Molecular orbital calculations rationalize this failure by revealing a rather low value for the sum of the excited-state free-valence indices at the appropriate *ortho* and *ortho'* carbon atoms in distyrylbenzene **246**.[181] Irradiation of **246** is complicated because photodimer **247** undergoes further photochemical transformation, leading in part to styrylphenanthrene **231**. This *indirect* pathway to 3-styrylphenanthrene (**231**) from *p*-distyrylbenzene (**246**) is illustrated in Scheme 22.[181]

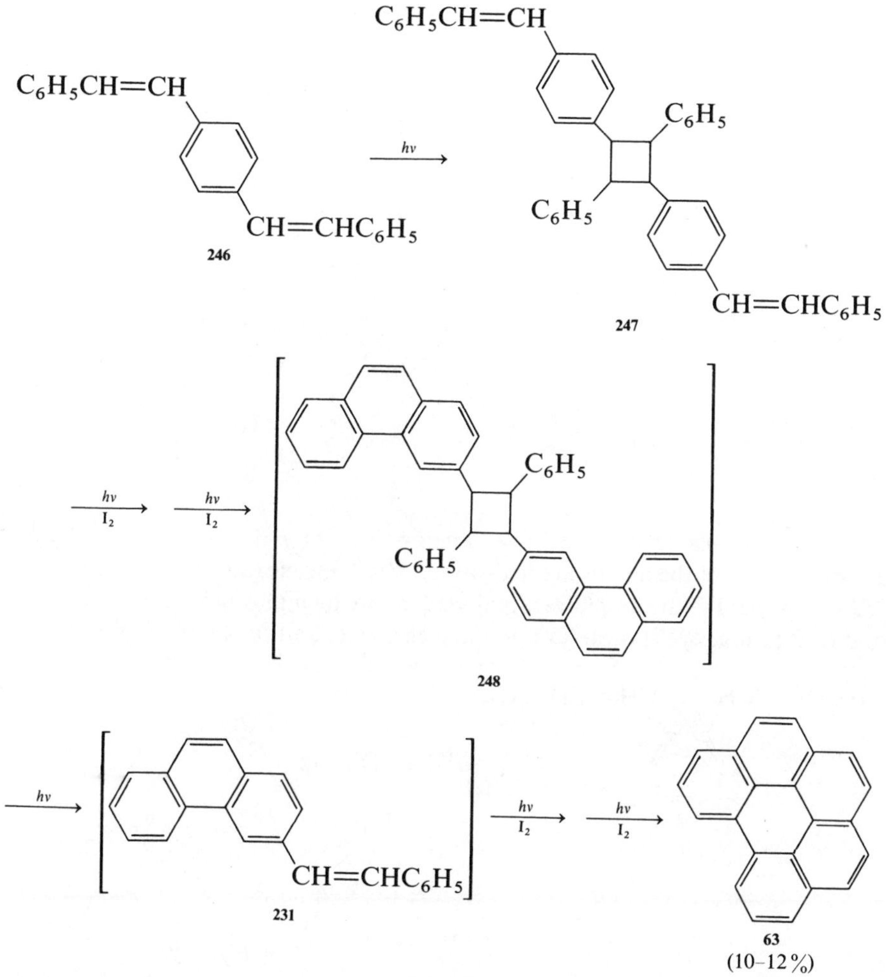

$C_6H_5CH{=}CH$ ... $CH{=}CHC_6H_5$ **246** $\xrightarrow{h\nu}$ **247** $\xrightarrow[I_2]{h\nu}$ $\xrightarrow[I_2]{h\nu}$ [**248**] $\xrightarrow{h\nu}$ [**231**] $\xrightarrow[I_2]{h\nu}$ $\xrightarrow[I_2]{h\nu}$ **63** (10–12%)

Scheme 22

A photodimer of type **247** is, in effect, an alkyl-substituted stilbene and, as such, undergoes two straightforward oxidative photocyclizations to give a phenanthrylcyclobutane derivative of type **248**. Photochemical cleavage of the cyclobutane ring in **248** leads to 3-styrylphenanthrene (**231**). Under the reaction conditions, **231** is further transformed by way of dibenzo[*c,g*]phenanthrene to benzo[*ghi*]perylene (**63**). The overall conversion of *p*-distyrylbenzene (**246**) to benzo[*ghi*]perylene (**63**) requires prolonged irradiation and proceeds in poor yield.[169,181]

The oxidative photocyclization of the dimethyl derivative **249**, which presumably proceeds by way of photodimers by analogy with *p*-distyrylbenzene itself (**246**), gives dibenz[*a,h*]anthracene **250** as the ultimate product in low yield.[183] In the apparent precursor to **250**, the 3-styrylphenanthrene derivative **251**, the normal photocyclization at C-4 on the phenanthrene ring is blocked by a methyl substituent; therefore, photocyclization takes place at C-2.

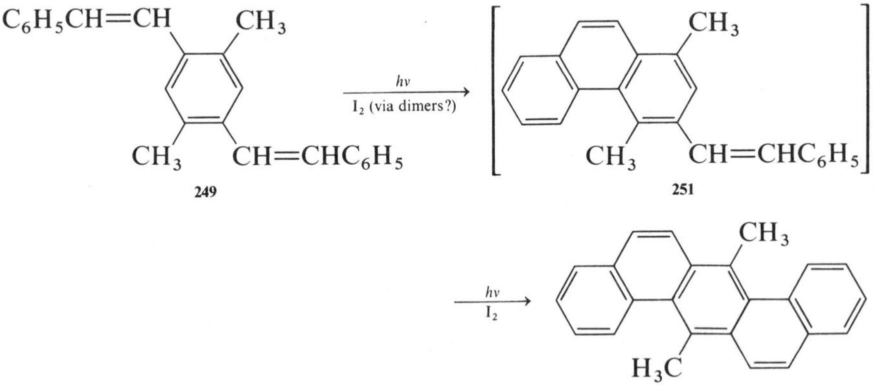

$C_6H_5CH{=}CH$ CH_3 CH_3 $CH{=}CHC_6H_5$

249

$\xrightarrow[I_2 \text{ (via dimers?)}]{h\nu}$

[CH_3 CH_3 $CH{=}CHC_6H_5$]

251

$\xrightarrow[I_2]{h\nu}$

CH_3 H_3C

250
(23%)

o-Distyrylbenzene (**252**) behaves analogously to *p*-distyrylbenzene (**246**), although some of the details are in dispute.[181,184] Picene (**200**) is obtained from **252** in low yield[169] by an indirect pathway involving a photochemical sequence of dimerization, cyclization, cyclobutane cleavage, and further cyclization.[181]

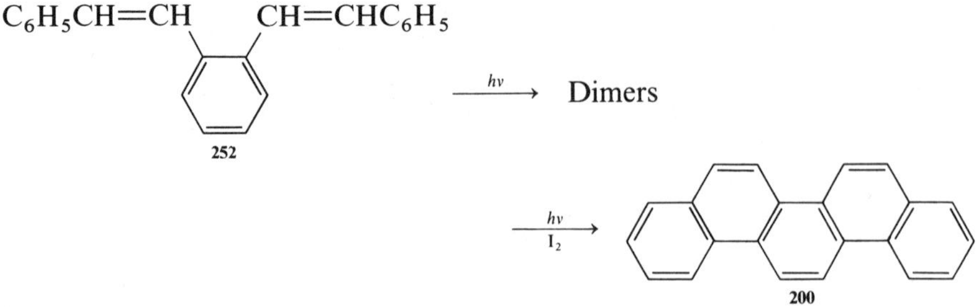

$C_6H_5CH{=}CH$ $CH{=}CHC_6H_5$

252

$\xrightarrow{h\nu}$ Dimers

$\xrightarrow[I_2]{h\nu}$

200
(15%)

Various helicenes, including [6]helicene,[185] [11]helicene,[186] [13]helicene,[187] and [14]helicene,[186] are available in 45–84% yield by oxidative photocyclization of polynuclear analogs of *p*-distyrylbenzene (**246**). The high yields suggest *direct* photocyclization pathways for these particular analogs.

Derivatives of *o*-divinylbenzene usually undergo photoreactions other than electrocyclic conversion to 2,3-dihydronaphthalene derivatives.[188–192] An exception is the transformation of **253** to **254**, presumably by photocyclization of **253** to a 2,3-dihydronaphthalene intermediate followed by a 1,5-hydrogen shift.[193]

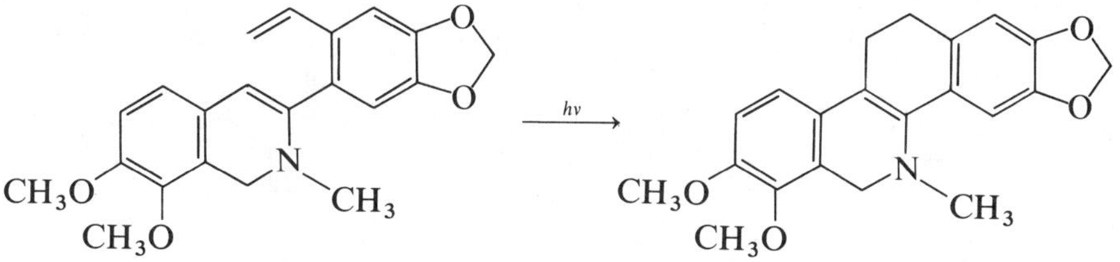

CH_3O CH_3O N CH_3 O O — $h\nu$ → CH_3O CH_3O N CH_3 O O

253 **254** (70%)

Irradiation of metacyclophane **255** in solution under nonoxidizing conditions produces dihydropyrene **256**, an [18]annulene of special interest.[194] On exposure of the resulting green solution to air, dihydropyrene **256** is oxidized quantitatively to pyrene (**257**).[194]

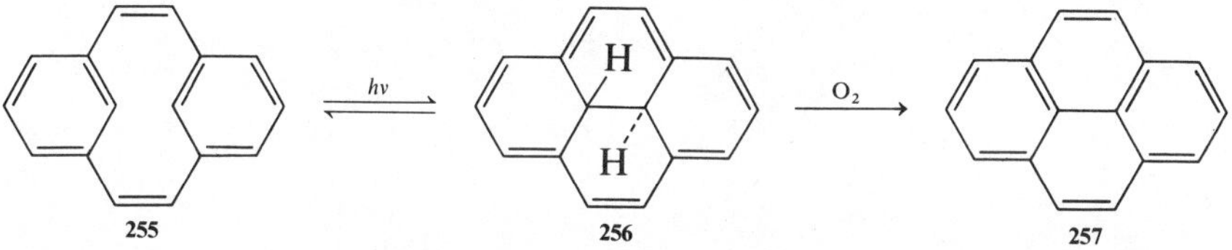

255 $\underset{}{\overset{h\nu}{\rightleftharpoons}}$ **256** (H, H) $\xrightarrow{O_2}$ **257**

Gram amounts of 2,7-dimethylpyrene can be prepared by this method.[195]

2,7-Distyrylnaphthalene (**258**) gives [6]helicene (**220**),[185] and the cyclic analog **259** gives circulene **260**.[196]

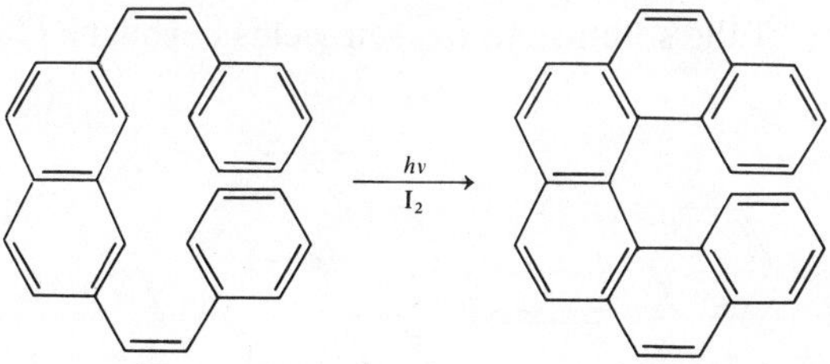

258 $\xrightarrow[I_2]{h\nu}$ **220** (60%)

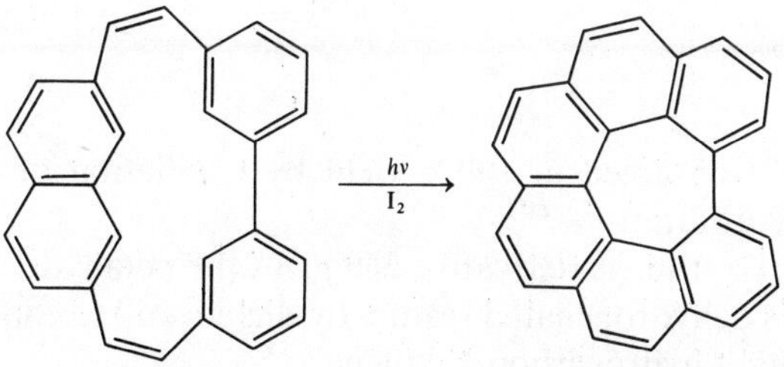

259 $\xrightarrow[I_2]{h\nu}$ **260** (52%)

Attempts to prepare other unusual circulene systems by analogous oxidative photocyclization of cyclophanes **261**,[197] **262**,[198] and **263**[199] have failed.

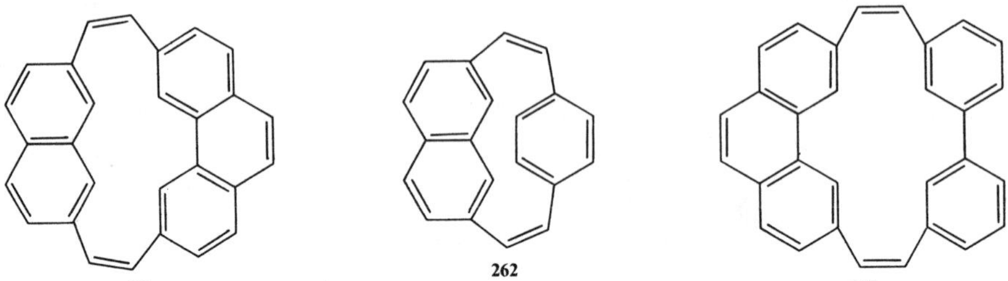

261 **262** **263**

Oxidative photocyclization of diene **264**, however, gives coronene (**265**).[200]

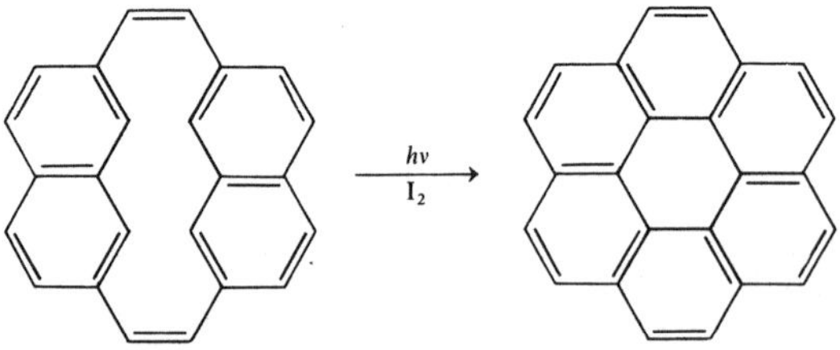

264 $\xrightarrow[I_2]{h\nu}$ **265** (100%)

Irradiation of triene **266** in degassed tetrahydrofuran-d_8 leads to a photostationary mixture containing about 20% of hexahydrocoronene **267**; subsequent exposure of the solution to oxygen yields coronene (**265**).[201]

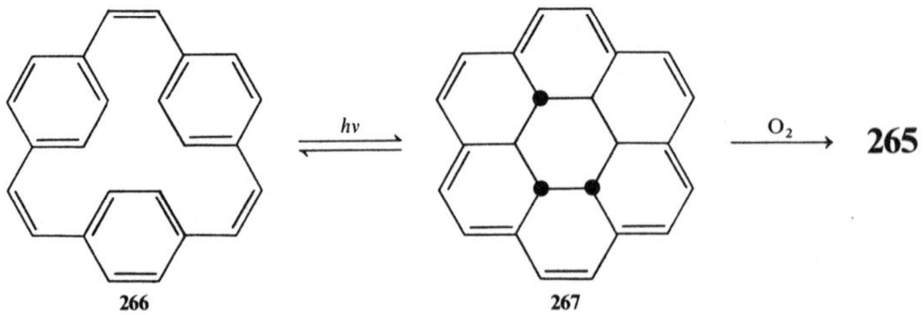

266 $\underset{}{\overset{h\nu}{\rightleftharpoons}}$ **267** $\xrightarrow{O_2}$ **265**

Coronene (**265**) is obtained in 100% yield by irradiation of **266** in oxygen-saturated tetrahydrofuran-d_8.[201]

The macrocyclic stilbene derivative **268** gives the octahydrokekulene **269**, a compound that is dehydrogenated readily by dichlorodicyanobenzoquinone to give the long-sought hydrocarbon kekulene.[202]

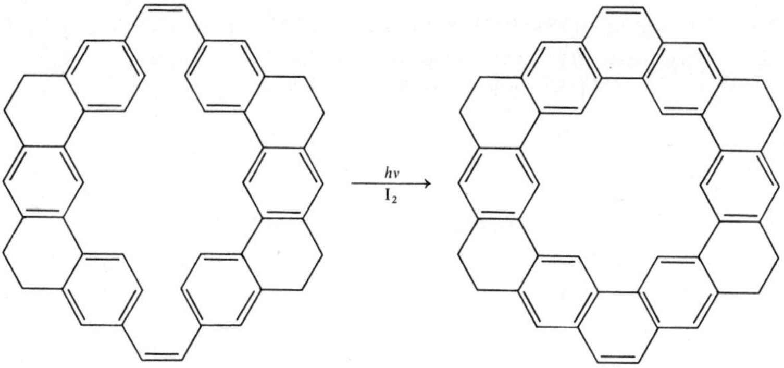

$\xrightarrow[I_2]{h\nu}$

268

269
(70%)

Another striking oxidative photocyclization produces dibenzo[*def,pqr*]tetraphenylene (**270**) from **271**, a stilbene derivative prepared by a double Wittig reaction starting with 3,3′-bis(bromomethyl)biphenyl and 3,3′-biphenyldicarboxaldehyde.[203,204]

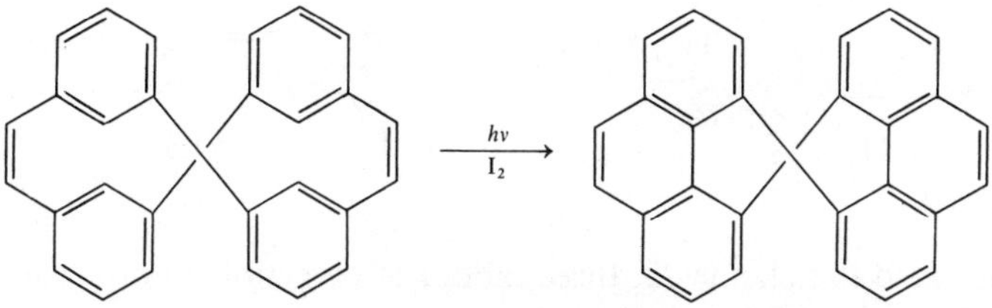

$\xrightarrow[I_2]{h\nu}$

271

270
(63–65%)

From 2-Vinylbiphenyls and 4-Vinylphenanthrenes. The product obtained from the photocyclization of a 2-vinylbiphenyl or a 4-vinylphenanthrene depends on whether the first-formed dihydroaromatic intermediate is allowed to isomerize by a 1,5-hydrogen shift or is trapped by an oxidant. For example, the irradiation of vinylbiphenyl **54** produces either dihydrophenanthrene **55** or phenanthrene **272**.[36,73]

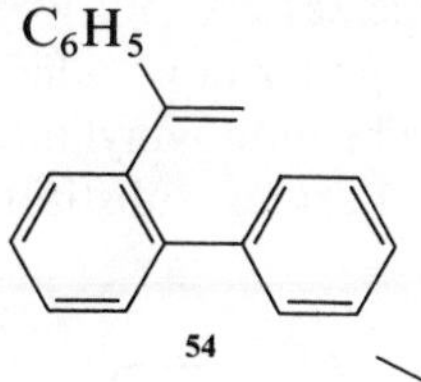

C_6H_5

54 $\xrightarrow[\text{Degassed}]{h\nu}$ 9,10-Dihydro-9-phenylphenanthrene (**55**)
(85–98%)

55 $\not\xrightarrow{h\nu,\ I_2}$ 9-Phenylphenanthrene (**272**)

54 $\xrightarrow[I_2]{h\nu}$ 9-Phenylphenanthrene (**272**)
(80%)

Key intermediates in two multistep syntheses of the natural product juncusol (**273**, R = vinyl) are produced by photocyclization of vinylbiphenyls of type **274** with R = CH_2OH[205] or R = CO_2CH_3.[206]

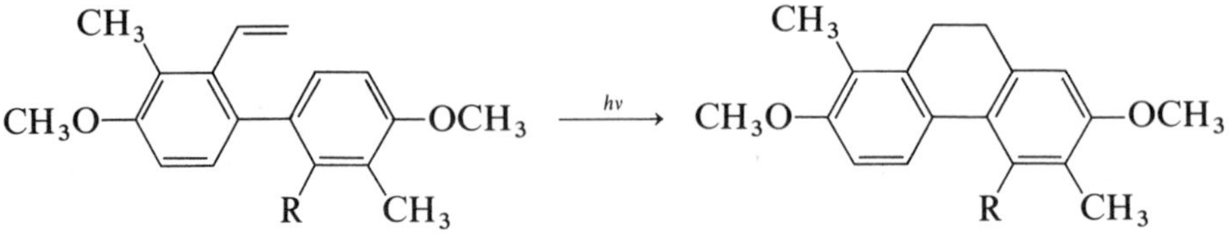

274 → (hν) 273

(65% for R = CH_2OH)
(85% for R = CO_2CH_3)

The conversion of vinylbiphenyl **275** to phenanthrene **276** apparently involves the elimination of water from the dihydroaromatic intermediate.[207]

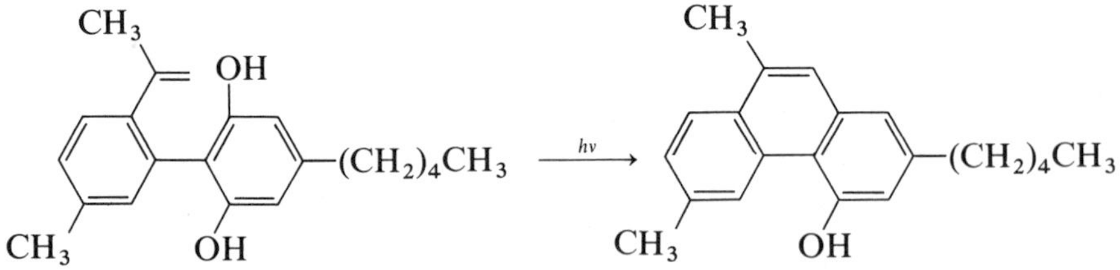

275 → (hν) 276

As illustrated in Scheme 23, three modes of reversible photocyclization are possible for 2-styrylbiphenyl **277**, giving the 8a,9-dihydrophenanthrenes **278** and **279** and the 4a,4b-dihydrophenanthrene **280**. Under nonoxidizing conditions, the only isolable product (80–90% yield) is 9,10-dihydrophenanthrene **281** formed by irreversible 1,5-hydrogen shifts in intermediates **278** and **279**.[208] In the presence of iodine, all the dihydro intermediates are trapped oxidatively, giving rise to 1-phenylphenanthrene (**282**) (up to 64% yield)[107,209] and 9-phenylphenanthrene (**272**).[157,208] Although oxygen traps the 4a,4b-dihydro intermediate **280** to give 1-phenylphenanthrene (**282**), it is not sufficiently reactive to trap the 8a,9-dihydro intermediates **278** and **279** in competition with the 1,5-hydrogen shifts that give 9,10-dihydrophenanthrene **281**.[208]

2,2′-Divinylbiphenyl (**283**) undergoes photocyclization under nonoxidizing conditions to give 4,5,9,10-tetrahydropyrene (**284**).[33,210] The monovinyl intermediate **285** can be isolated after brief irradiation.[33] Various substituted tetrahydropyrenes also are available by this method.[33]

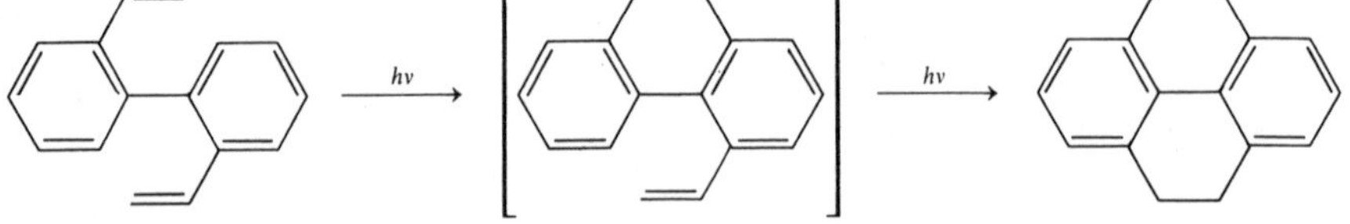

283 → (hν) [285] → (hν) 284 (85–90%)

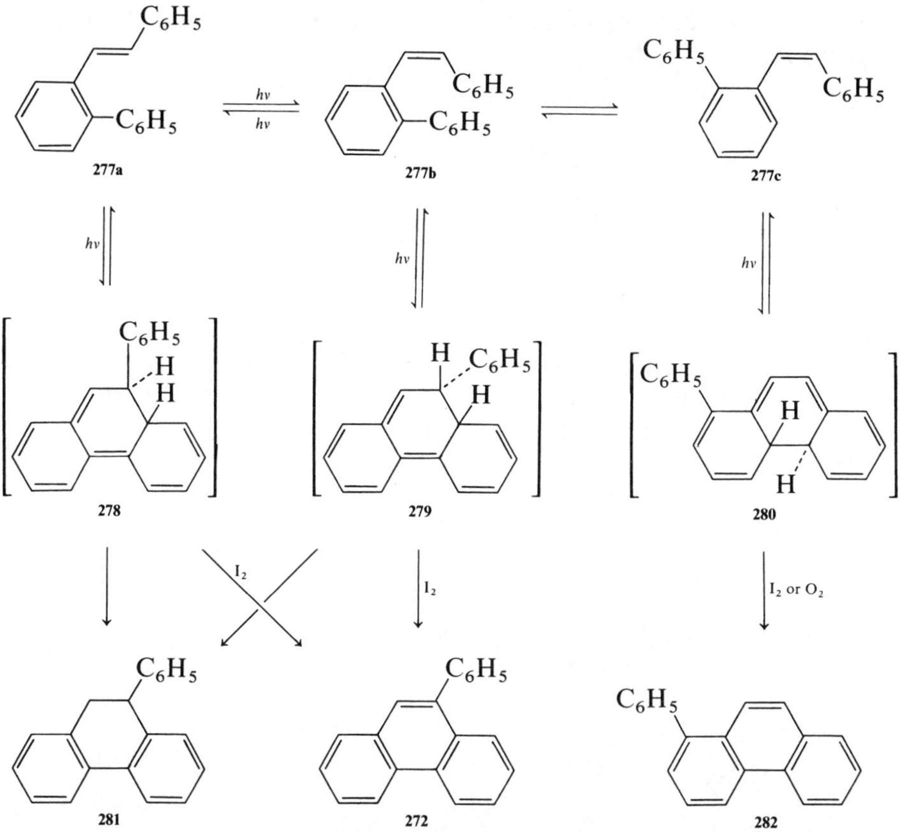

Scheme 23

The vinylbiphenyl **286** undergoes triplet-state photocyclization at two different ring carbon atoms to give a mixture of aldehydes **287** and **288**, the latter by way of a 1,5-acyl shift.[33] Ketone **92** (Scheme 10) behaves analogously.

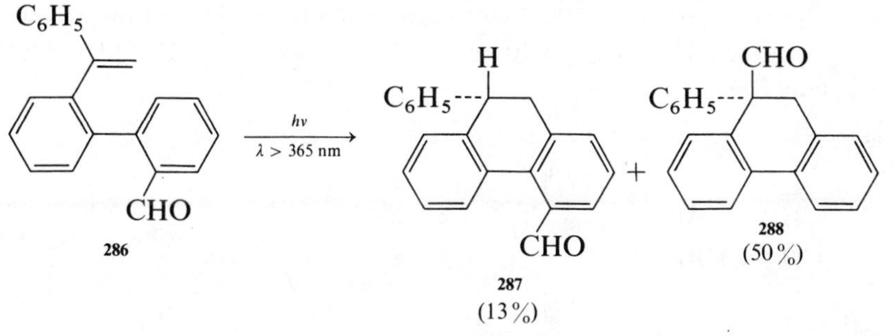

2,2′-Distyrylbiphenyl (**289**) shows particularly varied photochemical behavior. Depending on the irradiation conditions, **289** can give any of three products, each having a different carbon skeleton. Brief irradiation of **289** under nonoxidizing conditions gives predominantly the intramolecular cycloaddition product **290**.[33, 210–212]

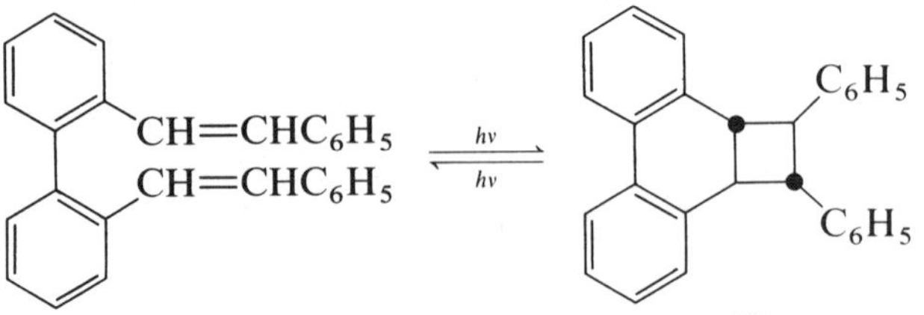

$CH{=}CHC_6H_5$ $CH{=}CHC_6H_5$ $\underset{h\nu}{\overset{h\nu}{\rightleftharpoons}}$ C_6H_5 C_6H_5

289 **290** (68–90%)

With wavelengths of less than about 300 nm, this cycloaddition is photochemically reversible, and prolonged irradiation of either **289** or **290** in the absence of an oxidant leads to the tetrahydropyrene derivative **291**.[33,210,211] Remarkably, irradiation of bis-olefin **289** in the presence of iodine gives benzochrysene **292** rather than the pyrene derivative corresponding to tetrahydropyrene **291**.[211]

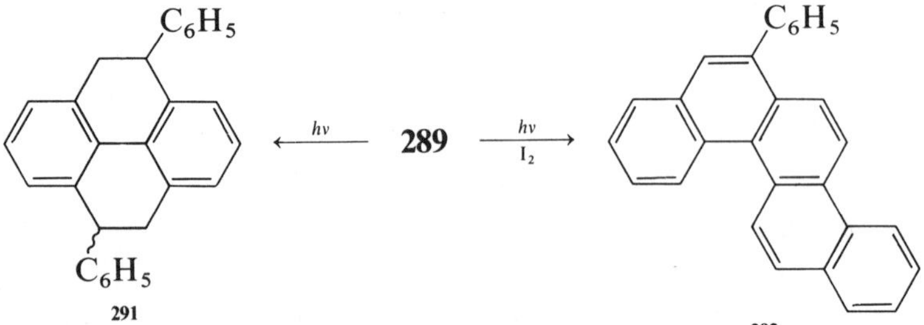

C_6H_5 $\xleftarrow{h\nu}$ **289** $\xrightarrow[I_2]{h\nu}$ C_6H_5

C_6H_5

291 (52–70%) **292** (60%)

As shown in Scheme 24, oxidative photocyclization of **289** gives 4-styrylphenanthrene **293**, and subsequent photocyclization of **293** takes place with aryl–aryl closure to give **294** (two aromatic rings are sacrificed) rather than with aryl–vinyl closure to give **295** (three aromatic rings would be sacrificed). Oxidative trapping of **294** then gives the observed product **292**.

The biphenyl derivative **296** fails to undergo either aryl–aryl or aryl–vinyl photocyclization; irradiation with or without added iodine leads only to cyclobutane **297**.[210]

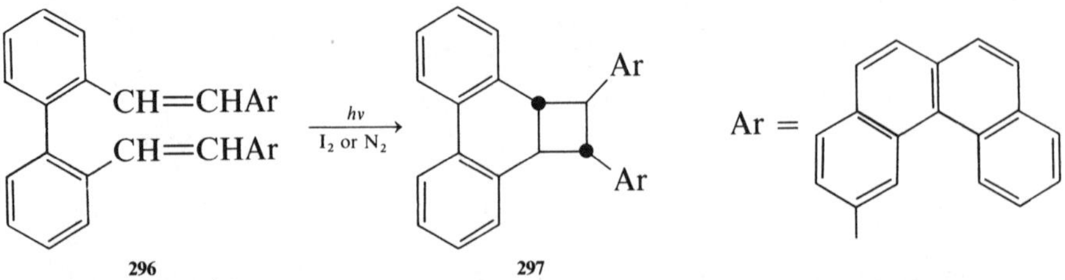

CH=CHAr CH=CHAr $\xrightarrow[I_2 \text{ or } N_2]{h\nu}$ Ar Ar Ar =

296 **297**

4-Styrylphenanthrene (**298**) also exhibits competing aryl–aryl and aryl–vinyl photocyclization pathways, giving benzo[*c*]chrysene (**210**) as the major

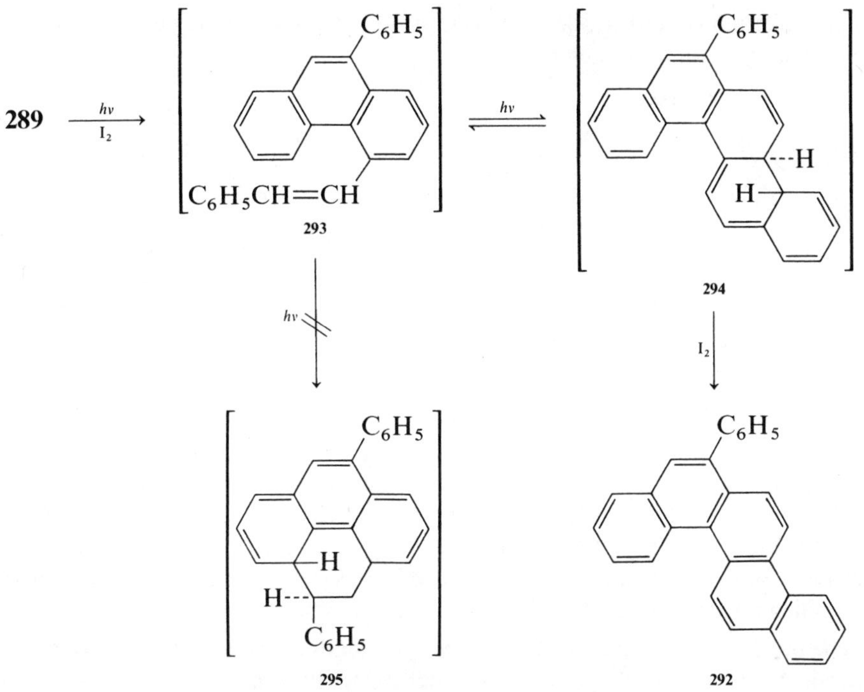

289 $\xrightarrow[I_2]{h\nu}$ [C_6H_5 ... $C_6H_5CH{=}CH$] 293 $\underset{}{\overset{h\nu}{\rightleftharpoons}}$ [C_6H_5 ... H ... H] 294

293 $\xrightarrow{h\nu}$ (not observed) [C_6H_5 ... H ... H ... C_6H_5] 295

294 $\xrightarrow{I_2}$ C_6H_5 292

Scheme 24

product with 4-phenylpyrene (**299**) as a minor product under oxidizing conditions. The yield of **299** is reported as 28% by one group[182] and 2% by another;[181] the reasons for this discrepancy are not clear.

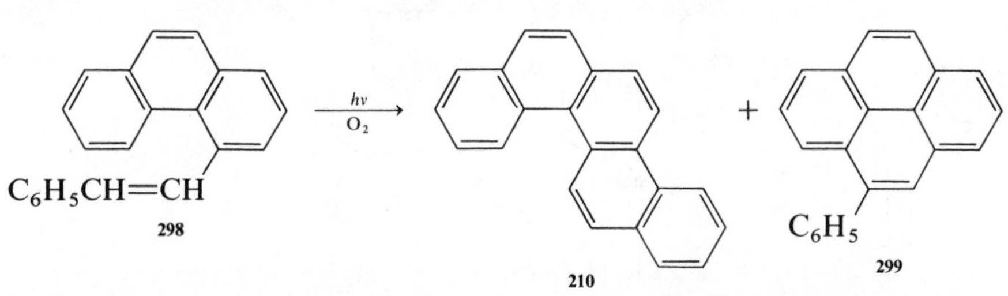

$C_6H_5CH{=}CH$ 298 $\xrightarrow[O_2]{h\nu}$ 210 + C_6H_5 299

Presumably it would be possible to obtain pyrene **299** in good yield by photocyclizing **298** under nonoxidizing conditions to give exclusively dihydropyrene **300** and then treating **300** with an appropriate dehydrogenation reagent.

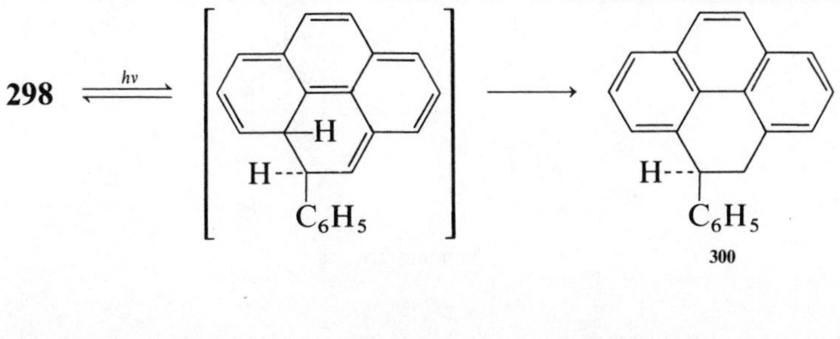

298 $\overset{h\nu}{\rightleftharpoons}$ [H ... H ... C_6H_5] $\longrightarrow$ H ... C_6H_5 300

The conversion of **301** to **302** is an example of the nonoxidative photocyclization of a 4-styrylphenanthrene derivative.[213]

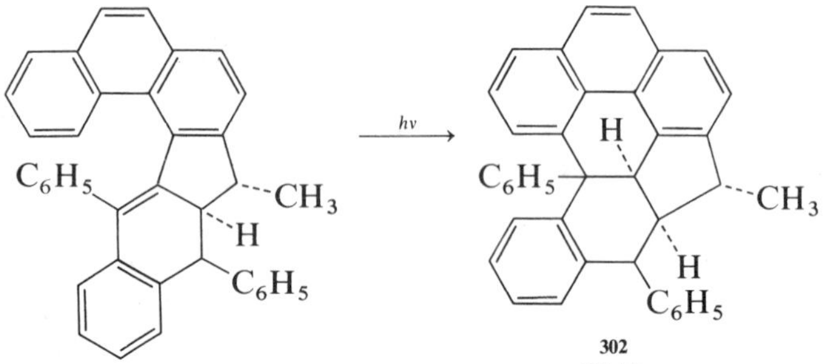

$\xrightarrow{h\nu}$

301

302 (85%)

From Aryl Polyenes. Aryl–vinyl photocyclizations are known for a great variety of aryl-substituted butadienes. For example, simple monosubstituted 1,4-diphenylbutadienes of type **303** (with X = CH_3, CH_3O, CHO, CN, F, or Cl) undergo oxidative photocyclization to give mixtures of products of types **304** and **305** in low yield along with photodimers (Scheme 25).[214] 1,4-Diphenylbutadiene itself gives 1-phenylnaphthalene.[88]

More highly substituted aryl dienes give better yields of photocyclization products,[215–221] perhaps because the competing photodimerizations are sterically hindered. For example, under nonoxidizing conditions diene **306** undergoes photocyclization to dihydro intermediate **307**, with a subsequent

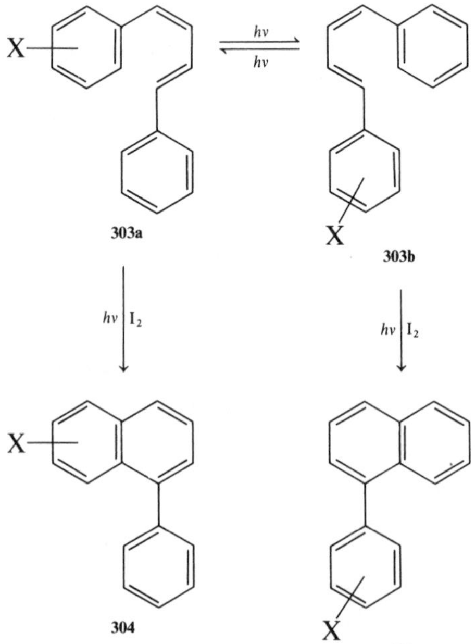

303a **303b**

304 **305**

Scheme 25

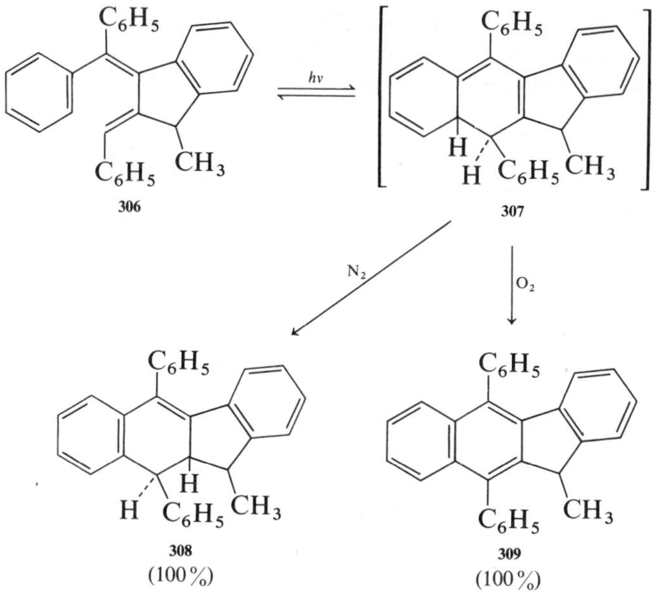

Scheme 26

1,5-hydrogen shift to give dihydronaphthalene **308**; under oxidizing conditions the irradiation of **306** leads to naphthalene **309** (Scheme 26).[218]

Substituent effects are important in the photocyclizations of aryl dienes **310–312** in methanol solution. The dimethoxy compound **310** gives the conjugated ester **313**, whereas the trifluoromethyl compound **311** gives the unconjugated ester **314** (and the corresponding carboxylic acid produced by hydrolysis of **314** during work-up).[222]

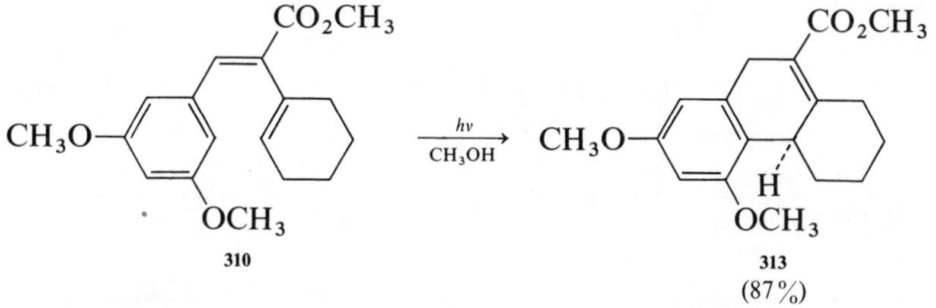

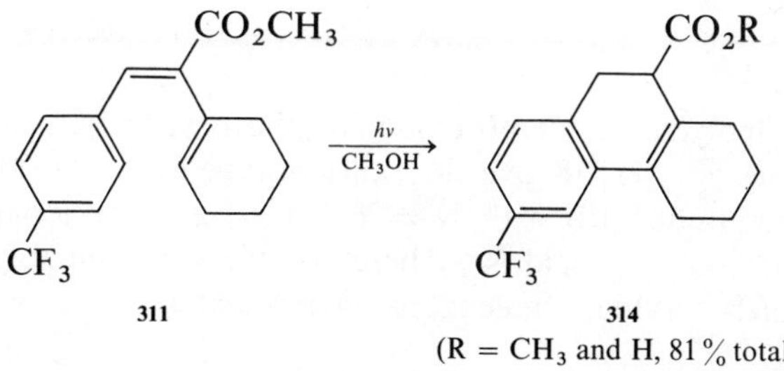

In addition to the unconjugated ester **315**, the elimination product **316** also is obtained from the photocyclization of the dichloro compound **312** in methanol containing potassium carbonate (Scheme 27).[222]

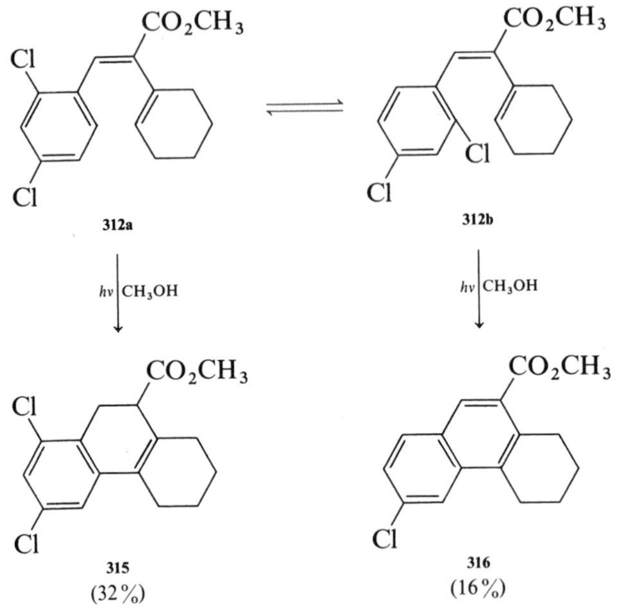

CO_2CH_3 Cl Cl **312a** ⇌ CO_2CH_3 Cl Cl **312b**

hν | CH_3OH ↓

CO_2CH_3 Cl Cl **315** (32%) CO_2CH_3 Cl **316** (16%)

Scheme 27

Remarkably, none of the isolated products from the nonoxidative photocyclizations of esters **310–312** arise from 1,5-hydrogen shifts. These three esters also undergo oxidative photocyclization; for example, ester **310** gives naphthalene **317**.[222]

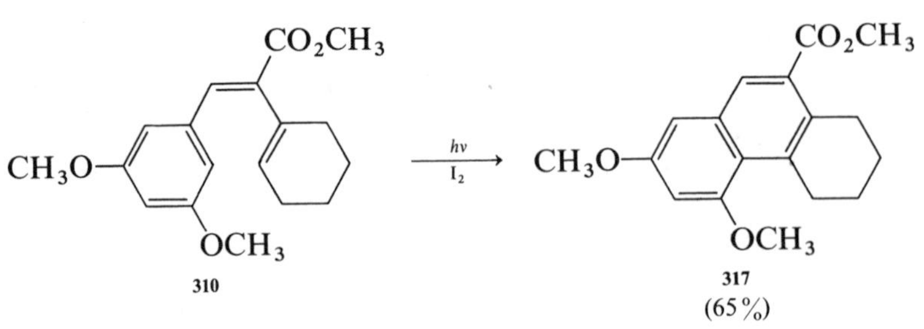

CO_2CH_3 CH_3O OCH_3 **310** $\xrightarrow[I_2]{h\nu}$ CO_2CH_3 CH_3O OCH_3 **317** (65%)

Higher diarylpolyenes undergo oxidative photocyclization in poor yields. A series of trienes of type **318** give chrysenes of type **319** in 11–14% yields;[214] the dicyano compound, **318** with X = Y = CN, is exceptional, giving 3,9-dicyanochrysene in 60% yield.[214] These reactions presumably proceed in stages by way of 1-styrylnaphthalenes as intermediates.[223]

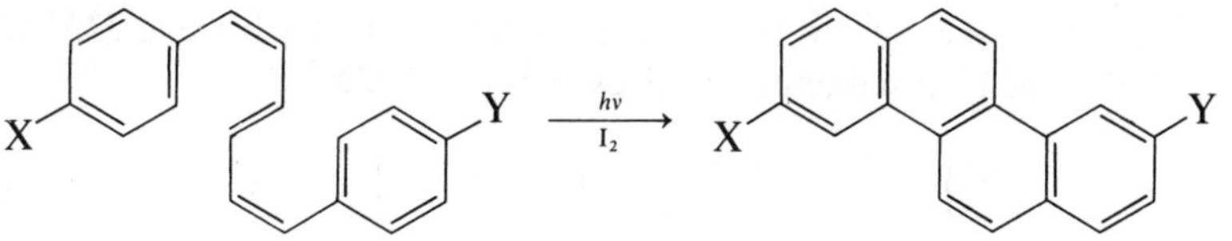

X Y $\xrightarrow[I_2]{h\nu}$ X Y

318 319

Triene **320** photocyclizes differently, giving 1,2,4-triphenylbenzene.[223]

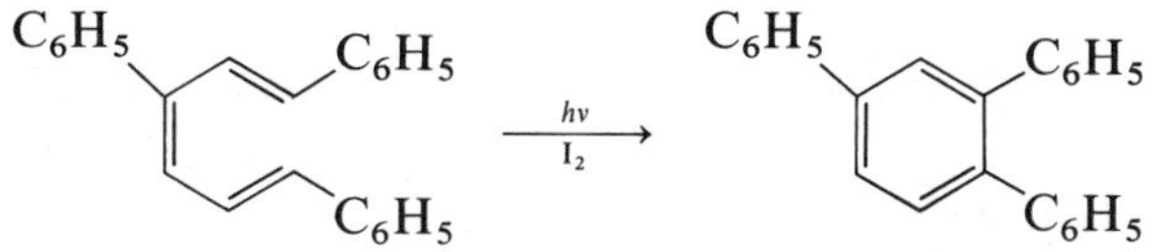

C_6H_5 C_6H_5 C_6H_5 $\xrightarrow[I_2]{h\nu}$ C_6H_5 C_6H_5 C_6H_5

320 (90%)

On irradiation in the presence of iodine, tetraene **321** gives 1-phenylphenanthrene (2%), and pentaene **322** gives picene (2%).[157]

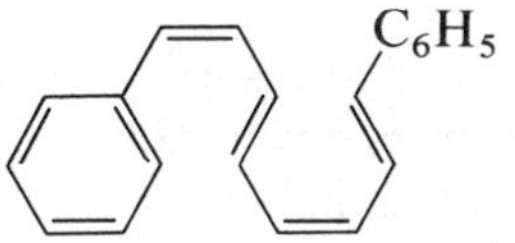

C_6H_5

321 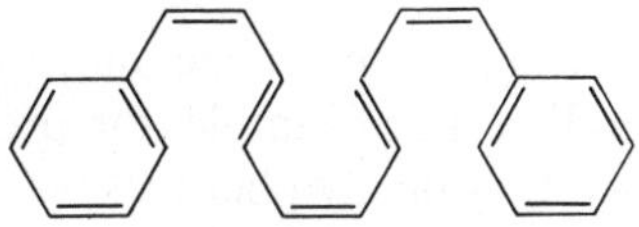 322

The syntheses of chrysenes, phenanthrenes, and picenes are more suitably carried out by high-yield oxidative photocyclizations of the appropriate diarylethylenes.

From Aryl Enynes. 1,4-Diarylbutenynes, readily available from Wittig reactions of $ArCH_2Br$ and Ar′C≡CCHO, undergo photochemically induced cyclization under nonoxidizing conditions; usually the yields are superior to those of the oxidative photocyclization of the corresponding 1,4-diarylbutadienes.[224,225] For example, 4-phenylphenanthrene is obtained in 50–55% yield from enyne **323**[225–227] but in 7% yield from diene **324**.[157]

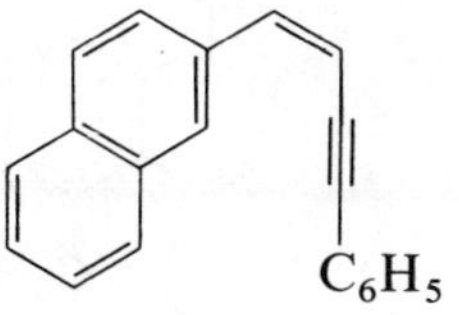

C_6H_5 C_6H_5

323 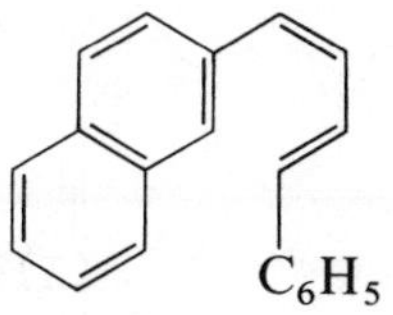 324

Although these enyne and diene photoreactions appear similar, they proceed by quite different mechanisms. The enyne cyclization, as illustrated for enyne **323**, seems best formulated as a free-radical process initiated by hydrogen abstraction from the solvent by the singlet-state excited acetylene **323*** to give radical **325**, which subsequently cyclizes to radical **326**.[225]

323 $\xrightarrow[\text{RH}]{h\nu}$ [**325**] → [**326**]

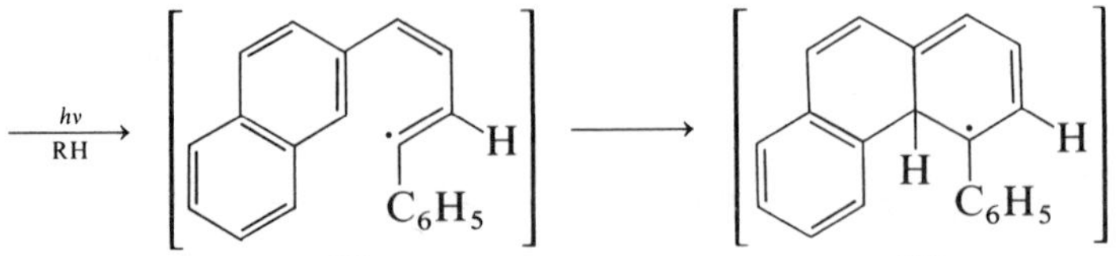

The cyclized radical **326** might be transformed to 4-phenylphenanthrene by transfer of its angular hydrogen to the starting acetylene **323**, thus producing radical **325** and constituting a chain reaction. Alternatively, a molecule of dissolved oxygen could serve as the hydrogen acceptor, reacting with radical **326** to give $HO_2\cdot$ and 4-phenylphenanthrene.

Enyne photocyclization apparently follows an ionic mechanism in methanol involving proton transfer from methanol to the photoexcited acetylene to give a vinyl cation that undergoes intramolecular electrophilic aromatic substitution.[225]

A useful variant of the enyne photoreaction involves irradiation in benzene containing iodine. For example, the product from enyne **323** under these conditions is 3-iodo-4-phenylphenanthrene,[227] presumably formed by way of a chain reaction involving addition of I· to the triple bond.

323 $\xrightarrow[I_2]{h\nu}$ (58%)

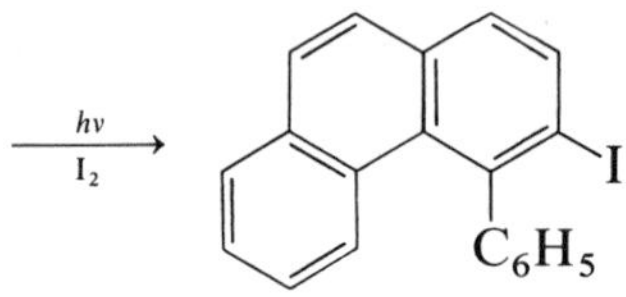

Similarly, enyne **327** gives 1-phenylphenanthrene on irradiation; with added iodine, the product is 2-iodo-1-phenylphenanthrene (50%).[225]

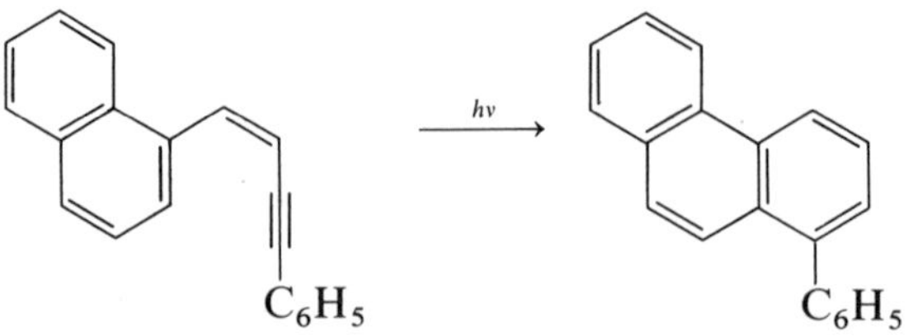

327 $\xrightarrow{h\nu}$ (55%)

Aromatic iodo compounds are susceptible to photolysis.[228] Thus prolonged irradiation of 2-iodo-1-phenylphenanthrene in benzene gives 1,2-diphenylphenanthrene in 88% yield.[225]

The irradiation of enyne **328** in benzene under nitrogen gives the expected 1-phenyltriphenylene (**329**);[229,230] in benzene containing dissolved oxygen, however, enyne **328** undergoes a triplet-state photocyclization to give lactone **330**.[229]

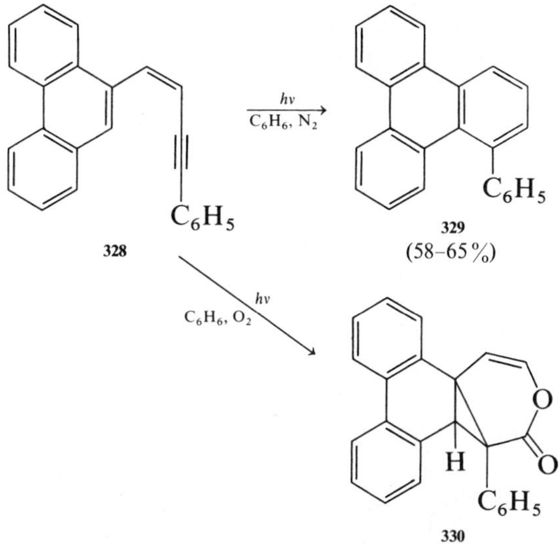

C_6H_5 **328** $\xrightarrow[C_6H_6,\ N_2]{h\nu}$ C_6H_5 **329** (58–65%)

328 $\xrightarrow[C_6H_6,\ O_2]{h\nu}$ H, O, O, C_6H_5 **330** (61%)

The enyne cyclization method is especially valuable for preparing certain highly crowded compounds. For example, enyne **331** gives 4,5-diphenylphenanthrene.[231]

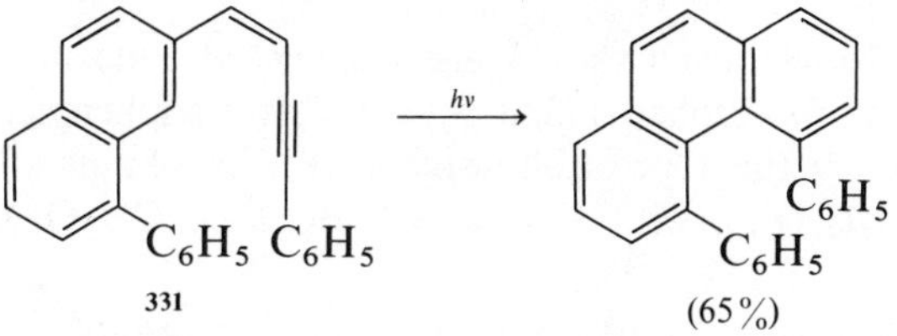

C_6H_5 C_6H_5 **331** $\xrightarrow{h\nu}$ C_6H_5 C_6H_5 (65%)

Apparently there are both upper and lower limits on the size of the aromatic system that can be annelated by the enyne photoreaction. Although phenanthryl enyne **332** cyclizes successfully to give 1-phenylbenzo[*c*]phenanthrene (**333**),[225]

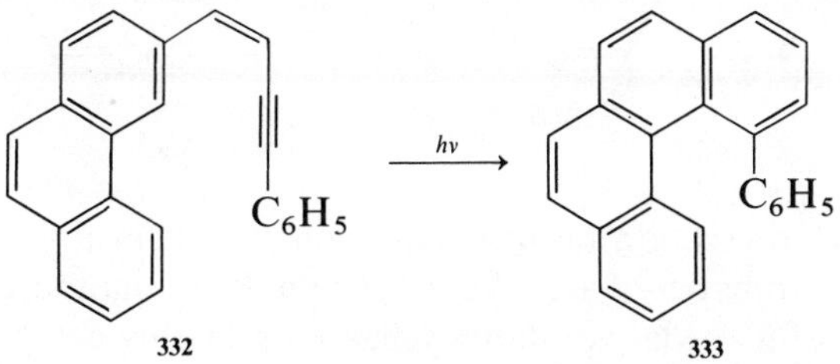

C_6H_5 **332** $\xrightarrow{h\nu}$ C_6H_5 **333** (60%)

the next-higher benzolog **334** fails to yield the corresponding dibenzo[*c*,*g*]phenanthrene derivative, giving instead a mixture of cyclic photodimers under all irradiation conditions examined.[225,232] Furthermore, although the naphthylphenyl compound **327** gives 1-phenylphenanthrene, the isomeric phenylnaphthyl compound **335** fails to give 1,1′-binaphthyl.[225]

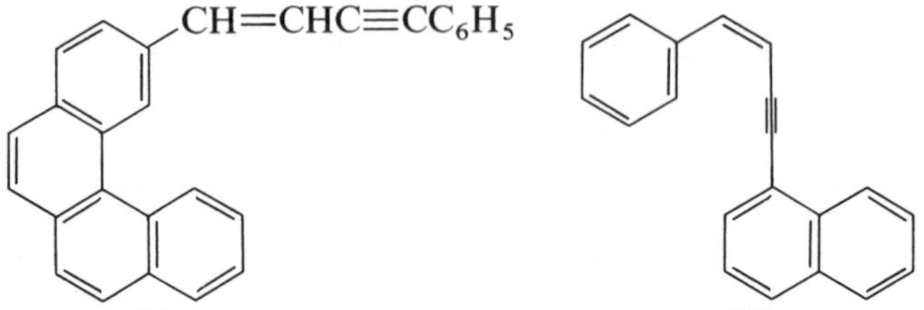

$CH{=}CHC{\equiv}CC_6H_5$

334 **335**

Similarly, the simple 1,4-diphenylbutenyne gives 1-phenylnaphthalene in less than 2% yield even under forcing conditions.[225] Suitably placed methyl substituents evidently enhance the reactivity of a benzene ring toward cyclization, however, as illustrated by the conversion of **336** to **337**.[225,226]

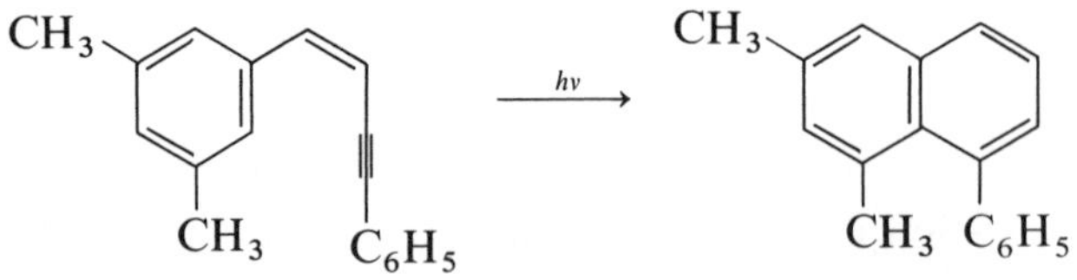

CH_3 CH_3 C_6H_5 $\xrightarrow{h\nu}$ CH_3 CH_3 C_6H_5

336 **337** (44%)

From *o*-Terphenyls and Other Arene Analogs of Stilbene. The oxidative photocyclization of *o*-terphenyl (**338**, X = H) gives triphenylene (**339**, X = H) in 88% yield;[20,233] this type of photoreaction also is known for many substituted *o*-terphenyls (e.g., with X = F, Cl, Br, CN, CH_3O, $CO_2C_2H_5$, and C_6H_5).[20,234,235]

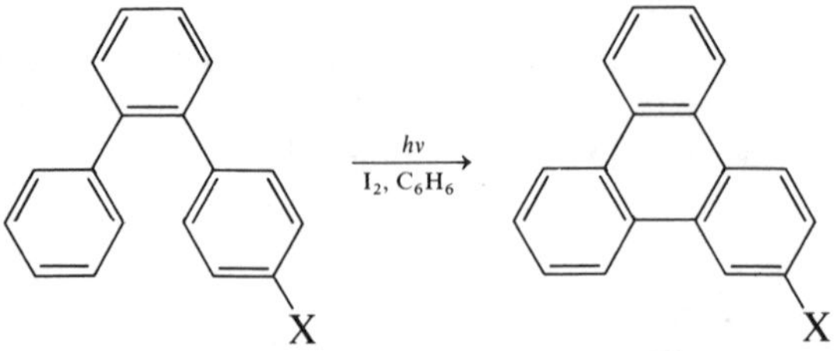

X $\xrightarrow[I_2,\ C_6H_6]{h\nu}$ X

338 **339**

As discussed in the section on mechanism, it is critical to use iodine as the oxidant and an aromatic solvent like benzene in these photocyclizations.

Other successful oxidative photocyclizations in this category include the conversions of 4-phenylphenanthrene to benzo[*e*]pyrene (**340**),[236] naphthylphenanthrene **341** to hydrocarbon **342**,[237] and benzopyrene **343** to hydrocarbon **344**.[231]

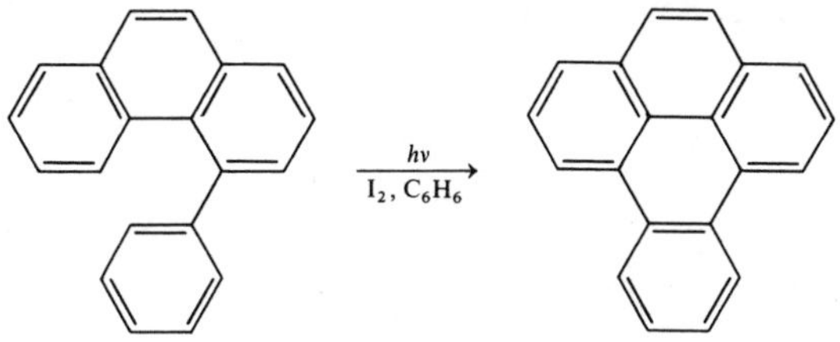

$\xrightarrow[I_2,\ C_6H_6]{h\nu}$

340
(46%)

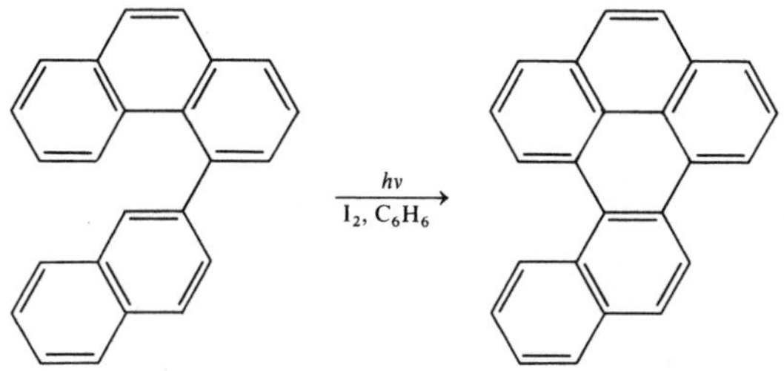

$\xrightarrow[I_2,\ C_6H_6]{h\nu}$

341 **342**
(35%)

$\xrightarrow[I_2,\ C_6H_6]{h\nu}$

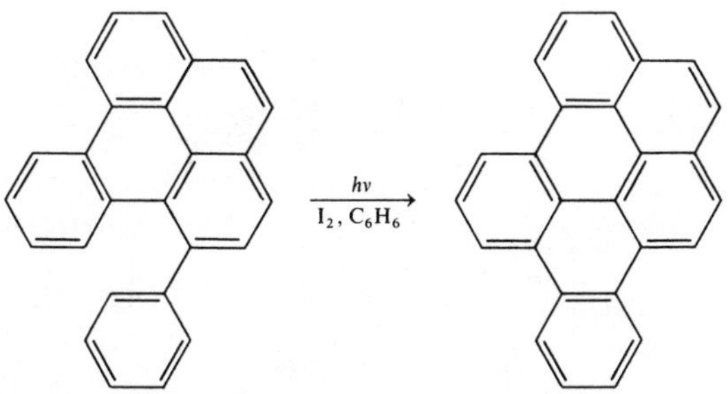

343 **344**
(85%)

In addition, 1,2,3-triphenylbenzene (**345**) is converted to dibenzonaphthacene **346** by two successive oxidative photocyclizations.[20]

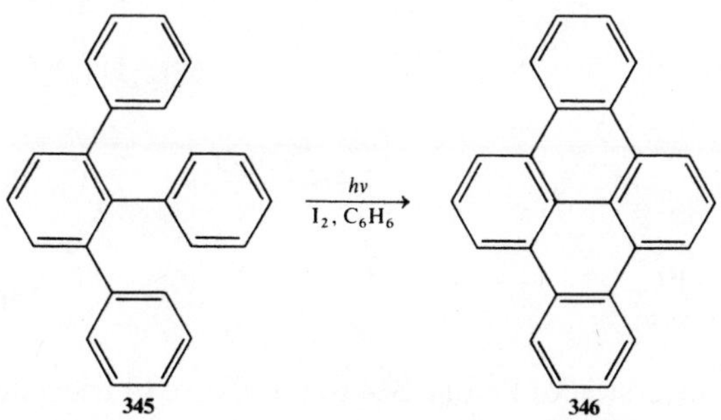

$\xrightarrow[I_2,\ C_6H_6]{h\nu}$

345 **346**
(21%)

Hydrocarbon **346** also is obtained by oxidative photocyclization of 2,2′-diphenylbiphenyl (**347**) (57% yield) and by eliminative photocyclization of dichloro compound **348** under nonoxidizing conditions (67% yield).[20]

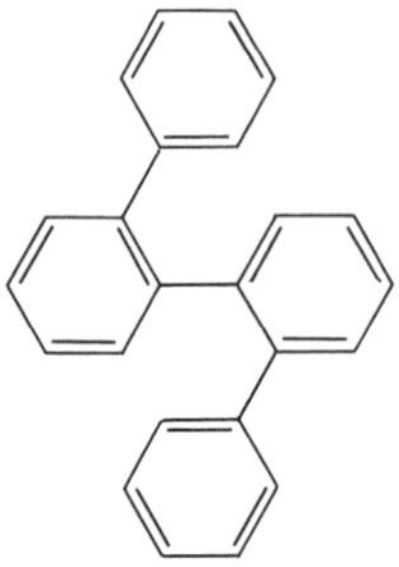

347

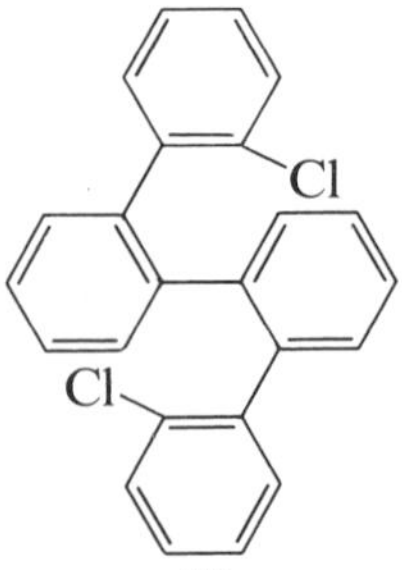

348

From Enolates. On irradiation in ethanol containing sodium ethoxide, ester **349** undergoes photocyclization by way of enolate anion **350**; naphthol **351** is isolated after acidification.[238]

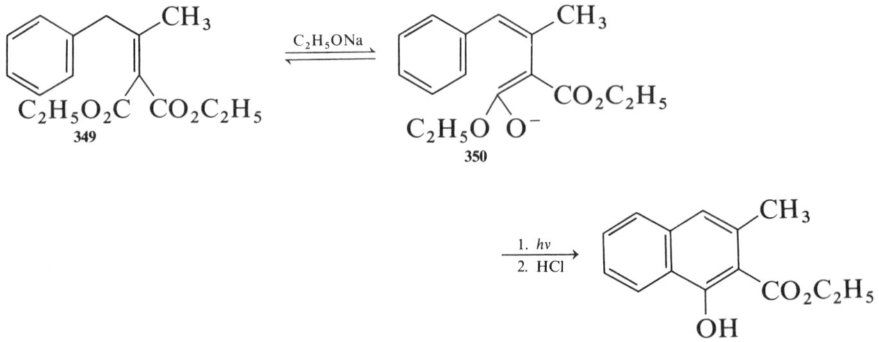

CH_3 $C_2H_5O_2C$ $CO_2C_2H_5$
349

C_2H_5ONa

CH_3 $CO_2C_2H_5$ C_2H_5O O^-
350

1. $h\nu$
2. HCl

CH_3 $CO_2C_2H_5$ OH
351
(43%)

The conversion of ester **352** to phenanthrol **353** also involves photocyclization of an enolate anion.[238]

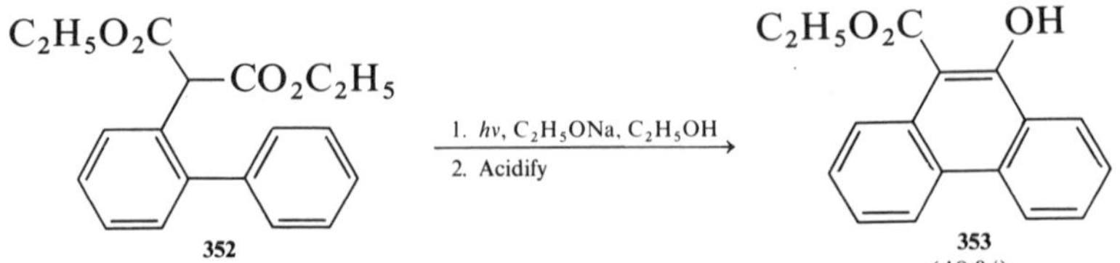

$C_2H_5O_2C$ $CO_2C_2H_5$
352

1. $h\nu$, C_2H_5ONa, C_2H_5OH
2. Acidify

$C_2H_5O_2C$ OH
353
(48%)

However, the conversion of ketone **354** to a mixture of tricyclic epimers **355** by irradiation under acidic conditions apparently proceeds by an intramolecular

Friedel–Crafts alkylation mechanism rather than an electrocyclic mechanism involving the corresponding enol.[239]

354 → ($h\nu$, $BF_3 \cdot (C_2H_5)_2O$) 355 (87%)

Heterocyclic Systems

From Azastilbene Derivatives. Stilbene analogs that have one or more nitrogen atoms in place of ring carbon atoms undergo oxidative photocyclization to give the corresponding azaphenanthrenes. Generally, the yields are lower than those for carbocyclic systems. 2-Azastilbene (**356**) gives 1-azaphenanthrene (**357**).[240,241]

356 → ($h\nu$, O_2) 357 (35%)

Unlike most simple *meta*-substituted stilbenes, 3-azastilbene (**358**) exhibits marked regioselectivity, giving predominantly 2-azaphenanthrene (**359**).[240,242] This regioselectivity is rationalized by molecular orbital calculations.[240,243,244]

358 → ($h\nu$, O_2) 359 (66%)

4-Azastilbene gives 3-azaphenanthrene (21%).[240]

In many azastilbene photocyclizations, iodine is less satisfactory than oxygen as the oxidant,[245] perhaps because charge-transfer complexes of iodine and azastilbenes[246] function as competitive light absorbers or singlet quenchers.

The yields of azaphenanthrenes in these reactions often depend strikingly on the solvent. For example, the yield of 9-cyano-1-azaphenanthrene from the oxidative photocyclization of α'-cyano-2-azastilbene is 70% in benzene, 62% in a 92.5:7.5 mixture of *tert*-butyl alcohol and benzene (preferred over pure benzene because much shorter irradiation times suffice for complete conversion of the azastilbene), 38% in cyclohexane, 30% in acetonitrile, 14% in ethanol, and 0% in 1,2-dimethoxyethane.[241] Irradiation of the azastilbene through a

Corex filter (to remove light of short wavelength) minimizes unwanted polymerization.[241]

The oxidative photocyclization of 1-styrylpyridinium ion (**360**) in ethanol gives cation **361**.[247]

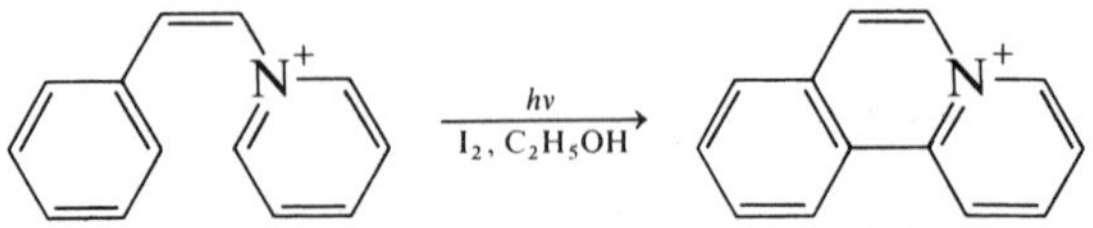

$h\nu$, I_2, C_2H_5OH

360 **361** (60%)

The photochemical conversion of styrylpyridone **362** to the light- and air-sensitive compound **363** in acetonitrile requires hydrochloric acid, suggesting that the photocyclization step involves 2-hydroxy-1-styrylpyridinium ion.[248]

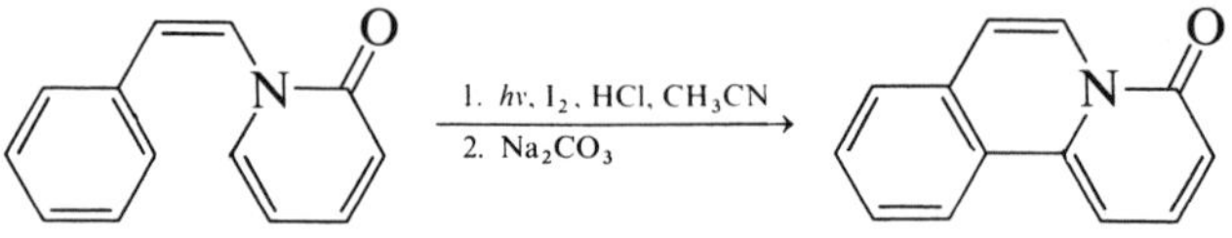

1. $h\nu$, I_2, HCl, CH_3CN
2. Na_2CO_3

362 **363** (25%)

Diazastilbenes give diazaphenanthrenes, as exemplified by the conversion of **364** to **365**.[249]

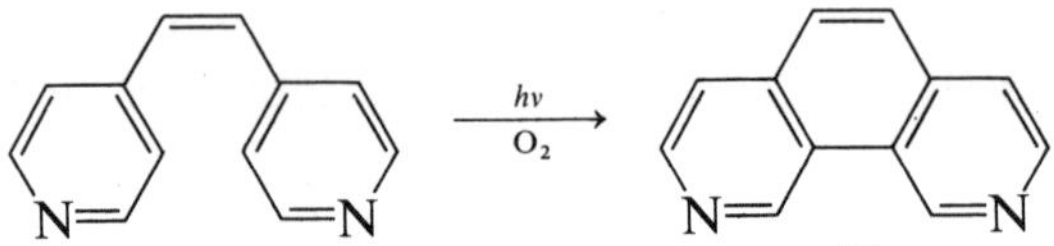

$h\nu$, O_2

364 **365** (47%)

Diazastilbene **366** exhibits the regioselectivity characteristic of 3-azastilbenes, giving predominantly 2,6-diazaphenanthrene (**367**).[249]

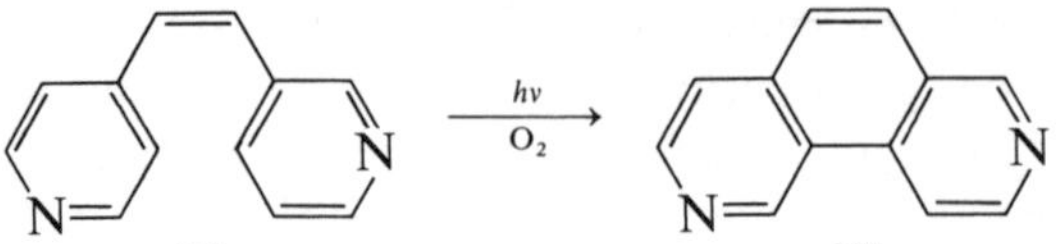

$h\nu$, O_2

366 **367** (40%)

3,3′-Diazastilbene gives a mixture of 2,5-diazaphenanthrene (20%), 2,7-diazaphenanthrene (12%), and 4,5-diazaphenanthrene (1%).[249] Molecular orbital calculations indicate that electrostatic repulsion between electron-rich nitrogen atoms causes the observed regioselectivity in the photocyclization of diazastilbenes.[250]

Diazastilbenes with two adjacent ring nitrogen atoms fail to undergo photocyclization under the usual conditions. In most azastilbenes and diazastilbenes, the lowest $^1n,\pi^*$ and $^1\pi,\pi^*$ excited states are sufficiently similar in energy[251–256] that $^1\pi,\pi^*$ photocyclization competes with the decay pathways characteristic of the $^1n,\pi^*$ state. But in a diaza compound such as **368**, for example, interaction between the adjacent lone pairs causes the lowest $^1n,\pi^*$ state to lie significantly lower in energy than the stilbene-like $^1\pi,\pi^*$ state,[253] and photocyclization fails. As with the azobenzenes, this difficulty can be circumvented by irradiating in sulfuric acid.[246]

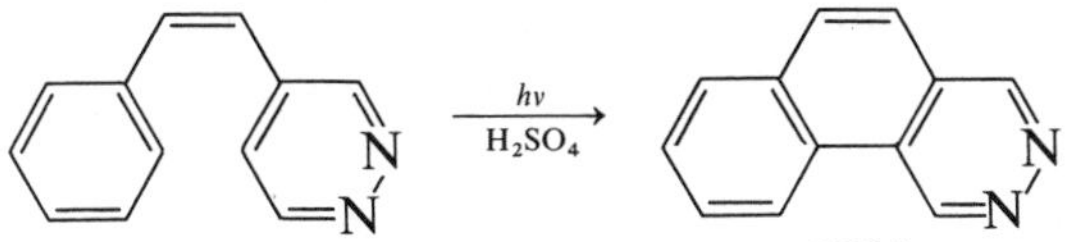

368 $\xrightarrow[H_2SO_4]{h\nu}$ (59%)

Under these conditions, the diazastilbenes are protonated, and the resulting cations have $^1\pi,\pi^*$ lowest excited singlet states that are capable of photocyclization.

Oxidative photocyclization has been used to synthesize many heteroaromatic systems larger than azaphenanthrenes;[257–262] an example is the conversion of **369** to **370**.[257]

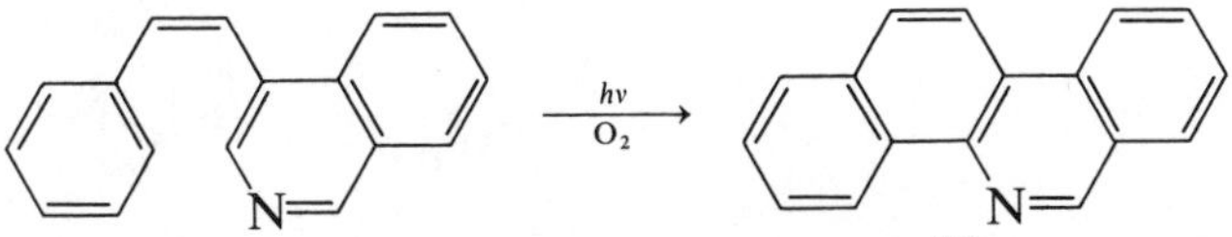

369 $\xrightarrow[O_2]{h\nu}$ **370** (43%)

Inexplicably, although tetramethoxy compound **371** undergoes oxidative photocyclization to give azachrysene **372**,[263] the closely related compounds **373** and **374** fail to give the analogous photocyclization products under the same conditions.[264]

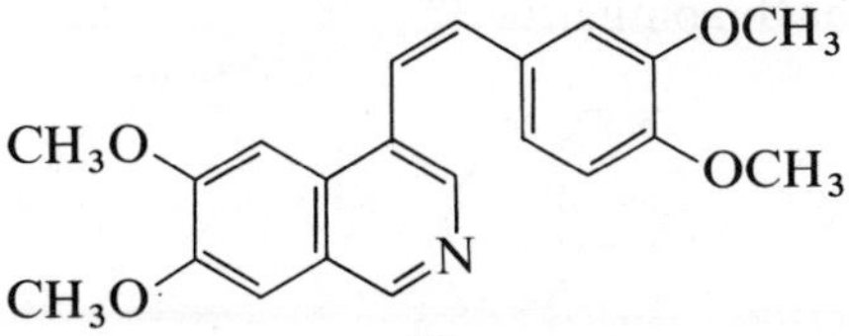

371 (OCH₃, OCH₃, CH₃O, CH₃O, N)

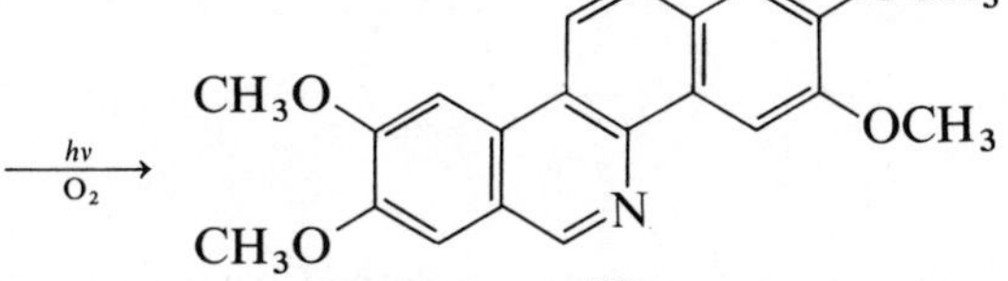

$\xrightarrow[O_2]{h\nu}$ **372** (27%)

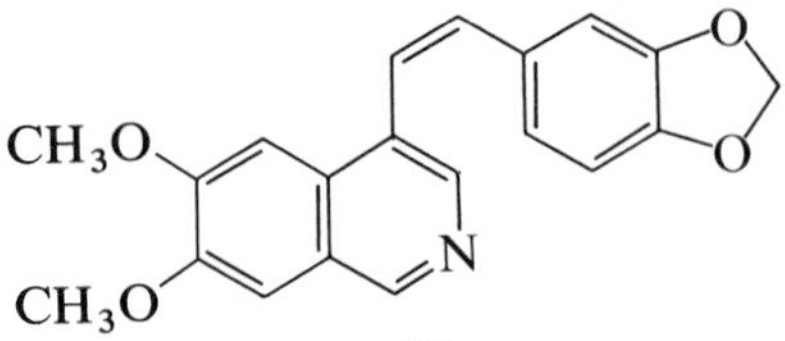

CH_3O CH_3O N 373

OCH_3 OCH_3 CH_3O CH_3O N 374

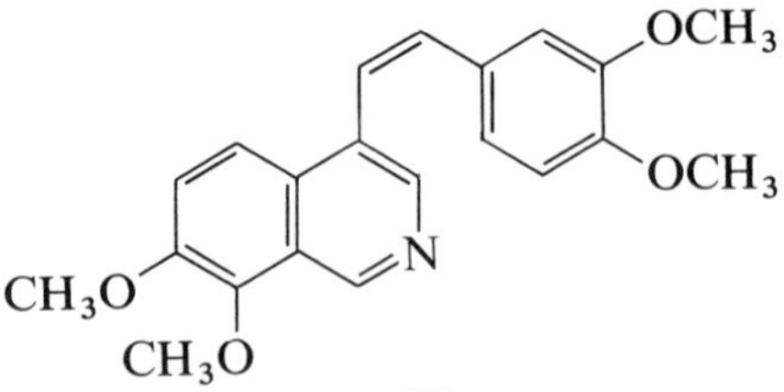

The conversion of vinyl compound **375** to azachrysene **376** is a heterocyclic example of a nonoxidative photocyclization involving the elimination of methanol.[265]

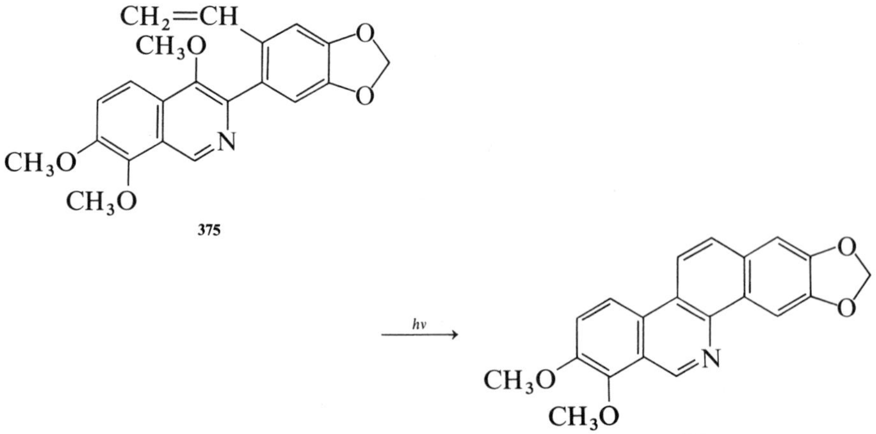

$CH_2{=}CH$ CH_3O CH_3O CH_3O N 375

$\xrightarrow{h\nu}$

CH_3O CH_3O N 376 (57%)

The oxidative photocyclization of 3,4-diphenylquinoline (**377**) succeeds only under acidic conditions, indicating that ring closure involves the corresponding quinolinium ion.[266] Oxygen suffices for the oxidative photocyclization of this quinolinium ion, whereas iodine is required for the oxidative photocyclization of its isoelectronic counterpart, 1,2-diphenylnaphthalene.[76]

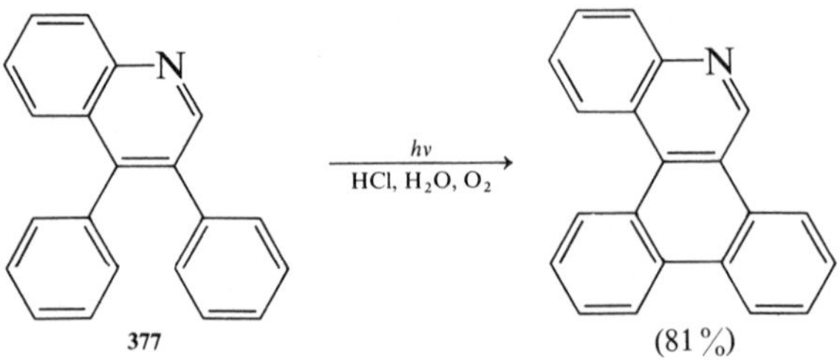

N 377 $\xrightarrow[\text{HCl, } H_2O\text{, } O_2]{h\nu}$ N (81%)

These observations are explained by the proposal that an irreversible proton transfer to the solvent traps the initially formed dihydroaromatic intermediate

378 to give the partially aromatized intermediate **379**. Because **379** does not undergo competing ring opening, its oxidation to give the fully aromatic product does not demand an especially reactive oxidant.

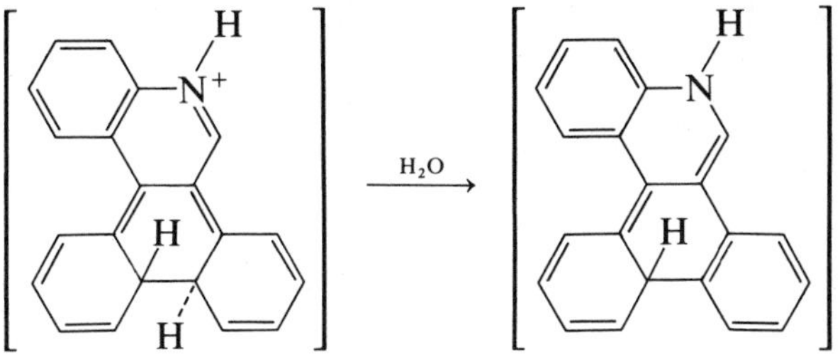

H_2O

378

379

The 4-phenylphenanthrene analog **380** also requires acid and tolerates oxygen for its oxidative photocyclization;[267] a similar explanation is suggested.

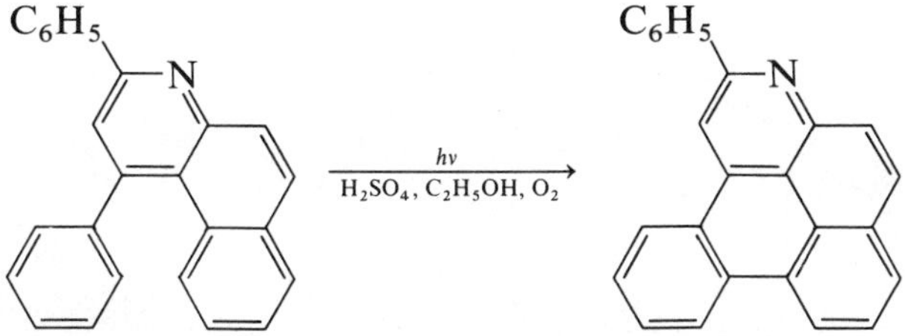

C_6H_5

$h\nu$

H_2SO_4, C_2H_5OH, O_2

380

(91%)

From Stilbene Analogs with Five-Membered Heteroaromatic Rings. Both 2-styrylindole (**381**)[268] and 3-styrylindole (**382**)[269] undergo oxidative photocyclization to give benzocarbazoles.

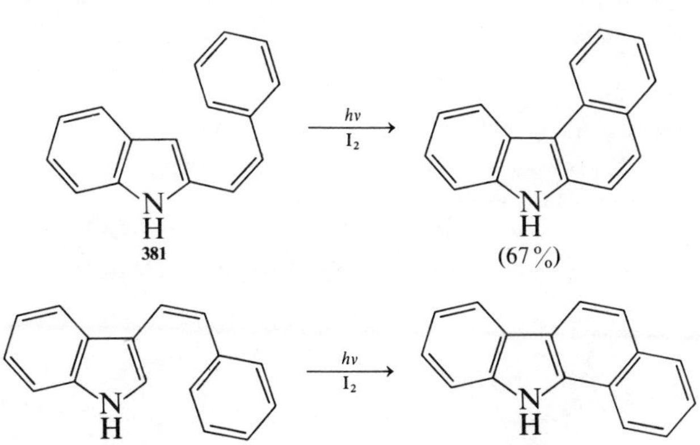

$h\nu$

I_2

381

(67%)

$h\nu$

I_2

382

(36%)

Peculiarly, although oxidative photocyclization succeeds for pyridyl compound **383**, it fails for the isomeric pyridyl compound **384**.[269]

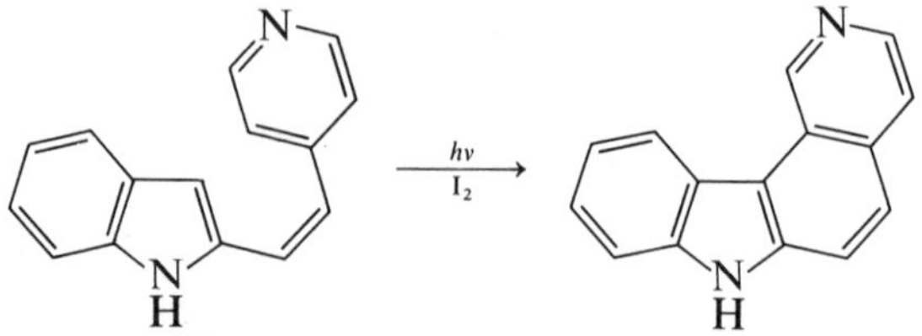

383 → (hν, I_2) (85%)

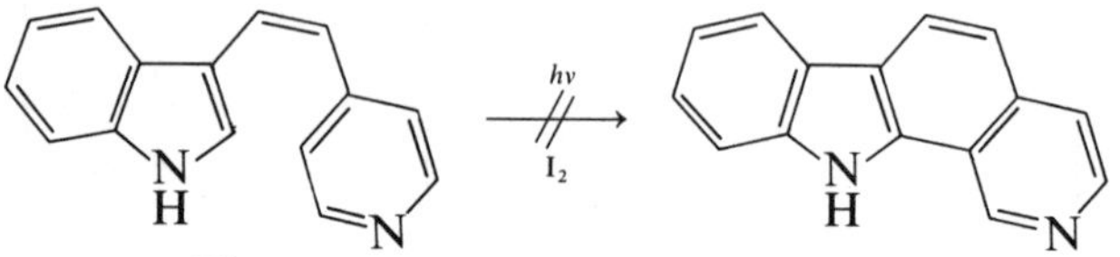

384 ⇸ (hν, I_2)

This failure is especially puzzling in view of the successful oxidative photocyclizations of the methyl, ethyl, and carbethoxy derivatives of type **385**.[269]

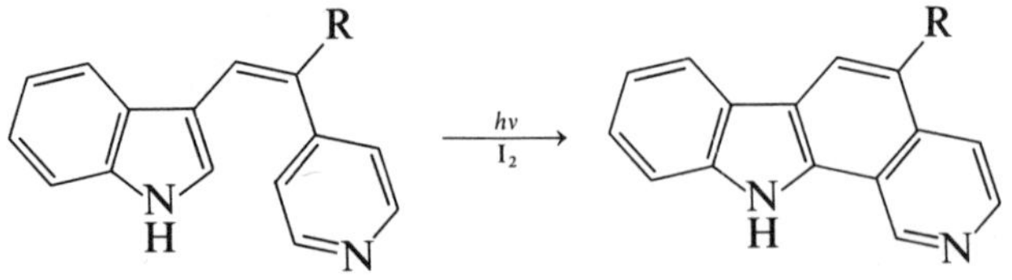

385 → (hν, I_2)

$R = CH_3$ (28%)
$R = C_2H_5$ (48%)
$R = CO_2C_2H_5$ (79%)

High yields are obtained in the oxidative photocyclizations of the heteroaromatic *o*-terphenyl analogs **386–388**.

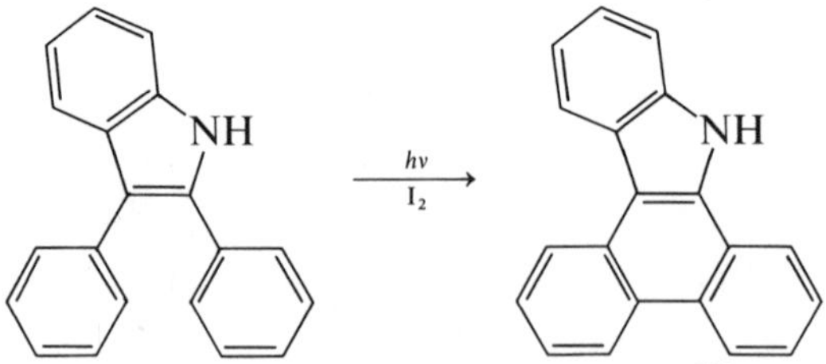

386 → (hν, I_2) (80%)[270]

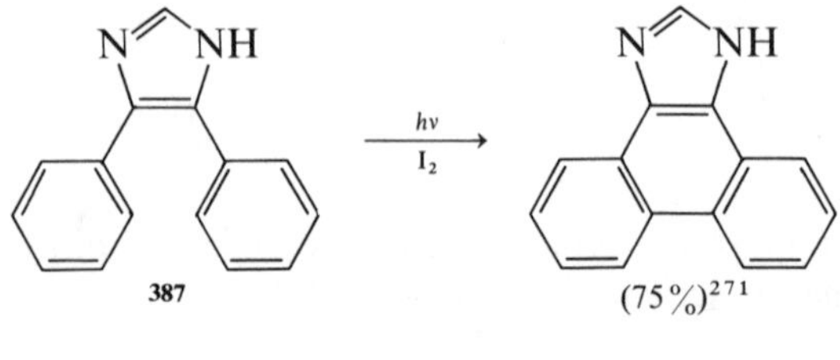

387 → (hν, I_2) (75%)[271]

CH_3 CH_3

N S → N S

$h\nu$, I_2

388 (87%)[271]

1,4,5-Triphenylpyrazole (**389**) undergoes oxidative photocyclization, but 1,3,4-triphenylpyrazole does not.[272]

N–C_6H_5 → N–C_6H_5

$h\nu$, I_2

389 (67%)

Although *o*-terphenyl requires iodine for its oxidative photocyclization, oxygen suffices for the oxidative photocyclization of 4,5-diphenyloxazole (**390**).[273]

N O → N O

$h\nu$, O_2

390 (42%)

The dihydro intermediate in this reaction is easier to trap than the analogous intermediate in the photocyclization of *o*-terphenyl because it has a smaller driving force for ring opening. (An oxazole ring has less resonance stabilization than a benzene ring.) A similar explanation accounts for the success of oxygen in the oxidative photocyclizations of diaryl heterocycles **391** and **392**.

S → S

$h\nu$, O_2

391 (34%)[274]

392 → ($h\nu$, O_2) (30%)[275]

Chloro compound **393** undergoes nonoxidative photocyclization,[276] presumably by photolysis of the carbon–chlorine bond assisted by complexation of the incipient chlorine atom by the adjacent phenyl group.[37,277]

393 → ($h\nu$) (92%)

1-Styrylimidazole (**394**) undergoes oxidative photocyclization exclusively at C-2 rather than C-5.[278] In the proposed dipolar intermediate **395**,[278,279] in contrast to the isomeric intermediate resulting from closure at C-5, a significant fraction of the negative charge is accommodated on a nitrogen atom. It is not known whether intermediate **395** is trapped directly by oxidation or indirectly by way of the partially aromatized intermediate **396** to give the isolated product **397**.

394 ⇌ ($h\nu$) [395] → (Proton transfers) [396] → (I_2) 397 (12%)

Similar oxidative photocyclization is found for benzimidazole **398**.[278]

398 → ($h\nu$, I_2, CH_3OH) (53%)

2-Styrylbenzimidazole (**399**) undergoes ring closure at a nitrogen atom.[278]

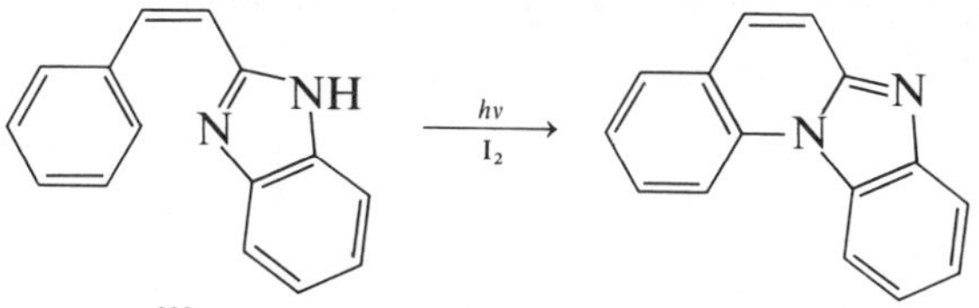

$\xrightarrow[I_2]{h\nu}$

399 (64%)

The diphenyltetrazolium ion **400** undergoes oxidative photocyclization in aqueous nitric acid.[280] Many substituted derivatives of **400** behave analogously.[281–283]

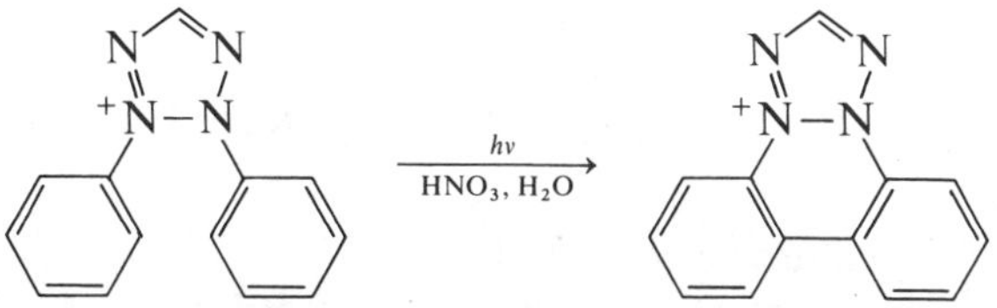

$\xrightarrow[HNO_3,\ H_2O]{h\nu}$

400 (69%)

Photocyclization of cation **400** by the usual mechanism would give **401**, a dihydro intermediate with an unfavorable distribution of positive charge.

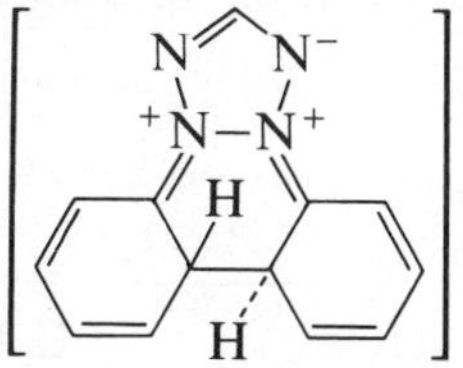

401

As a consequence, this reaction may proceed by a special mechanism involving cleavage and subsequent reformation of the central nitrogen–nitrogen bond.[284]

From Imine Analogs of Stilbene. Irradiation of imine **402** under normal oxidative photocyclization conditions in organic solvents fails to produce 9-azaphenanthrene (**403**),[108,285]

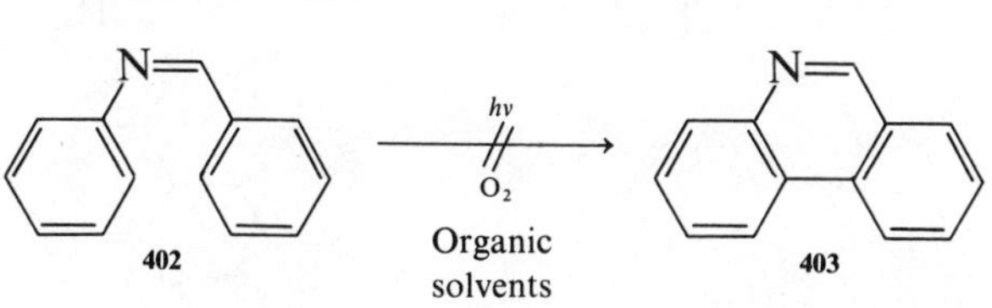

$h\nu$, O_2, Organic solvents

402 **403**

but the photocyclization of **402** to **403** succeeds in 98% sulfuric acid.[286] This behavior is typical of most diarylimines.[287,288] Photocyclization involves protonated imine **404**, and dihydro intermediate **405** is trapped oxidatively by **404**, sulfuric acid, or oxygen.

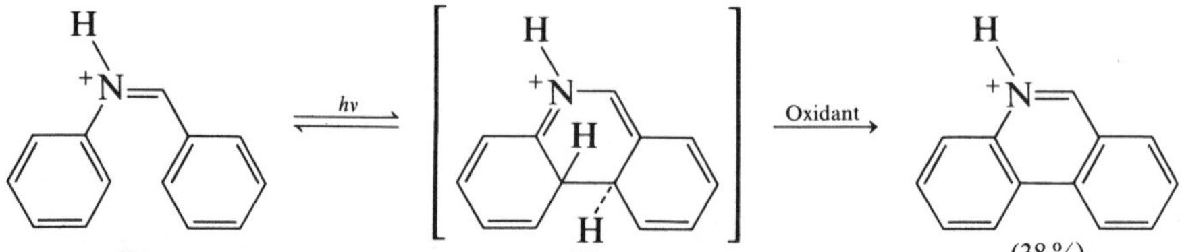

404 $\xrightleftharpoons{h\nu}$ [**405**] $\xrightarrow{\text{Oxidant}}$ (38%)

The amount of water present must be minimized to avoid acid-catalyzed hydrolysis of the imine.

$$\text{ArN=CHAr}' + H_2O \xrightarrow{H_2SO_4} \text{ArNH}_2 + \text{Ar}'\text{CHO}$$

Typically, 98–99% sulfuric acid is used.

The electronic absorption spectrum of imine **465** at $-100°$ [289] suggests an S_1 state of the $^1n,\pi^*$ type;[290] this would constitute an impediment to photocyclization that protonation would remove.

The failure of imine **402** to undergo photocyclization in organic solvents can also be attributed, at least in part, to its thermal conversion to the related *trans* isomer **406**[285,291] with a half-life of about 1 second at room temperature.[292,293]

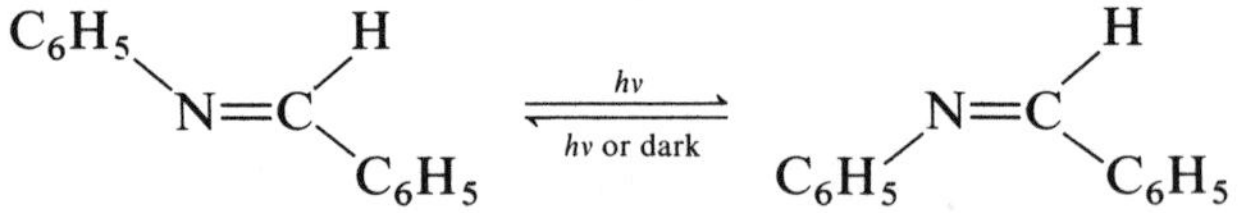

406 $\xrightleftharpoons[h\nu \text{ or dark}]{h\nu}$ **402**

No such problem exists in sulfuric acid solution since the thermal *cis* → *trans* isomerization of the protonated imine **404** is very slow.[289] Consistent with this explanation is the finding that imine **407**, which cannot escape having two phenyl groups *cis* to one another, undergoes oxidative photocyclization in cyclohexane.[285]

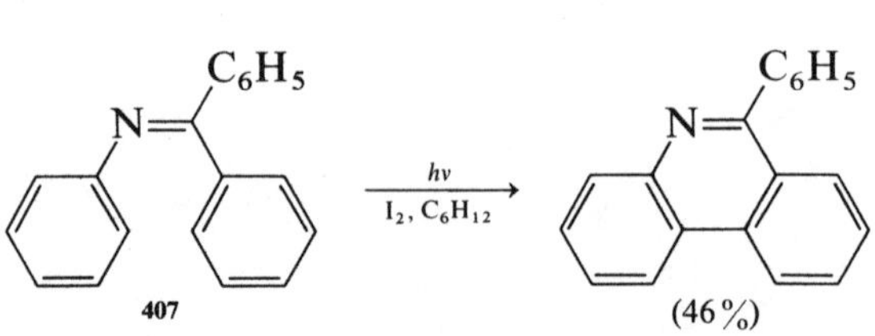

407 $\xrightarrow[I_2,\ C_6H_{12}]{h\nu}$ (46%)

As an exception to the general rule that acidic conditions are required, imine **408** is reported to undergo oxidative photocyclization in diethyl ether,[294a] although subsequent workers were unable to realize preparatively useful results from this photocyclization.[294b]

$$\mathbf{408} \xrightarrow[O_2,\ (C_2H_5)_2O]{h\nu} (56\%)$$

Imine **409** undergoes oxidative photocyclization successfully in ethanol;[295,296]

$$\mathbf{409} \xrightarrow[O_2,\ C_2H_5OH]{h\nu} (40\%)$$

the eight other imines of type ArN=CHAr′ in which the Ar and Ar′ groups are phenyl, 1-naphthyl, or 2-naphthyl all fail to give the corresponding oxidative photocyclization products on irradiation under nonacidic conditions, however.[296] Imine **410**, like its parent imine **407**, undergoes oxidative photocyclization under neutral conditions.[297]

$$\mathbf{410} \xrightarrow[O_2,\ CH_2Cl_2]{h\nu} (42\%)$$

Although imine **411** gives diazachrysene **412** in sulfuric acid,[298] imines of type **413** give unsatisfactory results under similar conditions.[299]

$$\mathbf{411} \xrightarrow[H_2SO_4]{h\nu} \mathbf{412}\ (53\%)$$

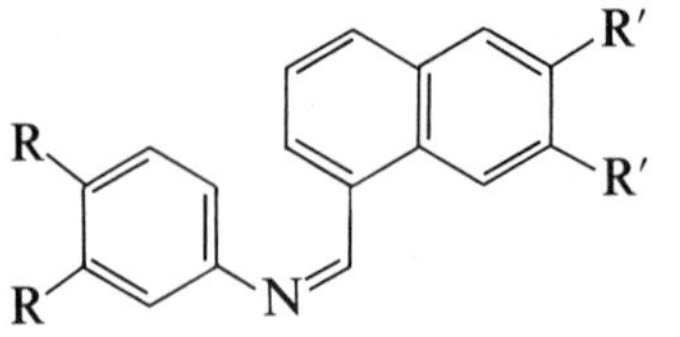

R R R' R' N 413

An alternative synthesis of 9-azaphenanthrenes that avoids the problems of imine photocyclization is illustrated by the three-step conversion of hydroxamic acid **414** to 9-azaphenanthrene **415**.[300]

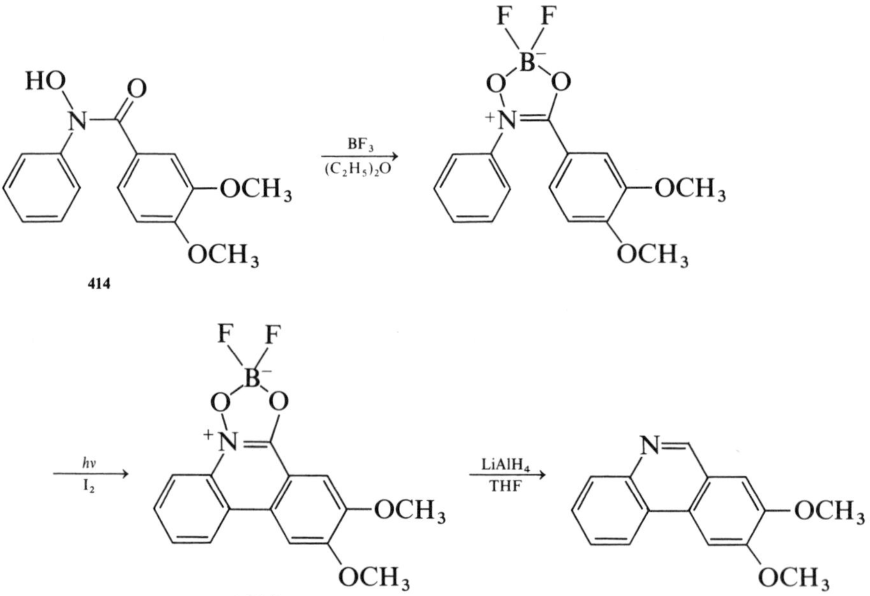

HO N O OCH_3 OCH_3 **414** $\xrightarrow[(C_2H_5)_2O]{BF_3}$ F F B O O N OCH_3 OCH_3

$\xrightarrow[I_2]{h\nu}$ F F B O O N OCH_3 OCH_3 (95%) $\xrightarrow[THF]{LiAlH_4}$ N OCH_3 OCH_3 **415** (66%)

The nonoxidative conversion of imine **416** to the dihydrophenanthridine **417** requires a triplet sensitizer,[301] and probably proceeds by a free-radical cyclization rather than an electrocyclic photoreaction.

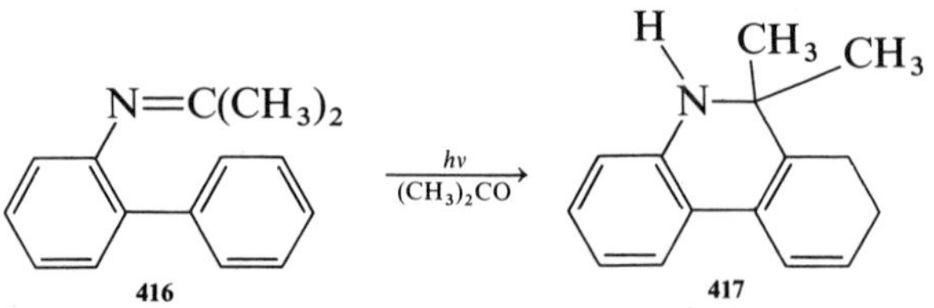

$N{=}C(CH_3)_2$ **416** $\xrightarrow[(CH_3)_2CO]{h\nu}$ H N CH_3 CH_3 **417** (66%)

The bis-imine **418** undergoes double photocyclization under nonoxidizing conditions to give the tetracyclic product **419**, a compound that is air-oxidized on work-up to give **420** as the isolated product.[302]

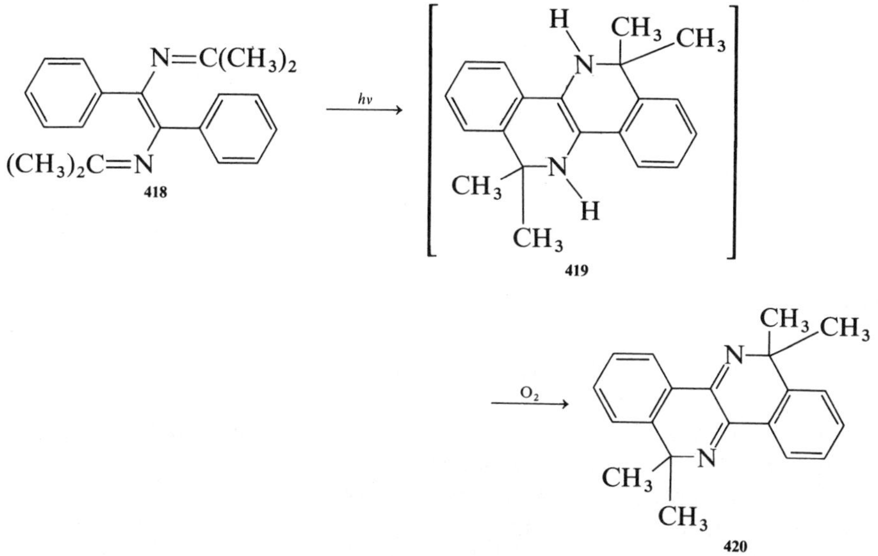

418 → hν → [**419**] → O_2 → **420** (53%)

Irradiation of imines of type **421** in air-saturated alcohols of type $R'CH_2CH_2OH$ gives products of type **422** resulting from incorporation of a fragment derived from the solvent.[303,304]

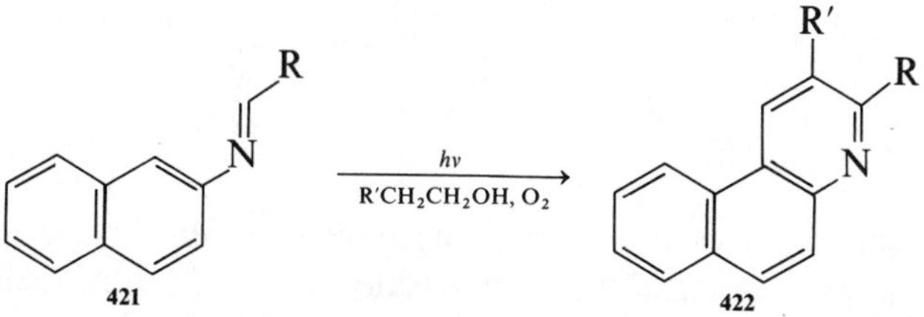

421 → hν, $R'CH_2CH_2OH$, O_2 → **422**

The photochemistry of imines is discussed in two recent reviews.[305,306]

From Azobenzenes. Azobenzenes undergo oxidative photocyclization to give 9,10-diazaphenanthrenes only under strongly acidic conditions.[307,308] For photocyclization to occur, the azobenzene must be converted to a species having an S_1 state that is $^1\pi,\pi^*$ rather than $^1n,\pi^*$.[49,50] This is achieved by protonation in sulfuric acid,[307,308] interaction with Lewis acids in 1,2-dichloroethane,[29] or intramolecular hydrogen bonding.[309] The 9,10-diazaphenanthrenes are obtained in limited yields in 22 *N* sulfuric acid (68% sulfuric acid) because up to half of the protonated azo starting material is consumed by functioning as the oxidant. For example, azobenzene gives comparable amounts of 9,10-diazaphenanthrene and benzidine.[308]

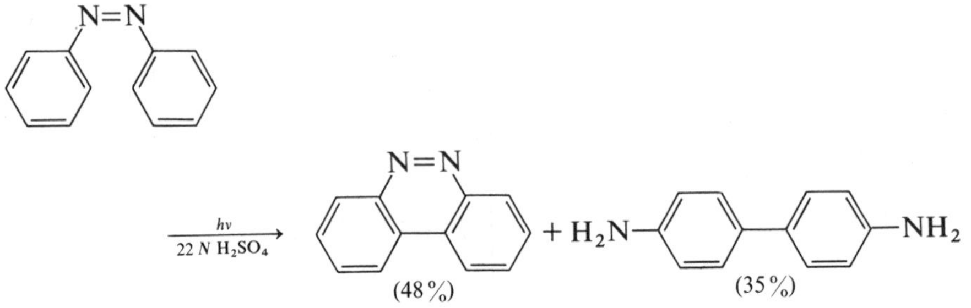

$\xrightarrow[22\,N\,H_2SO_4]{h\nu}$ (48%) + H_2N–⟨biphenyl⟩–NH_2 (35%)

Higher yields of 9,10-diazaphenanthrenes are obtained in 98% sulfuric acid because the solvent acts as the oxidant.[13,310] However, preparative-scale photocyclizations of azobenenes in 98% sulfuric acid can require excessively long periods of irradiation for complete conversion,[311] perhaps because the thermal *cis* → *trans* isomerization of the starting material is very much faster in 98% sulfuric acid than in 22 *N* sulfuric acid.[48]

Although stilbene photocyclization is thwarted by amino and nitro substituents, azobenzenes **423** and **424** photocyclize readily in 98% sulfuric acid,[310] presumably because the substituents become protonated.

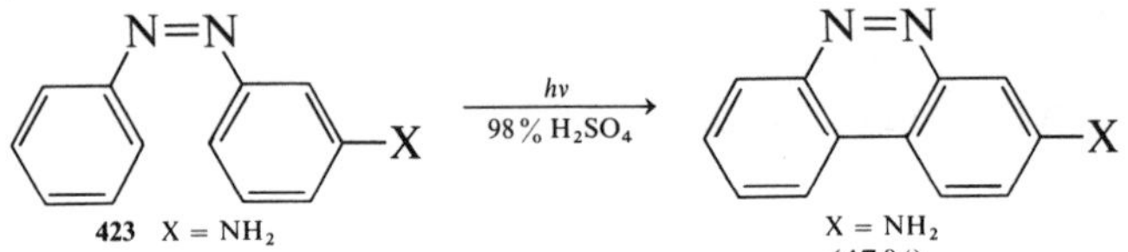

423 X = NH_2
424 X = NO_2

$\xrightarrow[98\%\,H_2SO_4]{h\nu}$

X = NH_2 (47%)
X = NO_2 (90%)

p-Nitroazobenzene gives 3-nitro-9,10-diazaphenanthrene in 30% yield, but *p*-aminoazobenzene (**425**) is photochemically inert,[310,312] probably because dication **426** lacks the stilbene-like electron distribution that is required for photocyclization.[310]

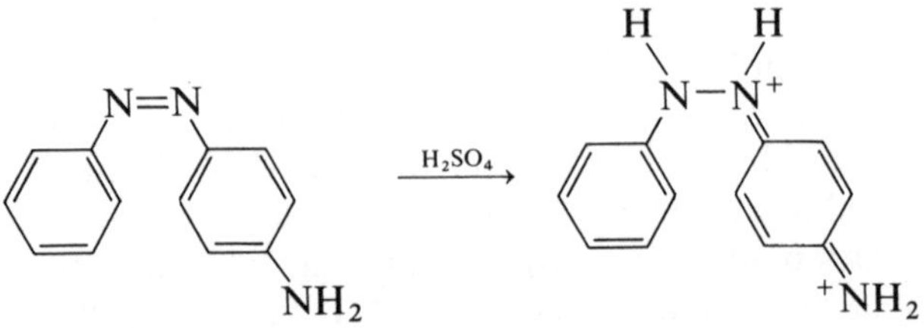

$\xrightarrow{H_2SO_4}$

425 **426**

The conversion of **425** to **427** can be achieved, however, by irradiating the *N*-benzylidene derivative **428**; the protecting *N*-benzylidene group is lost hydrolytically in the work-up procedure.[310]

N=N

N=CHC$_6$H$_5$

428

$h\nu$ / 98% H_2SO_4

N=N

NH$_2$

427
(97%)

In the oxidative photocyclization of iodoazobenzenes, in contrast with iodostilbenes, the iodo substituents are retained.[313] For example, **429** gives **430**.[313]

N=N I

429

$h\nu$ / 22 *N* H_2SO_4

N=N I

430
(29%)

Protonated azobenzenes and protonated 9,10-diazaphenanthrenes in sulfuric acid solution have $^1\pi,\pi^*$ S_1 states with excitation energies of about 67 kcal/mol,[13,48] whereas an excitation energy of at least 77 kcal/mol is required for population of the dissociative ^{3}n,σ^* excited triplet state in which homolysis of the aryl carbon–iodine bond occurs.[314]

Oxidative photocyclization of *meta*-substituted azobenzenes of type **431** with certain simple substituents such as chloro,[313] iodo,[313] methyl,[308] and phenyl,[315] gives mixtures of 2-substituted and 4-substituted 9,10-diazaphenanthrenes.

N=N

X

431

$h\nu$ / H_2SO_4

N=N

X +

(18–35%)

N=N

X

(9–13%)

X = Cl, I, CH_3, C_6H_5

For certain other substituents, however, such as acetyl, amino, and nitro, oxidative photocyclization of the *meta*-substituted azobenzene in sulfuric acid gives exclusively the 2-substituted product.[310] This high degree of regioselectivity may reflect a conformational bias stemming from electrostatic repulsion between the protonated substituent and the protonated nitrogen atom of the azo linkage.

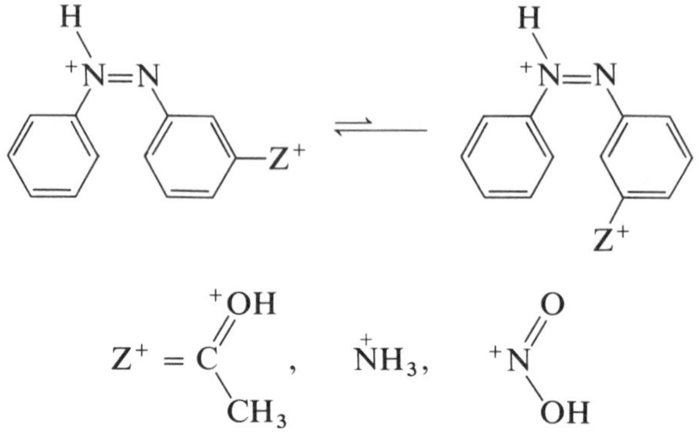

$Z^+ = C(=^+OH)CH_3$, $\overset{+}{N}H_3$, $^+N(=O)OH$

The predominant formation of the 2-aza product rather than the 4-aza product from irradiation of the pyridyl compound **432**[316] may have a similar explanation.

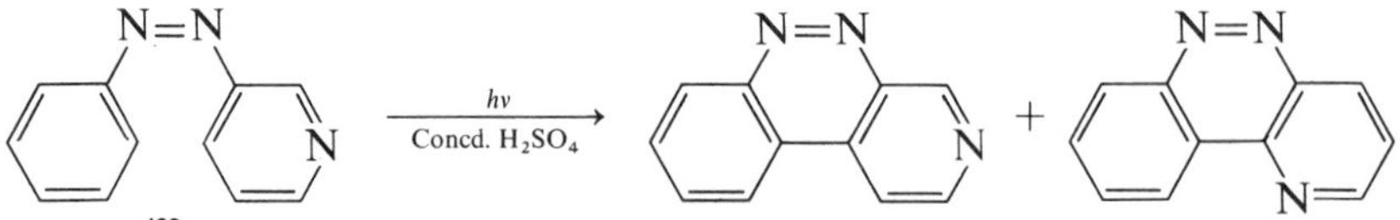

hv, Concd. H_2SO_4

432 → (19%) + (3%)

Oxidative photocyclization of each disubstituted azobenzene of type **433** with R = H, CH_3, and C_2H_5 gives exclusively the 2,7-disubstituted product of type **434**; the corresponding 2,5 or 4,5 isomers are not observed.[317]

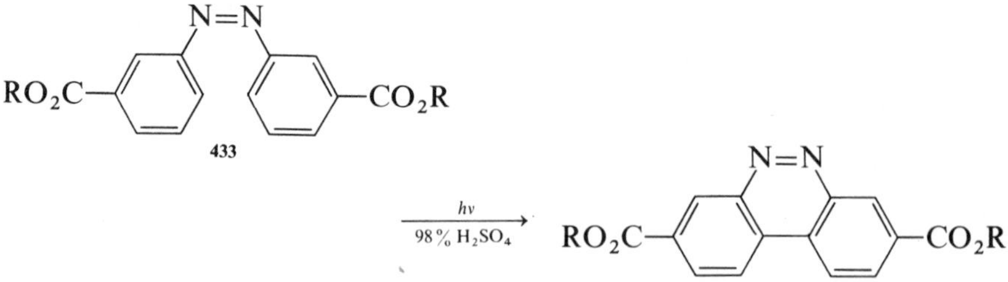

433 → (*hv*, 98% H_2SO_4) 434 (60–62%)

R = H, CH_3, C_2H_5

Although this striking regioselectivity may be due in part to a simple steric effect,[317] electrostatic repulsion between the protonated substituents probably is a more important factor. Consistent with this proposed electrostatic effect is the fact that marked regioselectivity is *not* observed in the oxidative photocyclization of *m,m'*-dimethylazobenzene (**435**).[308]

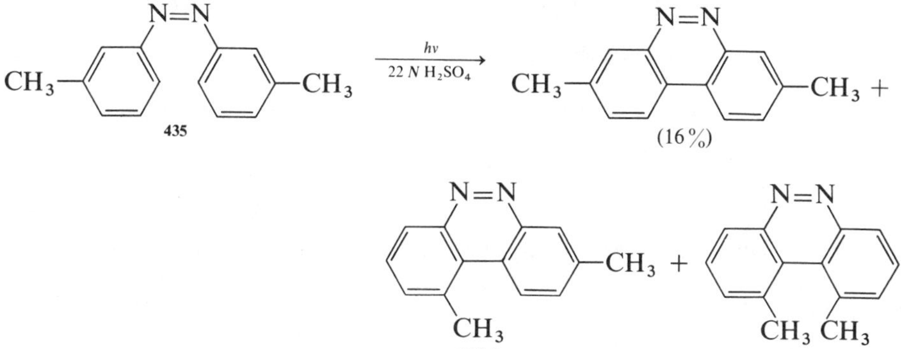

435

(16%)

(7%) (3%)

Mysteriously, the scope of the photocyclization of azobenzene analogs with polynuclear aryl groups appears severely limited, in contrast to the situation with diaryl olefins. For example, although 1-phenylazonaphthalene (**436**) gives diazachrysene **437**, the attempted photocyclizations of 2-phenylazonaphthalene, 1,1′-azonaphthalene, 1,2′-azonaphthalene, and 2,2′-azonaphthalene all fail.[318] 6,6′-Azochrysene also fails to photocyclize.[319]

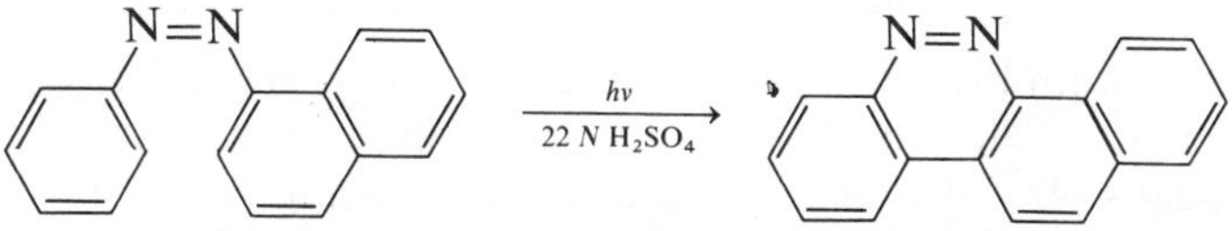

436 437 (42%)

The photochemistry of azobenzene and its derivatives has been reviewed.[320]

From Isocyanates. Isocyanate **438** undergoes a triplet-sensitized ring closure to give 6(5*H*)-phenanthridinone (**66**).[301]

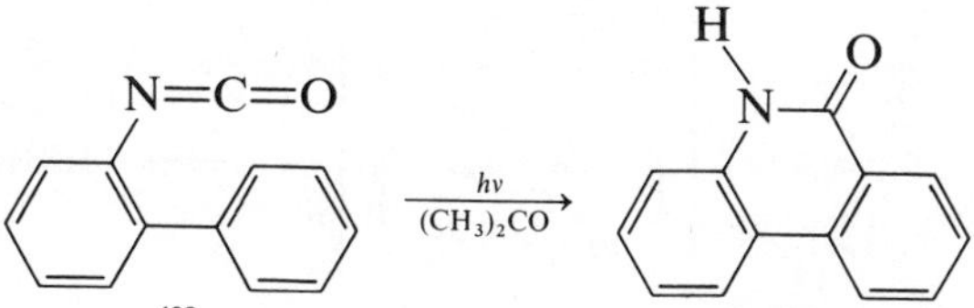

438 66 (70%)

As with imine **416**, this reaction might involve photocyclization of **438** to a dihydroaromatic intermediate followed by a 1,5-hydrogen shift to produce **66**, but it seems more likely that it proceeds by a radical chain mechanism.

From Anilinoboranes. Borazarophenanthrenes of type **439** with various common substituents (X = H, Br, Cl, CH_3, and CH_3O) are obtained by oxidative photocyclization of anilinoboranes of type **440**.[321]

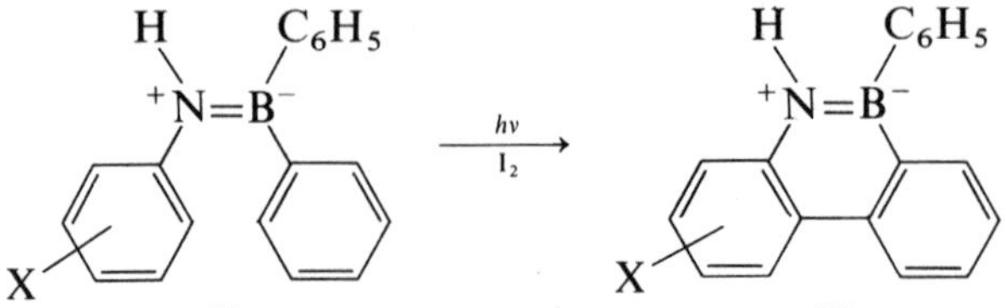

440 → ($h\nu$, I_2) → **439** (4–33%)

From Thiophene Derivatives. The oxidative photocyclization of di-2-thienylethylene **441** proceeds in excellent yield, whereas that of di-3-thienylethylene **442** fails.[322a,b]

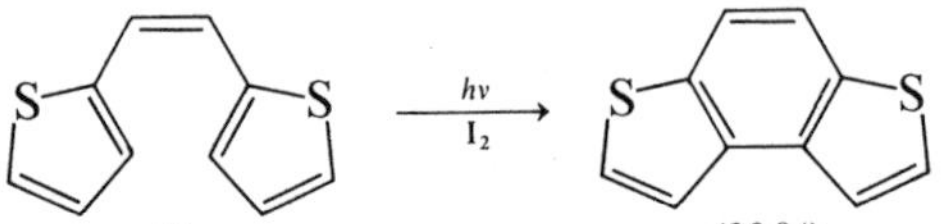

441 → ($h\nu$, I_2) → (90%)

442 ↛ ($h\nu$, I_2)

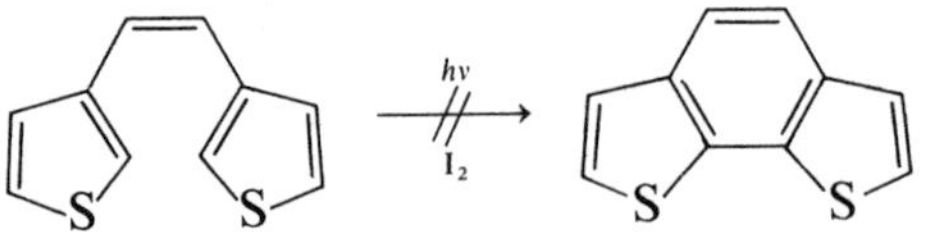

The problem with the latter photocyclization lies in the oxidative trapping of dihydro intermediate **443**. Abstraction of an allylic hydrogen atom from **443** gives radical **444**, a species that readily suffers carbon–sulfur bond cleavage because of the large driving force associated with the formation of a benzene ring; the resulting thiyl radical **445** gives intractable products.[322a]

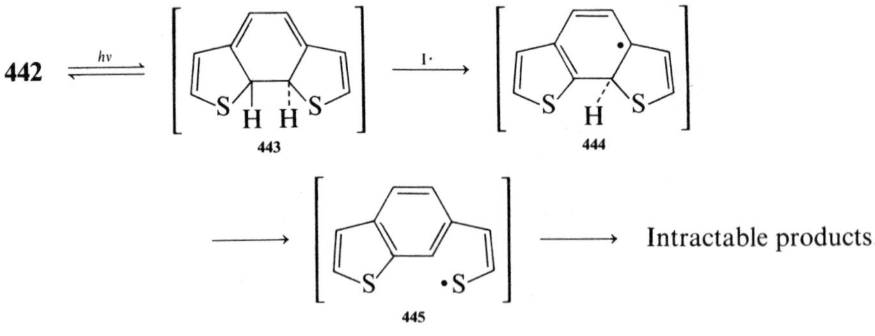

442 ⇌ ($h\nu$) [**443**] → (I·) [**444**] → [**445**] → Intractable products

The partial success in the oxidative photocyclization of the mixed dithienylethylene **446** to **447** can be rationalized by assuming that the two different allylic hydrogen atoms in the dihydro intermediate are abstracted with about equal probability, leading to comparable amounts of **447** and intractable material.[322a]

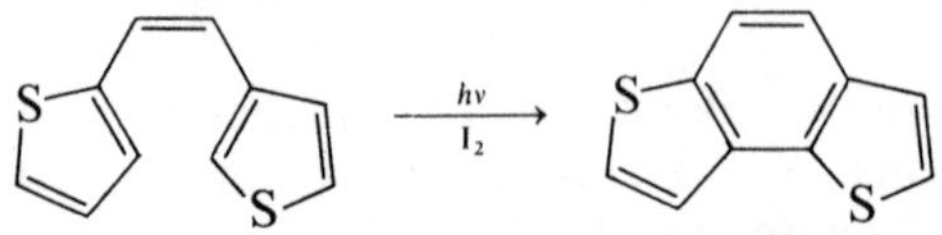

446 → ($h\nu$, I_2) → **447** (47%)

The modest yield in the oxidative photocyclization of 3-styrylthiophene (**448**)[323] can be accounted for by a similar assumption.

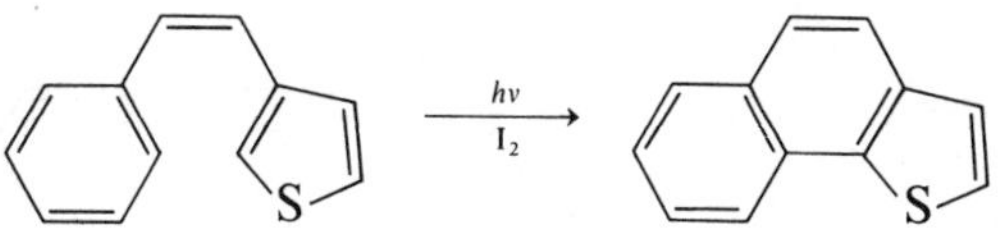

448 (40–50%)

2-Styrylthiophene (**449**) undergoes oxidative photocyclization without problems.[324]

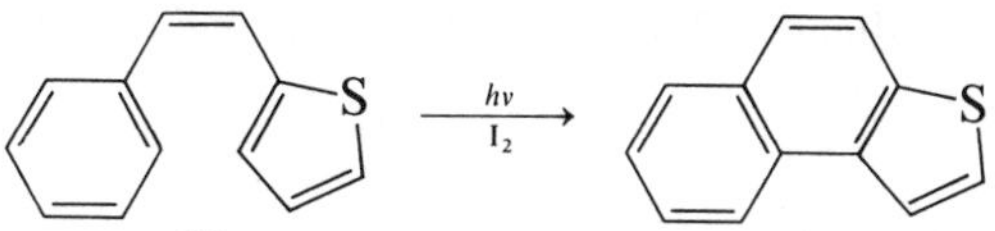

449 (100%)

Thiophene analogs of higher polynuclear aromatic systems are readily available by oxidative photocyclization, as illustrated by the conversion of olefins **450–452** to heterohelicenes **453–455**, respectively.[325]

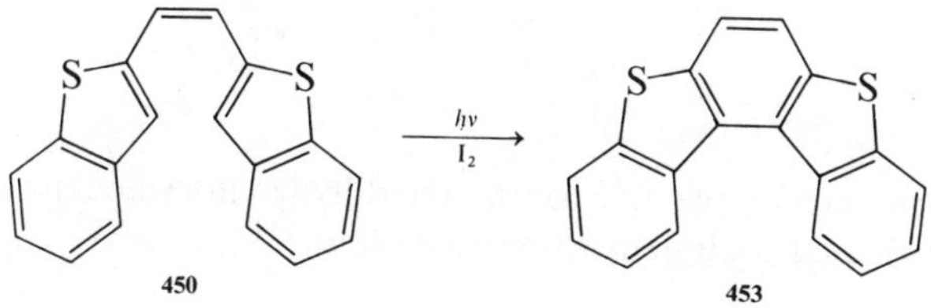

450 **453** (57%)[326]

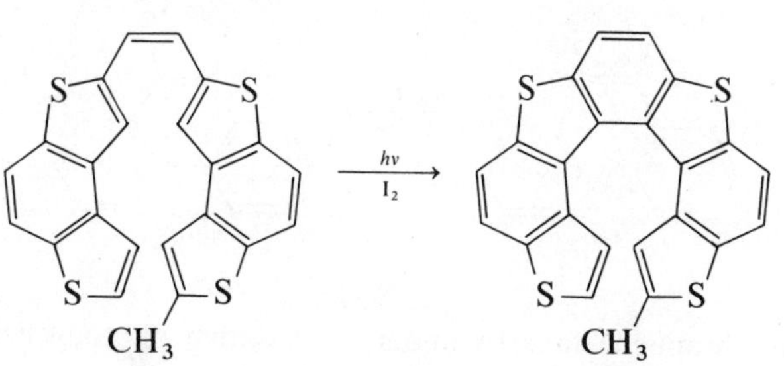

451 **454** (72%)[327]

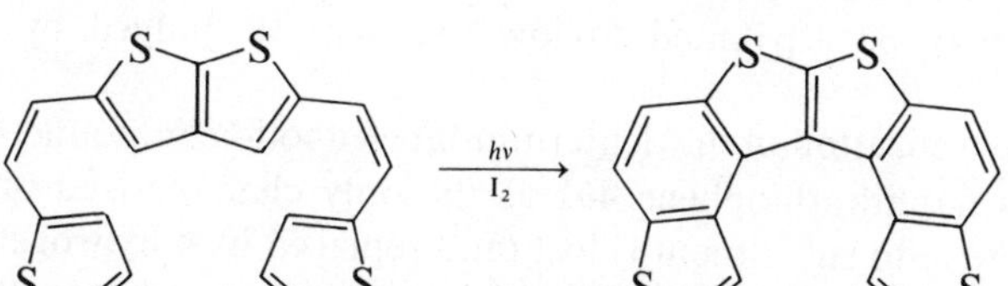

452 **455** (51%)[328]

The preliminary report that the 2-styrylnaphthalene analog **456** photocyclizes at C-3 rather than C-1 is unprecedented and merits further investigation.[329]

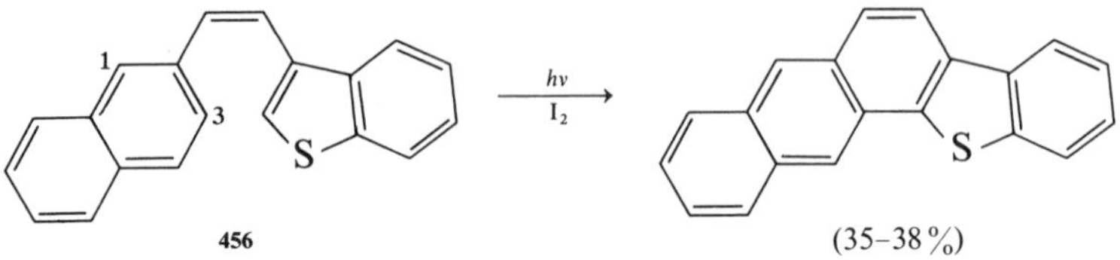

456 (35–38%)

The conversion of diene **457** to benzothiophene **458** is an example of a vinyl cyclization involving a thiophene ring.[330]

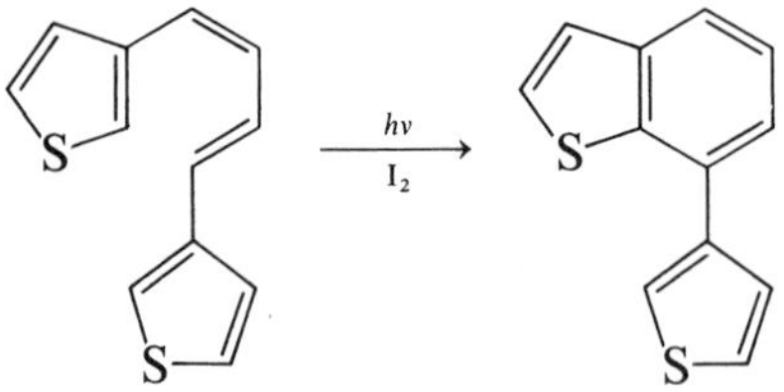

457 **458** (62%)

Oxygen is sufficiently reactive to trap the dihydro intermediate generated by photocyclization of 2,3-diphenylthiophene (**459**).[331]

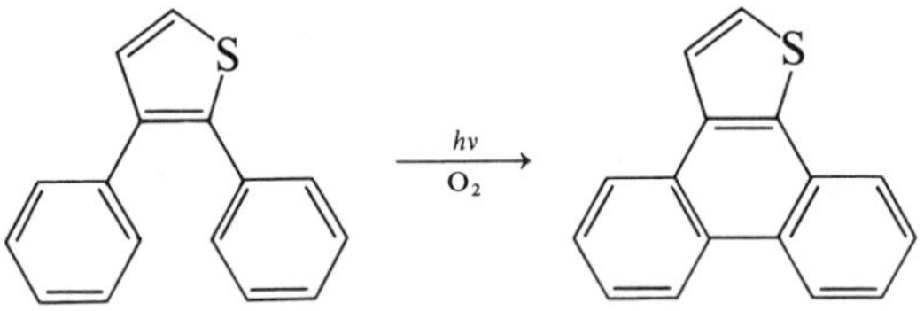

459 (70%)

Tetraphenylthiophene fails to undergo oxidative photocyclization,[160,331] perhaps because its phenyl groups at C-3 and C-4 are nearly orthogonal to the rest of the π system and hence not involved in the photoexcitation. Alternatively, this failure may be attributed to low reactivity as judged by free-valence indices.[160]

The photocyclization of the deuterium-labeled iodo compound **460** in cyclohexane gives naphthothiophene **461** as the only characterized product, indicating that the iodo substituent is lost (and replaced by a hydrogen atom from the solvent) in a process independent of the electrocyclic photocyclization.[323]

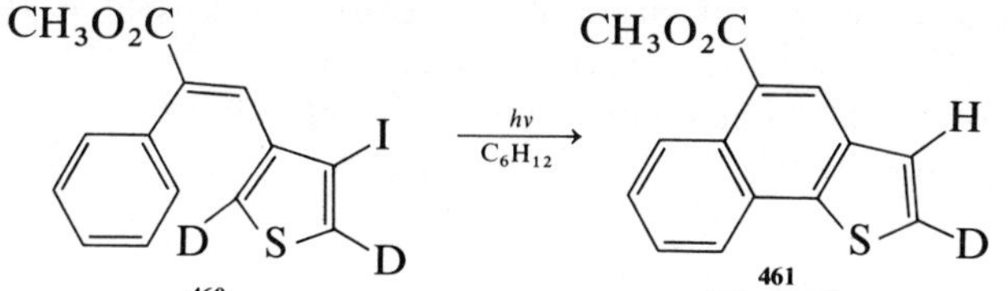

CH_3O_2C ... I, D, S, D — $\xrightarrow[C_6H_{12}]{h\nu}$ — CH_3O_2C ... H, S, D

460 **461** (50–60%)

The same product, **461**, is obtained in 40–50% yield by oxidative photocyclization of the analog of **460** lacking the iodo substituent.[323]

From Furan Derivatives. 2-Styrylfuran (**462**) undergoes oxidative photocyclization in poor yield.[332]

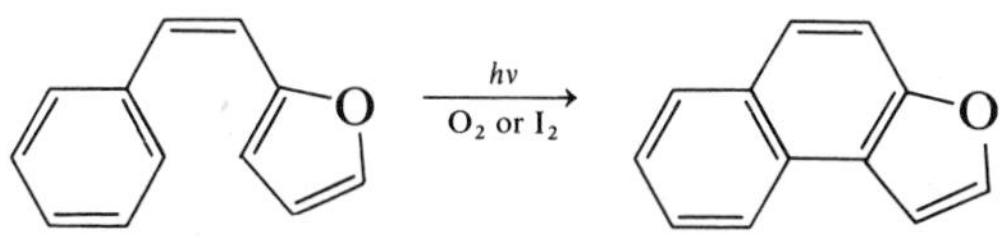

O $\xrightarrow[O_2 \text{ or } I_2]{h\nu}$ O

462 (9%)

Better results are obtained for 2,3-diphenylfuran (**463**).[333]

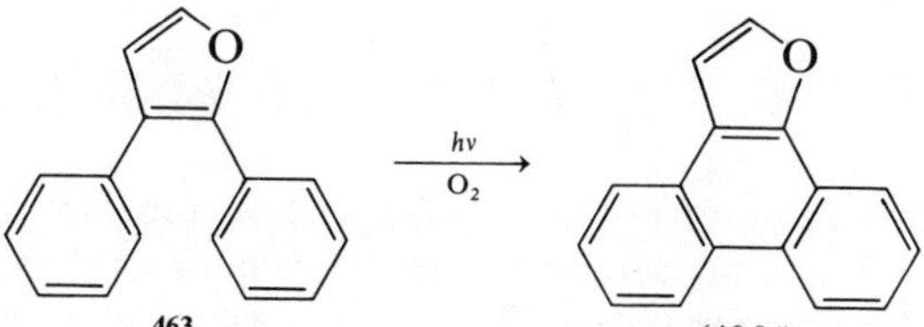

O $\xrightarrow[O_2]{h\nu}$ O

463 (40%)

The attempted conversion of tetraphenylfuran to phenanthrene **464** by oxidative photocyclization fails.[160] An indirect route from tetraphenylfuran to **464** involves eliminative photocyclization of **465**, the Diels–Alder adduct of tetraphenylfuran and tetrachloro-*o*-benzoquinone.[334]

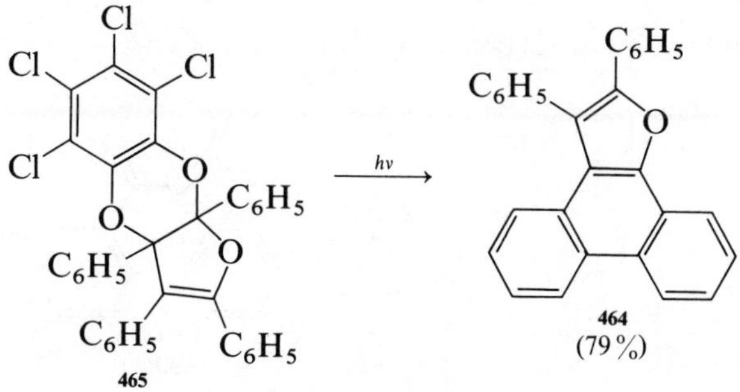

Cl, Cl, Cl, Cl, O, O, C_6H_5, O, C_6H_5, C_6H_5, C_6H_5 $\xrightarrow{h\nu}$ C_6H_5, C_6H_5, O

465 **464** (79%)

The oxidative photocyclization of the benzofuran derivative **466** in various organic solvents gives the expected product **467**.[335,336] The photoreaction of **466** follows a peculiar course in propylamine, however, giving dihydroaromatic compound **468**.[335–337]

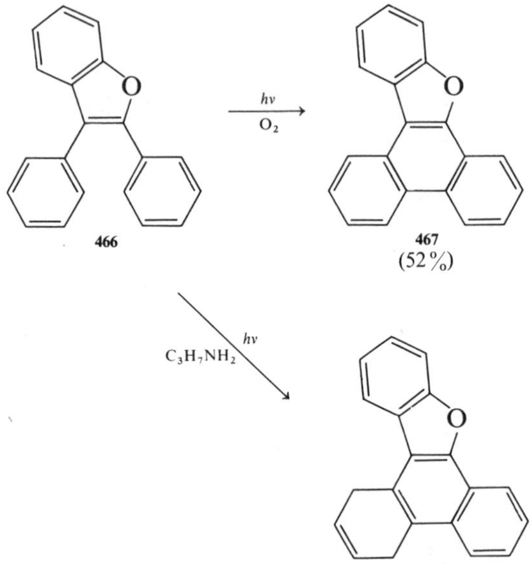

466 → ($h\nu$, O_2) → 467 (52%)

466 → ($h\nu$, $C_3H_7NH_2$) → 468 (65%)

From Pyrylium Ions. Irradiation of pyrylium ion **469** in acetic acid or acetic anhydride gives a mixture of the oxidative photocyclization product **470** and the reduced starting material **471**.[338,339] It seems plausible that the first-formed dihydroaromatic intermediate is trapped by proton transfer to a solvent molecule and hydride transfer to a molecule of the starting pyrylium ion **469** as illustrated in Scheme 28. Similar photochemical behavior is found for derivatives of cation **469**.[338,340]

Lactams and Other Heterocyclic Systems from Amides

Benzanilide (**65**) undergoes oxidative photocyclization to give 6(5*H*)-phenanthridinone (**66**).[77]

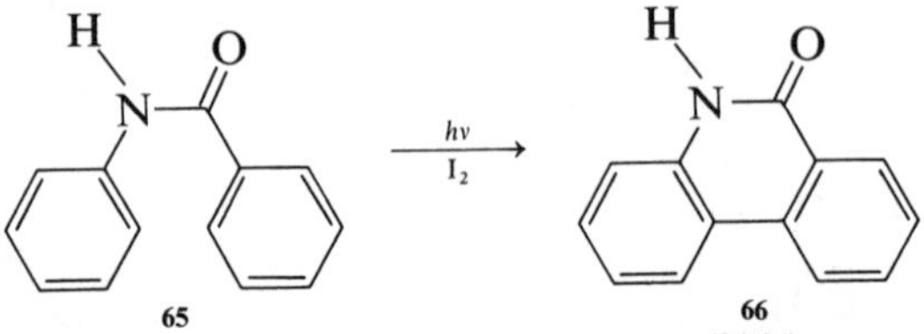

65 → ($h\nu$, I_2) → 66 (20%)

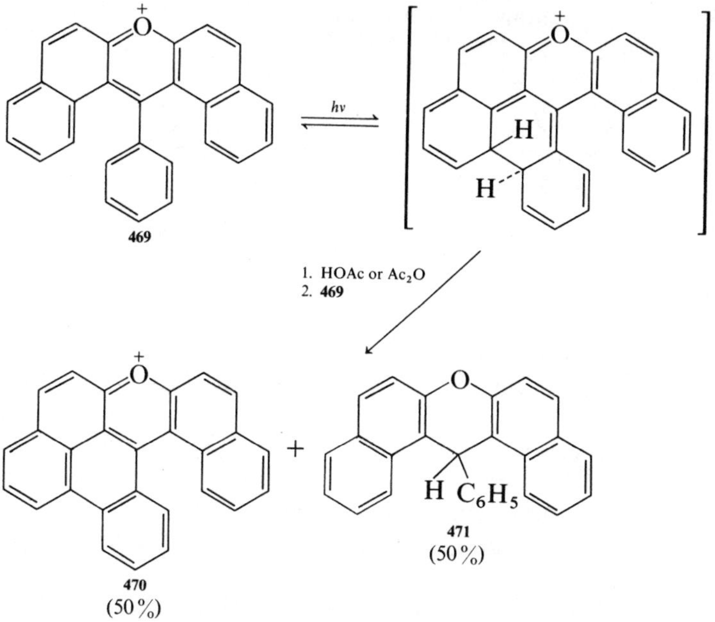

Scheme 28

Oxidative photocyclization also succeeds for the 2-pyridyl and 4-pyridyl compounds **472** and **473**; in contrast, the 3-pyridyl compound **474** gives only rearrangement products.[341]

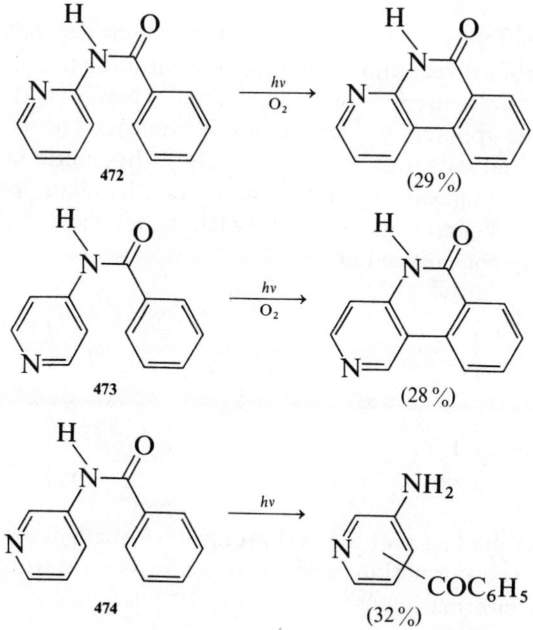

Other heterocyclic analogs of benzanilide that undergo oxidative photocyclization include indole **475**.[342]

475 $\xrightarrow[O_2]{h\nu}$ (77%)

As illustrated for the parent system **476**, *o*-chlorobenzanilides undergo nonoxidative photocyclization in cyclohexane to give the corresponding phenanthridinone in good yield.[343] This reaction is quenched by air and also by polar solvents. A triplet mechanism is proposed, involving carbon–chlorine photolysis assisted by complexation of the leaving chlorine atom with the *N*-phenyl ring; as a consequence of this assistance, the aryl radical is generated in a *syn* conformation suitable for an intramolecular arylation reaction.[343]

476 $\xrightarrow[C_6H_{12},\ N_2]{h\nu\ (254\ nm)}$ **66** (71%)

The outcome of the irradiation of *o*-iodobenzanilide depends on the solvent. In a 9:1 mixture of benzene and methanol, the cyclization product **66** is obtained in 57% yield.[344] In benzene[77] or cyclohexane,[343] however, the irradiation of *o*-iodobenzanilide apparently involves simple photolysis of the carbon–iodine bond followed predominantly[77] or exclusively[343] by intermolecular reactions of the resulting aryl radical with solvent molecules. Irradiation of the *N*-methyl derivative **477** in benzene gives a somewhat better yield of the cyclization product *N*-methylphenanthridinone.[345]

477 $\xrightarrow[I_2\ \text{or}\ O_2,\ C_6H_6]{h\nu}$ (30%)

This may reflect the fact that the aryl groups in *N*-methylbenzanilides such as **477** prefer a *syn* conformation, whereas those in *N*-unsubstituted benzanilides prefer an *anti* conformation.[345]

Certain *o*-bromobenzanilide derivatives give good yields of the corresponding phenanthridinones on irradiation in alcohol solvents;[299] an example is **478**.[346]

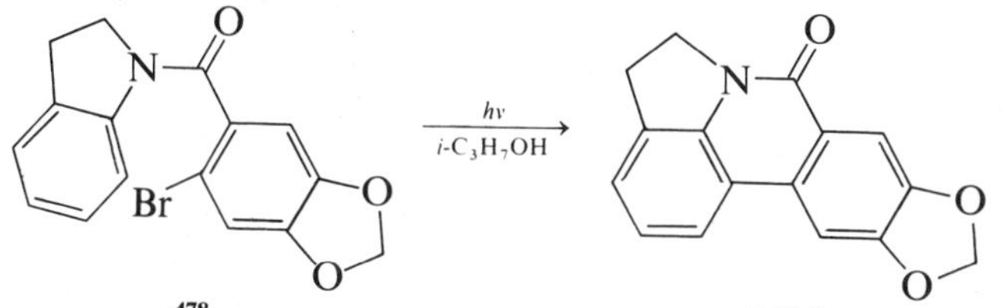

$h\nu$, i-C_3H_7OH

478

(67%)

It is not clear whether these reactions proceed by carbon–bromine photolysis, followed or accompanied by intramolecular radical arylation,[37a] or, alternatively, by an electrocyclic reaction to give a zwitterionic intermediate that subsequently undergoes solvent-assisted elimination of the elements of hydrogen bromide.

Other examples of eliminative photocyclization include the conversions of the four different isomeric dimethoxybenzanilides of type **479** to methoxyphenanthridinones of type **480**.[78]

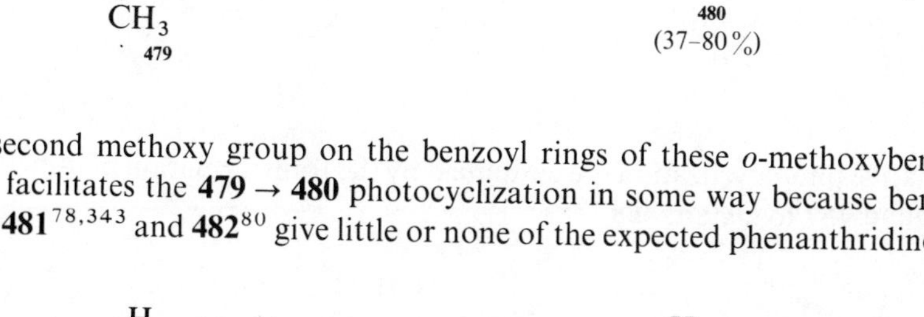

$h\nu$, C_6H_6, C_2H_5OH

479

480
(37–80%)

The second methoxy group on the benzoyl rings of these *o*-methoxybenzanilides facilitates the **479** → **480** photocyclization in some way because benzanilides **481**[78,343] and **482**[80] give little or none of the expected phenanthridinones.

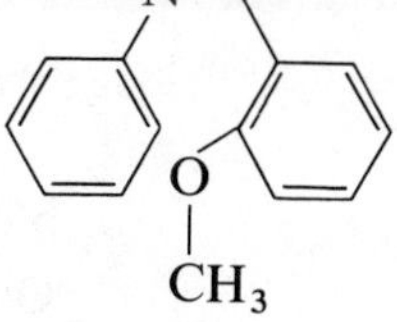

481

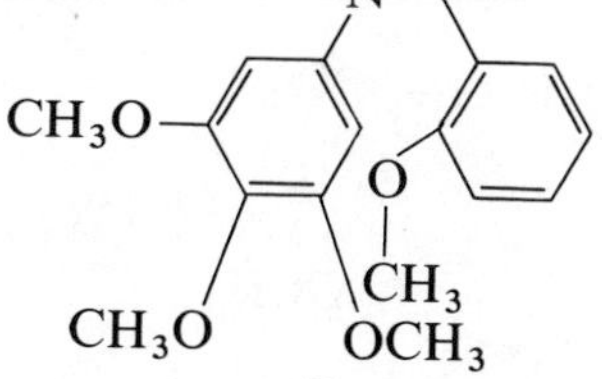

482

Vinylogous amides of type **483** with X = Br or X = I undergo nonoxidative photocyclization to give cation **484**; the analogous reaction of **483** with X = Cl is slow and not preparatively useful.[347]

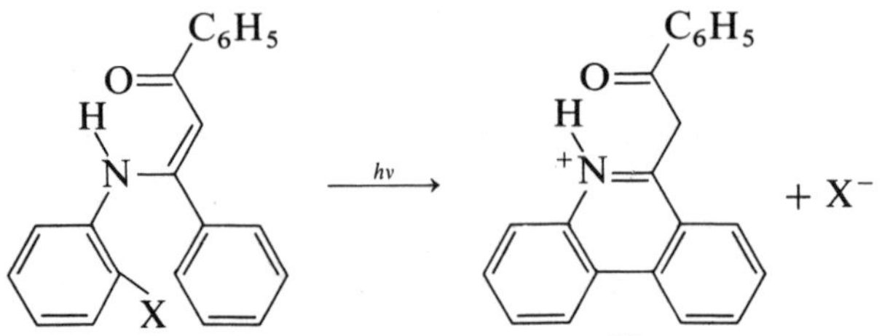

483
X = Br or I

484
(60%)

The widespread use of the photocyclization of enamides as a preparative method, particularly for the synthesis of lactams related to alkaloid natural products, is discussed thoroughly in two recent reviews.[348a,b] Several earlier extensive reviews also exist.[349–352] Some examples of oxidative photocyclization are illustrated for enamides **485** and **486**.

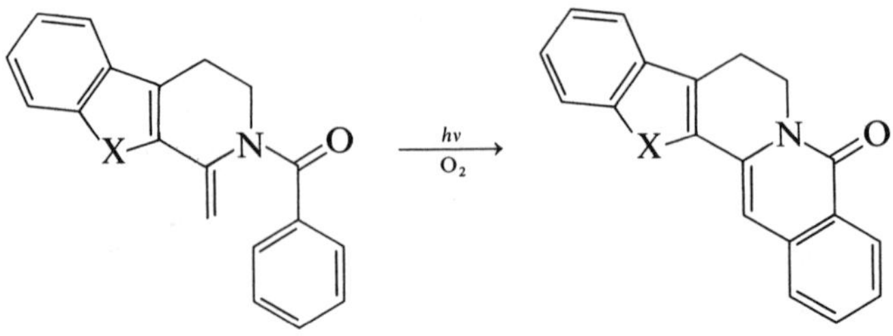

485
X = NH
486
X = S

X = NH (37%)[353,354]
X = S (72%)[355]

Eliminative photocyclization is exemplified by the transformation of *o*-methoxy compound **487** to lactam **488**.[80,356]

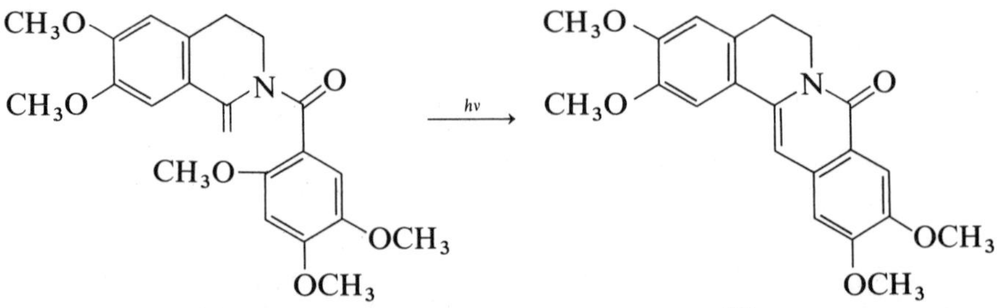

487

488
(85%)

The conversion of enamide **489** to lactam **490** illustrates eliminative photocyclization involving a methylenedioxy group to give a phenol by the loss of formaldehyde from the labile hemiacetal produced initially ($ArOCH_2OH \rightarrow ArOH + CH_2O$).[357,358]

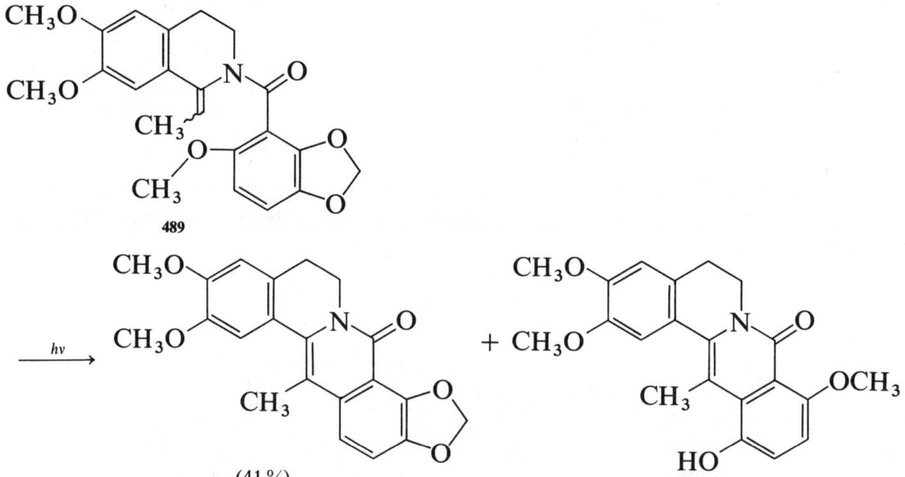

CH_3O CH_3O N O CH_3 O CH_3 O O

489

$\xrightarrow{h\nu}$ CH_3O CH_3O N O CH_3 O O (41 %) + CH_3O CH_3O N O OCH_3 CH_3 HO

490
(29 %)

The stereochemistry of the lactams produced by nonoxidative photocyclization of enamides **491** and **492** is consistent with a conrotatory ring closure followed by a concerted, suprafacial 1,5-hydrogen shift.

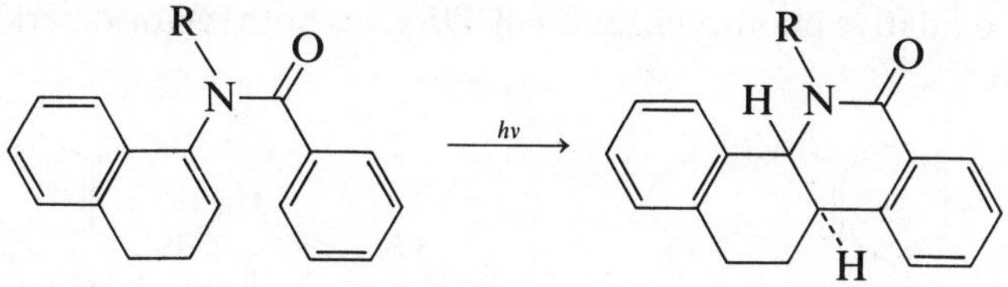

R N O $\xrightarrow{h\nu}$ R H N O H

491 (51–55 %)[350,359,360]

$R = CH_3$ or $C_6H_5CH_2$

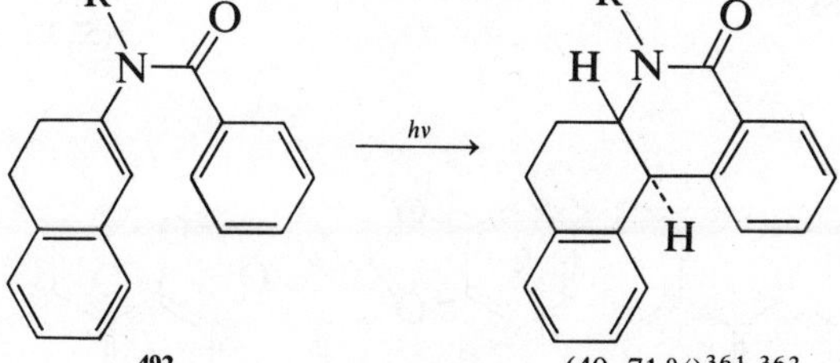

R N O $\xrightarrow{h\nu}$ R H N O H

492 (40–71 %)[361,362]

$R = CH_3$, n-C_4H_9, $C_6H_5CH_2$, or $CH_2{=}CHCH_2$

meta-Substituted enamides such as **493** undergo nonoxidative photocyclization with a high degree of regioselectivity.[81]

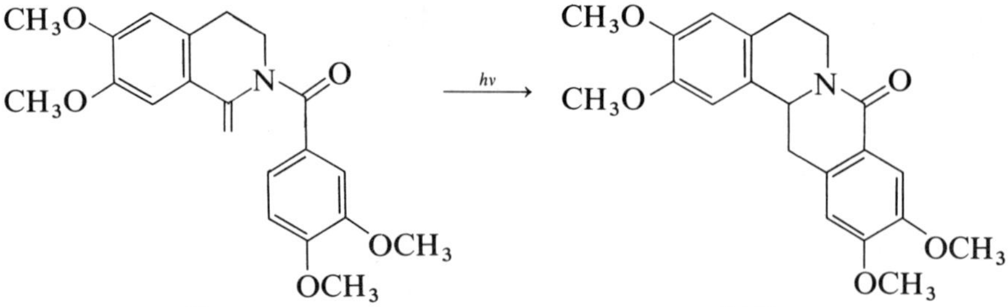

493 (94%)

The corresponding oxidative photocyclizations usually are less regioselective; for example, enamide **493** gives a mixture of unsaturated lactams **488** and **494**.[363,364]

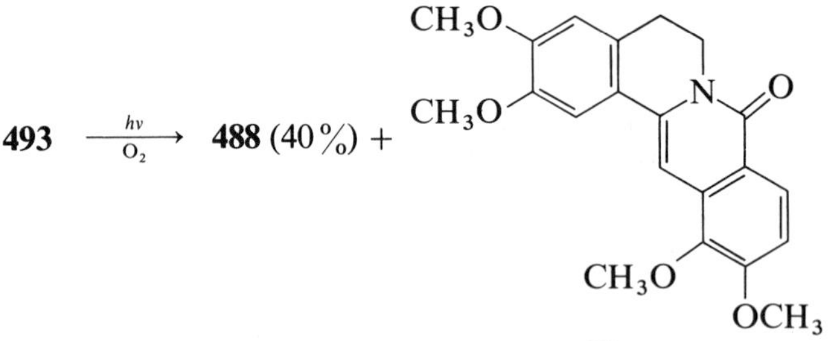

494 (5%)

Similarly, although nonoxidative photocyclization of enamide **495** gives only lactam **496**, oxidative photocyclization of **495** gives both regioisomeric products **497** and **498**.[350,365]

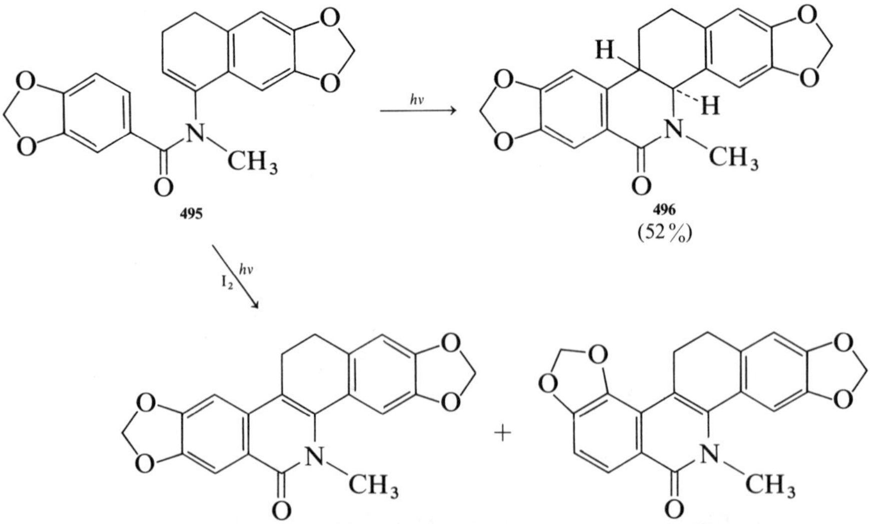

495 496 (52%)

497 (25%) 498 (6%)

As shown with abbreviated formulas in Scheme 29, two isomeric zwitterions, **499a** and **499b**, are produced by photocyclization of an enamide such as **493**.

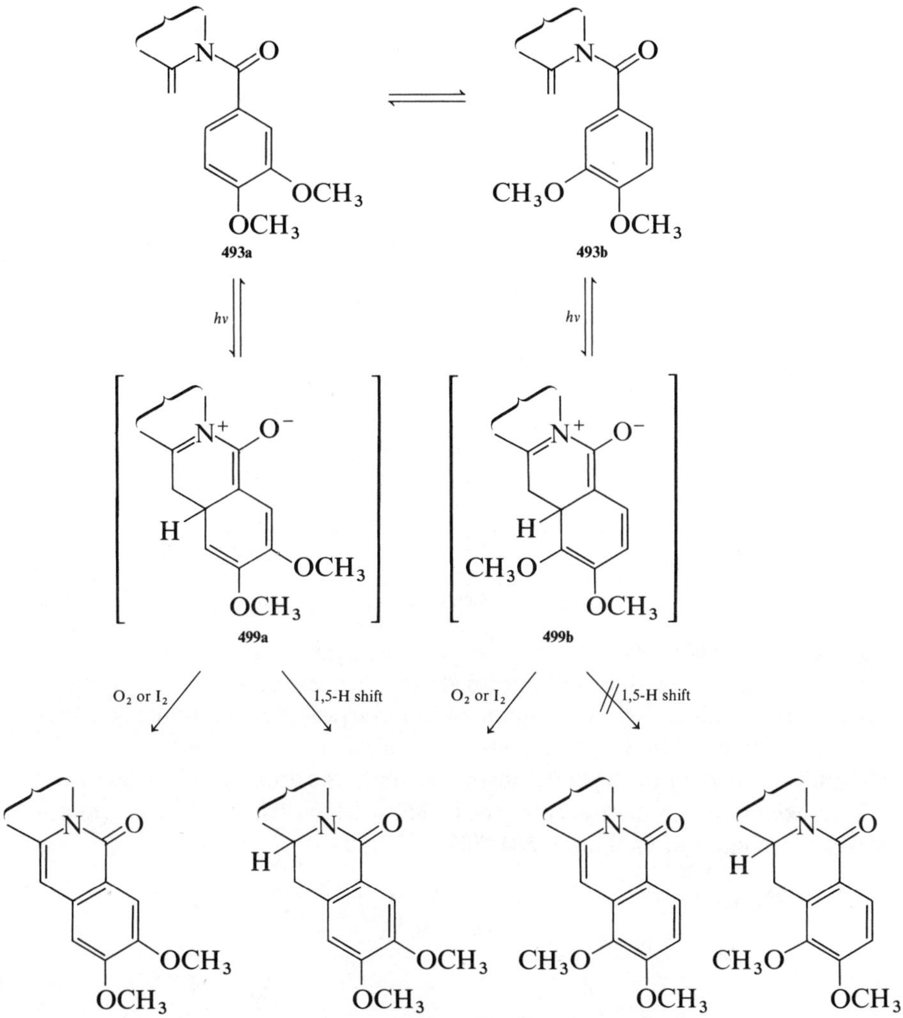

493a ⇌ 493b

499a 499b

O_2 or I_2 1,5-H shift O_2 or I_2 1,5-H shift

Scheme 29

Evidently oxidative trapping with oxygen or iodine competes with ring opening for *both* zwitterions, whereas trapping by a 1,5-hydrogen shift is too slow to compete effectively with the sterically accelerated ring opening of the more crowded zwitterion **499b**.

Remarkably, although irradiation of enamide **500** produces a mixture of isomers **500** and **501** by reversible *cis–trans* photoisomerization of the ethylidene group, only enamide **500** undergoes photocyclization (Scheme 30).[30a]

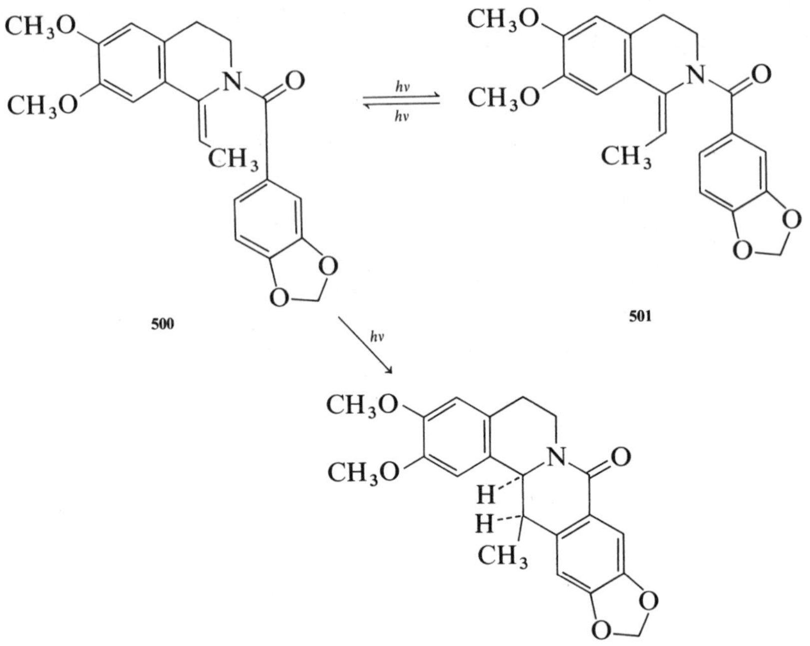

500 ⇌ (hν / hν) 501

500 → (hν) (98%)

Scheme 30

N-Arylamides of α,β-unsaturated carboxylic acids also undergo photocyclization by way of zwitterionic intermediates. For example, oxidative photocyclization of anilide **502** gives lactam **503**, and irradiation of anilide **502** under nonoxidizing conditions gives a mixture of lactams **504** and **505**.[366] The **504**:**505** product ratio depends on the solvent: the *trans* product **504** forms almost exclusively in diethyl ether (**504**:**505** = 94:6), whereas the *cis* product **505** predominates in methanol (**504**:**505** = 29:71).[366]

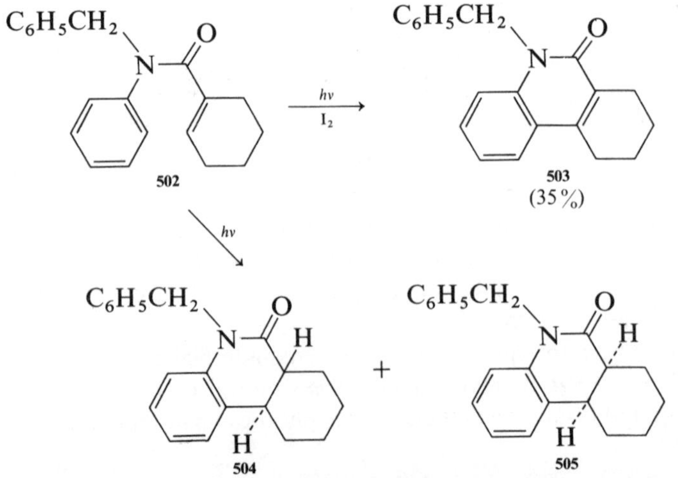

502 → ($h\nu$, I_2) 503 (35%)

502 → ($h\nu$) 504 + 505

In diethyl ether, the intermediate zwitterion **506** is trapped mainly by a concerted 1,5-hydrogen shift; the product in D_2O-saturated diethyl ether consists of almost pure *trans* isomer **504** (53% yield) with little incorporation of deuterium (only 12% of one deuterium atom) (Scheme 31).[366] In methanol, zwitterion **506**, which is especially acidic because proton loss creates an aromatic ring, is trapped mainly by a sequence of proton transfers to and from the solvent; the *cis* product in CH_3OD contains 90% of one deuterium atom (Scheme 31).[366]

502 $\xrightarrow{h\nu}$ [506] $\xrightarrow[\text{1,5-H shift}]{(C_2H_5)_2O,\ D_2O}$ 504

506 $\xrightarrow{CH_3OD}$ (Proton transfers) $\xrightarrow{CH_3OD}$ product

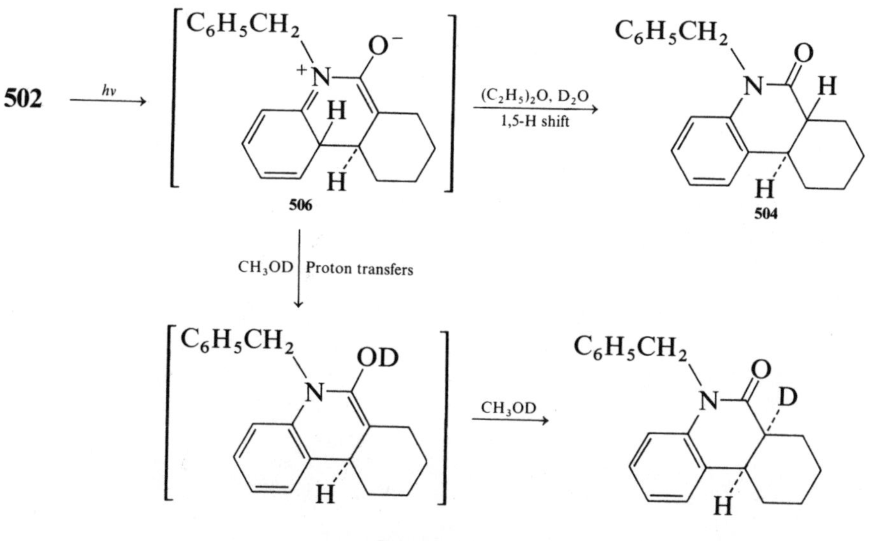

Scheme 31

The photocyclizations of certain *ortho*-substituted anilides involve 1,5 shifts of the substituent. For example, anilide **507** gives rearranged lactam **508**.[367,368]

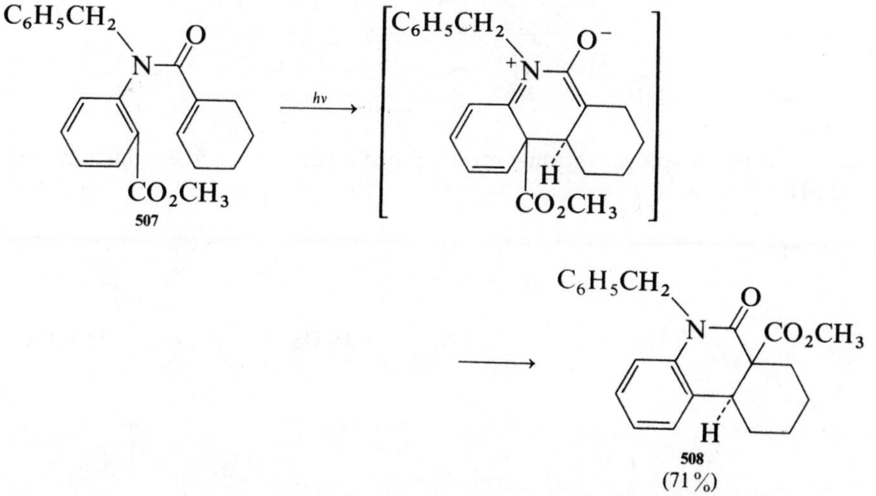

507 $\xrightarrow{h\nu}$ [intermediate] $\longrightarrow$ 508 (71%)

Analogous 1,5 shifts also are known for $COCH_3$, $CONH_2$, and CN groups.[367,368] In anilide **509**, photocyclization at the *ortho* carbon atom bearing the carbomethoxy group is prevented by intramolecular hydrogen bonding; the major product arises from a 1,5-hydrogen shift.[368]

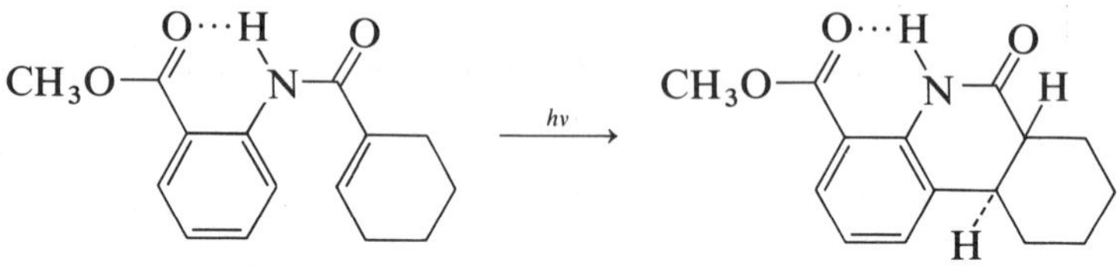

509 → (hν) (60%)

The pyridyl compound **510** gives unsaturated lactam **511** under oxidizing conditions and gives a nearly equimolar mixture of lactams **512** and **513** (total yield 73–88%) under nonoxidizing conditions in diethyl ether, benzene, or methanol.[369]

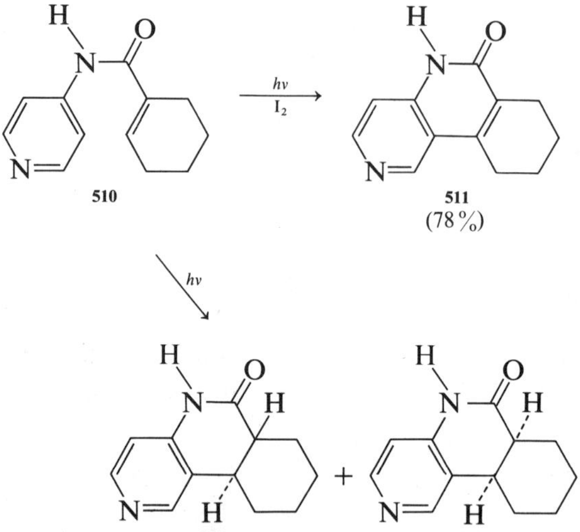

510 → (hν, I_2) **511** (78%)

510 → (hν) **512** + **513**

Nonoxidative photocyclization converts pyridine derivative **514** to lactams **515** and **516**.[370]

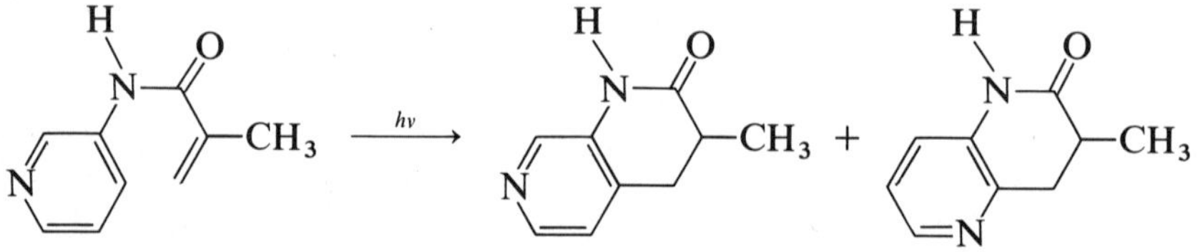

514 → (hν) **515** (22%) + **516** (53%)

Excited-state free-valence indices for amide **514** are in accordance with the preponderance of **516** over **515** as photocyclization products.[370]

Naphthyl compounds of type **517** undergo nonoxidative photocyclization to give lactams of type **518**.[371]

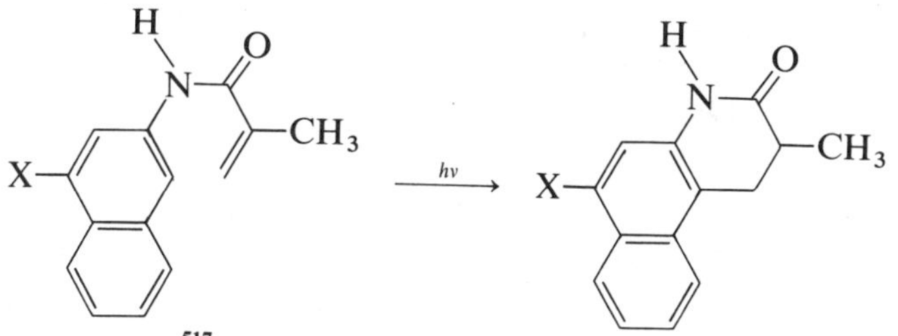

517

518
(43–59%)

X = H or CO_2CH_3

Irradiation of anilide **519** causes reversible *cis–trans* photoisomerization, and both stereoisomers, **519** and **520**, undergo photocyclization, as illustrated in Scheme 32.[372] This contrasts with the behavior shown in Scheme 30.

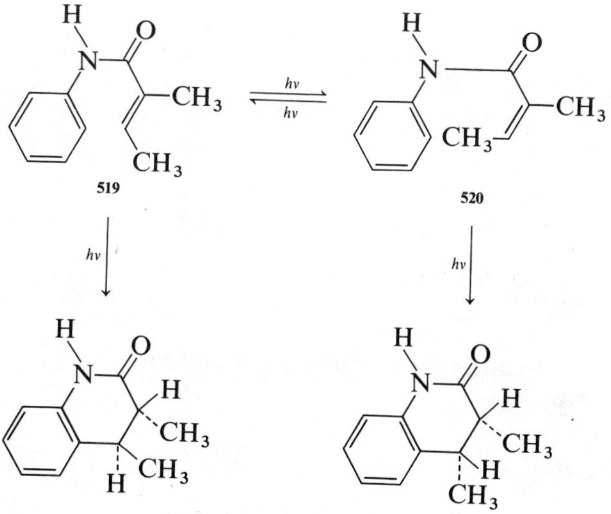

519

520

Scheme 32

Irradiation of formamide **521** in a mixture of dioxane and *tert*-butyl alcohol containing hydroiodic acid gives isoquinolinium ion **522**.[373] The reason for the regioselectivity observed in the photocyclization of this *meta*-substituted enamide is not known.[373]

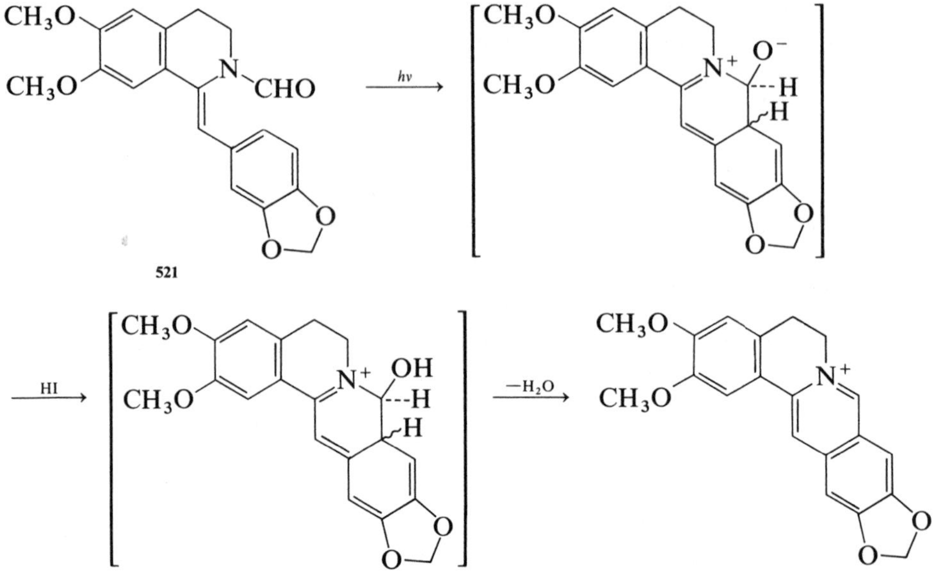

521

522
(100%)

The acetamide derivative **523** shows similar photochemical behavior.[374]

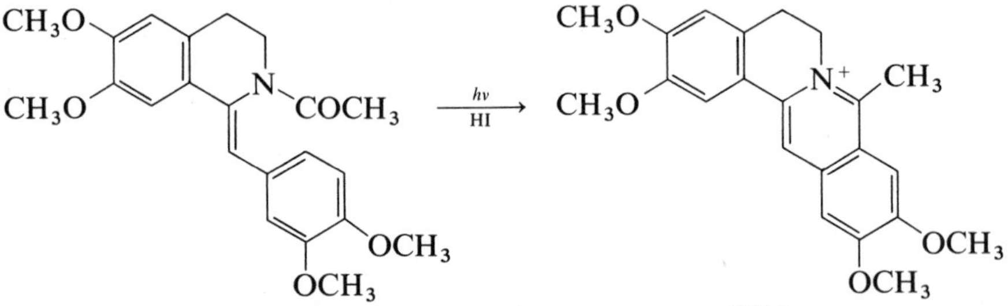

523

(75%)

Benzamide **524** undergoes oxidative photocyclization to give unsaturated lactam **525**.[375]

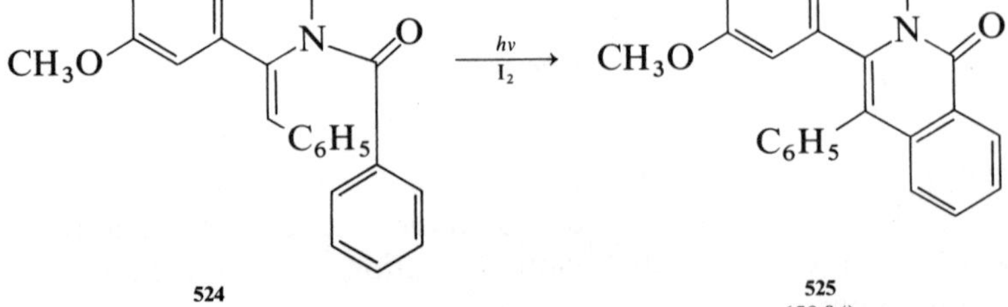

524

525
(50%)

The photocyclization of carbamate **526** gives phenanthridinone (**66**).[376]

$NHCO_2C_2H_5$

526

$h\nu$

H O⁻ OC_2H_5 ^+N H

$-C_2H_5OH$

H O N

66
(85%)

Carbamate **527** gives lactam **528** under nonoxidizing conditions but gives phenanthrene **529** as the major product under oxidizing conditions.[375]

CH_3O CH_3O N $CO_2C_2H_5$

527

$h\nu$

CH_3O CH_3O N O C_6H_5

528
(60%)

$h\nu$ I_2

CH_3O CH_3O N $CO_2C_2H_5$ + **528** (9%) C_6H_5

529
(60%)

The irradiation of carbamate **530** causes olefinic *cis–trans* photoisomerization but inexplicably fails to result in eliminative photocyclization to give either a lactam or a phenanthrene.[80]

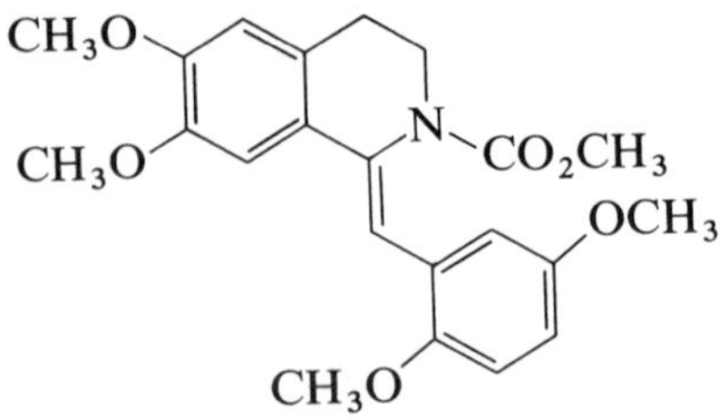

530

Photocyclizations Producing Five-Membered Rings

From 1-Vinylnaphthalenes. As shown in Scheme 33, the 1-vinylnaphthalene derivative **531** undergoes photocyclization to create a five-membered ring by a

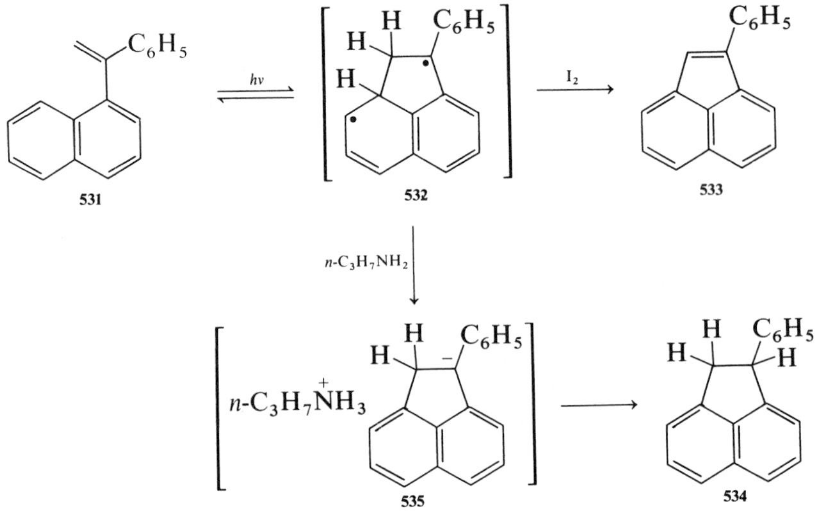

Scheme 33

mechanism analogous to that of stilbene photocyclization, except that dihydro intermediate **532** is formally a biradical.

Intermediate **532** can be trapped oxidatively with iodine in cyclohexane to give acenaphthylene derivative **533** in 70% yield[73] or trapped nonoxidatively by a proton-transfer sequence catalyzed by *n*-propylamine in acetonitrile to give acenaphthene derivative **534** in 95% yield.[377] Consistent with the postulated involvement of an anionic intermediate such as **535**, nonoxidative photocyclization succeeds with nitrile **536**[377] and ester **537**[378] but fails with 1-isopropenylnaphthalene,[377] a hydrocarbon that lacks an anion-stabilizing substituent.

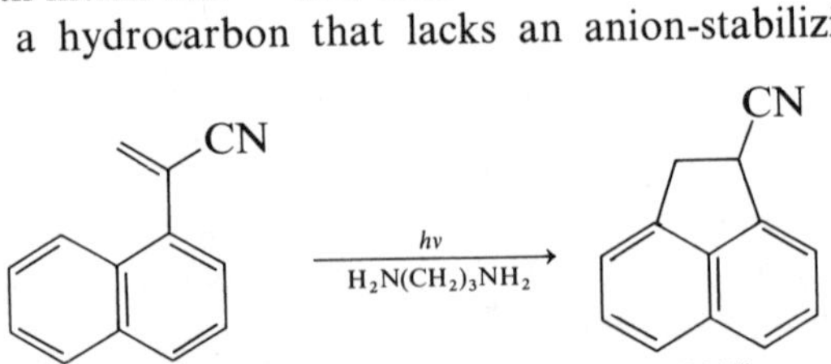

(95%)

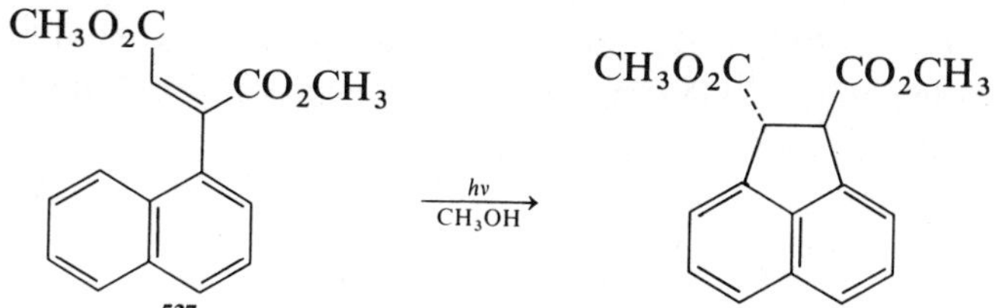

CH_3O_2C CO_2CH_3 537 $\xrightarrow[CH_3OH]{h\nu}$ CH_3O_2C CO_2CH_3 (74%)

Stilbene-like photocyclization of 1-styrylnaphthalene analog **538** to give an azachrysene derivative fails to compete successfully with the formation of compounds **539** and **540**.[379]

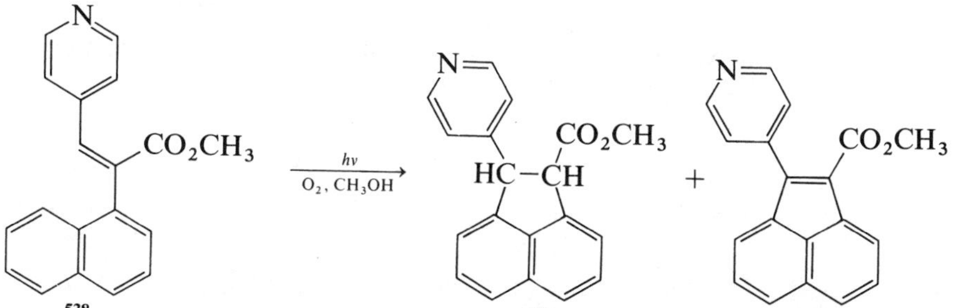

N CO_2CH_3 538 $\xrightarrow[O_2, CH_3OH]{h\nu}$ N CO_2CH_3 HC—CH 539 (56%) + N CO_2CH_3 540 (19%)

Although photocyclization succeeds for pyrene **541**, it fails, unaccountably, for chrysene **542** under similar conditions.[377]

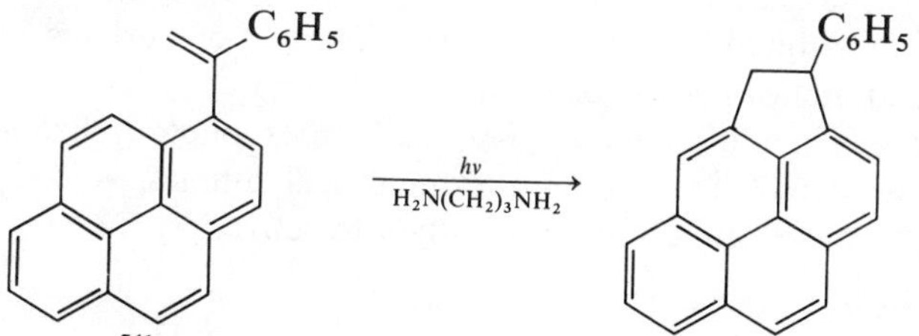

C_6H_5 541 $\xrightarrow[H_2N(CH_2)_3NH_2]{h\nu}$ C_6H_5 (60%)

C_6H_5 542

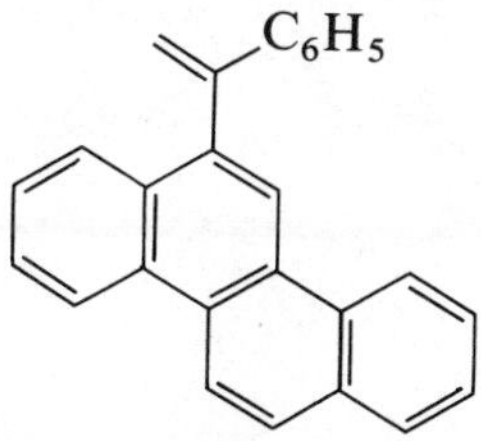

1-Arylnaphthalenes apparently do not photocyclize to give biradical intermediates analogous to **532**. Thus 1-phenylnaphthalene fails to undergo oxidative photocyclization to give fluoranthene.[158,380,381] The successful conversion of *o*-chloro derivative **543** to fluoranthene evidently involves photolysis of the carbon–chlorine bond.[381] This photolysis may be assisted by complexation of the incipient chlorine atom with the naphthalene ring.[37]

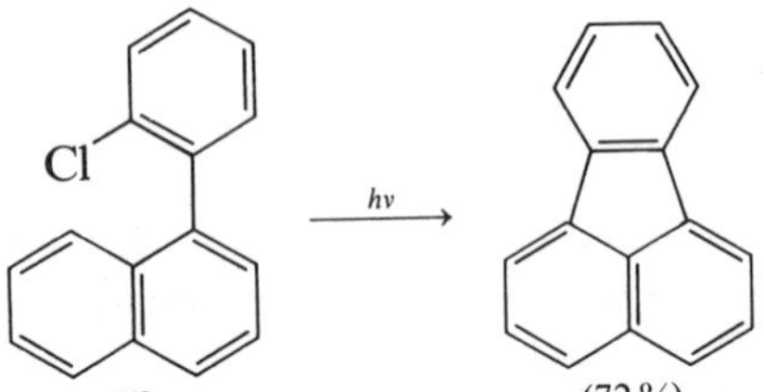

Cl

hν

543 (72%)

From Diaryl and Aryl Vinyl Amines, Ethers, and Sulfides. The irradiation of diphenylamine (**544**) produces carbazole (**545**).[382]

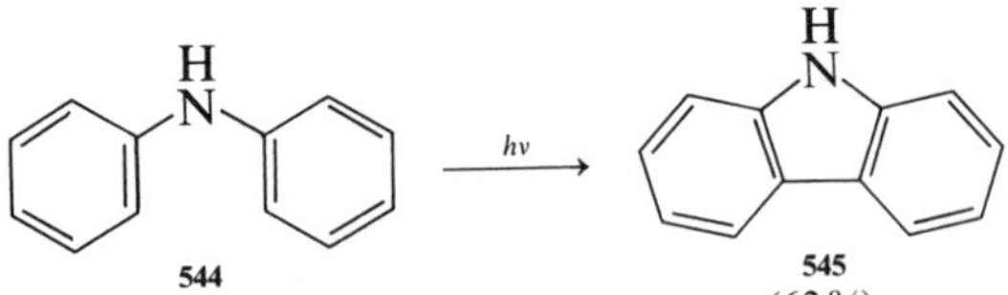

H
N

hν

544 **545** (62%)

The three isomeric *N*-pyridylanilines behave analogously, giving the expected azacarbazoles in 70–81% yield.[382] The photocyclizations of these four secondary amines are reported to take place under a nitrogen atmosphere.[382] Perhaps the carbazoles are produced by way of the elimination of molecular hydrogen from a dihydrocarbazole intermediate,[83b] or perhaps residual dissolved oxygen functions as the oxidant.

Although di-2-naphthylamine (**546**) undergoes photocyclization to give dibenzocarbazole **547**,[383] several unsymmetrical amines, including 1- and 2-anilinonaphthalene, are photochemically unreactive.[382–384]

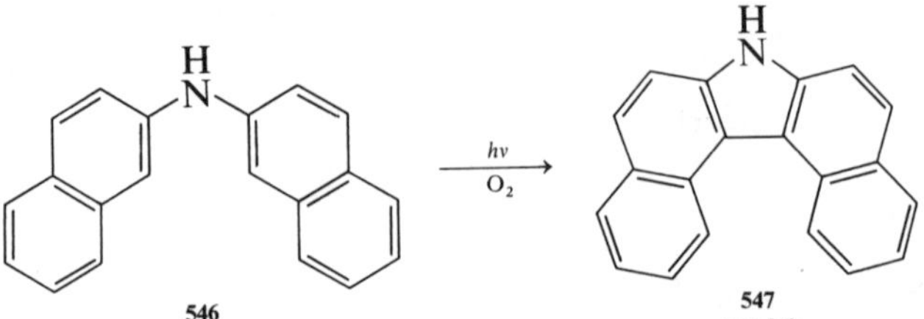

H
N

hν, O_2

546 **547** (18%)

An interesting explanation has been offered for these contrasting results.[383]

The syntheses of carbazoles by the photocyclization of tertiary amines such as *N*-methyldiphenylamine (**81**) and triphenylamine (**548**) are best carried out with the use of oxygen as the oxidant.[385]

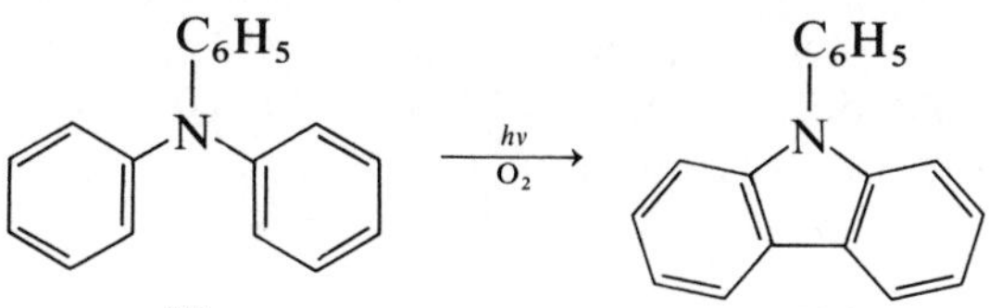

C_6H_5 N C_6H_5 N

548 — $h\nu$, O_2 → (65%)

Diphenyl ether and diphenyl sulfide undergo oxidative photocyclization with iodine as the trapping agent.[386]

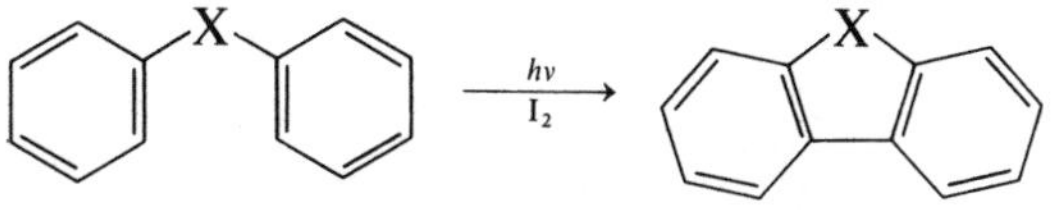

X — $h\nu$, I_2 → X

X = O or S (54–61%)

1-Phenoxynaphthalene fails to undergo oxidative photocyclization, but the *o*-chloro derivative **549** undergoes eliminative photocyclization.[387]

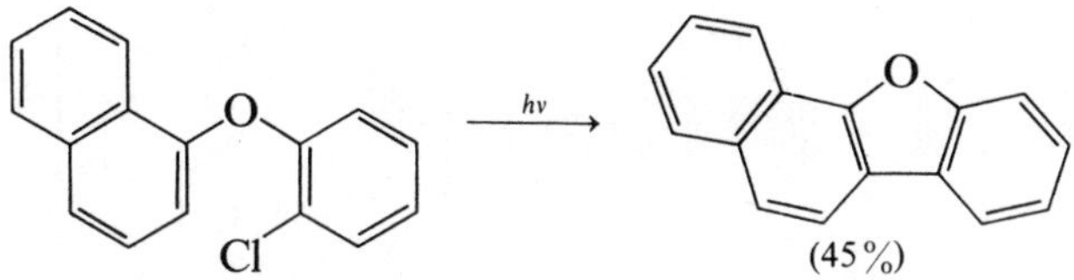

O Cl — $h\nu$ → O

549 (45%)

Ether **550** also photocyclizes with loss of hydrogen chloride.[388a,b]

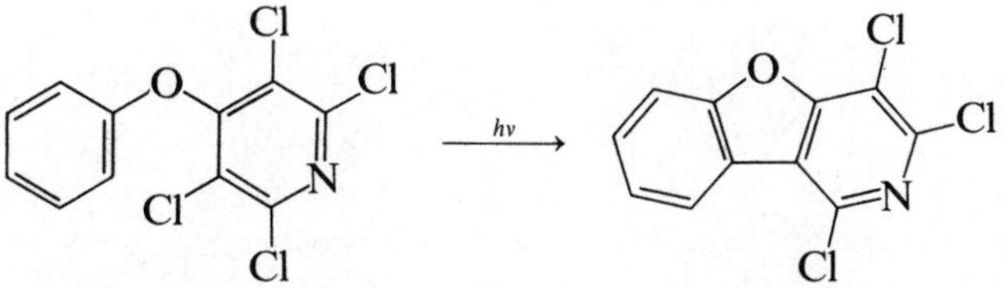

Cl O Cl N Cl Cl — $h\nu$ → Cl O Cl N Cl

550 (54%)

Photocyclization of diaryl ethers with elimination of an *o*-methoxy group proceeds in very low yield except in certain multisubstituted systems such as **551**.[389,390]

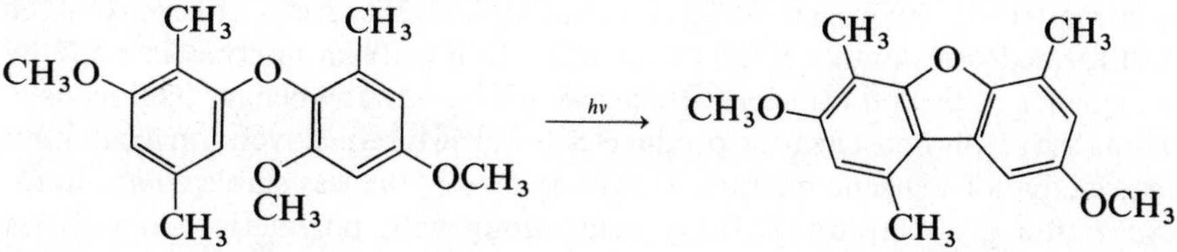

CH_3 CH_3 CH_3O O O CH_3 CH_3 OCH_3 — $h\nu$ → CH_3 CH_3 O CH_3O CH_3 OCH_3

551 (35%)

Nonoxidative photocyclization of enamine **552** gives the *trans*-indoline **553**;[19,391a] the photocyclization of **552** with circularly polarized light gives **553** with a slight enantiomeric excess.[391b]

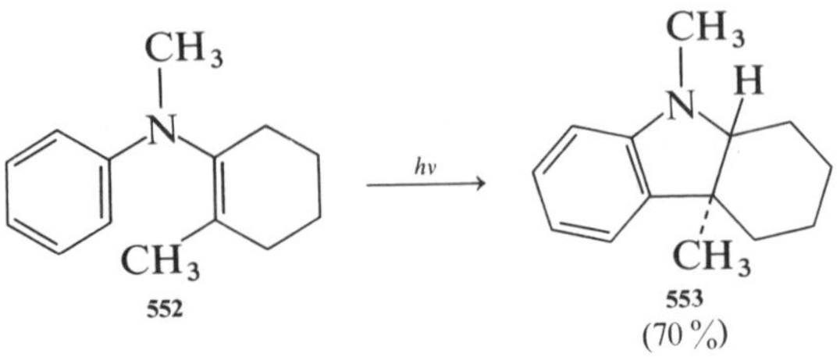

CH_3 $h\nu$ CH_3 H

CH_3 CH_3

552 553 (70%)

Irradiation of keto derivative **554**, however, gives a mixture of stereoisomeric indolines, **555** and **556**, as shown in Scheme 34.[392]

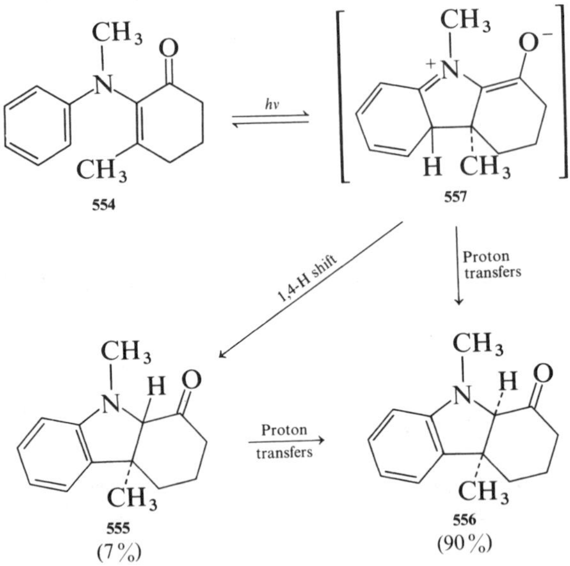

CH_3 O $h\nu$ CH_3 O^- $+N$

CH_3 H CH_3

554 557

1,4-H shift Proton transfers

CH_3 H O Proton transfers CH_3 H O

CH_3 CH_3

555 (7%) 556 (90%)

Scheme 34

Presumably, *trans*-indoline **555** is formed from dihydro intermediate **557** by a concerted 1,4-hydrogen shift, and *cis*-indoline **556** arises either from the first-formed *trans*-indoline **555** by epimerization or from intermediate **557** by a sequence of proton transfers. Ether **558** and sulfide **559** behave analogously, giving the *cis* photocyclization products **560** and **561**, respectively, on irradiation in a methanol–benzene mixture but giving mostly the less stable *trans* photocyclization products (in 27–30% yield, along with polymeric material) on irradiation in benzene.[34,35]

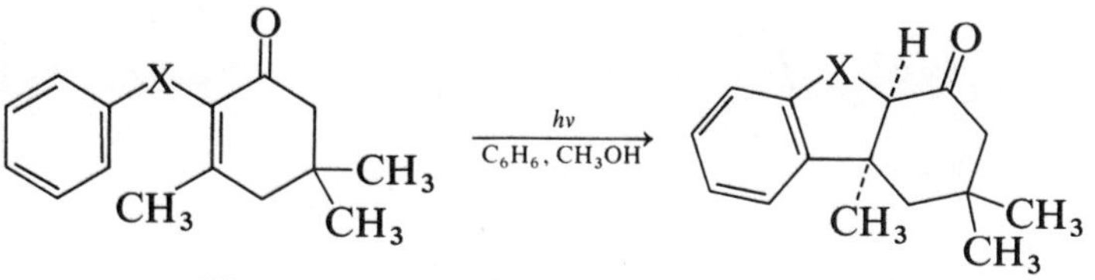

558
X = O
559
X = S

$h\nu$
C_6H_6, CH_3OH

560
X = O (88%)
561
X = S (91%)

meta-Substituted derivatives of **558** and **559** exhibit a remarkable regioselectivity in their photocyclizations, undergoing ring closure predominantly at the *ortho* carbon adjacent to the substituent.[34,35] For example, methyl-substituted ether **562** gives a 3:1 mixture of products **563** and **564**, and the methoxy analog **565** gives **566** as the only detected product.[35]

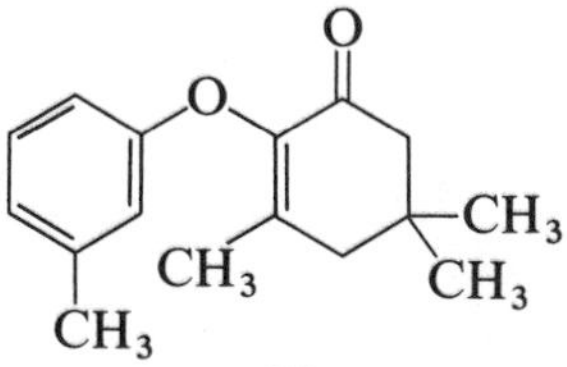

562

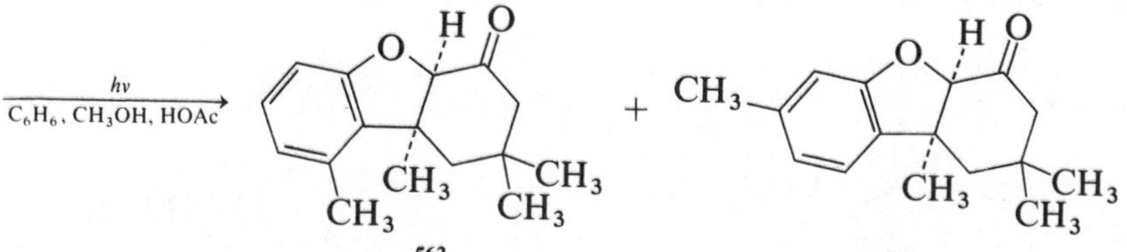

$h\nu$
$C_6H_6, CH_3OH, HOAc$

563
(75%)

564
(25%)

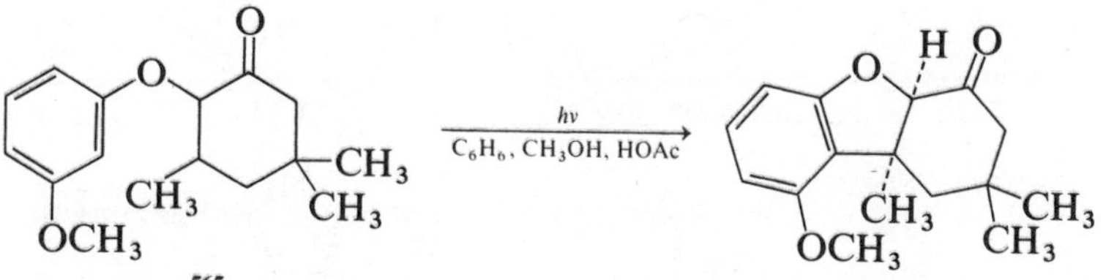

565

$h\nu$
$C_6H_6, CH_3OH, HOAc$

566
(90%)

When the irradiation of sulfide **567** is carried out in the presence of *N*-phenylmaleimide, the photochemically generated dihydro intermediate **568** is trapped, giving adduct **569** in 81% yield; otherwise, intermediate **568** rearranges by a 1,4-hydrogen shift to give *trans* product **570** in 78% yield, as depicted in Scheme 35.[74,393]

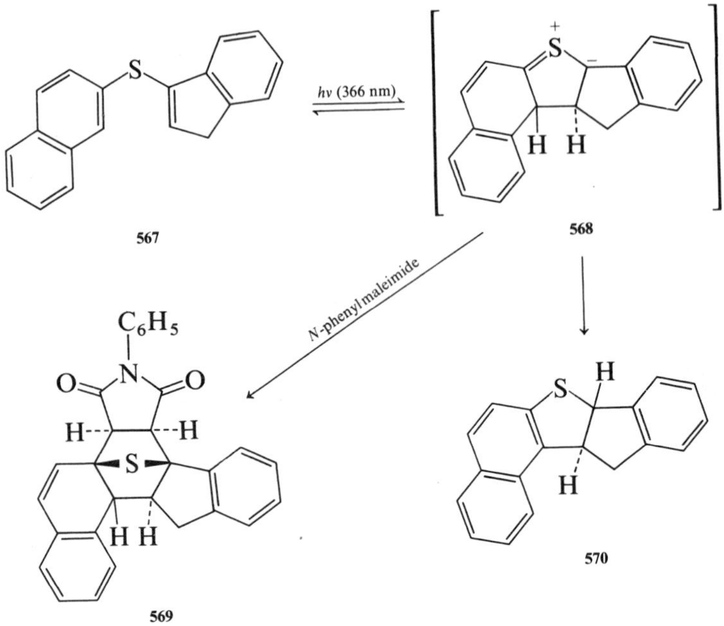

567 568 569 570

N-phenylmaleimide

$h\nu$ (366 nm)

Scheme 35

Irradiation of enol **571** under nonacidic conditions gives alcohol **572**;[394] under acidic conditions, dehydration product **573** is obtained.[395]

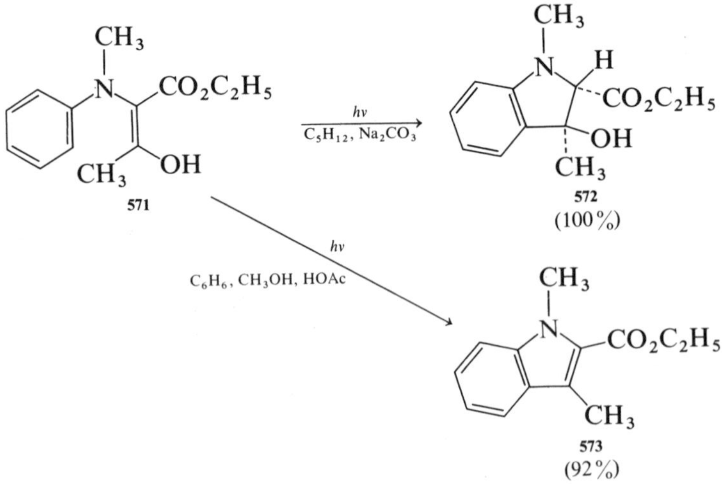

$h\nu$, C_5H_{12}, Na_2CO_3

$h\nu$, C_6H_6, CH_3OH, HOAc

571 572 (100%) 573 (92%)

The analog of **571** lacking the *N*-methyl substituent behaves similarly,[396] as does enol **574**, the only reported example of a selenide that undergoes photocyclization.[397]

Se COCH3 CH3 OH 574 → hv, C_6H_6, p-$CH_3C_6H_4SO_3H$ → Se COCH3 CH3 (60%)

Uracil derivatives such as **575** undergo photocyclization.[398]

575 → hv → (87%)

Photocyclization involving a carbon–nitrogen double bond is observed for conversions such as **576** to **577**.[399]

576 → hv, CH_3CN, HOAc → 577 (92%)

Amine **578** undergoes eliminative photocyclization to give indole **579**.[400]

578 → hv → 579 (80%)

Cyano compound **580** photocyclizes with loss of hydrogen cyanide.[401]

580 → hv → 545 (87%)

The conversion of amine **581** to carbazole **582** exemplifies a type of eliminative photocyclization that takes place in hydrocarbon solvents but is suppressed completely in ethanol, tetrahydrofuran, or carbon tetrachloride.[402]

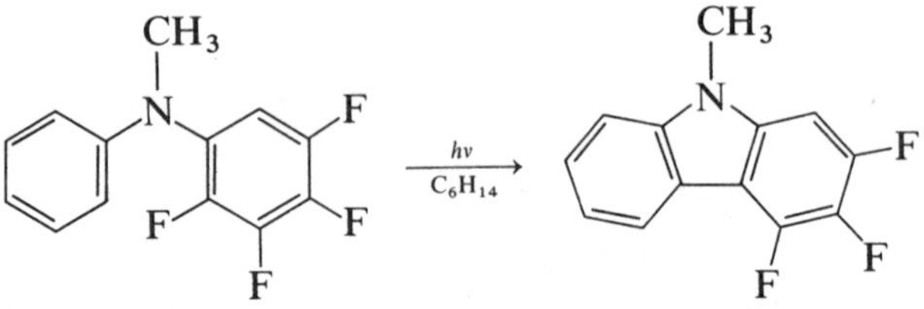

581 **582** (66%)

It remains uncertain whether the photochemical conversions of keto sulfides such as **583** to dihydrobenzothiophenes like **584** in acetonitrile[403] proceed by photocyclization of the corresponding enol or by intramolecular photo-addition of an *ortho* carbon atom to the carbonyl group of the ketone itself.

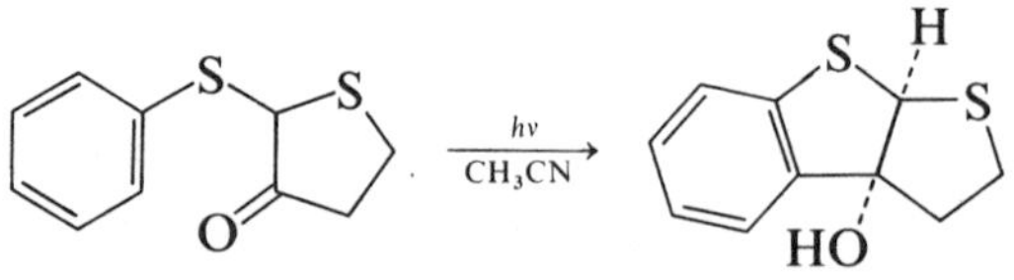

583 **584** (43%)

From Thiocarbonyl Compounds. Irradiation of thioketone **585** gives cyclized product **586**.[404]

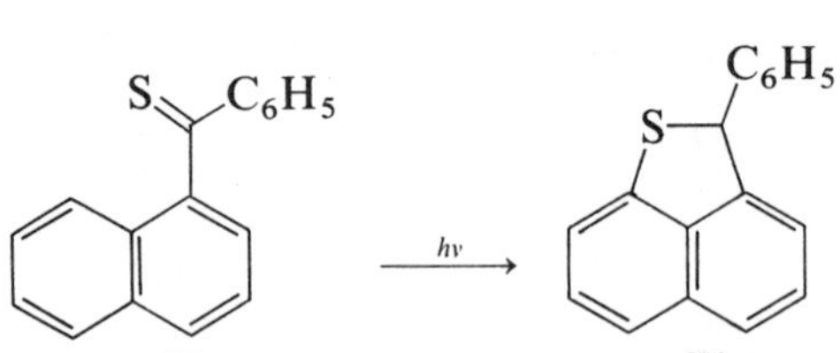

585 **586** (51%)

Oxidative photocyclization of thioamide **587** gives heterocycle **588**.[405]

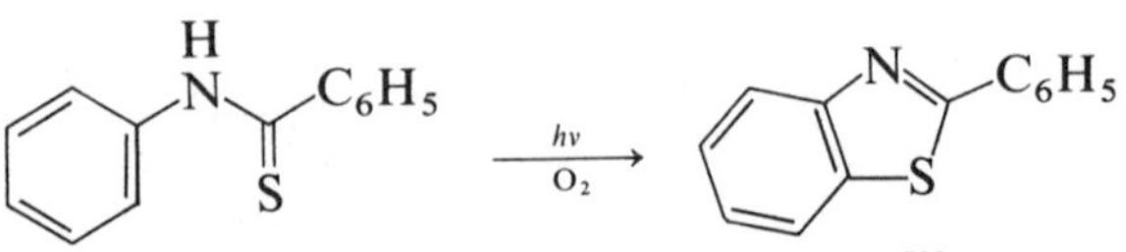

587 **588**

Rearrangements

Photocyclizations rarely involve rearrangements of substituents other than hydrogen. Migrations are known only for certain groups attached to one of the carbon atoms at which ring closure takes place.

For example, oxidative photocyclization of phenyl-substituted dinaphthylethylene **589** gives benzo[*a*]coronene (**590**) by way of a phenyl migration as indicated in Scheme 36.[232] This multistep conversion begins with oxidative

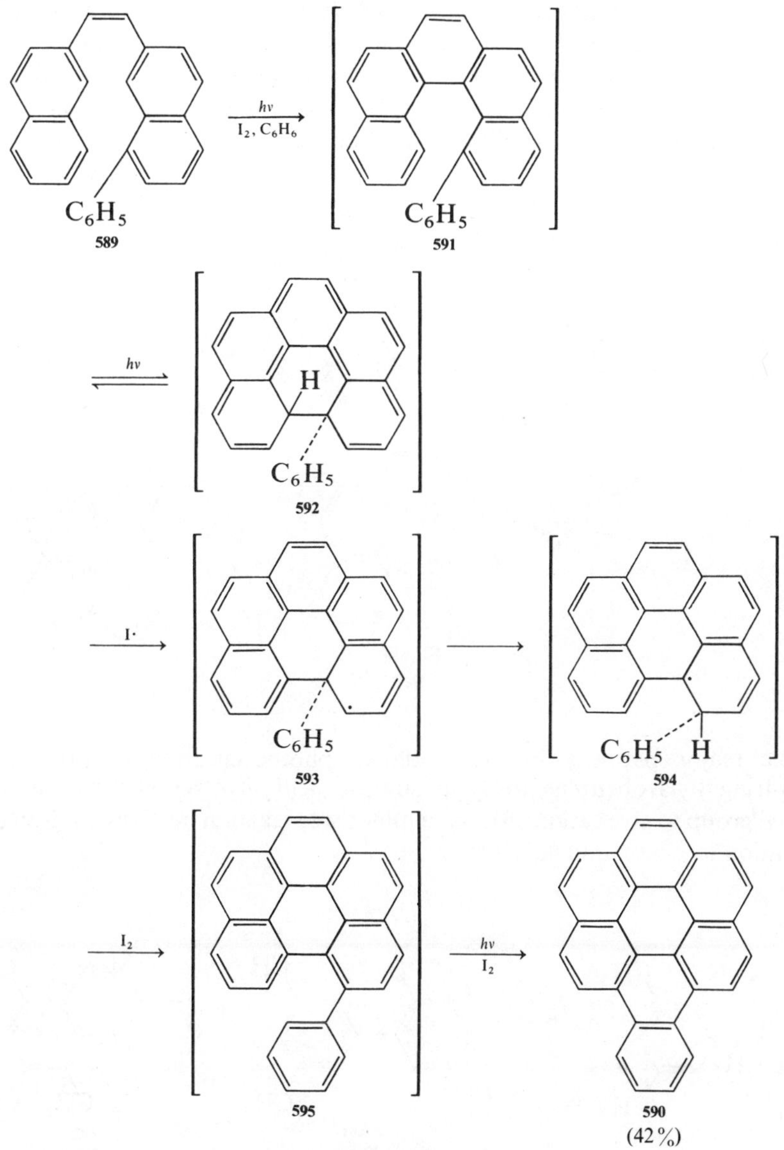

Scheme 36

photocyclization of **589** to 1-phenyl[5]helicene (**591**). Subsequent photocyclization of **591** to **592** followed by hydrogen abstraction gives **593**, a neophyl-type radical that undergoes 1,2-phenyl migration to give the less crowded radical **594**. Hydrogen transfer converts **594** to hydrocarbon **595**, which then undergoes a final oxidative photocyclization to give the isolated product **590**.

The irradiation of certain other overcrowded phenyl-substituted systems related to **591** leads in part to phenyl rearrangements and in part to some unusual photofragmentation reactions.[231,406]

In the oxidative photocyclizations of some 1-fluoro[5]helicenes, a remarkable rearrangement occurs in which the fluorine atom migrates to a different ring.[407] For example, [5]helicene derivative **596**, obtained by oxidative photocyclization of styrylphenanthrene **597**, gives rearranged benzo[*ghi*]perylene **598**; elimination product **599** also is obtained.[407]

F, Cl, **597** $\xrightarrow[I_2,\ C_6H_6]{h\nu}$ F, Cl, **596**

$\xrightarrow[I_2,\ C_6H_6]{h\nu}$ Cl, F, **598** + Cl, **599**

The major pathway in the oxidative photocyclization of protonated 2,4,6-trimethylazobenzene (**600**) in sulfuric acid involves elimination of a methyl group to give cation **601**, but a minor side reaction involves a 1,2-methyl migration to give cation **602**.[313]

H, $^{+}$N=N, CH_3, CH_3, CH_3, **600** $\xrightarrow[H_2SO_4]{h\nu}$ H, $^{+}$N=N, CH_3, CH_3, **601** (20%) + H, $^{+}$N=N, CH_3, CH_3, CH_3, **602** (2%)

In the oxidative photocyclization of anilinoborane **603** in cyclohexane with $[I_2] = 10^{-2}$ *M*, methyl migration to give product **604** greatly predominates over methyl loss to give product **605**;[321b,408] with $[I_2] = 5 \times 10^{-4}$ *M*, however, only the demethylated product **605** is observed.[408]

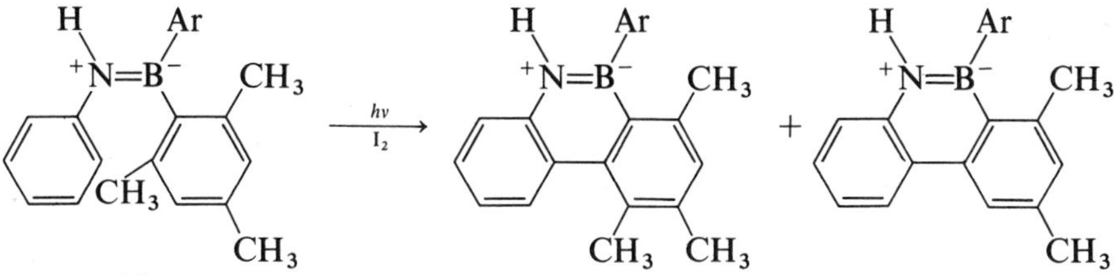

Ar = 2,4,6-trimethylphenyl

Suprafacial 1,5 shifts of substituents other than hydrogen occur in the dihydro intermediates produced by the photocyclization of certain amides. For example, irradiation of amide **606** leads to a 1,5-methoxy shift,[103,367] and irradiation of amides of type **607** leads to 1,5-X shifts.[367,368]

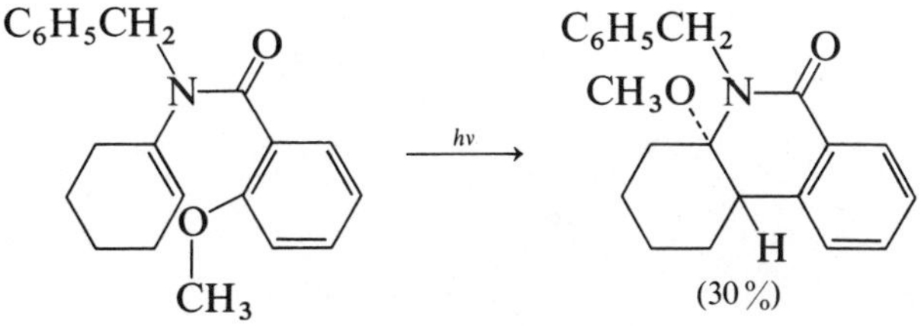

(30%)

(40–71%)

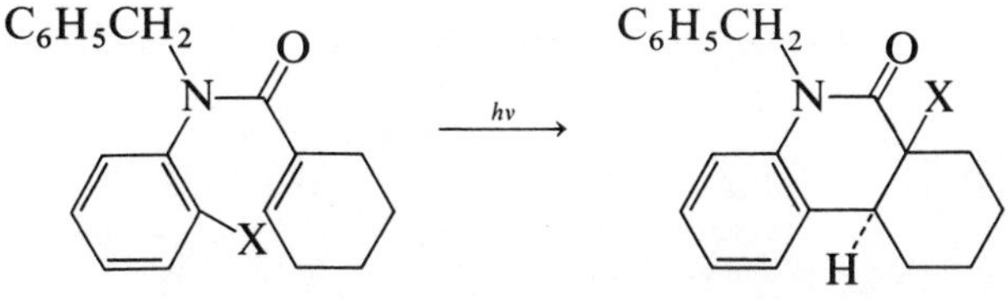

$X = CO_2CH_3, COCH_3, CONH_2$, or CN

Similarly, 1,5-acyl shifts are observed in the photocyclization of ketone **92** (Scheme 10) and aldehyde **286**.[33]

The irradiation of diester **608** is reported to give anthracene **609** as a minor product along with the expected phenanthrene **610** as the major product.[409] A plausible multistep pathway to anthracene **609** can be imagined, starting with the photochemical formation of a new bond between C-α′ and C-2 in the *trans* isomer of stilbene **608**.

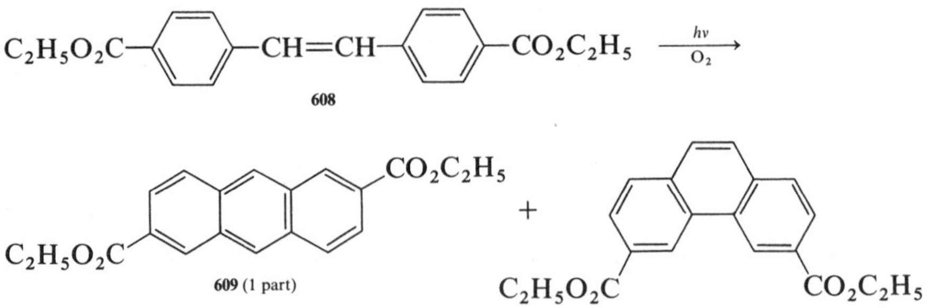

$C_2H_5O_2C-C_6H_4-CH{=}CH-C_6H_4-CO_2C_2H_5 \xrightarrow[O_2]{h\nu}$

608

609 (1 part) + **610** (4 parts)

COMPARISON WITH OTHER METHODS

As described in a recent review, many different nonphotochemical methods have been developed for the construction of the phenanthrene ring system.[410] These include intramolecular cyclizations of various derivatives of benzene, biphenyl, naphthalene (e.g., the Haworth synthesis illustrated in Scheme 37), and stilbene (e.g., the Pschorr synthesis illustrated in Scheme 38); intermolecular

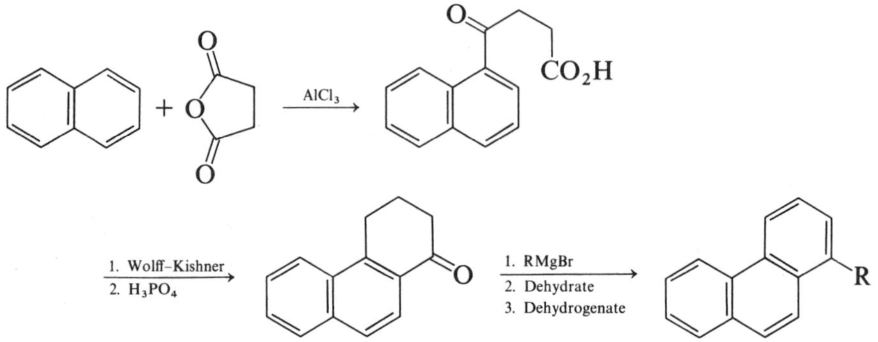

$\xrightarrow{AlCl_3}$ ... CO_2H

1. Wolff-Kishner
2. H_3PO_4

1. RMgBr
2. Dehydrate
3. Dehydrogenate

Scheme 37

CH_2CO_2H, NO_2 + CHO $\xrightarrow{-H_2O}$ HO_2C ... NO_2

$\xrightarrow[HCl]{Sn}$ HO_2C ... NH_2 $\xrightarrow[NaI]{C_5H_{11}ONO}$ HO_2C

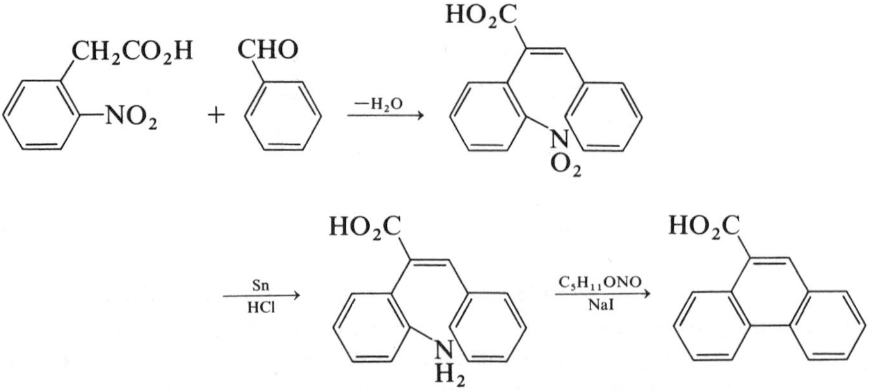

Scheme 38

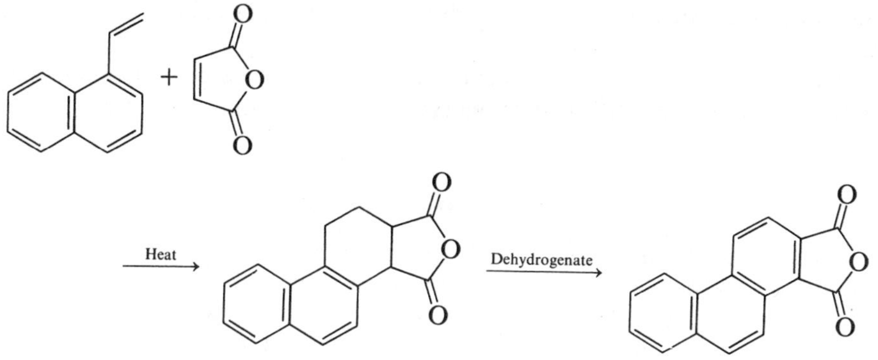

Heat

Dehydrogenate

Scheme 39

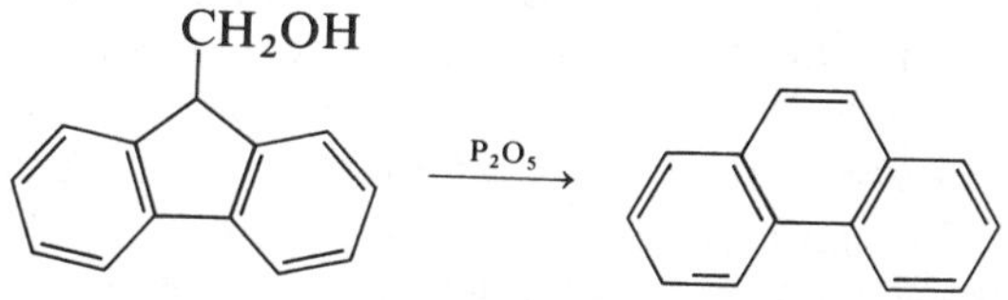

CH_2OH

P_2O_5

Scheme 40

cycloadditions (e.g., the Diels–Alder reaction illustrated in Scheme 39); and various ring expansions of fluorene derivatives (e.g., the Wagner–Meerwein rearrangement illustrated in Scheme 40). These methods often require lengthy reaction sequences to obtain the immediate precursors to the desired phenanthrenes.

Because of its brevity and simplicity, the approach culminating in the photocyclization of a stilbene derivative usually is the preferred synthetic route to the phenanthrene system. The method is reliable for the synthesis of substituted phenanthrenes with a great variety of specifically chosen substitution patterns; only with most *meta*-substituted stilbenes are there significant regiochemical complications. In contrast, direct electrophilic substitution on phenanthrenes usually proceeds without adequate regioselectivity for synthetic practicality.

The superiority of the photocyclization approach is especially prominent in the syntheses of polynuclear aromatic systems larger than phenanthrene.

Seen in broader perspective, the two-step sequence involving the preparation of a stilbene by an olefin-forming reaction (e.g., Wittig, Grignard, Perkin, Meerwein, or Siegrist) followed by the photocyclization of that stilbene constitutes a method for the construction of a six-membered ring by the assembly of two smaller precursor molecules. In this generalized sense, the photocyclization method can be compared with the Diels–Alder reaction. The photocyclization and the Diels–Alder approaches differ in synthetic strategy: in the former each precursor contributes three of the carbon atoms in the new ring, whereas in the latter one precursor contributes four and the other precursor contributes two of the ring carbon atoms. The two synthetic approaches also are complementary in their outcomes: the photocyclization method usually creates an aromatic ring, whereas the Diels–Alder reaction usually creates an alicyclic ring.

The easy access to various six-membered and five-membered heteroaromatic polycyclic systems by stilbene-like photocyclization is without parallel.

EXPERIMENTAL CONDITIONS

Many different commercially available mercury arc lamps can be used as sources of UV light for stilbene or stilbene-like photocyclizations on a preparative scale. Among the most popular of these are Hanovia medium-pressure immersion lamps (supplied by Ace Glass, Inc., P.O. Box 688, Vineland, New Jersey 08360) and Rayonet low-pressure lamps (supplied by Southern New England Ultraviolet Company, P.O. Box 4134, Hamden, Connecticut 06514). The fundamental requirement for the light source is emission in a wavelength region that corresponds to an electronic absorption band for the reactant. In addition, only sources of very high intensity are practical for preparative-scale work. Medium- and high-pressure mercury lamps emit light throughout the region of 200–400 nm and are widely used for photocyclizations. The emission from simple low-pressure mercury lamps is limited almost entirely to a narrow region around 254 nm; these lamps are not as well suited for stilbene photocyclizations because the products usually absorb more strongly than do the reactants in the region of 200–300 nm. Satisfactory results are often obtained, however, with low-pressure mercury lamps that are coated on the inside with a material that absorbs the 254-nm radiation and subsequently emits its own characteristic fluorescence at longer wavelengths. Lamps that have a main output at around either 300 or 350 nm are available.

Caution: The intense irradiation from the UV light sources commonly used for preparative-scale photocyclizations can cause serious damage to unprotected eyes and skin. In addition, small amounts of ozone are produced in the air around most such light sources. To avoid both of these hazards, one can carry out the irradiation in a hood with the door covered by an opaque material such as cardboard or aluminum foil. Manipulations of the apparatus while the light source is in operation should be minimized and should not be carried out unless goggles or safety glasses, long sleeves, and gloves are worn.

Some light sources are designed to be enclosed in a transparent well with a water-cooled jacket and immersed in the solution to be irradiated. Surrounding of the light source by the reaction mixture in this way maximizes the effectiveness with which the emitted photons are captured by the reactant molecules. Other light sources are designed to be mounted outside the reaction vessel.

The use of a quartz immersion well (or a quartz reaction vessel with an external light source) allows light of all wavelengths above about 200 nm to enter the reaction mixture. For some photocyclizations, higher chemical yields are obtained by employing a Pyrex immersion well (or a Pyrex reaction vessel).

This excludes from the reaction mixture light of wavelengths below about 300 nm and thereby protects the product from photochemical destruction.

The irradiation time required for optimal conversion of reactant to product in a photocyclization reaction depends on several factors: the type of light source and its positioning relative to the reaction mixture; the transparency of the material through which the light enters the reaction vessel; the purity of the reactant and the solvent; the effectiveness with which the reaction mixture is stirred; and the scale of the reaction. (Surprisingly, the importance of the last factor is often overlooked. Photons can be regarded as one of the reactants in a photoreaction; keeping other reaction conditions unchanged, a larger scale requires a greater number of photons and thus a longer irradiation time.) Because so many factors are operative, the reported duration of irradiation in a published description of a particular photoreaction is not a reliable guide for prediction of the irradiation time that would be required for the repetition of the work by an independent investigator. Therefore, it is essential to monitor the progress of a photocyclization to determine when to stop the irradiation. Any convenient analytical method can be employed, such as gas–liquid chromatography (GLC), thin-layer chromatography (TLC), or UV spectroscopy.

During the course of a photoreaction, the wall of the immersion well (or reaction vessel) through which the light enters the reaction mixture often becomes coated with opaque material, thereby interfering with the completion of the reaction (as revealed by monitoring). In this event the reaction should be interrupted and the immersion well and reaction vessel cleaned; it may also be advisable to filter the reaction mixture through a short column of alumina to remove highly colored or polymeric materials before returning the mixture to the cleaned reaction vessel for resumption of irradiation.

For oxidative photocyclizations employing iodine as the oxidant, it may be necessary to add more iodine during the course of the reaction if the reaction mixture loses its purple color.

For some photocyclizations, it is desirable to exclude dissolved oxygen. Three techniques have been employed for this purpose: degassing the solution under high vacuum by a series of freeze–thaw cycles and then sealing the reaction vessel before irradiation; heating the solution under reflux and then allowing it to cool under an atmosphere of an inert gas such as nitrogen or argon before irradiation; and continuously purging the solution with nitrogen or argon during irradiation. This last method is recommended for preparative-scale work because of its simplicity and effectiveness.

In most preparative-scale photoreactions, the absorptivity of the reactant is sufficiently high that the light penetrates only an extremely short distance into the reaction mixture. As a consequence, stirring is very important in order to remove the product from this very thin reaction zone and bring in fresh reactant.

A more detailed general discussion concerning experimental methods in photochemical syntheses is given elsewhere.[411]

EXPERIMENTAL PROCEDURES

3-Fluorophenanthrene.[412] A solution of 1.98 g (0.01 mol) of *trans-p*-fluorostilbene and 0.127 g (0.5 mmol) of iodine in 1 L of cyclohexane was stirred magnetically and irradiated with a modified General Electric H100A4/T 100-W medium-pressure mercury lamp in a water-cooled quartz immersion well for 10 hours until the reaction was judged complete by GLC. At intermediate stages in the conversion, analysis on SE-30, SE-52, or neopentyl glycol succinate columns indicated three peaks; in order of increasing retention time these corresponded to *cis-p*-fluorostilbene, *trans-p*-fluorostilbene, and 3-fluorophenanthrene. The reaction mixture was evaporated to dryness at reduced pressure, and the residue was chromatographed on alumina by using cyclohexane as the eluent. Evaporation of the solvent and recrystallization of the residue from methanol gave 1.48 g (76%) of 3-fluorophenanthrene, mp 88.2–89.0°.

***trans*-9,10-Dicyano-9,10-dihydrophenanthrene.**[69] A solution of 5.22 g (0.023 mol) of *trans*-α,α'-dicyanostilbene in 275 mL of purified benzene was degassed at $-78°$ under high vacuum. The solution was then irradiated with a Hanovia 400-W medium-pressure mercury lamp through a water-cooled Pyrex jacket for 15.75 hours. During irradiation the solution was maintained under vacuum and was stirred magnetically. The product crystallized from the solution during the reaction. After filtration, removal of the solvent, and recrystallization of the residue from acetone, *trans*-9,10-dicyano-9,10-dihydrophenanthrene (3.23 g, 85%) was obtained as prisms with mp 199–204°. Mass spectrum *m/e*: 230; IR (KBr) cm^{-1}: 3080 (aromatic CH), 3022 (aromatic CH), 2948 (saturated CH), 2886 (saturated CH), 2245 (unconjugated C≡N); ^{1}H NMR ($CHCl_3$) δ: 4.42 (singlet, 9,10-protons); ^{1}H NMR [$(CH_3)_2SO$] δ: 7.66–8.64 (multiplet, aromatic protons), 5.32 (singlet, 9,10-protons).

1,3-Dimethoxyphenanthrene.[28] A solution of 1.0 g (0.0037 mol) of 2,4,6-trimethoxystilbene in 600 mL of deoxygenated cyclohexane was stirred and irradiated under nitrogen for 48 hours with a Hanovia 500-W mercury lamp through a silica cooling jacket. The crude product obtained by evaporation of the solvent was chromatographed on silica gel with a mixture of 5% ethyl acetate and 95% petroleum ether as eluent. 1,3-Dimethoxyphenanthrene (0.75 g, 85%) was obtained as an oil, bp 130° (bath) at 0.05 mm. Mass spectrum *m/e*: 238; ^{1}H NMR (CCl_4) δ: 8.39 (multiplet, 1*H*, H-5), 8.05 (doublet, $J =$ 9.0 Hz, 1*H*, H-10), 7.25–7.80 (multiplet, 5*H*, H-4, H-6, H-7, H-8, H-9), 6.40 (doublet, $J = 2.0$ Hz, 1*H*, H-2), 3.75 (singlet, 6*H*, CH_3O protons).

[6]Helicene.[185] A magnetically stirred solution of 200 mg (0.6 mmol) of 2,7-distyrylnaphthalene and 6 mg (0.02 mmol) of iodine in 1 L of benzene was irradiated through Pyrex for 1.75 hours with a Hanovia 450-W mercury lamp. The reaction product was chromatographed on alumina by using 60–70°

petroleum ether as eluent to give 120 mg (60%) of [6]helicene as pale yellow crystals, mp 240–242°.

Triphenylene.[233] A solution of 235 mg (1.02 mmol) of *o*-terphenyl and 257 mg (1.01 mmol) of iodine in 60 mL of benzene in a quartz cell was irradiated for 20 hours under nitrogen with a 1000-W high-pressure mercury lamp in a water-cooled probe. Recrystallization of the product from ethanol gave 205 mg (88%) of triphenylene as colorless needles, mp 194–195°. Picrate: mp 219–220°.

4-Methylbenzo[*c*]cinnoline (9,10-Diaza-1-methylphenanthrene).[29] 2-Methylazobenzene (200 mg, 1.02 mmol) in 1,2-dichloroethane (850 mL) in the presence of anhydrous aluminum chloride (1 g) was irradiated in a Hanovia 1-L photochemical reactor for 70 hours. The mixture was added to water (50 mL) and the organic solvent removed by distillation. The aqueous solution was neutralized with sodium bicarbonate (5 g) and was then extracted with diethyl ether. The ether extract was washed, dried, and evaporated. The residue was chromatographed on alumina. Elution with petroleum ether gave unchanged 2-methylazobenzene (15 mg, 8%). Subsequent elution with benzene followed by evaporation of the benzene and recrystallization of the residue from ethanol gave 4-methylbenzo[*c*]cinnoline (95 mg, 48%) as yellow needles, mp 129°. Finally, elution with benzene–chloroform [9:1, (v/v)] gave benzo[*c*]cinnoline (20 mg, 10%), mp 156°. Similar results were obtained by irradiation of 2-methylazobenzene with anhydrous stannic chloride or anhydrous ferric chloride in place of anhydrous aluminum chloride.

3-Nitrobenzo[*c*]cinnoline (9,10-Diaza-2-nitrophenanthrene).[310] After a solution of 3-nitroazobenzene (1.0 g, 4.4 mmol) in 98% sulfuric acid (120 mL) was irradiated through Pyrex with a Philips 125-W mercury lamp for 250 hours, it was diluted with ice and the resulting precipitate was collected. The filtrate was made basic and extracted with benzene, and the extract was evaporated. The residue was combined with the original precipitate and treated with concentrated hydrochloric acid (10 mL). The resulting solution was filtered and the filtrate diluted with water to precipitate a yellow solid (0.89 g, 90%). Recrystallization from a 1:1 mixture of ethanol and benzene gave 3-nitrobenzo[*c*]cinnoline as shining yellow plates, mp 259°. The same product was obtained in similar yield following irradiation of 3-nitroazobenzene in sulfuric acid in sunlight for 26 days.

Benzo[1,2-*b*:4,3-*b'*]dithiophene.[322a] A vigorously stirred solution of 0.96 g (5 mmol) of 1,2-di(2-thienyl)ethylene and 13 mg (0.05 mmol) of iodine in 550 mL of benzene under an air atmosphere was irradiated at 350 nm in a Rayonet reactor. The progress of the reaction was followed by GLC (4-ft Apiezon L, 270°) or by UV spectroscopy. After irradiation for 2–3 hours, the solvent was evaporated and the product was purified by chromatography on alumina by using mixtures of 40–60° petroleum ether and benzene as eluent. The product was purified either by sublimation under reduced pressure or by recrystallization from petroleum ether to give 0.86 g (90%) of benzo[1,2-*b*:4,3-*b*′]dithiophene, mp 117–118°. Ultraviolet (C_6H_{12}) nm max (log ε): 219 (4.47), 234 (4.11) shoulder, 248 (4.09) shoulder, 252 (4.16), 257 (4.10), 268 (3.95), 277 (4.11), 288 (4.29), 300 (4.20), 317 (3.53); ^{1}H NMR (CCl_4) δ: 7.69 (singlet, 2*H*), 7.56 (doublet, J = 5.5 Hz, 2*H*), 7.38 (doublet, J = 5.5 Hz, 2*H*). Picrate: mp 148.5–149.5°.

[9,10-*b*]Phenanthrofuran.[333] A solution of 0.50 g (2.3 mmol) of 2,3-diphenylfuran in 100 mL of anhydrous benzene was irradiated through a Vycor filter with a Hanovia Type L 450-W mercury lamp in a water-cooled well. The reaction was followed by GLC (5% SE-30, 225°). After 20 hours of irradiation the peak for the starting material disappeared and a new peak appeared with a slightly longer retention time. The solution was evaporated to give an orange oil that was chromatographed on silica gel with benzene as the eluent. A middle fraction of the eluate was concentrated to dryness under reduced pressure to give 0.20 g (40%) of [9,10-*b*]phenanthrofuran as a white solid. Recrystallization from pentane and sublimation under reduced pressure gave material with mp 118.5–119°. Infrared cm^{-1}: 763; UV (95% C_2H_5OH) nm max (log ε): 237 (4.48), 249 (4.72), 254 (4.81), 280 (4.17), 290 (4.03), 302 (4.09), 320 (2.90), 335 (3.08), 352 (3.11).

α-Carboline (9*H*-Pyrido[2,3-*b*]indole, or 1-Azacarbazole).[382] A magnetically stirred solution of 102 mg (0.6 mmol) of 2-anilinopyridine in 300 mL of cyclohexane under nitrogen was irradiated through Pyrex with a Hanovia 507/7 250-W mercury arc. The progress of the reaction was followed by UV spectroscopy or TLC. After 9 hours of irradiation, the solvent was evaporated and the residue was extracted with two 100-mL portions of boiling benzene. Evaporation of the extract gave 91 mg of a brown solid. Recrystallization of this crude product from benzene and treatment with activated charcoal gave 81 mg (80%) of α-carboline, mp 215–216°. Ultraviolet (CH_3OH) nm max (log ε): 233 (4.28), 259 (4.09), 297 (4.21), 327 (3.61); UV (0.02 *N* methanolic hydrogen chloride) nm max: 244, 263, 268, 306. A benzene solution of the α-carboline exhibited violet fluorescence in UV light. Cyclization went equally well in tetrahydrofuran (80% yield in 8 hours), but no change in the UV spectrum was observed following irradiation in methanol or 98% sulfuric acid.

2,3-Dimethoxy-8-oxoberbine (5,6,13,13a-Tetrahydro-2,3-dimethoxy-8*H*-dibenzo[*a*,*g*]quinolizin-8-one).[81] A solution of 1.0 g (3.2 mmol) of 2-benzoyl-1,2,3,4-tetrahydro-6,7-dimethoxy-1-methyleneisoquinoline in 600 mL of *tert*-butyl alcohol in a quartz irradiation vessel was degassed by four freeze–thaw cycles. The degassed solution was irradiated at 300 nm in a Rayonet reactor for 1.5 hours. The solvent was removed under reduced pressure, and the residue was recrystallized from methanol to give, in two crops, 0.97 g (97%) of 2,3-dimethoxy-8-oxoberbine, mp 143–145°. Ultraviolet (CH_3OH) nm max (log ε): 229 (4.26), 254 (3.81), 264 (3.74), 279 (3.78); IR (KBr) cm^{-1}: 1660, 1520; ^1H NMR ($CDCl_3$) δ: 8.17 (multiplet, 1*H*), 7.34 (multiplet, 3*H*), 6.75 (singlet, 1*H*), 6.73 (singlet, 1*H*), 4.65–5.15 (multiplet, 2*H*), 3.91 (singlet, 6*H*), 2.7–3.5 (multiplet, 5*H*).

1-Carbomethoxy-8-methoxy-10-nitrophenanthrene.[121] A magnetically stirred solution of 50 mg (0.11 mmol) of *cis*-2-carbomethoxy-2′-iodo-6′-methoxy-α-nitrostilbene in 475 mL of cyclohexane was irradiated with a 300-W mercury lamp. The progress of the reaction was followed by UV spectroscopy. After irradiation for 2 hours, the reaction mixture was evaporated under reduced pressure to give 31 mg of a brown, oily residue. Preparative TLC on silica gel with 10% methanol in chloroform gave 20 mg (60%) of 1-carbomethoxy-8-methoxy-10-nitrophenanthrene, mp 173–175°. Recrystallization from ethanol gave 17 mg of yellow needles with mp 175–176°. Ultraviolet (C_2H_5OH) nm max (log ε): 230 (4.54), 253.5 (4.39), 279 (4.11); IR ($CHCl_3$) cm^{-1}: 1721, 1520, 1351.

***N*-Carbethoxydehydronorneolitsine (Ethyl 5,6-Dihydro-7*H*-bis[1,3]benzodioxolo[6,5,4-*de*:5′,6′-*g*]quinoline-7-carboxylate).**[25] A stirred mixture of 1.0 g (2.0 mmol) of *N*-carbethoxy-1-(6′-bromo-3′,4′-methylenedioxybenzylidene)-6,7-methylenedioxy-1,2,3,4-tetrahydroisoquinoline, 0.90 g (8.0 mmol) of potassium *tert*-butoxide, 50 mL of *tert*-butyl alcohol, and 200 mL of benzene under nitrogen gas was irradiated through a Corex filter with a Hanovia 450-W mercury lamp for 21 hours. Dilution with water and evaporation of the washed and dried benzene phase gave a solid that was recrystallized from methanol to give 0.60 g (72%) of *N*-carbethoxydehydronorneolitsine, mp 198–201°. Further purification by recrystallization from a mixture of chloroform and methanol gave material with mp 206–207°. Ultraviolet (dioxane) nm max (log ε): 264 (4.99), 270 (4.97), 295 (4.25), 327 (4.32), 340 (4.33), 363 (3.89), 382 (3.83); ^1H NMR ($CDCl_3$) δ:8.45 (singlet, 1*H*), 7.70 (singlet, 1*H*), 7.13 (singlet, 1*H*), 6.95 (singlet, 1*H*), 6.18 (singlet, 2*H*), 6.07 (singlet, 2*H*), 4.31 (quartet, $J = 7$ Hz, 2*H*), 4.05 (triplet, $J = 5$ Hz, 2*H*), 3.12 (triplet, $J = 5$ Hz, 2*H*), 1.35 (triplet, $J = 7$ Hz, 3*H*).

5,7-Dihydro-6*H*-indolo[2,3-*c*]quinolin-6-one.[342] A solution of 200 mg (0.85 mmol) of indole-2-carboxanilide in 180 mL of benzene and 20 mL of ethanol was irradiated with a 100-W high-pressure mercury lamp at room temperature for 1 hour. The solvent was removed under reduced pressure and the residue purified by chromatography on a silica gel column with a 3:1 mixture of benzene and ethyl acetate as eluent. Evaporation of the solvent and recrystallization from ethanol gave 154 mg (77%) of 5,7-dihydro-6*H*-indolo-[2,3-*c*]quinolin-6-one as pale yellow needles, mp 314–315°. Ultraviolet (C_2H_5OH) nm max (log ε): 246.5 (4.58), 269.5 (4.25), 292 (3.89), 305.5 (3.97), 320 (3.88), 332.5 (4.10), 348.5 (4.11); IR (Nujol) cm^{-1}: 3370, 3200, 1670–1655 (NHCO); mass spectrum *m*/*e*: 234, 216, 205, 189, 177, 164, 151, 117, 103.

Benzo[*f*]quinoline-6-carbonitrile.[413] A magnetically stirred solution of 8.00 g (0.039 mol) of 2-phenyl-3-(2-pyridyl)acrylonitrile in 4.4 L of *tert*-butyl alcohol and 475 mL of benzene was irradiated with a Hanovia 450-W mercury lamp contained in a water-jacketed immersion well fitted with a Corex 9700 sleeve as a filter. Oxygen was bubbled through the reaction mixture continuously. After the solution was irradiated for 12–13 hours, the solvents were evaporated under reduced pressure. The resulting red–brown residue was dissolved in 2 L of benzene and this solution was passed through a column containing 150 g of Woelm activity III alumina to remove polymeric material. The column was washed with an additional 500 mL of benzene. Reduced-pressure evaporation of the combined benzene solutions gave 6.07 g of crude product as tan needles. Recrystallization of this material from chloroform–hexane gave 4.35 g of off-white needles, mp 162–163°. Evaporation of the mother liquor under reduced pressure gave 1.58 g of a red–brown oil. Purification of this oil by chromatography on 50 g of Woelm activity III alumina with mixtures of hexane and benzene as eluents gave an additional 0.23 g of product with mp 161–163°. The combined 4.58 g of product was purified by two recrystallizations from chloroform–hexane to give, in two crops, 4.26 g (54%) of benzo[*f*]quinoline-6-carbonitrile, mp 163–164°.

***trans*-9-Methyl-1,2,3,4,4a,9a-hexahydrocarbazole.**[86] A solution of 4.20 g (23.6 mmol) of the moisture-sensitive enamine 1-(*N*-methylanilino)cyclohexene in 300 mL of diethyl ether from a freshly opened can was purged with argon for 30 minutes and then irradiated with a Hanovia 550-W mercury lamp in a Pyrex immersion well. The progress of the reaction was monitored by TLC on Silica Gel H with ethyl acetate in Skelly B petroleum ether. After 6 hours of irradiation, the solvent was evaporated from the reaction mixture. The resulting oil was crystallized from 95% ethanol to give 2.3 g (55%) of *trans*-9-methyl-1,2,3,4,4a,9a-hexahydrocarbazole, mp 58–60°. Picrate: mp 125–126°; methiodide: mp 233–234°.

TABULAR SURVEY

This review is based on information derived from a thorough search of the chemical literature through the end of 1980; in addition, some references from 1981 and 1982 are cited. The Tabular Survey is organized into the following eight tables:

I	Phenanthrenes and 9,10-Dihydrophenanthrenes from Simple Stilbenes
II	Higher Carbocyclic Aromatic Systems from Stilbene Analogs
III	Carbocyclic Systems from Reactants Other Than 1,2-Diarylethylenes
IV	Polycyclic Systems from Dianthrones and Other Doubly Bridged Tetraarylethylenes
V	Nitrogen Heterocyclic Systems

VI	Oxygen and Sulfur Heteroaromatic Systems from Furans, Thiophenes, and Pyrylium Ions
VII	Lactams and Other Heterocyclic Systems from Amides
VIII	Photocyclizations Producing Five-Membered Rings

The entries in each table are listed in order of increasing number of carbon atoms in the starting material. Within a given carbon number in each table, the starting materials are arranged in order of increasing number of heteroatoms (other than hydrogen) according to the following alphabetical priority sequence: boron, bromine, chlorine, fluorine, iodine, nitrogen, oxygen, and sulfur. In searching for information about the photocyclization of a particular heteroatom-substituted starting material, however, the reader is advised not to rely on this priority sequence but rather to scan all of the entries for the carbon number of interest before concluding that the information is not in the table. This is because a few heteroatom-substituted systems are tabulated out of sequence through the space-saving device of representing a series of related starting materials by a single generalized structure with substituents designated by symbols such as X or Y. Operationally, the actual starting material used in the photocyclization of a stilbene or a related olefinic system that has both *E* and *Z* isomers usually is the synthetically more accessible *E* isomer, from which the mechanistically required *Z* isomer is generated *in situ* by photoisomerization. In the tables, however, the starting material in such a system either is depicted arbitrarily as the *Z* isomer or is listed with a stereochemically noncommittal name that lacks the *E* or *Z* descriptor.

Solvents, gases, and reactants other than the starting material are tabulated for each photocyclization under the Reaction Conditions column. For entries listing EtOH as the solvent, the original sources do not specify whether absolute ethanol or 95% ethanol was used. The term "solvent" is used as a tabular entry if the original source fails to identify the solvent for the photocyclization. The following symbols and abbreviations are used:

Abs	Absolute
Ac	Acetyl
Me	Methyl
Et	Ethyl
n-Pr	*n*-Propyl
i-Pr	*i*-Propyl
n-Bu	*n*-Butyl
t-Bu	*tert*-Butyl
Ph	Phenyl
DDQ	2,3-Dichloro-5,6-dicyanobenzoquinone
C_5H_{12}	Pentane
C_6H_6	Benzene
C_6H_{12}	Cyclohexane
C_6H_{14}	Hexane

$CH_3C_6H_{11}$	Methylcyclohexane
DCE	1,2-Dichloroethane
Pet. ether	Petroleum ether
Py	Pyridine
THF	Tetrahydrofuran
p-TSA	*p*-Toluenesulfonic acid

Products usually are indicated in the tables by either name or structure. In some of the tables, however, the first tabular entry for certain polycyclic aromatic systems includes both the name and the structure; where relevant, the numbering scheme is given also. In subsequent entries for those same polycyclic products or simple derivatives thereof, only the name is given to save space and printing cost.

The names of some ring systems that are cited frequently as "starting materials" or "products" are designated by boldface letters. Each of these abbreviations is defined in the first footnote of the table in which it is used.

Percentage yields or product ratios are indicated (in parentheses) for most of the tabular entries; a dash (—) signifies that the yield is not reported in the cited reference. For most of the photocyclizations in the tables, the original reports provide no evidence of systematic attempts to optimize yields. Considerable amounts of recovered starting material are reported in many instances. In general, reported photocyclizations that were not carried out on a preparative scale or those from which products were not isolated and characterized are not included in the Tabular Survey.

TABLE I. Phenanthrenes and 9,10-Dihydrophenanthrenes from Simple Stilbenes

	Stilbene	Reaction Conditions	Phenanthrenes and Yields (%)	Refs.
C_{14}	Stilbene	C_6H_{12}, I_2, air	Phenanthrene (73)	10,55
		C_6H_{12}, air	" (good)	414
		C_6H_{12}, sulfur[a]	" (60)	415a
		C_6H_{12}, $(PhSe)_2$, air	" (86)	156
		C_6H_{12}, $(PhSe)_2$	" (21)	156
		C_6H_{14}, Al_2O_3, air	" (—)	415b
		C_6H_{14}, $(PhCH_2S)_2$[b]	" (—)	416
		C_6H_6, air	" (20)	417
		EtOH, air	" (—)	418
		Abs EtOH, sulfur, N_2[c]	" (—)	419
		EtOH, N_2[d,e]	" (10)	420,421
		CH_2Cl_2, $(NC)_2C{=}C(CN)_2$	" (very good)	31,65
		n-$PrNH_2$, N_2	1,4-Dihydro (70)	72a
	4-Deuterio	C_6H_{12}, I_2, air	3-Deuterio (30)	422
	3,5-Dideuterio	C_6H_{12}, I_2, O_2	2,4-Dideuterio (33)	423
	2,4,6-Trideuterio	"	1,3-Dideuterio (27)	423
	3,3′,5,5′-Tetradeuterio	"	2,4,5,7-Tetradeuterio (25)	423
	2,3,4,5,6-Pentadeuterio	"	1,2,3,4-Tetradeuterio (48)	423
	2,2′,4,4′,6,6′-Hexadeuterio	"	1,3,6,8-Tetradeuterio (39)	423
	α,α′-$(^{13}C)_2$	"	9,10-$(^{13}C)_2$ (—)	424
	2-Bromo	Solvent, air	1-Bromo (—)	425
	3-Bromo	C_6H_{12}, I_2	2-Bromo (74), 4-bromo (16)	115
	4-Bromo	C_6H_{12}, I_2, air	3-Bromo (76)	55
		Solvent, air	" (—)	425

TABLE I. PHENANTHRENES AND 9,10-DIHYDROPHENANTHRENES FROM SIMPLE STILBENES (*Continued*)

Stilbene	Reaction Conditions	Phenanthrenes and Yields (%)	Refs.
4,4′-Dibromo	C_6H_{12}, air	3,6-Dibromo (40)	64
4-Bromo-3′,5′-dideuterio	C_6H_{12}, I_2, air	3-Bromo-5,7-dideuterio (42)	422
4-Bromo-2′,4′,6′-trideuterio	"	3-Bromo-6,8-dideuterio (45)	422
2-Bromo-2′-fluoro	"	1-Bromo-8-fluoro (68)	426
2-Bromo-4′-fluoro	"	1-Bromo-6-fluoro (43)	427
3-Bromo-4′-fluoro	"	2-Bromo-6-fluoro (—), 4-bromo-6-fluoro (—)	76
4′-Bromo-2-fluoro	"	6-Bromo-1-fluoro (88)	76
4′-Bromo-3-fluoro	"	6-Bromo-2-fluoro + 6-bromo-4-fluoro (63:37)[f]	76
4-Bromo-4′-fluoro	"	3-Bromo-6-fluoro (64)	426
2-Chloro	"	1-Chloro (57)	55
3-Chloro	"	2-Chloro + 4-chloro (52:48)[f]	21
4-Chloro	"	3-Chloro (76)	55
α,α′-Dichloro	C_6H_{12}, I_2	9,10-Dichloro (12)	128
4-Chloro-4′-deuterio	C_6H_{12}, I_2, air	3-Chloro-6-deuterio (59)	422
4-Chloro-3′,5′-dideuterio	"	3-Chloro-5,7-dideuterio (55)	422
4-Chloro-2′,4′,6′-trideuterio	"	3-Chloro-6,8-dideuterio (60)	422
2-Chloro-2′-fluoro	"	1-Chloro-8-fluoro (87)	426
2-Chloro-3′-fluoro	"	1-Chloro-7-fluoro (—), 1-chloro-5-fluoro (—)	76
2-Chloro-4′-fluoro	"	1-Chloro-6-fluoro (67)	76
3′-Chloro-2-fluoro	"	7-Chloro-1-fluoro (—), 5-chloro-1-fluoro (—)	76
3-Chloro-4′-fluoro	"	2-Chloro-6-fluoro (—), 4-chloro-6-fluoro (—)	76
4′-Chloro-2-fluoro	"	6-Chloro-1-fluoro (72)	76
4′-Chloro-3-fluoro	"	6-Chloro-2-fluoro (—), 6-chloro-4-fluoro (—)	76
4-Chloro-4′-fluoro	"	3-Chloro-6-fluoro (76)	426
2-Chloro-4,4′-difluoro	"	1-Chloro-3,6-difluoro (40)	76
4-Chloro-2,2′-difluoro	"	3-Chloro-1,8-difluoro (63)	76
2-Chloro-3′,5-difluoro	"	1-Chloro-4,7-difluoro (—), 1-chloro-4,5-difluoro (—)	76
2-Fluoro	"	1-Fluoro (78)	427
3-Fluoro	"	2-Fluoro + 4-fluoro (59:41)[f]	21
4-Fluoro	"	3-Fluoro (76)	55,412

	2,2′-Difluoro	”	1,8-Difluoro (86)	426
	2,3′-Difluoro	”	1,7-Difluoro (—), 1,5-difluoro (—)	76
	2,4′-Difluoro	”	1,6-Difluoro (73)	76
	3,4′-Difluoro	”	2,6-Difluoro + 4,6-difluoro (67:33)[f]	76
	4,4′-Difluoro	”	3,6-Difluoro (84)	426
	4-Fluoro-3′,5′-dideuterio	”	3-Fluoro-5,7-dideuterio (25)	422
	4-Fluoro-2′,4′,6′-trideuterio	”	3-Fluoro-6,8-dideuterio (23)	422
	2-Iodo	C_6H_{12}	Phenanthrene (90)	121,428
	2-Iodo-α′-nitro	”	9-Nitro (40)	121,428
	3-Hydroxy	C_6H_{12}, I_2, air	2-Hydroxy + 4-hydroxy (3:1)	429
C_{15}	2-Methyl	1. *n*-$PrNH_2$, N_2 2. DDQ	1-Methyl + phenanthrene (9:1) (82 total)	72b
		C_6H_{12}, I_2, air	1-Methyl + phenanthrene (65:35)	72b
		”	1-Methyl (57)	55
		C_6H_{14}, I_2, air	” (50)	192
		C_6H_{14}, I_2	” (—)	430
	3-Methyl	C_6H_{12}, I_2, air	2-Methyl + 4-methyl (51:49)[f]	21
		C_6H_{14} or C_6H_6, I_2, N_2	” + ” (1:1)	159
	4-Methyl	C_6H_{12}, I_2, air	3-Methyl (67)	55
		C_6H_6, I_2	” (50)	431
	α-Methyl	EtOH, air	9-Methyl (39)	69
	4-Methyl-3′,5′-dideuterio	C_6H_{12}, I_2, air	3-Methyl-5,7-dideuterio (70)	422
	4-Methyl-2′,4′,6′-trideuterio	”	3-Methyl-6,8-dideuterio (63)	422
	2-Methyl-α-tritio	”	1-Methyl-10-tritio (—)	432
	2-Methyl-α′-tritio	”	1-Methyl-9-tritio (—)	432
	3-Methyl-α-tritio	”	2-Methyl-10-tritio (—), 4-methyl-10-tritio (—)	432
	3-Methyl-α′-tritio	”	2-Methyl-9-tritio (—), 4-methyl-9-tritio (—)	432
	4-Methyl-α-tritio	”	3-Methyl-10-tritio (—)	432
	4-Methyl-α′-tritio	”	3-Methyl-9-tritio (—)	432
	3-Bromo-4-methyl	”	2-Bromo-3-methyl (43)[g]	433
	2-Fluoro-2′-methyl	”	1-Fluoro-8-methyl (84)	426
	2-Fluoro-4′-methyl	”	1-Fluoro-6-methyl (68)	427
	3′-Fluoro-2-methyl	”	7-Fluoro-1-methyl (—), 5-fluoro-1-methyl (—)	76

TABLE I. PHENANTHRENES AND 9,10-DIHYDROPHENANTHRENES FROM SIMPLE STILBENES (*Continued*)

Stilbene	Reaction Conditions	Phenanthrenes and Yields (%)	Refs.
3-Fluoro-4′-methyl	C_6H_{12}, I_2, air	2-Fluoro-6-methyl (24), 4-fluoro-6-methyl (18)	76
4′-Fluoro-2-methyl	"	6-Fluoro-1-methyl (75)	427
4′-Fluoro-3-methyl	"	6-Fluoro-2-methyl (—), 6-fluoro-4-methyl (—)	76
4-Fluoro-4′-methyl	"	3-Fluoro-6-methyl (69)	426
3′,5-Difluoro-2-methyl	"	4,7-Difluoro-1-methyl (—), 4,5-difluoro-1-methyl (—)	76
3-Trifluoromethyl	"	2-Trifluoromethyl (46), 4-trifluoromethyl (43)	21,55
4-Trifluoromethyl	"	3-Trifluoromethyl (60)	55
4-Trifluoromethyl-3′,5′-dideuterio	"	3-Trifluoromethyl-5,7-dideuterio (35)	422
4-Trifluoromethyl-2′,4′,6′-trideuterio	"	3-Trifluoromethyl-6,8-dideuterio (35)	422
2-Fluoro-2′-methoxy	"	1-Fluoro-8-methoxy (40)	426
2-Fluoro-3′-methoxy	"	1-Fluoro-7-methoxy (—), 1-fluoro-5-methoxy (—)	76
2-Fluoro-4′-methoxy	"	1-Fluoro-6-methoxy (74)	427
3′-Fluoro-2-methoxy	"	7-Fluoro-1-methoxy (—), 5-fluoro-1-methoxy (—)[h]	76
3′-Fluoro-3-methoxy	"	7-Fluoro-2-methoxy + 5-fluoro-2-methoxy (1:1)[i]	76
3-Fluoro-4′-methoxy	"	2-Fluoro-6-methoxy (—), 4-fluoro-6-methoxy (—)	76
4′-Fluoro-2-methoxy	"	6-Fluoro-1-methoxy (63)	427
4′-Fluoro-3-methoxy	"	6-Fluoro-2-methoxy (—), 6-fluoro-4-methoxy (—)	76
4-Fluoro-4′-methoxy	"	3-Fluoro-6-methoxy (61)	426
2′-Iodo-2-methyl	C_6H_{12}	1-Methyl (96)	121,428
3-Cyano	C_6H_{12}, I_2	2-Cyano (71), 4-cyano (19)	115
4-Cyano	"	3-Cyano (48)	422
α-Cyano	EtOH, air	9-Cyano (17)	69
	MeOH, H_2O	9-Cyano-9,10-dihydro (80)	32
4-Cyano-3′,5′-dideuterio	C_6H_{12}, I_2, air	3-Cyano-5,7-dideuterio (11)	422
4-Cyano-2′,4′,6′-trideuterio	"	3-Cyano-6,8-dideuterio (34)	422
α-Carbamido	EtOH, air	9-Carbamido (41)	69
	Solvent, I_2 (or O_2)	" (—)	434
3-Formyl	C_6H_6, I_2, N_2	2-Formyl (23), 4-formyl (2)	104
4-Formyl	"	3-Formyl (40)	104
2-Methoxy	C_6H_{12}, I_2, air	1-Methoxy (46)	55

		C_6H_{12}, N_2	Phenanthrene (58), 1-methoxy (15)[e]	28,435
	3-Methoxy	C_6H_{12}, I_2, air	2-Methoxy (—), 4-methoxy (—)	21
		Solvent, air	2-Methoxy (—)	436
	4-Methoxy	C_6H_{12}, I_2, air	3-Methoxy (42)	55
		"	" (31)	60
		"	" (38)	46
		Abs EtOH, $CuCl_2$, I_2, air	" (32–39)	60,436
	α-Methoxy	C_6H_6 or C_6H_{14}, I_2	9-Methoxy (7–20)	437
	4-Methoxy-4′-deuterio	C_6H_{12}, I_2, air	3-Methoxy-6-deuterio (33)	422
	4-Methoxy-3′,5′-dideuterio	"	3-Methoxy-5,7-dideuterio (23)	422
	4-Methoxy-2′,4′,6′-trideuterio	"	3-Methoxy-6,8-dideuterio (36)	422
	α-Carboxy	C_6H_6, I_2, air	9-Carboxy (72)	55
		Solvent, I_2 (or O_2)	" (—)	434
	α′,α- (cyclic carbamate: O–C(=O)–NH)	95% EtOH, $CuCl_2$, I_2, N_2	9,10- (cyclic carbamate: O–C(=O)–NH) (39)	132
		MeOH, $CuCl_2$, I_2, air	" (8)	133
	α′,α- (cyclic carbonate: O–C(=O)–O)	C_6H_{12}, I_2	9,10- (cyclic carbonate: O–C(=O)–O) (77)	129
		EtOH, O_2	" (13)	130
C_{16}	2,2′-Dimethyl	1. *n*-$PrNH_2$, N_2 2. DDQ	1,8-Dimethyl + 1-methyl (95:5) (70 total)	72b
		C_6H_{12}, I_2, air	" + " (7:3)	72b
		C_6H_{12}, I_2, O_2	1,8-Dimethyl (49)	438
		C_6H_{12}, I_2	" (45)	72a
		n-$PrNH_2$, N_2	1,4-Dihydro-1,8-dimethyl (50)	72a
	3,3′-Dimethyl	C_6H_{12}, I_2 (5×10^{-2} *M*), air	2,5-Dimethyl (54), 2,7-dimethyl (28), 4,5-dimethyl (18)[f]	21
		C_6H_{12}, I_2 (5×10^{-4} *M*), air	2,5-Dimethyl (63), 2,7-dimethyl (33), 4,5-dimethyl (4)[f]	21
	4,4′-Dimethyl	C_6H_{12}, I_2	3,6-Dimethyl (81)	72a
		C_6H_{14}, I_2	" (70)	439

TABLE I. PHENANTHRENES AND 9,10-DIHYDROPHENANTHRENES FROM SIMPLE STILBENES (*Continued*)

Stilbene	Reaction Conditions	Phenanthrenes and Yields (%)	Refs.
	C_6H_{12}, C_6H_6, I_2, air	" (61)	440,441
	C_6H_{12}, I_2	" (46)	442
	Solvent, air	" (—)	443
	n-$PrNH_2$, N_2	1,4-Dihydro-3,6-dimethyl (75)	72a
α,α′-Dimethyl	EtOH, air	9,10-Dimethyl (53)	69
	Abs EtOH, $CuCl_2$, I_2, air	" (55)	60
	C_6H_{12}, I_2, air	" (28)	60
2-Vinyl	C_6H_{14}, N_2[e]	1-Vinyl (15)[j]	192,444
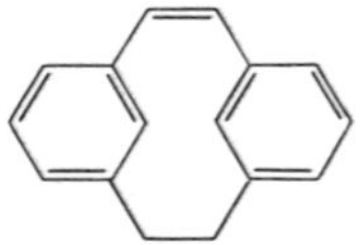	C_6H_{14}, O_2	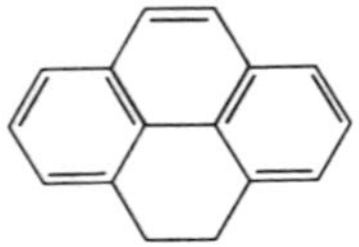 (good)	445–447
	$CH_3C_6H_{11}$, I_2	" (100)	448
4-Bromo-α-ethyl	C_6H_{12}, I_2	3-Bromo-10-ethyl (56)	449
4-Bromo-α-carbomethoxy	C_6H_{12}, I_2, O_2	3-Bromo-10-carbomethoxy (44)	450
4′-Chloro-2,6-dimethoxy	C_6H_{12}, N_2	6-Chloro-1-methoxy (56)	28,435
4-Chloro-α,α′-dicyano	CH_2Cl_2, I_2	3-Chloro-9,10-dicyano (—)	71a
	EtOH, argon	3-Chloro-9,10-dicyano-9,10-dihydro (—)	71a
5,5′-Difluoro-2,2′-dimethyl	C_6H_{12}, I_2, air	4,5-Difluoro-1,8-dimethyl (49), 4,7-difluoro-1-methyl (13)	451,452
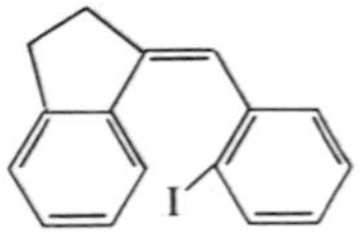 I	MeOH	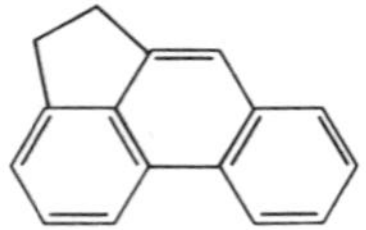 (25)	122
α′-Carbomethoxy-2-iodo	C_6H_6	9-Carbomethoxy (71)	121,428
2-Carbomethoxy-2′-iodo-α-nitro	"	1-Carbomethoxy-10-nitro (40)	121,428
4,4′-Dicyano	C_6H_6, air	3,6-Dicyano (40)	64
4,α-Dicyano	CH_2Cl_2, I_2	3,10-Dicyano (—)	71a

	EtOH, argon	3,10-Dicyano-9,10-dihydro (—)	71a
4,α′-Dicyano	C_6H_6, I_2	3,9-Dicyano (85)	453
α,α′-Dicyano	$CHCl_3$, I_2, air	9,10-Dicyano (60)	69
	CH_2Cl_2, I_2	" (—)	71a
	CCl_4, air	" (—)	454
	C_6H_6, N_2[e,k]	" (23)	455
	$CHCl_3$, air	9,10-Dicyano (major), *trans*-9,10-dicyano-9,10-dihydro (7)	456
	H_2O (suspension), O_2	9,10-Dicyano (60), *trans*-9,10-dicyano-9,10-dihydro (22)[l]	41,457
	C_6H_6, degassed	*trans*-9,10-Dicyano-9,10-dihydro (85)	69
	EtOH, degassed	" (—)	456
	EtOH, argon	" (—)	71
α-Cyano-4-methoxy	C_6H_6, I_2	10-Cyano-3-methoxy (50)	453
α′-Cyano-4-methoxy	"	9-Cyano-3-methoxy (60)	453
α′-Cyano-α-methoxy	Et_2O, air	9-Cyano-10-methoxy (65)	458a
α′,α- O–C(=O)–N–CH_3	95% EtOH, $CuCl_2$, I_2, O_2	9,10- O–C(=O)–N–CH_3 (70)	132
	MeOH, $CuCl_2$, I_2, air	" (65)	133
	MeOH, $CuCl_2$, I_2, N_2	" (84)	458b
α′,α- O=C–NH–C=O	EtOH, air	*cis*-9,10-Dihydro-9,10- O=C–NH–C=O (40)[m]	69,456
3-Acetyl	C_6H_{12}, I_2	2-Acetyl (57), 4-acetyl (13)	115
4-Acetyl	C_6H_{12}, I_2, air	3-Acetyl (67)	76
α-Acetoxy	Dry HOAc, N_2[e]	9-Acetoxy (—)	421
4-Carbomethoxy	C_6H_{12}, C_6H_6, I_2, air	3-Carbomethoxy (63)	422
α-Carbomethoxy	MeOH, argon	9-Carbomethoxy-9,10-dihydro (90)	32
	MeOH, N_2	" (72)	459
	C_6H_6, dry HCl, argon	" (50)	32
	MeOD	9-Carbomethoxy-9-deuterio-9,10-dihydro (—)	32
	CD_3OD	9-Carbomethoxy-9,10-dideuterio-9,10-dihydro (—)	459

TABLE I. PHENANTHRENES AND 9,10-DIHYDROPHENANTHRENES FROM SIMPLE STILBENES (*Continued*)

	Stilbene	Reaction Conditions	Phenanthrenes and Yields (%)	Refs.
	4-Carbomethoxy-3′,5′-dideuterio	C_6H_{12}, C_6H_6, I_2, air	3-Carbomethoxy-5,7-dideuterio (31)	422
	4-Carbomethoxy-2′,4′,6′-trideuterio	"	3-Carbomethoxy-6,8-dideuterio (49)	422
	4,4′-Diformyl	C_6H_6, air	3,6-Diformyl (40)	105
	α′,α- 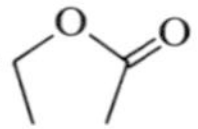	C_6H_6, O_2[n]	9,10- 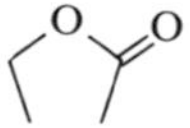 (—)	460
	3,5-Dimethoxy	C_6H_{12}, I_2	2,4-Dimethoxy (35)	461
	3,3′-Dimethoxy	C_5H_{12}, air	2,7-Dimethoxy (45), 2,5-dimethoxy (28)	108
	2,5-Dimethoxy	C_6H_{14}, I_2	1,4-Dimethoxy (71)	462
	2,6-Dimethoxy	C_6H_{12}, N_2	1-Methoxy (90)	28,435
	4,4′-Dimethoxy	CH_2Cl_2, $(NC)_2C{=}C(CN)_2$	3,6-Dimethoxy (very good)	31,65
		C_6H_{12}, I_2	" (79)	72a
		n-$PrNH_2$, N_2	1,4-Dihydro-1,8-dimethoxy (60)	72a
	3′-Hydroxy-3,5-dimethoxy	Solvent, air	7-Hydroxy-2,4-dimethoxy (—)	113
	α′,α- 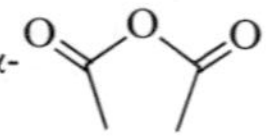	H_2O, Na_2CO_3, air	9,10- 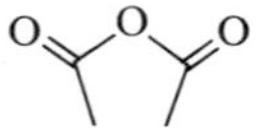 (43)	69
		C_6H_6, degassed[e]	" (20), 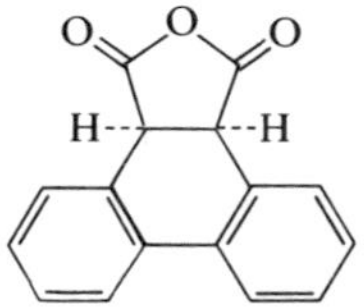 (15)	69
		C_6H_6 or MeCN, O_2[o]	" (—), " (—)	463
C_{17}	2,2′,3-Trimethyl	1. n-$PrNH_2$, N_2 2. DDQ	1,2,8-Trimethyl + 1,2- and 1,5-dimethyl (95:5) (>63 total)	72b
		C_6H_{12}, I_2, air	" + " (65:35)	72b
	2,4,6-Trimethyl	C_6H_{12}, air	1,3-Dimethyl (3)	64

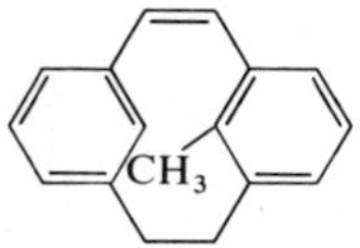 CH_3	C_6H_{14}, O_2	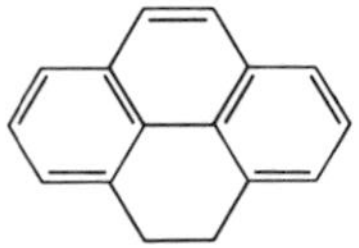 (good)	446,447
	$CH_3C_6H_{11}$, I_2	" (100)	448
α,α′-Trimethylene	C_6H_{12}, I_2, air	9,10-Trimethylene (82)	55,56
2-Propenyl	C_6H_{14}, degassed[e]	1-Propenyl (10)	192
2-Bromo-4′-carbomethoxy-3-methyl	C_6H_{12}, I_2	1-Bromo-6-carbomethoxy-2-methyl (73)	117
3-Bromo-6′-iodo-3′-methoxy-2′,4-dimethyl-5-nitro	*t*-BuOH, CH_2Cl_2, N_2	2-Bromo-7-methoxy-3,8-dimethyl-4-nitro (12)	464
4,4′-Dibromo-α,α′-trimethylene	C_6H_{12}, I_2, air	3,6-Dibromo-9,10-trimethylene (82)	56
4-Chloro-2,2′-dimethoxy-3-methyl	C_6H_{12}, I_2	3-Chloro-1,8-dimethoxy-2-methyl (20)	465
	C_6H_{12}, N_2[e]	3-Chloro-1-methoxy-2-methyl (12), 3-chloro-1,8-dimethoxy-2-methyl (7)	28
4,4′-Dichloro-α,α′-trimethylene	C_6H_{12}, I_2, air	3,6-Dichloro-9,10-trimethylene (61)	56
5-Fluoro-2,2′,5′-trimethyl	Solvent, air	4-Fluoro-1,5,8-trimethyl (—)	466
4,4′-Difluoro-α,α′-trimethylene	C_6H_{12}, I_2, air	3,6-Difluoro-9,10-trimethylene (84)	56
α-Carbomethoxy-2′-iodo-2-methyl	C_6H_{12}	10-Carbomethoxy-1-methyl (62)	121,428
2-Carbomethoxy-2′-iodo-6′-methoxy-α-nitro	"	1-Carbomethoxy-8-methoxy-10-nitro (60)	121
2-Carbomethoxy-2′-iodo-4,5-methylenedioxy-α-nitro	"	1-Carbomethoxy-3,4-methylenedioxy-10-nitro (26)	121,428
2-Carbomethoxy-2′-iodo-4′,5′-methylenedioxy-α-nitro	"	1-Carbomethoxy-6,7-methylenedioxy-10-nitro (39)	121,428
3′-Acetoxy-2′-iodo-3-methoxy	"	7-Acetoxy-2-methoxy (24), 7-acetoxy-4-methoxy (9), 5-acetoxy-2-methoxy (6)	114
α′,α- 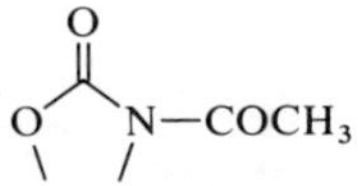 (O, N—$COCH_3$ cyclic carbamate)	MeOH, $CuCl_2$, I_2, air	9,10- 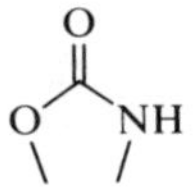 (O, NH cyclic carbamate) (5)	133
α,α′-Dicyano-4-methoxy	CH_2Cl_2, I_2	9,10-Dicyano-3-methoxy (—)	71a
	EtOH, argon	9,10-Dicyano-9,10-dihydro-3-methoxy (—)	71a
4,α,α′-Tricyano	CH_2Cl_2, I_2	3,9,10-Tricyano (—)	71a

TABLE I. PHENANTHRENES AND 9,10-DIHYDROPHENANTHRENES FROM SIMPLE STILBENES (*Continued*)

Stilbene	Reaction Conditions	Phenanthrenes and Yields (%)	Refs.
	EtOH, argon	3,9,10-Tricyano-9,10-dihydro (—)	71a
α′,α- 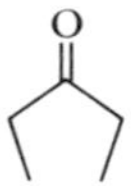	MeOH, N_2[e]	9,10- 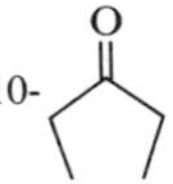 (65),[p] 9,10-  (7)[p]	135
2-Methoxy-4,5-dimethyl	C_6H_{12}, I_2	1-Methoxy-3,4-dimethyl (30), 2,3-dimethyl (10)	467
α-Carbethoxy	Solvent, air	9-Carbethoxy (—)	468
α′-Carbomethoxy-4-methyl	C_6H_{14}, I_2	9-Carbomethoxy-3-methyl (48)	439
4′-Carbomethoxy-2-methyl	C_6H_{12}, I_2	6-Carbomethoxy-1-methyl (74)	117
4′-Ethoxy-2-methoxy	"	6-Ethoxy-1-methoxy (—)	461
4′-Ethoxy-4-methoxy	"	6-Ethoxy-3-methoxy (—)	461
4,4′-Dimethoxy-α-methyl	Abs EtOH, $CuCl_2$, I_2, air	3,6-Dimethoxy-9-methyl (33)	60,469
	C_6H_{12}, I_2, air	" (32)	60,469
4,6-Dimethoxy-2-methyl	C_6H_{12}, N_2	3-Methoxy-1-methyl (38)	28,435
2,4-Dimethoxy-2′-methyl	Solvent, air	1,3-Dimethoxy-8-methyl (44)	470
3,5-Dimethoxy-2′-methyl	C_6H_{14}, air	2,4-Dimethoxy-8-methyl (33)	470
2-Acetoxy-3-methoxy	C_6H_{12}, I_2, O_2	1-Acetoxy-2-methoxy (65)	471
	C_6H_6, I_2, O_2	" (54)	471
4′-Carbomethoxy-2-methoxy	Pet. ether, I_2	6-Carbomethoxy-1-methoxy (17), 3-carbomethoxy (6)	465
	C_6H_{12}, I_2, air	" (—), " (—)	435
4′-Ethoxy-3,4-methylenedioxy	C_6H_{12}, I_2	6-Ethoxy-2,3-methylenedioxy (—)	461
2,2′,6-Trimethoxy	C_6H_{12}, N_2	1,8-Dimethoxy (91)	28,435
2,4,6-Trimethoxy	"	1,3-Dimethoxy (85)	28,435
2,3-Dimethoxy-3′,4′-methylenedioxy	C_6H_{12}, I_2	1,2-Dimethoxy-6,7-methylenedioxy (40), 1,2-dimethoxy-5,6-methylenedioxy (10)	461
3-Hydroxy-3′,4,5′-trimethoxy	Solvent, air	2-Hydroxy-3,5,7-trimethoxy (—)	356
4,4′-Dimethoxy-α′,α- 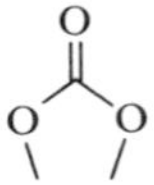	C_6H_{12}, I_2, O_2	3,6-Dimethoxy-9,10- 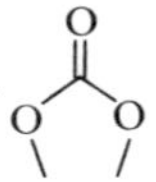 (50)	131

C_{18} 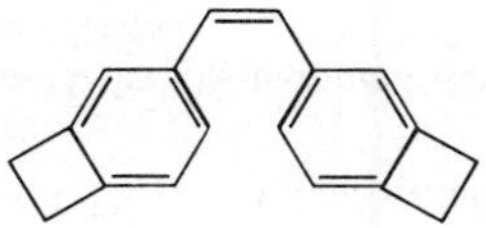	Solvent, air	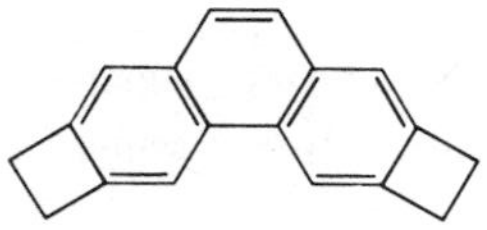 (18), 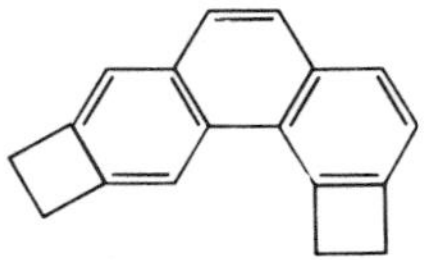 (36)	472
3,3′,5,5′-Tetramethyl	C_6H_{12}, I_2, air	2,4,5,7-Tetramethyl (60)	64
	C_6H_{12}, $(PhSe)_2$	" (51)	156
2,2′,3,3′-Tetramethyl	Pet. ether, I_2, air	1,2,5-Trimethyl (49), 1,2,7,8-tetramethyl (24)	473
4-*tert*-Butyl	Solvent, air	3-*tert*-Butyl (—)	474
α-*tert*-Butyl	C_6H_{12}, air[a]	9-*tert*-Butyl (6)	475
2-Carbomethoxy-2′-iodo-6′-methoxy-4,5-methylenedioxy-α-nitro	C_6H_{12}	1-Carbomethoxy-8-methoxy-3,4-methylenedioxy-10-nitro (54)	121,428
3′-Acetoxy-2′-iodo-3,5-dimethoxy	C_6H_{12}, air (or N_2)	7-Acetoxy-2,4-dimethoxy (57), 5-acetoxy-2,4-dimethoxy (14)	114
α′-Carbomethoxy-2-iodo-4,5-dimethoxy	C_6H_{12}	9-Carbomethoxy-2,3-dimethoxy (61)	476
α′,α- 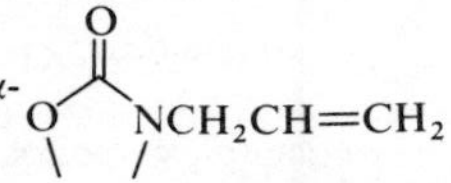	MeOH, $CuCl_2$, I_2, N_2	9,10- 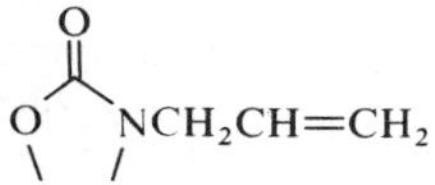 (89)	458b
4-Cyano-2,2′-dimethoxy-3-methyl	C_6H_{12}, I_2, air	3-Cyano-1,8-dimethoxy-2-methyl (31), 3-cyano-1-methoxy-2-methyl (11)	435,465
2-Carbethoxy-4′-methyl	C_6H_{12}, I_2	1-Carbethoxy-6-methyl (49)	117
3-Carbethoxy-4′-methyl	"	2-Carbethoxy-6-methyl (60)	117
4′-Carbomethoxy-3,5-dimethyl	C_6H_6, I_2, N_2	6-Carbomethoxy-2,4-dimethyl (68)	232
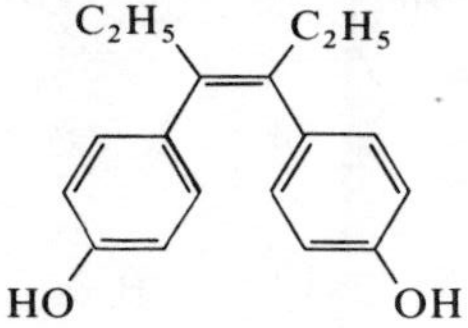	EtOH, H_2O, K_2HPO_4	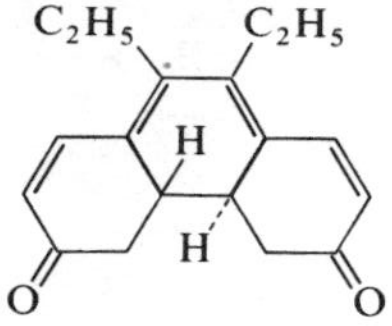 (70)	477a
	"	" (—)	477b
	MeOH, H_2O, K_2HPO_4	" (25)	17
	"	" (39)	39
	C_5H_{12}, air	9,10-Diethyl-3,6-dihydroxy (14)	108

TABLE I. PHENANTHRENES AND 9,10-DIHYDROPHENANTHRENES FROM SIMPLE STILBENES (*Continued*)

	Stilbene	Reaction Conditions	Phenanthrenes and Yields (%)	Refs.
	C_2H_5, $CH{=}CH_2$, HO, OH (structure)	MeOH, H_2O, K_2HPO_4, KH_2PO_4^{r}	C_2H_5, $CH{=}CH_2$, H, H, O, O (structure) (—)	478
	4,4′-Dimethoxy-α,α′-dimethyl	Abs EtOH, $CuCl_2$, I_2, air	3,6-Dimethoxy-9,10-dimethyl (47)	60,469
		C_6H_{12}, I_2, air	" (11)	60,469
	4′-Ethoxy-3,4-dimethoxy	C_6H_{12}, I_2	6-Ethoxy-2,3-dimethoxy (35), 6-ethoxy-3,4-dimethoxy (5)	461
	4′-Ethoxy-2,3-dimethoxy	"	6-Ethoxy-1,2-dimethoxy (40)	461
	2′,4,6-Trimethoxy-2-methyl	C_6H_{12}, N_2	3,8-Dimethoxy-1-methyl (63)	28,435
	3′-Acetoxy-3,5-dimethoxy	EtOH, I_2	5-Acetoxy-2,4-dimethoxy (31), 7-acetoxy-2,4-dimethoxy (28)	113
		C_6H_{12}, I_2	5-Acetoxy-2,4-dimethoxy (36), 7-acetoxy-2,4-dimethoxy (35)	114
	2,2′-Dicarbomethoxy	C_6H_{12}, I_2, O_2	1,8-Dicarbomethoxy (51)	438
		Solvent, air	" (—)	479
	4,4′-Dicarbomethoxy	C_6H_6, I_2	3,6-Dicarbomethoxy (51)	187
		Solvent, air	" (—)	479
	4′-Carbomethoxy-2,6-dimethoxy	C_6H_{12}, N_2	6-Carbomethoxy-1-methoxy (48)	28,435
	2,2′,4,6-Tetramethoxy	"	1,3,8-Trimethoxy (74)	28,435
	3,3′,4,4′-Tetramethoxy	Abs EtOH, I_2	2,3,6,7-Tetramethoxy (52)	480
	3,3′,4,5′-Tetramethoxy	EtOH, I_2	2,3,5,7-Tetramethoxy (50)	112
		Solvent, air	" (—), 3,4,5,7-tetramethoxy (—)	481
	2,3,5′-Trimethoxy-3′,4′-methylenedioxy	C_6H_{12}, I_2	1,2,5-Trimethoxy-6,7-methylenedioxy (20), 1,2,7-trimethoxy-5,6-methylenedioxy (10)	461
	3′,4′,5′-Trimethoxy-2,3-methylenedioxy	"	5,6,7-Trimethoxy-1,2-methylenedioxy (60)	461
	3′,4′-Diacetoxy-2,3,5′-trimethoxy	"	6,7-Diacetoxy-1,2,5-trimethoxy (10), 5,6-diacetoxy-1,2,7-trimethoxy (10)	461
C_{19}	α-Methyl-3,4-tetramethylene	Pet. ether, I_2, air	10-Methyl-2,3-tetramethylene (30), 10-methyl-3,4-tetramethylene (30)	158

4,4′-Dimethyl-α,α′-trimethylene	C_6H_{12}, I_2, air	3,6-Dimethyl-9,10-trimethylene (63)	56
	MeCN, I_2, N_2	 (66)	482,483
2-Bromo-3′,4′,5,5′-tetramethoxy-3,4-methylenedioxy	C_6H_{12}, THF, I_2	1-Bromo-4,5,6,7-tetramethoxy-2,3-methylenedioxy (84)	120
2-Carbomethoxy-2′-iodo-4′,6′-dimethoxy-4,5-methylenedioxy-α-nitro	C_6H_{12}	1-Carbomethoxy-6,8-dimethoxy-3,4-methylenedioxy-10-nitro (27)	484
3′-Cyano-3,5′-dimethoxy-2,4′-dimethyl	C_6H_6, I_2	7-Cyano-2,5-dimethoxy-1,6-dimethyl (61), 5-cyano-2,7-dimethoxy-1,6-dimethyl (9)	116
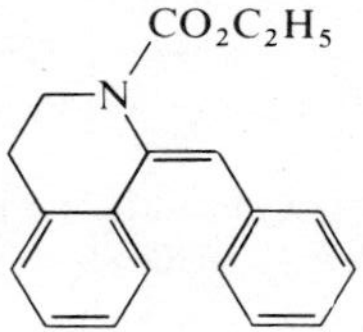 $CO_2C_2H_5$	MeOH, I_2	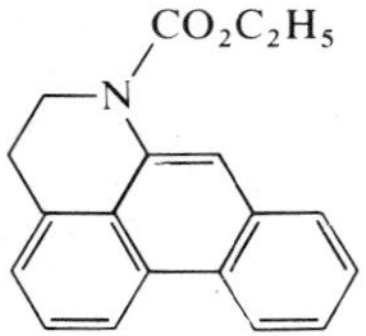 $CO_2C_2H_5$ (65), 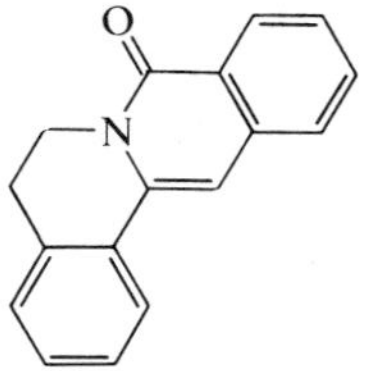 (10–21)	109,376
α′,α- 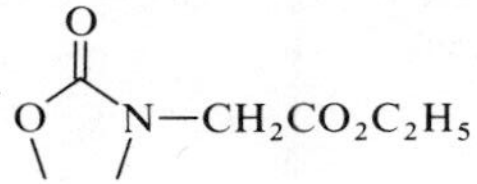 O, N—$CH_2CO_2C_2H_5$	MeOH, $CuCl_2$, I_2, N_2	9,10- 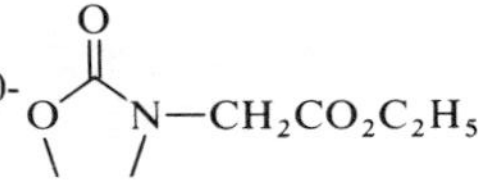 O, N—$CH_2CO_2C_2H_5$ (85)	458b
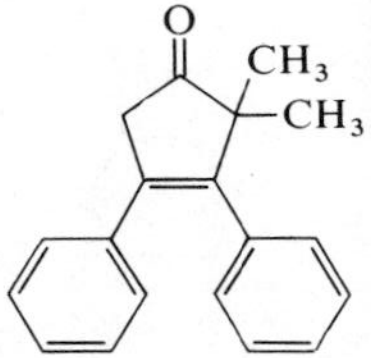 CH_3, CH_3	*i*-PrOH, N_2[e]	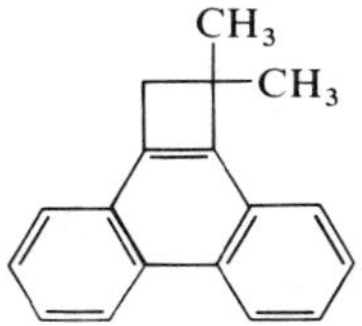 CH_3, CH_3 (34)	135

TABLE I. PHENANTHRENES AND 9,10-DIHYDROPHENANTHRENES FROM SIMPLE STILBENES (*Continued*)

Stilbene	Reaction Conditions	Phenanthrenes and Yields (%)	Refs.
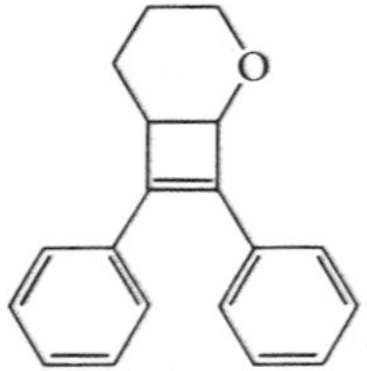	C_6H_{12}, air	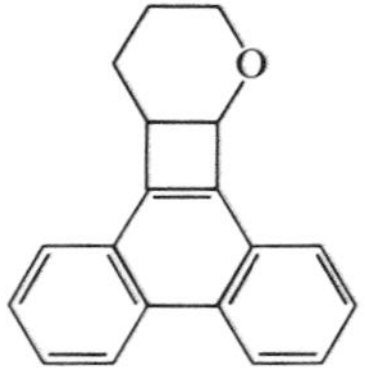 (27)	54
4,4′-Dimethoxy-α,α′-trimethylene	C_6H_{12}, I_2, air	3,6-Dimethoxy-9,10-trimethylene (50)	56
α-Ethyl-4,4′-dimethoxy-α′-methyl	Abs EtOH, $CuCl_2$, I_2, air	10-Ethyl-3,6-dimethoxy-9-methyl (55)	60,469
	C_6H_{12}, I_2, air	" (16)	60,469
4′-Isopropyl-2,6-dimethoxy	C_6H_{12}, N_2	6-Isopropyl-1-methoxy (87)	28,435
4-Carbomethoxy-2,2′-dimethoxy-3-methyl	C_6H_{12}, I_2, air	3-Carbomethoxy-1,8-dimethoxy-2-methyl (35), 3-carbomethoxy-1-methoxy-2-methyl (16)	465,485
4-Acetoxy-3,3′,5′-trimethoxy	EtOH, I_2	3-Acetoxy-2,5,7-trimethoxy (30)	486
3-Acetoxy-3′,4,5′-trimethoxy	"	2-Acetoxy-3,5,7-trimethoxy (35–40)	487
3,4,5-Trimethoxy-α′-methyl-3′,4′-methylenedioxy	C_6H_{12}, I_2, air	2,3,4-Trimethoxy-9-methyl-6,7-methylenedioxy (36)	111
2,3,3′,4′,5′-Pentamethoxy	C_6H_{12}, I_2, O_2	1,2,5,6,7-Pentamethoxy (—)	488
	Solvent, air	" (—)	481
3,3′,4,5-Tetramethoxy-4′,5′-methylenedioxy	C_6H_{12}, THF, I_2, N_2	2,3,4,7-Tetramethoxy-5,6-methylenedioxy (40), 2,3,4,5-tetramethoxy-6,7-methylenedioxy (40)	120
α′-Carboxy-3,4,5-trimethoxy-3′,4′-methylenedioxy	C_6H_6, I_2	9-Carboxy-2,3,4-trimethoxy-6,7-methylenedioxy (81)	437
C_{20} 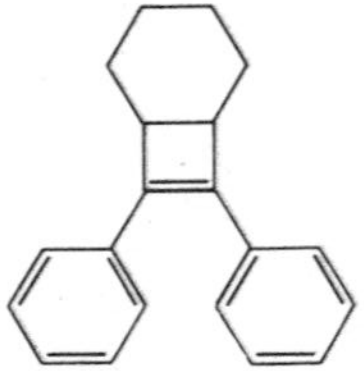	C_6H_{12}, air	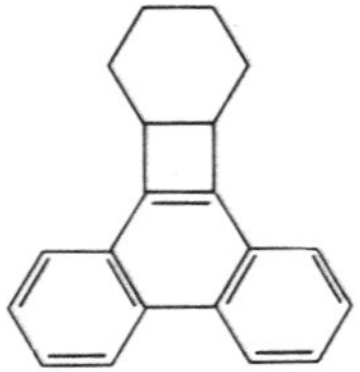 (11)	54

4,4′-Diisopropyl	C_6H_{12}, I_2	3,6-Diisopropyl (69)	489
2-Phenyl	C_6H_6, C_6H_{12}, I_2, air	1-Phenyl (64)	107,209
	C_6H_6, I_2, N_2	1-Phenyl (18), 9-phenyl (18)	157
	C_6H_6, C_6H_{14}, or alcohols; I_2	" (—), " (—)	208
	C_6H_6, C_6H_{14}, or alcohols; O_2	1-Phenyl + 9,10-dihydro-9-phenyl (1:1)	208
3-Phenyl	C_6H_6, C_6H_{12}, I_2, air	2-Phenyl (51), 4-phenyl (38)	107,209
4-Phenyl	"	3-Phenyl (86)	107,209
α-Phenyl	C_6H_{12}, I_2, air	9-Phenyl (85)	55,59
	"	" (82)	60
	Abs EtOH, $CuCl_2$, I_2, air	" (82)	60
	EtOH, air	" (57)	69
	Solvent, air	" (—)	209
	"	" (—)	490
4-Bromo-α′-phenyl	C_6H_{12}, I_2, air	3-Bromo-9-phenyl (90)	491a
3-Bromo-5-[(ethoxycarbonyl)amino]-3′-methoxy-2′,4-dimethyl	*t*-BuOH, CH_2Cl_2, I_2	2-Bromo-4-[(ethoxycarbonyl)amino]-7-methoxy-3,8-dimethyl (23), 2-[(ethoxycarbonyl)amino]-4-hydroxy-7-methoxy-3,8-dimethyl (14), 4-*tert*-butoxy-2-[(ethoxycarbonyl)amino]-7-methoxy-3,8-dimethyl (6)	464
	C_6H_6, I_2	2-Bromo-4-[(ethoxycarbonyl)amino]-7-methoxy-3,8-dimethyl (—), 2-[(ethoxycarbonyl)amino]-7-methoxy-3,8-dimethyl (—)	464
3-Bromo-5-[(ethoxycarbonyl)amino]-6′-iodo-3′-methoxy-2′,4-dimethyl	*t*-BuOH, CH_2Cl_2, Et_3N, N_2	2-Bromo-4-[(ethoxycarbonyl)amino]-7-methoxy-3,8-dimethyl (15), 2-[(ethoxycarbonyl)amino]-7-methoxy-3,8-dimethyl (8–10)	464
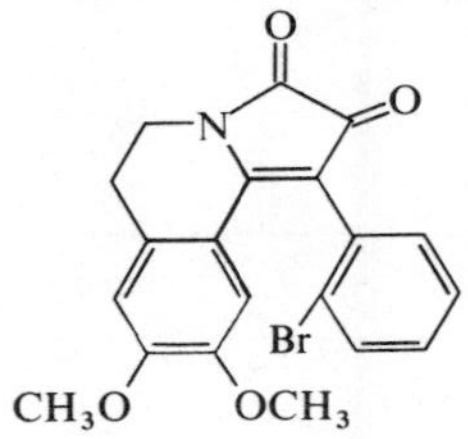	C_6H_6, *t*-$BuNH_2$, argon	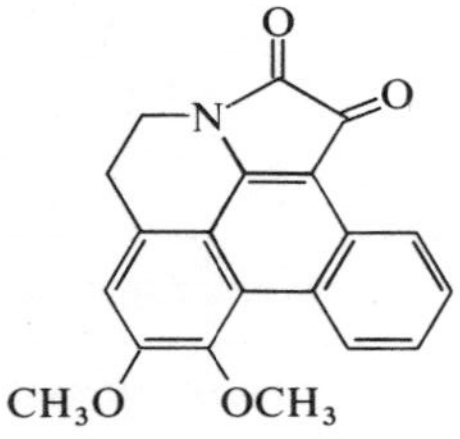 (66)	491b

TABLE I. Phenanthrenes and 9,10-Dihydrophenanthrenes from Simple Stilbenes (*Continued*)

Stilbene	Reaction Conditions	Phenanthrenes and Yields (%)	Refs.
	C_6H_6, Et_3N, argon	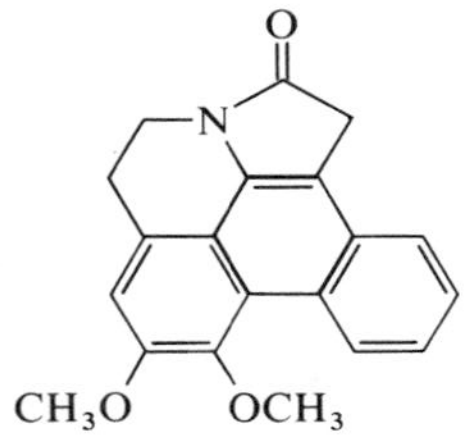 O, N, CH_3O, OCH_3 (84)	491b
4-Chloro-α′-phenyl	C_6H_{12}, I_2, air	3-Chloro-9-phenyl (91)	491a
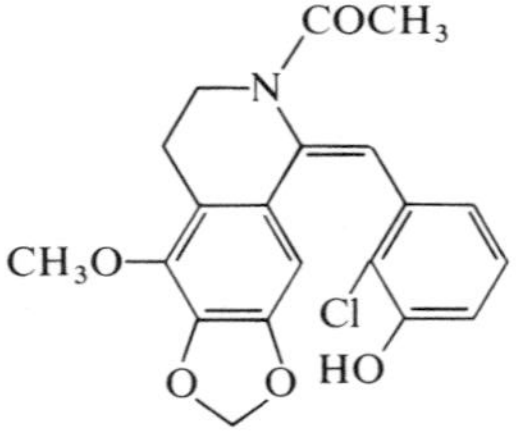 $COCH_3$, N, CH_3O, Cl, O, O, HO	*t*-BuOH, C_6H_6, *t*-BuOK, N_2	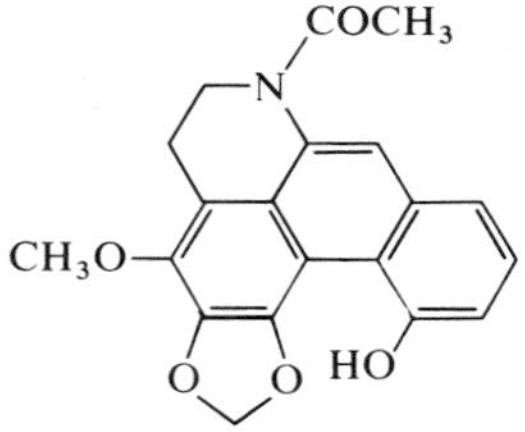 $COCH_3$, N, CH_3O, O, O, HO (20)	491c
4-Fluoro-α′-phenyl	C_6H_{12}, I_2, air	3-Fluoro-9-phenyl (87)	491a
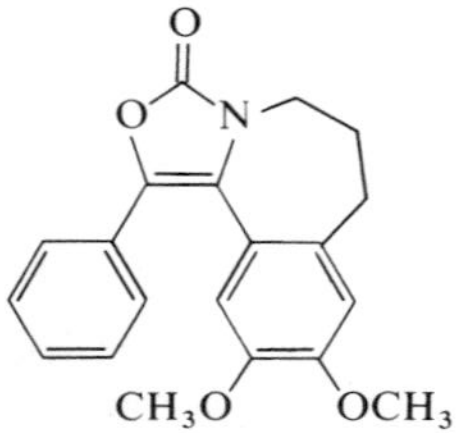 O, O, N, CH_3O, OCH_3	MeOH, $CuCl_2$, I_2, N_2	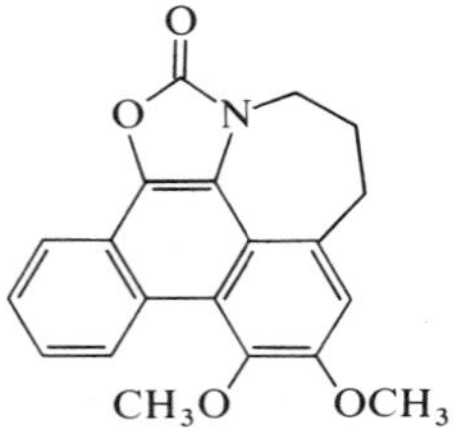 O, O, N, CH_3O, OCH_3 (20)	491d

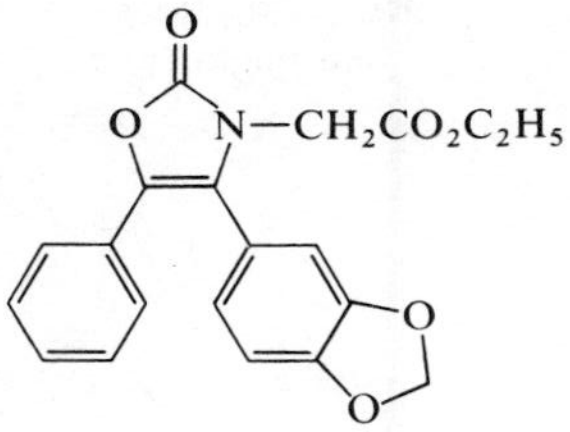

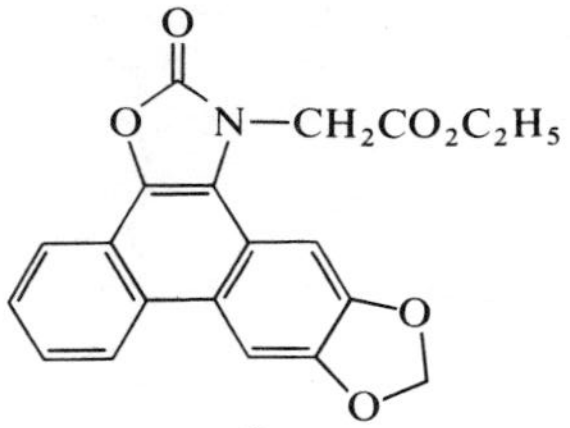

Substituents	Conditions	Product (yield %)	Ref.
(structure: 3-($CH_2CO_2C_2H_5$)-4-(3,4-methylenedioxyphenyl)-5-phenyl-oxazol-2-one)	"	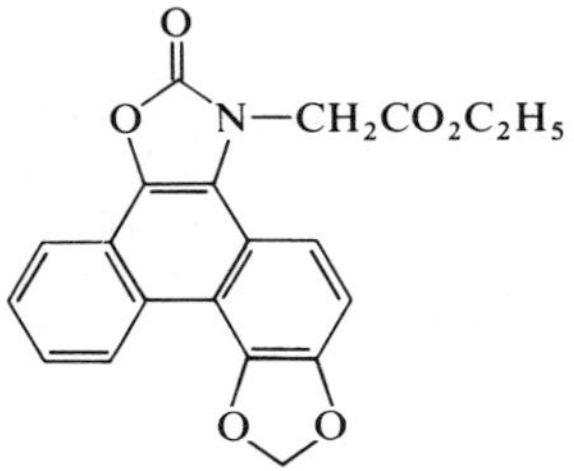 (structure: N—$CH_2CO_2C_2H_5$ oxazolone fused phenanthrene with methylenedioxy) (30), (structure: isomeric N—$CH_2CO_2C_2H_5$ oxazolone fused phenanthrene with methylenedioxy) (20)	458b
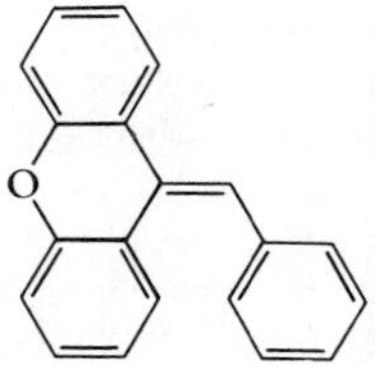 (structure: 9-benzylidenexanthene)	C_6H_6, N_2[e]	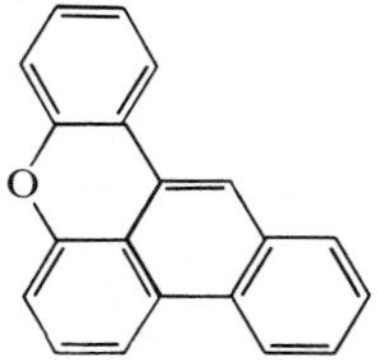 (structure) (20)	152
	Me_2CO, N_2[e]	" (19)	152
α,α′-Diethyl-4,4′-dimethoxy	Abs EtOH, $CuCl_2$, I_2, air	9,10-Diethyl-3,6-dimethoxy (63)	60,469
	C_6H_{12}, I_2, air	" (42)	60,469
	C_5H_{12}, air	" (45)	108
5,5′-Dimethoxy-2,2′,4,4′-tetramethyl	C_6H_{12}, I_2, O_2	4,5-Dimethoxy-1,3,6,8-tetramethyl (4), 4,7-dimethoxy-1,3,6-trimethyl (16)	118
2,2′-Dicarbethoxy	Solvent, air	1,8-Dicarbethoxy (—)	479
4,4′-Dicarbethoxy	C_6H_6, I_2, air	3,6-Dicarbethoxy (53)	492

TABLE I. PHENANTHRENES AND 9,10-DIHYDROPHENANTHRENES FROM SIMPLE STILBENES (*Continued*)

	Stilbene	Reaction Conditions	Phenanthrenes and Yields (%)	Refs.
		Solvent, air	" (—)	479
		n-BuOH, air	3,6-Dicarbethoxy + 2,6-dicarbethoxyanthracene (4:1)	409
	3′,4-Diacetoxy-3,5-dimethoxy	C_6H_{12}, I_2, O_2	3,7-Diacetoxy-2,4-dimethoxy (—)	488
	4-Acetoxy-3,5-dimethoxy-2′,3′-methylenedioxy	C_6H_6, I_2, O_2	3-Acetoxy-2,4-dimethoxy-7,8-methylenedioxy (—)	488
	2-Acetoxy-3,3′,4′,5′-tetramethoxy	C_6H_{12}, I_2	1-Acetoxy-2,5,6,7-tetramethoxy (50)	461
	3′-Acetoxy-3,4,4′,5-tetramethoxy	EtOH, I_2, air	7-Acetoxy-2,3,4,6-tetramethoxy (35–40)	487
	4′-Acetoxy-2,3,3′,5′-tetramethoxy	Solvent[e]	6-Acetoxy-1,2,5,7-tetramethoxy (—)	481
	α′-Carbomethoxy-3,3′,4,4′-tetramethoxy	C_6H_{12}, I_2	9-Carbomethoxy-2,3,6,7-tetramethoxy (31)	493
	3,3′,4,4′,5,5′-Hexamethoxy	C_6H_{12}, THF, I_2	2,3,4,5,6,7-Hexamethoxy (82)	119
	α′-Acetoxy-3,4,5-trimethoxy-3′,4′-methylenedioxy	C_6H_6, I_2, air	9-Acetoxy-2,3,4-trimethoxy-6,7-methylenedioxy (49)	111
	α′-Carboxy-3,4,5-trimethoxy-3′,4′-methylenedioxy	C_6H_6, I_2	9-Carboxy-2,3,4-trimethoxy-6,7-methylenedioxy (81)	494
	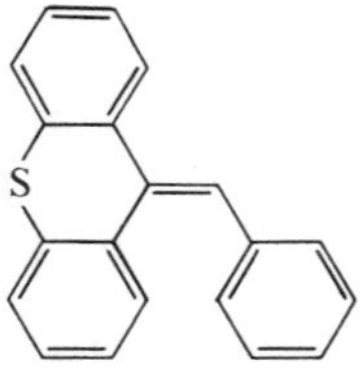 [structure: 9-benzylidenethioxanthene]	Me_2CO, N_2[e]	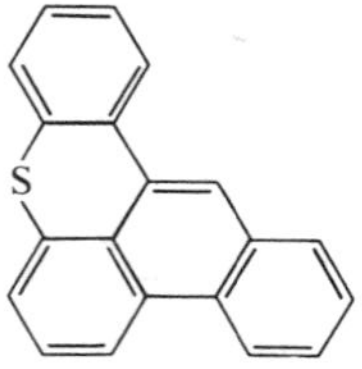 [structure] (7)	152
C_{21}	4-Methyl-α′-phenyl	C_6H_{12}, I_2, air	3-Methyl-9-phenyl (96)	491a
	4,4′-Heptamethylene	C_6H_{12}, I_2	3,6-Heptamethylene (good)	106

Reactant	Conditions	Product(s) (Yield, %)	Refs.
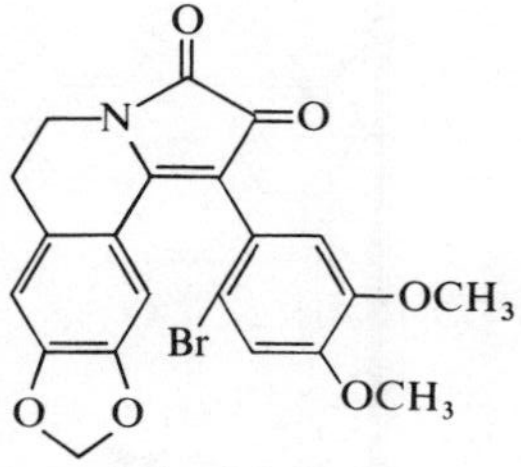	C_6H_6, Et_3N, argon	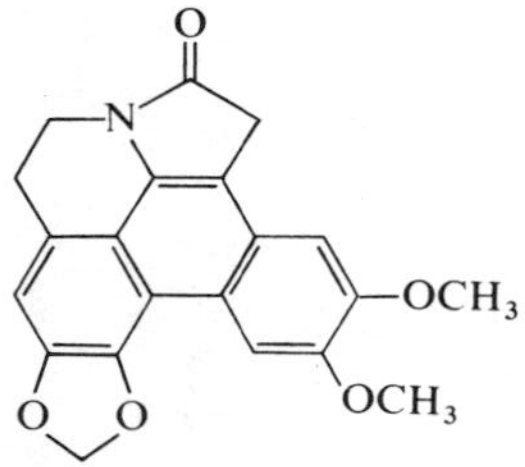 (70)	491b
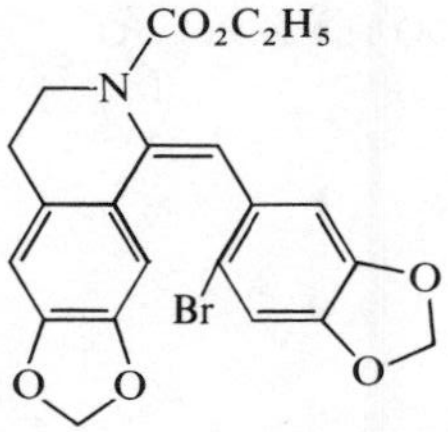	*t*-BuOH, C_6H_6, *t*-BuOK, N_2	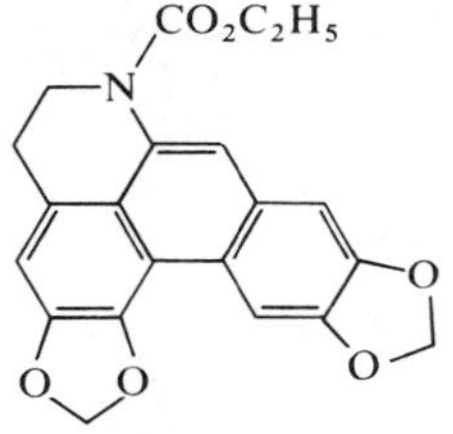 (72)	25
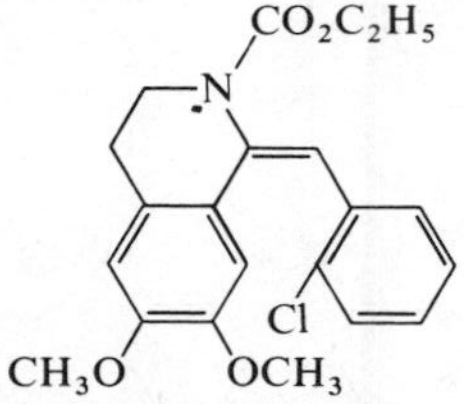	MeOH, $CaCO_3$, N_2	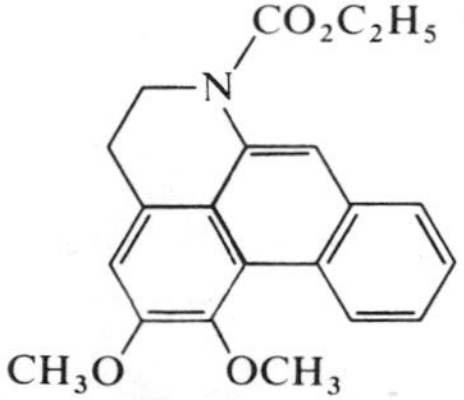 (32)	110
4-Cyano-α′-phenyl	C_6H_{12}, I_2, air	3-Cyano-9-phenyl (91)	491a
α′,α- 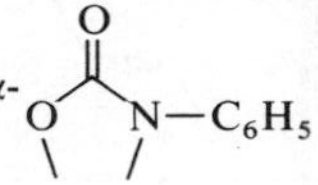	C_6H_6, I_2	9,10- 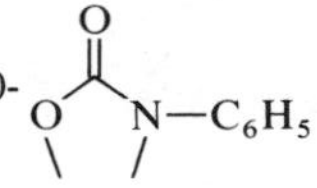 (52)	134

TABLE I. PHENANTHRENES AND 9,10-DIHYDROPHENANTHRENES FROM SIMPLE STILBENES (*Continued*)

Stilbene	Reaction Conditions	Phenanthrenes and Yields (%)	Refs.
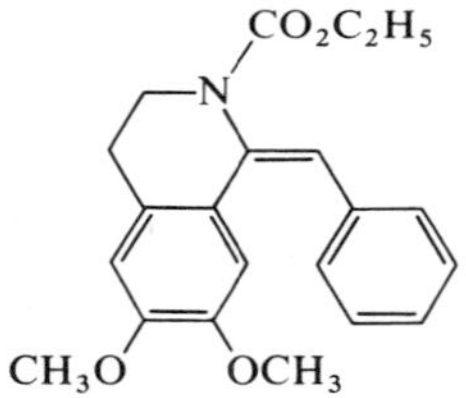 $CO_2C_2H_5$ N; CH_3O OCH_3	EtOH, $Cu(OAc)_2$, I_2, N_2	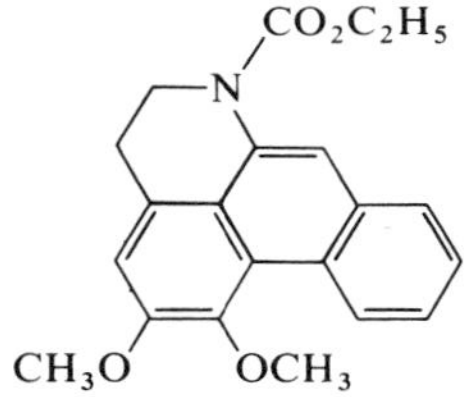 $CO_2C_2H_5$ N; CH_3O OCH_3 (35)	110
	EtOH, I_2	" (15)	495
	Abs EtOH, N_2[e]	" (20), 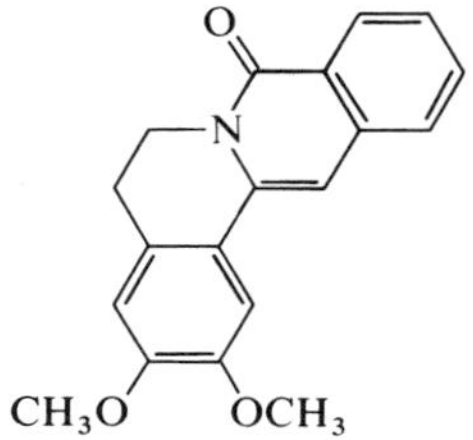 O N; CH_3O OCH_3 (55)	80
α′,α-  O; O N$CH_2CH(OC_2H_5)_2$	MeOH, $CuCl_2$, I_2, N_2	9,10-  O; O N$CH_2CH(OC_2H_5)_2$ (60)	458b
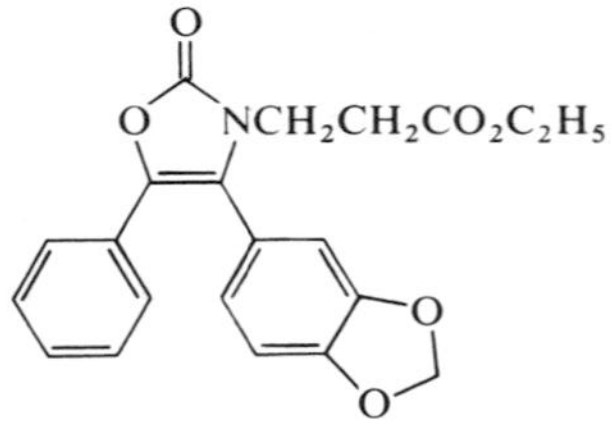 O; O N$CH_2CH_2CO_2C_2H_5$; O O	"	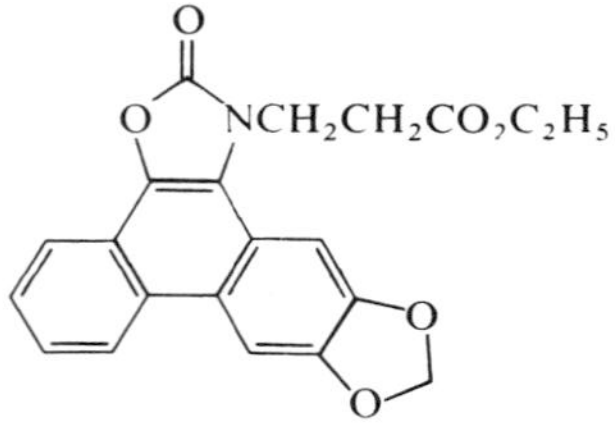 O; O N$CH_2CH_2CO_2C_2H_5$; O O (63),	491d

Reactant	Conditions	Product(s) (Yield, %)	Refs.
		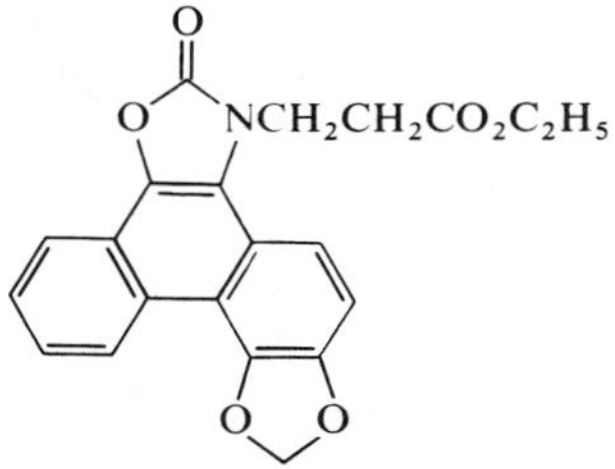 $NCH_2CH_2CO_2C_2H_5$ (7)	
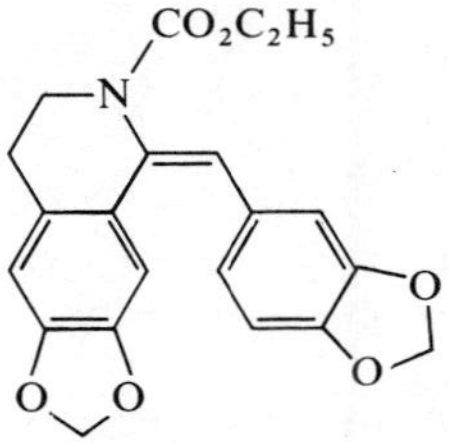 $CO_2C_2H_5$	EtOH, $Cu(OAc)_2$, I_2	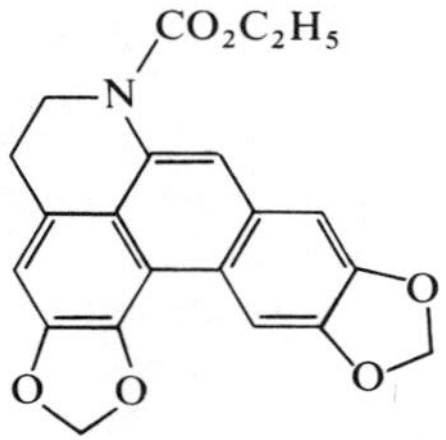 $CO_2C_2H_5$ (40)	496
α′,α- 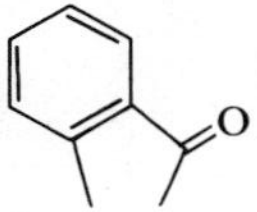	*i*-PrOH, N_2[e]	9,10- 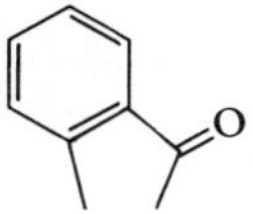 (3)	66a
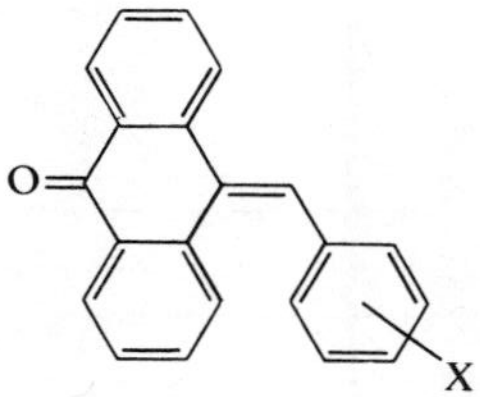 X = H, *p*-Br, *o*-Cl, or *p*-Cl	C_6H_6 or C_6H_{12}, I_2	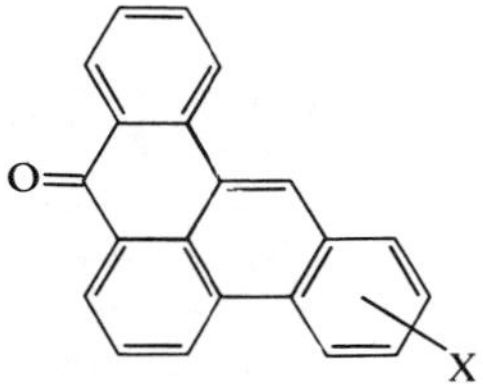 (20–90)	58
4-Methoxy-α′-phenyl	C_6H_{12}, I_2, air	3-Methoxy-9-phenyl (90)	491a

TABLE I. PHENANTHRENES AND 9,10-DIHYDROPHENANTHRENES FROM SIMPLE STILBENES (*Continued*)

	Stilbene	Reaction Conditions	Phenanthrenes and Yields (%)	Refs.
	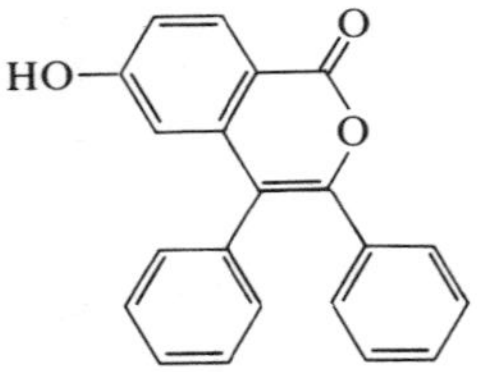	MeOH, I_2, N_2[s]	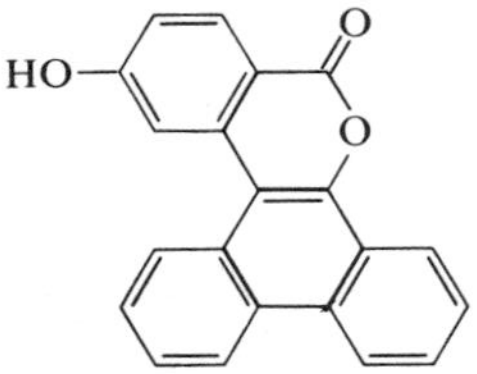 (38)	497,498
		MeOH, air	" (—)	497,498
	3,5-Dimethoxy-3′-trimethylacetoxy	Solvent, air	2,4-Dimethoxy-5-trimethylacetoxy + 2,4-dimethoxy-7-trimethylacetoxy (1:1)	113
	2,3-Diacetoxy-3′,4′,5′-trimethoxy	C_6H_{12}, I_2	1,2-Diacetoxy-5,6,7-trimethoxy (20)	461
	3′,4′-Diacetoxy-3,4,5-trimethoxy	EtOH, I_2, air	6,7-Diacetoxy-2,3,4-trimethoxy (15)	499
C_{22}	4,4′-Di-*tert*-butyl	Solvent, air	3,6-Di-*tert*-butyl (—)	474
	 C_6H_5	C_6H_{12}, I_2, air	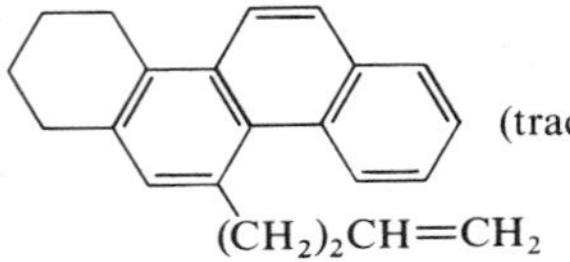 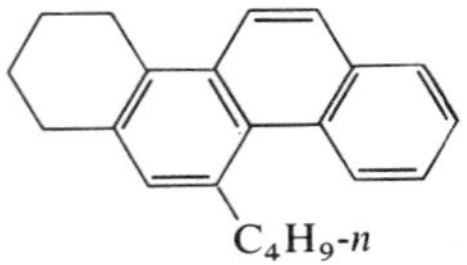 [$(CH_2)_2CH{=}CH_2$] (trace), [C_4H_9-*n*] (trace)	64
	4,4′-Octamethylene	C_6H_{12}, I_2	3,6-Octamethylene (good)	106
	α′-Phenyl-2-vinyl	C_6H_{14}, degassed[e]	9-Phenyl-1-vinyl (10)	500

	C_6H_6, t-$BuNH_2$, argon	(60)	491b
	C_6H_6, Et_3N, argon	(82)	491b
X = Br	t-BuOH, C_6H_6, t-BuOK, N_2	” (56)	25
X = Cl	”	” (44)	25

TABLE I. PHENANTHRENES AND 9,10-DIHYDROPHENANTHRENES FROM SIMPLE STILBENES (*Continued*)

Stilbene	Reaction Conditions	Phenanthrenes and Yields (%)	Refs.
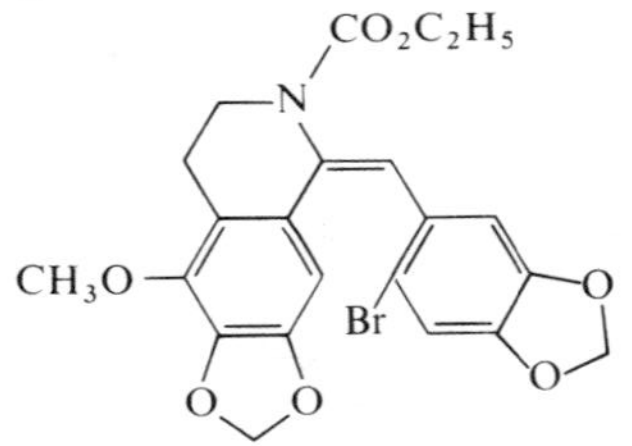 $CO_2C_2H_5$, N, CH_3O, Br, O, O, O, O	t-BuOH, C_6H_6, t-BuOK, N_2	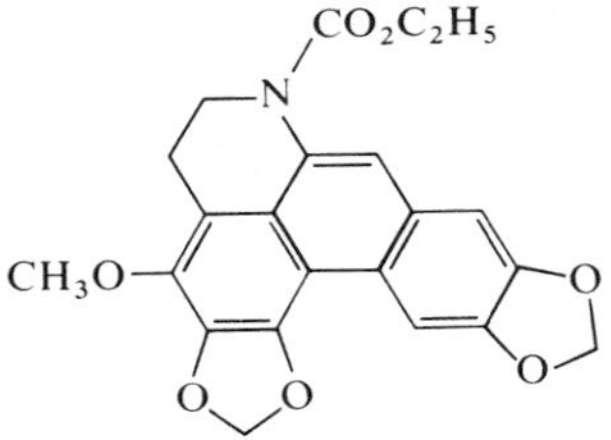 $CO_2C_2H_5$, N, CH_3O, O, O, O, O (32)	501
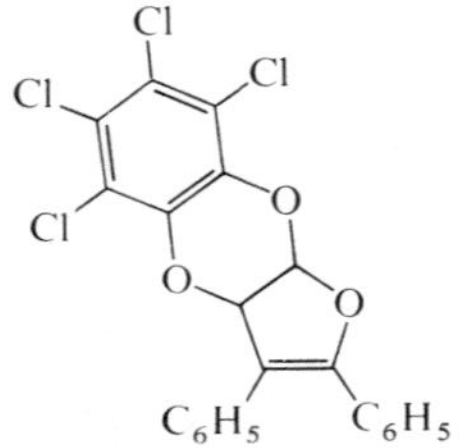 Cl, Cl, Cl, Cl, O, O, O, C_6H_5, C_6H_5	i-PrOH, N_2	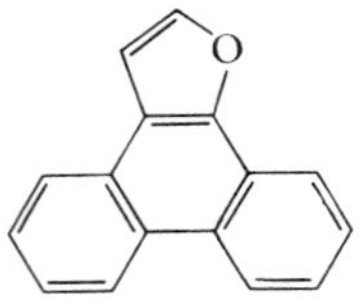 O (73)	334
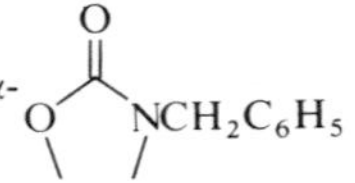 α',α- O, $NCH_2C_6H_5$, O	MeOH, $CuCl_2$, I_2, air	9,10- O, $NCH_2C_6H_5$, O (80)	133
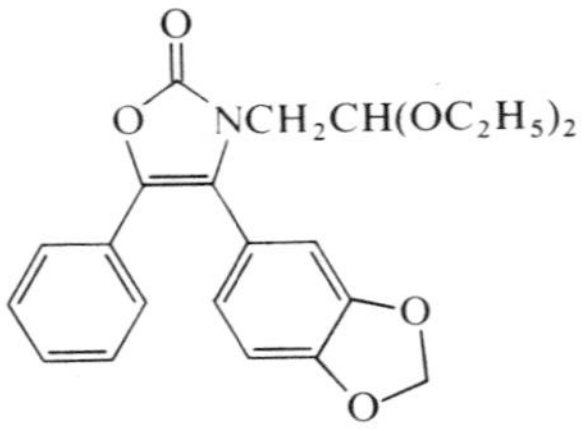 O, O, $NCH_2CH(OC_2H_5)_2$, O, O	MeOH, $CuCl_2$, I_2, N_2	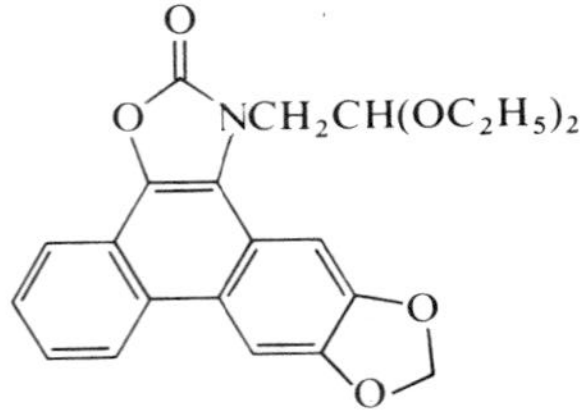 O, O, $NCH_2CH(OC_2H_5)_2$, O, O (33),	458b

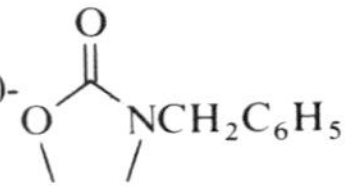

Reactant	Conditions	Product (yield %)	Ref.
		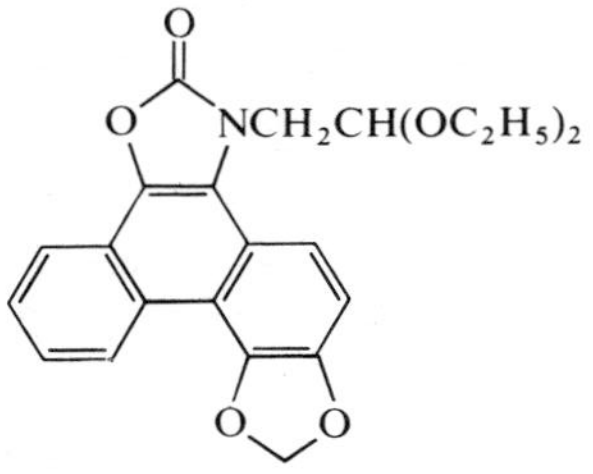 [structure] (17)	
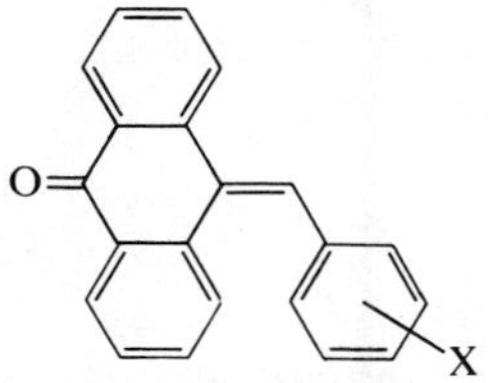 [structure] $X = o\text{-}CH_3$, $p\text{-}CH_3$, or $p\text{-}CH_3O$	C_6H_6 or C_6H_{12}, I_2	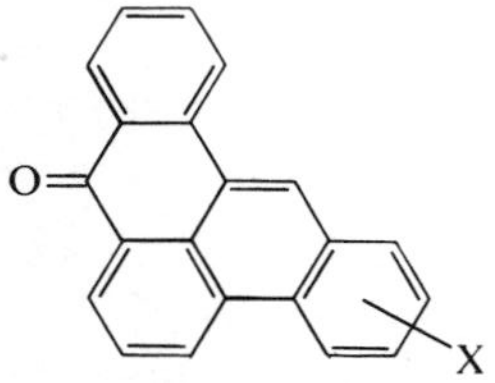 [structure] (20–90)	58
α′,α- 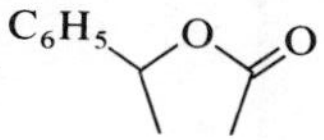 [structure: C_6H_5, O, O]	C_6H_6, air	9,10- 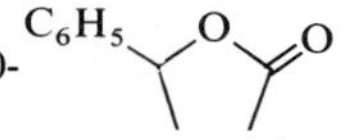 [structure: C_6H_5, O, O] (65)	67
	i-PrOH, degassed[e]	" (53)	67
	C_6H_6, O_2[f]	" (high)	460
α′,α- 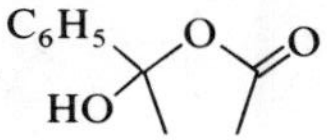 [structure: C_6H_5, HO, O, O]	C_6H_6, O_2	9,10- 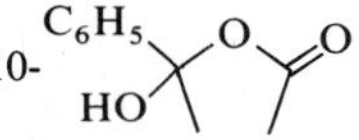 [structure: C_6H_5, HO, O, O] (67)	67,502
	Py, H_2O (1:1), air	*trans*-9-Benzoyl-10-carboxy-9,10-dihydro (68)	67,502
	Py, D_2O (1:1), air	*trans*-9-Benzoyl-10-carboxy-9,10-dideuterio (—)	67,502

TABLE I. PHENANTHRENES AND 9,10-DIHYDROPHENANTHRENES FROM SIMPLE STILBENES (*Continued*)

	Stilbene	Reaction Conditions	Phenanthrenes and Yields (%)	Refs.
	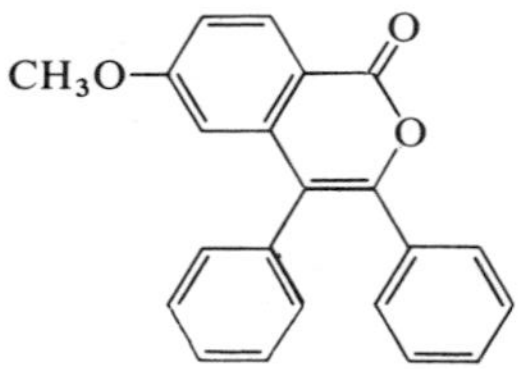 CH_3O (structure)	MeOH, N_2[e,u]	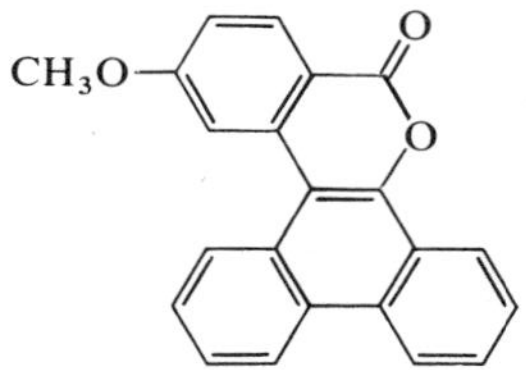 CH_3O (structure) (51)	498
		MeOH, air[u]	" (42)	497
	2,2′-Di-*n*-propoxycarbonyl	Solvent, air	1,8-Di-*n*-propoxycarbonyl (—)	479
	4,4′-Di-*n*-propoxycarbonyl	"	3,6-Di-*n*-propoxycarbonyl (—)	479
	α′,α- 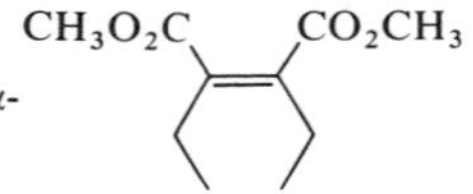 CH_3O_2C CO_2CH_3 (structure)	C_6H_6, I_2, O_2	9,10- 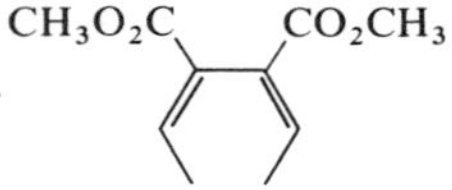 CH_3O_2C CO_2CH_3 (structure) (29), 1,2-dicarbomethoxy-4,5-diphenylbenzene (62)	482,483
	α′,α- 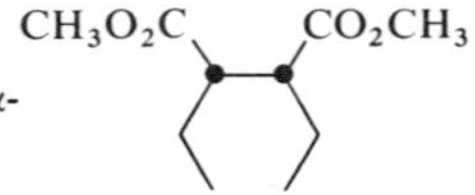 CH_3O_2C CO_2CH_3 (structure)	"	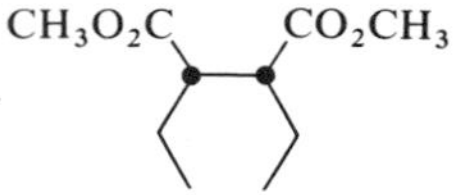 9,10- CH_3O_2C CO_2CH_3 (structure) (54)	483
	3,3′,4,5′-Tetraacetoxy	EtOH, I_2	2,3,5,7-Tetraacetoxy (48)	112
	α′,α- 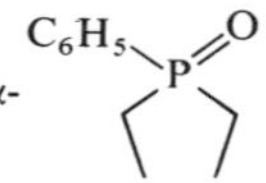 C_6H_5 P=O (structure)	C_6H_6, I_2, O_2	9,10- 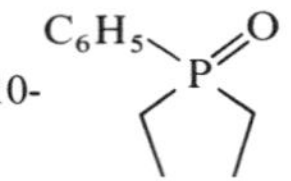 C_6H_5 P=O (structure) (56)	503
C_{23}	4,4′-Nonamethylene	C_6H_{12}, I_2	3,6-Nonamethylene (good)	106
	α′-Phenyl-2-propenyl	C_6H_{14}, degassed[e]	9-Phenyl-1-propenyl (25)	500

Reactant	Conditions	Product(s) and Yield(s) (%)	Refs.
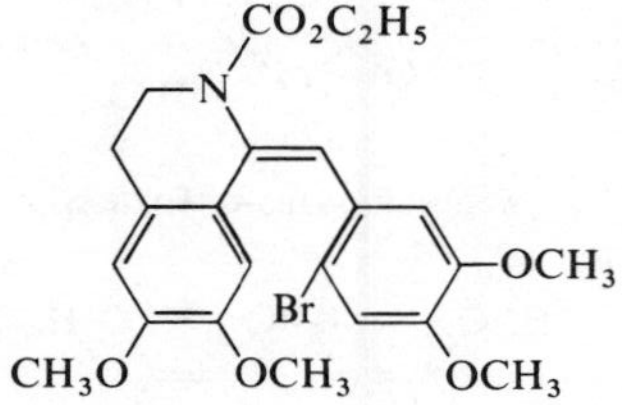 $CO_2C_2H_5$, CH_3O, OCH_3, OCH_3, OCH_3, Br	t-BuOH, C_6H_6, t-BuOK, N_2	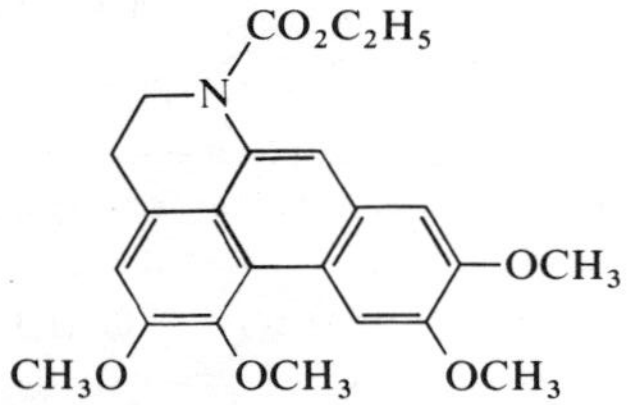 $CO_2C_2H_5$, CH_3O, OCH_3, OCH_3, OCH_3 (59)	25
	MeOH, $CaCO_3$, N_2	" (24)	110
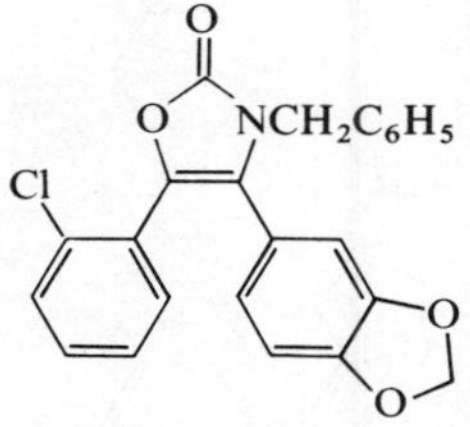 O, $NCH_2C_6H_5$, Cl, O	MeOH, $CuCl_2$, I_2, air	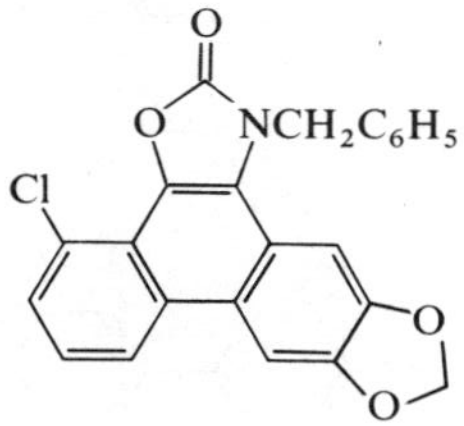 O, $NCH_2C_6H_5$, Cl, O (85)[v]	133
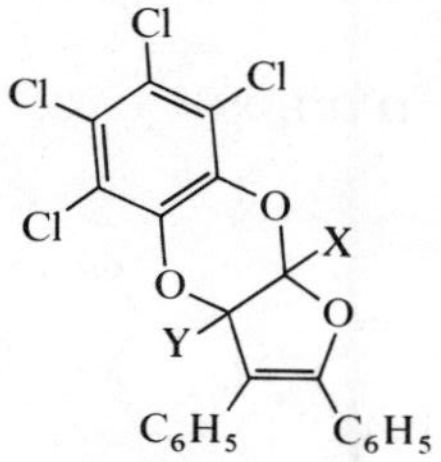 Cl, Cl, Cl, Cl, O, X, O, Y, C_6H_5, C_6H_5 (I)		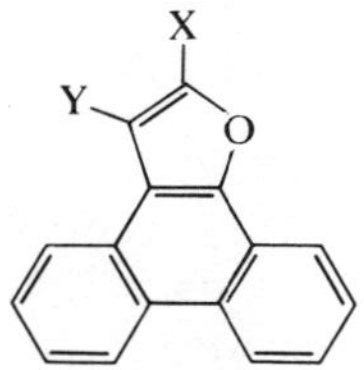 X, Y, O (II)	
I, X = CH_3, Y = H	i-PrOH, N_2	II, X = CH_3, Y = H (88)	334
I, X = H, Y = CH_3	"	II, X = H, Y = CH_3 (78)	334
	MeOH, N_2	" (68)	334

TABLE I. PHENANTHRENES AND 9,10-DIHYDROPHENANTHRENES FROM SIMPLE STILBENES (*Continued*)

Stilbene	Reaction Conditions	Phenanthrenes and Yields (%)	Refs.
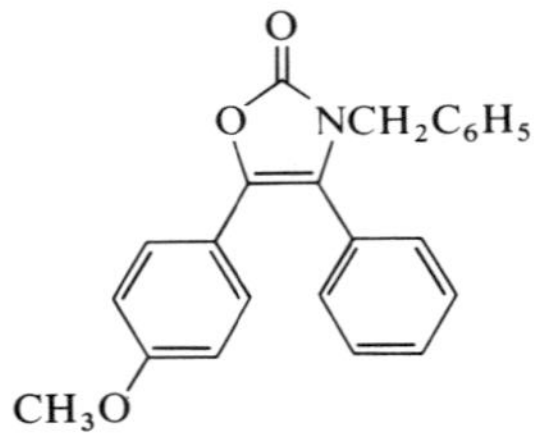 O; O, $NCH_2C_6H_5$; CH_3O	MeOH, $CuCl_2$, I_2, air[v]	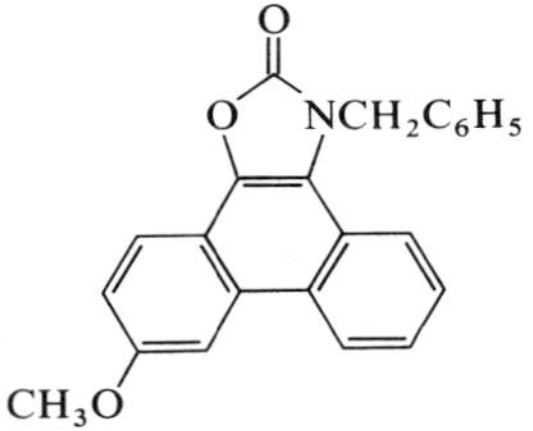 O; O, $NCH_2C_6H_5$; CH_3O (86)	133
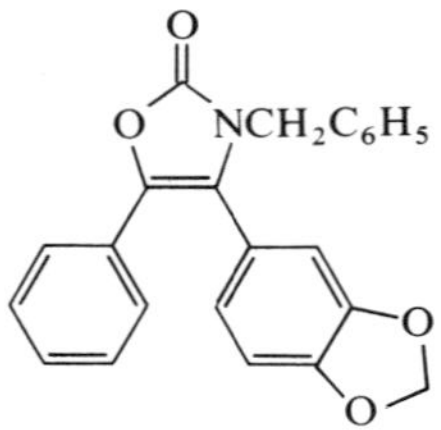 O; O, $NCH_2C_6H_5$; O; O	"	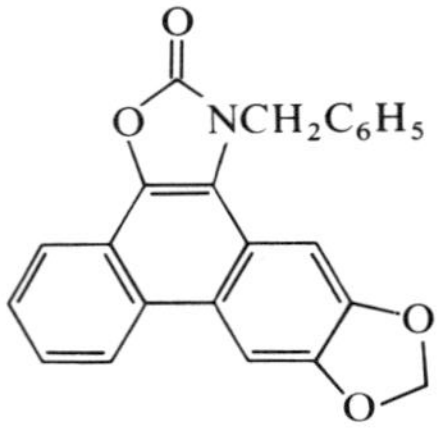 O; O, $NCH_2C_6H_5$; O; O (67)	133
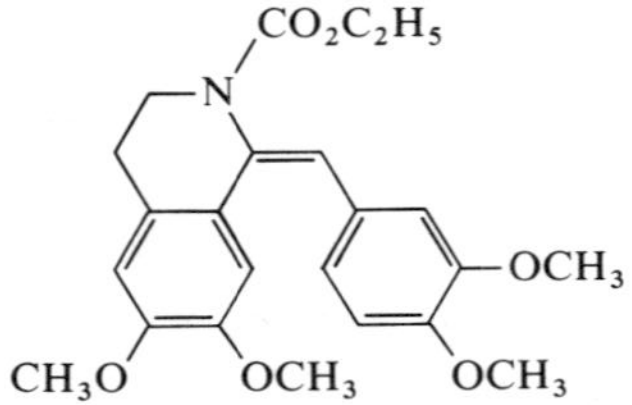 $CO_2C_2H_5$; N; OCH_3; CH_3O, OCH_3, OCH_3	MeOH, I_2	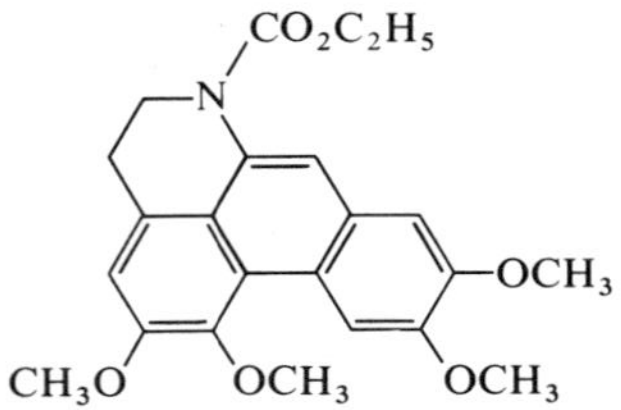 $CO_2C_2H_5$; N; OCH_3; CH_3O, OCH_3, OCH_3 (25)	109
	EtOH, $Cu(OAc)_2$, I_2, N_2	" (19)	110
α′-Benzoyl-α-carbomethoxy	C_6H_6, O_2	9-Benzoyl-10-carbomethoxy (43)	67
	Py, H_2O (1:1)	*trans*-9-Benzoyl-10-carbomethoxy-9,10-dihydro (55)	67
α′,α- 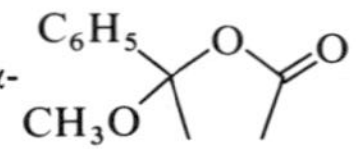 C_6H_5, O, O; CH_3O	C_6H_6, O_2	9,10- 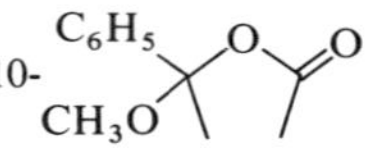 C_6H_5, O, O; CH_3O (70)	67

		i-PrOH [or Py, H_2O (1:1)], degassed	” (50)	67
	3′-Benzyloxy-3,5-dimethoxy	PhMe, I_2	7-Benzyloxy-2,4-dimethoxy (38), 5-benzyloxy-2,4-dimethoxy (17)	113
C_{24}	4-*tert*-Butyl-α′-phenyl	C_6H_{12}, I_2, air	3-*tert*-Butyl-9-phenyl (76)	491a
	4,4′-Decamethylene	C_6H_{12}, I_2	3,6-Decamethylene (good)	106
	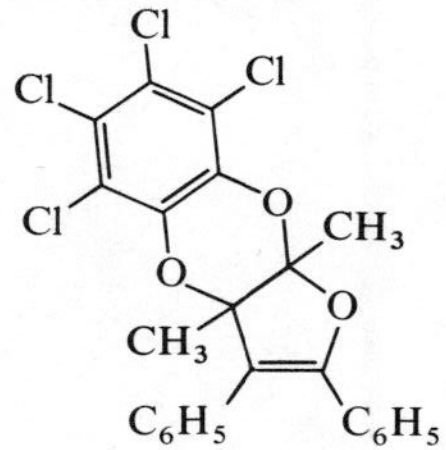	i-PrOH, N_2	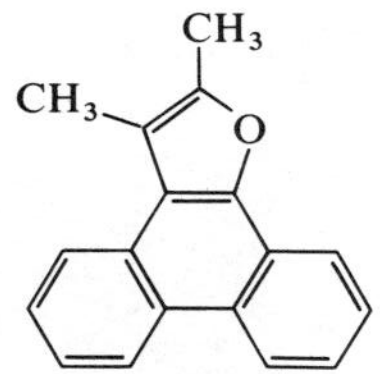 (69)	334
	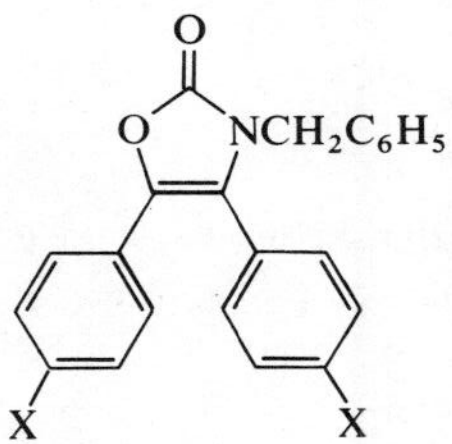 (I)		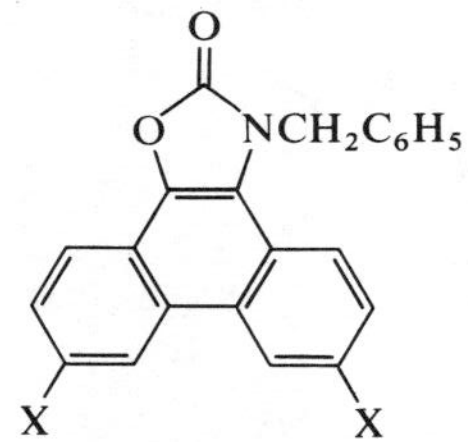 (II)	
	I, X = CH_3	MeOH, $CuCl_2$, I_2, air	II, X = CH_3 (84)	133
	I, X = CH_3O	”	II, X = CH_3O (90)	133
	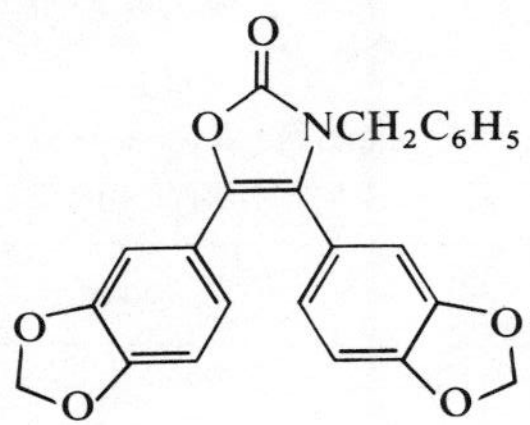	”	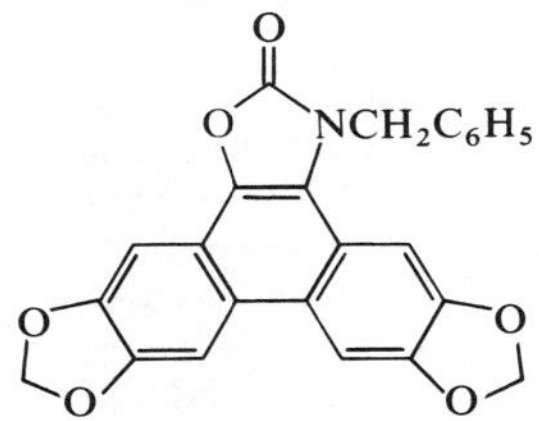 (83)	133

TABLE I. PHENANTHRENES AND 9,10-DIHYDROPHENANTHRENES FROM SIMPLE STILBENES (*Continued*)

	Stilbene	Reaction Conditions	Phenanthrenes and Yields (%)	Refs.
	2,2′-Di-*n*-butoxycarbonyl	Solvent, air	1,8-Di-*n*-butoxycarbonyl (—)	479
	4,4′-Di-*n*-butoxycarbonyl	"	3,6-Di-*n*-butoxycarbonyl (—)	479
	α′,α- 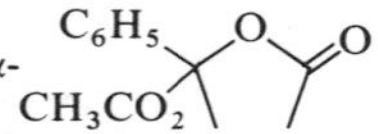 C_6H_5, CH_3CO_2 [structure]	C_6H_6, O_2	9,10- 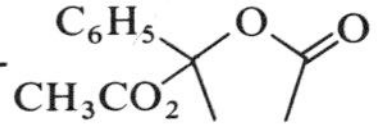 C_6H_5, CH_3CO_2 [structure] (68)	67
		i-PrOH, degassed	9,10- 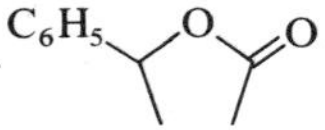 C_6H_5 [structure] (70)	67
		Py, H_2O (1:1), air	" (60)	67
	4-Benzyloxy-3,3′,5′-trimethoxy	EtOH, I_2	3-Benzyloxy-2,5,7-trimethoxy (30)	504
C_{25}	4,4′-Di-*tert*-butyl-α,α′-trimethylene	C_6H_{12}, I_2, air	3,6-Di-*tert*-butyl-9,10-trimethylene (82)	56
	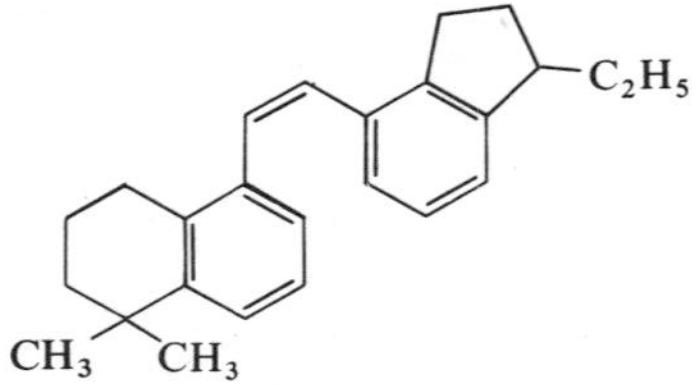 [structure: C_2H_5, CH_3, CH_3]	C_6H_{12}, I_2	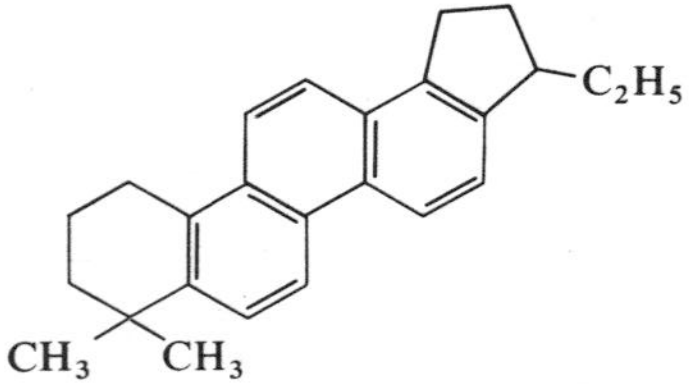 [structure: C_2H_5, CH_3, CH_3] (84)	505
	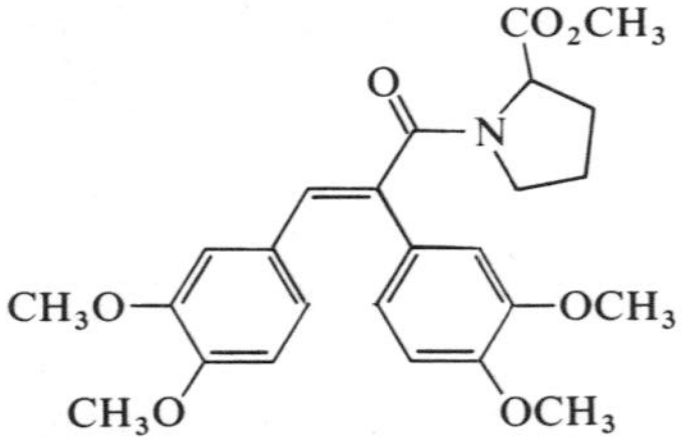 [structure: CO_2CH_3, O, N, CH_3O, CH_3O, OCH_3, OCH_3]	"	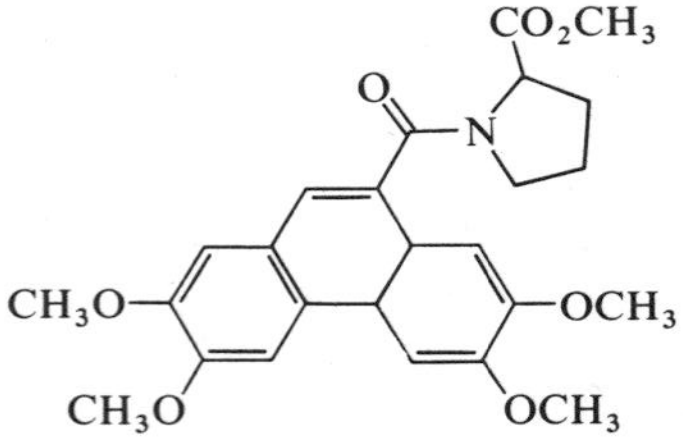 [structure: CO_2CH_3, O, N, CH_3O, CH_3O, OCH_3, OCH_3] (–)	493
C_{26}	α,α′-Diphenyl	C_6H_{12}, I_2, air	9,10-Diphenyl (90)	76
		"	" (88)	60

	Abs EtOH, $CuCl_2$, I_2, air	" (83)	60
	C_6H_6, I_2	" (50–70)	506
	EtOH, I_2, air	" (65)	69
	MeOH, air	" (—)	507
	Cyclopentene, air[w]	" (80)	508a
	Solvent, air	" (—)	508b
α,α′-Diphenyl-α,α′-(^{13}C)$_2$	C_6H_{12}, I_2, O_2	9,10-Diphenyl-9,10-(^{13}C)$_2$ (—)	424
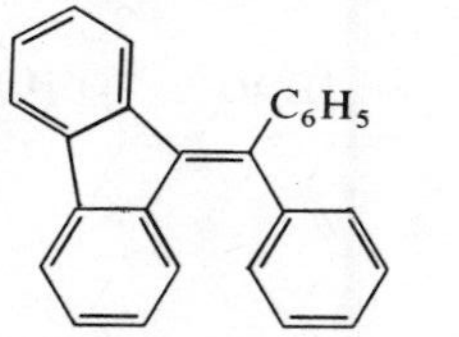 C_6H_5	Solvent, air	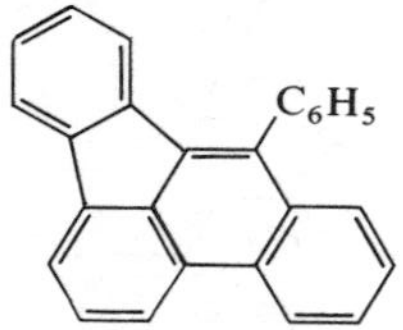 C_6H_5 (1)	58
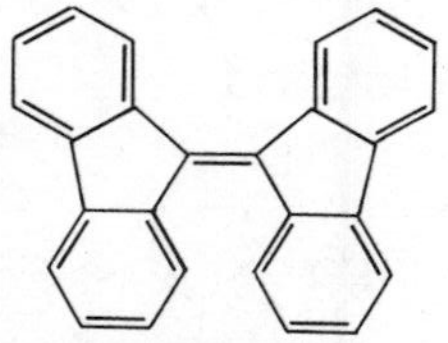	C_6H_{14}, N_2	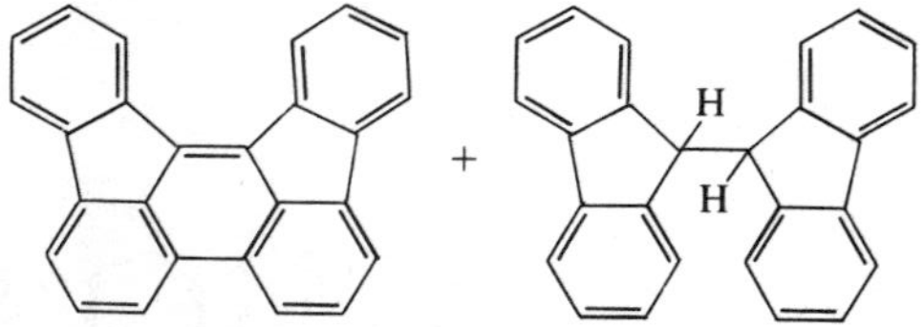 + H H (1:1)	509
4-Chloro-α,α′-diphenyl	C_6H_6, I_2	3-Chloro-9,10-diphenyl (26–37), 9-(*p*-chlorophenyl)-10-phenyl (24–33)	506
α′-Phenoxy-α-phenyl	Solvent, O_2	9-Phenoxy-10-phenyl (—)	510
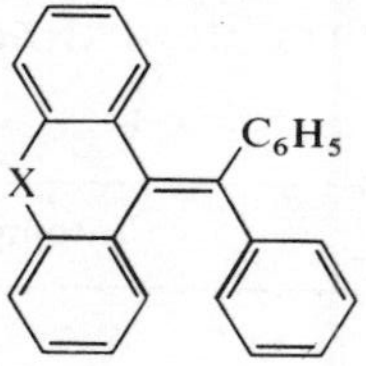 X C_6H_5 (I)		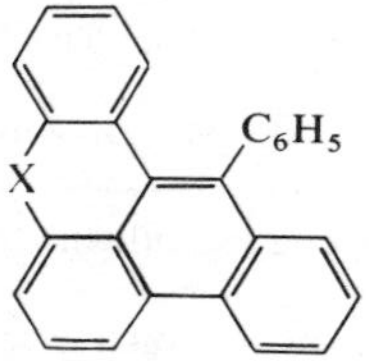 X C_6H_5 (II)	
I, X = O	C_6H_6, N_2[e]	II, X = O (14)	152
	Me_2CO, N_2[e]	" (11)	152

TABLE I. PHENANTHRENES AND 9,10-DIHYDROPHENANTHRENES FROM SIMPLE STILBENES (*Continued*)

	Stilbene	Reaction Conditions	Phenanthrenes and Yields (%)	Refs.
	I, X = S	C_6H_6, N_2[e]	II, X = S (7)	152
		Me_2CO, air	" (5)	152
	2,2′-Di-[$(CH_2)_5CO_2H$]	MeOH, I_2, O_2	1,8-Di-[$(CH_2)_5CO_2H$] (58)	438
C_{27}	4-Methyl-α,α′-diphenyl	C_6H_6, I_2	3-Methyl-9,10-diphenyl (29–41), 9-(*p*-methylphenyl)-10-phenyl (21–29)	506
	α′,α- 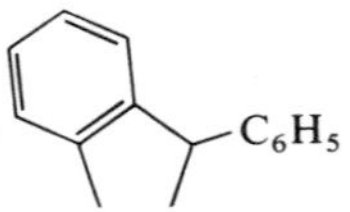 [structure] C_6H_5	C_6H_{14}, air	9,10- 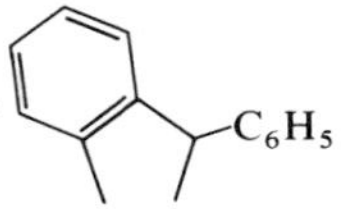 [structure] C_6H_5 (20)	511a
		C_6H_{12} or C_6H_6, argon[e]	" (–)	511b
	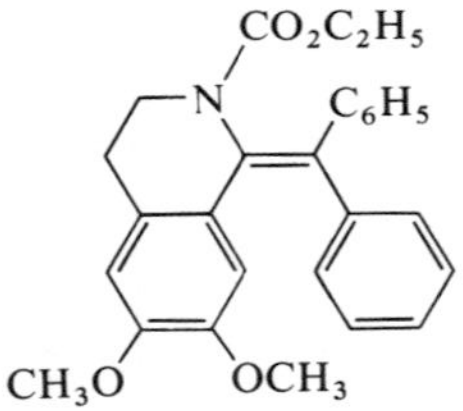 [structure: $CO_2C_2H_5$, C_6H_5, CH_3O, OCH_3]	EtOH, I_2	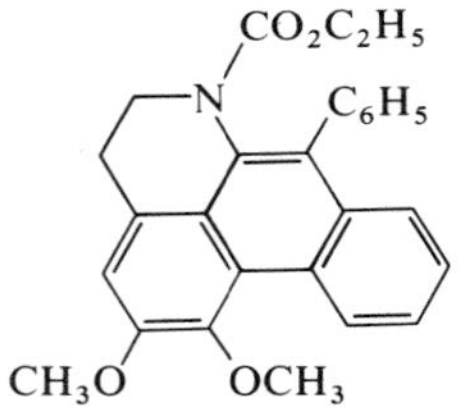 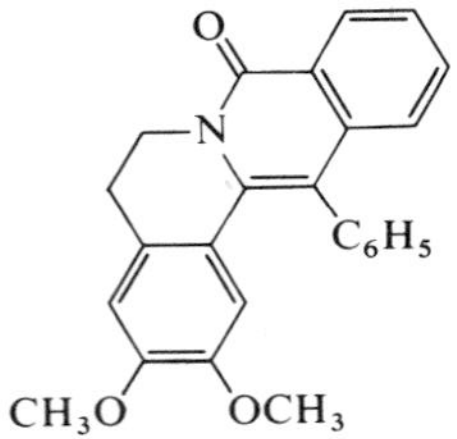 [structure: $CO_2C_2H_5$, C_6H_5, CH_3O, OCH_3] (60), [structure: O, N, C_6H_5, CH_3O, OCH_3] (9)	375

	Reactant	Conditions	Product (Yield %)	Refs.
	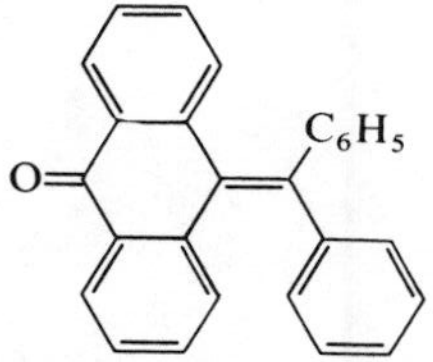 (structure: C_6H_5)	Xylene, air	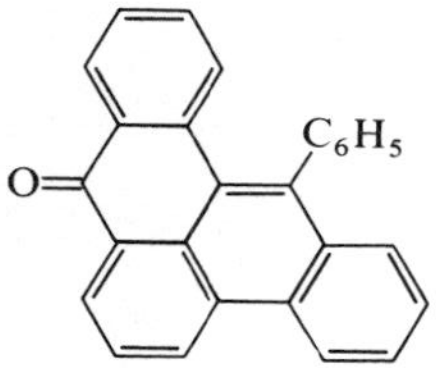 (structure: C_6H_5) (—)	512
	4-Methoxy-α,α′-diphenyl	C_6H_6, I_2	3-Methoxy-9,10-diphenyl (45–63), 9-(*p*-methoxyphenyl)-10-phenyl (5–7)	506
C_{28}	4-Chloro-4′-methoxy-α-(*p*-chlorophenyl)-α′-(*p*-methoxyphenyl)	C_6H_6, I_2, air	3-Chloro-6-methoxy-10-(*p*-chlorophenyl)-9-(*p*-methoxyphenyl) (20–40)	513
	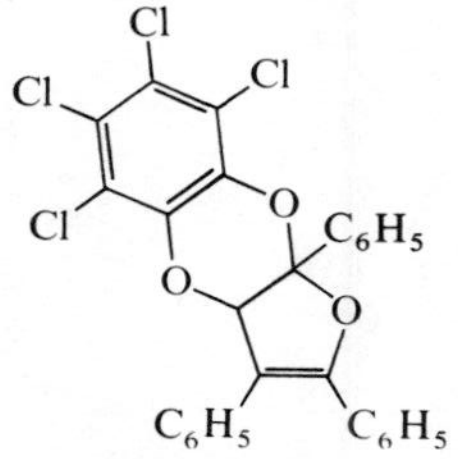 (structure: Cl, Cl, Cl, Cl, O, O, O, C_6H_5, C_6H_5, C_6H_5)	*i*-PrOH, N_2	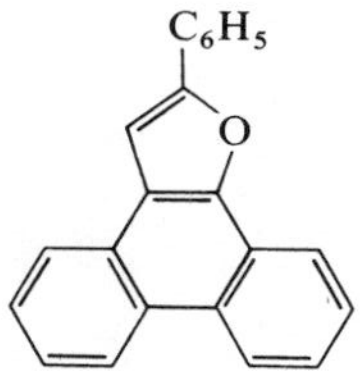 (structure: C_6H_5, O) (85)	334
	α′,α- 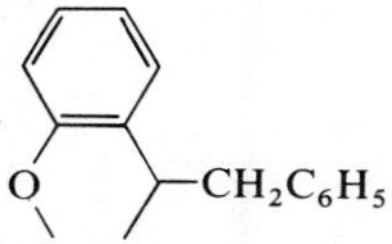 (structure: O, $CH_2C_6H_5$)	C_6H_{12}, N_2[e]	9,10- 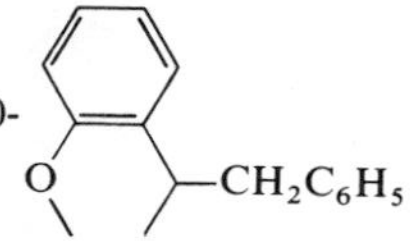 (structure: O, $CH_2C_6H_5$) (60)	514
	4,4′-Dimethoxy-α,α′-diphenyl	C_6H_6, I_2	3-Methoxy-9-(*p*-methoxyphenyl)-10-phenyl (44–62), 3,6-dimethoxy-9,10-diphenyl (6–8)	506
	2,2′-Di-[$(CH_2)_5CO_2CH_3$]	C_6H_{12}, I_2, O_2	1,8-Di-[$(CH_2)_5CO_2CH_3$] (70)	438
C_{29}	α′,α- 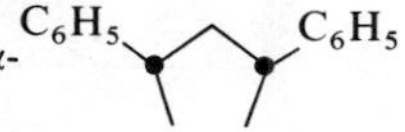 (structure: C_6H_5, C_6H_5)	C_6H_{12}, air	9,10- 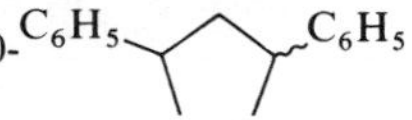 (structure: C_6H_5, C_6H_5) (10)	515a

TABLE I. PHENANTHRENES AND 9,10-DIHYDROPHENANTHRENES FROM SIMPLE STILBENES (*Continued*)

Stilbene	Reaction Conditions	Phenanthrenes and Yields (%)	Refs.
α′,α-C_6H_5, 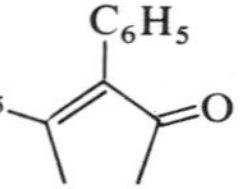 C_6H_5, O	*i*-PrOH, chloranil, N_2	9,10-C_6H_5, 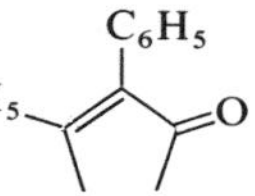 C_6H_5, O (18)	66a
	HOAc, N_2	9,10-C_6H_5, 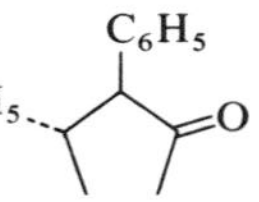 C_6H_5, O (40), 9,10-C_6H_5, C_6H_5, O, C_6H_5, C_6H_5 (19) 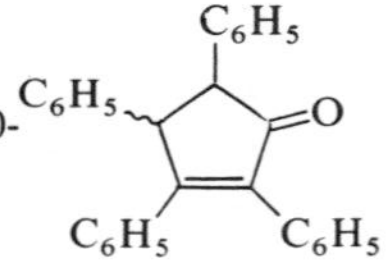	515b
	i-PrOH, N_2	9,10-C_6H_5, 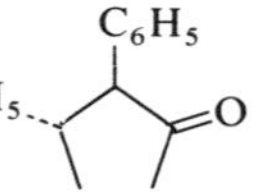 C_6H_5, O (45), 9,10-C_6H_5, C_6H_5 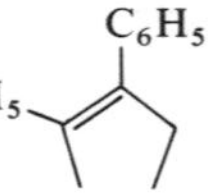 (10)	136,516
	MeOH or *t*-BuOH, N_2	9,10-C_6H_5, 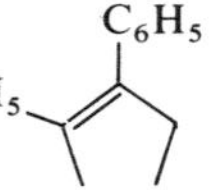 C_6H_5 (12–17)	515b
α′,α-C_6H_5, 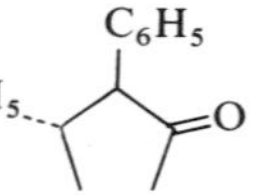 C_6H_5, O	*i*-PrOH, O_2 (or N_2[e])	9,10-C_6H_5, 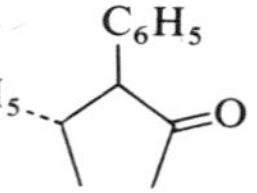 C_6H_5, O (54–55), 9,10-C_6H_5, 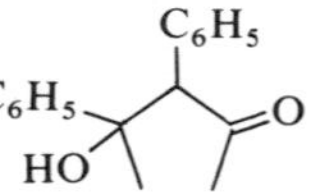 C_6H_5, O, HO (18–23)	135
α′,α- 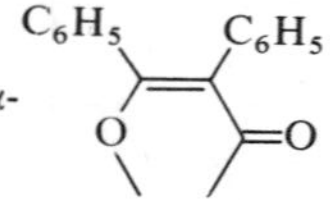 C_6H_5, C_6H_5, O, O	MeCN, I_2, N_2	9,10- 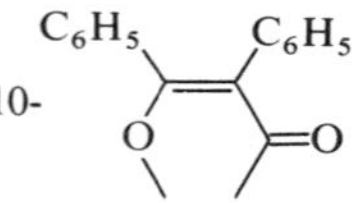 C_6H_5, C_6H_5, O, O (50)	137
	MeCN, air	" (35)	137
	CH_2Cl_2, N_2[e]	" (7), tetraphenyl-α-pyrone (42)	517

	Reactant	Conditions	Product (Yield %)	Refs.
C_{30}	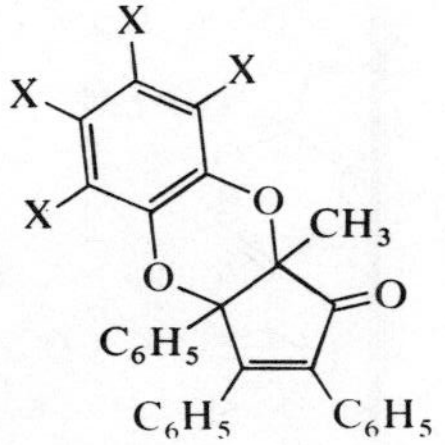		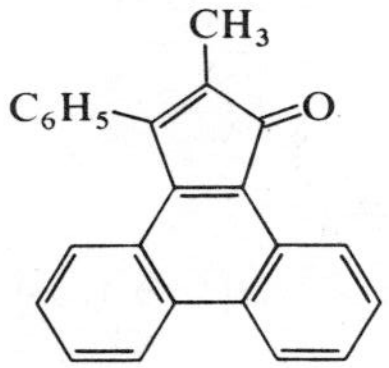	
	$X = Br$	*i*-PrOH, N_2	" (13)	66a
	$X = Cl$	"	" (49)	66a
	4,4′-Dimethoxy-α,α′-di-(*p*-methoxyphenyl)	C_6H_6, I_2, air	3,6-Dimethoxy-9,10-di-(*p*-methoxyphenyl) (57)	513
		C_6H_6, I_2	" (50–70)	506
C_{31}	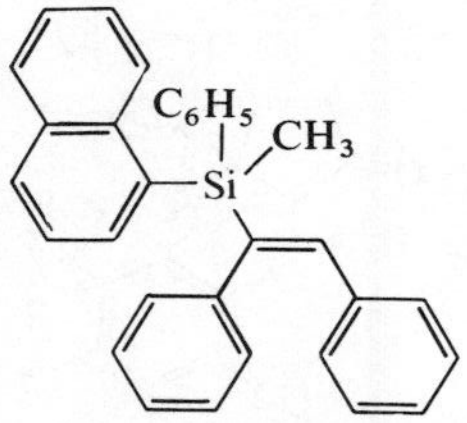	C_6H_{12}, air	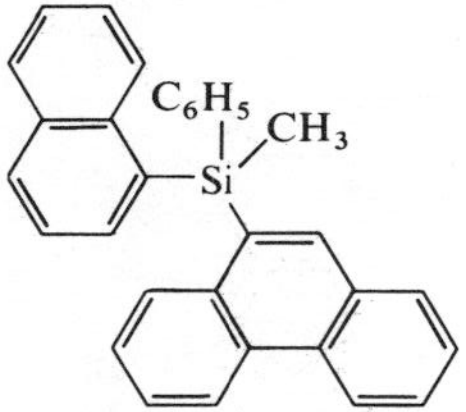 (40)	518
C_{32}	4,α,α′-Triphenyl	C_6H_6, I_2	3,9,10-Triphenyl (50–70)	506
	α-Triphenylsilyl	C_6H_{12}, air	9-Triphenylsilyl (28)	518

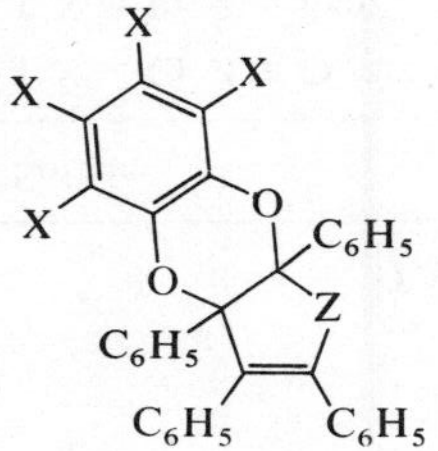

(I)

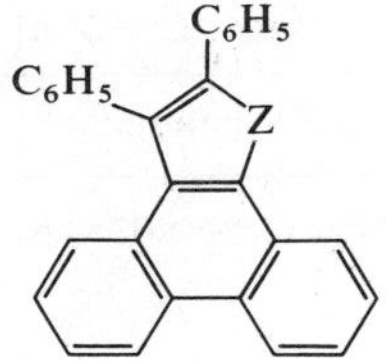

(II)

TABLE I. PHENANTHRENES AND 9,10-DIHYDROPHENANTHRENES FROM SIMPLE STILBENES (*Continued*)

	Stilbene	Reaction Conditions	Phenanthrenes and Yields (%)	Refs.
C_{34}	I, X = Cl, Z = O	*i*-PrOH, N_2	II, Z = O (79)	334
C_{35}	I, X = Br, Z = CO	"	II, Z = CO (13)	66a
	I, X = Cl, Z = CO	"	" (50)	66a,b
	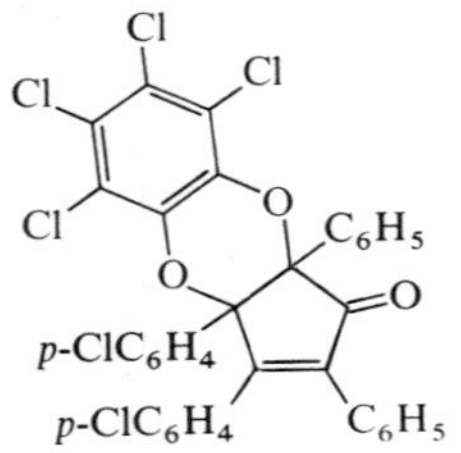 [structure: Cl, Cl, Cl, Cl; O; O; C_6H_5; =O; *p*-ClC_6H_4; *p*-ClC_6H_4; C_6H_5]	"	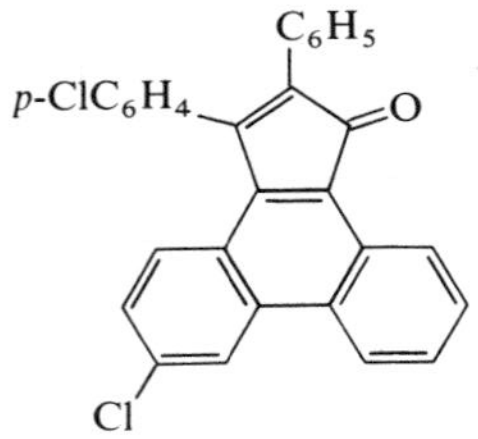 [structure: C_6H_5; *p*-ClC_6H_4; =O; Cl] (53)	66a
C_{36}	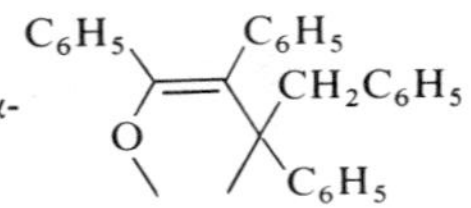 α',α- [structure: C_6H_5; C_6H_5; $CH_2C_6H_5$; O; C_6H_5]	C_6H_{12}, N_2[e]	9,10- 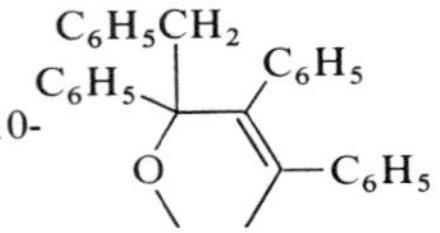 [structure: $C_6H_5CH_2$; C_6H_5; C_6H_5; O; C_6H_5] (61)	514
C_{37}	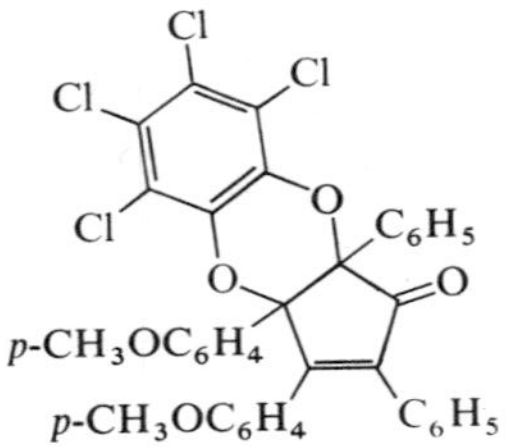 [structure: Cl, Cl, Cl, Cl; O; O; C_6H_5; =O; *p*-$CH_3OC_6H_4$; *p*-$CH_3OC_6H_4$; C_6H_5]	*i*-PrOH, N_2	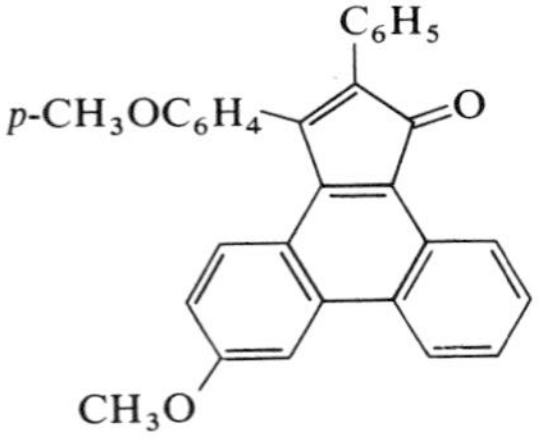 [structure: C_6H_5; *p*-$CH_3OC_6H_4$; =O; CH_3O] (62)	66a

[a] The starting material and the presumed oxidant, elemental sulfur, were generated *in situ* by irradiation of *cis*-stilbene episulfide.

[b] The starting material and the presumed oxidant, dibenzyl disulfide, were generated *in situ* by irradiation of dibenzyl sulfide.

[c] The starting material and the presumed oxidant, elemental sulfur, were generated *in situ* by irradiation of either *cis*- or *trans*-stilbene trithiocarbonate.

[d] The starting material was generated *in situ* by photoreduction of diphenylacetylene.

[e] The identity of the oxidant in this reaction is obscure. One possibility is residual dissolved oxygen.

[f] These are relative yields as determined by GLC.

[g] An additional 20% yield was obtained as a 3:1 mixture of 4-bromo-3-methylphenanthrene and 2-bromo-3-methylphenanthrene.

[h] In addition, 2-fluorophenanthrene and 4-fluorophenanthrene were formed, representing about half of the total product.

[i] The crude product mixture also contained small amounts of 2-fluoro-5-methoxyphenanthrene and 4-fluoro-5-methoxyphenanthrene.

[j] The major product, in 70% yield, was *exo*-5-phenylbenzobicyclo[2.1.1]hex-2-ene.

[k] The starting material was generated *in situ* by irradiation of 1-azido-2-bromo-2-iodo-1-phenylethylene.

[l] There are discrepancies between the yields reported in Refs. 41 and 457.

[m] In addition, phenanthrene-9,10-dicarboximide was obtained in 10% yield.

[n] The 2,3-diphenyl starting material was generated *in situ* by irradiation of the 2,4-diphenyl isomer.

[o] The starting diphenylmaleic anhydride was generated *in situ* by irradiation of diphenylcyclobutendione.

[p] The respective yields of the ketone and the alcohol depend on the solvent; reported values are 33% and 11% in isopropyl alcohol and 20% and 35% in tetrahydrofuran, respectively.

[q] The starting material was generated *in situ* by irradiation of 3,3-dimethyl-1,1-diphenyl-1-butene.

[r] The starting material was generated *in situ* by irradiation of 3,4-di-(*p*-hydroxyphenyl)-2,4-hexadiene.

[s] The starting material was generated *in situ* by irradiation of 7-hydroxy-2-phenylisoflavone.

[t] The starting material was generated *in situ* by irradiation of the lactone of 4-hydroxy-2,4,4-triphenyl-2-butenoic acid.

[u] The starting material was generated *in situ* by irradiation of 7-methoxy-2-phenylisoflavone.

[v] The starting material is a mixture of the indicated 4-aryl-5-aryl′-oxazol-2-one and the isomeric 5-aryl-4-aryl′-oxazol-2-one; correspondingly, the product also is a mixture of isomers.

[w] Irradiation of tetraphenylethylene episulfide also gives 9,10-diphenylphenanthrene (Refs. 508a and 519).

[x] The starting material was generated *in situ* by irradiation of tetraphenylcyclopropene.

TABLE II. HIGHER CARBOCYCLIC AROMATIC SYSTEMS FROM STILBENE ANALOGS[a]

	Starting Material	Reaction Conditions	Products and Yields (%)	Refs.
C_{16}	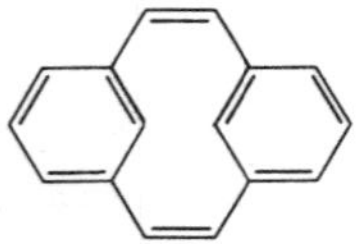	C_6H_{12}, degassed[b]	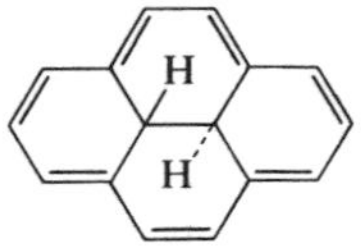 (60),	194
			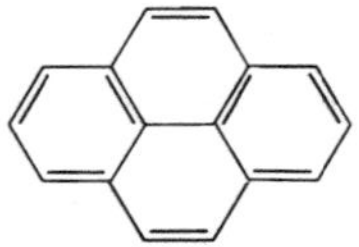 (35) (Pyrene)	
C_{18}	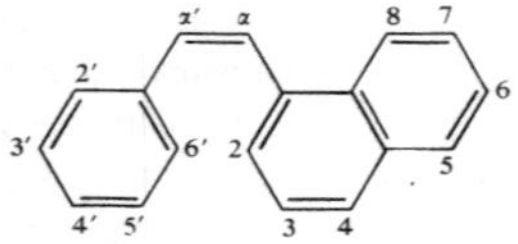 **A** (1-Styrylnaphthalene)	C_6H_{12}, I_2, air	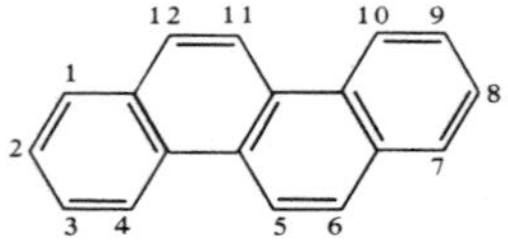 (77) **B** (Chrysene)	55
		HOAc, $FeCl_3$, air	" (38)	108
		CH_2Cl_2, $(NC)_2C{=}C(CN)_2$	" (very good)	31,65
	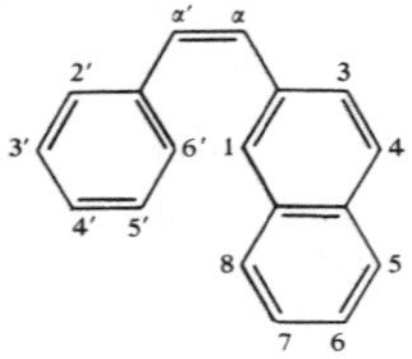 **C** (2-Styrylnaphthalene)	C_6H_{12}, I_2, air	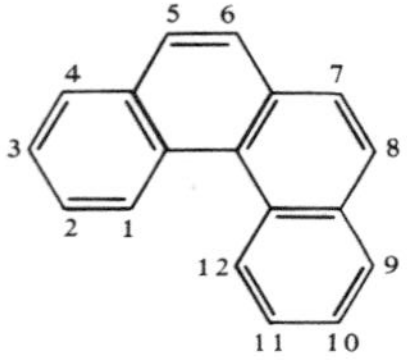 (74) **D** (Benzo[*c*]phenanthrene)	163
		C_6H_{12}, air	" (64)	165
		Pet. ether, I_2, air	" (50)	158

	CH_2Cl_2, $(NC)_2C{=}C(CN)_2$	" (very good)	31,65
	Solvent, air	" (—)	520
	Solvent, I_2, air	" (—)	72b
	n-$PrNH_2$, N_2	5,6-Dihydro-**D** (86)	72b
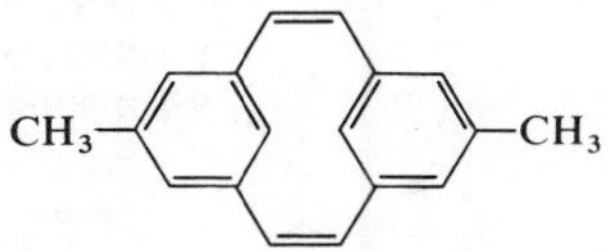 CH_3—[structure]—CH_3	C_6H_{12}, O_2	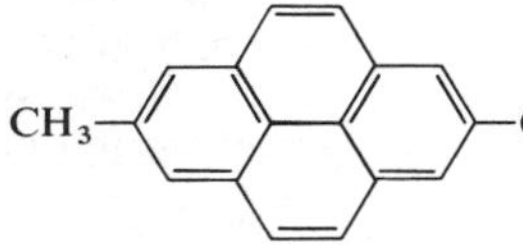 CH_3—[structure]—CH_3 (100) (2,7-Dimethylpyrene)	195
2′-Tritio-**C**	C_6H_6, I_2	4-Tritio-**D** (35–50)	521
3′-Tritio-**C**	"	1-Tritio-**D** + 3-tritio-**D** (35–50 total)	521
4′-Tritio-**C**	"	2-Tritio-**D** (35–50)	521
α-Tritio-**C**	"	6-Tritio-**D** (35–50)	521
α′-Tritio-**C**	"	5-Tritio-**D** (35–50)	521
1-Bromo-**C**	C_6H_{12}, I_2, air	**D** (18)	89
3′-Bromo-**C**	C_6H_6, I_2	3-Bromo-**D** (42)	521
2′-Bromo-5′-fluoro-**C**	C_6H_{12}, I_2, air	4-Bromo-1-fluoro-**D** (43)	76
2′-Bromo-8-fluoro-**C**	"	9-Bromo-1-fluoro-**D** (50), 1-fluoro-**D** (13)	76
2′-Bromo-5′,8-difluoro-**C**	"	4-Bromo-1,12-difluoro-**D** (39)	76
4′-Chloro-**A**	C_6H_6, I_2, N_2	3-Chloro-**B** (70)	214
2′,6′-Dichloro-**A**	C_6H_{12}, I_2, air	1-Chloro-**B** (53)	89
4′-Fluoro-**A**	C_6H_6, I_2, N_2	3-Fluoro-**B** (82)	214
3-Fluoro-**C**	C_6H_{12}, I_2, air	6-Fluoro-**D** (82)	126
4-Fluoro-**C**	"	5-Fluoro-**D** (poor)	126
8-Fluoro-**C**	"	1-Fluoro-**D** (40)	76
4-Fluoro-2′-iodo-**C**	C_6H_{12}, N_2	5-Fluoro-**D** (80)	126
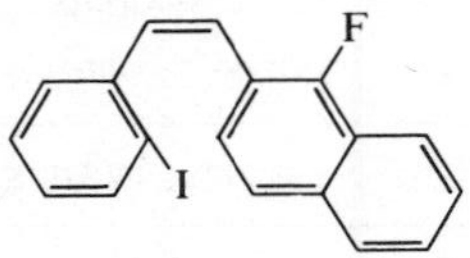 [structure: F, I]	"	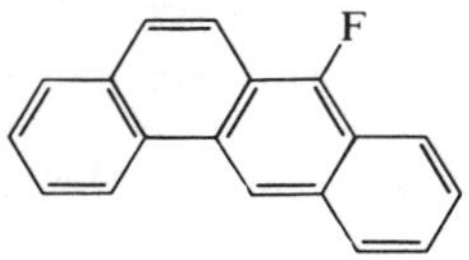 [structure: F] (52) (7-Fluorobenz[a]anthracene)	123

TABLE II. HIGHER CARBOCYCLIC AROMATIC SYSTEMS FROM STILBENE ANALOGS (*Continued*)[a]

	Starting Material	Reaction Conditions	Products and Yields (%)	Refs.
	3-Fluoro-2′-iodo-**C**	C_6H_{12}, N_2	6-Fluoro-**D** (91)	126
C_{19}	2′-Methyl-**C**	1. *n*-$PrNH_2$, N_2 2. DDQ	4-Methyl-**D** (85)	72b
		C_6H_{12}, I_2, air	" (68)	522
		Solvent, I_2, air	4-Methyl-**D** + **D** (65:35)	72b
		n-$PrNH_2$, N_2	5,6-Dihydro-4-methyl-**D** (84)	72b
	3′-Methyl-**C**	C_6H_{12}, I_2, air	3-Methyl-**D** (53), 1-methyl-**D** (28)	21
	4′-Methyl-**C**	Solvent, air	2-Methyl-**D** (70)	523
		"	" (51)	178
		C_6H_{12}, I_2, air	" (52)	522
		C_6H_6, I_2	" (—)	524
	6-Methyl-**C**	C_6H_{12}, I_2, air	3-Methyl-**D** (66)	522
	8-Methyl-**C**	"	1-Methyl-**D** (89)	522
	α-Methyl-**C**	"	6-Methyl-**D** (72)	522
		Pet. ether, I_2, air	" (—)	158
	α′-Methyl-**C**	C_6H_{12}, I_2, air	5-Methyl-**D** (72)	522
	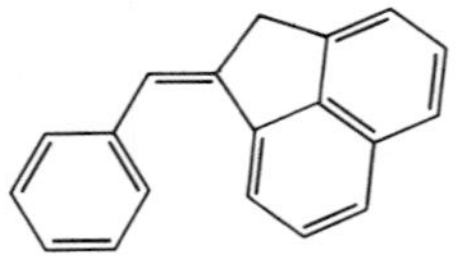	"	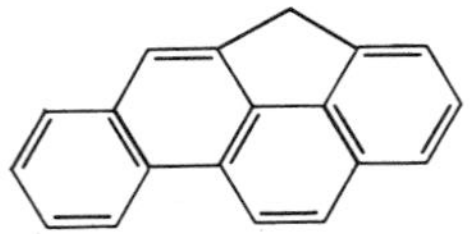 (8) (4,5-Methylenechrysene)	522
	2-Methyl-**A**	"	**B** (34)	89
	4-Methyl-**A**	C_6H_6, I_2, N_2	6-Methyl-**B** (70)	159
	4′-Methyl-**A**	"	3-Methyl-**B** (69)	214
	α-Methyl-**A**	C_6H_{12}, I_2, air	5-Methyl-**B** (65)	522
		C_6H_6, I_2, air	" (31, 29)	525,526
	3′-Bromo-4′-methyl-**C**	C_6H_{12}, I_2	3-Bromo-2-methyl-**D** (78)	433,527
	α-Carboxy-2′,6′-dichloro-**A**	"	11-Carboxy-1-chloro-**B** (82)	528

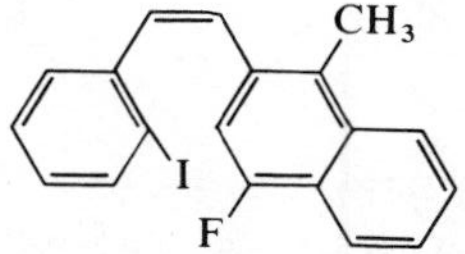	C_6H_{12}, N_2	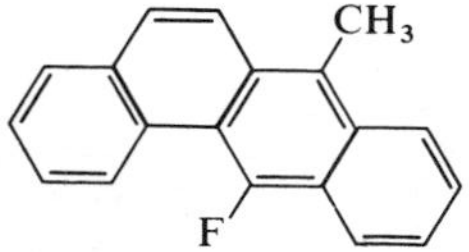 (11)	123
2′-Fluoro-3-methyl-**A**	C_6H_6, I_2, air	1-Fluoro-5-methyl-**B** (6)	529
2′-Fluoro-4′-methyl-**A**	”	1-Fluoro-4-methyl-**B** (77)	529
2′-Fluoro-α-methyl-**A**	”	1-Fluoro-11-methyl-**B** (9)	529
3-Fluoro-α-methyl-**A**	”	5-Fluoro-11-methyl-**B** (7)	529
4-Fluoro-α-methyl-**A**	”	6-Fluoro-11-methyl-**B** (30)	529
4′-Fluoro-3-methyl-**A**	”	3-Fluoro-5-methyl-**B** (19)	529
4′-Fluoro-α-methyl-**A**	”	3-Fluoro-11-methyl-**B** (13)	529
5′-Fluoro-2′-methoxy-**C**	C_6H_{12}, I_2, air	1-Fluoro-4-methoxy-**D** (43)	76
8-Fluoro-2′-methoxy-**C**	”	1-Fluoro-9-methoxy-**D** (35), 1-fluoro-**D** (15)	76
3′,8-Difluoro-α′-methyl-**C**	C_6H_6, I_2, air	1,12-Difluoro-5-methyl-**D** + 1,10-difluoro-8-methyl-**D** (1:1)	171
5′,8-Difluoro-2′-methyl-**C**	C_6H_{12}, I_2, air	1,12-Difluoro-4-methyl-**D** (48)	76
5′,8-Difluoro-2′-methoxy-**C**	”	1,12-Difluoro-4-methoxy-**D** (48)	76
2′-Cyano-**A**	C_6H_{12}, I_2	1-Cyano-**B** (95)	528
4′-Cyano-**A**	C_6H_6, I_2, N_2	3-Cyano-**B** (80)	214
	C_6H_{12}, I_2	” (94)	528
α-Cyano-**A**	”	5-Cyano-**B** (76)	528
α′-Cyano-**A**	”	6-Cyano-**B** (94)	528
4′-Formyl-**C**	”	2-Formyl-**D** (61)	530
4′-Methoxy-**A**	C_6H_6, I_2, N_2	3-Methoxy-**B** (51)	214
α′-Methoxy-**A**	C_6H_6, I_2, air	6-Methoxy-**B** (15)	529
4′-Methoxy-**C**	C_6H_{12}, I_2, air	2-Methoxy-**D** (94)	522
α-Carboxy-**A**	C_6H_{12}, I_2	5-Carboxy-**B** (60)	528
C_{20} 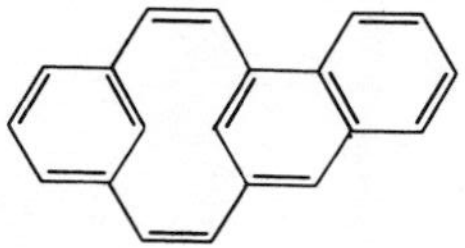	THF-d_8, degassed	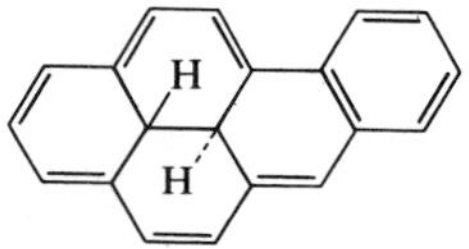 (—)	531

TABLE II. Higher Carbocyclic Aromatic Systems from Stilbene Analogs (*Continued*)[a]

	Starting Material	Reaction Conditions	Products and Yields (%)	Refs.
	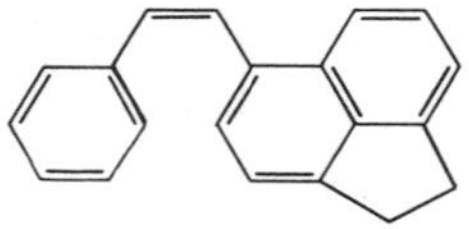	C_6H_{12}, I_2	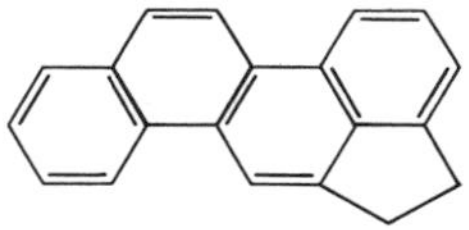 (74) (4,5-Dihydrocyclopenta[*hi*]chrysene)	532
	α-Ethyl-**A**	C_6H_{12}, I_2, air	5-Ethyl-**B** (80)	522
	2′,5-Dimethyl-**A**	C_6H_{14}, I_2, air	1,7-Dimethyl-**B** (41), 1-methyl-**B** (small)	533
	2′-Acetyl-**A**	C_6H_{12}, I_2	1-Acetyl-**B** (52), **B** (26)	528
	3′-Acetyl-**A**	"	2-Acetyl-**B** (50), 4-acetyl-**B** (24)	528
	4′-Acetyl-**A**	"	3-Acetyl-**B** (78)	528
	4-Methoxy-α-methyl-**A**	C_6H_6, I_2, air	12-Methoxy-5-methyl-**B** (37)	529
	α′-Methoxy-α-methyl-**A**	"	6-Methoxy-5-methyl-**B** (3)	529
	4′-Carbomethoxy-**C**	C_6H_6, I_2	2-Carbomethoxy-**D** (—)	524
C_{21}	2′,4′,6′-Trimethyl-**A**	C_6H_{12}, I_2, air	1,3-Dimethyl-**B** (40)	89
	2′,4′,6′-Trimethyl-**C**	"	2,4-Dimethyl-**D** (40)	89
C_{22}	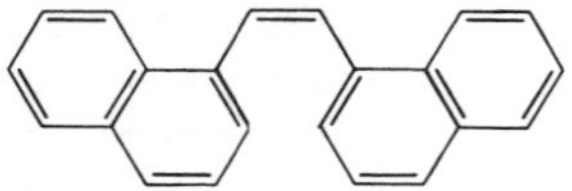	C_6H_6, I_2, air	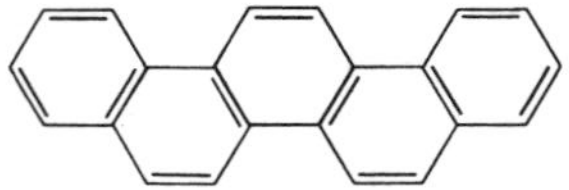 (91) (Picene)	59
		C_6H_{12}, air	" (100)	156
		CH_2Cl_2, $(NC)_2C{=}C(CN)_2$	" (very good)	31,65
	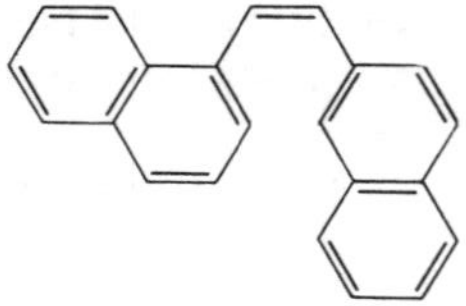	C_6H_6, I_2, air	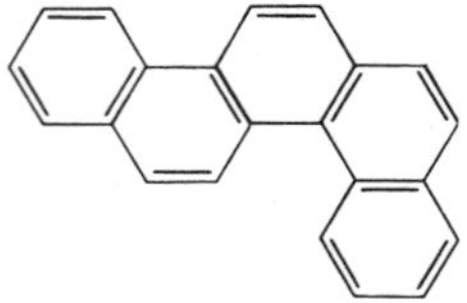 (76) (Benzo[*c*]chrysene)	163

	C_6H_6, air	" (61)	165
	Solvent, air	" (—)	38a
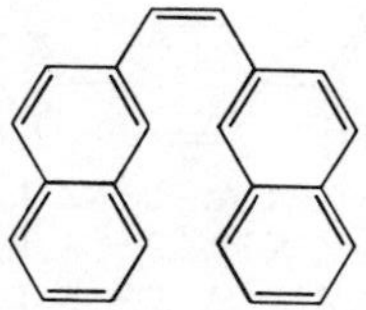	C_6H_6, air	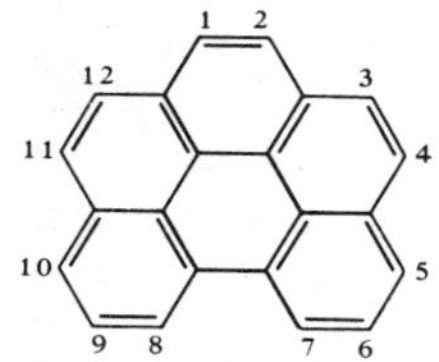 (89) (Benzo[*ghi*]perylene)	165
	Et_2O, air	" (16)	89
	Pet. ether, I_2, air	" (23)	158
	C_6H_6, I_2, air	" (37)[c], 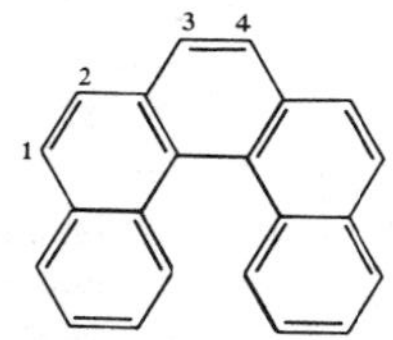 (25)[c]; (Dibenzo[*c,g*]phenanthrene) 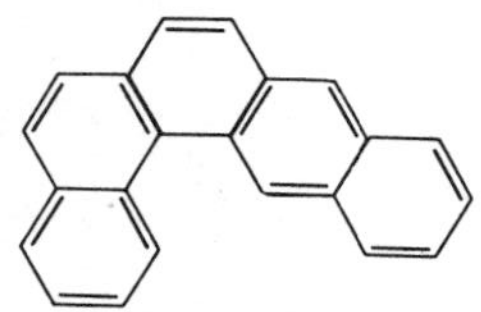 (38)[c] (Dibenzo[*b,g*]phenanthrene)	76,163
	CH_2Cl_2, $(NC)_2C{=}C(CN)_2$	Dibenzo[c,*g*]phenanthrene (—)[d]	65
	n-$PrNH_2$, N_2	3,4-Dihydrodibenzo[*c,g*]phenanthrene (60)	72b

TABLE II. Higher Carbocyclic Aromatic Systems from Stilbene Analogs (*Continued*)[a]

Starting Material	Reaction Conditions	Products and Yields (%)	Refs.
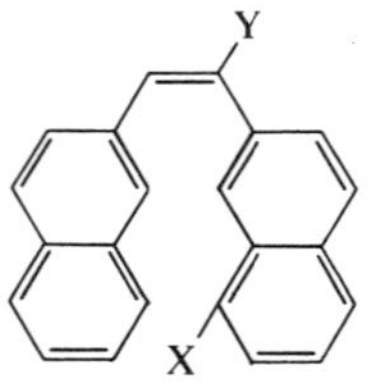 (I)		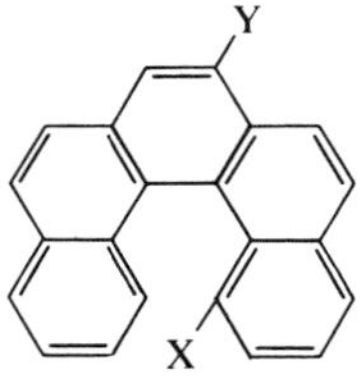 (II)	
I, X = T, Y = H	C_6H_6, I_2	II, X = T, Y = H (50)	431
I, X = H, Y = T	"	II, X = H, Y = T (50)	431
1-Styrylphenanthrene	C_6H_6, I_2, N_2	Picene (61)[e]	157
2-Styrylphenanthrene	C_6H_6, I_2	Benzo[*c*]chrysene (81)[e]	164
3-Styrylphenanthrene	C_6H_6, I_2, air	Benzo[*ghi*]perylene (88)[e]	169
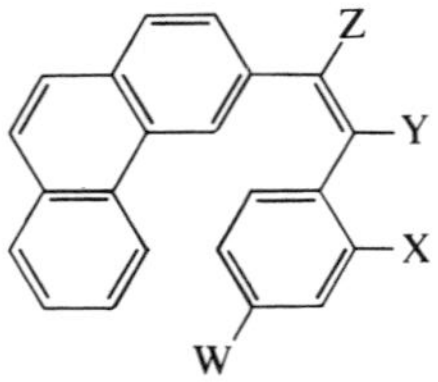 (I)		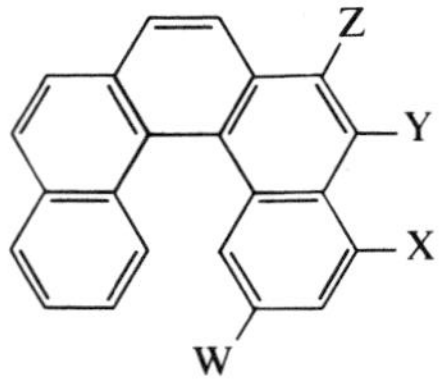 (II)	
I, W = T; X, Y, Z = H	C_6H_6, I_2	II, W = T; X, Y, Z = H (—)	431
I, X = T; W, Y, Z = H	"	II, X = T; W, Y, Z = H (—)	431
I, Y = T; W, X, Z = H	"	II, Y = T; W, X, Z = H (—)	431
I, Z = T; W, X, Y = H	"	II, Z = T; W, X, Y = H (—)	431
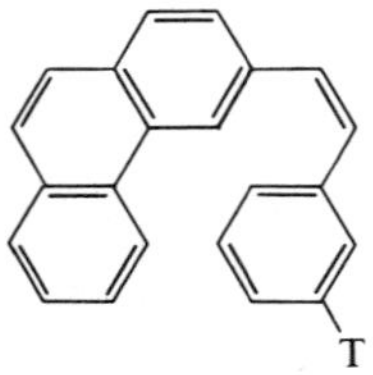	"	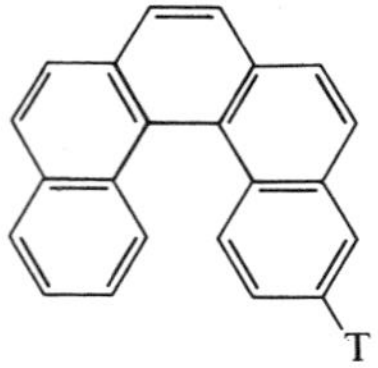 (—), 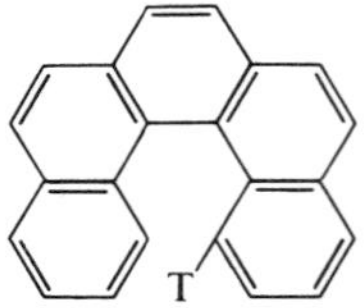 (—)	431

4-Styrylphenanthrene	C_6H_6, air	Benzo[*c*]chrysene (72),[c,e] (28)[c] C_6H_5 (4-Phenylpyrene)	182
	Solvent, air	Benzo[*c*]chrysene (main),[e] 4-phenylpyrene (2)	181
9-Styrylphenanthrene	Pet. ether, air	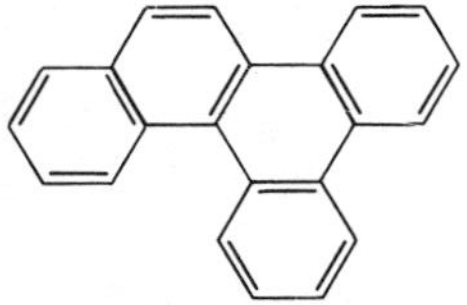 (50) (Benzo[*g*]chrysene)	158
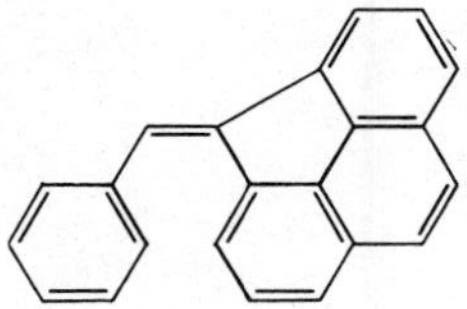	Solvent, I_2	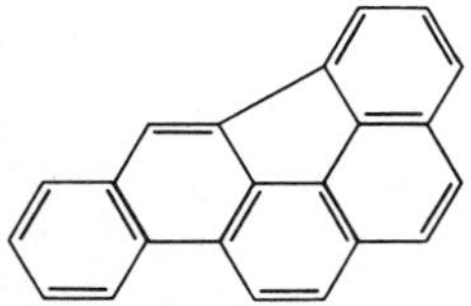 (—) (Dibenzo[*b*,*ghi*]fluoranthene)	38a
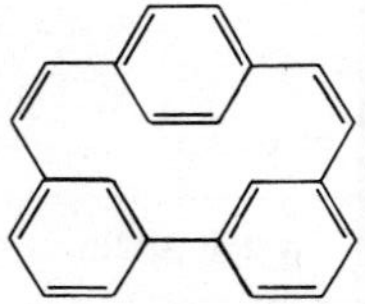	C_6H_{12}, I_2	Benzo[*ghi*]perylene (—)[e]	534
o-Distyrylbenzene	C_6H_{12}, I_2, air	Picene (15)[e]	169,181
	C_6H_6, I_2	" (—)[e]	181

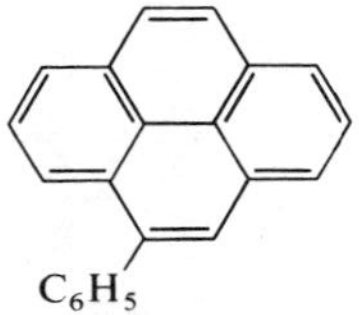

TABLE II. HIGHER CARBOCYCLIC AROMATIC SYSTEMS FROM STILBENE ANALOGS (*Continued*)[a]

Starting Material	Reaction Conditions	Products and Yields (%)	Refs.
m-Distyrylbenzene	C_6H_6, I_2, air	Benzo[*c*]chrysene (80)[e]	169
	Solvent, air	" (70)[e]	181
	C_6H_6, air	" (99),[c,e] 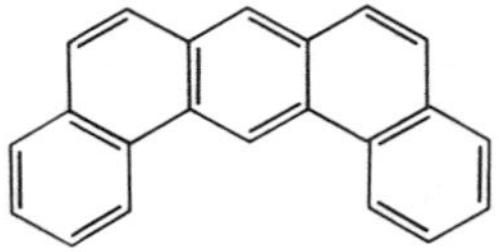 (1)[c] (Dibenz[*a,j*]anthracene)	182
p-Distyrylbenzene	C_6H_6, I_2, air	Benzo[*ghi*]perylene (12)[e]	169
	Solvent, I_2	" (10)[e]	181
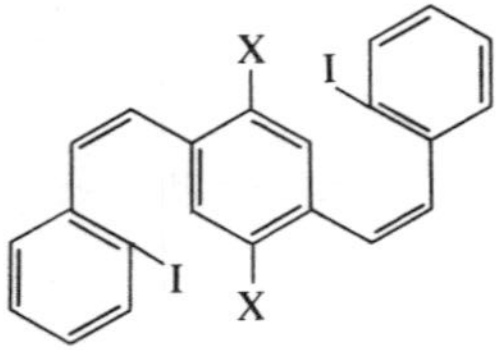 (I)		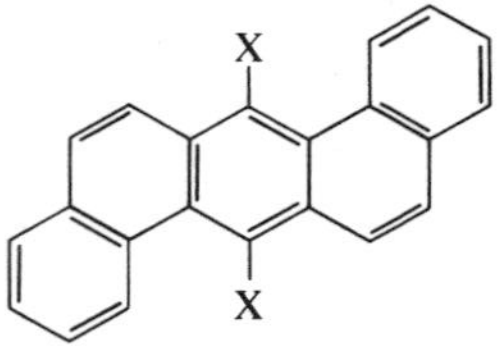 (II)	
I, X = Cl	C_6H_{12}, N_2	II, X = Cl (33)	183
I, X = F	"	II, X = F (31)	183
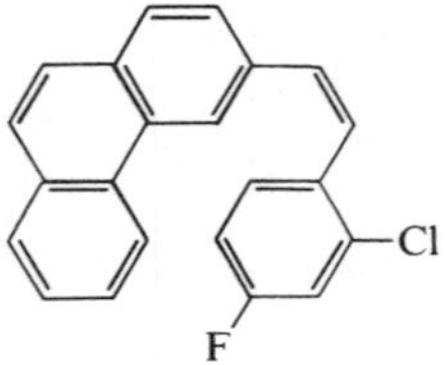	C_6H_6, I_2	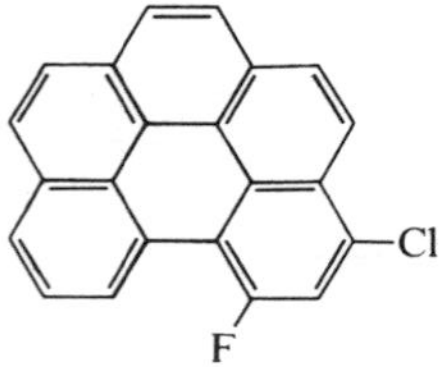 (44)	535

Starting material	Conditions	Product (yield, %)	Ref.
F, Cl	C_6H_6, I_2, air	F, Cl + Cl (1:2)	535
F, Cl	”	F, Cl (20)	535
F, Cl, Cl	”	F, Cl, Cl (55)	76
F	”	F + F (1:1)	535

TABLE II. HIGHER CARBOCYCLIC AROMATIC SYSTEMS FROM STILBENE ANALOGS (*Continued*)[a]

	Starting Material	Reaction Conditions	Products and Yields (%)	Refs.
	F	C_6H_6, I_2, air	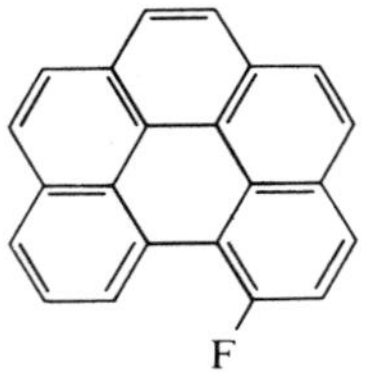 F (70)	535
C_{23} 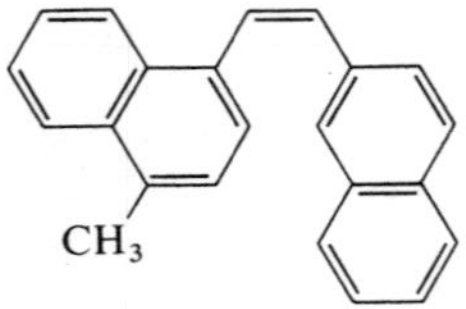	CH_3	C_6H_6, I_2, N_2	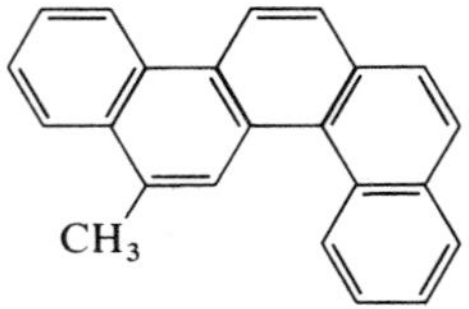 CH_3 (80)	159
	CH_3	"	CH_3 (80)	159
	CH_3	"	CH_3 (95)	159

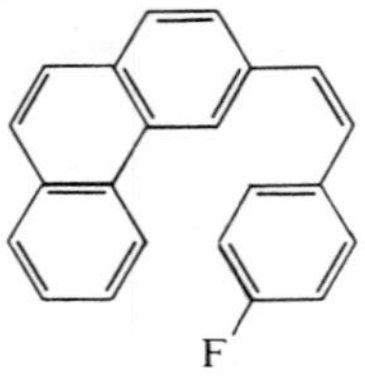

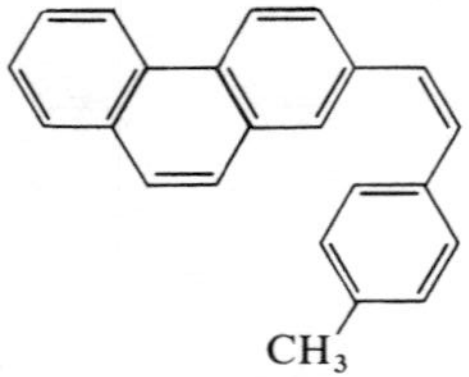

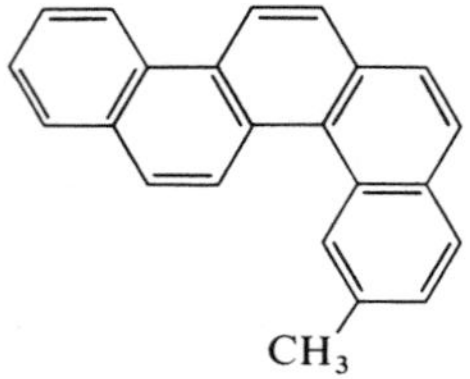

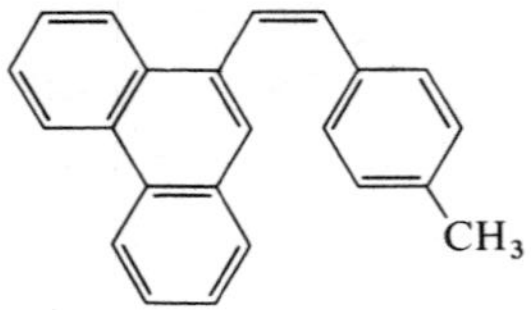

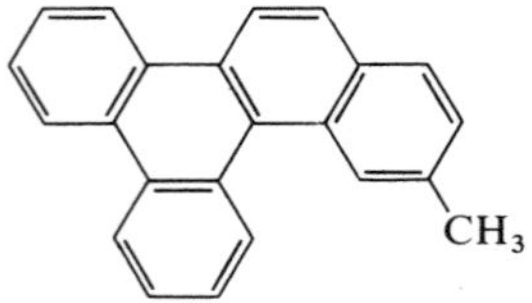

	Starting material	Conditions	Product (yield, %)	Ref.
		Pet. ether, air	(—)	158,536
		Xylene, I_2, N_2	(15)	159
		C_6H_6, I_2, air	(64)	171
		”	(5)	171
C_{24}		Isooctane, I_2	(—)	38a

TABLE II. Higher Carbocyclic Aromatic Systems from Stilbene Analogs (*Continued*)[a]

Starting Material	Reaction Conditions	Products and Yields (%)	Refs.
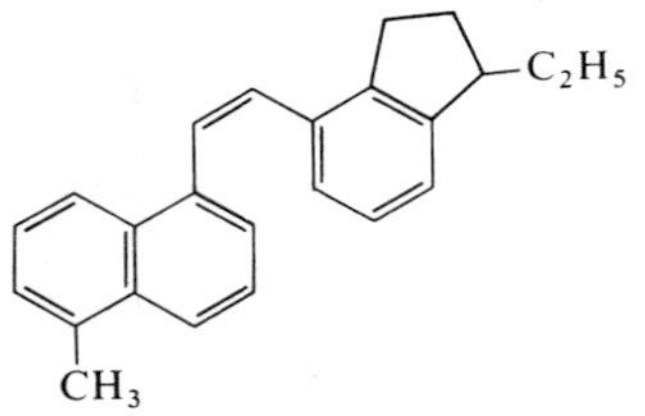 CH_3, C_2H_5	C_6H_{12}, I_2	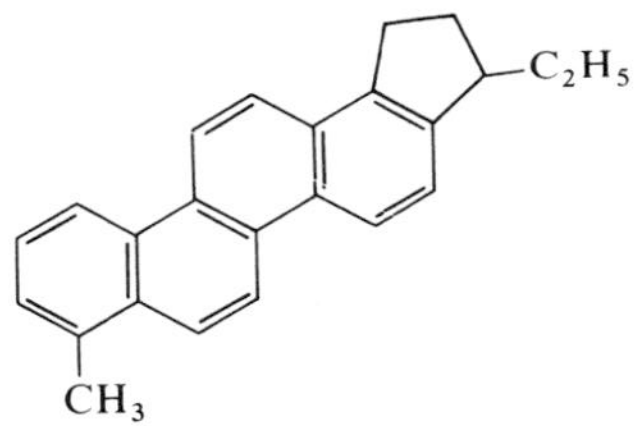 CH_3, C_2H_5 (95)	505
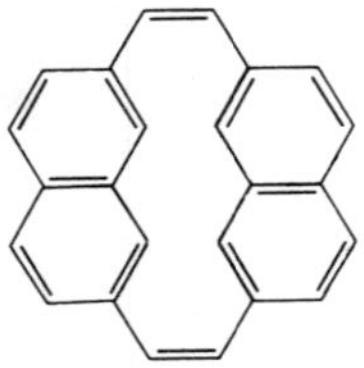	Pet. ether, I_2	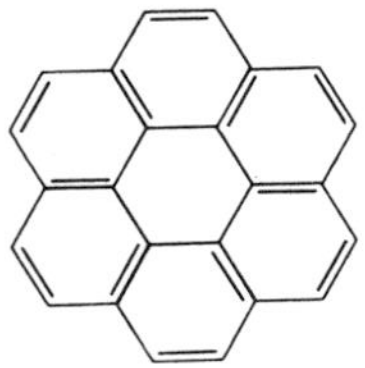 (100) (Coronene)	200
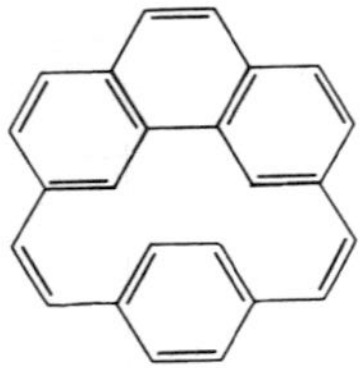	C_6H_{12}, I_2	" (—)	537
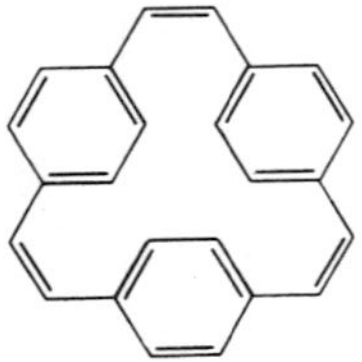	THF-d_8, O_2	" (100)	201

Reactant	Conditions	Product (yield)	Ref.
α′-Phenyl-**A**	C_6H_{12}, I_2, air	6-Phenyl-**B** (90)	76
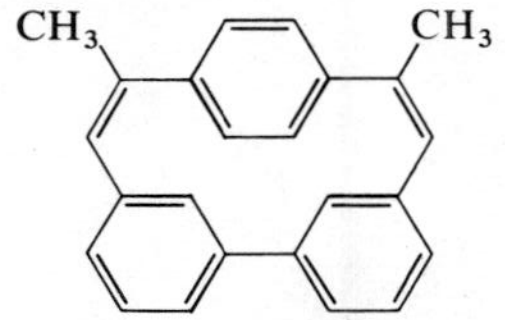 CH_3 CH_3	C_6H_{12}, I_2	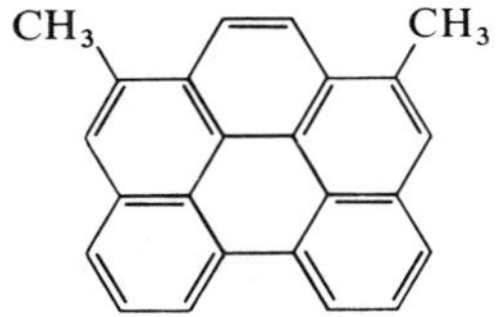 CH_3 CH_3 (—)	534
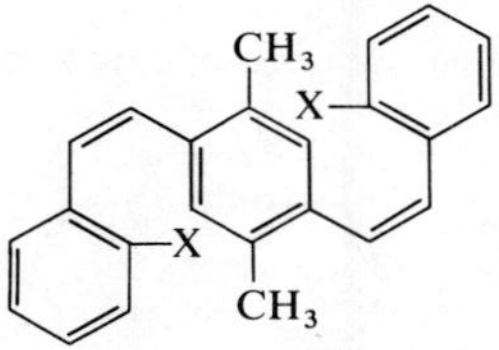 CH_3 X X CH_3		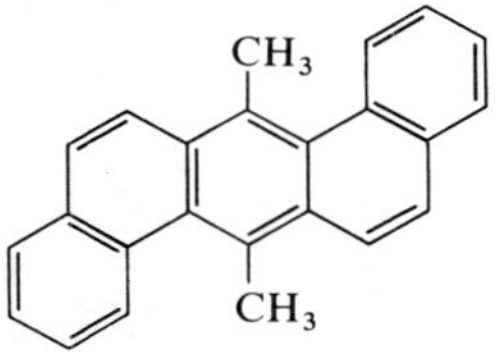 CH_3 CH_3	
X = H	C_6H_{12}, I_2, N_2	” (23)	183
X = I	C_6H_{12}, N_2	” (23)	183
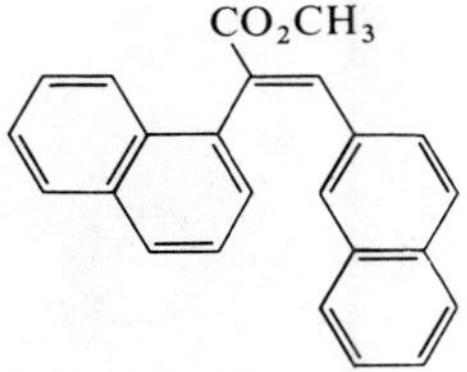 CO_2CH_3	C_6H_6, I_2, air	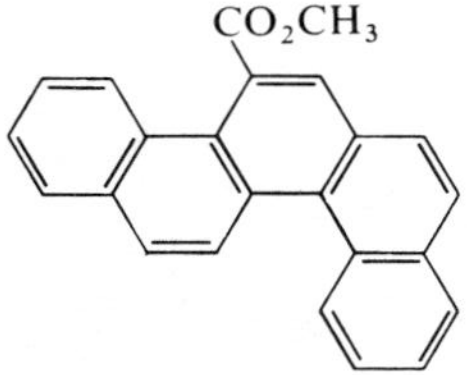 CO_2CH_3 (80)	538
C_{25} 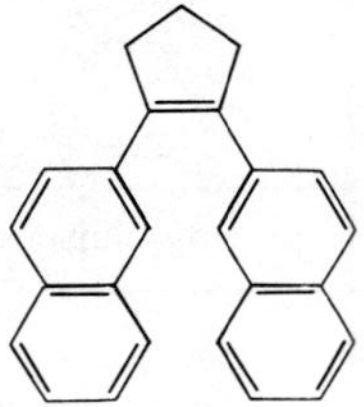	Solvent, I_2	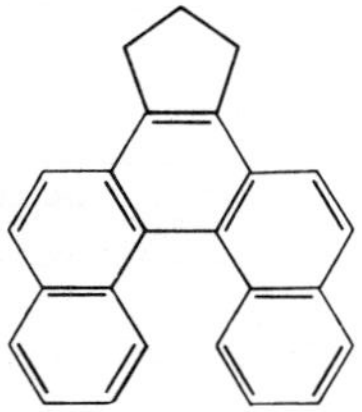 (—), 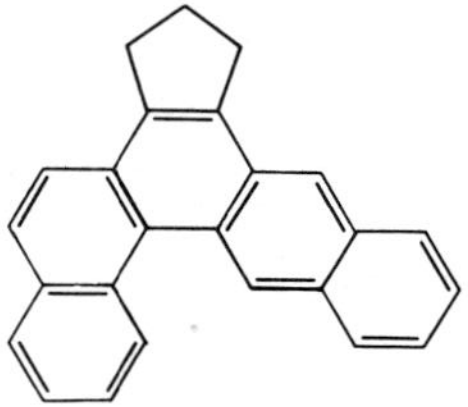 (—)	539

TABLE II. HIGHER CARBOCYCLIC AROMATIC SYSTEMS FROM STILBENE ANALOGS (*Continued*)[a]

Starting Material	Reaction Conditions	Products and Yields (%)	Refs.
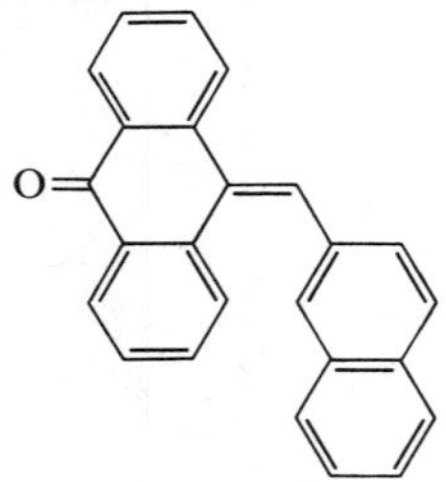	C_6H_{12} or C_6H_6, I_2	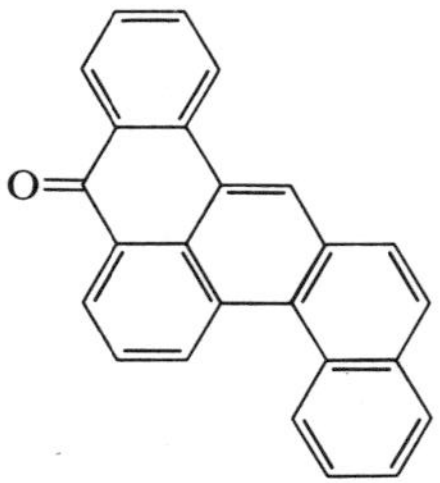 (20–90)	58
C_{26} 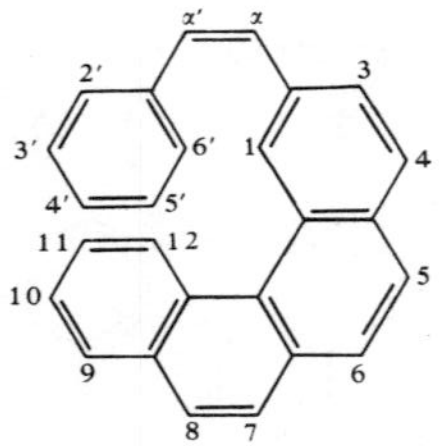 **E** (2-Styrylbenzo[*c*]phenanthrene)	C_6H_6, I_2	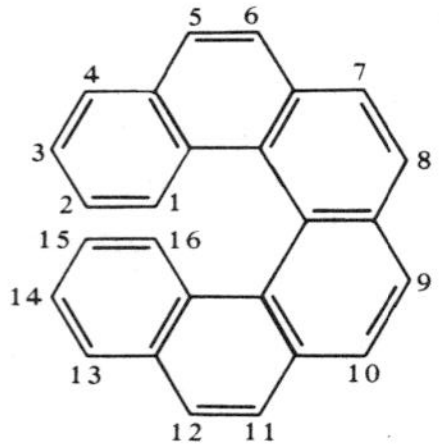 (85) **F** ([6]Helicene)[f]	168,170
	"	" (80)	524
	C_6H_6, I_2, air	" (30)	540,541
	C_6H_6, I_2 (0.5–2 mol %)	**F** (50), 5,6-dihydro-**F** (30)[g]	68a
	Cholesteryl nonanoate, cholesteryl chloride (3:2), I_2	**F** (75)	180a
	Cholesteryl benzoate, I_2	" (72)	180a
	p-$C_7H_{15}C_6H_4CO_2C_6H_4CN$-$p$, p-$C_4H_9C_6H_4CO_2C_6H_4CN$-p (1:1), I_2	" (70)	180b

Reactant	Conditions	Product(s) (Yield %)	Refs.
	C_6H_6, argon	5,6-Dihydro-**F** (60)	68b
2′,3′,4′,5′,6′-Pentadeuterio-**E**	”	1,2,3,4,5-Pentadeuterio-5,6-dihydro-**F** (—)	68b
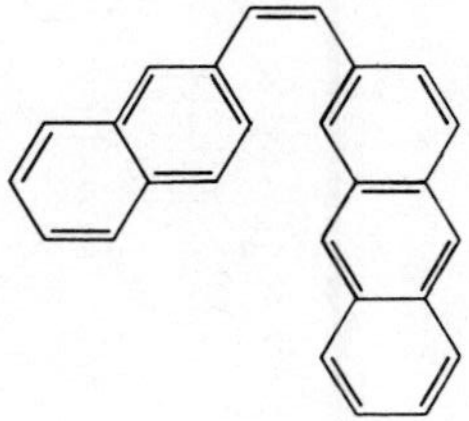	C_6H_6, I_2, N_2	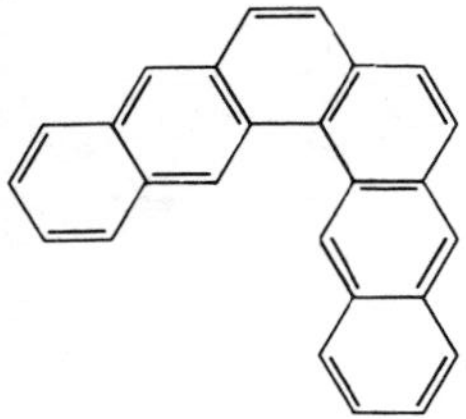 (5) (Anthra[1,2-*a*]anthracene)	159
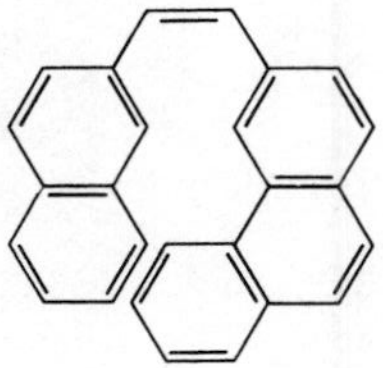	C_6H_{14}, C_6H_6, or xylene; I_2; N_2	**F** (22),[f]	159
		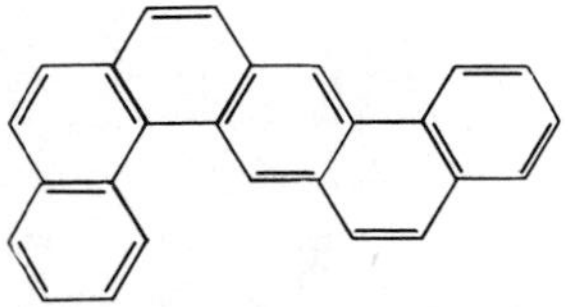 (50) (Benzo[*a*]naphth[1,2-*h*]anthracene)	
	C_6H_6, I_2	**F** (25)[f]	170
	C_6H_6, I_2, air	” (—)[f]	178

TABLE II. HIGHER CARBOCYCLIC AROMATIC SYSTEMS FROM STILBENE ANALOGS (*Continued*)[a]

Starting Material	Reaction Conditions	Products and Yields (%)	Refs.
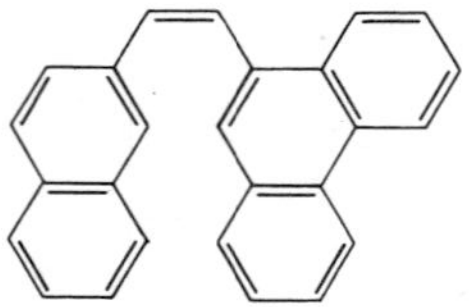	C_6H_6, I_2	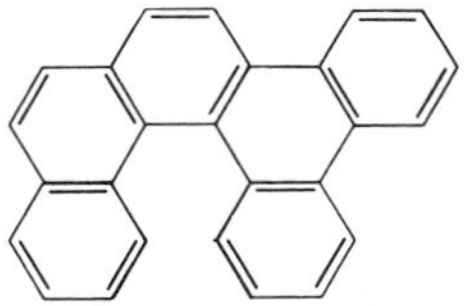 (26), (Dibenzo[*c,g*]chrysene) 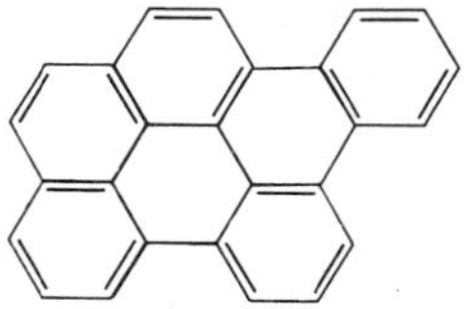 (12), (Dibenzo[*b,pqr*]perylene) 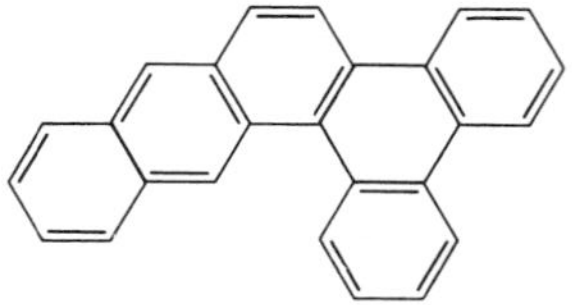 (28) (Dibenzo[*b,g*]chrysene)	231
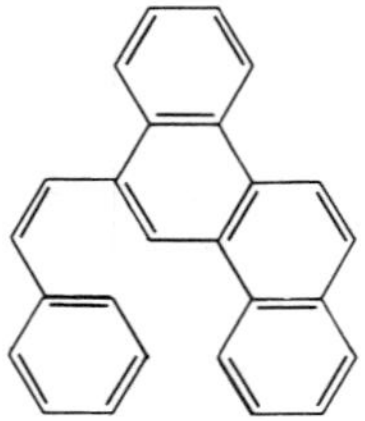	"	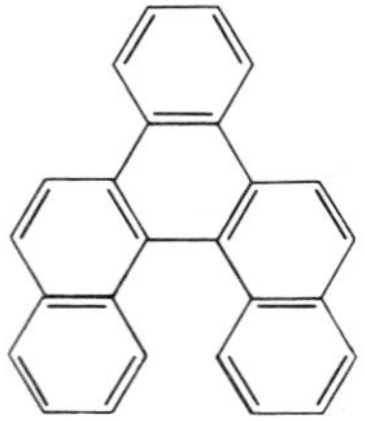 (72) (Naphtho[1,2-*g*]chrysene)	75

	Reactant	Conditions	Product (yield %)	Ref.
	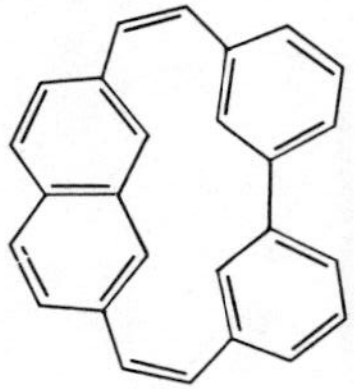 [structure]	C_6H_{12}, I_2	 [structure] (52) (Cyclohepta[1,2,3,4,5,6,7-*p*,*q*,*r*,*s*,*t*,*u*]-phenanthro[3,4-*c*]phenanthrene)	196,542
	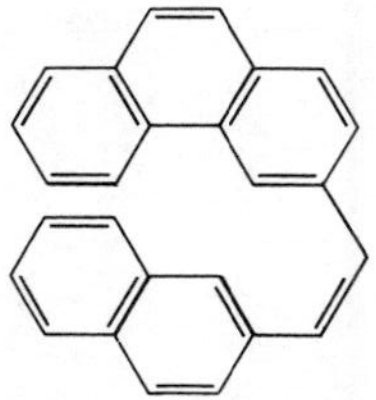 [structure]	C_6H_6, I_2	**F** (55)	185
	2,7-Distyrylnaphthalene	"	**F** (60)	185
		C_6H_6, argon	5,6-Dihydro-**F** (60)	68b
	3-Deuterio-**E**	C_6H_6, I_2	7-Deuterio-**F** (87)	433,527
	2′-Bromo-**E**	C_6H_6, I_2, air	4-Bromo-**F** (17)	541
	3-Bromo-**E**	C_6H_6, I_2	7-Bromo-**F** (67)	527
	4′-Bromo-**E**	C_6H_6, I_2, air	2-Bromo-**F** (23)	541
	2′-Chloro-**E**	"	4-Chloro-**F** (25)	541
	4′-Chloro-**E**	"	2-Chloro-**F** (36)	541
	2′-Fluoro-**E**	"	4-Fluoro-**F** (17)	541
	4′-Fluoro-**E**	"	2-Fluoro-**F** (25)	541
C_{27}	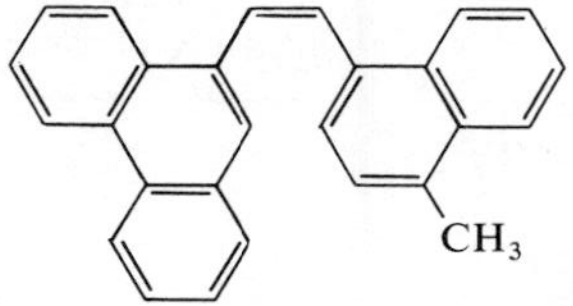 [structure] CH_3	C_6H_6, I_2	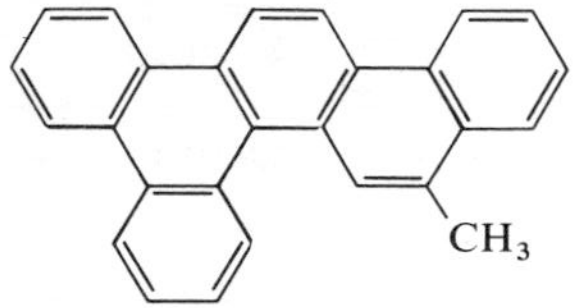 [structure] CH_3 (80) (10-Methylbenzo[*f*]picene)	543

TABLE II. Higher Carbocyclic Aromatic Systems from Stilbene Analogs (*Continued*)[a]

	Starting Material	Reaction Conditions	Products and Yields (%)	Refs.
	2′-Methyl-**E**	C_6H_6, I_2	4-Methyl-**F** (82)	544
		"	" (—)	545
		C_6H_6, I_2, air	" (20)	541
	3′-Methyl-**E**	C_6H_6, I_2	3-Methyl-**F** (72), 1-methyl-**F** (8)	523
	4′-Methyl-**E**	"	2-Methyl-**F** (70)	523
		C_6H_6, I_2, air	" (27)	541
	4′-Formyl-**E**	C_6H_{12}, I_2	2-Formyl-**F** (57)	530
C_{28}	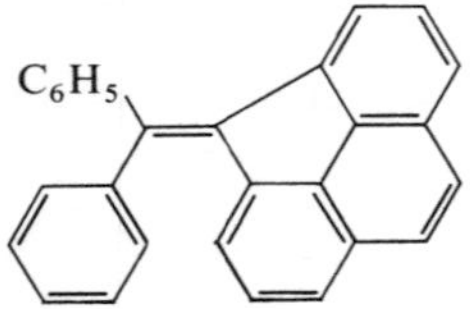 [structure, C_6H_5]	Solvent, I_2	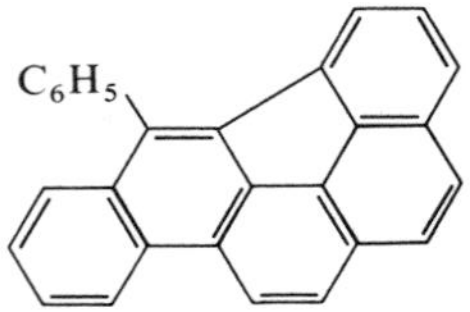 [structure, C_6H_5] (—)	38a
	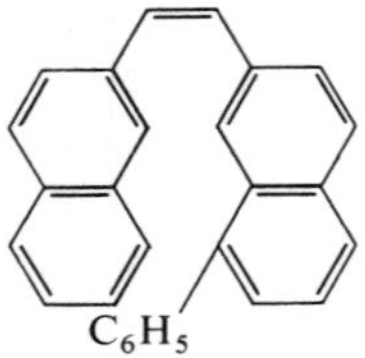 [structure, C_6H_5]	C_6H_6, I_2, N_2	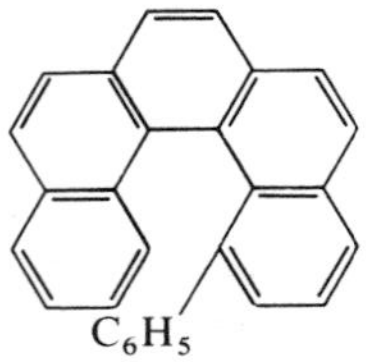 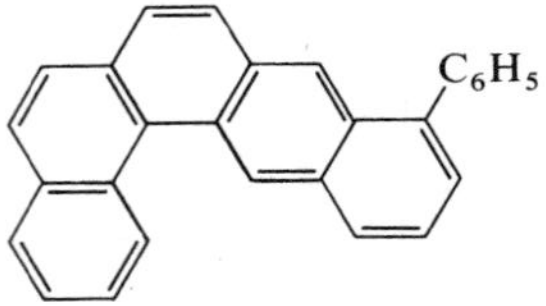 [structure, C_6H_5] (7), [structure, C_6H_5] (10),	232

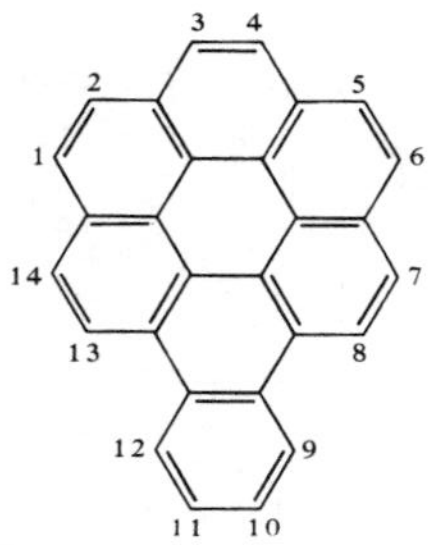

Reactant	Conditions	Product(s) (Yield, %)	Ref.
		(42)	
		(Benzo[a]coronene)	
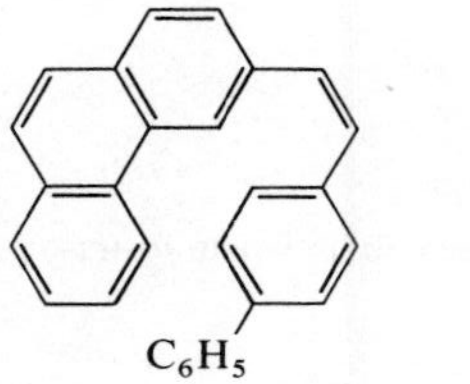 C_6H_5	C_6H_6, I_2, N_2	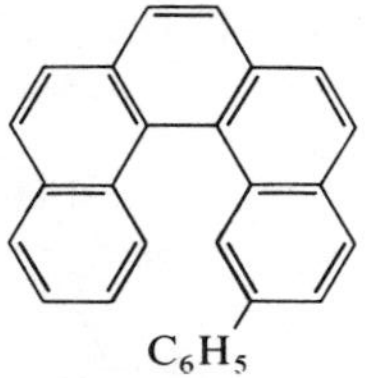 C_6H_5 (54), Benzo[a]coronene (29)[e]	232
C_6H_5	C_6H_6, I_2	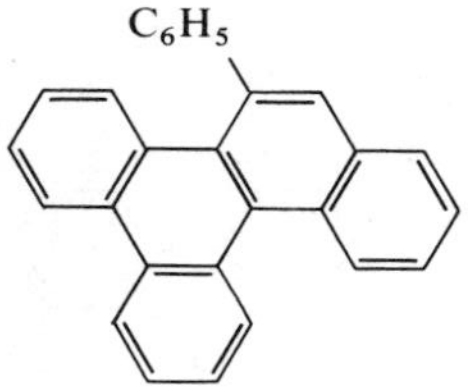 C_6H_5 (80)	122
3′,5′-Dimethyl-**E**	"	1,3-Dimethyl-**F** (80)	523

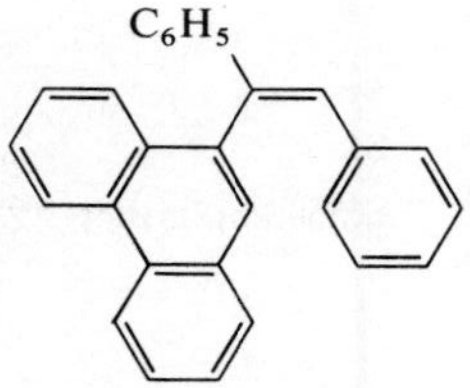

TABLE II. Higher Carbocyclic Aromatic Systems from Stilbene Analogs (*Continued*)[a]

Starting Material	Reaction Conditions	Products and Yields (%)	Refs.
2,2′-Distyrylbiphenyl	C_6H_{14}, I_2, N_2	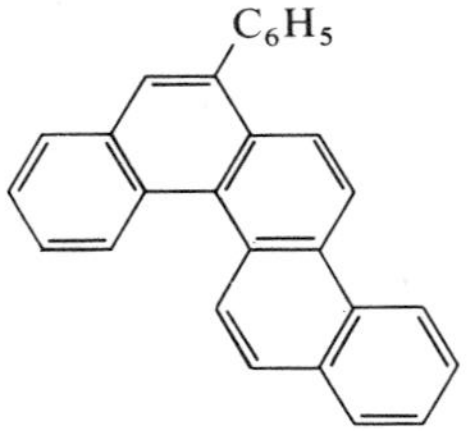 (60)	211
2,6-Di-(*p*-methylstyryl)naphthalene	Xylene, I_2, N_2	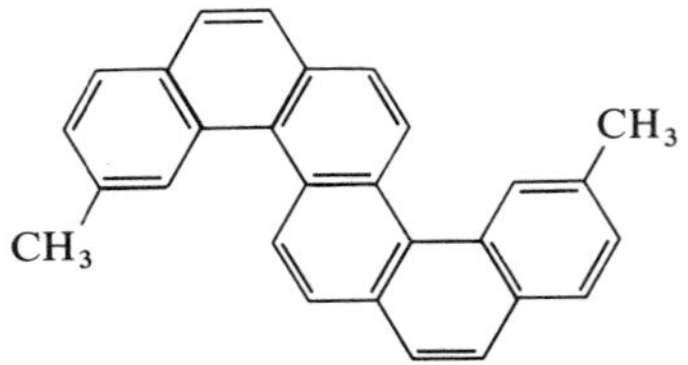 (60) (3,11-Dimethyldibenzo[*c*,*l*]chrysene)	172
2,7-Di-(*p*-methylstyryl)naphthalene	C_6H_6, I_2	2,15-Dimethyl-**F** (60)	546
2,7-Di-(*m*-methylstyryl)naphthalene	"	1,14-Dimethyl-**F** (40), 1,16-dimethyl-**F** (8), 3,14-dimethyl-**F** (30)	545
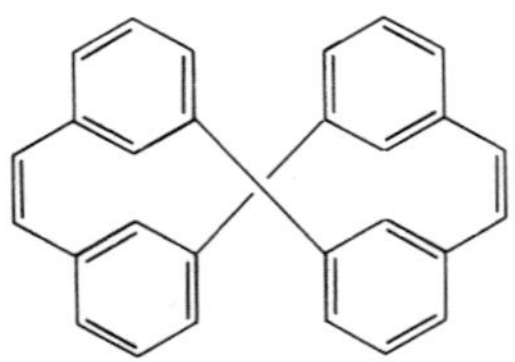	C_6H_{12}, I_2	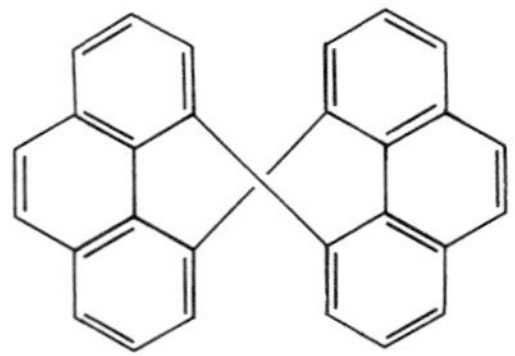 (65) (Dibenzo[*def*,*pqr*]tetraphenylene)	203
	"	" (63)	204

	2′-Bromo-5′-carbomethoxy-**E**	C_6H_6, I_2	4-Bromo-1-carbomethoxy-**F** (—)	175a
	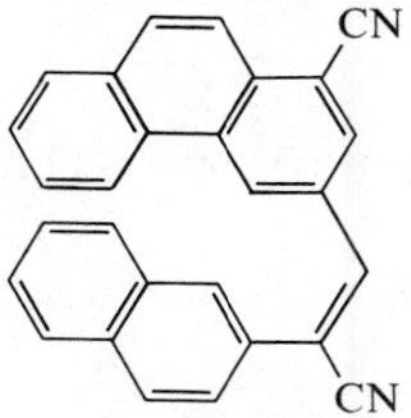 CN, CN	C_6H_6, I_2, air	7,10-Dicyano-**F** (96)	547
	3′-Carbomethoxy-**E**	C_6H_6, I_2	3-Carbomethoxy-**F** (major)	175a
	4′-Carbomethoxy-**E**	”	2-Carbomethoxy-**F** (55)	548
C_{29}	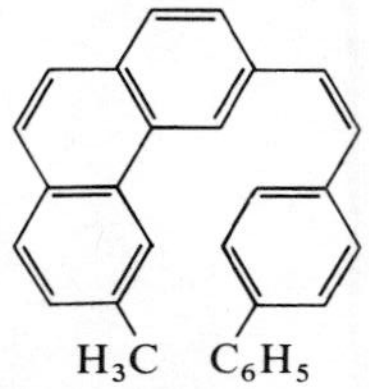 H_3C C_6H_5	C_6H_6, I_2, N_2	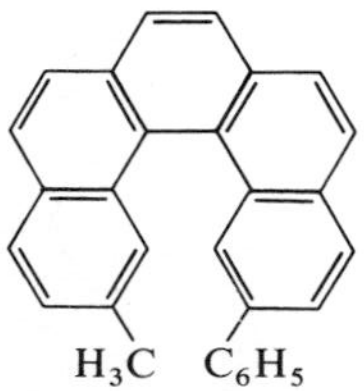 H_3C C_6H_5 (66)	232
	4′-Isopropyl-**E**	C_6H_6, I_2	2-Isopropyl-**F** (80)	523
C_{30}	1,2-Di-(3-phenanthryl)ethylene	C_6H_6 or C_6H_{14}, I_2, N_2	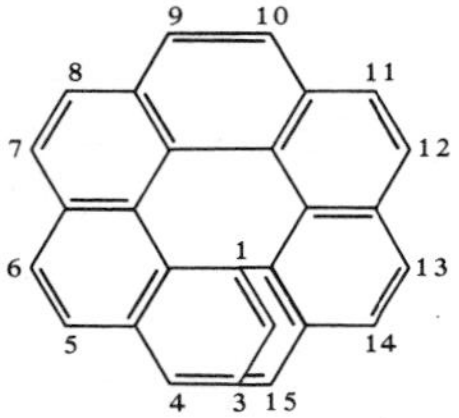 **G** ([7]Helicene)[h] (50)	159,166
		C_6H_6, I_2	” (12)	549
	1,2-Di-(2-deuterio-3-phenanthryl)ethylene	”	8,11-Dideuterio-**G** (61)	433

TABLE II. HIGHER CARBOCYCLIC AROMATIC SYSTEMS FROM STILBENE ANALOGS (*Continued*)[a]

Starting Material	Reaction Conditions	Products and Yields (%)	Refs.
1-(1-Phenanthryl)-2-(3-phenanthryl)ethylene	C_6H_6, I_2, N_2	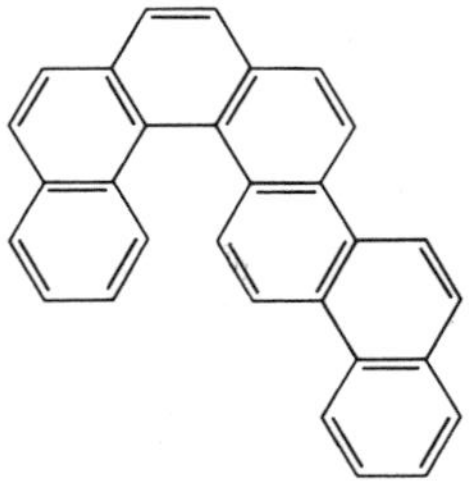 (12), (Naphtho[1,2-*a*]picene)  (68) (Benzo[*ghi*]naphtho[1,2-*b*]perylene)	159
1-(2-Phenanthryl)-2-(3-phenanthryl)ethylene	"	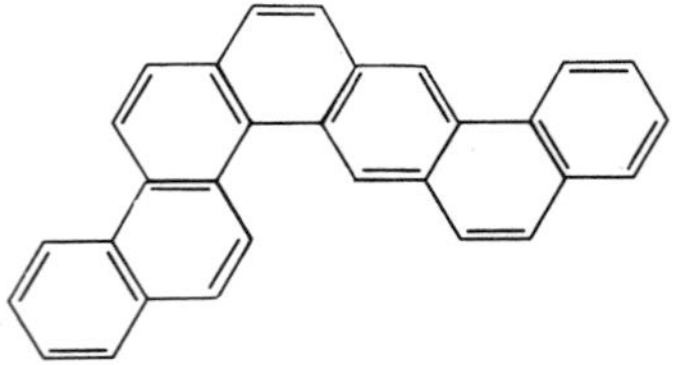 (75) (Phenanthro[3,2-*c*]chrysene)	159

1-(9-Phenanthryl)-2-(3-phenanthryl)ethylene	"	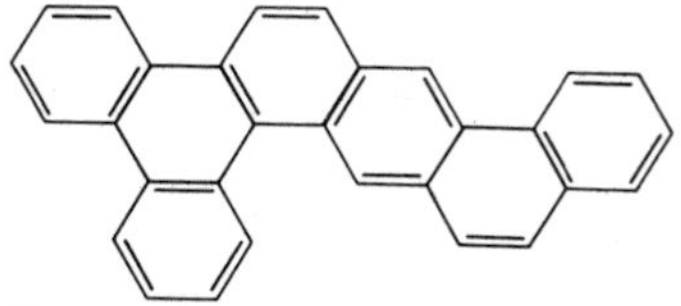 (Benzo[*g*]naphtho[1,2-*b*]chrysene) (75)	159
1,2-Di-(9-phenanthryl)ethylene	C_6H_6, I_2	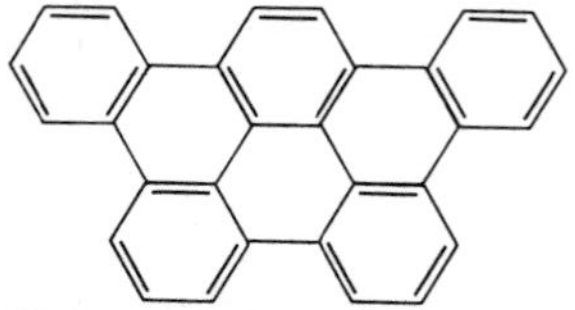 (Tribenzo[*b*,*n*,*pqr*]perylene) (30)	160
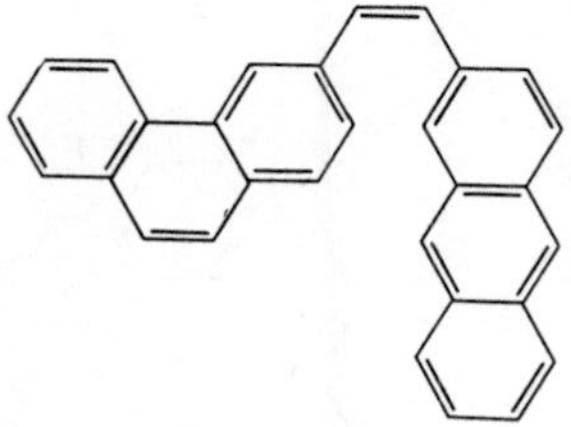 [structure]	Xylene, I_2, N_2	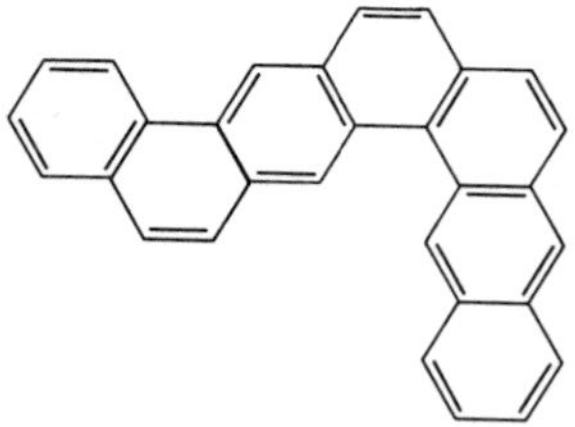 (Anthra[1,2-*a*]benz[*h*]anthracene) (60), 2,3-benzo-**F** (22)[i]	159,166
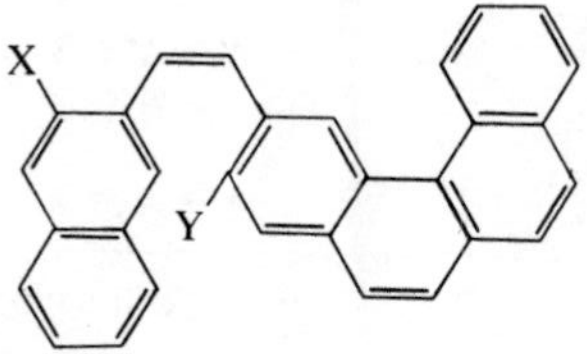 X, Y substituted (I)		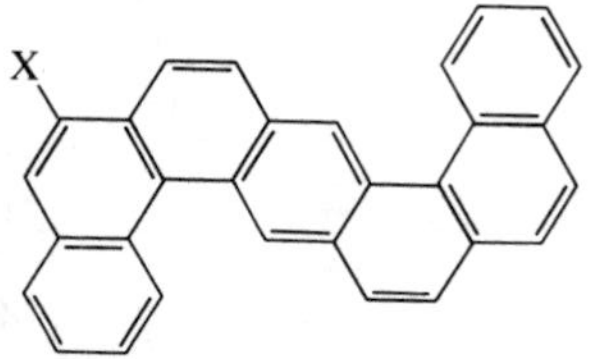 X substituted (II) (Dinaphth[1,2-*a*:1′,2′-*h*]anthracene)	

TABLE II. HIGHER CARBOCYCLIC AROMATIC SYSTEMS FROM STILBENE ANALOGS (*Continued*)[a]

Starting Material	Reaction Conditions	Products and Yields (%)	Refs.
I, X = Y = H	C_6H_6, I_2	II, X = H (67), **G** (20)	524
	"	**G** (—)	178
I, X = H, Y = D	"	II, X = H (—), 9-deuterio-**G** (14)	433
I, X = Y = D	"	II, X = D (67), 6,9-dideuterio-**G** (20)	433
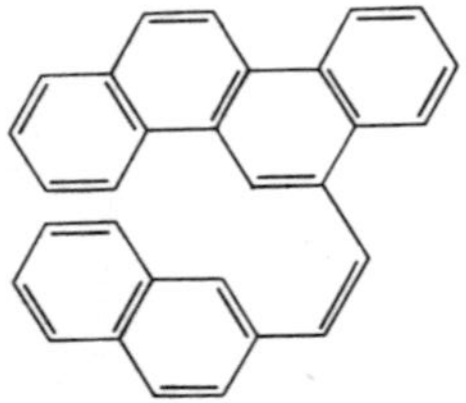	C_6H_6, I_2, N_2	7,8-Benzo-**F** (90)[j]	159,166
	C_6H_6 or C_6H_{14}, air	 (60) (Benzo[*c*]naphtho[2,1-*p*]chrysene)	538
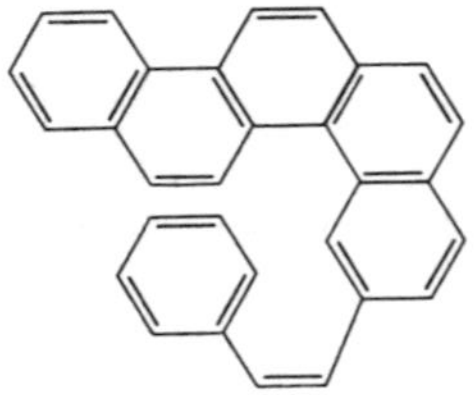	C_6H_6, I_2, N_2	3,4-Benzo-**F** (65)[k]	159,166

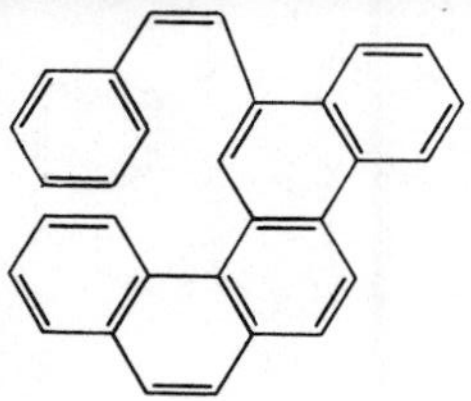	C_6H_6 or C_6H_{14}, I_2, N_2	7,8-Benzo-**F** (66)[j]	159,166
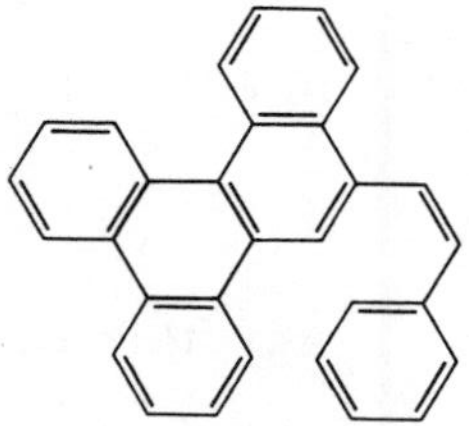	C_6H_6, I_2, N_2	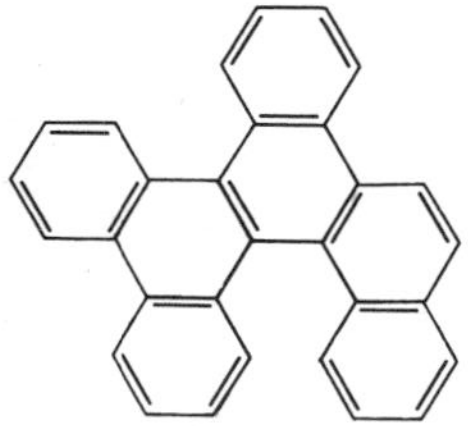 (52) (Tribenzo[*c,g,p*]chrysene)	232
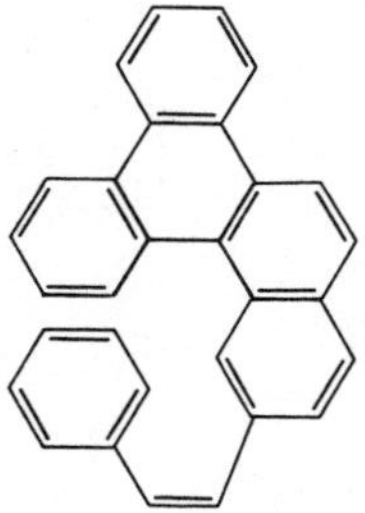	”	5,6-Benzo-**F** (60)[l]	159,166
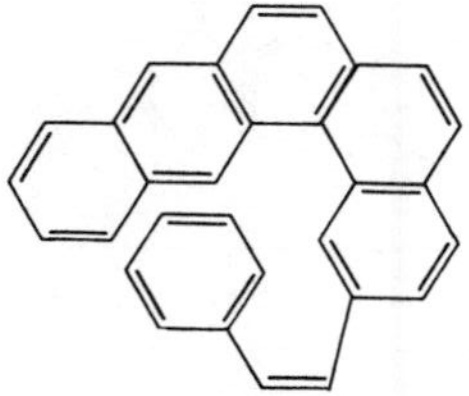	”	2,3-Benzo-**F** (10)[i]	159

TABLE II. Higher Carbocyclic Aromatic Systems from Stilbene Analogs (*Continued*)[a]

Starting Material	Reaction Conditions	Products and Yields (%)	Refs.
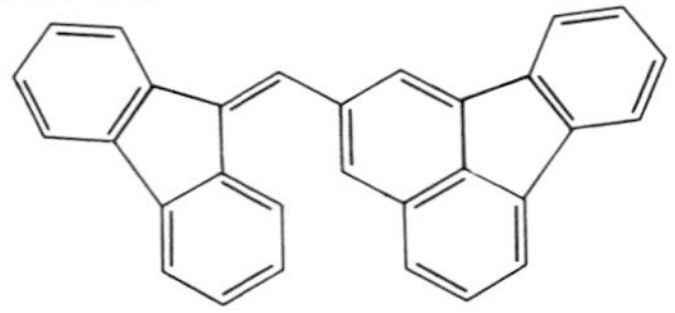 [structure]	C_6H_6, I_2, air	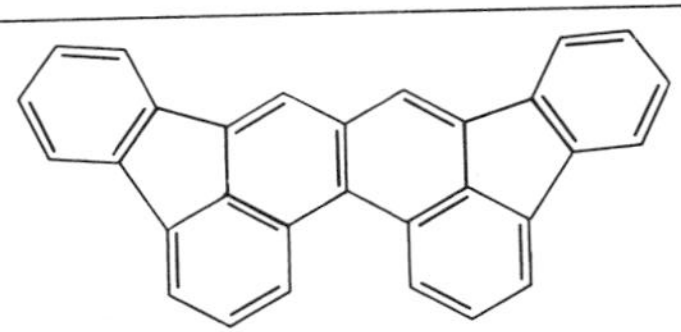 [structure] (50) (Benzo[*e*]fluoreno[9,1-*kl*]acephenanthrylene)	547
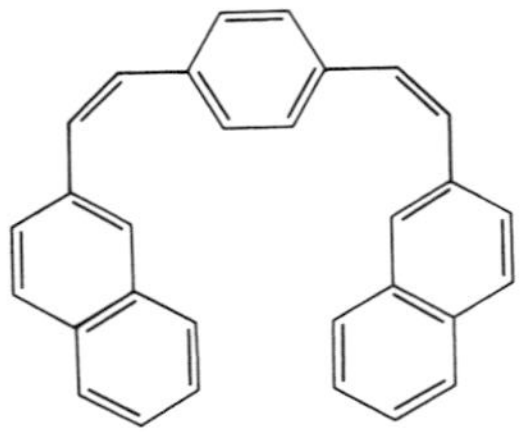 [structure]	C_6H_6, I_2	**G** (20), dinaphth[1,2-*a*:1′,2′-*h*]anthracene (20)[e]	185
4′-*tert*-Butyl-**E**	"	2-*tert*-Butyl-**F** (80)	523
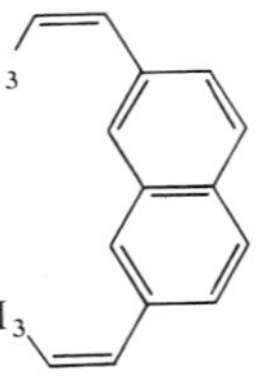 [structure with 3,5-$(CH_3)_2C_6H_3$ groups]	"	1,3,14,16-Tetramethyl-**F** (80)	546
[structure with Br, X]	Solvent, I_2	" (—), 5,6-dihydro-1,3,14,16-tetramethyl-**F** (—) [structure with Br]	68b

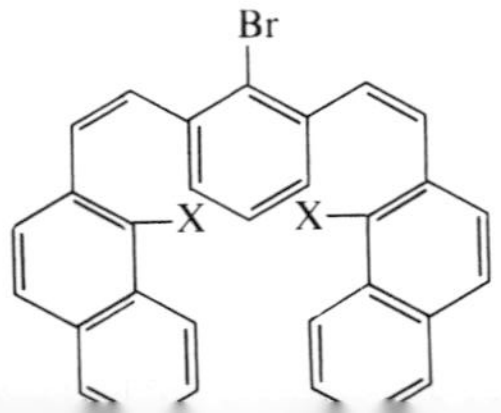

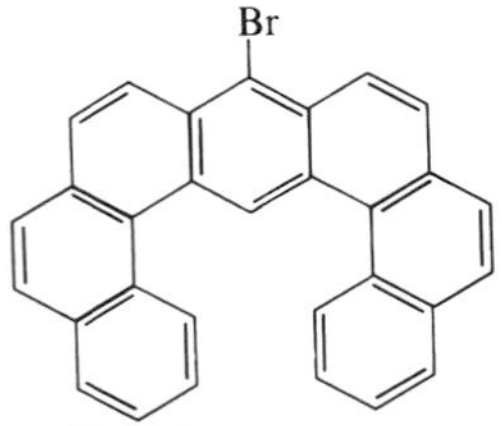

Reactant	Conditions	Product(s) (Yield, %)	Refs.
X = H	C_6H_6, I_2	" (62)	550
X = D	"	" (—)	550
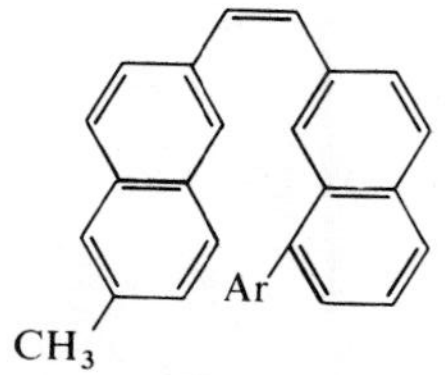 CH_3; Ar; Ar = *p*-Methoxyphenyl	C_6H_6, I_2, N_2	10-Methoxy-8-methylbenzo[*a*]coronene (25),[c]; 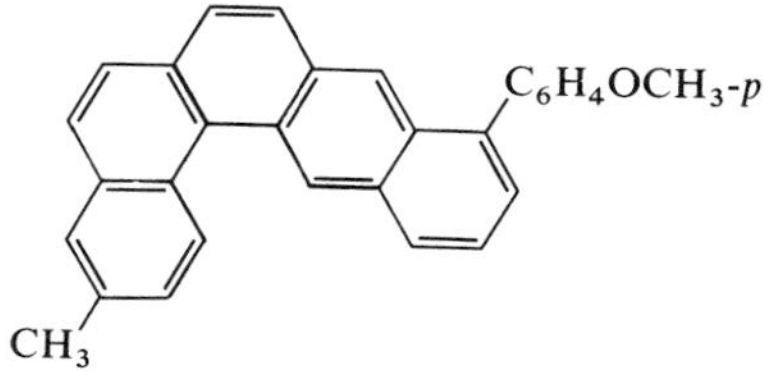 $C_6H_4OCH_3$-*p*; CH_3 (15)	232
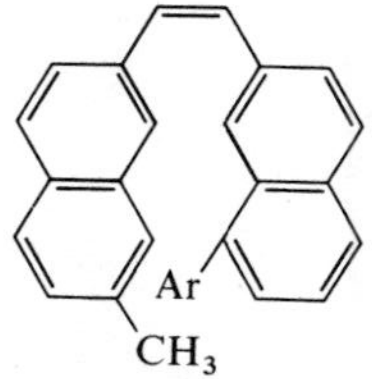 Ar; CH_3; Ar = *p*-Methoxyphenyl	"	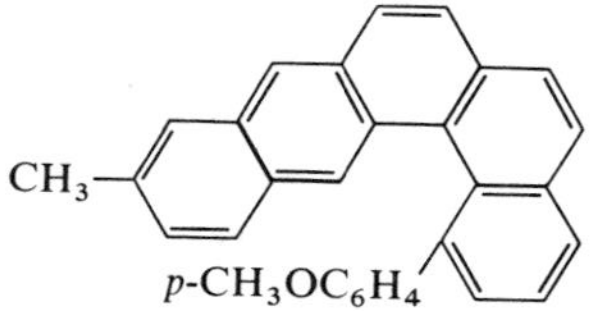 10-Methoxybenzo[*a*]coronene (8),[e]; 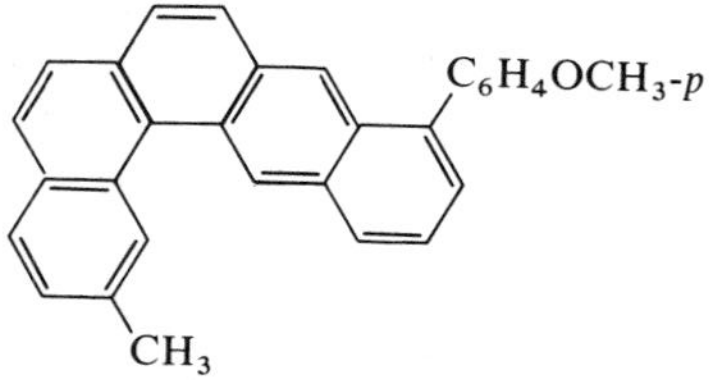 $C_6H_4OCH_3$-*p*; CH_3 (25),; CH_3; *p*-$CH_3OC_6H_4$ (25)	232

TABLE II. HIGHER CARBOCYCLIC AROMATIC SYSTEMS FROM STILBENE ANALOGS (*Continued*)[a]

	Starting Material	Reaction Conditions	Products and Yields (%)	Refs.
C_{31}	CH_3	C_6H_6, I_2	CH_3 (55),[k]	551
			CH_3 (7) (4-Methylphenanthro[4,3-*b*]chrysene)	
	CH_3	"	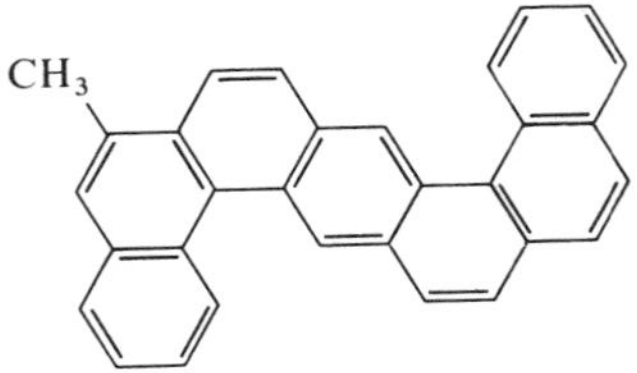 CH_3 (54), 6-methyl-**G** (30)	544
		EtOH, I_2	3,4-Benzo-**F** (60)[k]	552

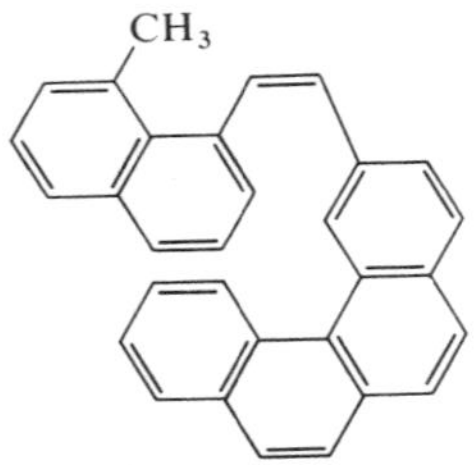

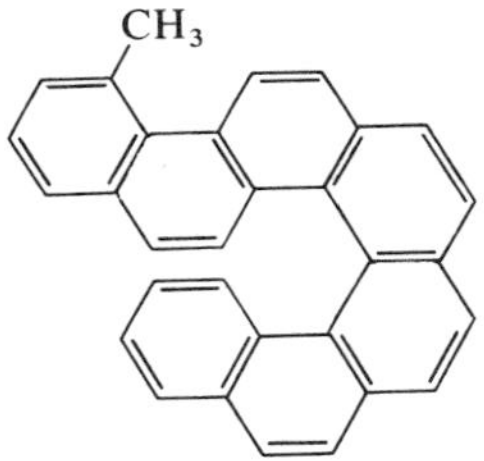

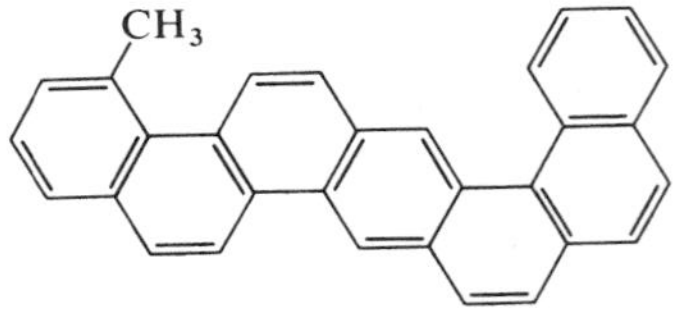

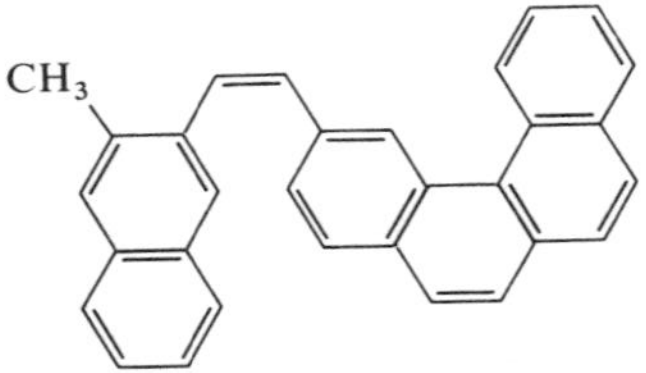

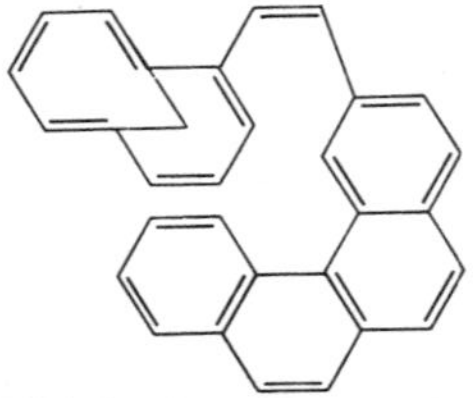

	Reactant	Conditions	Product(s) and Yield(s) (%)	Refs.
	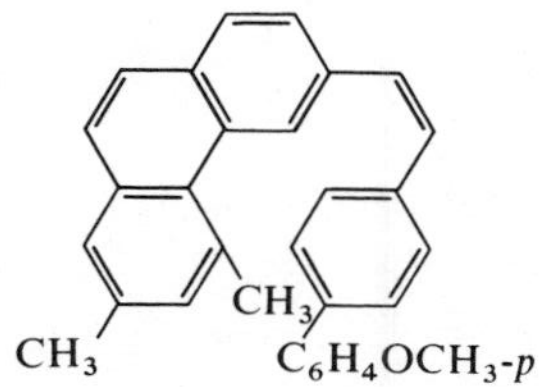 CH_3, CH_3, $C_6H_4OCH_3$-*p*	C_6H_6, I_2, N_2	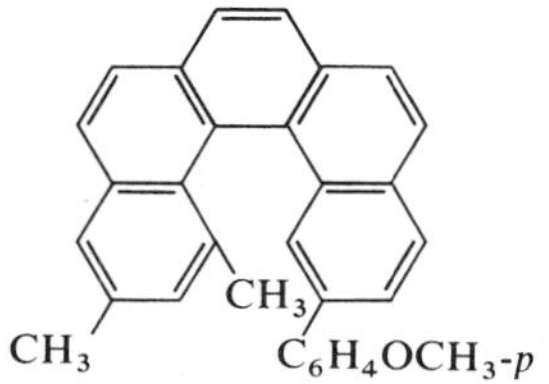 CH_3, CH_3, $C_6H_4OCH_3$-*p* (52), 10-Methoxy-8-methylbenzo[*a*]coronene (11)[e]	232
C_{32}	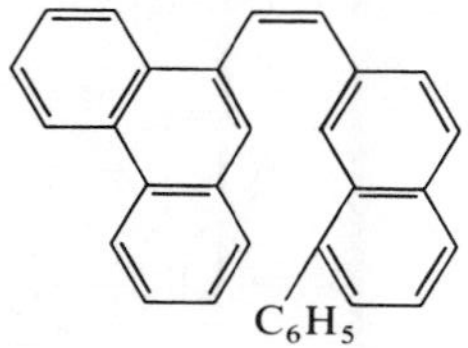 C_6H_5	"	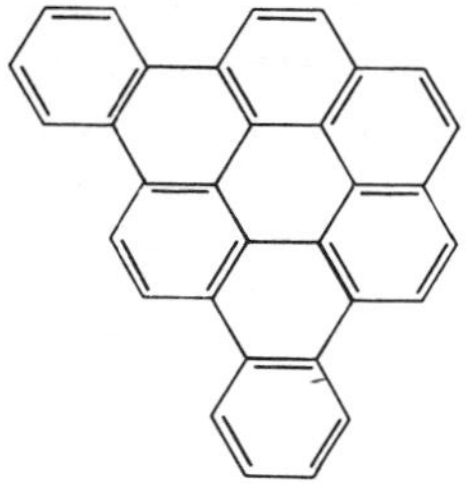 (30), (Dibenzo[*a,g*]coronene) 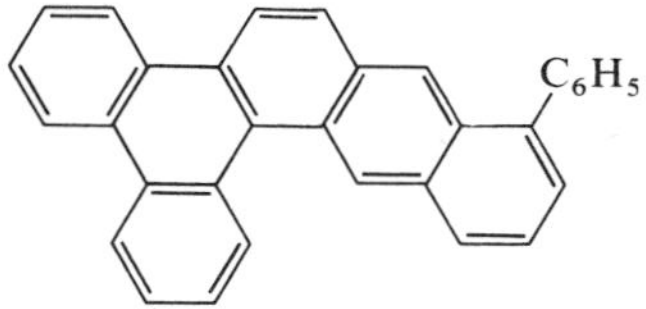 C_6H_5 (16)	232
	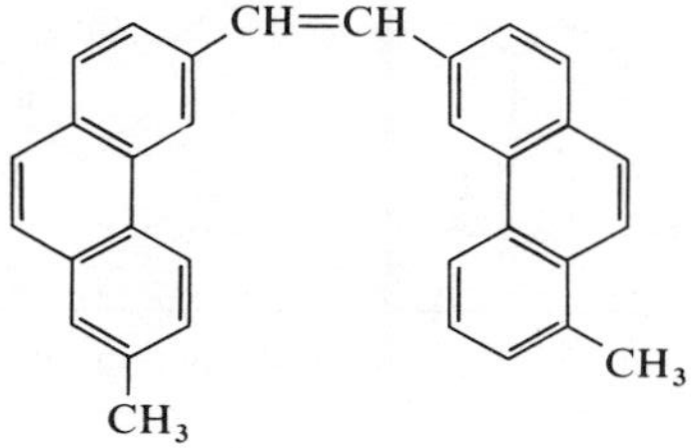 CH=CH, CH_3, CH_3	C_6H_{12}, I_2	3,15-Dimethyl-**G** (15)	117

TABLE II. HIGHER CARBOCYCLIC AROMATIC SYSTEMS FROM STILBENE ANALOGS (*Continued*)[a]

	Starting Material	Reaction Conditions	Products and Yields (%)	Refs.
	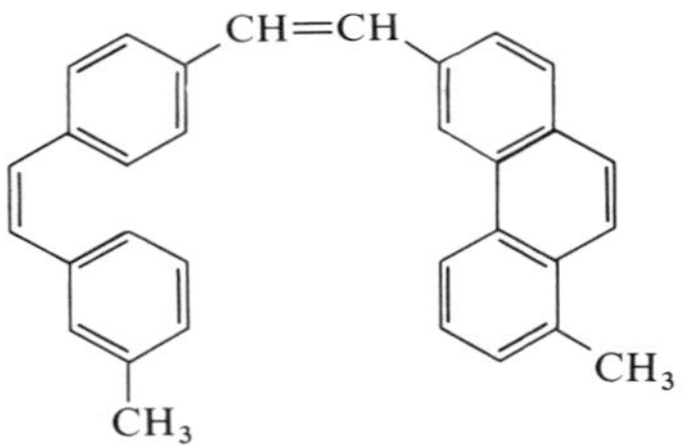 CH=CH, CH_3, CH_3	C_6H_6, I_2	1,15-Dimethyl-**G** (7), 3,15-dimethyl-**G** (7)	117
	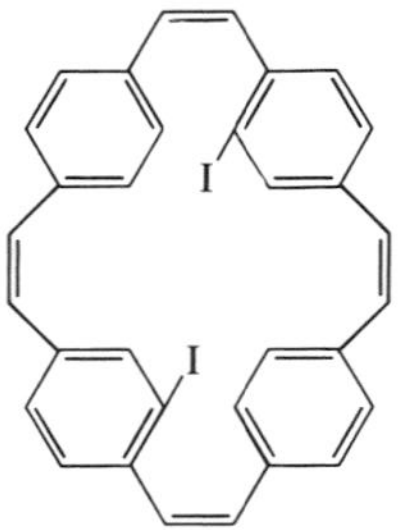 I, I	C_6H_6, N_2	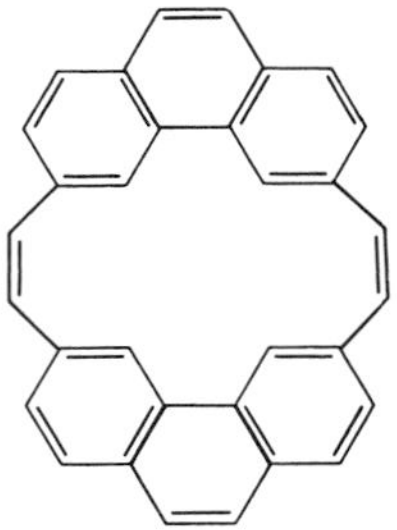 (15) (1,19:4,6:9,11:14,16-Tetraethenodibenzo[*a,i*]cyclohexadecene)	124
C_{33}	4′-(*p*-Methylphenyl)-**E**	C_6H_6, I_2	2-(*p*-Methylphenyl)-F (70)	523
	2′-Bromo-5-$CO_2CH(CH_3)C_4H_9$-*t*-**E**	C_6H_{12}, I_2	4-Bromo-1-$CO_2CH(CH_3)C_4H_9$-*t*-**F** (36), 3-$CO_2CH(CH_3)C_4H_9$-*t*-**F** (25)	553
	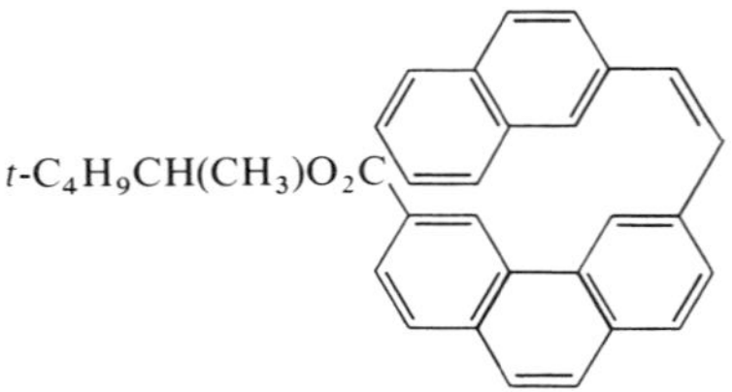 *t*-$C_4H_9CH(CH_3)O_2C$	"	2-$CO_2CH(CH_3)C_4H_9$-*t*-**F** (14)	553

2′-$CO_2CH(CH_3)C_4H_9$-*t*-**E**	”	4-$CO_2CH(CH_3)C_4H_9$-*t*-**F** (40)	553
3′-$CO_2CH(CH_3)C_4H_9$-*t*-**E**	”	3-$CO_2CH(CH_3)C_4H_9$-*t*-**F** (35)	553
4′-$CO_2CH(CH_3)C_4H_9$-*t*-**E**	”	2-$CO_2CH(CH_3)C_4H_9$-*t*-**F** (62)	553
11-$CO_2CH(CH_3)C_4H_9$-*t*-**E**	”	2-$CO_2CH(CH_3)C_4H_9$-*t*-**F** (54)	553
α′-$CO_2CH(CH_3)C_4H_9$-*t*-**E**	”	5-$CO_2CH(CH_3)C_4H_9$-*t*-**F** (48)	553
4′-$C(OCH_3)(CH_3)C_4H_9$-*t*-**E**	”	2-$C(OCH_3)(CH_3)C_4H_9$-*t*-**F** (43)	553
C_{34} 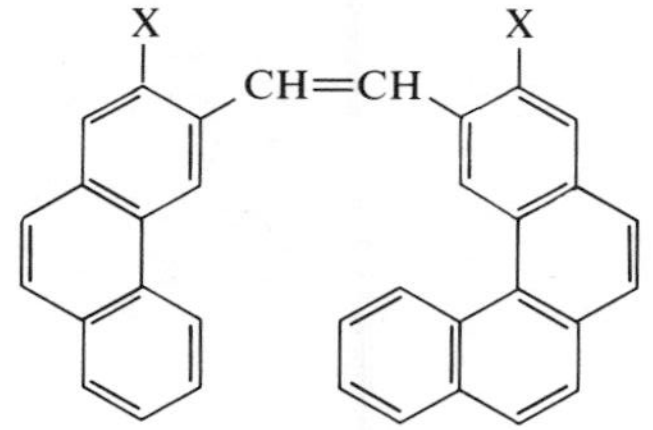 [structure: X-substituted bis(phenanthryl)ethylene, CH=CH]			
X = H	C_6H_6, I_2	[8]Helicene (85)[m]	168
	”	” (80)	167
	”	” (62)	166,524
	C_6H_6, I_2, air	” (—)	178
	p-$C_7H_{15}C_6H_4CO_2C_6H_4CN$-*p*, *p*-$C_4H_9C_6H_4CO_2C_6H_4CN$-*p* (1:1), I_2	” (40)	180b
X = D	C_6H_6, I_2	8,11-Dideuterio[8]helicene (68)	433
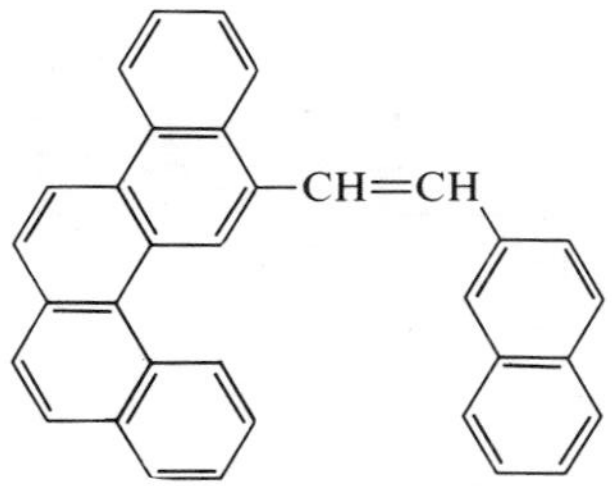 [structure: CH=CH]	”	9,10-Benzo-**G** (45)[n]	543

TABLE II. HIGHER CARBOCYCLIC AROMATIC SYSTEMS FROM STILBENE ANALOGS (*Continued*)[a]

	Starting Material	Reaction Conditions	Products and Yields (%)	Refs.
	2-Styryl-**F**	C_6H_6, I_2	[8]Helicene (80)[m]	167
		"	" (40)	548
	CH=CH	"	[8]Helicene (30)	168
	3′,5′-Di-*tert*-butyl-**E**	"	1,3-Di-*tert*-butyl-**F** (1)	523
C_{35} 	CH_3	C_6H_6, I_2, N_2	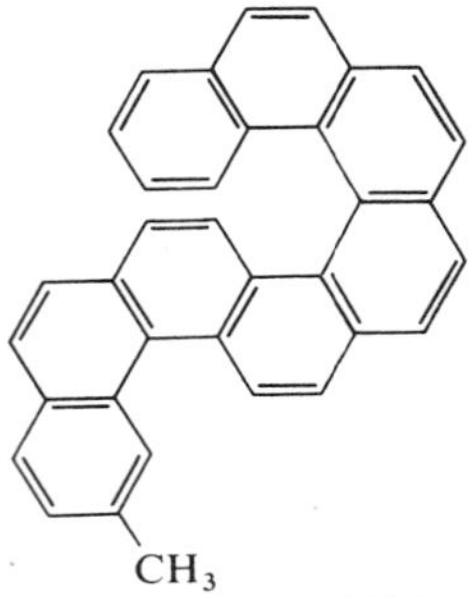 CH_3 (60) (15-Methylbenzo[*c*]phenanthro[3,4-*l*]chrysene)	172,554
	4′-$CO_2CH(CH_3)C_6H_5$-**E**	C_6H_{12}, I_2	2-$CO_2CH(CH_3)C_6H_5$-**F** (56)	553

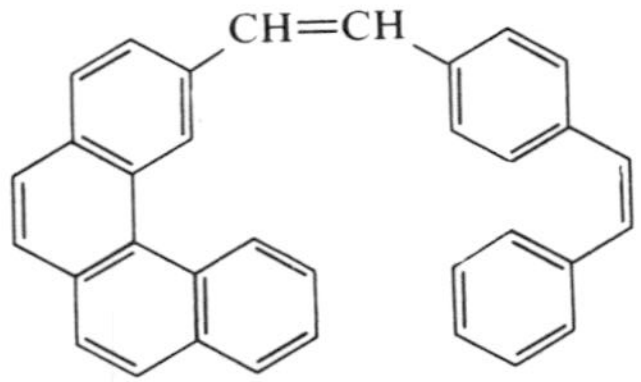

	Compound	Solvent	Product	Ref.
C_{36}	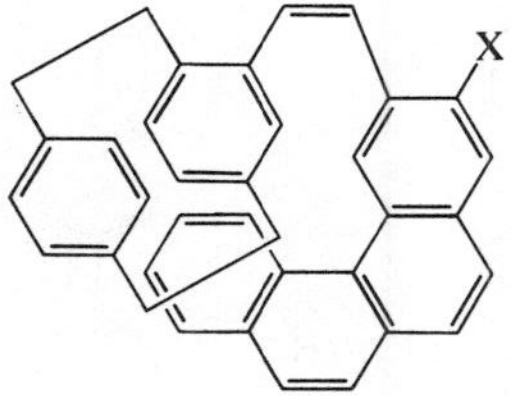 (I)		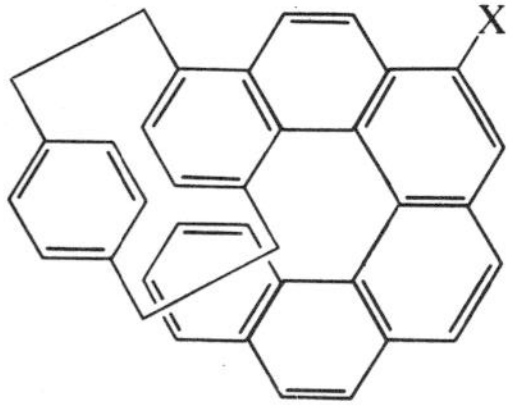 (II) (6-X-10,11,16,17-Tetrahydro-9,18:12,15-diethenocyclododeca[*c*]naphtho[1,2-*g*]phenanthrene)	
	I, X = H	C_6H_6, I_2	II, X = H (—)	176
	I, X = D	”	II, X = D (—)	176
	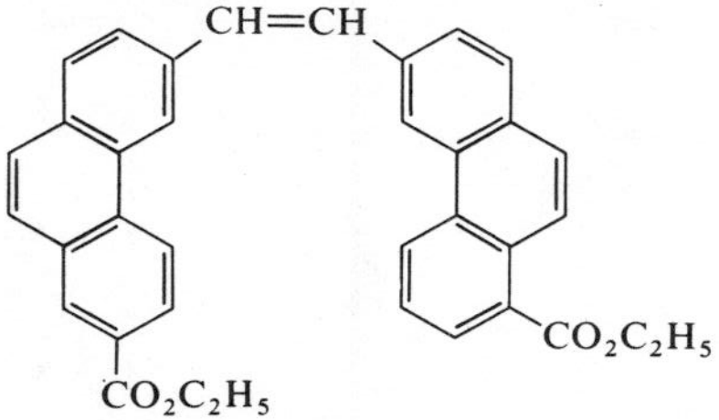 CH=CH; $CO_2C_2H_5$; $CO_2C_2H_5$	C_6H_{12}, I_2	3,15-Dicarbethoxy-**G** (35)	117
	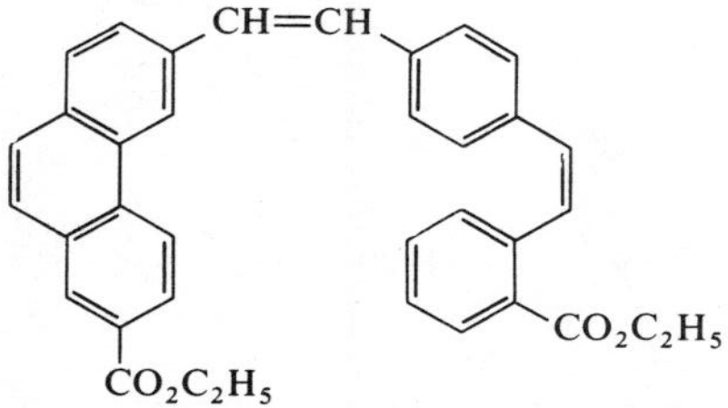 CH=CH; $CO_2C_2H_5$; $CO_2C_2H_5$	”	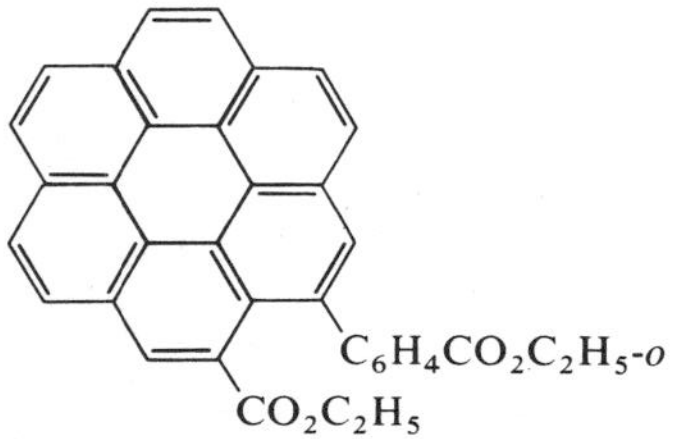 $C_6H_4CO_2C_2H_5$-*o*; $CO_2C_2H_5$ (10), 3,15-dicarbethoxy-**G** (20)	117
C_{37}	2′-Bromo-5-carbomenthoxy-**E**	”	4-Bromo-1-carbomenthoxy-**F** (18)	553
	3′-Carbomenthoxy-**E**	”	3-Carbomenthoxy-**F** (—)	177a
	4′-Carbomenthoxy-**E**	”	2-Carbomenthoxy-**F** (—)	177a

TABLE II. HIGHER CARBOCYCLIC AROMATIC SYSTEMS FROM STILBENE ANALOGS (*Continued*)[a]

Starting Material	Reaction Conditions	Products and Yields (%)	Refs.
C_{38} 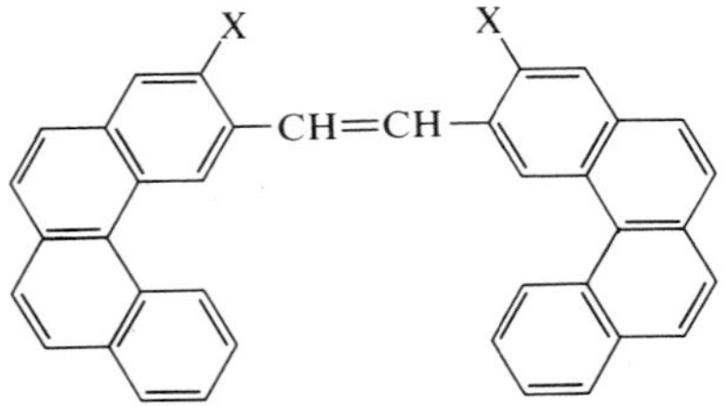 (2-X-3-phenanthryl)–CH=CH–(2-X-3-phenanthryl)			
X = H	C_6H_6, I_2	[9]Helicene (50)[o]	167
	"	" (48)	166,524
	C_6H_6, I_2, air	" (—)	178
X = D	C_6H_6, I_2	10,13-Dideuterio[9]helicene (61)	433
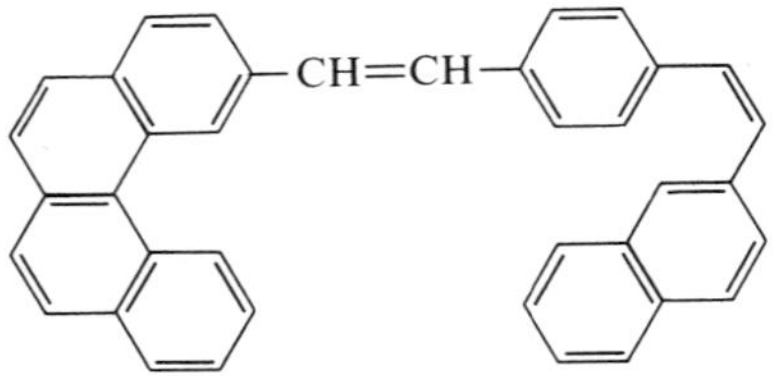 3-Phenanthryl–CH=CH–C₆H₄–CH=CH–2-naphthyl	"	[9]Helicene (50)	168
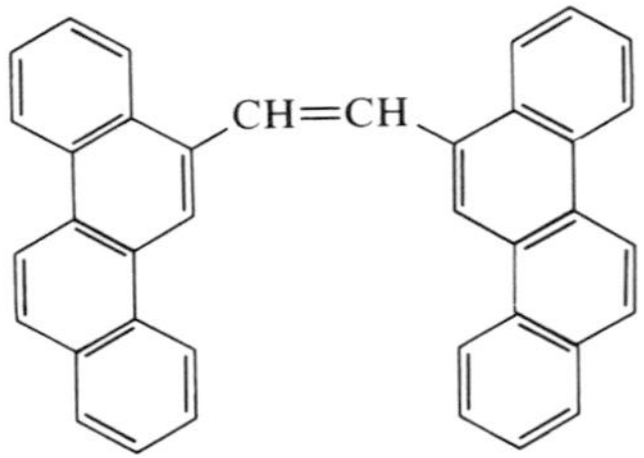 Chrysenyl–CH=CH–chrysenyl	"	7,8:11,12-Dibenzo-**G** (50)[p]	543

Reactant	Solvent, oxidant	Product(s) (yield, %)	Ref.
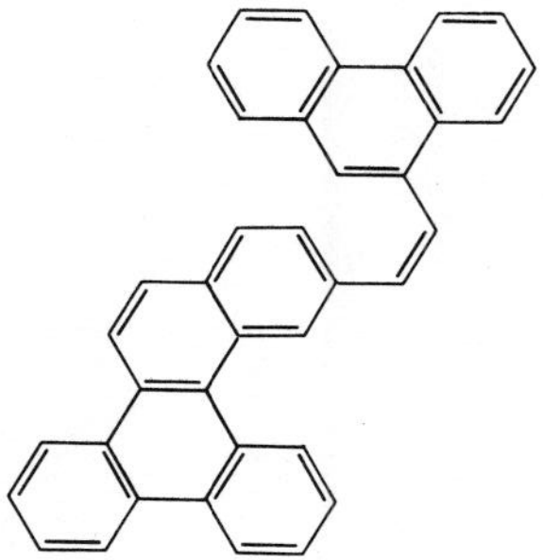	”	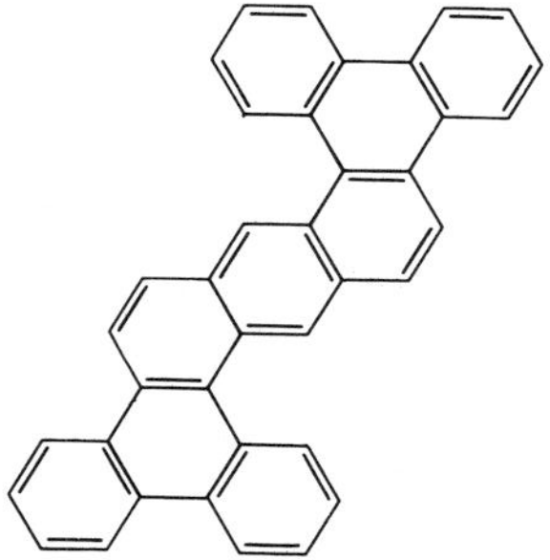 (60), (Benzo[*g*]triphenyleno[1,2-*b*]chrysene) 5,6:13,14-dibenzo-**G** (20)[q]	543
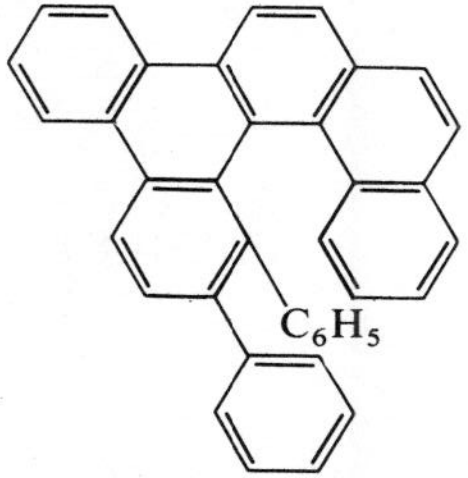 C_6H_5	EtOAc, I_2	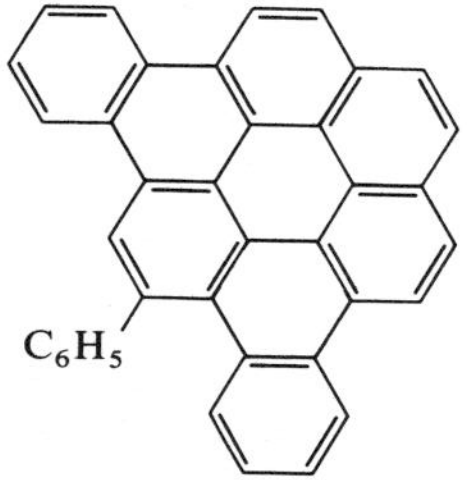 C_6H_5 (60) (15-Phenyldibenzo[*a,g*]coronene)	406
	C_6H_6, I_2	” (9), dibenzo[*a,g*]coronene (49)	406
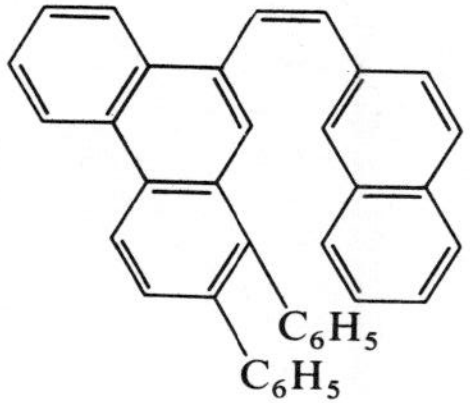 C_6H_5 C_6H_5	C_6H_{14}, C_6H_6, I_2	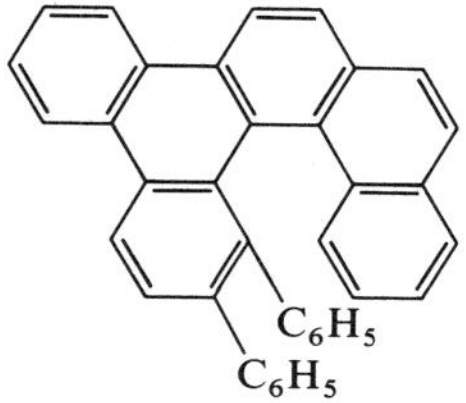 C_6H_5 C_6H_5 (70)	406

TABLE II. HIGHER CARBOCYCLIC AROMATIC SYSTEMS FROM STILBENE ANALOGS (*Continued*)[a]

	Starting Material	Reaction Conditions	Products and Yields (%)	Refs.
C_{40}	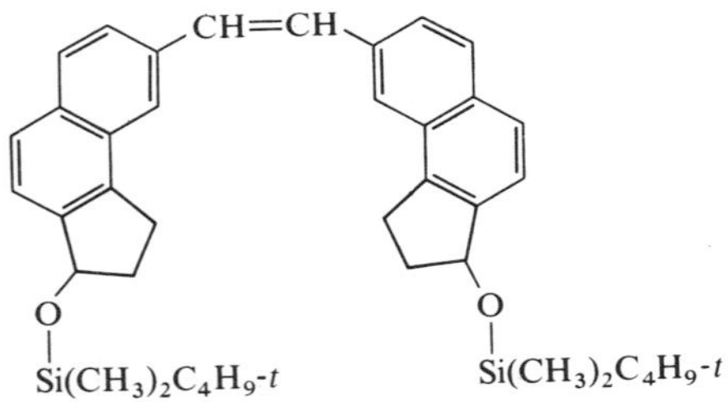 CH=CH; $Si(CH_3)_2C_4H_9$-*t*; $Si(CH_3)_2C_4H_9$-*t*	1. C_6H_6, I_2 2. *p*-TSA, dark	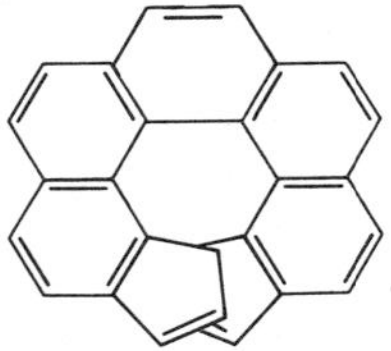 (53)	175b
C_{42}	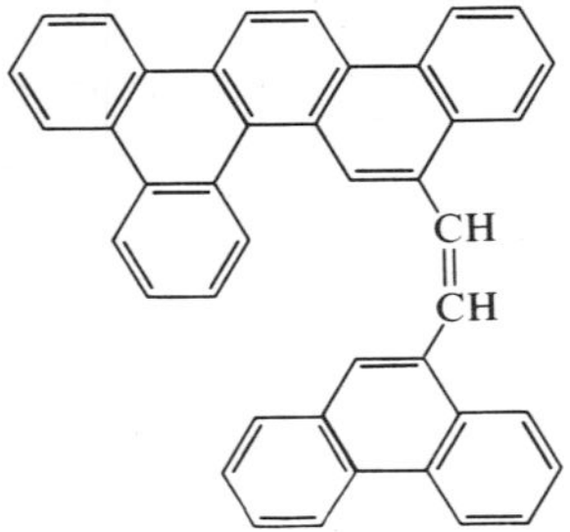 CH=CH	C_6H_6, I_2	5,6:9,10:13,14-Tribenzo-**G** (90)[r]	543
	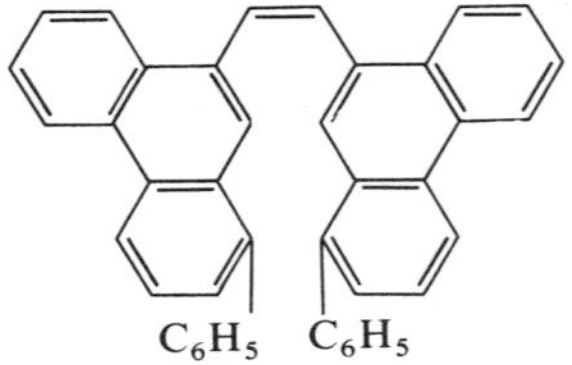 C_6H_5 C_6H_5	"	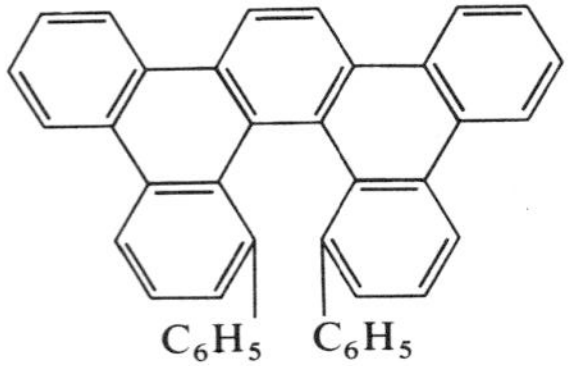 C_6H_5 C_6H_5 (1)	406

Starting material	Conditions	Product	Ref.
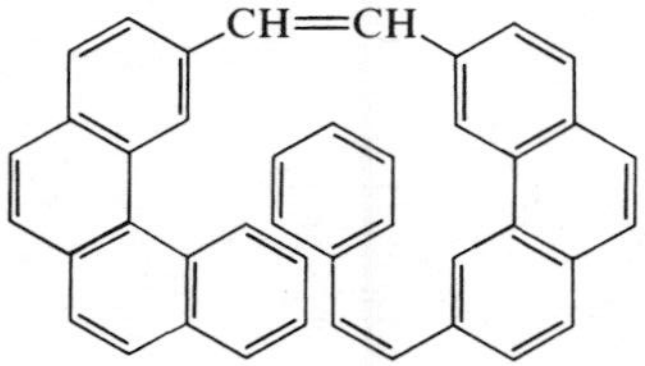 CH=CH … CH=CH	”	[10]Helicene (20)	168
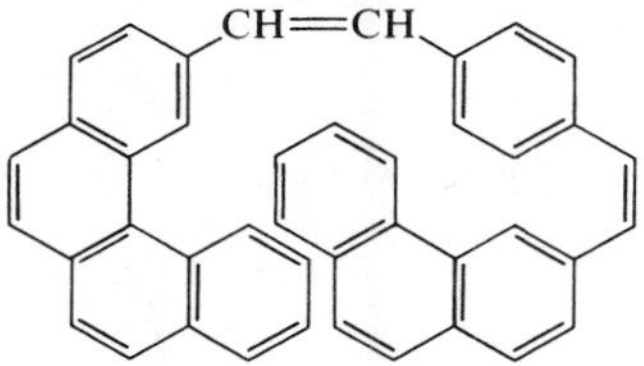 CH=CH … CH=CH	”	[10]Helicene (30)[s]	168
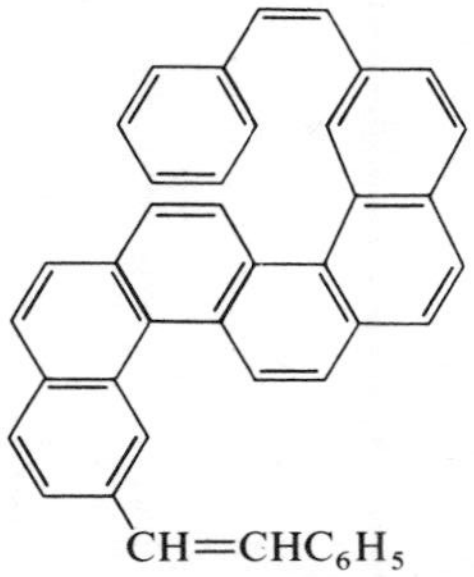 $CH{=}CHC_6H_5$	C_6H_6, I_2, N_2	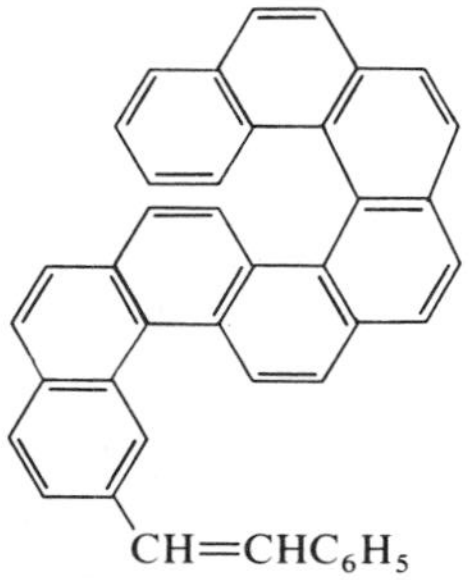 $CH{=}CHC_6H_5$ (−) (15-Styrylbenzo[*c*]phenanthro[3,4-*l*]chrysene)	172
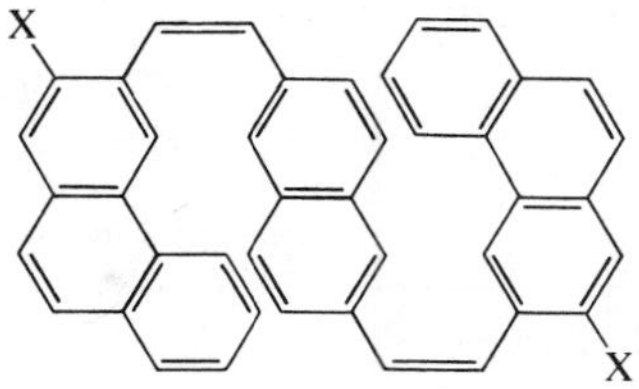 X, X (I)		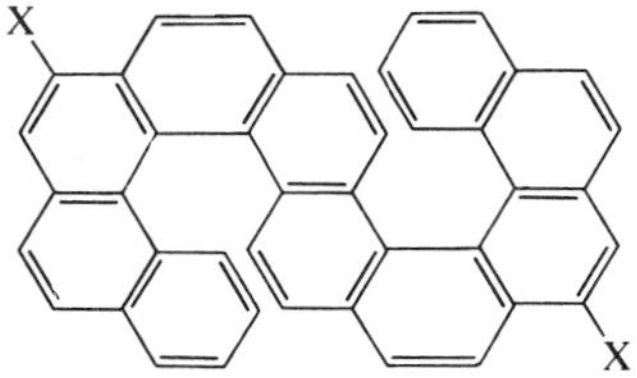 X, X (II) (4,16-Di-X-diphenanthro[3,4-*c*:3′,4′-*l*]chrysene)	

TABLE II. HIGHER CARBOCYCLIC AROMATIC SYSTEMS FROM STILBENE ANALOGS (*Continued*)[a]

Starting Material	Reaction Conditions	Products and Yields (%)	Refs.
I, X = H	C_6H_6, I_2, N_2	II, X = H (70)	172,554
I, X = Br	PhMe, I_2	II, X = Br (89)	173
C_{44} 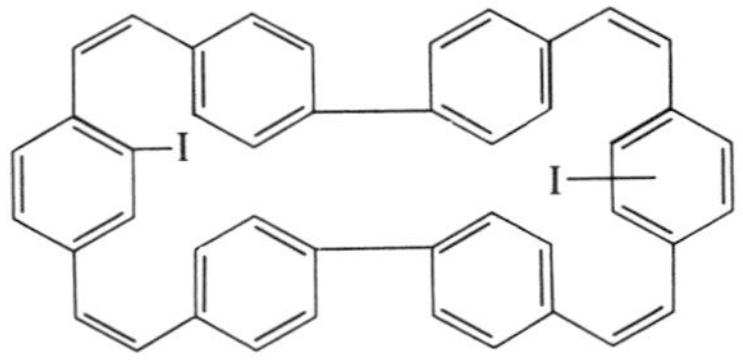	C_6H_6	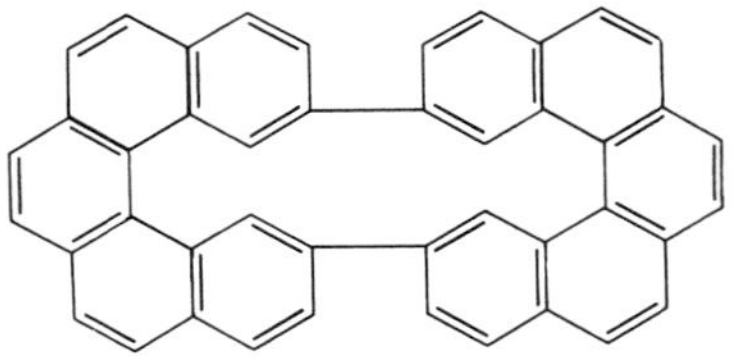 (70) (1,23:8,10:11,13:20,22-Tetraethenocyclohexadeca[1,2,3,4-*def*:9,10,11,12-*d′e′f′*]diphenanthrene)	555
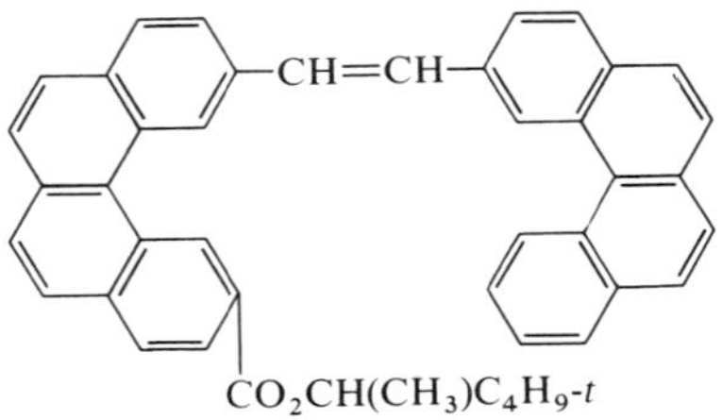 CH=CH; $CO_2CH(CH_3)C_4H_9$-*t*	C_6H_{12}, I_2	2-$CO_2CH(CH_3)C_4H_9$-*t*-[9]Helicene (8)[o]	553
C_{46} 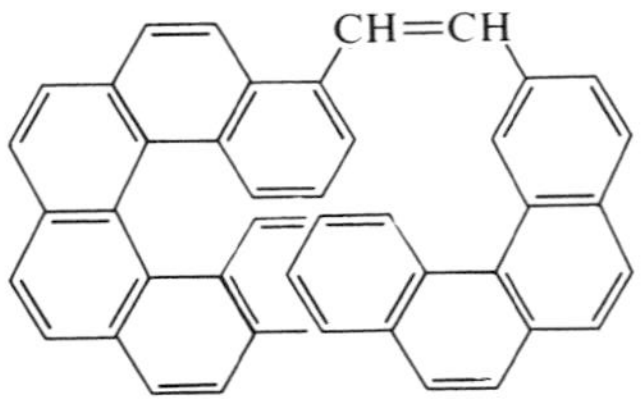 CH=CH	C_6H_6, I_2, N_2	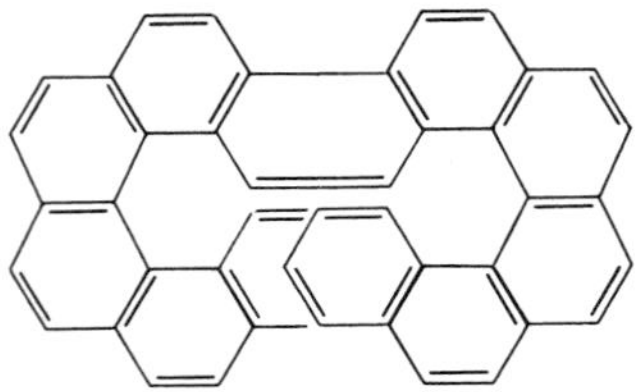 (90) (Dinaphtho[1,2-*g*:1′,2′-*g′*]benzo[2,1-*c*:3,4-*c′*]diphenanthrene)	174

Formula	Substrate	Solvent, oxidant	Product (yield %)	Refs.
		PhMe, I_2	" (73)	544
	Ar = Ar′ = 3-Phenanthryl	C_6H_6, I_2	[11]Helicene (54)[t]	186
	Ar = 2-Naphthyl, Ar′ = 2-Benzo[*c*]phenanthryl	"	" (80)	168
	Ar = 2-Benzo[*c*]phenanthryl	"	[11]Helicene (84)	186
		"	" (60)	168
C_{48}		"	(70)	202
	4′,11-Dicarbomenthoxy-**E**	C_6H_{12}, I_2	(1,2,4,5,10,11,13,14-Octahydro-15,23:16,22-dimethenobenzo[1,2-*a*:5,4-*a*′]dipentaphene) 2,15-Dicarbomenthoxy-**F** (30)	553

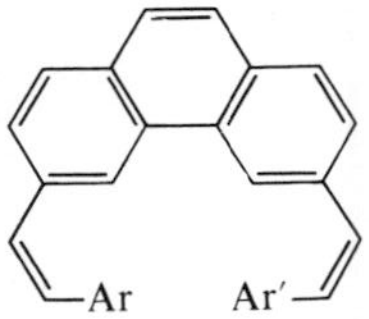

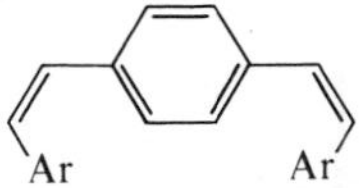

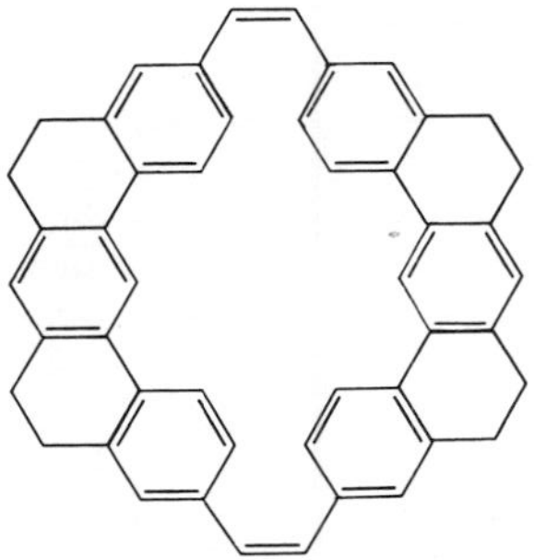

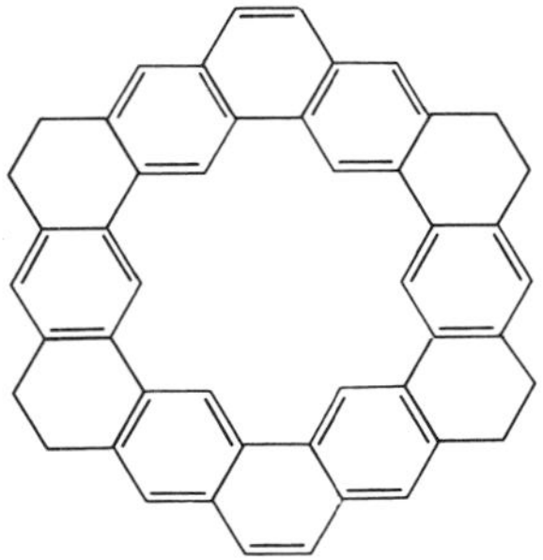

TABLE II. HIGHER CARBOCYCLIC AROMATIC SYSTEMS FROM STILBENE ANALOGS (*Continued*)[a]

	Starting Material	Reaction Conditions	Products and Yields (%)	Refs.
C_{50}		C_6H_6, I_2	(15), (Naphtho[1′,2′:5,6]phenanthro[4,3-*c*]-phenanthro[3,4-*l*]chrysene)	551
			Ar (70) Ar = 2-benzo[*c*]phenanthryl (8-Ar-benzo[*mno*]phenanthro[3,4-*c*]chrysene)	

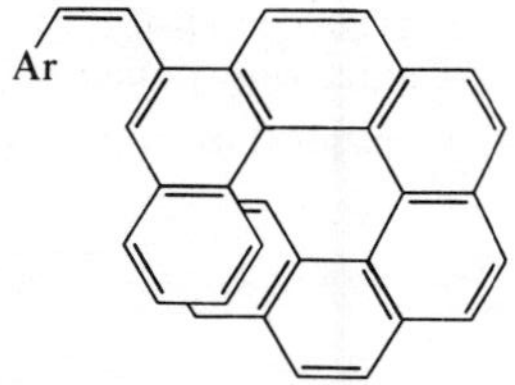 Ar = 2-Benzo[*c*]phenanthryl	PhMe, I_2	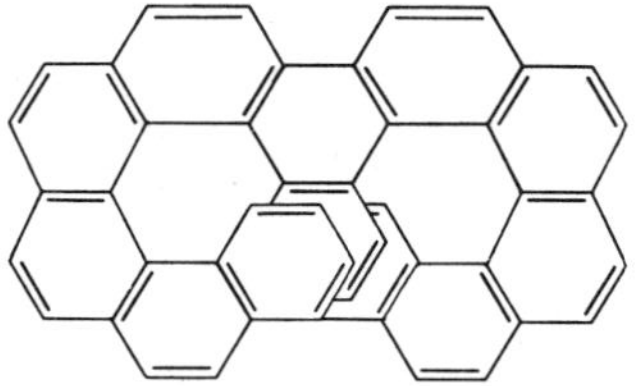 (Dinaphtho[1,2-*g*:1′,2′-*g*′]naphtho-[2,1-*c*:3,4-*c*′]diphenanthrene) (26)	544
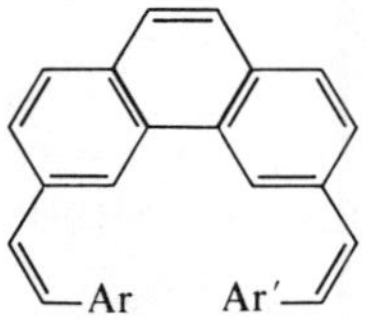 Ar = 3-Phenanthryl, Ar′ = 2-Benzo[*c*]phenanthryl	C_6H_6, I_2	[12]Helicene (30)[a]	168
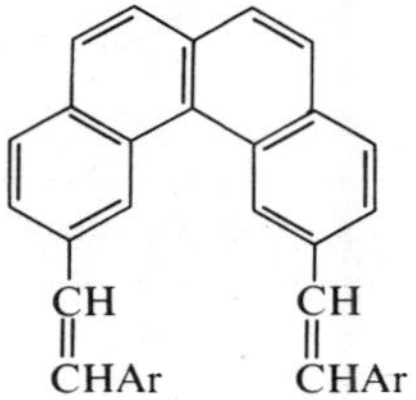 CH=CHAr, CH=CHAr; Ar = 3-Phenanthryl	"	[12]Helicene (32)	186

TABLE II. HIGHER CARBOCYCLIC AROMATIC SYSTEMS FROM STILBENE ANALOGS (*Continued*)[a]

	Starting Material	Reaction Conditions	Products and Yields (%)	Refs.
	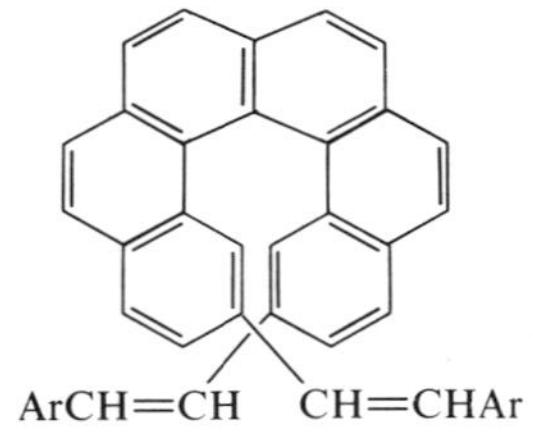 ArCH=CH CH=CHAr Ar = 2-Naphthyl	C_6H_6, I_2	[12]Helicene (42)	186
C_{52}	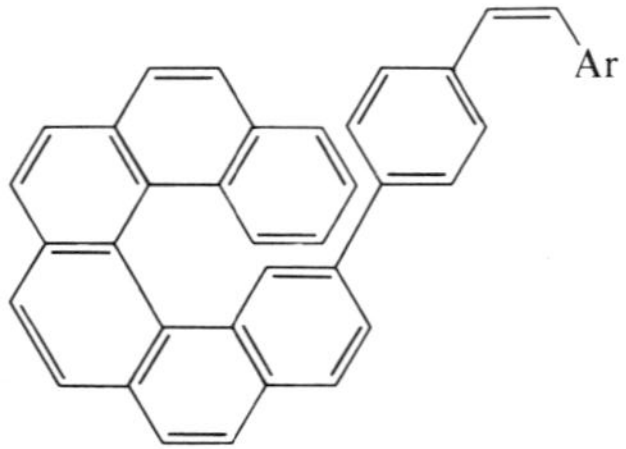 Ar = 2-Benzo[*c*]phenanthryl	"	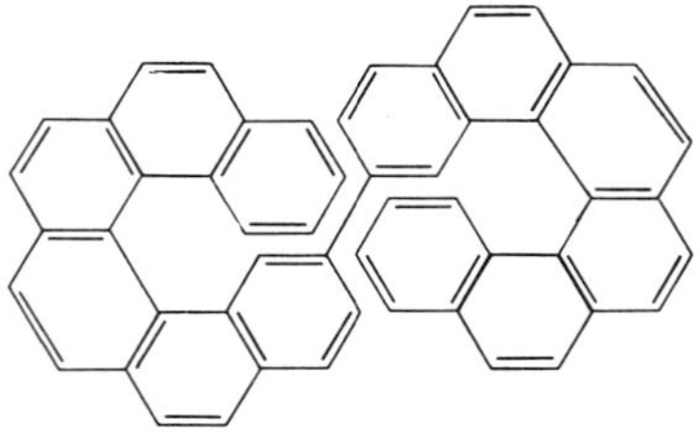 (50) (11,11′-Biphenanthro[3,4-*c*]phenanthrene)	556
C_{54}	1-(2-Benzo[*c*]phenanthryl)-2-(2-[8]helicenyl)ethylene	"	[13]Helicene (29)[v]	187

	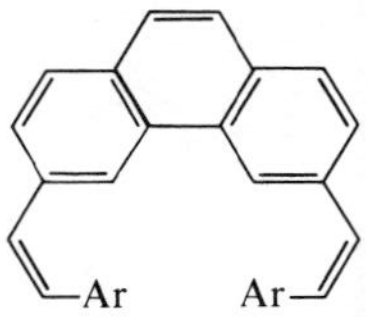 Ar = 3-X-2-Benzo[*c*]phenanthryl			
	X = H	”	[13]Helicene (52)	187
		”	” (40)	168
	X = D	”	10,21-Dideuterio[13]helicene (—)	187
	4′-CO_2CH_2(2-[6]Helicenyl)-**E**	C_6H_{12}, I_2	2-CO_2CH_2(2-[6]Helicenyl)-**F** (42)	553
C_{58}	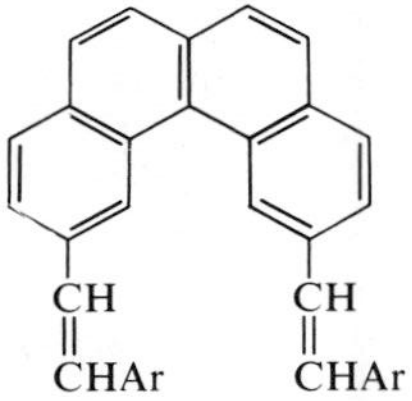 Ar = 2-Benzo[*c*]phenanthryl	C_6H_6, I_2	[14]Helicene (10)[w]	186
	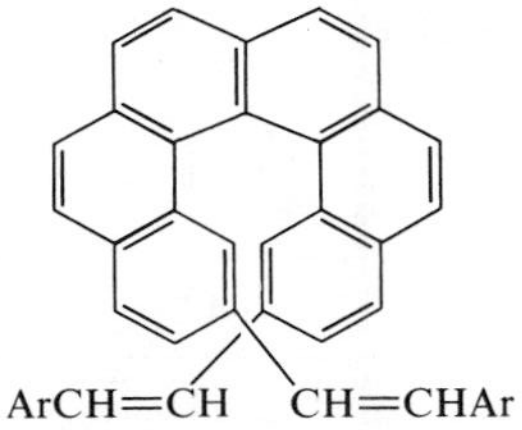 Ar = 3-Phenanthryl	”	[14]Helicene (45)	186

TABLE II. HIGHER CARBOCYCLIC AROMATIC SYSTEMS FROM STILBENE ANALOGS (*Continued*)[a]

Starting Material	Reaction Conditions	Products and Yields (%)	Refs.
C_n 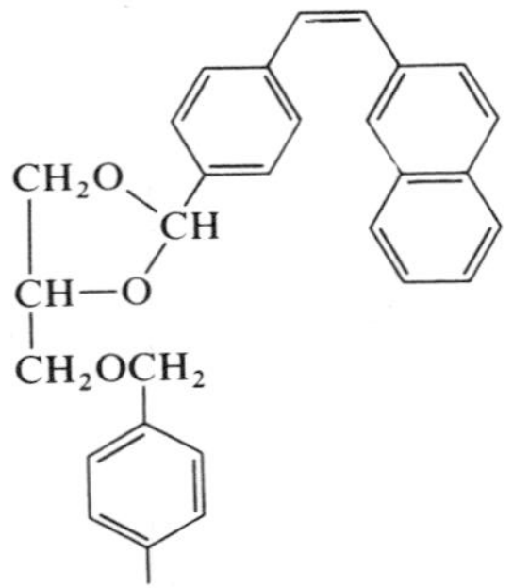 CH_2O, CH, CH—O, CH_2OCH_2; Styrene–divinylbenzene copolymer	1. C_6H_6, I_2 2. Acid hydrolysis	2-Formyl-**D** (24)	530
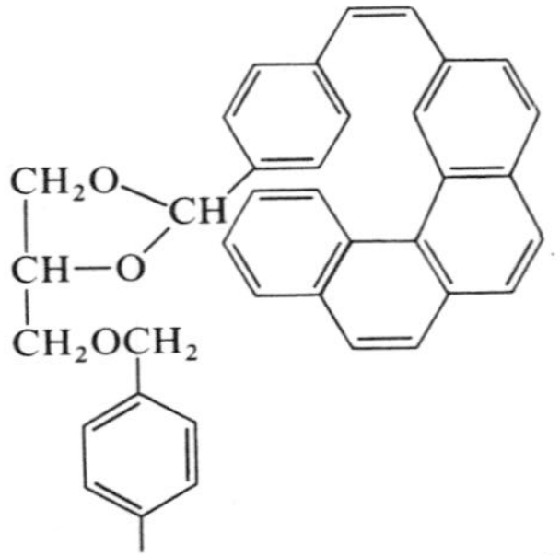 CH_2O, CH, CH—O, CH_2OCH_2; Styrene–divinylbenzene copolymer	1. C_6H_6, I_2 2. Acid hydrolysis	2-Formyl-**F** (5)	530

[a] The following symbols are used as abbreviations in Table II: **A** = 1-styrylnaphthalene, **B** = chrysene, **C** = 2-styrylnaphthalene, **D** = benzo[*c*]phenanthrene, **E** = 2-styrylbenzo[*c*]phenanthrene, **F** = [6]helicene, and **G** = [7]helicene. The numbering scheme for each of these systems is indicated in the entry for the parent compound (under C_{18} for **A**, **B**, **C**, and **D**; under C_{26} for **E** and **F**; and under C_{30} for **G**).

[b] The identity of the oxidant in this reaction is obscure. One possibility is residual dissolved oxygen.

[c] These are relative yields.

[d] The product was incorrectly reported as pentaphene in Ref. 31.

[e] The structure of the product is indicated in a previous entry in this table.

[f] [6]Helicene is phenanthro[3,4-*c*]phenanthrene.

[g] Unpublished results of W. H. Laarhoven cited in Ref. 68a.

[h] [7]Helicene is naphtho[2,1-*c*]phenanthro[4,3-*g*]phenanthrene.

[i] 2,3-Benzo[6]helicene is benzo[5,6]phenanthro[4,3-*a*]anthracene.

[j] 7,8-Benzo[6]helicene is benzo[*c*]naphtho[1,2-*g*]chrysene.

[k] 3,4-Benzo[6]helicene is phenanthro[3,4-*c*]chrysene.

[l] 5,6-Benzo[6]helicene is benzo[*g*]naphtho[2,1-*c*]chrysene.

[m] [8]Helicene is naphtho[2,1-*c*]phenanthro[4,3-*g*]phenanthrene.

[n] 9,10-Benzo[7]helicene is benzo[*c*]phenanthro[4,3-*g*]chrysene.

[o] [9]Helicene is diphenanthro[3,4-*c*:4′,3′-*g*]phenanthrene.

[p] 7,8:11,12-Dibenzo[7]helicene is benzo[2,1-*g*:3,4-*g*′]dichrysene.

[q] 5,6:13,14-Dibenzo[7]helicene is benzo[*g*]triphenyleno[2,1-*c*]chrysene.

[r] 5,6:9,10:13,14-Tribenzo[7]helicene is benzo[*f*]triphenyleno[1,2-*j*]picene.

[s] [10]Helicene is benzo[*g*]phenanthro[3,4-*c*:6,5-*c*′]diphenanthrene.

[t] [11]Helicene is bisbenzo[5,6]phenanthro[3,4-*c*:4′,3′-*g*]phenanthrene.

[u] [12]Helicene is dinaphtho[1,2-*g*:1′,2′-*g*′]naphtho[2,1-*c*:7,8-*c*′]diphenanthrene.

[v] [13]Helicene is bisnaphtho[1′,2′:5,6]phenanthro[3,4-*c*:4′,3′-*g*]phenanthrene.

[w] [14]Helicene is diphenanthro[4,3-*g*:4′,3′-*g*′]naphtho[2,1-*c*:7,8-*c*′]diphenanthrene.

TABLE III. Carbocyclic Systems from Reactants Other Than 1,2-Diarylethylenes

	Starting Material	Reaction Conditions	Products and Yields (%)	Refs.
C_{13}	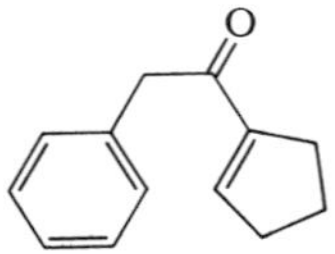 O	C_6H_6, $BF_3 \cdot Et_2O$, N_2 (or argon)	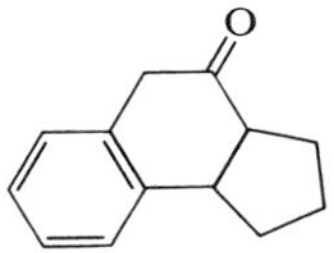 O (79)	239
		MeOH	" (90)	239
C_{14}	2-Vinylbiphenyl	C_6H_6, air	Phenanthrene (—)	557
		n-$PrNH_2$, N_2	9,10-Dihydrophenanthrene (70)	421
	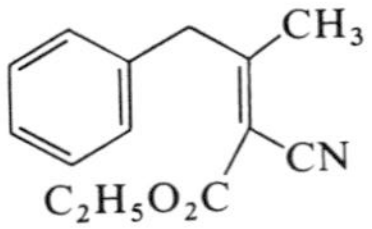 CH_3, CN, $C_2H_5O_2C$	EtOH, NaOEt	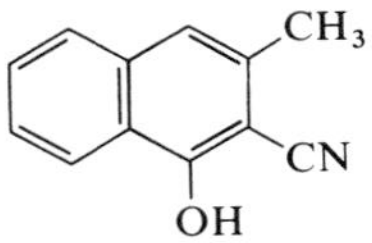 CH_3, CN, OH (37)	238
	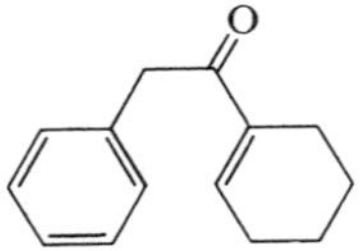 O	C_6H_6, $BF_3 \cdot Et_2O$, N_2 (or argon)	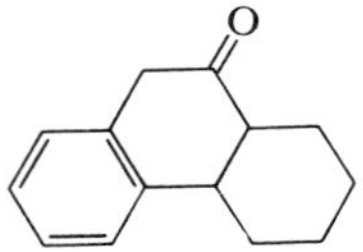 O (87)	239
		MeOH	" (80)	239
C_{15}	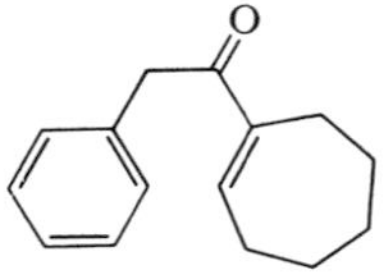 O	C_6H_6, $BF_3 \cdot Et_2O$, N_2 (or argon)	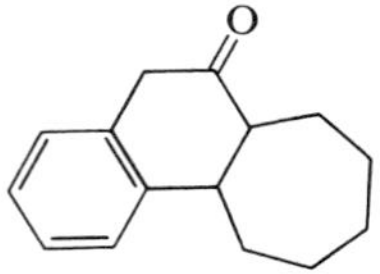 O (90)	239
	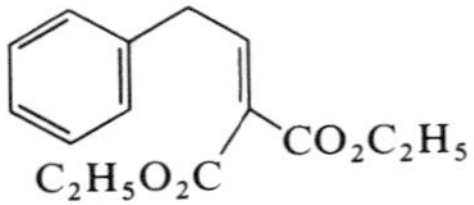 $CO_2C_2H_5$, $C_2H_5O_2C$	EtOH, NaOEt	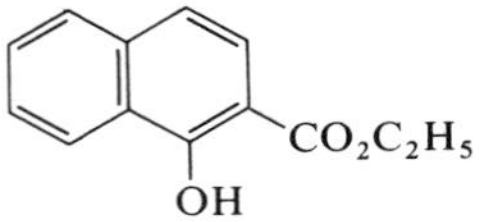 $CO_2C_2H_5$, OH (9)	238

C_{16}	4-Vinylphenanthrene	C_6H_6, air	4,5-Dihydropyrene (74), pyrene (26)	557
	2,2′-Divinylbiphenyl	Solvent	9,10-Dihydro-4-vinylphenanthrene (good)	558
		EtOH, N_2	4,5,9,10-Tetrahydropyrene (90)	210
		C_6H_6	" (85)	33
		"	" (80)	558
		Solvent, N_2	" (—)	559
		Solvent, I_2	4,5-Dihydropyrene (—), pyrene (—)	559
	9,10-Dihydro-4-vinylphenanthrene	Solvent	4,5,9,10-Tetrahydropyrene (—)	558
	$C_6H_5CH{=}CHCH{=}CHC_6H_5$	Et_2O, air	1-Phenylnaphthalene (50)	88
		C_6H_6, I_2, N_2	" (6–12)	560
	$C_6H_5CH{=}CHC{\equiv}CC_6H_5$	Solvent, I_2	1-Phenylnaphthalene (2)	225
	m-$ClC_6H_4CH{=}CHCH{=}CHC_6H_5$	C_6H_6, I_2, N_2	1-(*m*-Chlorophenyl)naphthalene + 6-chloro-1-phenylnaphthalene + 8-chloro-1-phenylnaphthalene (62:25:13) (6–12 total)	560
	p-$ClC_6H_4CH{=}CHCH{=}CHC_6H_5$	"	1-(*p*-Chlorophenyl)naphthalene + 7-chloro-1-phenylnaphthalene (50:50) (6–12 total)	560
	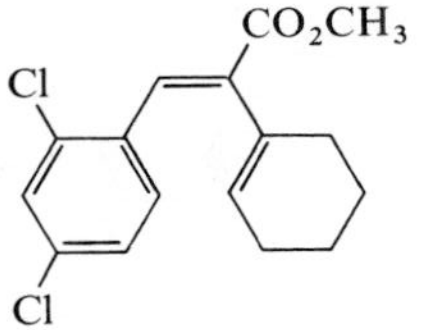 Cl, Cl, CO_2CH_3	MeOH, I_2	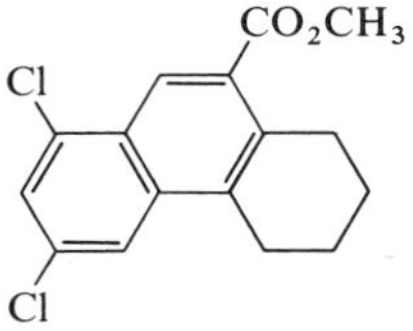 Cl, Cl, CO_2CH_3 (62)	222
		MeOH, N_2	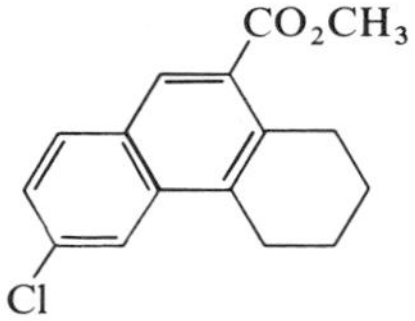 Cl, CO_2CH_3 (34)	222
		MeOH, K_2CO_3, N_2	" (16), 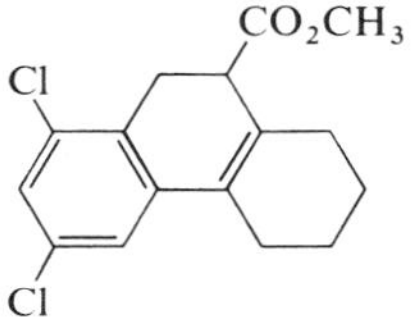 Cl, Cl, CO_2CH_3 (32)	222

TABLE III. CARBOCYCLIC SYSTEMS FROM REACTANTS OTHER THAN 1,2-DIARYLETHYLENES (*Continued*)

	Starting Material	Reaction Conditions	Products and Yields (%)	Refs.
	m-$FC_6H_4CH{=}CHCH{=}CHC_6H_5$	C_6H_6, I_2, N_2	1-(*m*-Fluorophenyl)naphthalene + 6-fluoro-1-phenylnaphthalene + 8-fluoro-1-phenylnaphthalene (87:8:5) (6–12 total)	560
	p-$FC_6H_4CH{=}CHCH{=}CHC_6H_5$	"	1-(*p*-Fluorophenyl)naphthalene + 7-fluoro-1-phenylnaphthalene (92:8) (6–12 total)	560
	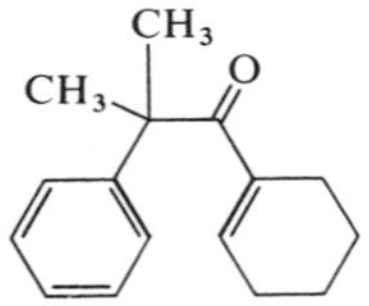 CH_3, CH_3, O (structure)	C_6H_6, $BF_3 \cdot Et_2O$, N_2 (or argon)	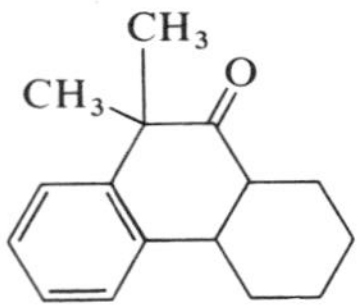 CH_3, CH_3, O (structure) (poor)	239
	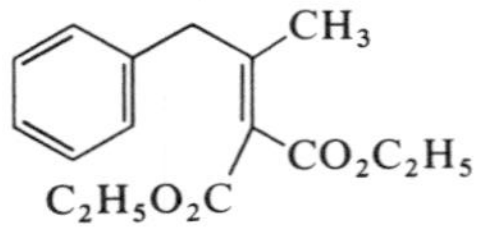 CH_3, $CO_2C_2H_5$, $C_2H_5O_2C$ (structure)	EtOH, NaOEt	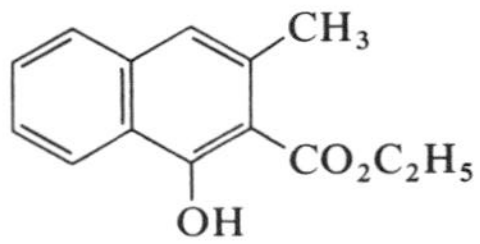 CH_3, $CO_2C_2H_5$, OH (structure) (43)	238
	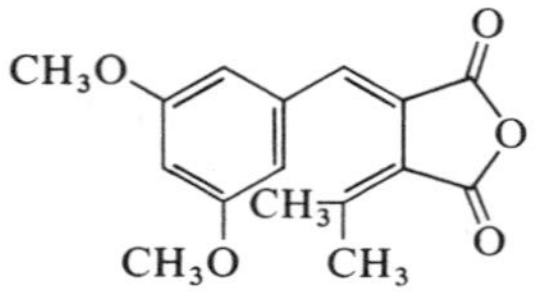 CH_3O, CH_3O, CH_3, CH_3, O (structure)	PhMe, 54°	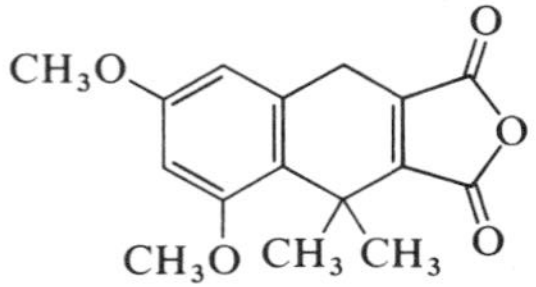 CH_3O, CH_3O, CH_3, CH_3, O (structure) (—)	71b
		$CDCl_3$, 12°	" (—)	71b
C_{17}	*m*-$CH_3C_6H_4CH{=}CHCH{=}CHC_6H_5$	C_6H_6, I_2, N_2	1-(*m*-Methylphenyl)naphthalene + 6-methyl-1-phenylnaphthalene + 8-methyl-1-phenylnaphthalene (52:38:10) (6–12 total)	560
	p-$CH_3C_6H_4CH{=}CHCH{=}CHC_6H_5$	"	1-(*p*-Methylphenyl)naphthalene + 7-methyl-1-phenylnaphthalene (41:59) (6–12 total)	560

Reactant	Conditions	Product(s) and Yield(s) (%)	Refs.
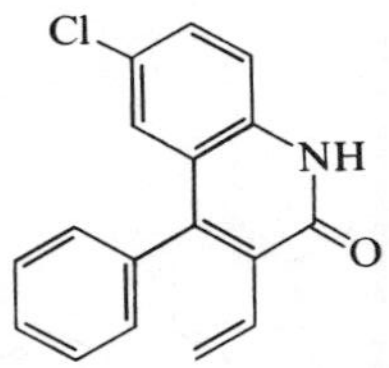	C_6H_6, air	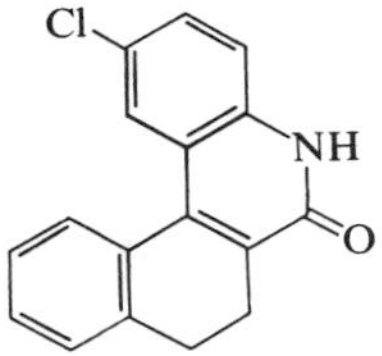 (65)	561a
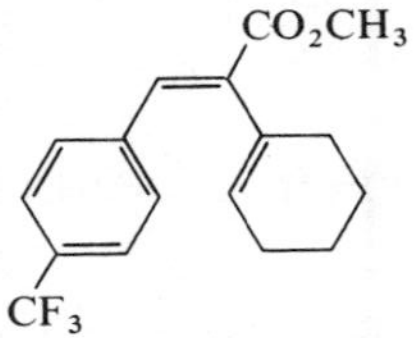	MeOH, I_2	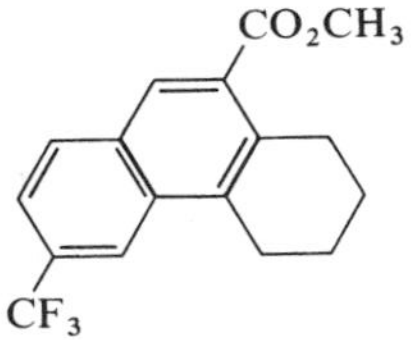 (71)	222
	MeOH, K_2CO_3, N_2	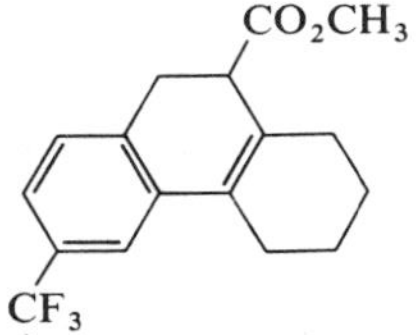 (57), 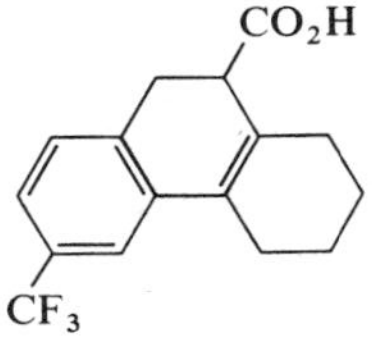 (24)	222
m-$NCC_6H_4CH{=}CHCH{=}CHC_6H_5$	C_6H_6, I_2, N_2	1-(m-Cyanophenyl)naphthalene + 6-cyano-1-phenylnaphthalene + 8-cyano-1-phenylnaphthalene (52:42:6) (6–12 total)	560
p-$NCC_6H_4CH{=}CHCH{=}CHC_6H_5$	"	1-(p-Cyanophenyl)naphthalene + 7-cyano-1-phenylnaphthalene (33:67) (6–12 total)	560
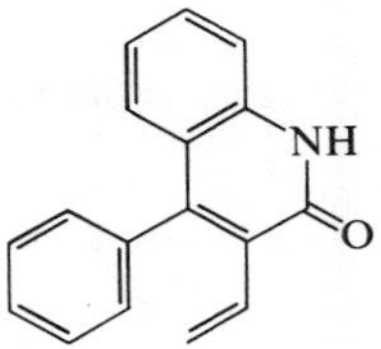	C_6H_6, air	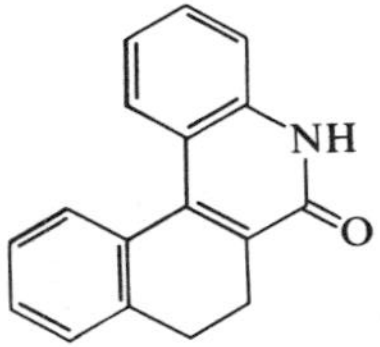 (80)	561a

TABLE III. CARBOCYCLIC SYSTEMS FROM REACTANTS OTHER THAN 1,2-DIARYLETHYLENES (*Continued*)

Starting Material	Reaction Conditions	Products and Yields (%)	Refs.
	C_6H_6, I_2	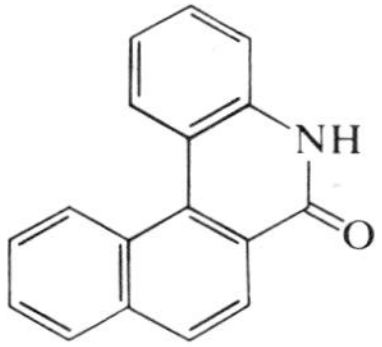 (58)	561a
m-$OHCC_6H_4CH{=}CHCH{=}CHC_6H_5$	C_6H_6, I_2, N_2	1-(*m*-Formylphenyl)naphthalene + 6-formyl-1-phenylnaphthalene + 8-formyl-1-phenylnaphthalene (40 total)	104
p-$OHCC_6H_4CH{=}CHCH{=}CHC_6H_5$	"	1-(*p*-Formylphenyl)naphthalene + 7-formyl-1-phenylnaphthalene (4:1) (35 total)	104
m-$CH_3OC_6H_4CH{=}CHCH{=}CHC_6H_5$	"	1-(*m*-Methoxyphenyl)naphthalene + 6-methoxy-1-phenylnaphthalene + 8-methoxy-1-phenylnaphthalene (55:33:12) (6–12 total)	560
p-$CH_3OC_6H_4CH{=}CHCH{=}CHC_6H_5$	"	1-(*p*-Methoxyphenyl)naphthalene + 7-methoxy-1-phenylnaphthalene (62:38) (6–12 total)	560
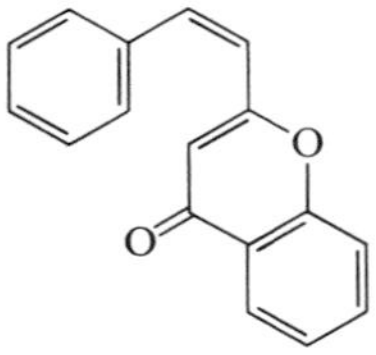	C_6H_6, I_2	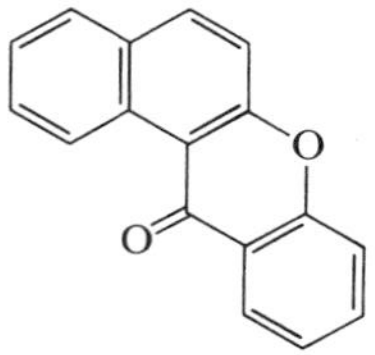 (12)	561b
	C_6H_6, O_2	" (15)	561b
	C_6H_6, air	" (8)	561b

	Reactant	Conditions	Product(s) (Yield %)	Refs.
	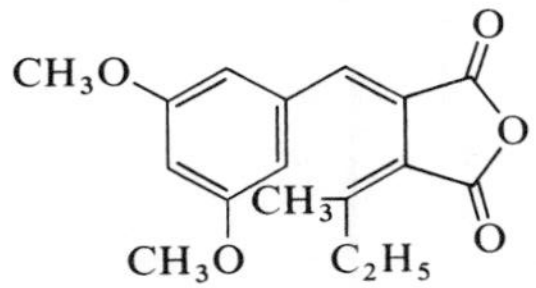 CH_3O, CH_3O, CH_3, C_2H_5 (structure)	C_6H_6, 54°	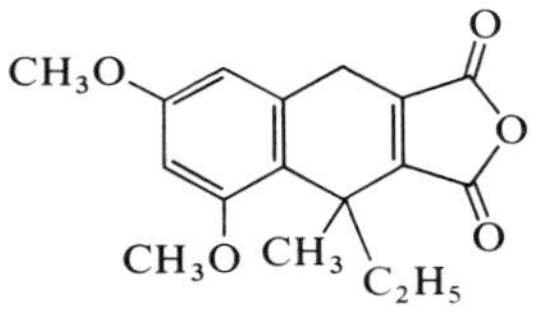 CH_3O, CH_3O, CH_3, C_2H_5 (structure) (—)	71b
		$CDCl_3$, 20°	" (—)	71b
C_{18}	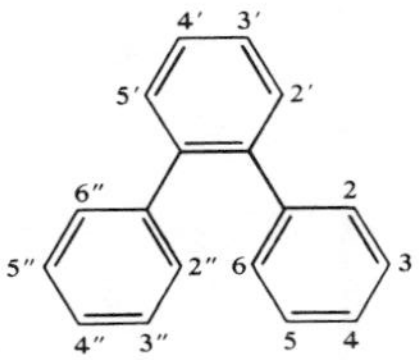 (*o*-Terphenyl)	C_6H_6, I_2, N_2	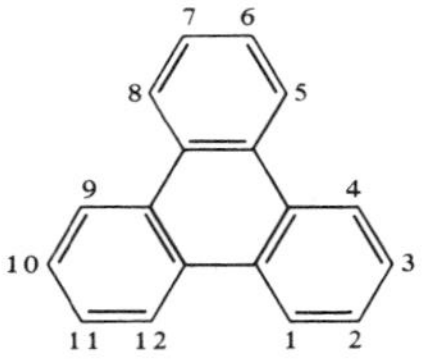 (Triphenylene) (88)	20,233
		C_6H_6, I_2	" (71)	234
		"	" (72)	76
	$C_6H_5(CH{=}CH)_3C_6H_5$	Et_2O, air	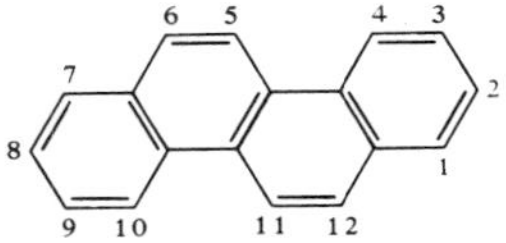 (Chrysene) (15—25)[a]	88
		C_6H_6, I_2, air	" (12)	223,562,563a
	$3,5\text{-}(CH_3)_2C_6H_3CH{=}CHC{\equiv}CC_6H_5$	C_6H_6, N_2	1,3-Dimethyl-8-phenylnaphthalene (44)	226
		Solvent, I_2	" (44)	225
	3′-Bromo-*o*-terphenyl	C_6H_6, I_2, N_2	2-Bromotriphenylene (66),[b] 2-phenyltriphenylene (11), triphenylene (14)	235
	4-Bromo-*o*-terphenyl	"	2-Bromotriphenylene (26),[b] 2-phenyltriphenylene (28), triphenylene (6)	235
	4,4″-Dibromo-*o*-terphenyl	"	3,6-Dibromotriphenylene (6),[b] 3-bromo-6-phenyltriphenylene (16)	235

TABLE III. Carbocyclic Systems from Reactants Other Than 1,2-Diarylethylenes (*Continued*)

Starting Material	Reaction Conditions	Products and Yields (%)	Refs.
4-Chloro-*o*-terphenyl	C_6H_6, I_2, N_2	2-Chlorotriphenylene (45)[b]	235
p-$ClC_6H_4(CH{=}CH)_3C_6H_5$	"	3-Chlorochrysene (13)[b]	214
2,5-Dichloro-*o*-terphenyl	C_6H_{12} or MeOH	2-Chlorotriphenylene (—)	563b
p-$ClC_6H_4(CH{=}CH)_3C_6H_4Cl$-*p*	C_6H_6, I_2, N_2	3,9-Dichlorochrysene (12)[b]	214
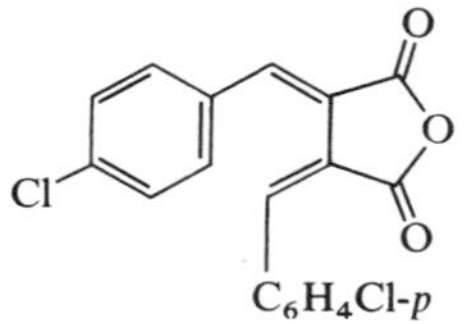 Cl, O, O, O, C_6H_4Cl-*p*	C_6H_6, I_2	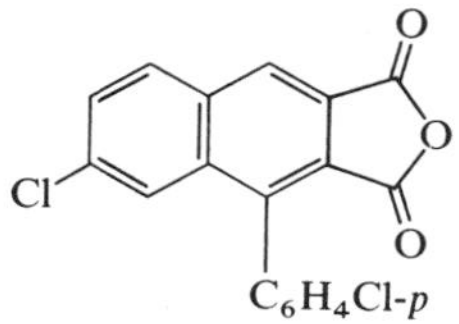 Cl, O, O, O, C_6H_4Cl-*p* (70)	215
4-Fluoro-*o*-terphenyl	C_6H_6, I_2, N_2	2-Fluorotriphenylene (77)[b]	235
p-$FC_6H_4(CH{=}CH)_3C_6H_5$	"	3-Fluorochrysene (14)[b]	214
p-$FC_6H_4(CH{=}CH)_3C_6H_4F$-*p*	"	3,9-Difluorochrysene (11)[b]	214
4-Iodo-*o*-terphenyl	"	2-Phenyltriphenylene (45),[b] triphenylene (36)	235
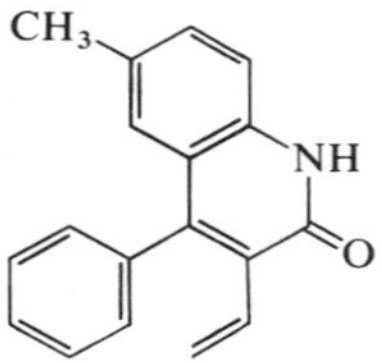 CH_3, NH, O	C_6H_6, air	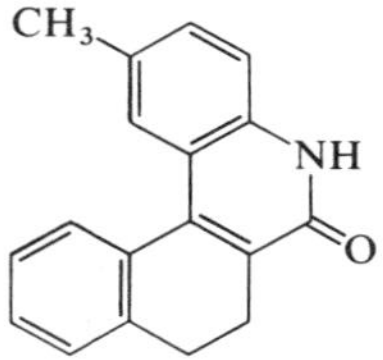 CH_3, NH, O (78)	561a
	C_6H_6, I_2	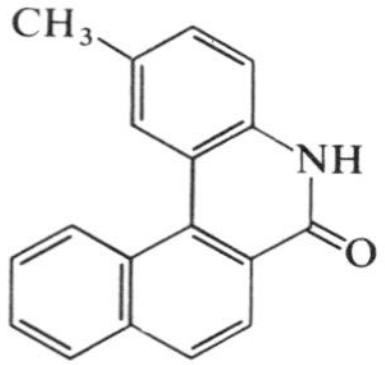 CH_3, NH, O (55)	561a

Reactant	Conditions	Product(s) (% yield)	Ref.
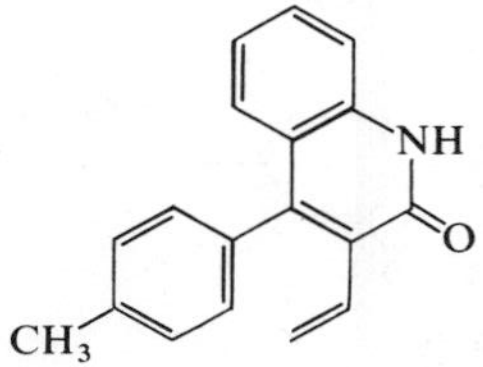 CH$_3$, NH, O	C_6H_6, air	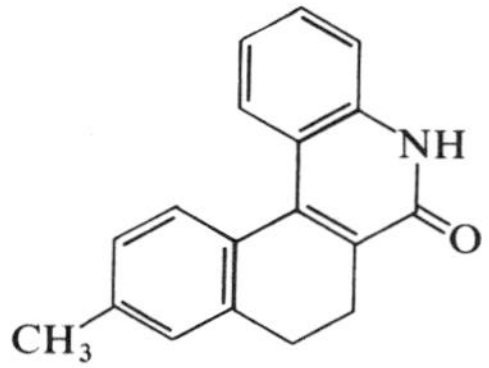 CH$_3$, NH, O (76)	561a
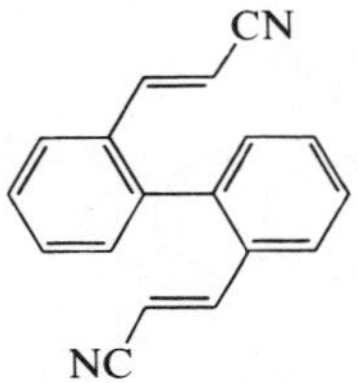 CN, NC	C_6H_6	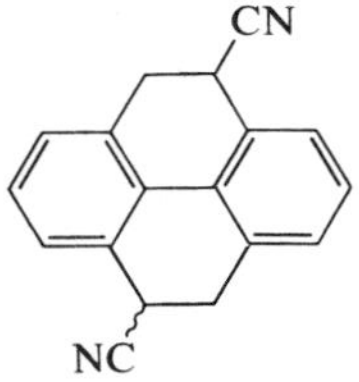 CN, NC (71)	33,558
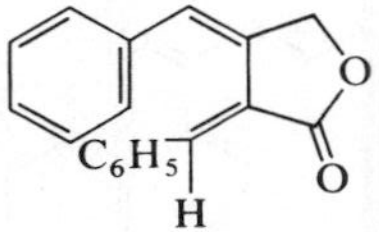 C_6H_5, H, O, O	C_6H_6, degassed	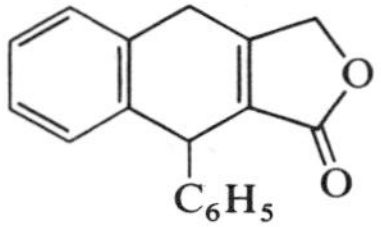 C_6H_5, O, O (100)	564
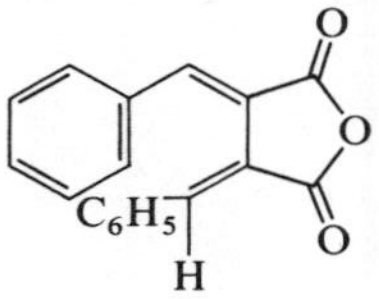 C_6H_5, H, O, O, O	$CDCl_3$, 100°	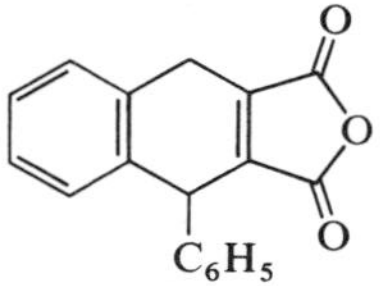 C_6H_5, O, O, O (—)	565
	C_6H_6, 25° or 80°	" (—)	565
	$CDCl_3$, CCl_3CO_2H, 25° or 100°	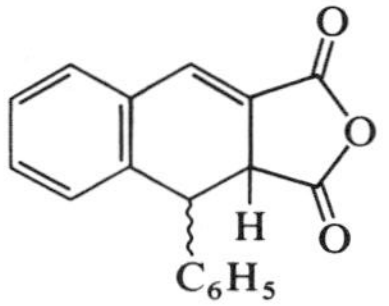 " (—), C_6H_5, H, O, O, O (—)	565

TABLE III. CARBOCYCLIC SYSTEMS FROM REACTANTS OTHER THAN 1,2-DIARYLETHYLENES (*Continued*)

Starting Material	Reaction Conditions	Products and Yields (%)	Refs.
	PhMe, Et_3N, 70° or 25°[c]	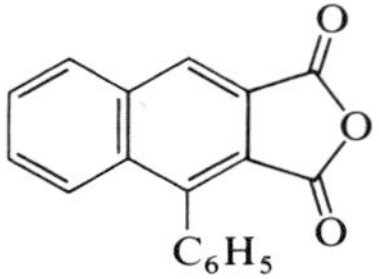 C_6H_5 (—)	565
	C_6H_6 or $CHCl_3$, I_2	" (—)	566
	$CHCl_3$, CO_2[d]	" (—)	567
	1. MeOH, −77° 2. O_2	" (—)	568
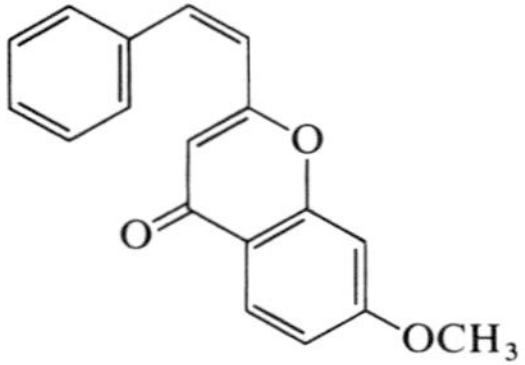 OCH_3	C_6H_6, I_2	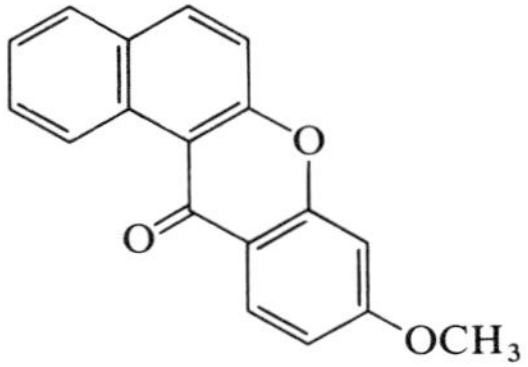 OCH_3 (12)	561b
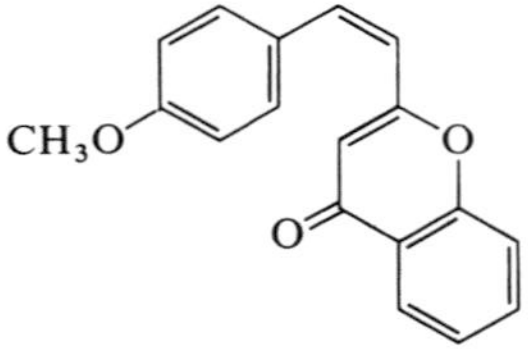 CH_3O	"	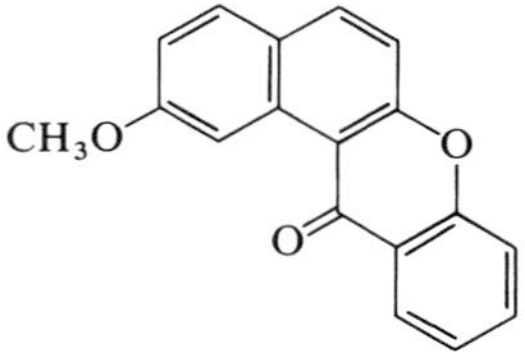 CH_3O (12)	561b
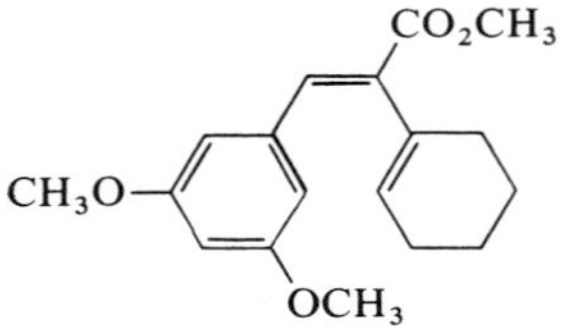 CO_2CH_3, CH_3O, OCH_3	MeOH, I_2	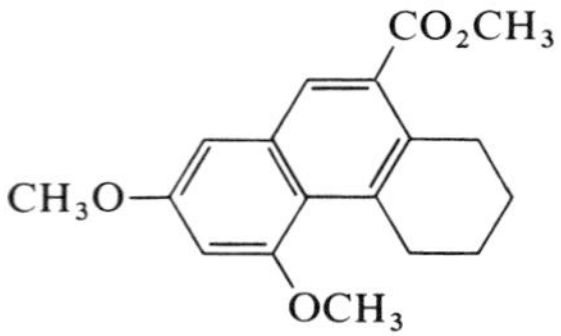 CO_2CH_3, CH_3O, OCH_3 (65)	222

	Reactant	Conditions	Product(s) (Yield, %)	Refs.
		MeOH, N_2	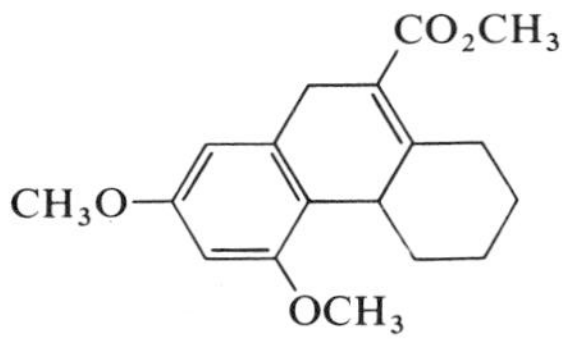 CO_2CH_3, CH_3O, OCH_3 (87)	222
	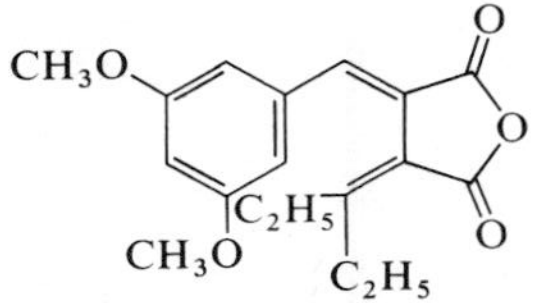 CH_3O, C_2H_5, CH_3O, C_2H_5	PhMe, 12°	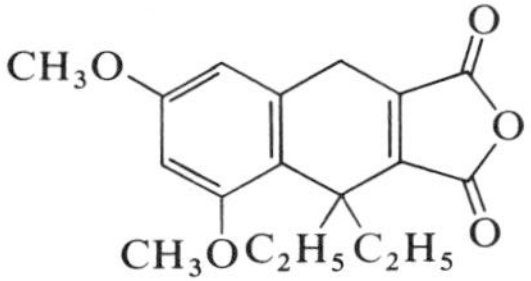 CH_3O, CH_3O C_2H_5 C_2H_5 (—)	71b
C_{19}	*p*-$CH_3C_6H_4(CH{=}CH)_3C_6H_5$	C_6H_6, I_2, N_2	3-Methylchrysene (13)[b]	214
	$C_6H_5CH{=}CHC(CH_3){=}CHCH{=}CHC_6H_5$	"	Chrysene (4)	223
	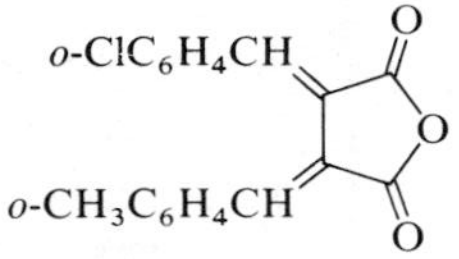 *o*-ClC_6H_4CH, *o*-$CH_3C_6H_4CH$	C_6H_6, I_2	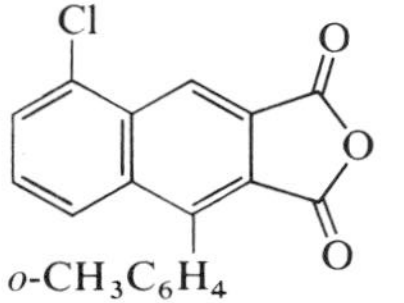 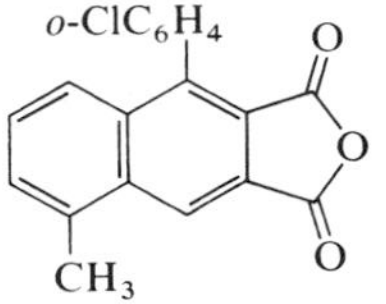 Cl, *o*-$CH_3C_6H_4$ (—), *o*-ClC_6H_4, CH_3 (—)	215
	4-Cyano-*o*-terphenyl	C_6H_6, I_2, N_2	2-Cyanotriphenylene (11)[b]	20
	p-$NCC_6H_4(CH{=}CH)_3C_6H_5$	"	3-Cyanochrysene (12)[b]	214
	4-Methoxy-*o*-terphenyl	"	2-Methoxytriphenylene (65),[b] triphenylene (1)	235
	p-$CH_3OC_6H_4(CH{=}CH)_3C_6H_5$	"	3-Methoxychrysene (13)[b]	214
	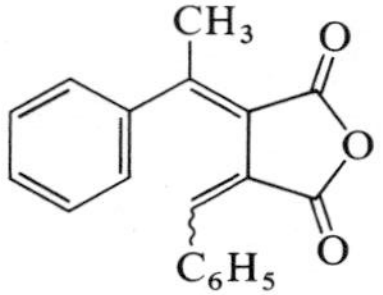 CH_3, C_6H_5		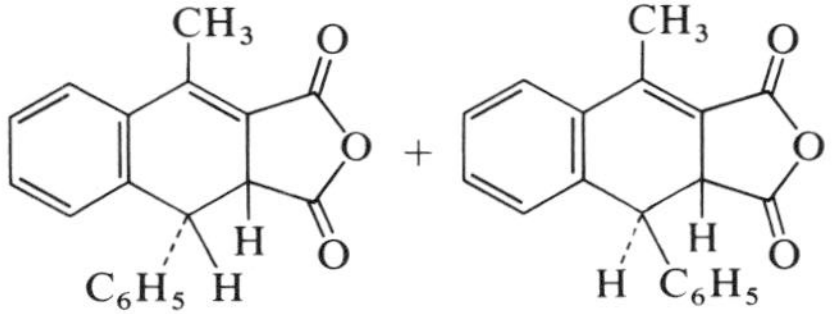 CH_3, C_6H_5 H, H + CH_3, H C_6H_5, H	
		$CHCl_3$, N_2, −8°	" (100), " (0)	569
		$CHCl_3$, N_2, 19°	" (75), " (25)	569
		C_6H_6, N_2, 54° or 80°	" (65), " (35)	569
		PhMe, N_2, 110°	" (35), " (65)	569

TABLE III. CARBOCYCLIC SYSTEMS FROM REACTANTS OTHER THAN 1,2-DIARYLETHYLENES (*Continued*)

Starting Material	Reaction Conditions	Products and Yields (%)	Refs.
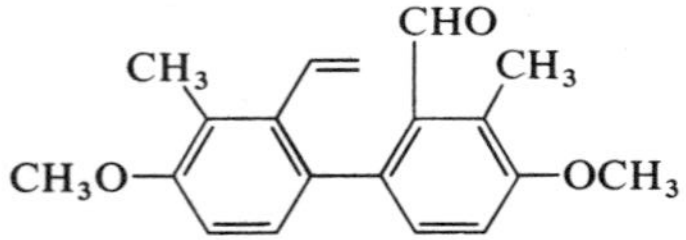 CHO, CH_3, CH_3, CH_3O, OCH_3	Solvent	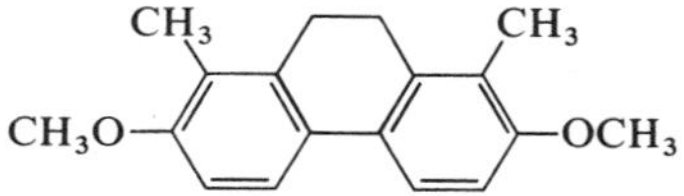 CH_3, CH_3, CH_3O, OCH_3 (major)	116
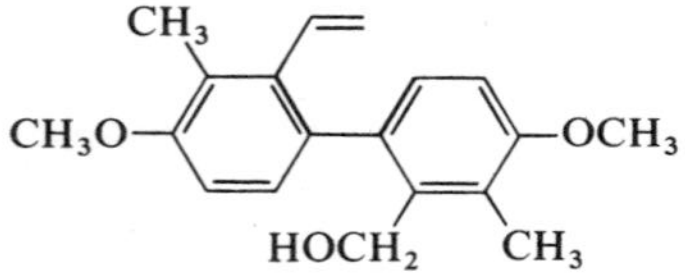 CH_3, CH_3O, OCH_3, $HOCH_2$, CH_3	C_6H_6, N_2	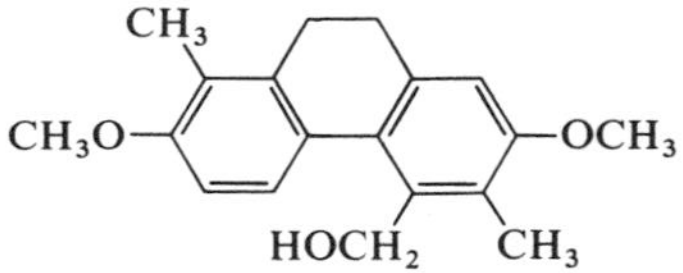 CH_3, CH_3O, OCH_3, $HOCH_2$, CH_3 (65)	116
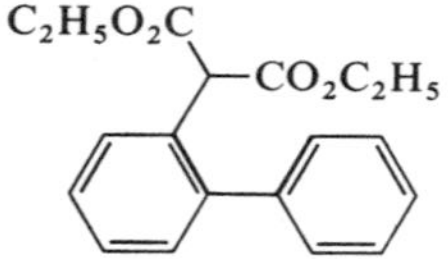 $C_2H_5O_2C$, $CO_2C_2H_5$	EtOH, NaOEt	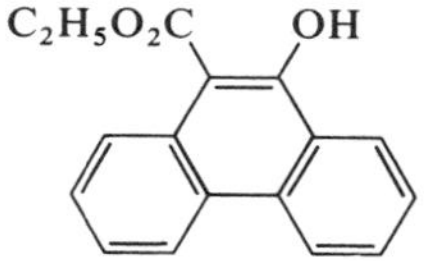 $C_2H_5O_2C$, OH (48)	238
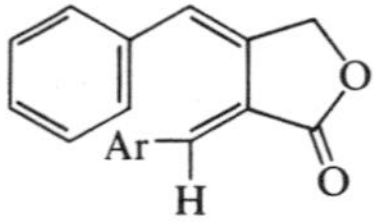 O, O, Ar, H Ar = 3,4-Methylenedioxyphenyl	C_6H_6, degassed	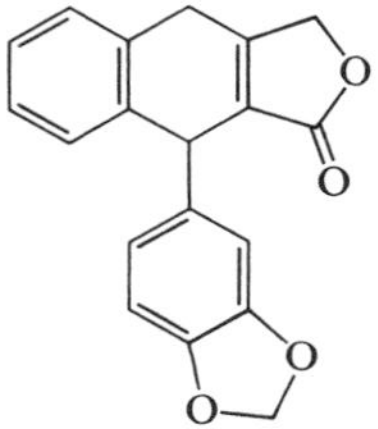 O, O, O, O (—)	564
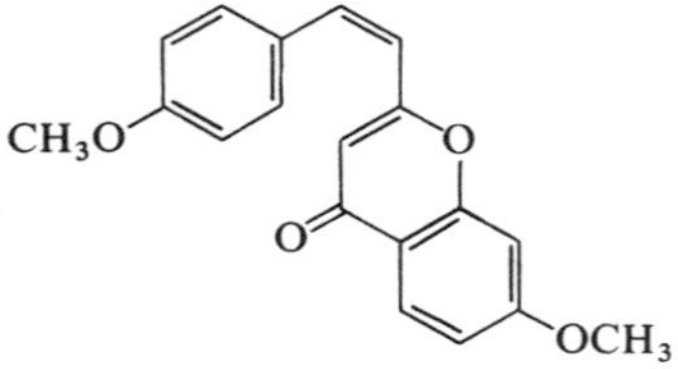 CH_3O, O, O, OCH_3	C_6H_6, I_2	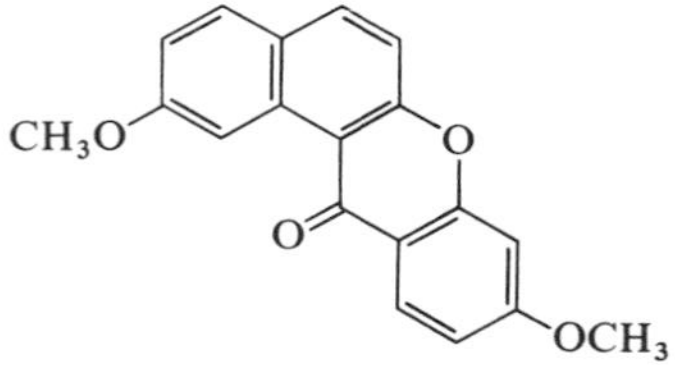 CH_3O, O, O, OCH_3 (12)	561b

	Reactant	Conditions	Product(s) (yield %)	Ref.
C_{20}	4-Phenylphenanthrene	C_6H_6, I_2, N_2	 (Benzo[*e*]pyrene) (46)	236
		Solvent, I_2	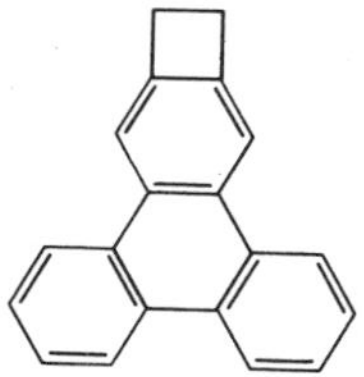 (10,11-Dihydrocyclobuta[*b*]triphenylene) (82)	570
	2-Styrylbiphenyl	C_6H_6, I_2, N_2	9-Phenylphenanthrene (18), 1-phenylphenanthrene (18)	157
		C_6H_6, C_6H_{14}, or alcohols; I_2	9-Phenylphenanthrene (—), 1-phenylphenanthrene (—)	208
		C_6H_6, C_6H_{14}, or alcohols; O_2	9,10-Dihydro-9-phenylphenanthrene + 1-phenylphenanthrene (1:1)	208
		C_6H_6, C_6H_{14}, or alcohols; degassed	9,10-Dihydro-9-phenylphenanthrene (80–90)	208
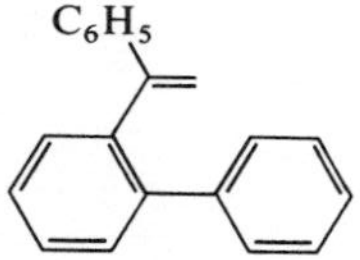	C_6H_5 (1-phenylvinyl)biphenyl	C_6H_{12}, degassed	9,10-Dihydro-9-phenylphenanthrene (98)	36a
		"	" (85)	73
		$CDCl_3$, degassed	" (—)	36a,73
		C_6H_{12}, O_2	" (—)	36a

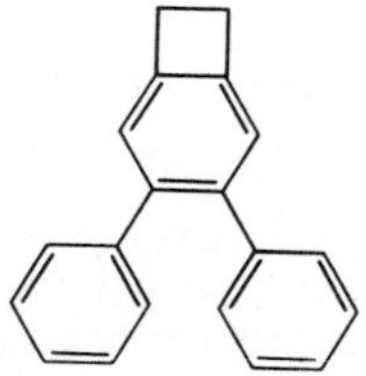

TABLE III. Carbocyclic Systems from Reactants Other Than 1,2-Diarylethylenes (*Continued*)

Starting Material	Reaction Conditions	Products and Yields (%)	Refs.
	Solvent, I_2, O_2	9-Phenylphenanthrene (80)	73
	C_6H_{12}, I_2, O_2	" (—)	36a
$CH{=}CHCH{=}CHC_6H_5$	C_6H_6, I_2, N_2	1-Phenylphenanthrene (9)	157
$CH{=}CHCH{=}CHC_6H_5$	"	4-Phenylphenanthrene (7)	157
$C_6H_5(CH{=}CH)_4C_6H_5$	"	1-Phenylphenanthrene (2)	157
o-$CH_3C_6H_4(CH{=}CH)_3C_6H_4CH_3$-$o$	"	1,7-Dimethylchrysene (7)[b]	223
p-$CH_3C_6H_4(CH{=}CH)_3C_6H_4CH_3$-$p$	"	3,9-Dimethylchrysene (14)[b]	214
$C_6H_5C(CH_3){=}CHCH{=}CHCH{=}C(CH_3)C_6H_5$	C_6H_6, I_2, air	6,12-Dimethylchrysene (20)[b]	223
$CH{=}CHC{\equiv}CC_6H_5$	C_6H_6, N_2	1-Phenylphenanthrene (45)	226
	MeOH, C_6H_{14}, or C_6H_6	" (55)	225
	C_6H_6, I_2	2-Iodo-1-phenylphenanthrene (50)	225
$CH{=}CHC{\equiv}CC_6H_5$, X			

X = H	C_6H_6, N_2	4-Phenylphenanthrene (55)	225,226
	MeOH	" (50)	227
	C_6H_6, I_2	3-Iodo-4-phenylphenanthrene (58)	227
	"	" (44)	225
X = Br	Solvent, I_2	4-Phenylphenanthrene (10)	225
2-(*m*-Chlorostyryl)biphenyl	C_6H_6, C_6H_{14}, or alcohols; degassed	9-(*m*-Chlorophenyl)-9,10-dihydrophenanthrene (80–90)	208
2-(*p*-Chlorostyryl)biphenyl	"	9-(*p*-Chlorophenyl)-9,10-dihydrophenanthrene (80–90)	208
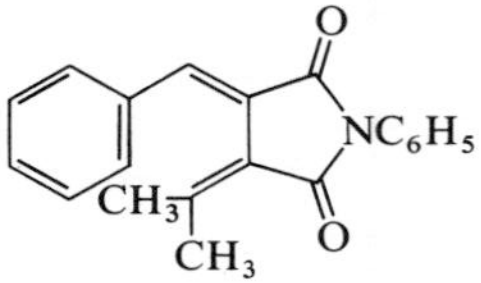	C_6H_6, pet. ether	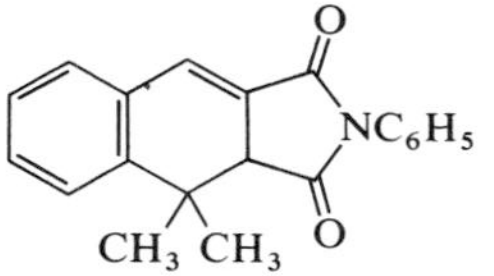 (100)	221
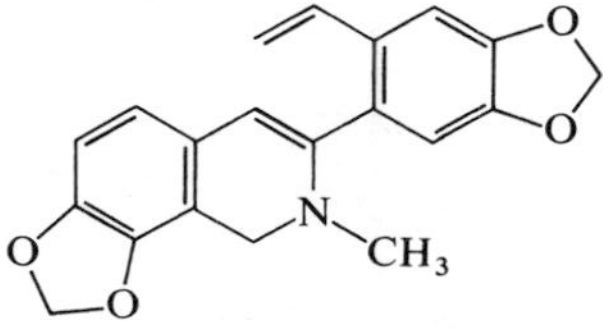	C_6H_6, N_2	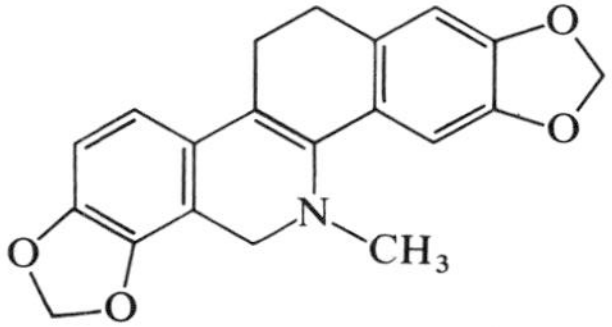 (67)	193
p-$NCC_6H_4(CH{=}CH)_3C_6H_4CN$-*p*	C_6H_6, I_2, N_2	3,9-Dicyanochrysene (60)[b]	214
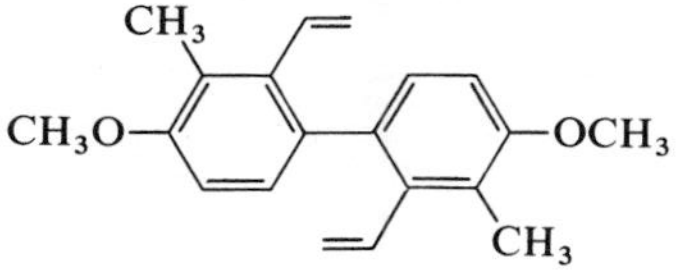	C_6H_6	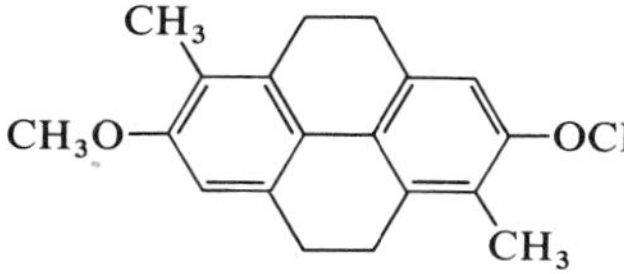 (—)	116
p-$CH_3OC_6H_4(CH{=}CH)_3C_6H_4OCH_3$-*p*	C_6H_6, I_2, N_2	3,9-Dimethoxychrysene (12)[b]	214
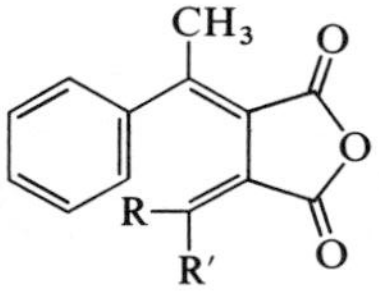		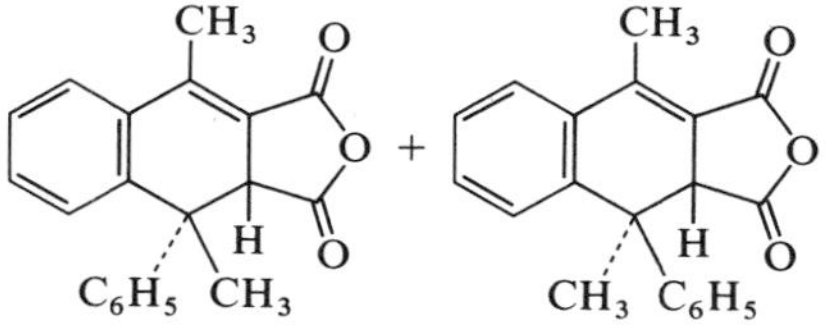	

TABLE III. CARBOCYCLIC SYSTEMS FROM REACTANTS OTHER THAN 1,2-DIARYLETHYLENES (*Continued*)

Starting Material	Reaction Conditions	Products and Yields (%)	Refs.
R = C_6H_5, R′ = CH_3	C_6H_6, N_2, 9°	" (100), " (0)	569
	$CHCl_3$, N_2, 54°	" (100), " (0)	569
	C_6H_6, N_2, 80°	" (80), " (20)	569
R = CH_3, R′ = C_6H_5	C_6H_6, N_2, 9°	" (90), " (10)	569
	C_6H_6, N_2, 54° or 80°	" (55), " (45)	569
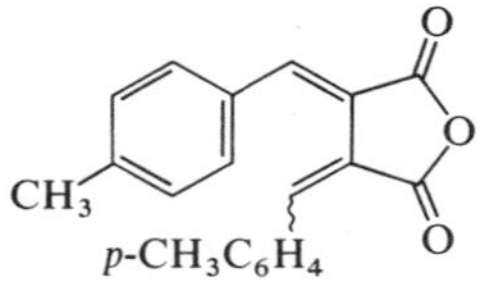 CH_3; p-$CH_3C_6H_4$; O; O; O	C_6H_6, air	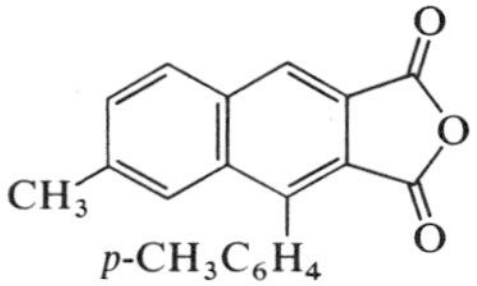 CH_3; p-$CH_3C_6H_4$; O; O; O (—)	571
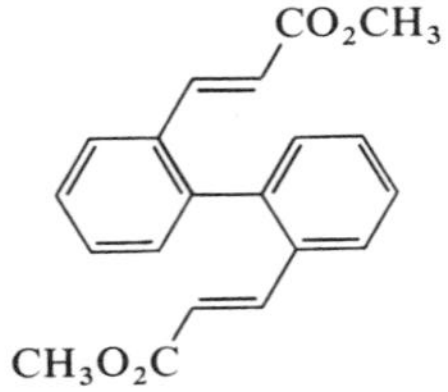 CO_2CH_3; CH_3O_2C	C_6H_6	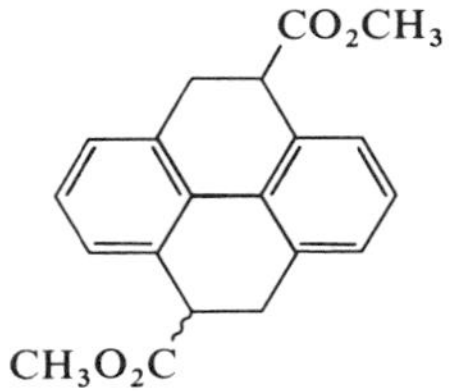 CO_2CH_3; CH_3O_2C (60)	33,558
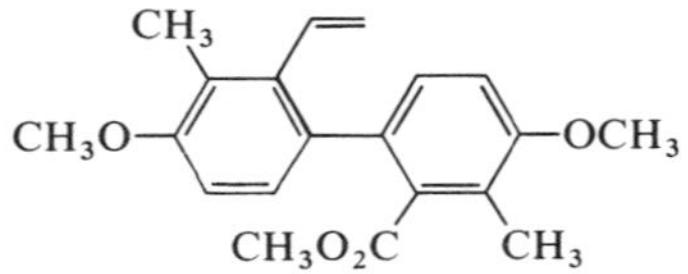 CH_3; CH_3O; OCH_3; CH_3O_2C; CH_3	C_6H_6, degassed	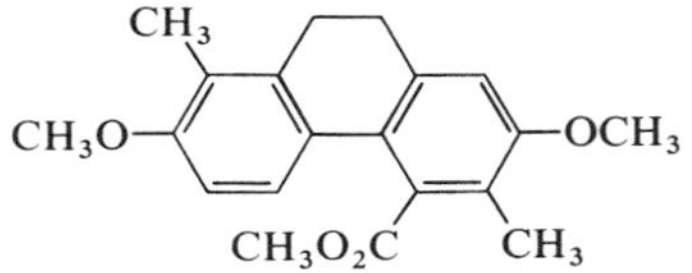 CH_3; CH_3O; OCH_3; CH_3O_2C; CH_3 (85)	206
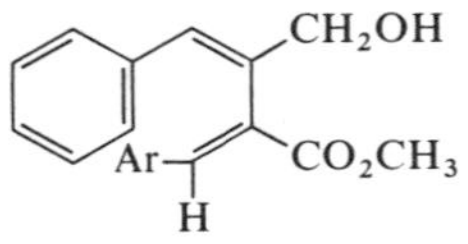 CH_2OH; Ar; H; CO_2CH_3 Ar = 3,4-Methylenedioxyphenyl	"	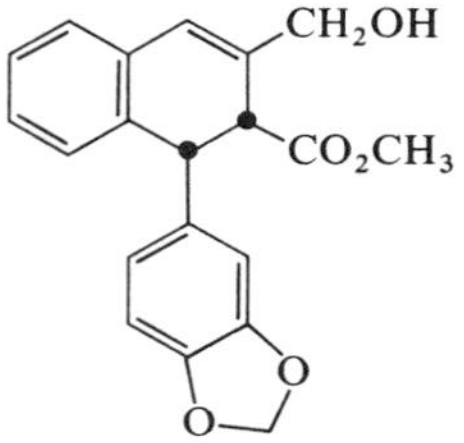 CH_2OH; CO_2CH_3; O; O (—)	564

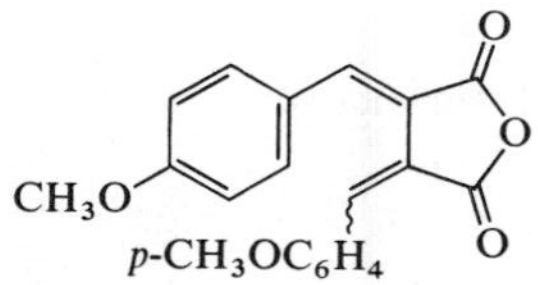 CH_3O, p-$CH_3OC_6H_4$	C_6H_6 or $CHCl_3$; I_2	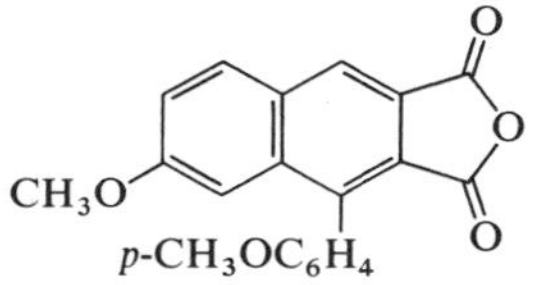 CH_3O, p-$CH_3OC_6H_4$ (100)	571
CH_3O, o-$CH_3OC_6H_4$	"	CH_3O, o-$CH_3OC_6H_4$ (100)	571
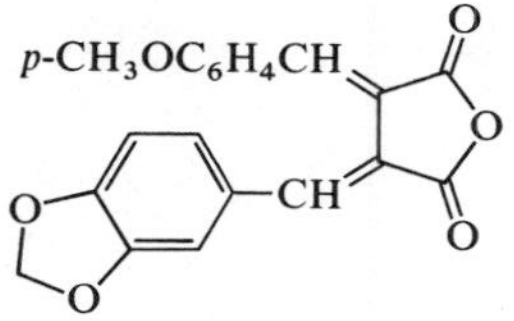 p-$CH_3OC_6H_4CH$, CH	C_6H_6, I_2	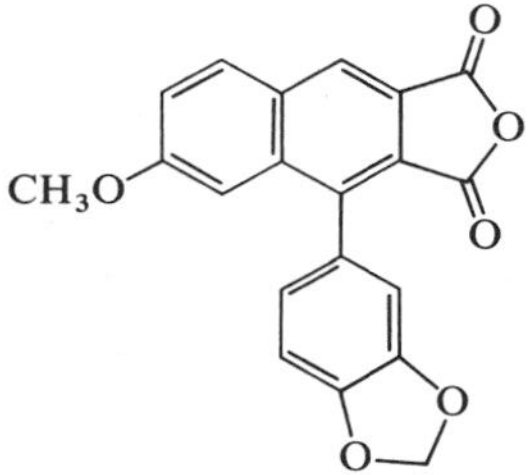 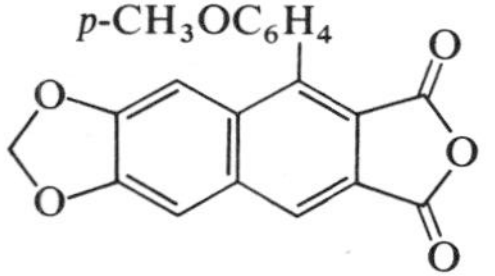 CH_3O (—), p-$CH_3OC_6H_4$ (—)	215
C_{21} $CH{=}CHC{\equiv}CC_6H_5$, CH_3	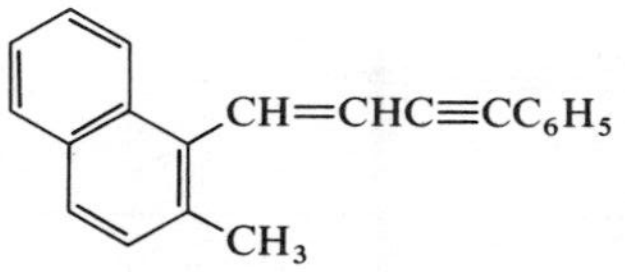 Solvent	1-Phenylphenanthrene (50)	225

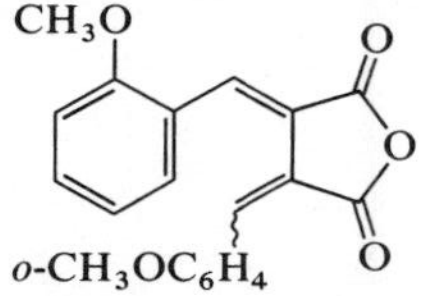

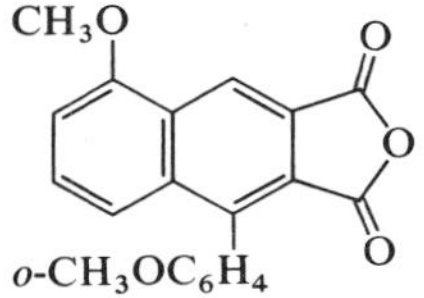

TABLE III. CARBOCYCLIC SYSTEMS FROM REACTANTS OTHER THAN 1,2-DIARYLETHYLENES (*Continued*)

Starting Material	Reaction Conditions	Products and Yields (%)	Refs.
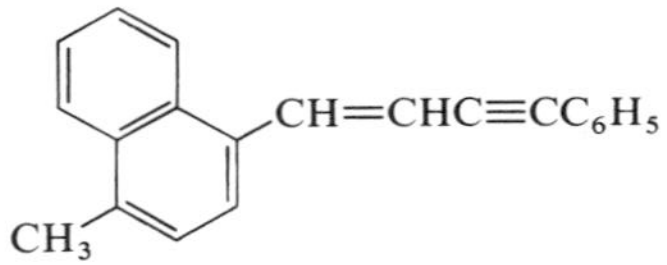 CH_3, $CH{=}CHC{\equiv}CC_6H_5$	C_6H_6, N_2	9-Methyl-1-phenylphenanthrene (54)	225,226
	C_6H_{14}	" (54)	406
	Solvent, I_2	2-Iodo-9-methyl-1-phenylphenanthrene (48)	225
	C_6H_6, I_2, N_2	" (19)	226
	C_6H_6, I_2	9-Methyl-1,2-diphenylphenanthrene (30)	406
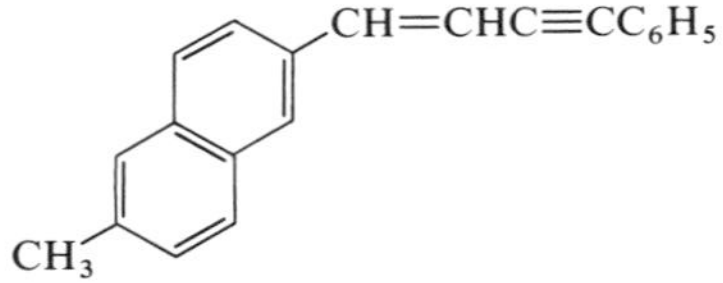 CH_3, $CH{=}CHC{\equiv}CC_6H_5$	C_6H_{14}	2-Methyl-5-phenylphenanthrene (50)	439
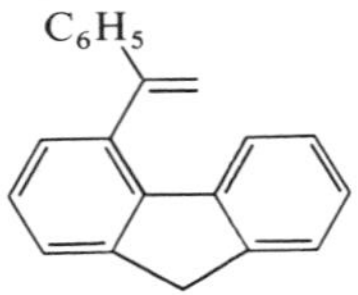 C_6H_5	C_6H_{12}, degassed	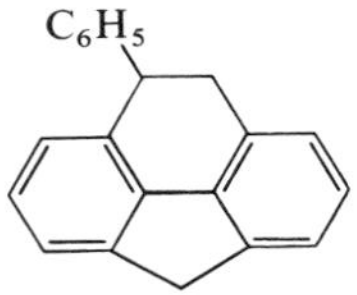 C_6H_5 (98)	36a,73
	C_6H_{12}, I_2 (or O_2)	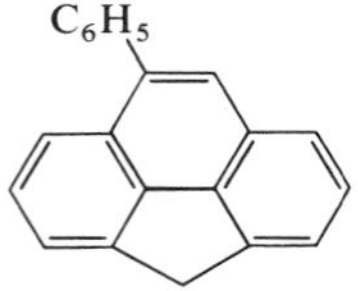 C_6H_5 (—)	36a
2-(*m*-Methylstyryl)biphenyl	C_6H_6, C_6H_{14}, or alcohols; degassed	9-(*m*-Methylphenyl)-9,10-dihydrophenanthrene (80–90)	208

Reactant	Conditions	Product(s) (Yield %)	Ref.
2-(*p*-Methylstyryl)biphenyl	"	9-(*p*-Methylphenyl)-9,10-dihydrophenanthrene (80–90)	208
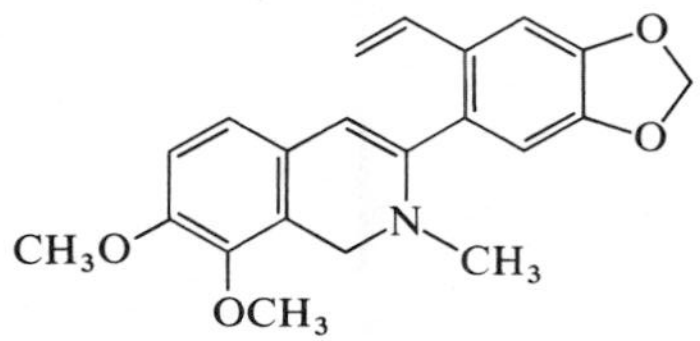	C_6H_6, N_2	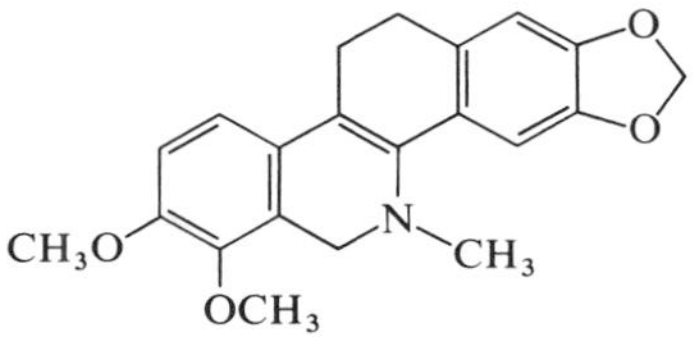 (70)	193
	C_6H_6, N_2, $CH_3O_2CC{\equiv}CCO_2CH_3$	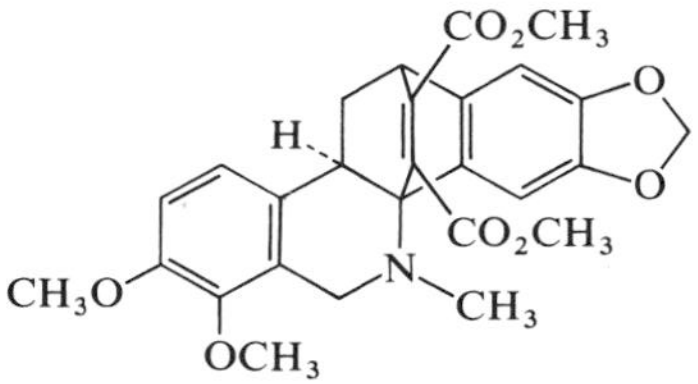 (45)	193
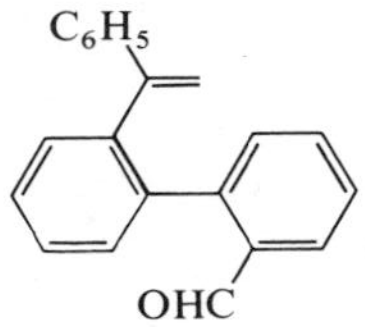	C_6H_6	9,10-Dihydro-9-phenylphenanthrene (58), 9-phenylphenanthrene (12), 4-formyl-9,10-dihydro-9-phenylphenanthrene (10)	33
	C_6H_6, $\lambda > 365$ nm	9-Formyl-9,10-dihydro-9-phenylphenanthrene (50), 4-formyl-9,10-dihydro-9-phenylphenanthrene (13)	33
2′-Benzoyl-2-vinylbiphenyl	C_6H_6	4-Benzoyl-9,10-dihydrophenanthrene (75)	33
2-(*m*-Methoxystyryl)biphenyl	C_6H_6, C_6H_{14}, or alcohols; degassed	9-(*m*-Methoxyphenyl)-9,10-dihydrophenanthrene (80–90)	208
2-(*p*-Methoxystyryl)biphenyl	"	9-(*p*-Methoxyphenyl)-9,10-dihydrophenanthrene (80–90)	208
4-Methoxy-2-styrylbiphenyl	"	9,10-Dihydro-2-methoxy-9-phenylphenanthrene (80–90)	208
5-Methoxy-2-styrylbiphenyl	"	9,10-Dihydro-3-methoxy-9-phenylphenanthrene (80–90)	208

TABLE III. Carbocyclic Systems from Reactants Other Than 1,2-Diarylethylenes (*Continued*)

Starting Material	Reaction Conditions	Products and Yields (%)	Refs.
3′-Methoxy-2-styrylbiphenyl	C_6H_6, C_6H_{14}, or alcohols; degassed	9,10-Dihydro-6-methoxy-9-phenylphenanthrene + 9,10-dihydro-8-methoxy-9-phenylphenanthrene (2:1) (80–90 total)	208
4′-Methoxy-2-styrylbiphenyl	"	9,10-Dihydro-7-methoxy-9-phenylphenanthrene (80–90)	208
4-Carbethoxy-*o*-terphenyl	C_6H_6, I_2, N_2	2-Carbethoxytriphenylene (—)[b]	20
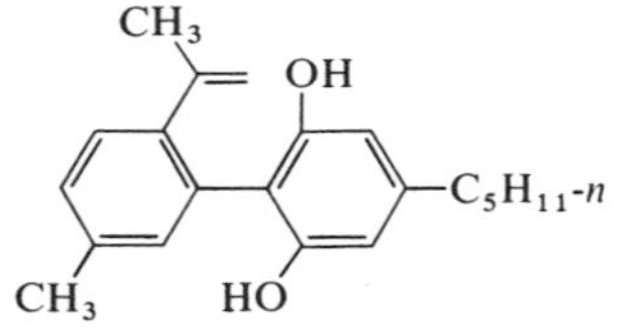 CH_3, OH, C_5H_{11}-*n*, CH_3, HO	EtOH	4-Hydroxy-6,9-dimethyl-2-(*n*-pentyl)phenanthrene (—)	207
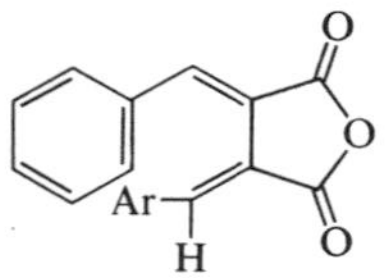 O, O, O, Ar, H; Ar = 2,4,6-Trimethylphenyl	PhMe, Et_3N[c]	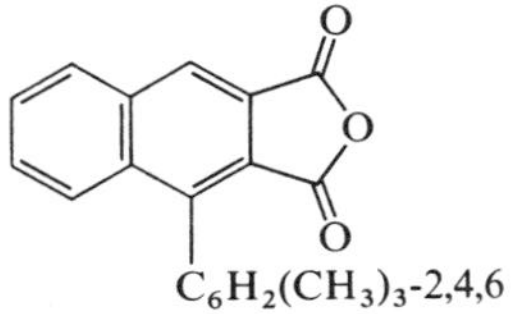 O, O, O, $C_6H_2(CH_3)_3$-2,4,6 (—)	565
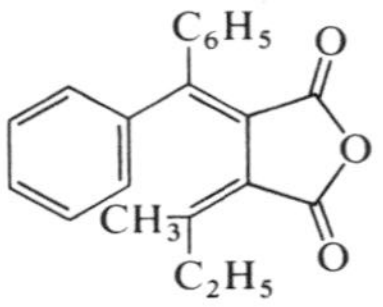 C_6H_5, O, O, O, CH_3, C_2H_5	PhMe, 60°[c]	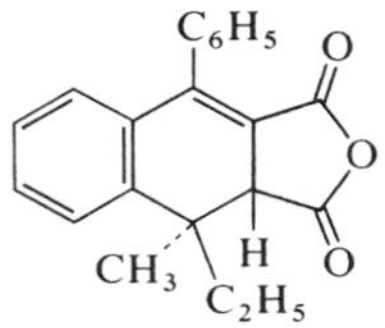 C_6H_5, O, O, O, CH_3, H, C_2H_5 (—)	71b
	PhMe, 18°	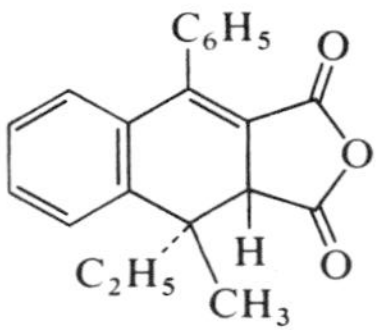 " + C_6H_5, O, O, O, C_2H_5, H, CH_3 (85:15)	71b

C_{22} 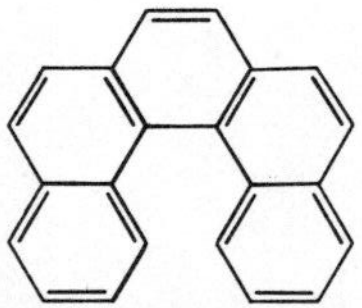 (Dibenzo[*c,g*]phenanthrene)	C_6H_{12}, I_2	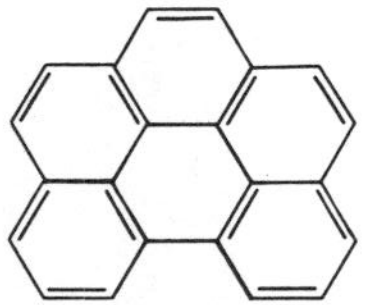 (59) (Benzo[*ghi*]perylene)	76,89
1,2-Diphenylnaphthalene	C_6H_6, I_2	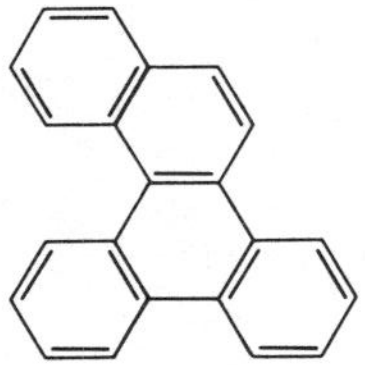 (—) (Benzo[*g*]chrysene)	76
	C_6H_{14}, I_2	" (—)	572
2-(1-Naphthyl)biphenyl	C_6H_{12}, I_2	Benzo[*g*]chrysene (11)	380,381
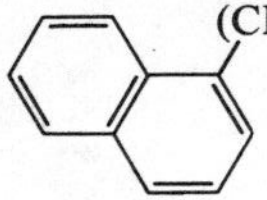 $(CH{=}CH)_3C_6H_5$	C_6H_6, I_2, N_2	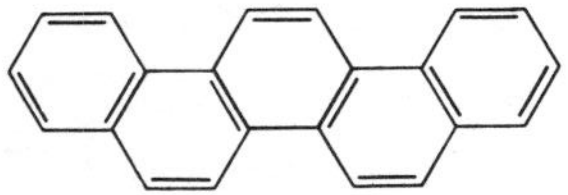 (2) (Picene)	157
$C_6H_5(CH{=}CH)_5C_6H_5$	C_6H_6, I_2	Picene (2)	157
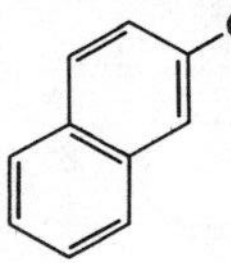 $CH{=}CHC{\equiv}CC_6H_3(CH_3)_2$-3,5	Solvent	4-(3,5-Dimethylphenyl)phenanthrene (65)	225
	Solvent, I_2	4-(3,5-Dimethyl)-3-iodophenanthrene (45)	225

TABLE III. CARBOCYCLIC SYSTEMS FROM REACTANTS OTHER THAN 1,2-DIARYLETHYLENES (*Continued*)

Starting Material	Reaction Conditions	Products and Yields (%)	Refs.
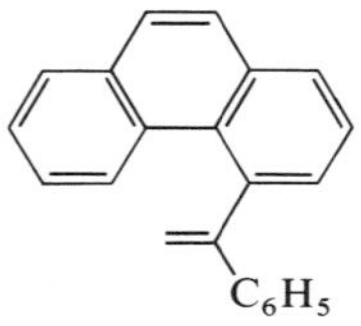 C_6H_5	C_6H_{12}, degassed	4,5-Dihydro-4-phenylpyrene (100)	36a
	C_6H_{12}, O_2	4,5-Dihydro-4-phenylpyrene + 4-phenylpyrene (41:59)	36a
	C_6H_{12}, I_2, O_2	4-Phenylpyrene (—)	36a
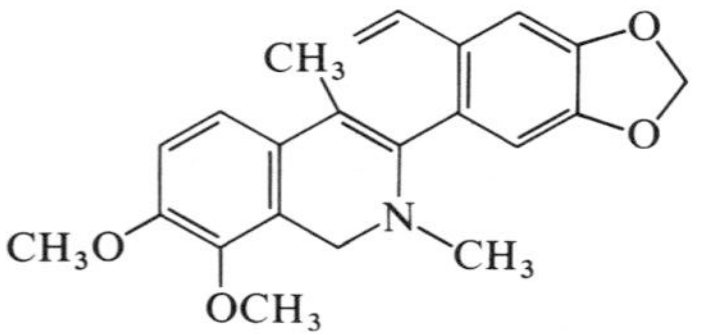 CH_3, CH_3O, OCH_3, N, CH_3, O, O	C_6H_6, N_2[e]	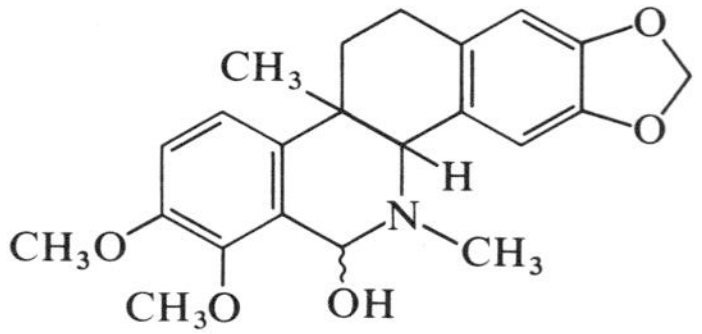 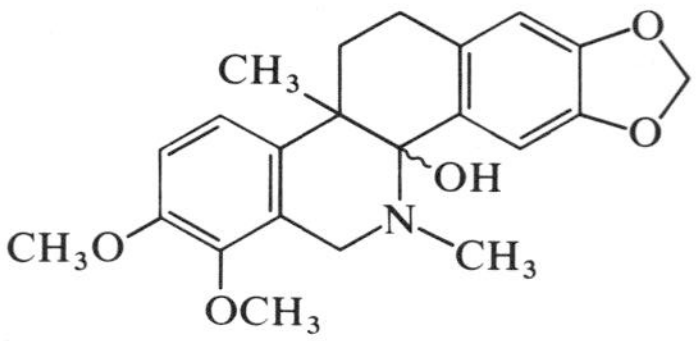 CH_3, H, N, CH_3, CH_3O, CH_3O, OH, O, O (19),	573
		CH_3, OH, N, CH_3, CH_3O, OCH_3, O, O (10)	
	C_6H_6, N_2, $CH_3O_2CC{\equiv}CCO_2CH_3$	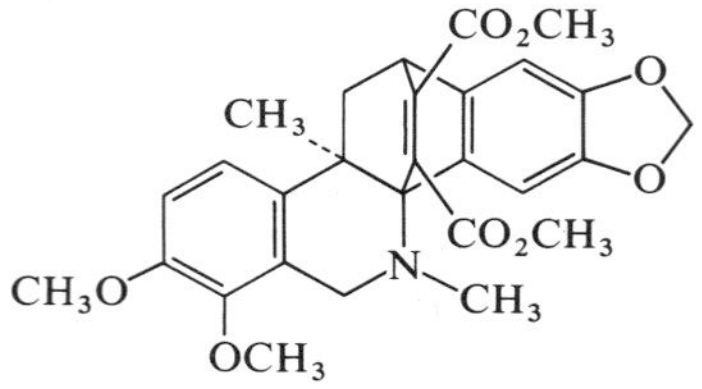 CO_2CH_3, CH_3, CO_2CH_3, N, CH_3, CH_3O, OCH_3, O, O (5)	573

Reactant	Conditions	Product(s) (Yield, %)	Refs.
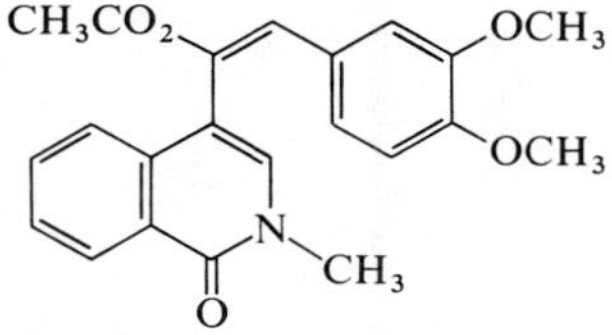 CH_3CO_2, OCH_3, OCH_3, N, CH_3, O	C_6H_6, I_2	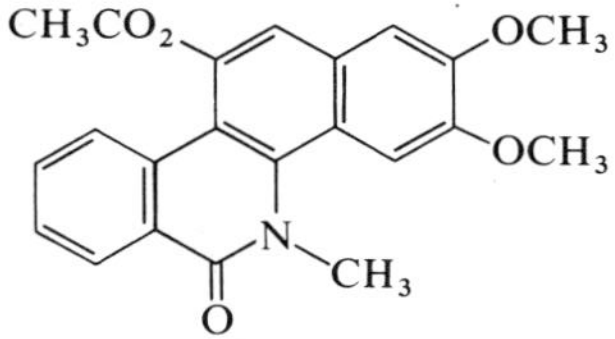 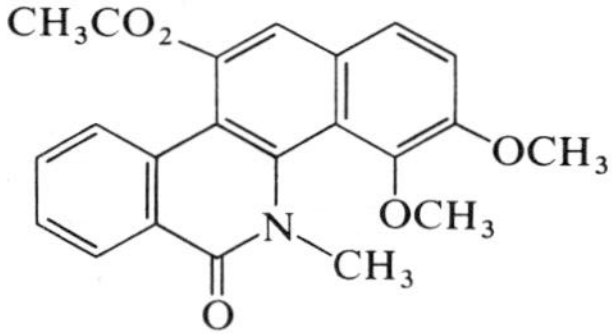 CH_3CO_2, OCH_3, OCH_3, N, CH_3, O (17–27), CH_3CO_2, OCH_3, OCH_3, N, CH_3, O (14–18)	574,575
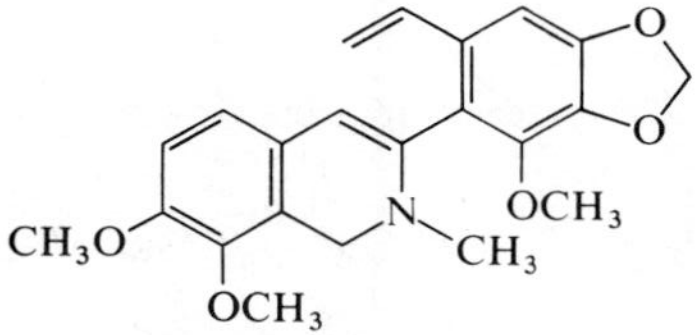 O, O, OCH_3, N, CH_3, CH_3O, OCH_3	C_6H_6, N_2	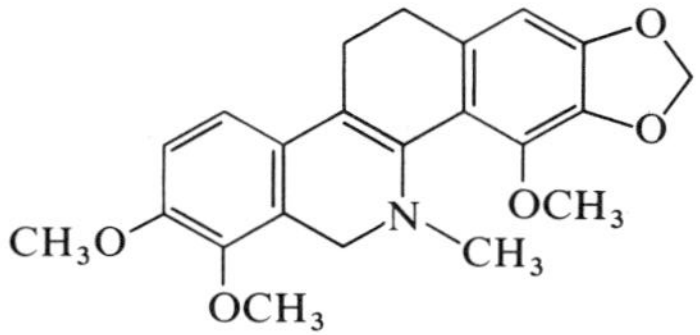 O, O, OCH_3, N, CH_3, CH_3O, OCH_3 (58)	576
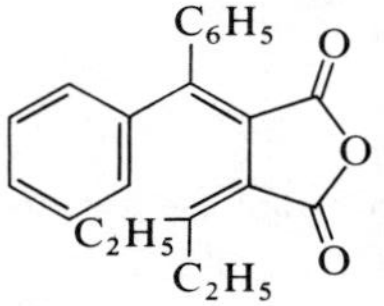 C_6H_5, O, O, O, C_2H_5, C_2H_5	$CDCl_3$, 20°	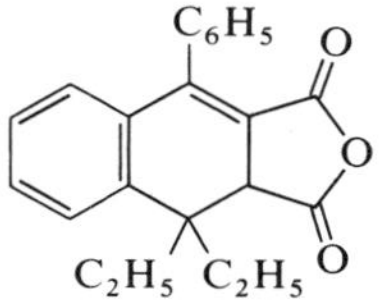 C_6H_5, O, O, O, C_2H_5, C_2H_5 (—)	71b
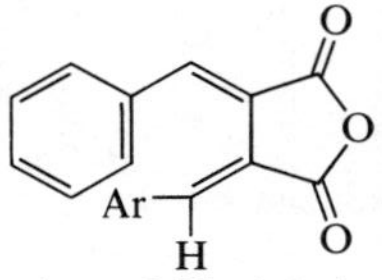 O, O, O, Ar, H; Ar = 2-Naphthyl	C_6H_6, degassed	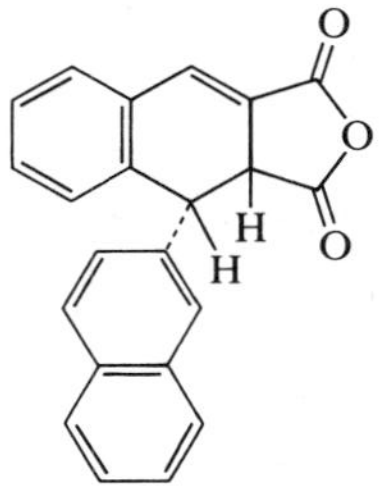 O, O, O, H, H (100)	577

TABLE III. CARBOCYCLIC SYSTEMS FROM REACTANTS OTHER THAN 1, 2-DIARYLETHYLENES (*Continued*)

	Starting Material	Reaction Conditions	Products and Yields (%)	Refs.
	3′,4′-Dicarbomethoxy-*o*-terphenyl	C_6H_6, I_2	2,3-Dicarbomethoxytriphenylene (27)[b]	483
	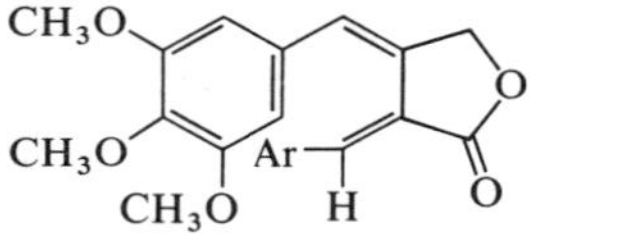 Ar = 3,4-Methylenedioxyphenyl	C_6H_6, degassed	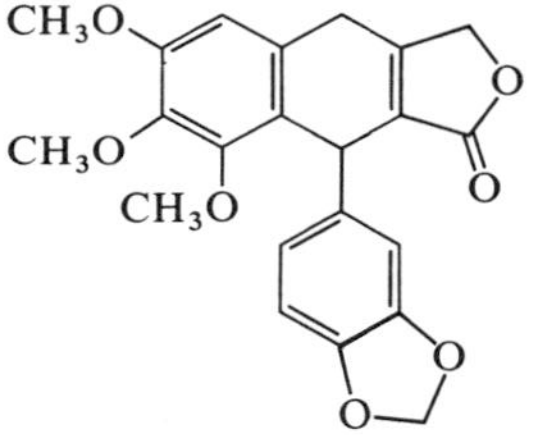 (—)	564
	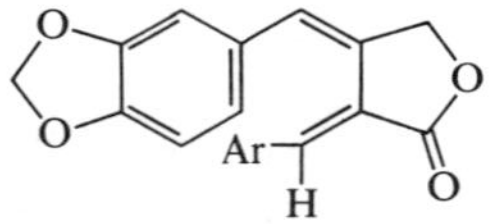 Ar = 3,4,5-Trimethoxyphenyl	"	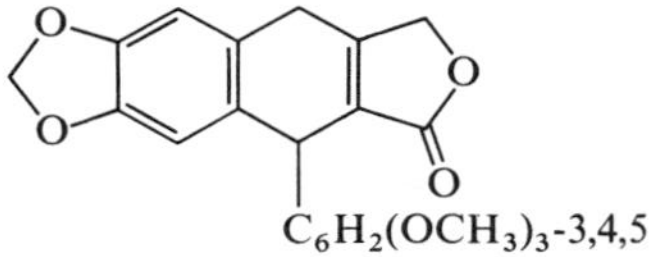 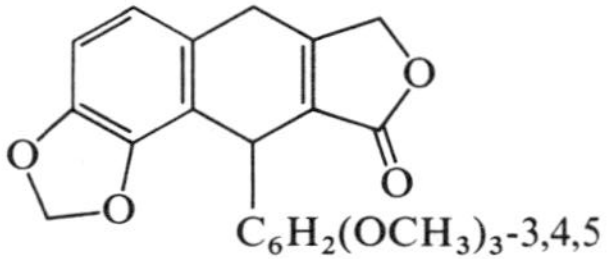 $C_6H_2(OCH_3)_3$-3,4,5 (—), $C_6H_2(OCH_3)_3$-3,4,5 (—)	564
C_{23}	9-Phenyl-10-propenylphenanthrene	C_6H_{14}, N_2	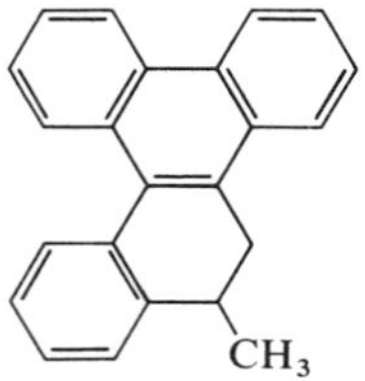 (100) (9,10-Dihydro-10-methylbenzo[*g*]chrysene)	231
		Solvent. I_2. N_2	" (—), 10-methylbenzo[*g*]chrysene (—)	231

	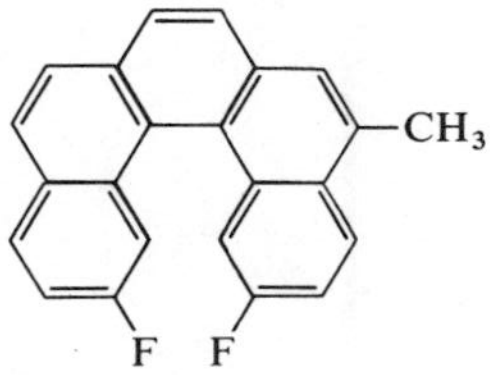 (difluoro-methyl compound; structure)	C_6H_6, I_2, air	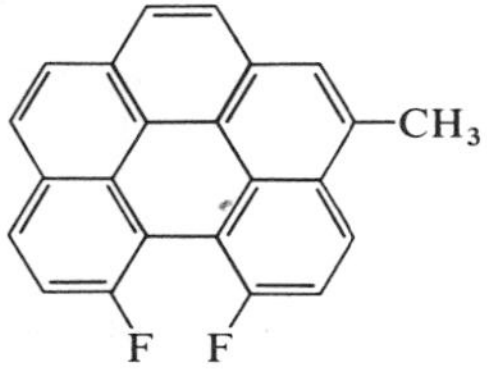 (cyclized difluoro-methyl product; structure) (—)	171
	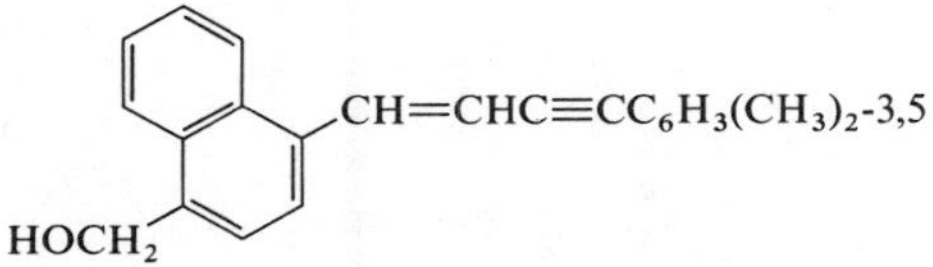 (1-naphthyl compound with 4-$HOCH_2$ group; structure) $CH{=}CHC{\equiv}CC_6H_3(CH_3)_2$-3,5	Solvent	1-(3,5-Dimethylphenyl)-9-hydroxymethylphenanthrene (62)	225
C_{24}	1,2,3-Triphenylbenzene	C_6H_6, I_2, N_2	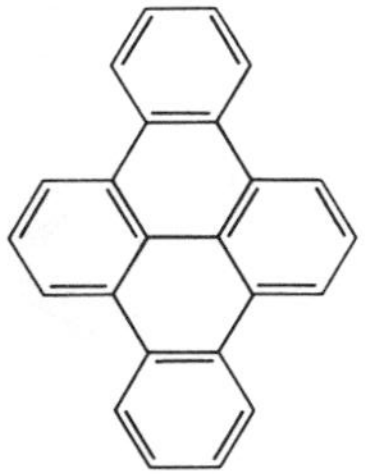 (21) (Dibenzo[*fg,op*]naphthacene)	20
	1,2,4-Triphenylbenzene[f]	"	2-Phenyltriphenylene (—)[b]	20
	2,2′-Diphenylbiphenyl	"	Dibenzo[*fg,op*]naphthacene (57)[g]	20
	2,4′-Diphenylbiphenyl	C_6H_6, I_2	2-Phenyltriphenylene (—)[b]	234
	4-(2-Naphthyl)phenanthrene	"	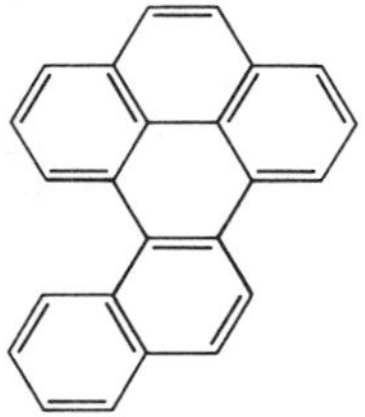 (35) (Naphtho[8,1,2-*ghi*]chrysene)	237

TABLE III. Carbocyclic Systems from Reactants Other Than 1, 2-Diarylethylenes (*Continued*)

Starting Material	Reaction Conditions	Products and Yields (%)	Refs.
1,4-Bis-(1-naphthyl)-1,3-butadiene	C_6H_6, I_2, N_2	1-(1-Naphthyl)phenanthrene (10)	237
1-(1-Naphthyl)-4-(2-naphthyl)-1,3-butadiene	"	4-(1-Naphthyl)phenanthrene (11)	237
1,4-Bis-(2-naphthyl)-1,3-butadiene	"	4-(2-Naphthyl)phenanthrene (2), naphtho[8,1,2-*ghi*]chrysene (9)[g]	237
1-(9-Phenanthryl)-4-phenyl-1,3-butadiene	"	1-Phenyltriphenylene (14)[b]	157
1-(9-Anthryl)-4-phenylbutadiene	"	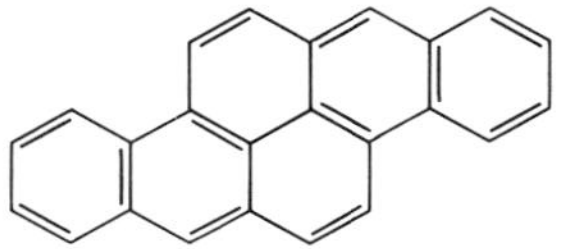 (7) (Dibenzo[*b,def*]chrysene)	236
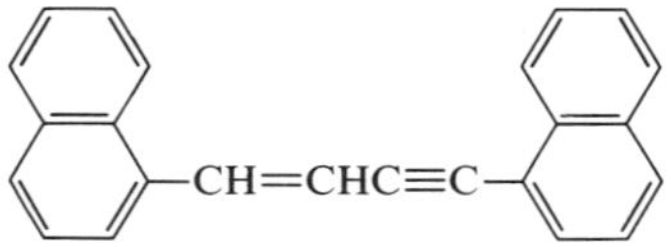 —CH=CHC≡C—	C_6H_6, N_2	1-(1-Naphthyl)phenanthrene (22)	226
	Solvent, I_2	" (22)	225
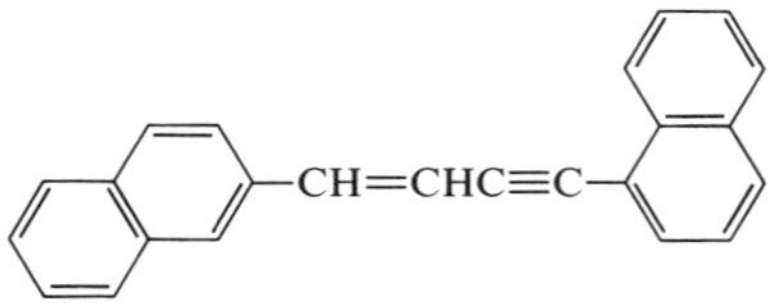 —CH=CHC≡C—	Solvent	4-(1-Naphthyl)phenanthrene (58)	225
	C_6H_6, N_2	" (38)	226
	C_6H_6, I_2, N_2	3-Iodo-4-(1-naphthyl)phenanthrene (38)	226

Reactant	Conditions	Product(s) (Yield, %)	Refs.
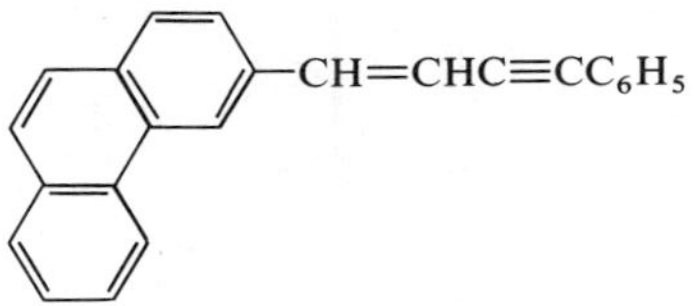 $-CH{=}CHC{\equiv}CC_6H_5$	Solvent, I_2	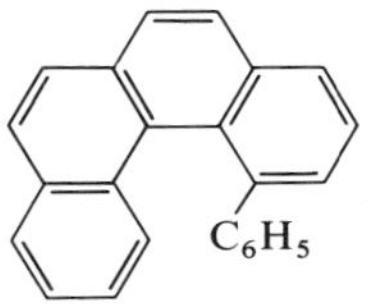 C_6H_5 (60)	225
		(1-Phenylbenzo[*c*]phenanthrene)	
	C_6H_6, N_2	" (50)	226
	C_6H_6, I_2	2-Iodo-1-phenylbenzo[*c*]phenanthrene (32)	225,226
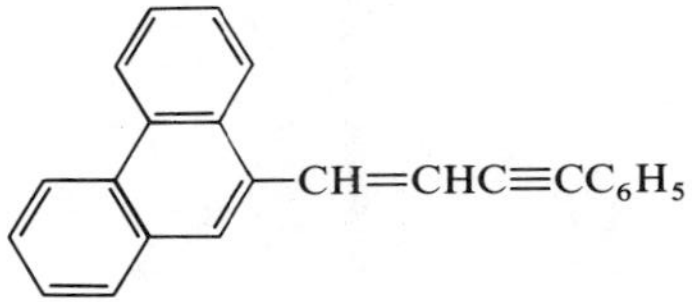 $-CH{=}CHC{\equiv}CC_6H_5$	MeOH, O_2 (or N_2)	1-Phenyltriphenylene (60–65)[b]	229,230
	C_6H_6, N_2	" (58–65)	229,230
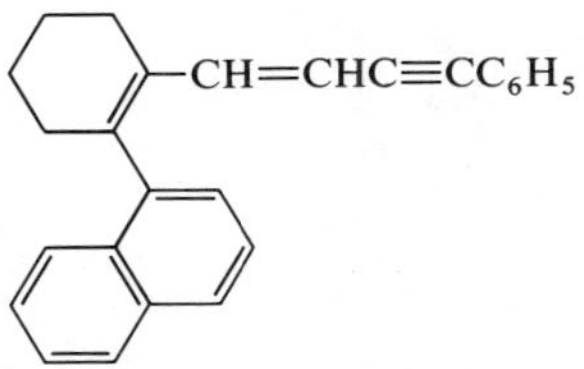 $-CH{=}CHC{\equiv}CC_6H_5$	C_6H_{14}	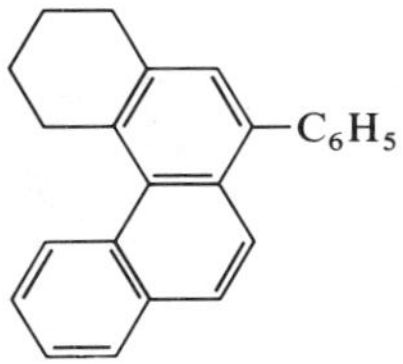 $-C_6H_5$ (45)	225
2-(*o*-Phenylstyryl)naphthalene	C_6H_6, C_6H_{14}, or alcohols; degassed	9,10-Dihydro-9-(2-naphthyl)phenanthrene (80–90)	208
2,2′-Bis-(*o*-chlorophenyl)biphenyl	C_6H_6	Dibenzo[*fg,op*]naphthacene (67)[g]	20
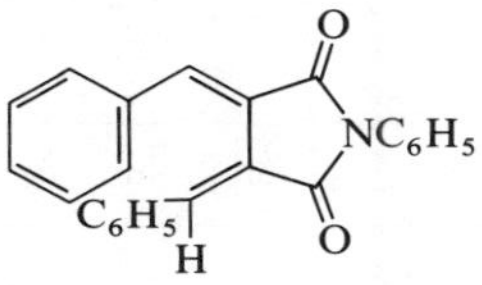 O, NC_6H_5, O, C_6H_5, H	C_6H_6, degassed, 18°	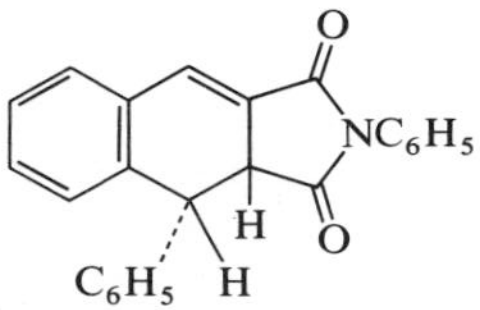 O, NC_6H_5 (70), H, O, C_6H_5, H	577

TABLE III. CARBOCYCLIC SYSTEMS FROM REACTANTS OTHER THAN 1,2-DIARYLETHYLENES (*Continued*)

Starting Material	Reaction Conditions	Products and Yields (%)	Refs.
	C_6H_6, degassed, 80°	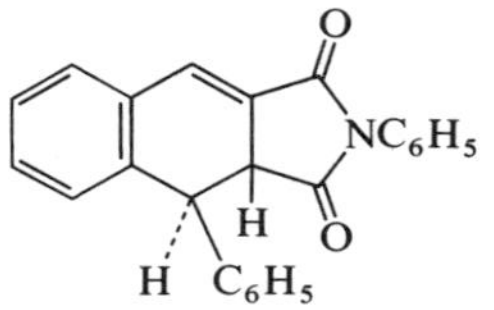 O, O, NC_6H_5, H, H, C_6H_5 (80)	577
	$CDCl_3$, 100°	" (—)	565
	C_6H_{14}, air	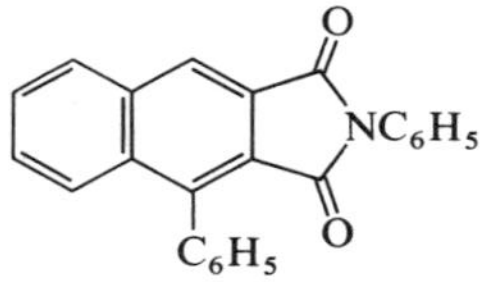 O, O, NC_6H_5, C_6H_5 (—)	217
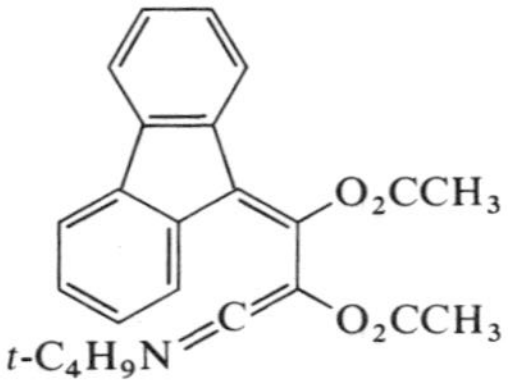 t-C_4H_9N=C=, O_2CCH_3, O_2CCH_3	THF, N_2	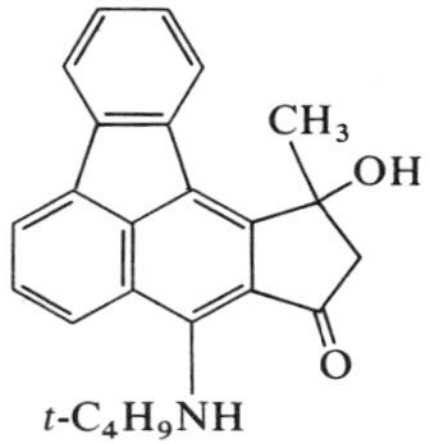 CH_3, OH, O, t-C_4H_9NH (11)	578
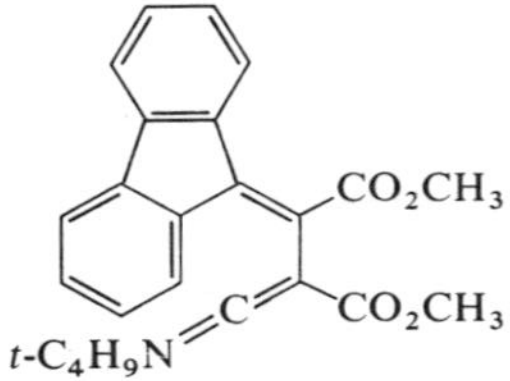 t-C_4H_9N=C=, CO_2CH_3, CO_2CH_3	"	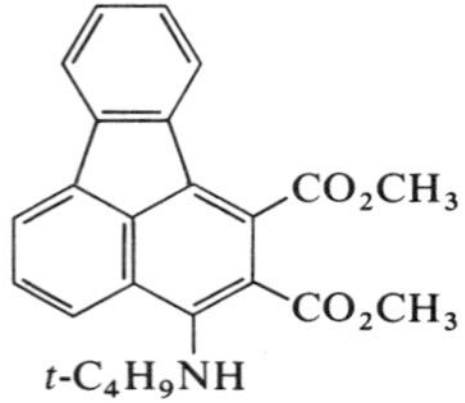 CO_2CH_3, CO_2CH_3, t-C_4H_9NH (32)	578

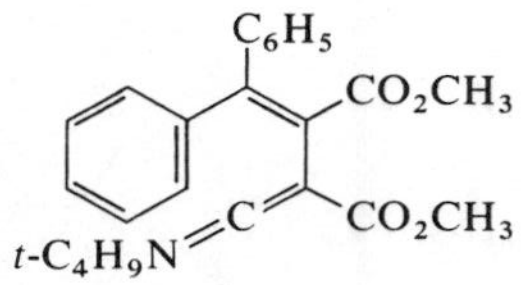

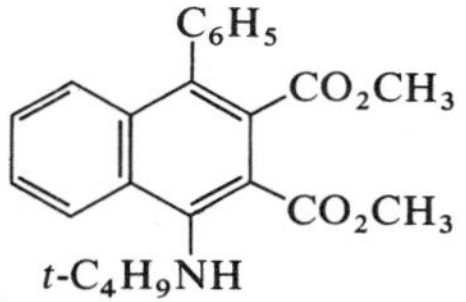

Reactant	Conditions	Product(s) and Yield(s) (%)		Refs.
C_6H_5, CO_2CH_3, CO_2CH_3, t-$C_4H_9N{=}C$	"	C_6H_5, CO_2CH_3, CO_2CH_3, t-C_4H_9NH	(63)	578
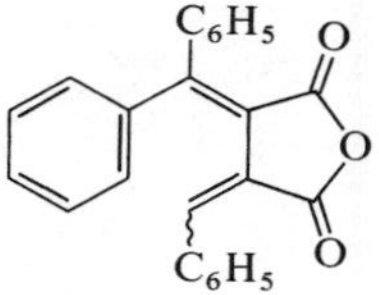 C_6H_5, O, O, O, C_6H_5	C_6H_6 or xylene, N_2	C_6H_5, O, O, O, H, C_6H_5	(—)	220
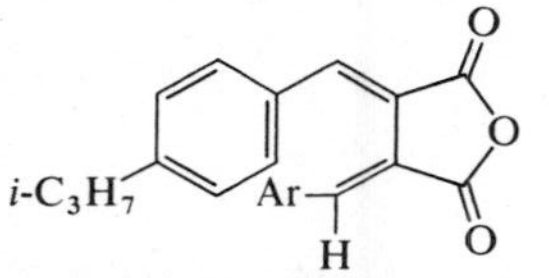 O, O, O, i-C_3H_7, Ar, H Ar = p-Isopropylphenyl	$CH_3C_6H_{11}$, I_2, air	O, O, O, i-C_3H_7, p-i-$C_3H_7C_6H_4$	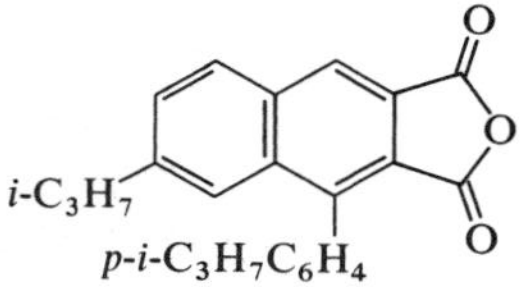 (16)	579
C_{25} 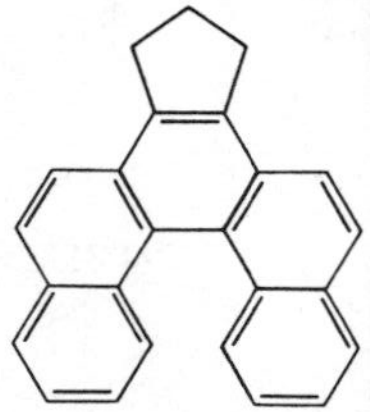	Solvent, I_2		(—)	539
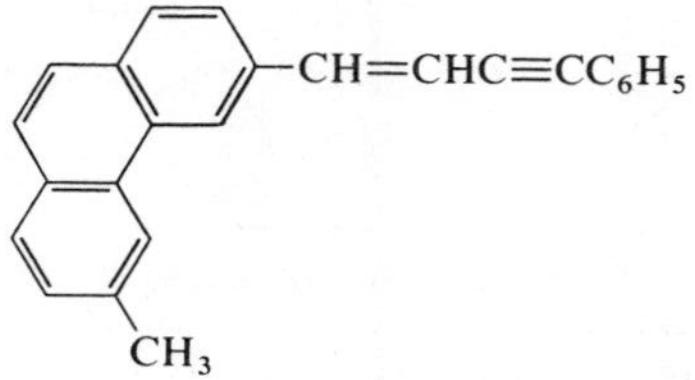 $-CH{=}CHC{\equiv}CC_6H_5$, CH_3	C_6H_6	C_6H_5, CH_3	(—)	439

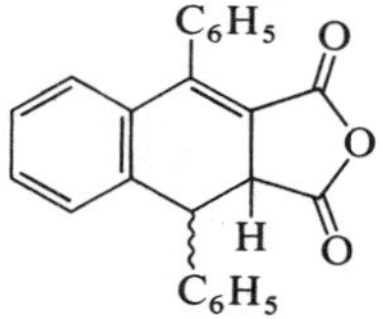

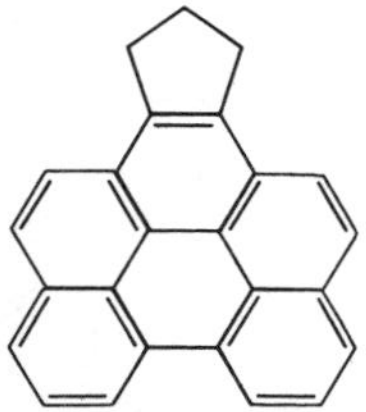

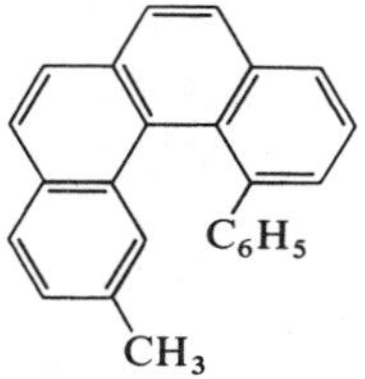

TABLE III. CARBOCYCLIC SYSTEMS FROM REACTANTS OTHER THAN 1,2-DIARYLETHYLENES (*Continued*)

	Starting Material	Reaction Conditions	Products and Yields (%)	Refs.
	O; CO_2CH_3; CO_2CH_3; $t\text{-}C_4H_9N{=}C$	THF, N_2	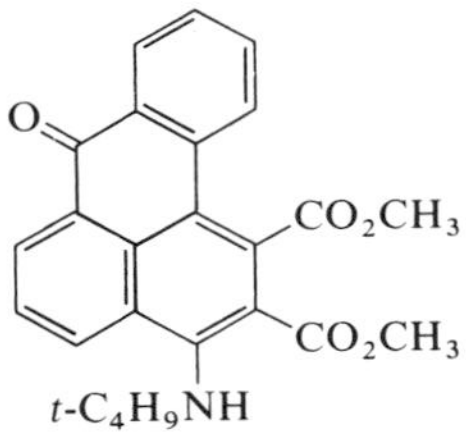 O; CO_2CH_3; CO_2CH_3; $t\text{-}C_4H_9NH$ (20)	578
	C_6H_5; O; O; O; C_6H_5; CH_3	PhMe, N_2, 54°	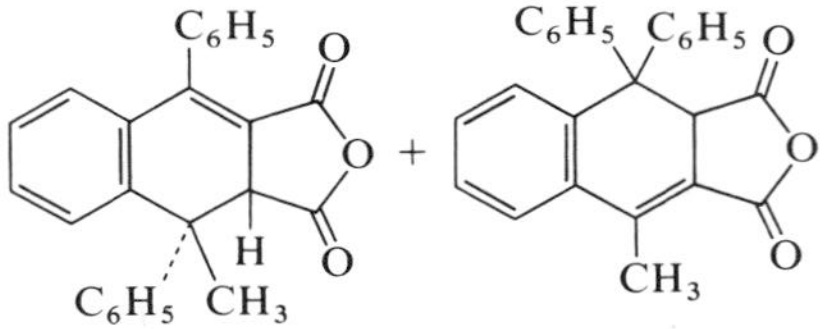 C_6H_5; O; O; O; H; C_6H_5; CH_3 + C_6H_5; C_6H_5; O; O; O; CH_3 (3:1)	569
C_{26}	4,5-Diphenylphenanthrene	C_6H_6, I_2, N_2	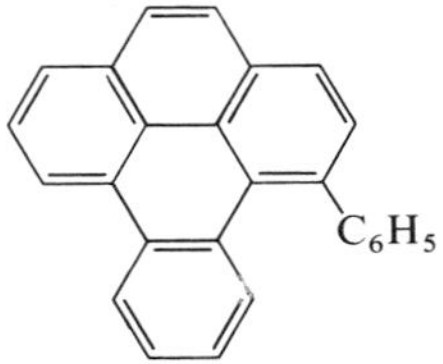 C_6H_5 (40) (1-Phenylbenzo[*e*]pyrene)	231
		C_6H_6, I_2	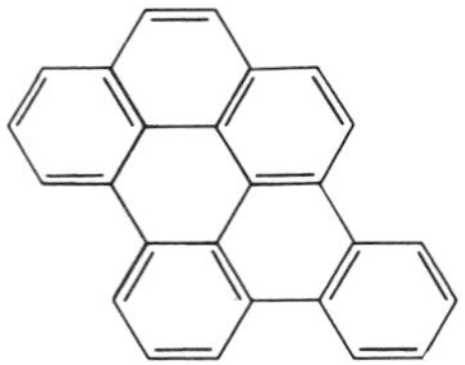 (85) (Dibenzo[*b,pqr*]perylene)	231

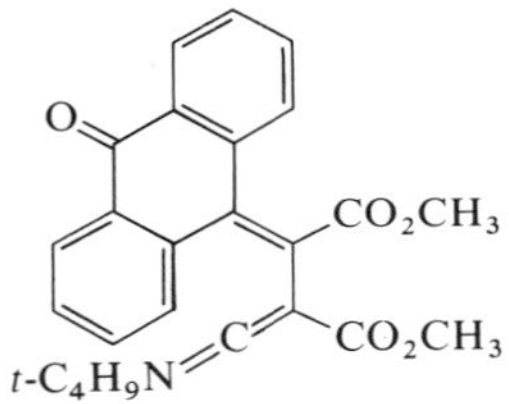

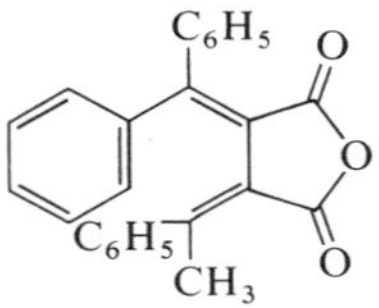

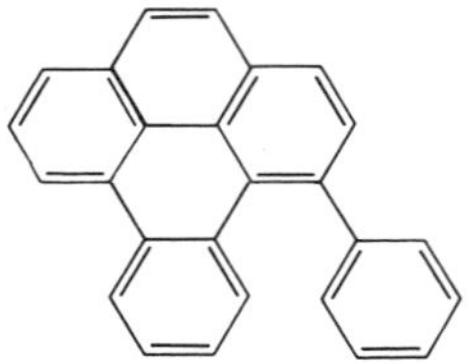

Reactant	Conditions	Product(s) and Yield(s) (%)	Refs.
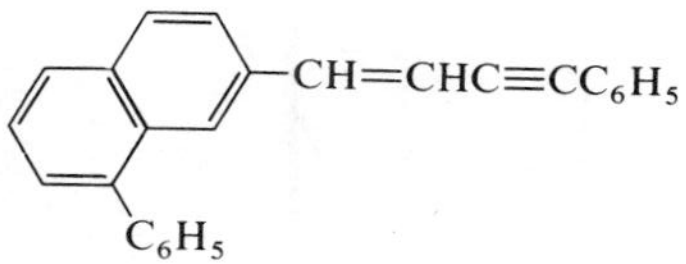 [structure]–$CH{=}CHC{\equiv}CC_6H_5$, C_6H_5	C_6H_{14}, N_2	4,5-Diphenylphenanthrene (65)	231
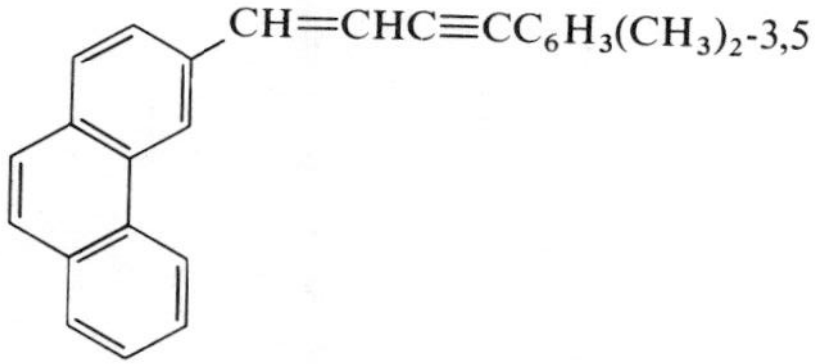 [structure]–$CH{=}CHC{\equiv}CC_6H_3(CH_3)_2$-3,5	Solvent, I_2 C_6H_6, I_2	3-Iodo-4,5-diphenylphenanthrene (26) " (16)	225 226
	Solvent, I_2 (0.05 eq)	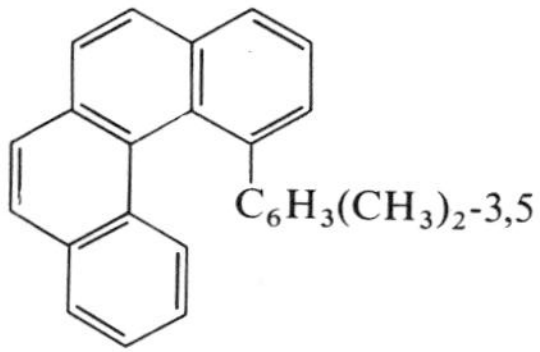 [structure] $C_6H_3(CH_3)_2$-3,5 (45)	225
	Solvent, I_2 (1.0 eq)	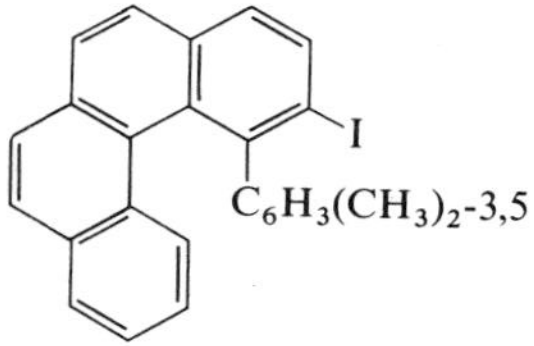 [structure] I, $C_6H_3(CH_3)_2$-3,5 (36)	225
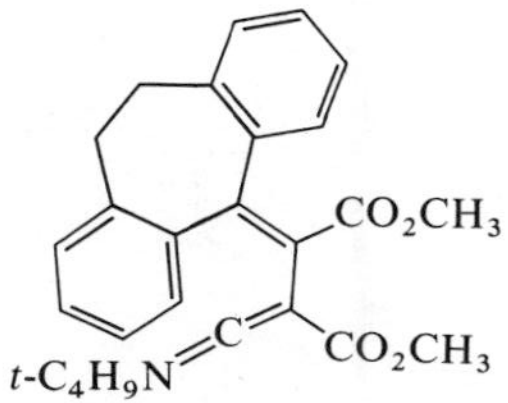 [structure] CO_2CH_3, CO_2CH_3, t-$C_4H_9N{=}C{=}$	THF, N_2	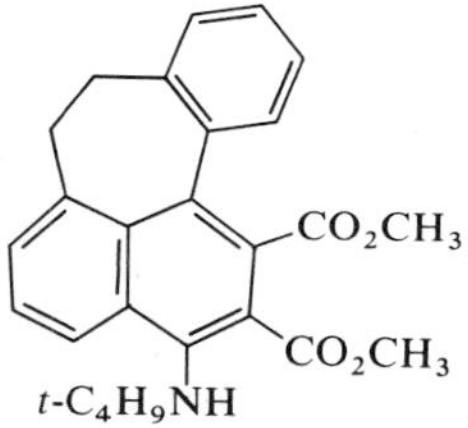 [structure] CO_2CH_3, CO_2CH_3, t-C_4H_9NH (58)	578

TABLE III. CARBOCYCLIC SYSTEMS FROM REACTANTS OTHER THAN 1,2-DIARYLETHYLENES (*Continued*)

	Starting Material	Reaction Conditions	Products and Yields (%)	Refs.
	$CH=CHC\equiv CC_6H_5$; CH_3O_2C	C_6H_6	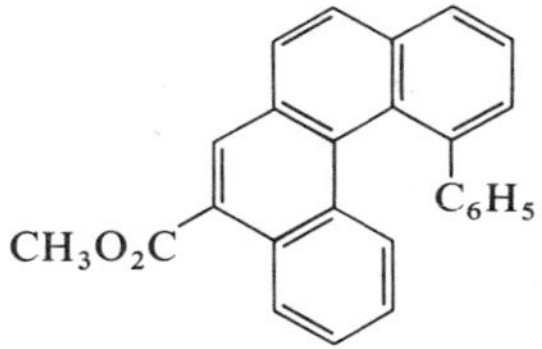 CH_3O_2C; C_6H_5 (—)	439
C_{27} 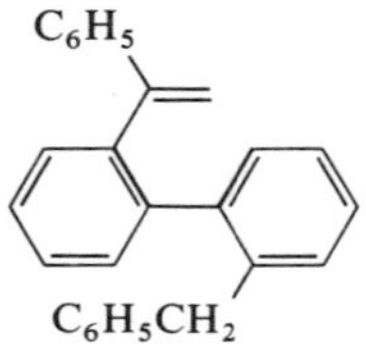	C_6H_5; $C_6H_5CH_2$	"	4-Benzyl-9,10-dihydro-9-phenylphenanthrene (—)	33
	CH_3; CH_3; CO_2CH_3; CO_2CH_3; t-$\dot{C}_4H_9N$=C	THF, N_2	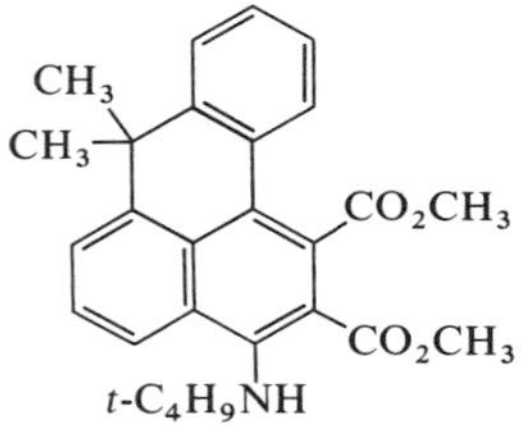 CH_3; CH_3; CO_2CH_3 (33); CO_2CH_3; t-C_4H_9NH	578
	C_6H_5; COC_6H_5	C_6H_6	4-Benzoyl-9,10-dihydro-9-phenylphenanthrene (60), 9-phenylphenanthrene (10)	33

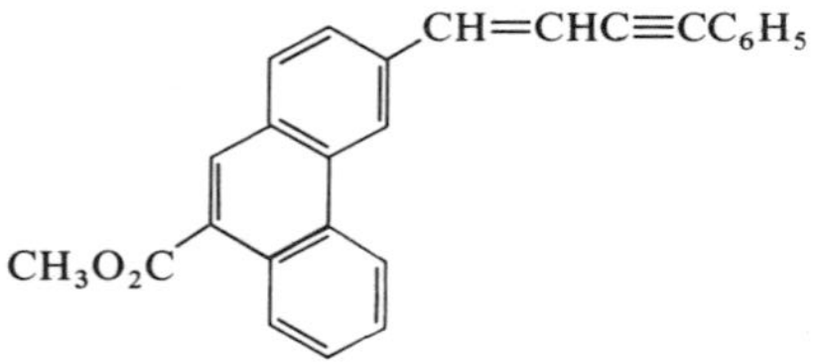

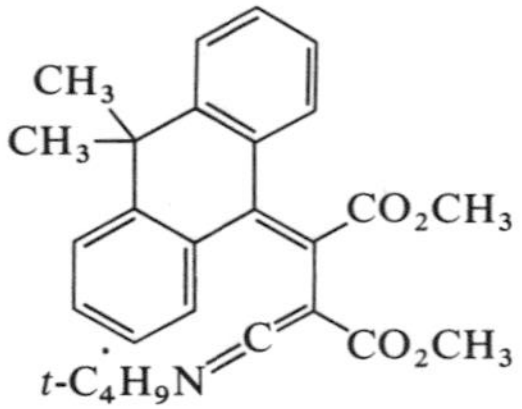

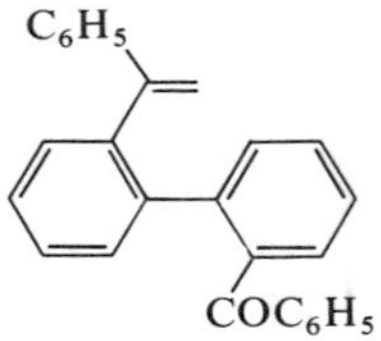

C_{28} [structure], C_6H_5	C_6H_6, $\lambda > 365$ nm	4-Benzoyl-9,10-dihydro-9-phenylphenanthrene (62), 9-benzoyl-9,10-dihydro-9-phenylphenanthrene (20)	33
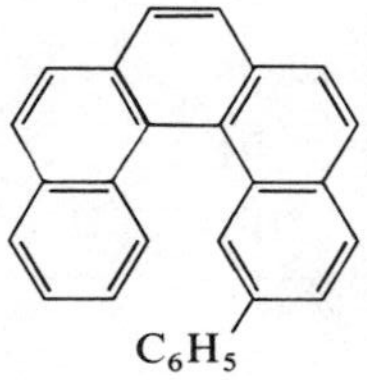	C_6H_6, I_2, N_2	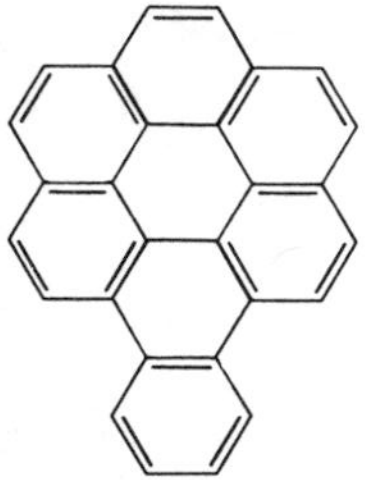 [structure] (29) (Benzo[*a*]coronene)	232
2,2′-Distyrylbiphenyl	C_6H_{14}, I_2, N_2	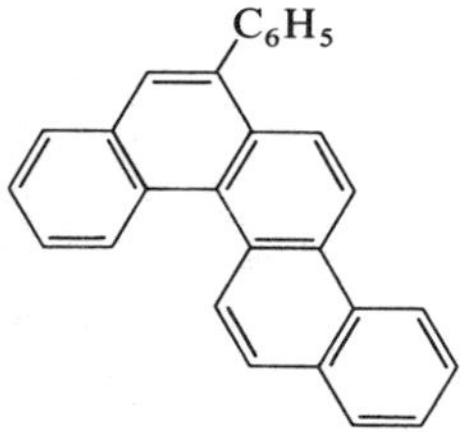 [structure, C_6H_5] (60) (7-Phenylbenzo[*c*]chrysene)	211
	C_6H_{14}, N_2	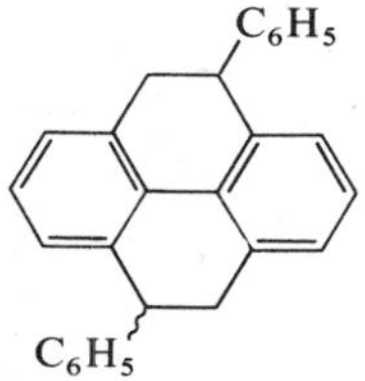 [structure, C_6H_5, C_6H_5] (70)	211
	EtOH, N_2	" (70)	210
	C_6H_6	" (52)	33

TABLE III. CARBOCYCLIC SYSTEMS FROM REACTANTS OTHER THAN 1,2-DIARYLETHYLENES (*Continued*)

Starting Material	Reaction Conditions	Products and Yields (%)	Refs.
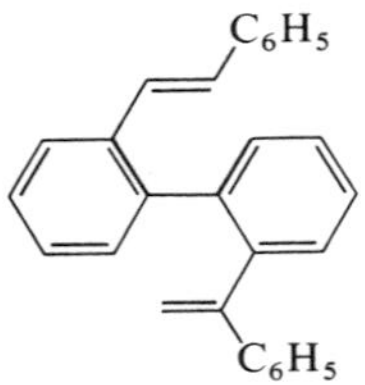 C_6H_5, C_6H_5	C_6H_6	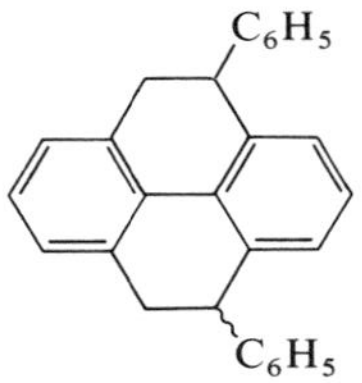 C_6H_5, C_6H_5 (80)	33
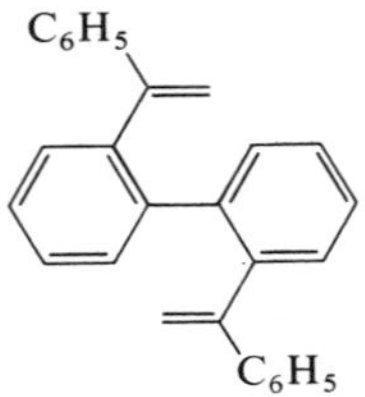 C_6H_5, C_6H_5	"	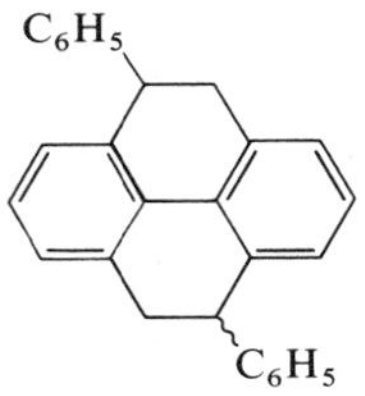 C_6H_5, C_6H_5 (38)	33,558
C_{29} 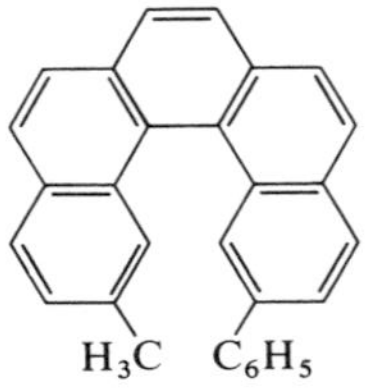 H_3C, C_6H_5	C_6H_6, I_2, N_2	Benzo[*a*]coronene (23)[g]	232
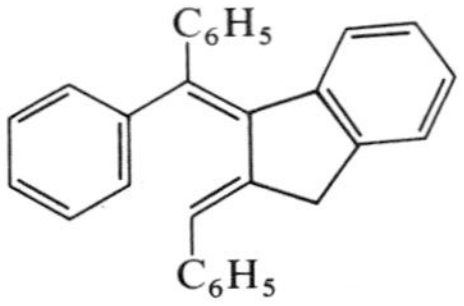 C_6H_5, C_6H_5	Pet. ether, O_2	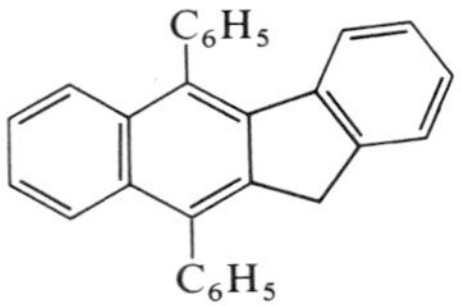 C_6H_5, C_6H_5 (100) (5,10-Diphenyl-11*H*-benzo[*b*]fluorene)	218

Reactant	Conditions	Product (yield %)	Ref.
	C_6H_{14}, air	" (—)	580
	Pet. ether, N_2	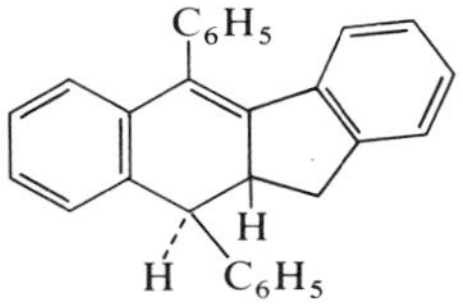 (*trans*-10a,11-Dihydro-5,10-diphenyl-10*H*-benzo[*b*]fluorene) (70)	218
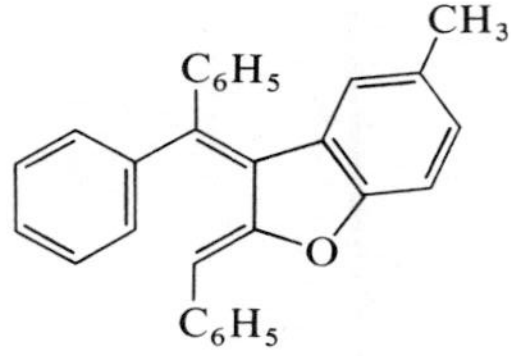 CH_3, C_6H_5, O, C_6H_5	$CHCl_3$, air	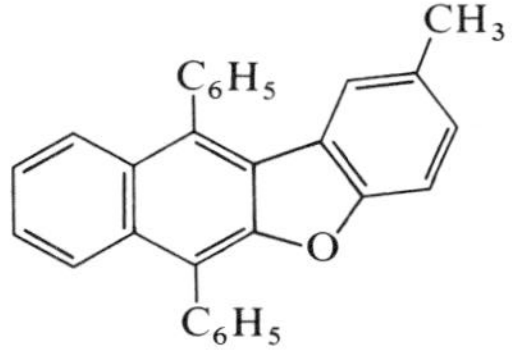 (2-Methyl-6,11-diphenylbenzo[*b*]naphtho[2,3-*d*]furan) (—)	581
C_{30} 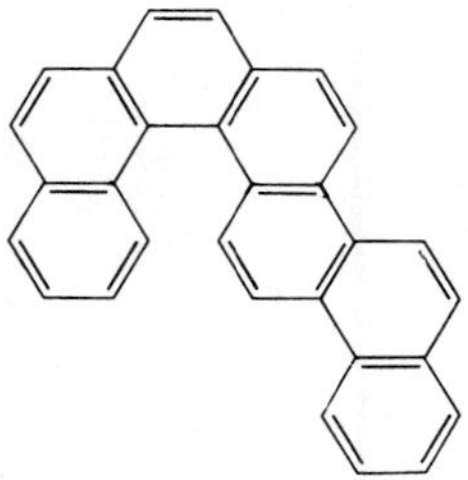	C_6H_6, I_2, N_2	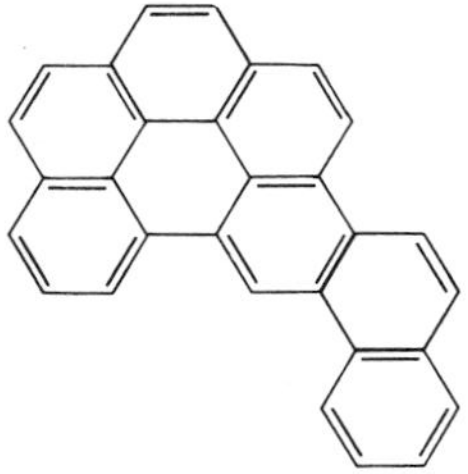 (Benzo[*ghi*]naphtho[1,2-*b*]perylene) (80)	75,159

TABLE III. CARBOCYCLIC SYSTEMS FROM REACTANTS OTHER THAN 1,2-DIARYLETHYLENES (*Continued*)

Starting Material	Reaction Conditions	Products and Yields (%)	Refs.
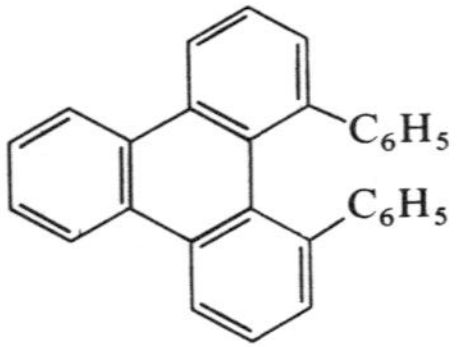 [structure: C_6H_5, C_6H_5]	C_6H_6, I_2, N_2	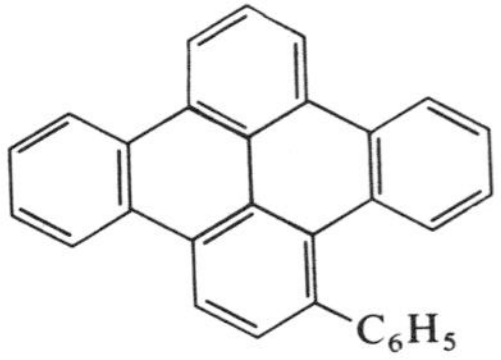 [structure: C_6H_5] (33), (1-Phenyldibenzo[*fg,op*]naphthacene) 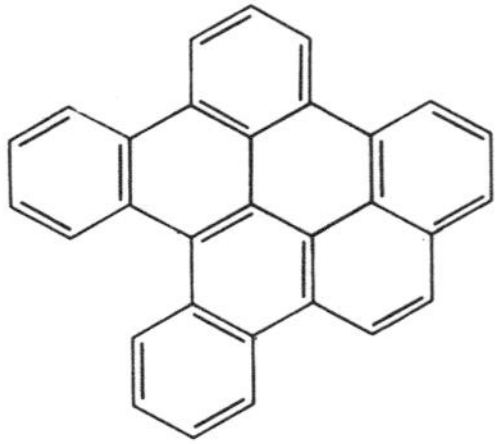 [structure] (29) (Benzo[*b*]naphtho[1,2,3,4-*pqr*]perylene)	231
4,4″-Diphenyl-*o*-terphenyl	C_6H_6, I_2	3,6-Diphenyltriphenylene (9)[b]	235
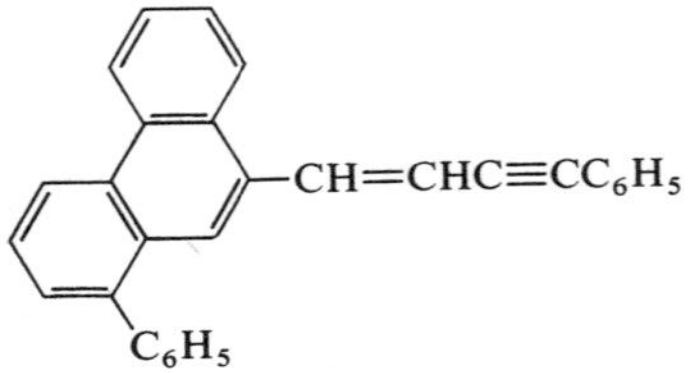 [structure: $-CH{=}CHC{\equiv}CC_6H_5$, C_6H_5]	C_6H_6, N_2	4,5-Diphenyltriphenylene (60)[b]	231
	C_6H_6, I_2	3-Iodo-4,5-diphenyltriphenylene (44)[b]	225,226

Reactant	Conditions	Product(s) (Yield, %)	Refs.
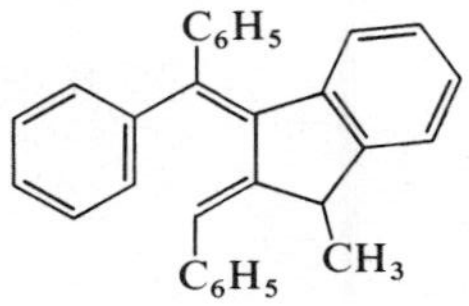 C_6H_5; C_6H_5; CH_3	Pet. ether, O_2	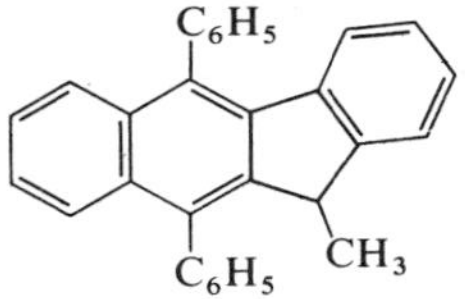 C_6H_5; C_6H_5; CH_3 (100) (11-Methyl-5,10-diphenyl-11*H*-benzo[*b*]fluorene)	218
	Pet. ether, N_2	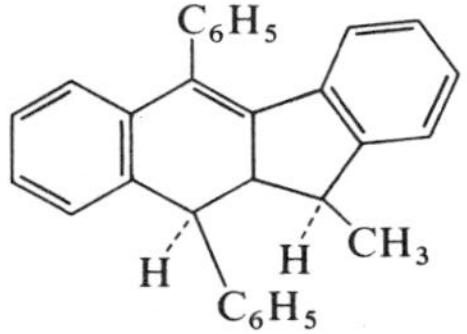 C_6H_5; H; H; CH_3; C_6H_5 (100) (10a,11-Dihydro-11-methyl-5,10-diphenyl-10*H*-benzo[*b*]fluorene)	218
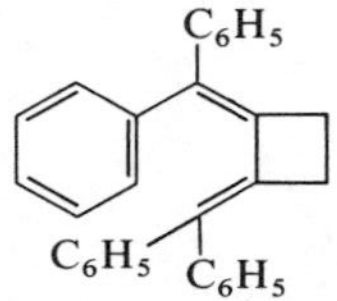 C_6H_5; C_6H_5; C_6H_5	C_6H_6, N_2	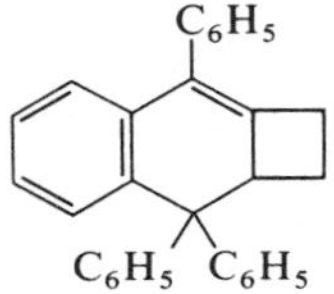 C_6H_5; C_6H_5; C_6H_5 (95)	582
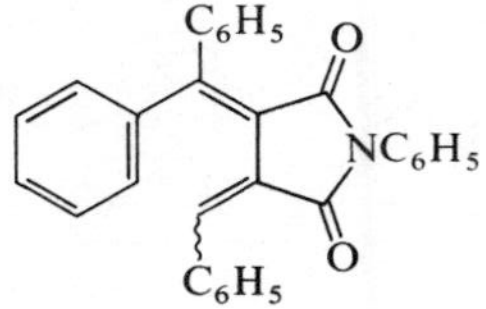 C_6H_5; O; NC_6H_5; O; C_6H_5	C_6H_6, C_6H_{14}, or xylene; N_2	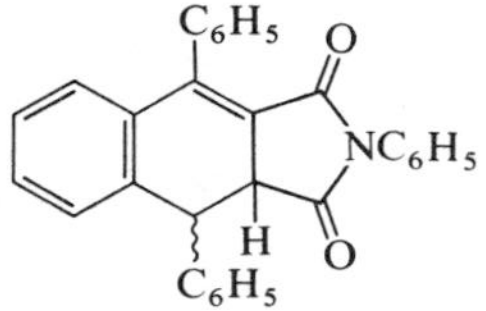 C_6H_5; O; NC_6H_5; H; O; C_6H_5 (—)	217,220
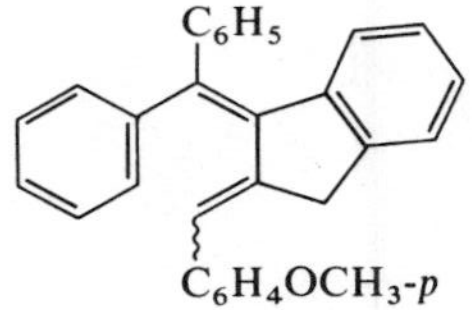 C_6H_5; $C_6H_4OCH_3$-*p*	C_6H_6, pet. ether, air	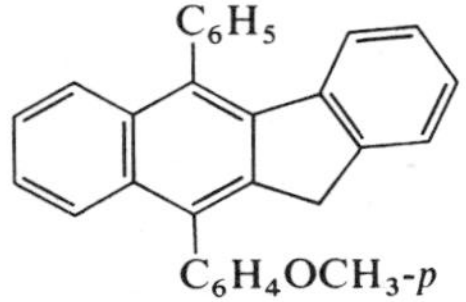 C_6H_5; $C_6H_4OCH_3$-*p* (—)	583

TABLE III. CARBOCYCLIC SYSTEMS FROM REACTANTS OTHER THAN 1,2-DIARYLETHYLENES (*Continued*)

Starting Material	Reaction Conditions	Products and Yields (%)	Refs.
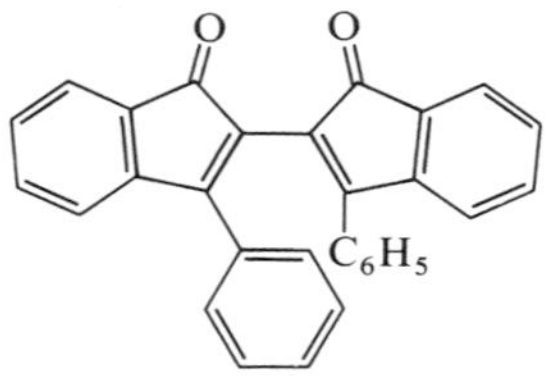 O O C_6H_5	C_6H_6, N_2	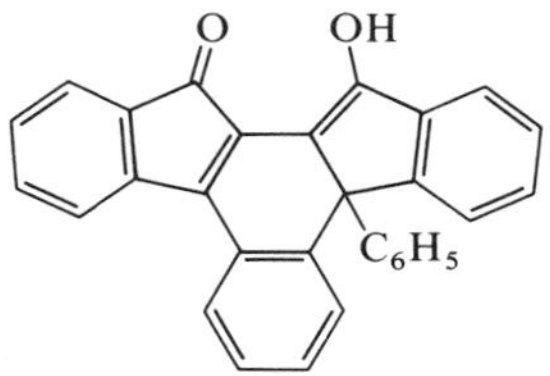 O OH C_6H_5 (70)	584a
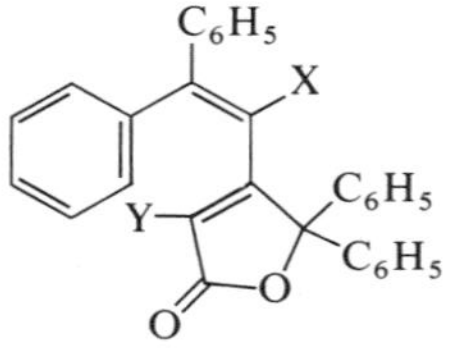 C_6H_5 X Y C_6H_5 C_6H_5 O O (I)		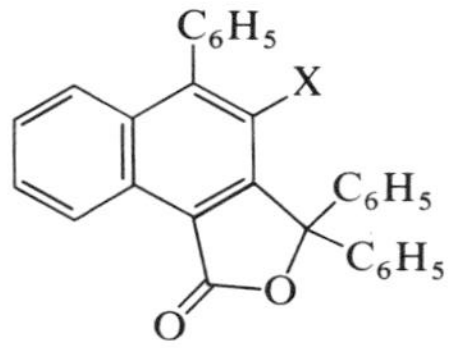 C_6H_5 X C_6H_5 C_6H_5 O O (II)	
I, X = Y = H	C_6H_6, N_2[d]	II, X = H (76)	584b
I, X = H, Y = Br	"	" (84)	584b
	MeOH, N_2	" (57)	584b
I, X = Br, Y = H	C_6H_6, N_2[d]	II, X = Br (60)	584b
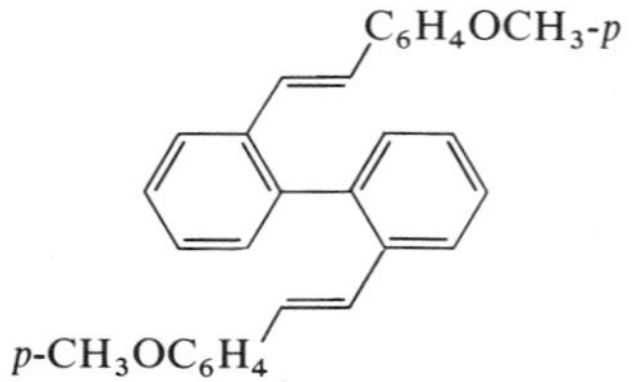 $C_6H_4OCH_3$-*p* *p*-$CH_3OC_6H_4$	EtOH, N_2	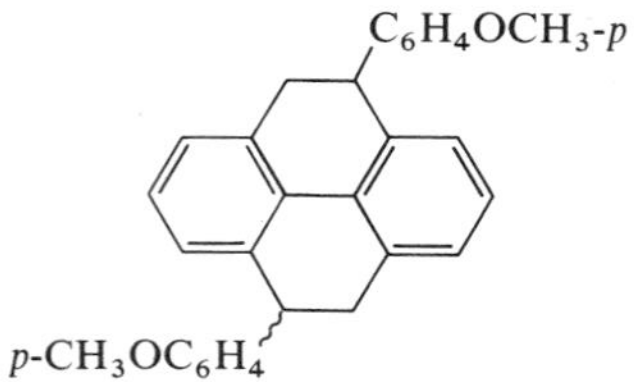 $C_6H_4OCH_3$-*p* *p*-$CH_3OC_6H_4$ (20–25),	210

	Reactant	Conditions	Product(s) (Yield, %)	Refs.
			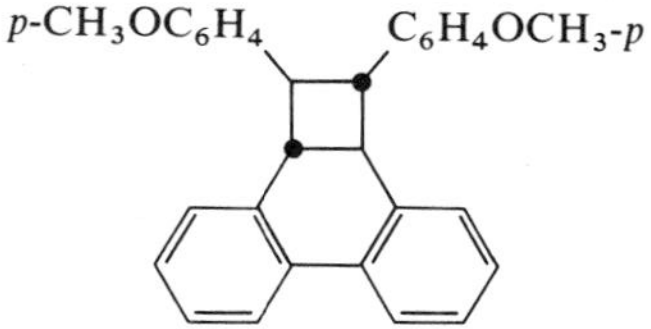 p-$CH_3OC_6H_4$, $C_6H_4OCH_3$-p (30–35)	
	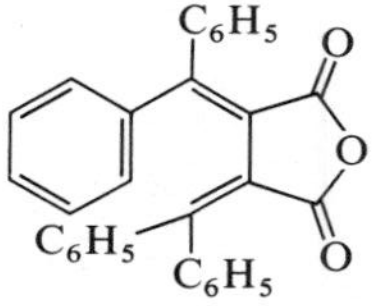 C_6H_5, C_6H_5, C_6H_5	C_6H_6, N_2	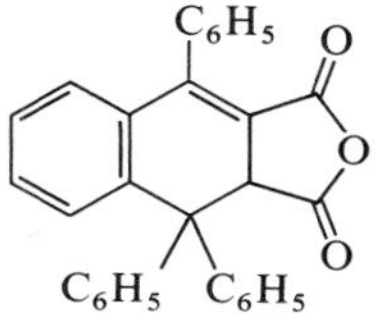 C_6H_5, C_6H_5, C_6H_5 (80)	217,220
	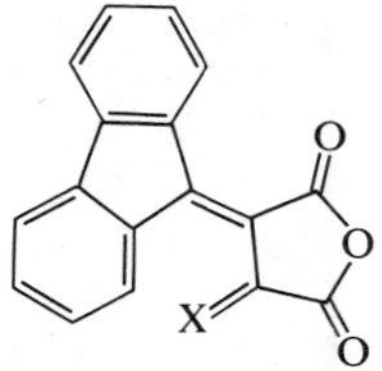 X = 9-Fluorenylidene	"	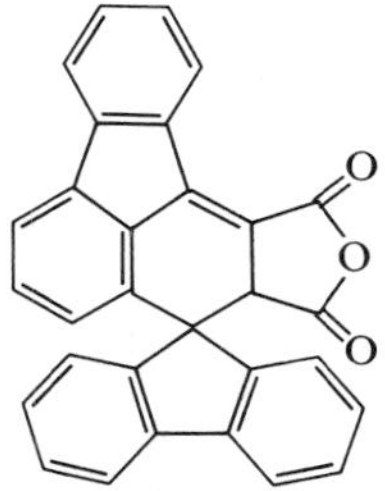 (—) (Spiro{fluorantheno[1,2-*c*]furan-8(11*H*),9′-[9*H*]fluorene}-9,11(8a*H*)-dione)	220
C_{31}	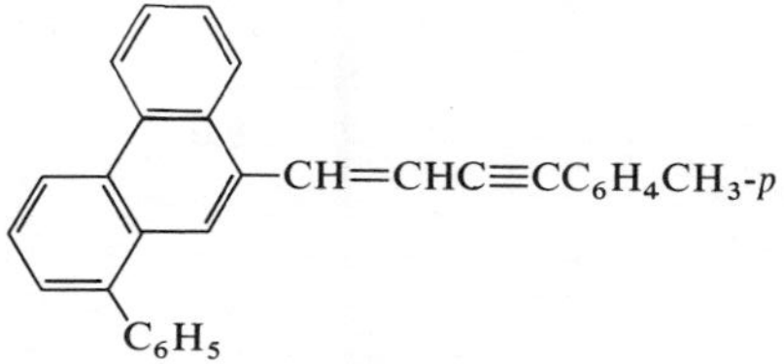 —CH=CHC≡$CC_6H_4CH_3$-p, C_6H_5	MeOH, deaerated	4-(*p*-Methylphenyl)-5-phenyltriphenylene (52)[b]	585

TABLE III. CARBOCYCLIC SYSTEMS FROM REACTANTS OTHER THAN 1,2-DIARYLETHYLENES (*Continued*)

	Starting Material	Reaction Conditions	Products and Yields (%)	Refs.
	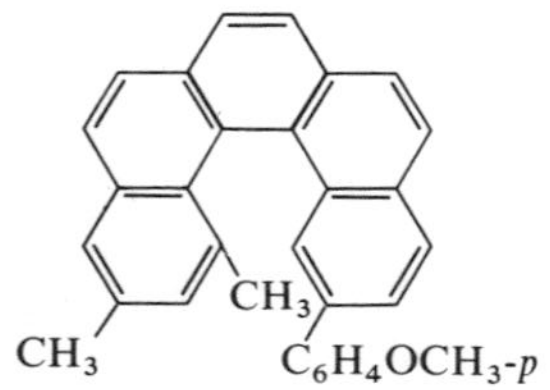 CH_3, CH_3, $C_6H_4OCH_3$-*p*	C_6H_6, I_2, N_2	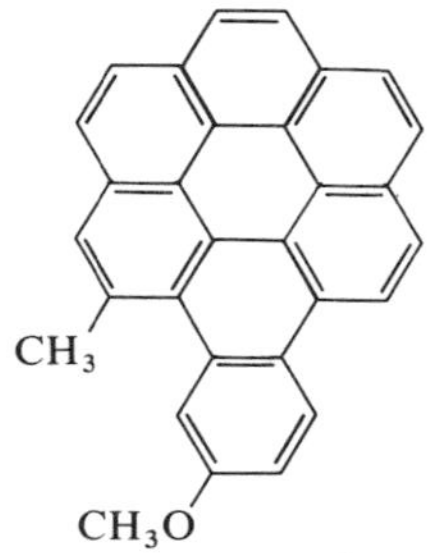 CH_3, CH_3O (37)	232
	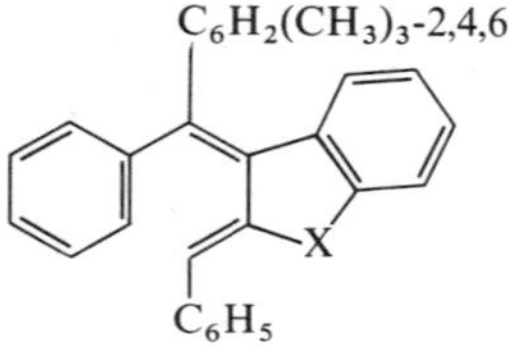 $C_6H_2(CH_3)_3$-2,4,6, X, C_6H_5 (I)		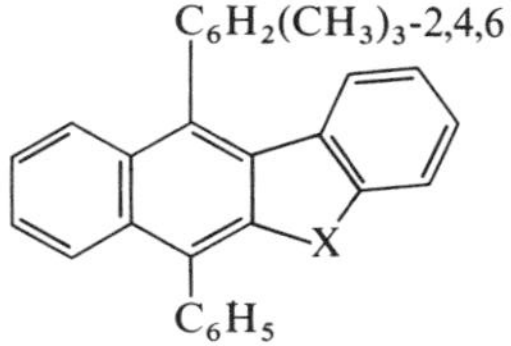 $C_6H_2(CH_3)_3$-2,4,6, X, C_6H_5 (II)	
	I, X = O	Pet. ether, O_2	II, X = O (100)	586
C_{32}	I, X = CH_2	C_6H_{14}, air	II, X = CH_2 (100)	587
	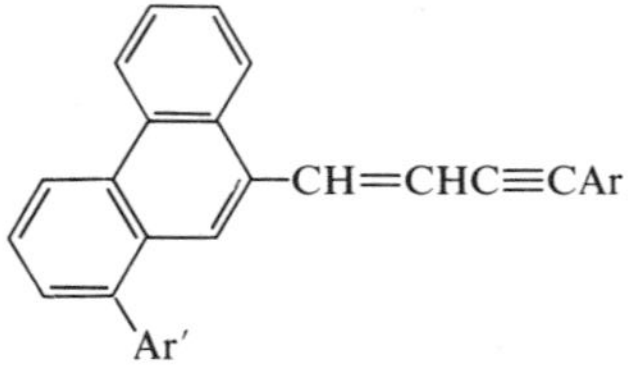 —CH=CHC≡CAr, Ar′			
	Ar = 3,5-Dimethylphenyl, Ar′ = Phenyl	MeOH, deaerated	4-(3,5-Dimethylphenyl)-5-phenyltriphenylene (65)[b]	585
C_{33}	Ar = 2,4,6-Trimethylphenyl, Ar′ = Phenyl	”	4-(2,4,6-Trimethylphenyl)-5-phenyltriphenylene (—)[b]	585

	Reactant	Solvent	Product(s) and Yield(s) (%)	Refs.
C_{34}	Ar = Ar′ = 3,5-Dimethylphenyl	Solvent	4,5-Bis-(3,5-dimethylphenyl)triphenylene (65)[b]	225
	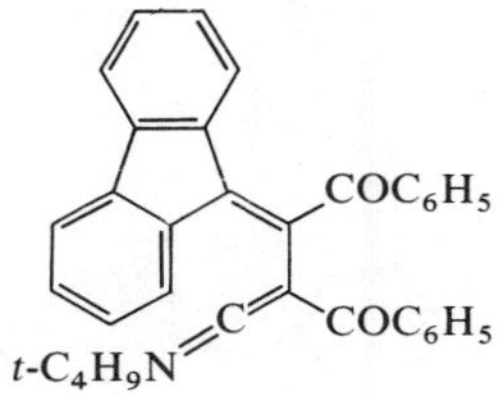 COC_6H_5, COC_6H_5, $t\text{-}C_4H_9N{=}C{=}$	THF, N_2	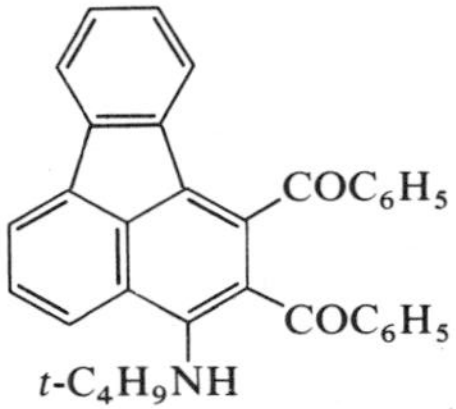 COC_6H_5, COC_6H_5, $t\text{-}C_4H_9NH$ (49)	578
C_{35}	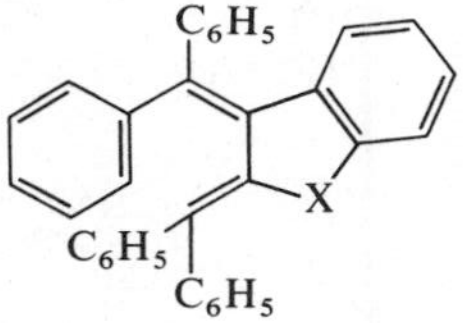 C_6H_5, C_6H_5, C_6H_5, X (I)		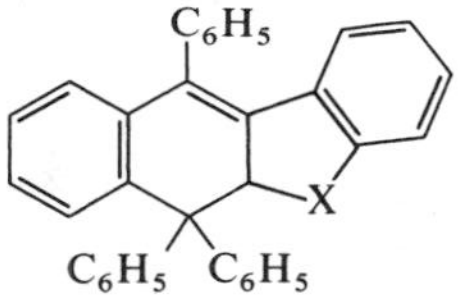 C_6H_5, X, C_6H_5, C_6H_5 (II)	
	I, X = CH_2	Pet. ether	II, X = CH_2 (20), 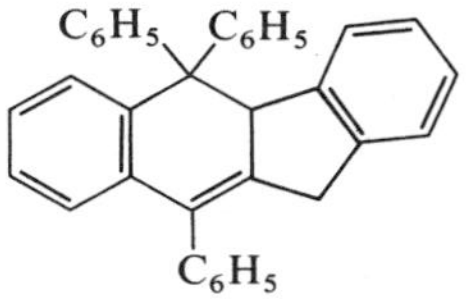 C_6H_5, C_6H_5, C_6H_5 (10)	216
	I, X = CO	Pet. ether, C_6H_6	II, X = CO (—)	588
C_{36}	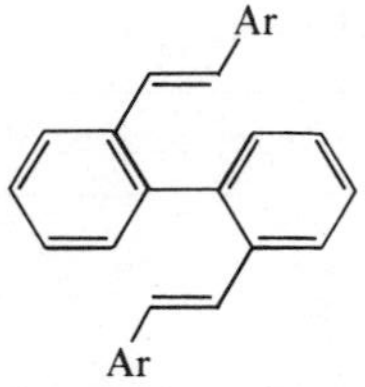 Ar, Ar; Ar = 2-Naphthyl	EtOH, N_2	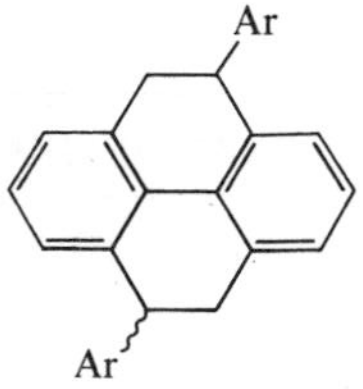 Ar, Ar (60); Ar = 2-Naphthyl	210

TABLE III. CARBOCYCLIC SYSTEMS FROM REACTANTS OTHER THAN 1,2-DIARYLETHYLENES (*Continued*)

	Starting Material	Reaction Conditions	Products and Yields (%)	Refs.
	C_6H_5, CH_3, C_6H_5, C_6H_5	Pet. ether	C_6H_5, CH_3, C_6H_5, C_6H_5 (—)	219
	C_6H_5, O, NC_6H_5, O, C_6H_5, C_6H_5	C_6H_6, C_6H_{14}, or $CHCl_3$; N_2	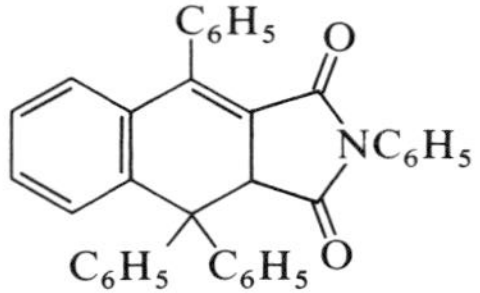 C_6H_5, O, NC_6H_5, O, C_6H_5, C_6H_5 (100)	217
C_{37} 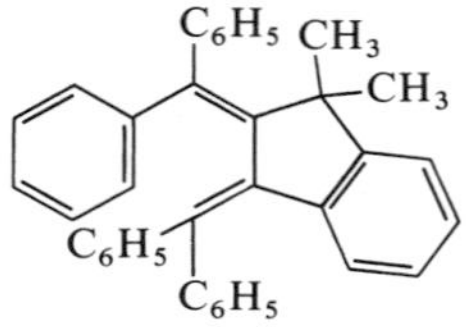	C_6H_5, CH_3, CH_3, C_6H_5, C_6H_5	Pet. ether	C_6H_5, CH_3, CH_3, C_6H_5, C_6H_5 (70), C_6H_5, C_6H_5, CH_3, CH_3, C_6H_5 (20)	216, 219

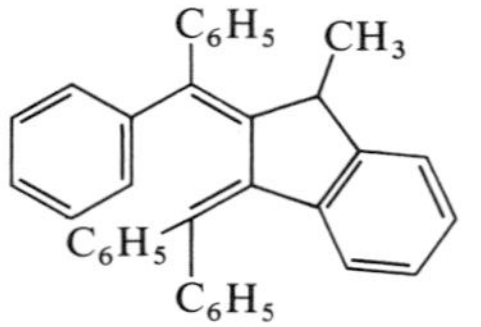

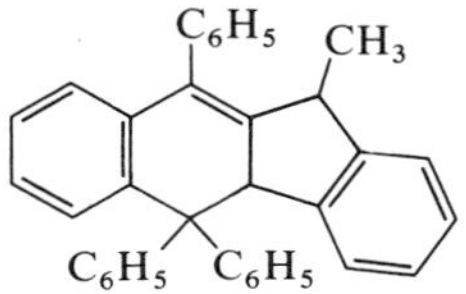

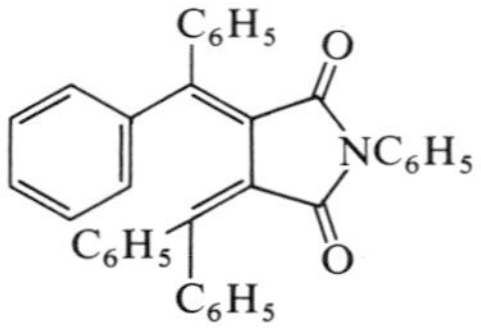

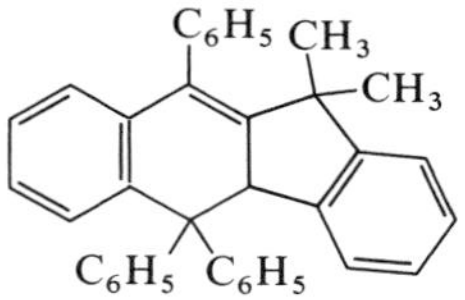

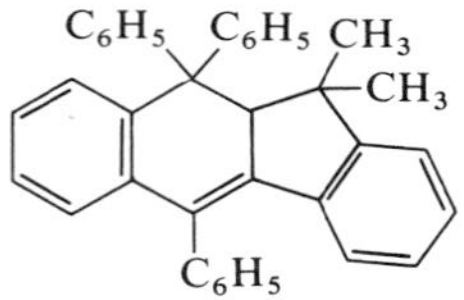

C_{38}	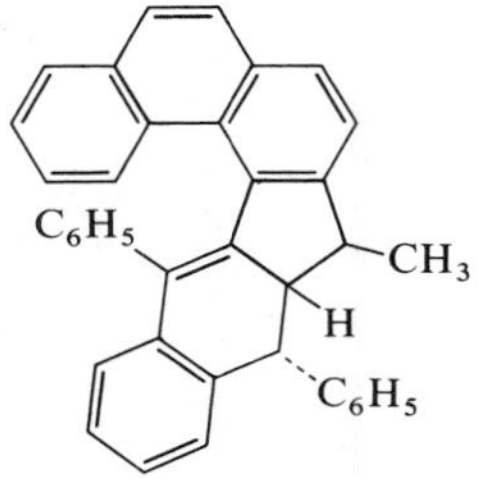 [structure: C_6H_5, CH_3, H, C_6H_5]	"	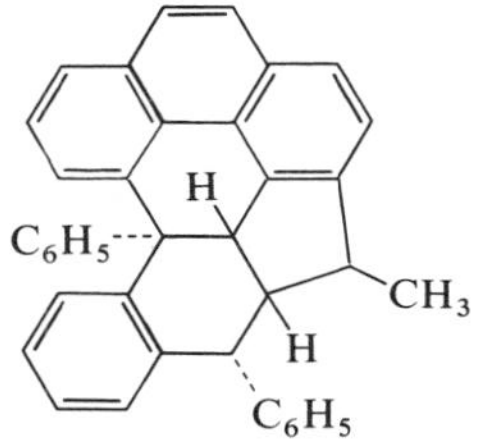 [structure: H, C_6H_5, CH_3, H, C_6H_5] (5a,6,10b,13e-Tetrahydro-5-methyl-6,10b-diphenyl-5*H*-cyclopenta[*def*]naphtho[8,1,2-*pqr*]chrysene) (85)	213,589
C_{48}	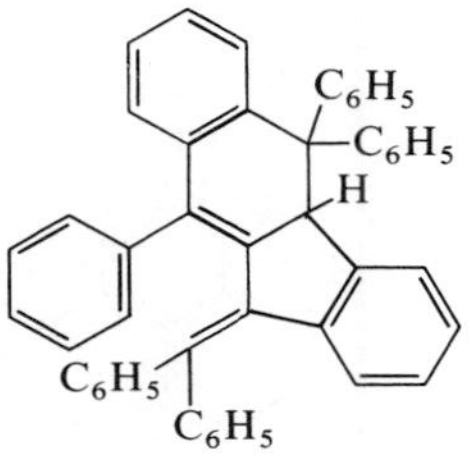 [structure: C_6H_5, C_6H_5, H, C_6H_5, C_6H_5]	"	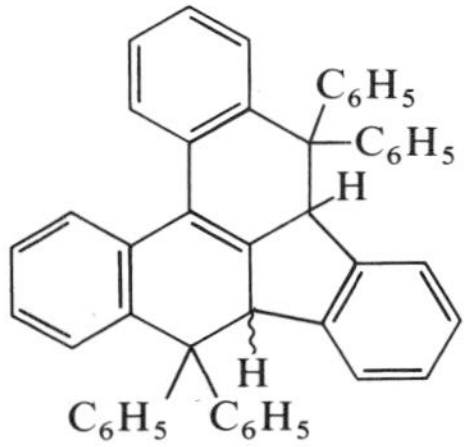 [structure: C_6H_5, C_6H_5, H, H, C_6H_5, C_6H_5] (9,9a,13b,14-Tetrahydro-9,9,14,14-tetraphenyldibenz[*a,e*]acephenanthrylene) (14)	588

[a] Subsequent workers were unable to detect chrysene under these conditions.[562]
[b] The numbering schemes for *o*-terphenyl, triphenylene, and chrysene are indicated in the first entries under C_{18}.
[c] The starting diene functions as an oxidant, undergoing 1,4-reduction to give a bis-(arylmethyl)maleic anhydride.
[d] The identity of the oxidant in this reaction is obscure. One possibility is residual dissolved oxygen.
[e] Adventitious water may be the source of the hydroxyl groups in the products.
[f] The starting material was generated *in situ* by UV irradiation of a benzene solution of 1-iodo-2,4-diphenylbenzene.
[g] The structure of the product is indicated in a previous entry in this table.

TABLE IV. POLYCYCLIC SYSTEMS FROM DIANTHRONES AND OTHER DOUBLY BRIDGED TETRAARYLETHYLENES[a]

	Starting Material	Reaction Conditions	Products and Yields (%)	Refs.
C_{26}	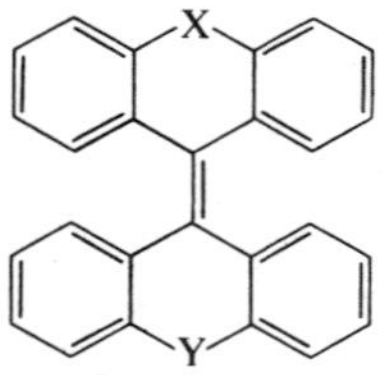 (I)		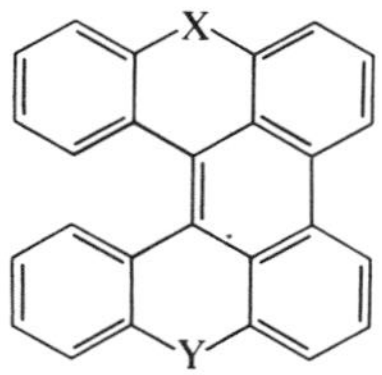 (II)	
	I, X = Y = O	C_6H_6, N_2[b]	II, X = Y = O (80)	142
	I, X = O, Y = S	"[b]	II, X = O, Y = S (80)	142
	I, X = Y = S	"[b]	II, X = Y = S (60)	142
C_{27}	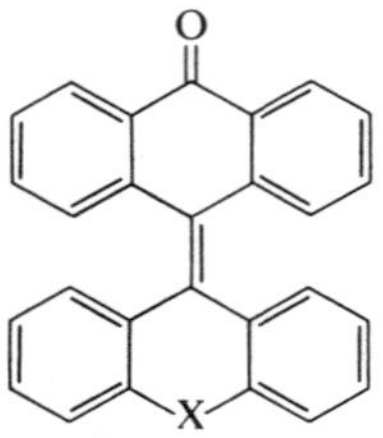 (I)		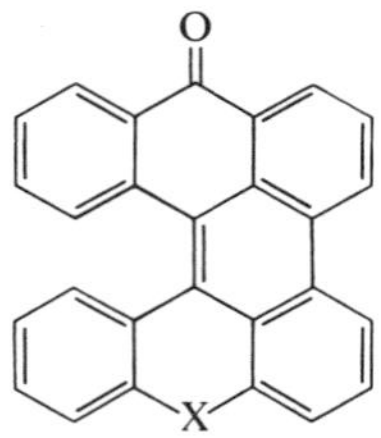 (II)	
	I, X = O	C_6H_6, CO_2[b]	II, X = O (50)	590
	I, X = S	"[b]	II, X = S (80)	150
		C_6H_6, air	" (60)	150

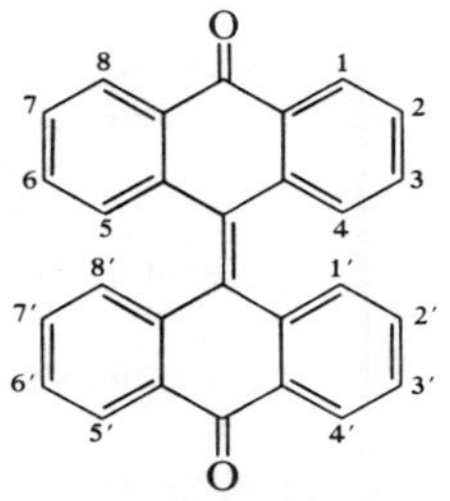 C_{28} H (Dianthrone)	Py, O_2	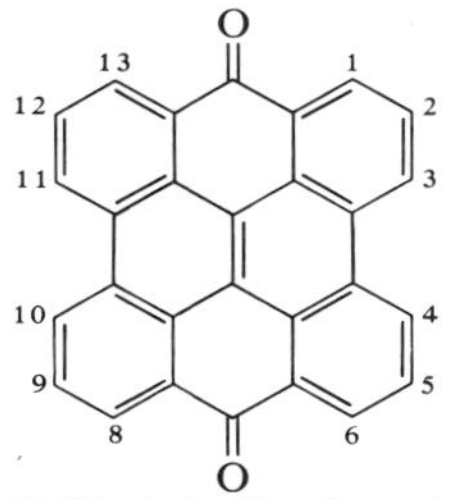 **K** (Naphthodianthrone) (55–60)	144
	H_2SO_4	" (55)	591
	Boiling Ac_2O, NaOAc, O_2	" (42)	144
	HOAc	" (—)	140
	Solvent, air[c]	" (—)	592
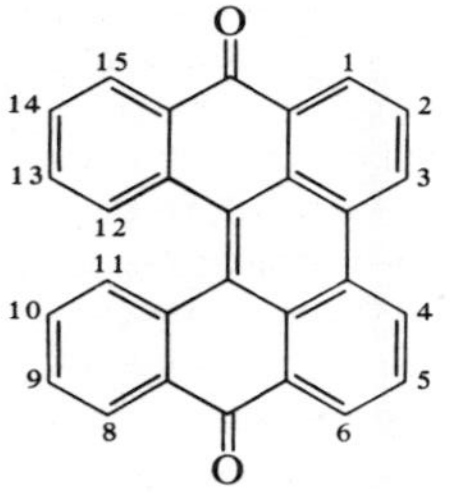 **J** (Helianthrone)	Py, O_2	**K** (98–99)	144
	C_6H_6, air	" (97)	593
	HOAc, air	" (90)	140
1′,5-Dibromo-**H**	Py or xylene	**K** (100)	145
	Ac_2O, NaOAc	" (60–90)	145
2,6′-Dibromo-**H**	Py, air	2,9-Dibromo-**K** (39)	145
	Ac_2O, NaOAc, air	2,9(or 2,5)-Dibromo-**K** (34)	145

TABLE IV. POLYCYCLIC SYSTEMS FROM DIANTHRONES AND OTHER DOUBLY BRIDGED TETRAARYLETHYLENES (*Continued*)[a]

	Starting Material	Reaction Conditions	Products and Yields (%)	Refs.
	3,7′-Dibromo-**H**	Ac_2O, NaOAc, air	3,10(or 3,4)-Dibromo-**K** (19)	145
	2,5-Dibromo-**J**	Py, air	2,5-Dibromo-**K** (100)	145
		Xylene, air	" (—)	594
	1,4′-Dichloro-**H**	"	1,6(or 1,8)-Dichloro-**K** (—)	594
	1,6-Dichloro-**J**	$PhNO_2$, air	1,6-Dichloro-**K** (—)	594
	2,6′(or 2,7′)-Dichloro-**H**	C_6H_6, air	2,9(or 2,4 or 2,6)-Dichloro-**K** (—)	594
	2,5-Dichloro-**J**	Xylene, air	2,5-Dichloro-**K** (—)	594
	1,1′,5,5′-Tetrachloro-**H**	Concd H_2SO_4	1,8(or 1,6)-Dichloro-**K** (—)	594
	1,4,5′,8′-Tetrachloro-**H**	"	1,8(or 1,6)-Dichloro-**K** (—)	594
	1,4′,5′,8-Tetrachloro-**H**	$PhNO_2$, air	1,6,8,13-Tetrachloro-**K** (—)	594
	1,6,8,11,12,15-Hexachloro-**J**[d]	H_2SO_4	1,6,8,13-Tetrachloro-**K** (—)[d]	595
	3,4-Dihydroxy-2-iodo-**J**	C_6H_6, air	3,4-Dihydroxy-2-iodo-**K** (—)	596
	3,4-Dihydroxy-**J**	Py, Ac_2O, air	3,4-Diacetoxy-**K** (87)	597
	1,2,5,6-Tetrahydroxy-**J**	H_2SO_4, air	1,2,5,6-Tetrahydroxy-**K** (—)	598
		Concd H_2SO_4, air	" (—)	599
		Py, Ac_2O, air	1,2,5,6-Tetraacetoxy-**K** (—)	599
	1,6,8,15-Tetrahydroxy-**J**	Py, air	1,6,8,15-Tetrahydroxy-**K** (100)	600
		Py, Ac_2O, air	1,6,8,15-Tetraacetoxy-**K** (80)	600
C_{30}	1,6-Dimethyl-**J**	Py, air	1,6-Dimethyl-**K** (94)	601
	2,3′-Dimethyl-**H**	"	2,5-Dimethyl-**K** (50)	145
		Ac_2O, NaOAc, air	" (39)	145
	2,5-Dimethyl-**J**	Py, air	2,5-Dimethyl-**K** (100)	145
		"	" (80)	601
	3,4-Dimethyl-**J**	Concd H_2SO_4, air	3,4-Dimethyl-**K** (80)	601
		C_6H_6, air	" (60)	601
		HOAc, air	" (—)	602
	2,5-Dibromo-1,6-dimethoxy-**J**	Solvent, air	2,5-Dibromo-1,6-dimethoxy-**K** (—)	598
	1,6-Dihydroxy-3,4-dimethyl-**J**	Py, Ac_2O, air	1,6-Diacetoxy-3,4-dimethyl-**K** (78)	603
		Concd H_2SO_4, air	1,6-Dihydroxy-3,4-dimethyl-**K** (—)	603

	1,6-Dihydroxy-2,5-dimethyl-**J**	Py, Ac_2O, air	1,6-Diacetoxy-2,5-dimethyl-**K** (70)	603
	1,4′-Dimethoxy-**H**	Me_2CO, air	1,6 (or 1,8)-Dimethoxy-**K** (70)	604
	1,6-Dimethoxy-**J**	PhCl, air	1,6-Dimethoxy-**K** (—)	598
	2′,3-Dimethoxy-**H**	Py, air	3,4 (or 3,10)-Dimethoxy-**K** (—)	604
	1,2′,3,4′,5′,8-Hexahydroxy-6,7′-dimethyl-**H**	Solvent, air	1,3,4,6,8,15-Hexahydroxy-10,13-dimethyl-**J** (Protohypericin) (—)	146
		Py or dioxane, air	" (—)	605
	1,3,4,6,8,15-Hexahydroxy-10,13-dimethyl-**J** (Protohypericin)	Py, Me_2CO, or concd H_2SO_4; O_2	1,3,4,6,8,13-Hexahydroxy-10,11-dimethyl-**K** (Hypericin) (—)	606
		Py or dioxane, air	" (—)	605
		Solvent, air	" (100)	146
		MeOH, dioxane, HOAc, air[e]	" (84)	606
C_{32}	1,3,5′,7′-Tetramethyl-**H**	Py, air	1,3,8,10(or 1,3,4,6)-Tetramethyl-**K** (100)	607
	1,6-Dimethoxy-2,5-dimethyl-**J**	Concd H_2SO_4, air	1,6-Dimethoxy-2,5-dimethyl-**K** (—)	603
	1,6-Dimethoxy-3,4-dimethyl-**J**	"	1,6-Dimethoxy-3,4-dimethyl-**K** (26)	603
	3,4-Diacetoxy-**J**	C_6H_6 or HOAc, air	3,4-Diacetoxy-**K** (100)	596,608
	2,3′,6,7′-Tetramethoxy-**H**	C_6H_6, air	2,5,10,11-Tetramethoxy-**K** (—)	609
	2,2′,3,3′-Tetramethoxy-**H**	C_6H_6, EtOH, air, 3 weeks	9,10,13,14 (or 2,3,9,10)-Tetramethoxy-**J** (75)	604
		C_6H_6, EtOH, air, 6 weeks	2,3,4,5 (or 2,3,9,10)-Tetramethoxy-**K** (—)	604
	1,6,8,15-Tetramethoxy-**J**	Py, air	1,6,8,13-Tetramethoxy-**K** (100)	600
C_{33}	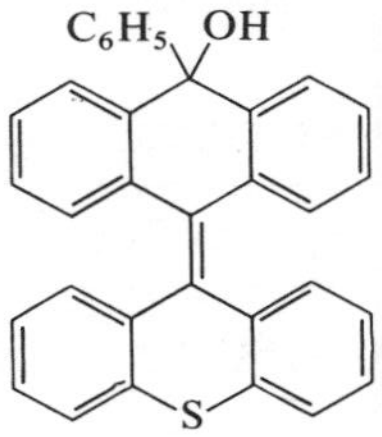 [structure: C_6H_5, OH; S]	C_6H_6, CO_2[b]	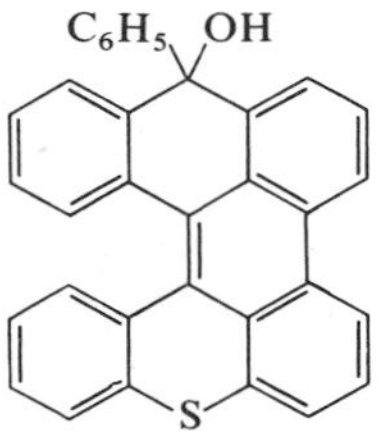 [structure: C_6H_5, OH; S] (70)	150
		C_6H_6, air	" (50)	150

TABLE IV. POLYCYCLIC SYSTEMS FROM DIANTHRONES AND OTHER DOUBLY BRIDGED TETRAARYLETHYLENES (*Continued*)[a]

	Starting Material	Reaction Conditions	Products and Yields (%)	Refs.
C_{36}	1,6,8,10,13,15-Hexamethoxy-3,4-dimethyl-**J**	C_6H_6, Me_2CO, or concd H_2SO_4; O_2	1,6,8,10,11,13-Hexamethoxy-3,4-dimethyl-**K** (100)	610
C_{40}	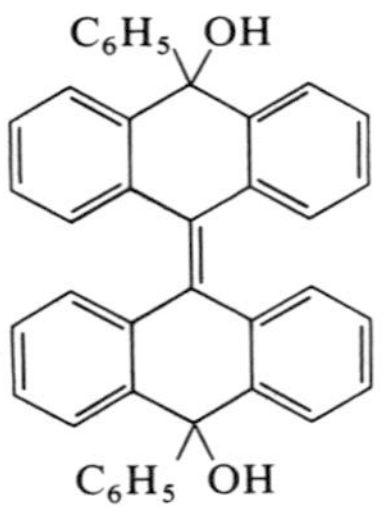 C_6H_5 OH / C_6H_5 OH	C_6H_6, I_2, air	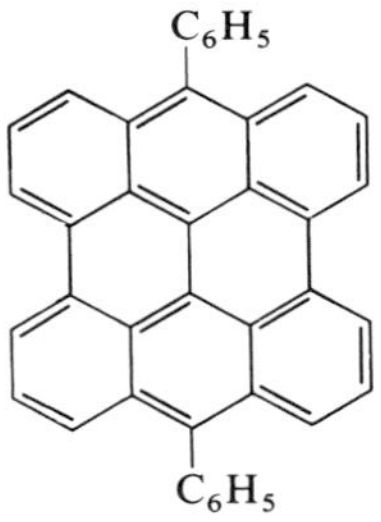 C_6H_5 / C_6H_5 (20)	139
		C_6H_6, I_2, N_2	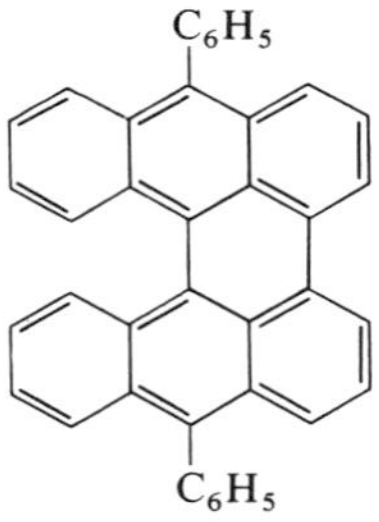 C_6H_5 / C_6H_5 (60)	139

	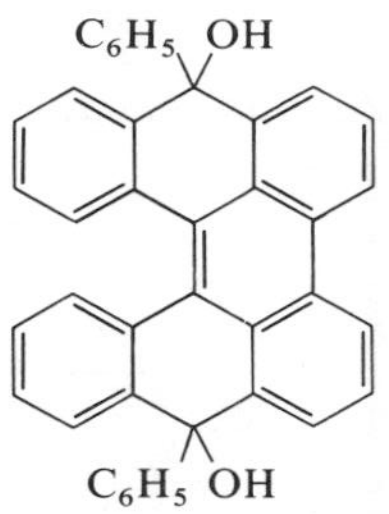 C_6H_5 OH / C_6H_5 OH (structure)	Et_2O, air	C_6H_5 OH / C_6H_5 OH (structure)	(—)	611
C_{72}	1,3,4,6,8,15-Hexabenzoyloxy-10,13-dimethyl-**J**	C_6H_6, H_2O, air	1,3,4,6,8,13-Hexabenzoyloxy-10,11-dimethyl-**K**	(—)	605

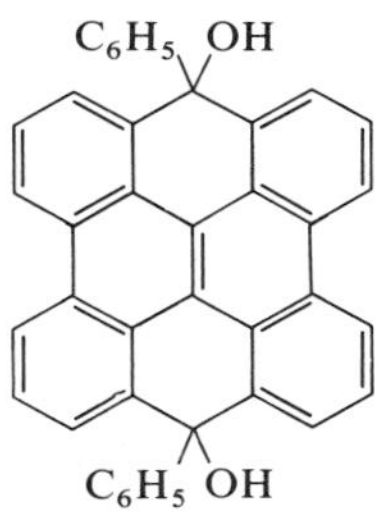

[a] The following symbols are used as abbreviations in Table IV: **H** = dianthrone, **J** = helianthrone, and **K** = naphthodianthrone. The numbering schemes for these systems are indicated in the first entries under C_{18}. Dianthrone is 10-(10-oxo-9(10*H*)-anthracenylidene)-9(10*H*)-anthracene. Helianthrone is dibenzo[*a,o*]-perylene-7,16-dione. Naphthodianthrone, also known as *meso*-naphthodianthrone, is phenanthro[1,10,9,8-*opqra*]perylene-7,14-dione.

[b] The identity of the oxidant in this reaction is obscure. One possibility is residual dissolved oxygen.

[c] The starting material was generated *in situ* by irradiation of 1,10-dibromoanthrone.

[d] The structural assignments for the starting material and the product are based on questionable evidence.

[e] The starting material was generated *in situ* by air oxidation of penicilliopsin in methanol.

TABLE V. NITROGEN HETEROCYCLIC SYSTEMS[a]

A. Heteroaromatic Systems with One Ring Nitrogen

	Starting Material	Reaction Conditions	Products and Yields (%)	Refs.
C_{13}	2-Styrylpyridine	C_6H_{12}, air	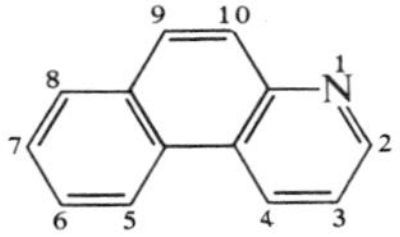 (35) 1-Aza-**L** (1-Azaphenanthrene)	240,245
		C_6H_{12}, O_2	" (35)	241
		Solvent, air	" (—)	612
	3-Styrylpyridine	C_6H_{14}, O_2	2-Aza-**L** + 4-Aza-**L** (4:1)	242
		C_6H_{12}, air	2-Aza-**L** (66)	240,245
	4-Styrylpyridine	"	3-Aza-**L** (21)	240,245
		Solvent, air	" (—)	612
	$C_6H_5N{=}CHC_6H_5$	98% H_2SO_4	9-Aza-**L** (38)	286
	1-Styrylpyridinium ion	EtOH, I_2	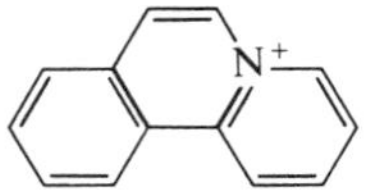 (60) (Benzo[*a*]quinolizinium ion)	247,613
	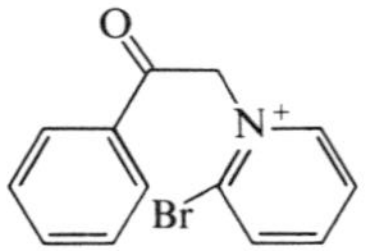	H_2O	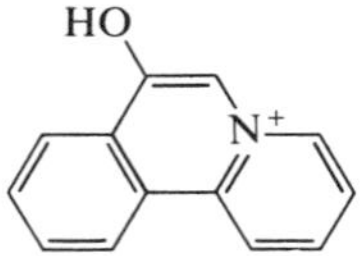 (—)	400
	1-(*p*-Chlorostyryl)pyridinium ion	EtOH, I_2	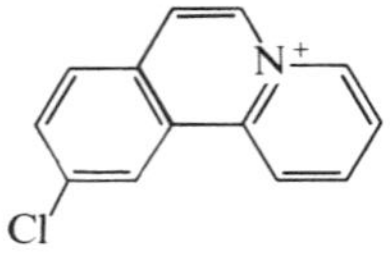 (60)	247,613

	Reactant	Conditions	Product (yield %)	Refs.
	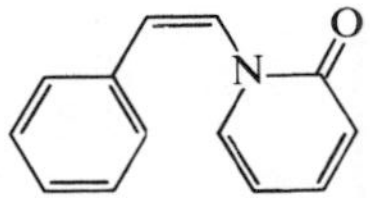	EtOH or MeCN, concd H_2SO_4, I_2, O_2	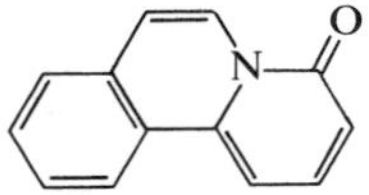 (25)	248
C_{14}	6-Methyl-2-styrylpyridine	C_6H_{12}, air	2-Methyl-1-aza-**L** (31)	240,245
		Solvent, air	" (—)	612
	5-Methyl-2-styrylpyridine	"	3-Methyl-1-aza-**L** (—)	612
	4-Methyl-2-styrylpyridine	"	4-Methyl-1-aza-**L** (—)	612
	3-Methyl-2-styrylpyridine	"	1-Aza-**L** (—)	612
	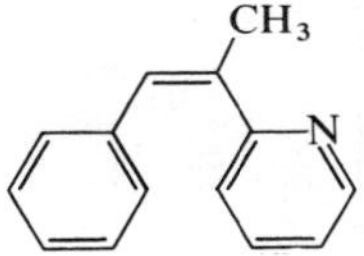	C_6H_{12}, air	10-Methyl-1-aza-**L** (32)	240,245
		Solvent, air	" (—)	614
	3-(*p*-Methylstyryl)pyridine	C_6H_{12}, air	6-Methyl-2-aza-**L** (57)	240,245
		C_6H_{14}, O_2	" (—), 6-methyl-4-aza-**L** (—)	612
	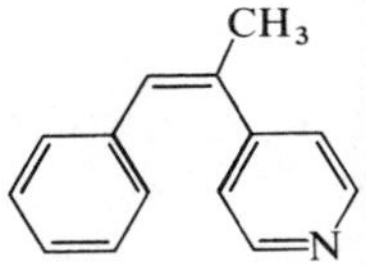	C_6H_{12}, air	10-Methyl-3-aza-**L** (13)	240,245
	1-(*o*-Methylstyryl)pyridinium ion	EtOH, I_2	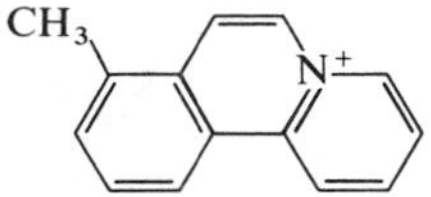 (56)	247
	1-(*p*-Methylstyryl)pyridinium ion	"	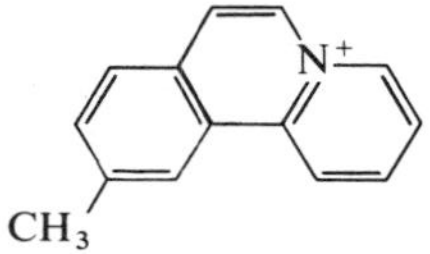 (66)	247

TABLE V. NITROGEN HETEROCYCLIC SYSTEMS (*Continued*)[a]

A. Heteroaromatic Systems with One Ring Nitrogen (*Continued*)

Starting Material	Reaction Conditions	Products and Yields (%)	Refs.
3-Methyl-1-styrylpyridinium ion	EtOH, I_2	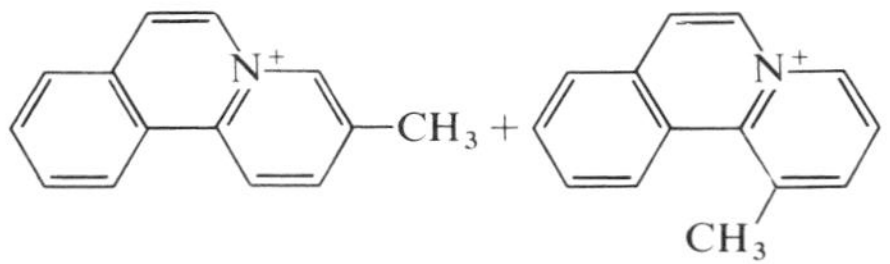 [structures] (56 total)	247
$C_6H_5N{=}C(CH_3)C_6H_5$	Concd H_2SO_4, argon	10-Methyl-9-aza-**L** (11)	287
	CF_3CO_2H	" (trace)	287
	HOAc, H_2SO_4	" (trace)	287
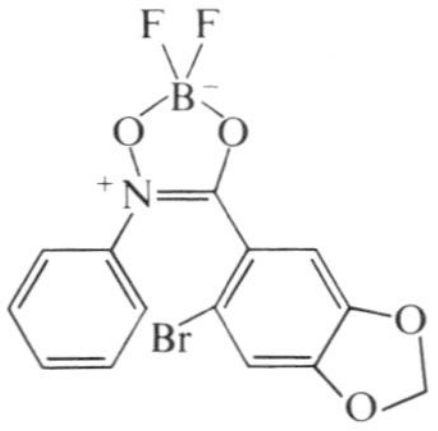 [structure]	C_6H_6	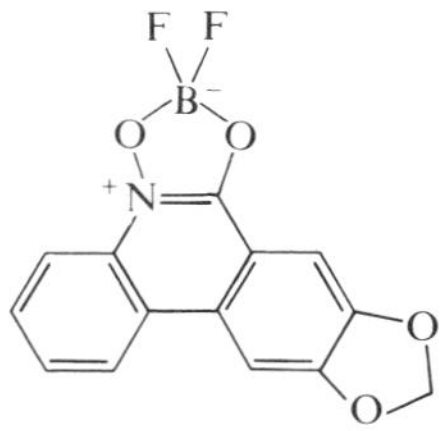 [structure] (99)	300
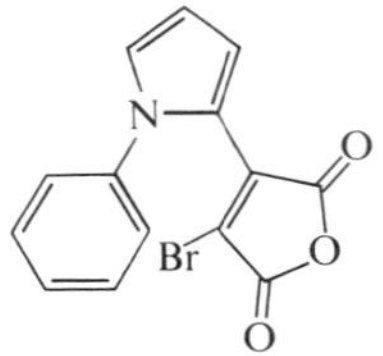 [structure]	CCl_4	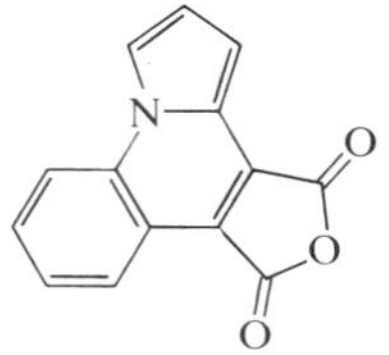 [structure] (100)	615
$C_6H_5C(CN){=}CHAr$ Ar = 2-Pyridyl	C_6H_6, *t*-BuOH, O_2	9-Cyano-1-aza-**L** (62)	241

		”	” (54)	413
		C_6H_6, O_2	” (70)	241
		C_6H_{12}, O_2	” (38)	241
		MeCN, O_2	” (30)	241
		EtOH, O_2	” (14)	241
	Ar = 3-Pyridyl	C_6H_{12}, air	9-Cyano-2-aza-**L** (41)	240,245
	Ar = 4-Pyridyl	*t*-BuOH, O_2	9-Cyano-3-aza-**L** (58)	57
	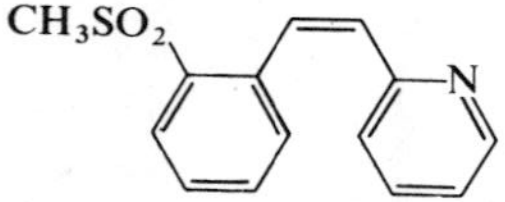 CH_3SO_2	*t*-BuOH, C_6H_6, O_2	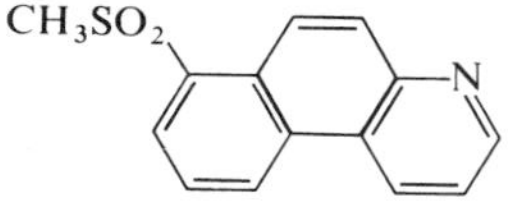 CH_3SO_2; N (21)	616
C_{15}	3,6-Dimethyl-2-styrylpyridine	Solvent, air	2-Methyl-1-aza-**L** (—)	612
	3,5-Dimethyl-1-styrylpyridinium ion	EtOH, I_2	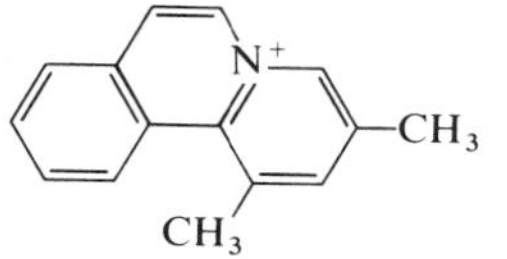 N^+; $-CH_3$; CH_3 (47)	247,613
	m-$CH_3C_6H_4N{=}C(CH_3)C_6H_5$	Concd H_2SO_4, argon	7,10-Dimethyl-9-aza-**L** (8)	287
	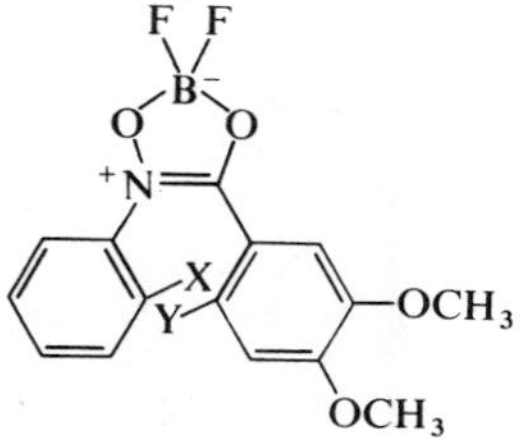 F F; B⁻; O O; ⁺N; X; Y; $-OCH_3$; OCH_3		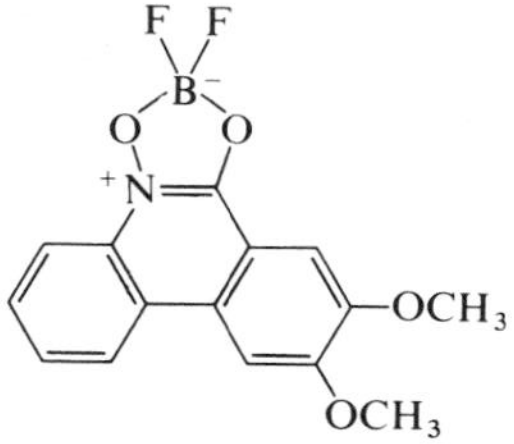 F F; B⁻; O O; ⁺N; $-OCH_3$; OCH_3	
	X = H, Y = Br	C_6H_6	” (99)	300
	X = Cl, Y = H	”	” (76)	300
	X = Y = H	C_6H_6, I_2	” (95)	300

TABLE V. Nitrogen Heterocyclic Systems (*Continued*)[a]

A. Heteroaromatic Systems with One Ring Nitrogen (*Continued*)

Starting Material	Reaction Conditions	Products and Yields (%)	Refs.
(structure: 2-X-phenyl, NC, 5-methoxy-2-pyridyl) $X = Cl$	C_6H_6, *t*-BuOH, O_2	8-Chloro-9-cyano-3-methoxy-1-aza-**L** (11), 9-cyano-3-methoxy-1-aza-**L** (9)	57
$X = H$	C_6H_6, O_2	9-Cyano-3-methoxy-1-aza-**L** (70)	241
$C_6H_5N{=}CHC_6H_4N(CH_3)_2$-*p*	Et_2O, air	3-Dimethylamino-9-aza-**L** (15)[b]	617
(I) I, X = O	EtOH, air	(II) II, X = O (42)	273
I, X = S	”	II, X = S (—)	619a
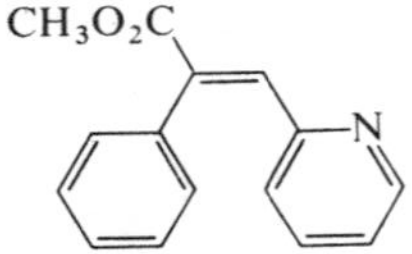 CH_3O_2C (structure)	C_6H_{12}, O_2	9-Carbomethoxy-1-aza-**L** (25)	241
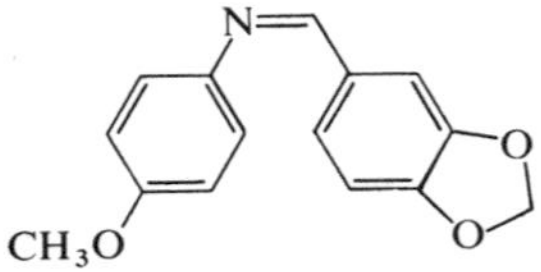 CH_3O (structure, N=)	Et_2O, air	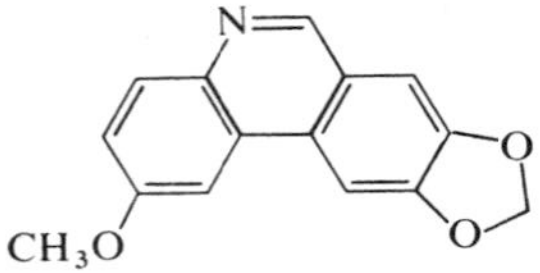 CH_3O (structure) (56)	294a

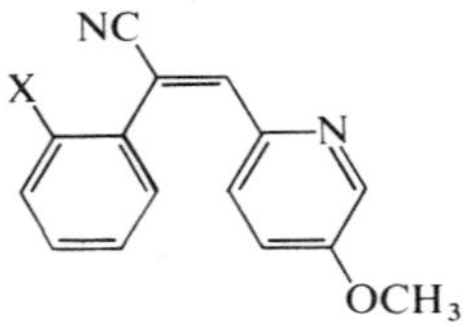

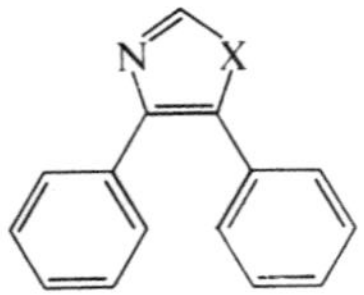

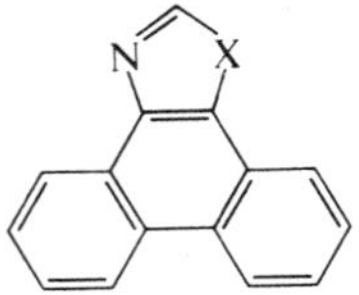

	Reactant	Conditions	Product (% yield)	Ref.
C_{16}	2-Styrylindole	EtOH, I_2	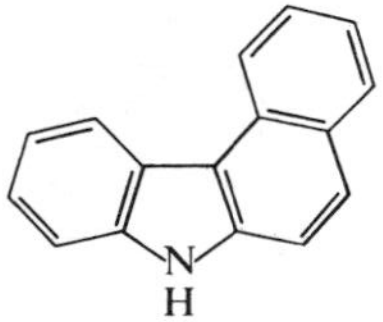 (67)	268
	3-Styrylindole	”	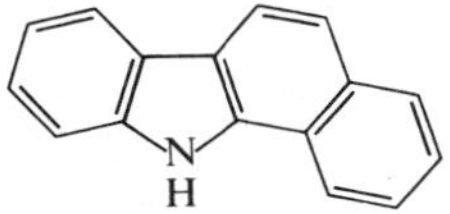 (36)	269
		C_6H_6, air	” (25)	619b
	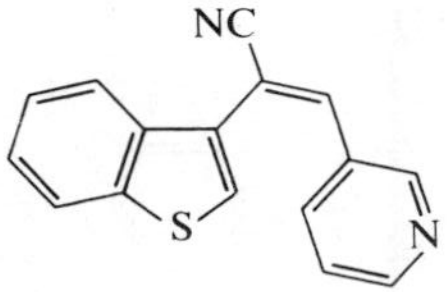 NC	EtOH, O_2	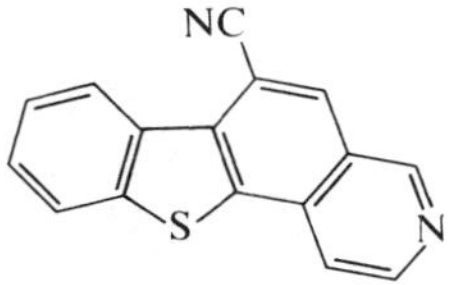 NC (62)[c]	620
	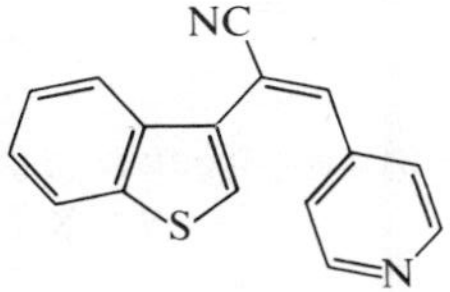 NC	”	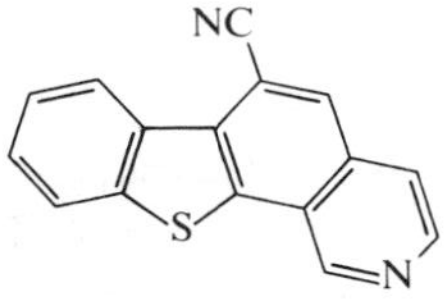 NC (78)	620
	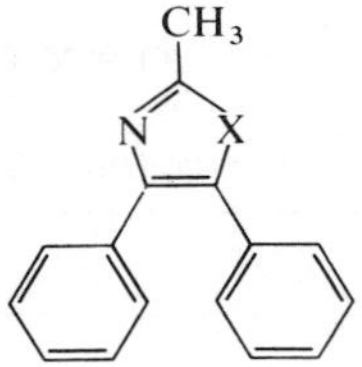 CH_3 (I)		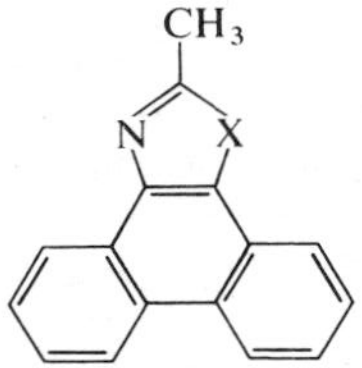 CH_3 (II)	

TABLE V. NITROGEN HETEROCYCLIC SYSTEMS (*Continued*)[a]

A. Heteroaromatic Systems with One Ring Nitrogen (*Continued*)

	Starting Material	Reaction Conditions	Products and Yields (%)	Refs.
	I, X = O	EtOH, I_2, air	II, X = O (55)	271
		EtOH, I_2, degassed	" (47)	271
		EtOH, I_2	" (—)	621
	I, X = S	EtOH, I_2, air	II, X = S (87)	271
	5-Carbethoxy-2-styrylpyridine	*t*-BuOH, C_6H_6, O_2	3-Carbethoxy-1-aza-**L** (19)	616
	CH_3O	EtOH, air	CH_3O (22)	273b
	CH_3	EtOH, O_2	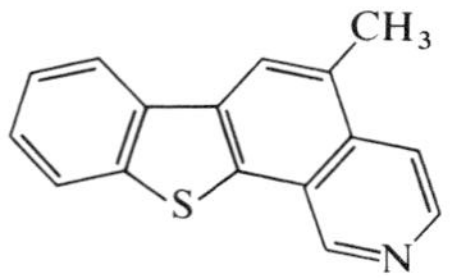 CH_3 (50)	620
C_{17}	4-Styrylquinoline	C_6H_{12}, air	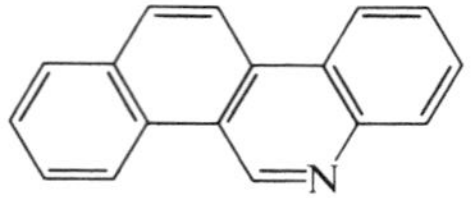 (25)	245,258
	4-Styrylisoquinoline	C_6H_{14}, air	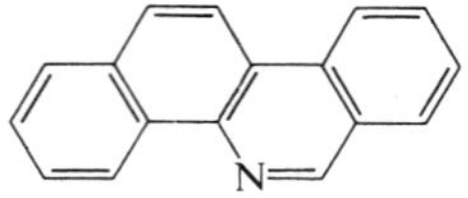 (43)	257

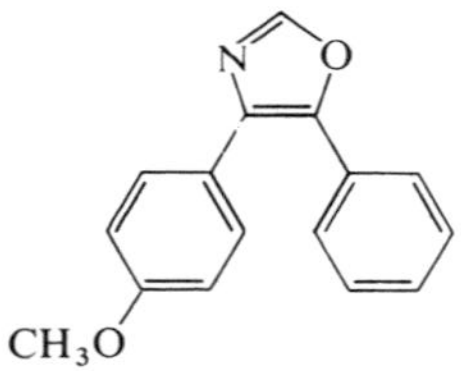

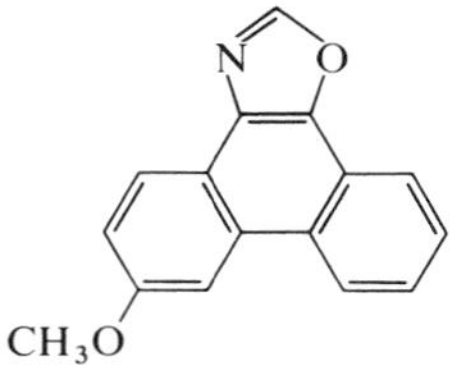

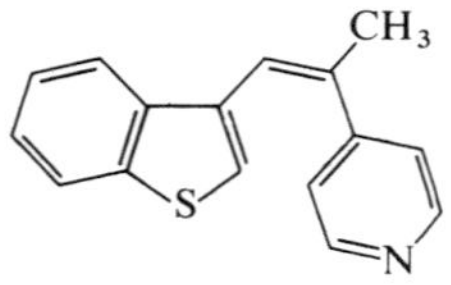

3-Styrylisoquinoline	”	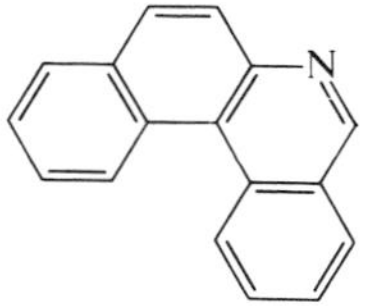 (34)	257
3-(*o*-Methylstyryl)indole	C_6H_6, air	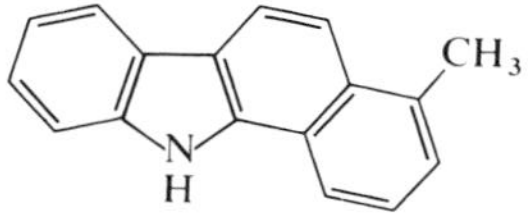 (16)	404
3-(*m*-Methylstyryl)indole	”	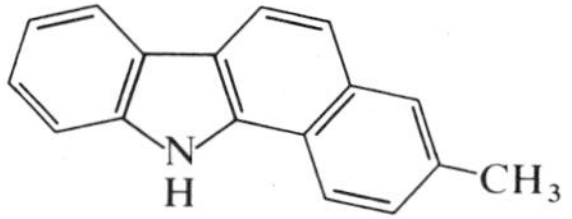 (8), 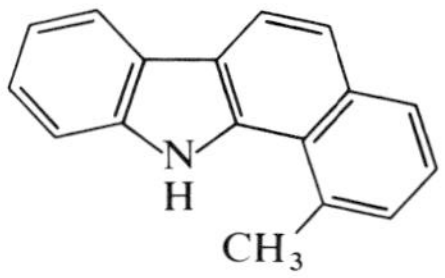 (7)	404
3-(*p*-Methylstyryl)indole	”	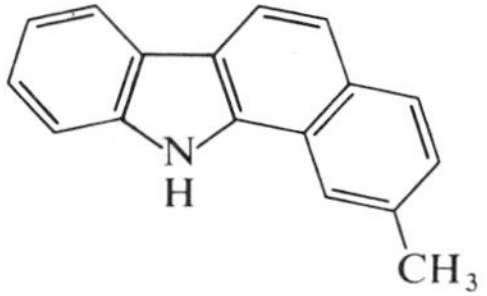 (17)	404
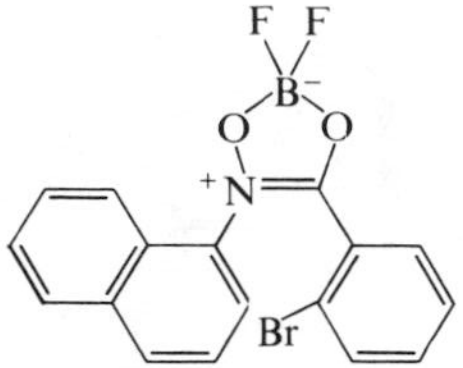	C_6H_6	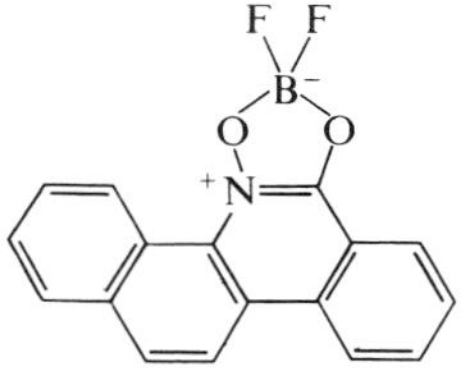 (—)	300

TABLE V. NITROGEN HETEROCYCLIC SYSTEMS (*Continued*)[a]

A. Heteroaromatic Systems with One Ring Nitrogen (*Continued*)

Starting Material	Reaction Conditions	Products and Yields (%)	Refs.
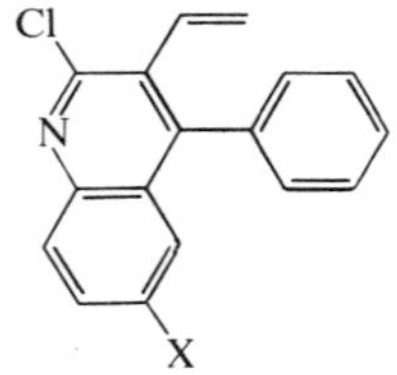 (I)		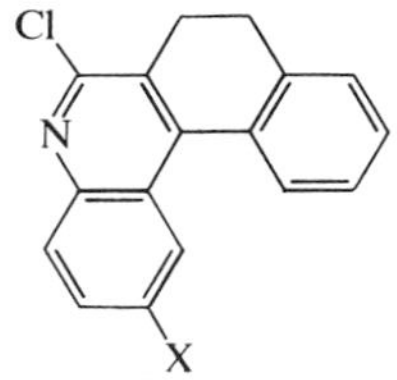 (II)	
I, X = H	C_6H_6, air	II, X = H (78)	561a
I, X = Cl	"	II, X = Cl (70)	561a
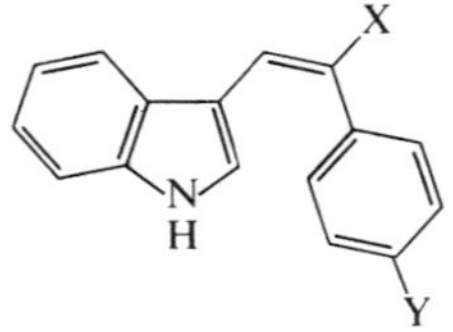 (I)		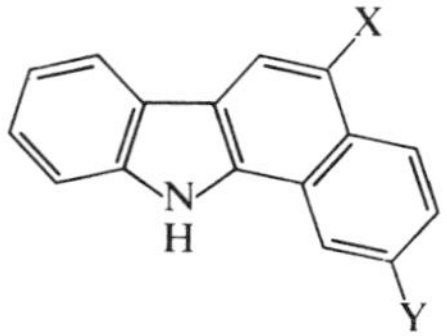 (II)	
I, X = CN, Y = H	EtOH, I_2	II, X = CN, Y = H (79)	269
I, X = H, Y = CH_3O	"	II, X = H, Y = CH_3O (24)	269
I, X = CO_2H, Y = H	"	II, X = CO_2H, Y = H (56)	269
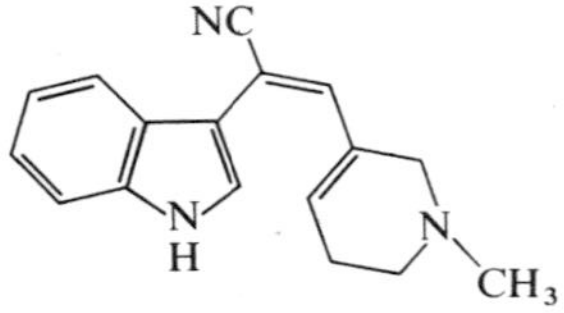	EtOH, I_2 (or $FeCl_3$)	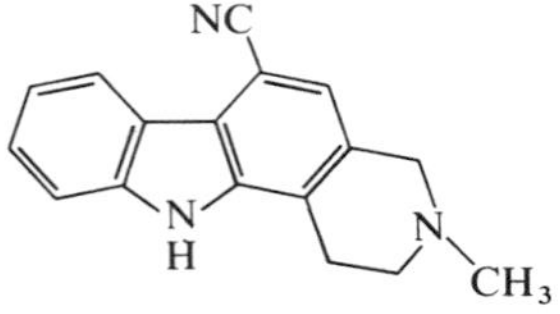 (—)	622
	EtOH, I_2, N_2	" (12)	623a

		EtOH	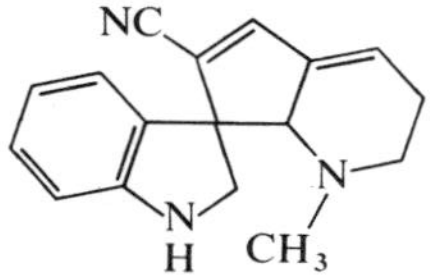 $(-)^d$	622
	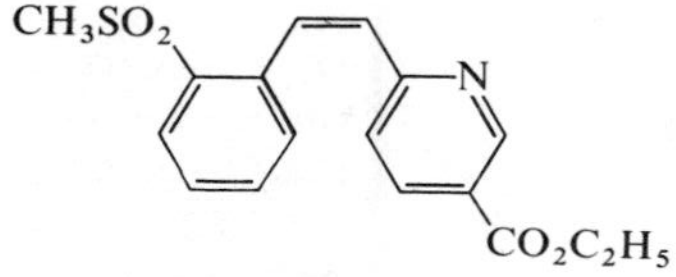	t-BuOH, C_6H_6, O_2	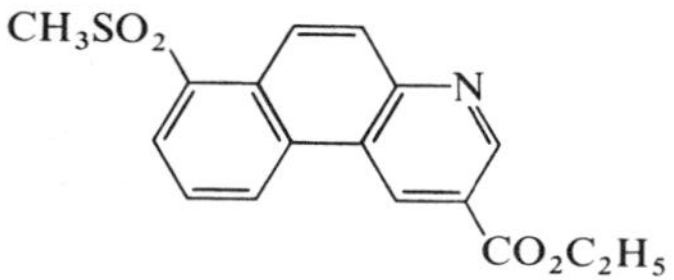 (26)	616
	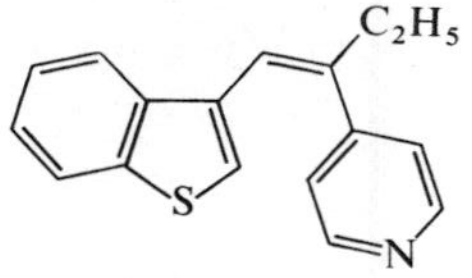	EtOH, O_2	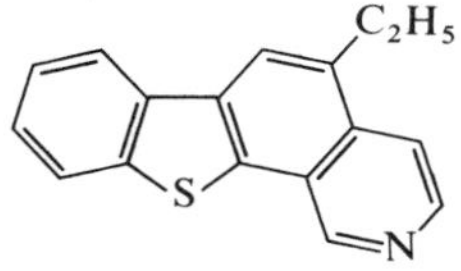 (46)	620
C_{18}	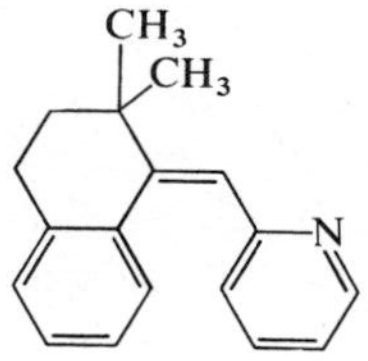	C_6H_6, t-BuOH, O_2	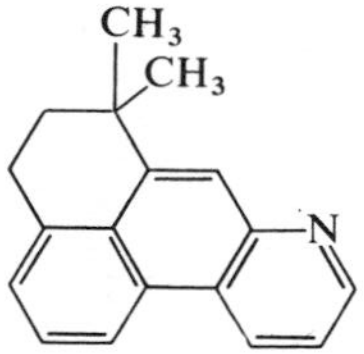 (53)	57
	$C_6H_5\overset{+}{N}H{=}\overset{-}{B}(C_6H_5)_2$	C_6H_{12}, I_2, N_2	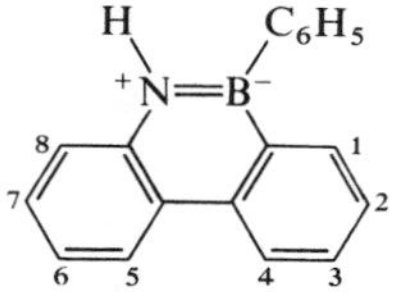 (30) 10-Phenyl-**M** (10-phenyl-10,9-borazarophenanthrene)	321a
	p-$BrC_6H_4\overset{+}{N}H{=}\overset{-}{B}(C_6H_5)_2$	"	6-Bromo-10-phenyl-**M** (4)	321a

TABLE V. NITROGEN HETEROCYCLIC SYSTEMS (*Continued*)[a]

A. Heteroaromatic Systems with One Ring Nitrogen (*Continued*)

	Starting Material	Reaction Conditions	Products and Yields (%)	Refs.
	o-$ClC_6H_4\overset{+}{N}H{=}\overset{-}{B}(C_6H_5)_2$	C_6H_{12}, I_2, N_2	8-Chloro-10-phenyl-**M** (20)	321a
	m-$ClC_6H_4\overset{+}{N}H{=}\overset{-}{B}(C_6H_5)_2$	"	7-Chloro-10-phenyl-**M** (17)	321a
	p-$ClC_6H_4\overset{+}{N}H{=}\overset{-}{B}(C_6H_5)_2$	"	6-Chloro-10-phenyl-**M** (15)	321a
	o-$IC_6H_4\overset{+}{N}H{=}\overset{-}{B}(C_6H_5)_2$	C_6H_{12}, N_2	10-Phenyl-**M** (51)	623b
	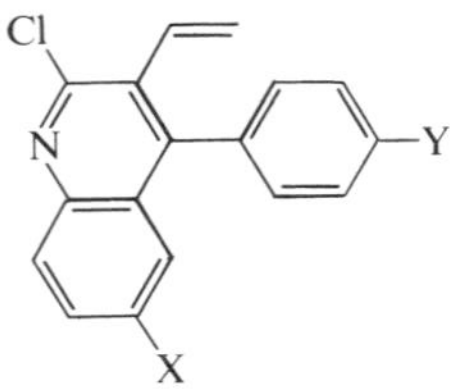 (I) [structure: Cl, N, X, Y]		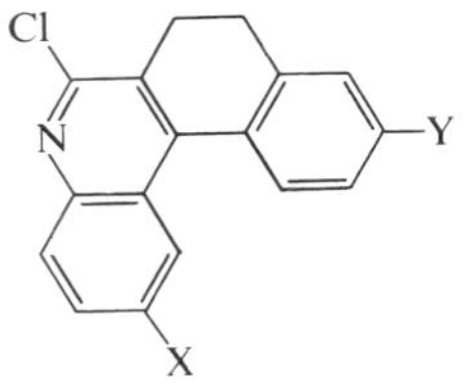 (II) [structure: Cl, N, X, Y]	
	I, X = CH_3, Y = H	C_6H_6, air	II, X = CH_3, Y = H (73)	561a
	I, X = H, Y = CH_3	"	II, X = H, Y = CH_3 (75)	561a
C_{19}	$C_6H_5CH{=}C(Ar)C_6H_5$			
	Ar = 2-Pyridyl	C_6H_{14}, O_2	10-Phenyl-1-aza-**L** (34), 9-(2-pyridyl)-**L** (12)	614
	Ar = 3-Pyridyl	C_6H_{14}, air	10-Phenyl-2-aza-**L** + 10-phenyl-4-aza-**L** (7:1)	624
	Ar = 4-Pyridyl	C_6H_{14}, O_2	10-Phenyl-3-aza-**L** (—), 9-(4-pyridyl)-**L** (—)	614
	$C_6H_5N{=}C(C_6H_5)_2$	C_6H_{12}, I_2, air	10-Phenyl-9-aza-**L** (46)	285
	4-Phenyl-1-aza-**L**	EtOH, air	 [structure: N] (51)	625

$C_6H_5\overset{+}{N}(CH_3)=\overset{-}{B}(C_6H_5)_2$	C_6H_{12}, I_2, N_2	9-Methyl-10-phenyl-**M** (18)	321a
o-$CH_3C_6H_4\overset{+}{N}H=\overset{-}{B}(C_6H_5)_2$	"	8-Methyl-10-phenyl-**M** (33)	321a
m-$CH_3C_6H_4\overset{+}{N}H=\overset{-}{B}(C_6H_5)_2$	"	7-Methyl-10-phenyl-**M** (33)	321a
p-$CH_3C_6H_4\overset{+}{N}H=\overset{-}{B}(C_6H_5)_2$	"	6-Methyl-10-phenyl-**M** (30)	321a
o-$CH_3OC_6H_4\overset{+}{N}H=\overset{-}{B}(C_6H_5)_2$	"	8-Methoxy-10-phenyl-**M** (20)	321a
m-$CH_3OC_6H_4\overset{+}{N}H=\overset{-}{B}(C_6H_5)_2$	"	7-Methoxy-10-phenyl-**M** (14)	321a
p-$CH_3OC_6H_4\overset{+}{N}H=\overset{-}{B}(C_6H_5)_2$	"	6-Methoxy-10-phenyl-**M** (10)	321a
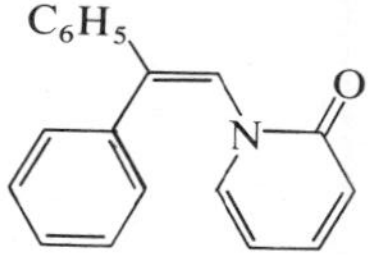 [structure: C_6H_5, N, O]	MeCN, HCl, degassed[e]	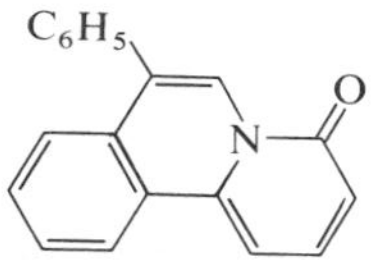 [structure: C_6H_5, N, O] (60)	248
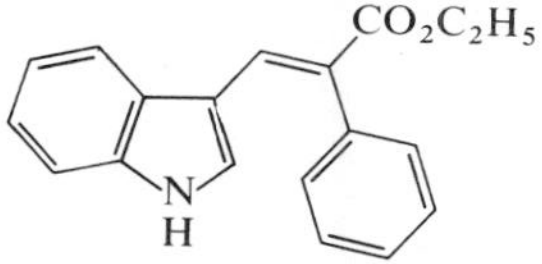 [structure: $CO_2C_2H_5$, N, H]	EtOH, I_2	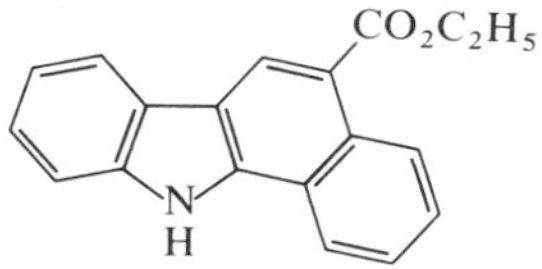 [structure: $CO_2C_2H_5$, N, H] (79)	269
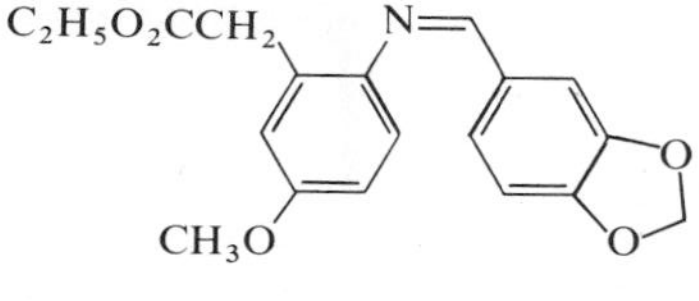 [structure: $C_2H_5O_2CCH_2$, N=, CH_3O, O, O]	Et_2O, air	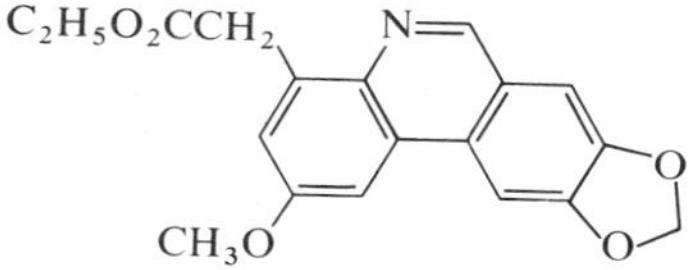 [structure: $C_2H_5O_2CCH_2$, N=, CH_3O, O, O] (21)	294a

TABLE V. NITROGEN HETEROCYCLIC SYSTEMS (*Continued*)[a]

A. Heteroaromatic Systems with One Ring Nitrogen (*Continued*)

	Starting Material	Reaction Conditions	Products and Yields (%)	Refs.
C_{20}	2,3-Diphenylindole	HOAc, I_2	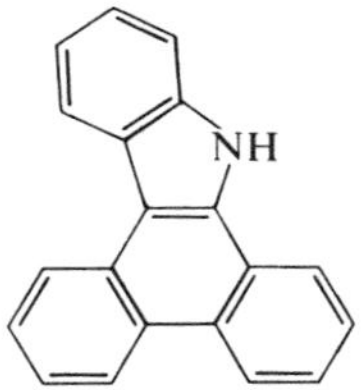 NH (80)	270
	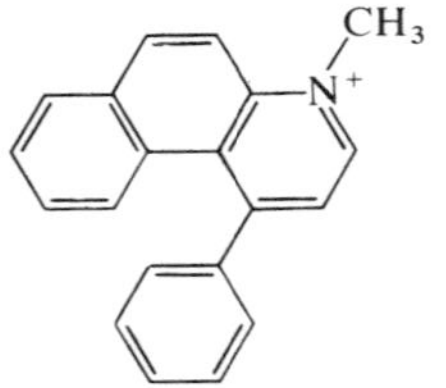 CH_3, N^+	EtOH, H_2O, air	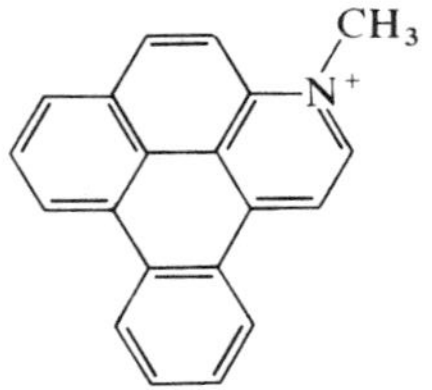 CH_3, N^+ (66)	625
	$2,3\text{-}(CH_3)_2C_6H_3\overset{+}{N}H{=}\overset{-}{B}(C_6H_5)_2$	C_6H_{12}, I_2, N_2	7,8-Dimethyl-10-phenyl-**M** (18)	321a
	$2,4\text{-}(CH_3)_2C_6H_3\overset{+}{N}H{=}\overset{-}{B}(C_6H_5)_2$	"	6,8-Dimethyl-10-phenyl-**M** (27)	321a
	$2,6\text{-}(CH_3)_2C_6H_3\overset{+}{N}H{=}\overset{-}{B}(C_6H_5)_2$	"	8-Methyl-10-phenyl-**M** (—)	321a,b
	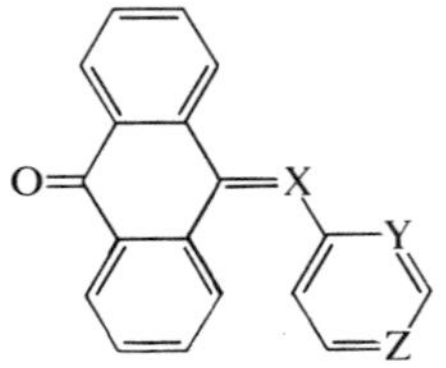 O=, =X, Y, Z (I)		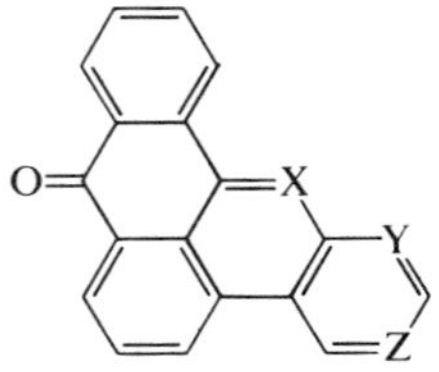 O=, =X, Y, Z (II)	

	Reactant	Conditions	Product (Yield, %)	Ref.
	I, X = N, Y = Z = CH	Concd H_2SO_4	II, X = N, Y = Z = CH (82)	58
	I, Y = N, X = Z = CH	C_6H_6 or C_6H_{12}, I_2	II, Y = N, X = Z = CH (20–90)	58
	I, Z = N, X = Y = CH	”	II, Z = N, X = Y = CH (20–90)	58
	1-(*o*-Benzoyloxystyryl)pyridinium ion	EtOH, I_2	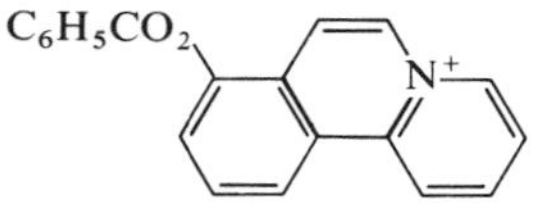 $C_6H_5CO_2$ (43)	247
C_{21}	3,4-Diphenylquinoline	2 *N* HCl, air	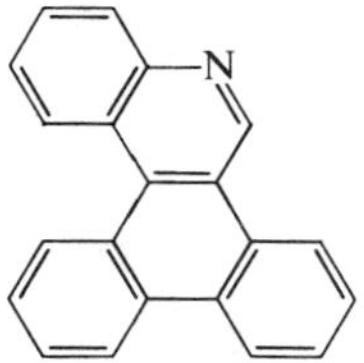 N (81)	266
	1,2-Diphenylquinolinium ion	MeCN, I_2	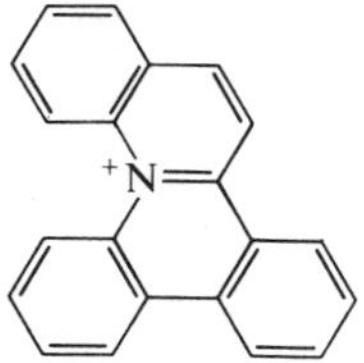 ^+N (50)	261
	1,2-Diphenylisoquinolinium ion	1. MeCN, I_2 2. $NaBH_4$	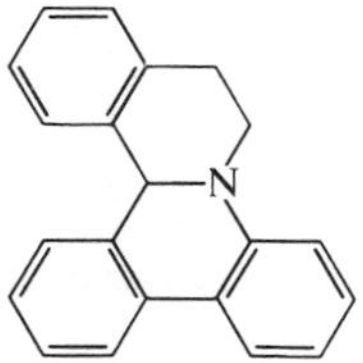 N (43)	262

TABLE V. NITROGEN HETEROCYCLIC SYSTEMS (*Continued*)[a]

A. Heteroaromatic Systems with One Ring Nitrogen (*Continued*)

Starting Material	Reaction Conditions	Products and Yields (%)	Refs.
4-Styryl-1-aza-**L**	EtOH, air	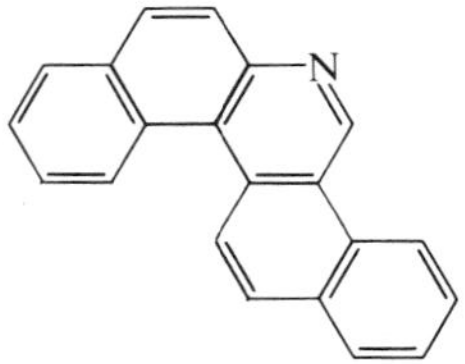 (4), 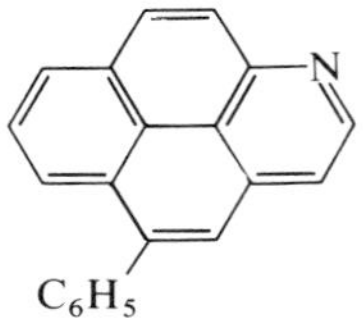 C_6H_5 (5)	625
1-Styryl-4-aza-**L**	”	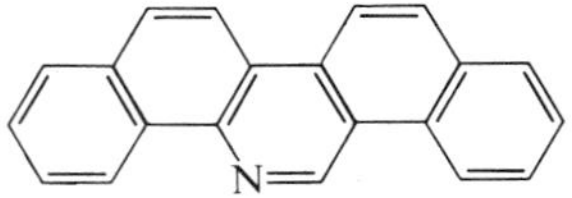 (18)	259
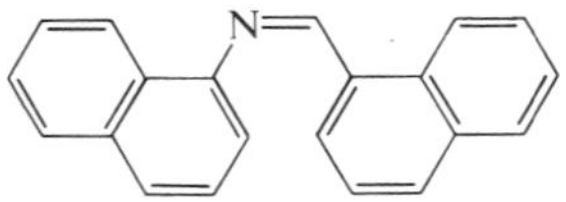	”	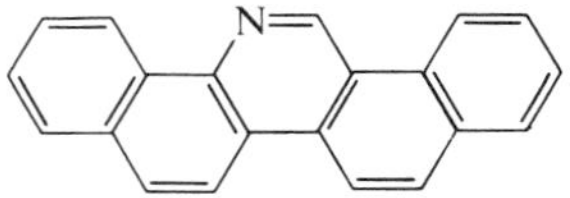 (40)	295
	C_6H_6, O_2	” (15)	295
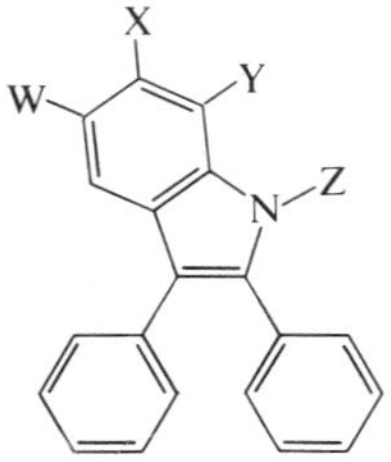 W X Y N Z (I)		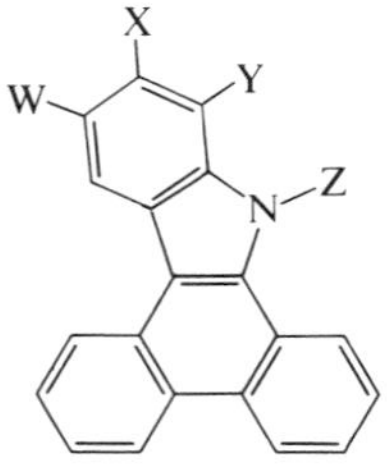 W X Y N Z (II)	

I, W = CH_3, X = Y = Z = H	HOAc, I_2	II, W = CH_3, X = Y = Z = H (72)	270
I, X = CH_3, W = Y = Z = H	”	II, X = CH_3, W = Y = Z = H (57)	270
I, Y = CH_3, W = X = Z = H	”	II, Y = CH_3, W = X = Z = H (64)	270
I, Z = CH_3, W = X = Y = H	”	II, Z = CH_3, W = X = Y = H (65)	270
o-$C_2H_5O_2CC_6H_4\overset{+}{N}H{=}\overset{-}{B}(C_6H_5)_2$	C_6H_{12}, I_2, N_2	8-Carbethoxy-10-phenyl-**M** (24)	321a
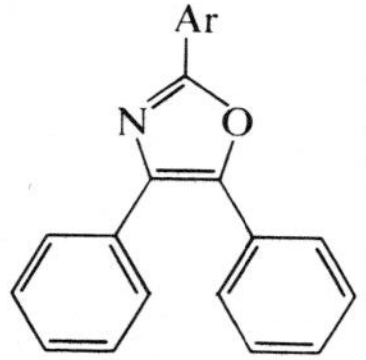 (I)		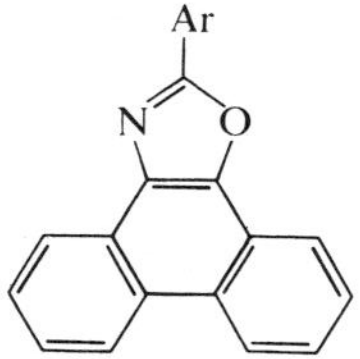 (II)	
I, Ar = Phenyl	EtOH, I_2, degassed	II, Ar = Phenyl (47)	271
	EtOH, I_2	” (50)	621
I, Ar = p-Chlorophenyl	”	II, Ar = p-Chlorophenyl (—)	621
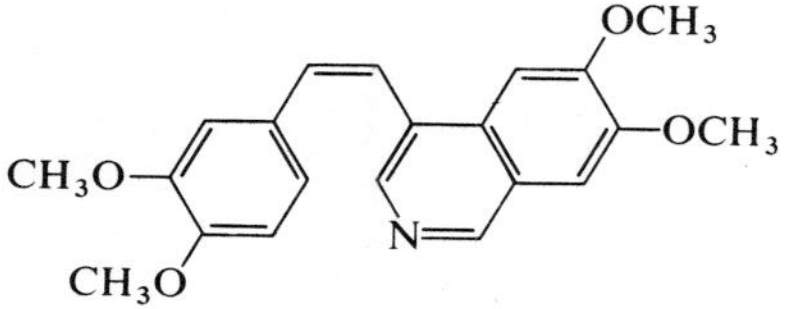 [structure: CH_3O, CH_3O, OCH_3, OCH_3, N]	EtOH, air	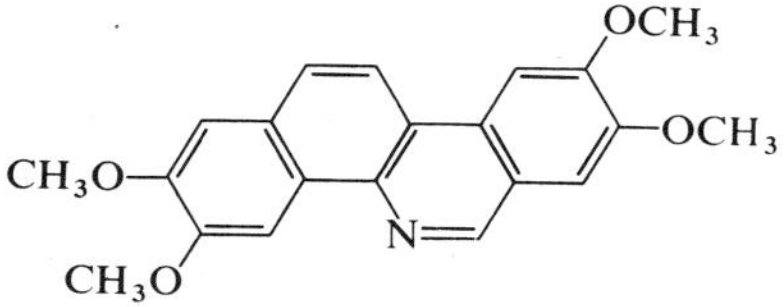 [structure: CH_3O, CH_3O, OCH_3, OCH_3, N] (27)	263
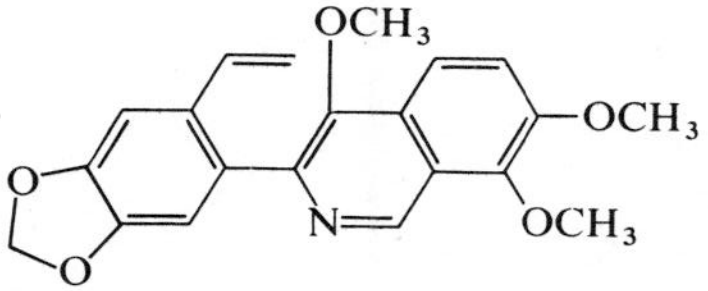 [structure: OCH_3, OCH_3, OCH_3, O, O, N]	C_6H_6, N_2	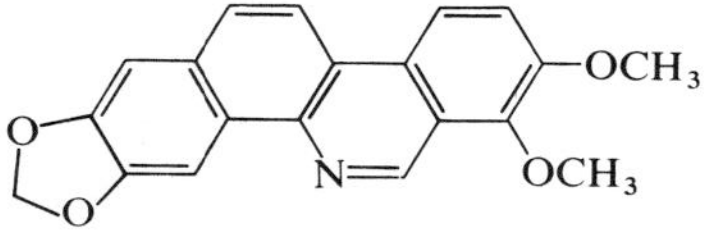 [structure: OCH_3, OCH_3, O, O, N] (57)	265

TABLE V. NITROGEN HETEROCYCLIC SYSTEMS (*Continued*)[a]

A. Heteroaromatic Systems with One Ring Nitrogen (*Continued*)

	Starting Material	Reaction Conditions	Products and Yields (%)	Refs.
C_{22}	(I)		(II)	
	I, X = CH_3, Y = H	HOAc, I_2	II, X = CH_3, Y = H (34)	270
	I, X = H, Y = CH_3O	C_6H_6, I_2	II, X = H, Y = CH_3O (22)	626
C_{24}	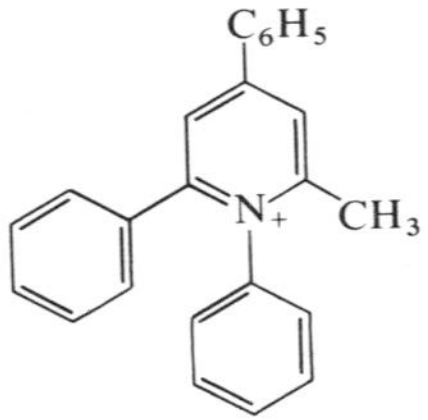 C_6H_5, N^+, CH_3	MeOH, O_2	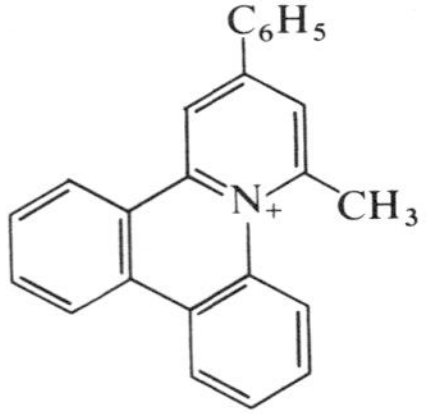 C_6H_5, N^+, CH_3 (16)	627,628a
	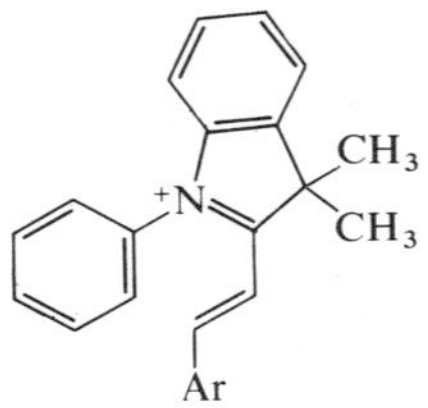 ^+N, CH_3, CH_3, Ar (I)		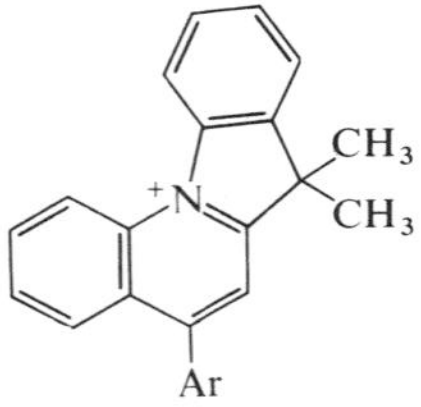 ^+N, CH_3, CH_3, Ar (II)	

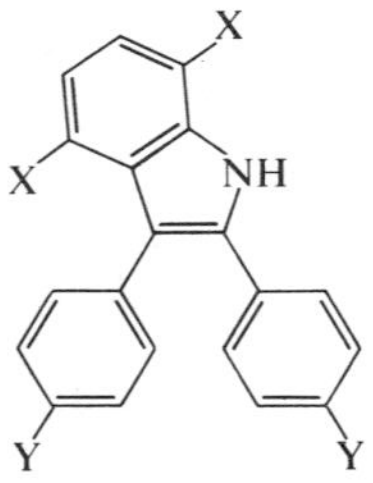

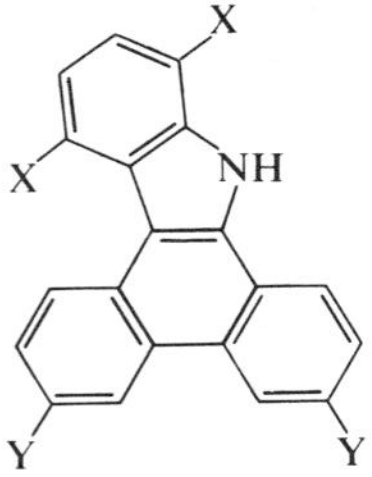

I, Ar = Phenyl	MeOH, air	II, Ar = Phenyl (very good)	628b
I, Ar = *p*-Chlorophenyl	"	II, Ar = *p*-Chlorophenyl (very good)	628b
I, Ar = *p*-Fluorophenyl	"	II, Ar = *p*-Fluorophenyl (very good)	628b
$C_6H_5\overset{+}{N}H{=}\overset{-}{B}[C_6H_2(CH_3)_3\text{-}2,4,6]_2$	C_6H_{12}, I_2	1,3,4-Trimethyl-10-(2,4,6-trimethylphenyl)-**M** (58)	408
	C_6H_{12}, I_2, N_2	1,3,4-Trimethyl-10-(2,4,6-trimethylphenyl)-**M** + 1,3-dimethyl-10-(2,4,6-trimethylphenyl)-**M** (88:12)	321b
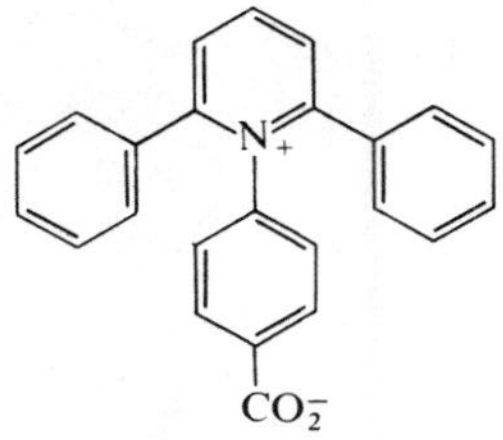	Solvent, air	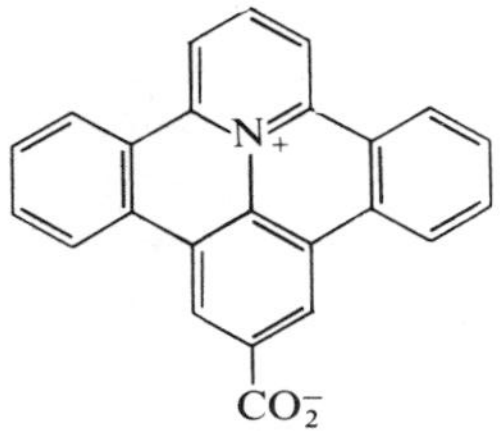 (—)	628a
C_{25} 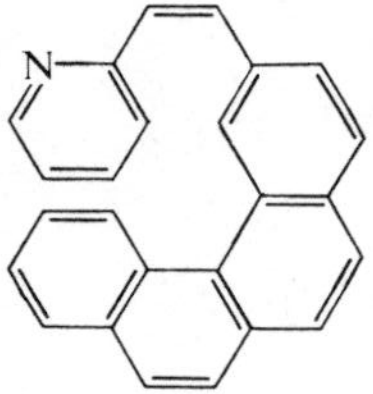	C_6H_6, I_2, air	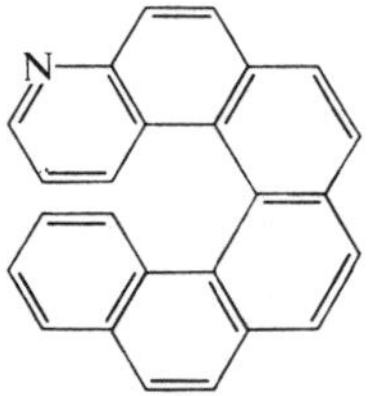 (18)	260
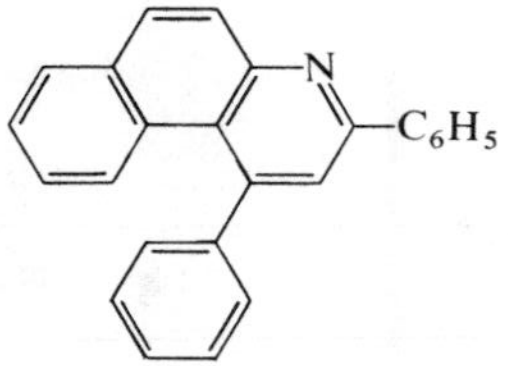	EtOH, H_2SO_4, air	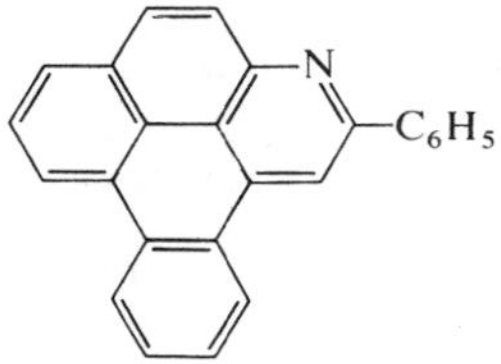 (91)	267

TABLE V. NITROGEN HETEROCYCLIC SYSTEMS (*Continued*)[a]

A. Heteroaromatic Systems with One Ring Nitrogen (*Continued*)

Starting Material	Reaction Conditions	Products and Yields (%)	Refs.
3,5-Diphenyl-1-styrylpyridinium ion	EtOH, I_2	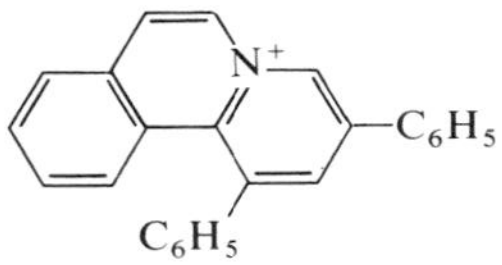 [structure: N^+, C_6H_5, C_6H_5] (50)	247
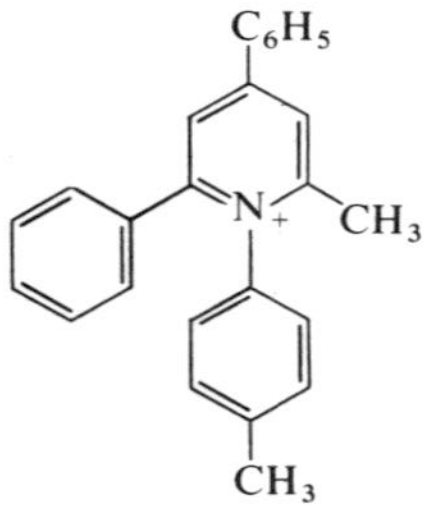 [structure: C_6H_5, N^+, CH_3, CH_3]	MeOH, O_2	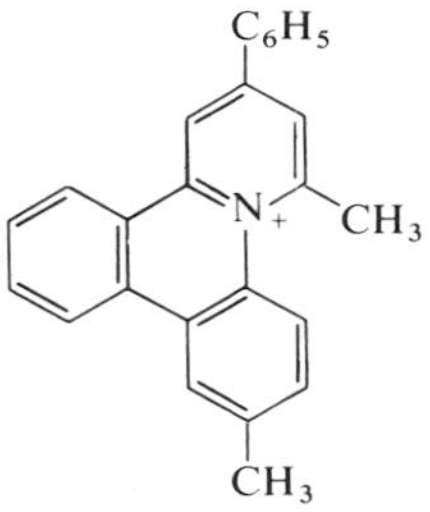 [structure: C_6H_5, N^+, CH_3, CH_3] (15)	627
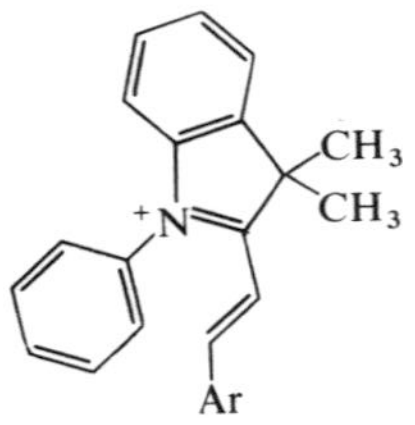 [structure: ^+N, CH_3, CH_3, Ar] (I)		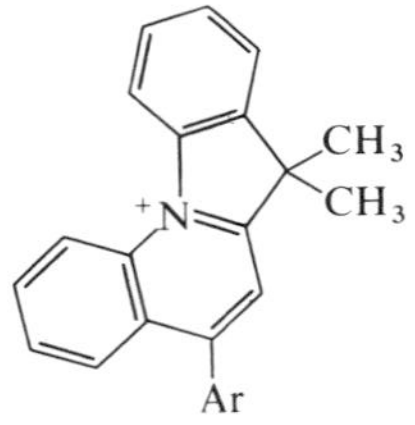 [structure: ^+N, CH_3, CH_3, Ar] (II)	
I, Ar = *p*-Methylphenyl	MeOH, air	II, Ar = *p*-Methylphenyl (very good)	628b
I, Ar = *o*-Methoxyphenyl	"	II, Ar = *o*-Methoxyphenyl (very good)	628b
I, Ar = *m*-Methoxyphenyl	"	II, Ar = *m*-Methoxyphenyl (very good)	628b
I, Ar = *p*-Methoxyphenyl	"	II, Ar = *p*-Methoxyphenyl (very good)	628b

Reactant	Conditions	Product (Yield %)		Ref.
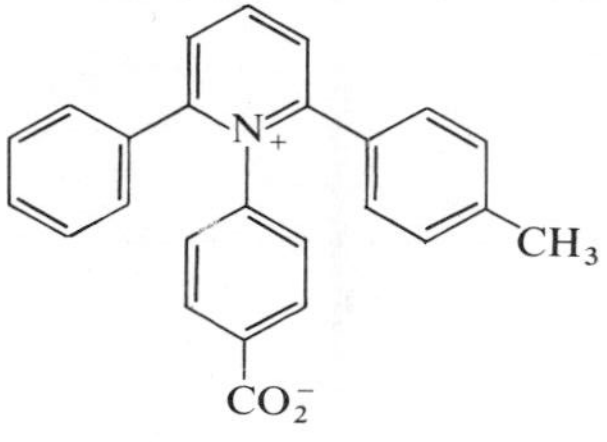 CH_3, CO_2^-	Solvent, air	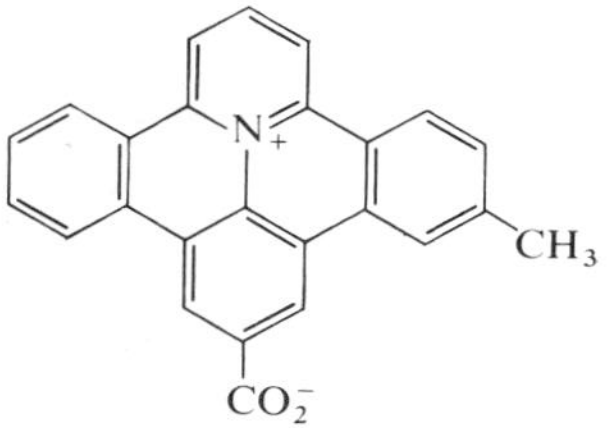 CH_3, CO_2^-	(—)	628a
C_{26} 1,2,3-Triphenylindole	$HOAc, I_2$	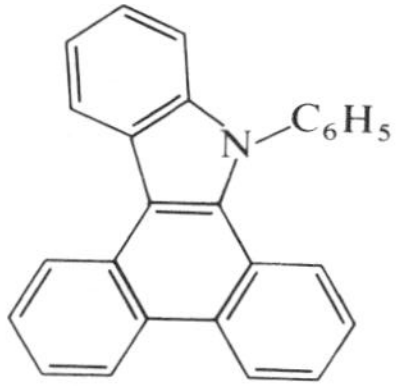 C_6H_5	(77)	270
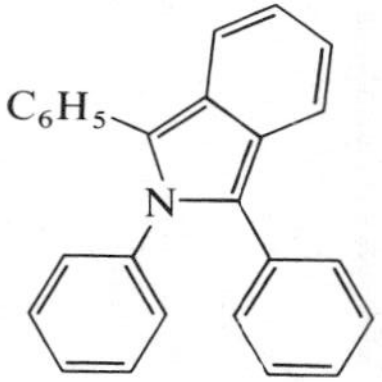 C_6H_5	CH_2Cl_2, O_2	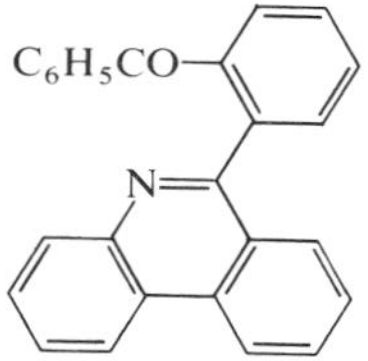 C_6H_5CO	(—)	297
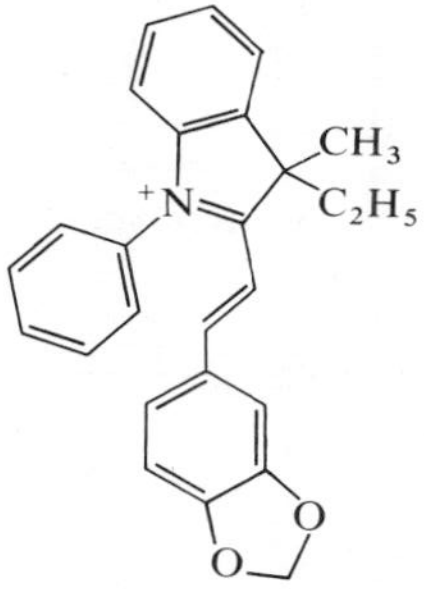 CH_3, C_2H_5	MeOH, air	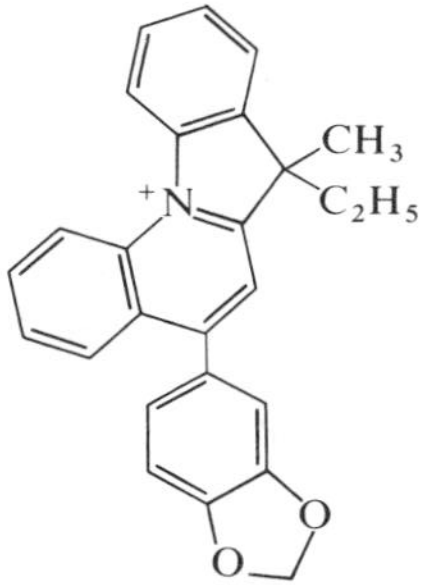 CH_3, C_2H_5	(very good)	628b

TABLE V. NITROGEN HETEROCYCLIC SYSTEMS (*Continued*)[a]

A. Heteroaromatic Systems with One Ring Nitrogen (*Continued*)

	Starting Material	Reaction Conditions	Products and Yields (%)	Refs.
C_{27}	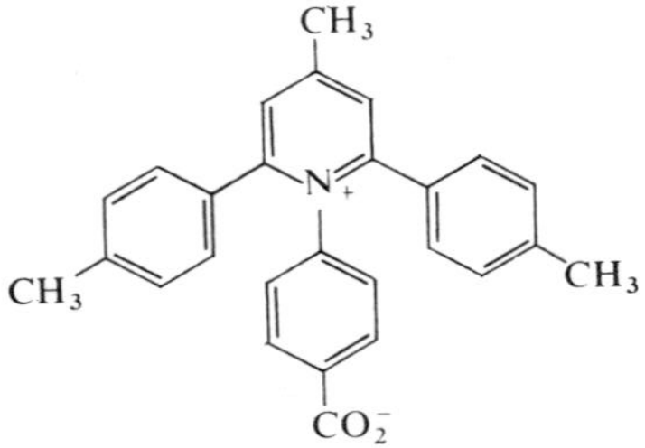 CH_3, CH_3, CH_3, N^+, CO_2^-	Solvent, air	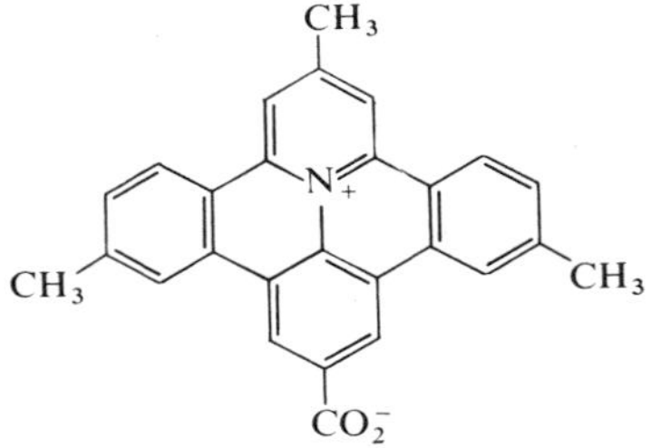 CH_3, N^+, CH_3, CH_3, CO_2^- (—)	628a
	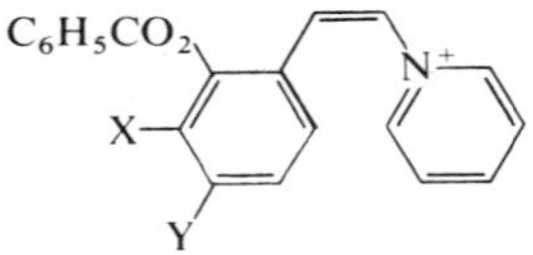 $C_6H_5CO_2$, X, Y, N^+ (I)		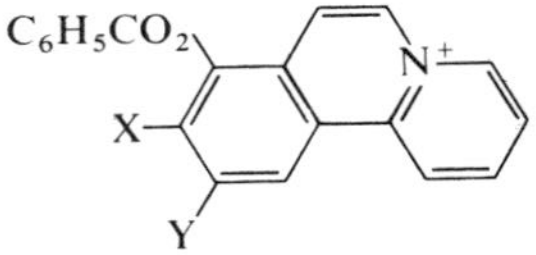 $C_6H_5CO_2$, X, Y, N^+ (II)	
	I, X = $C_6H_5CO_2$, Y = H	EtOH, I_2	II, X = $C_6H_5CO_2$, Y = H (25)	247
	I, X = H, Y = $C_6H_5CO_2$	"	II, X = H, Y = $C_6H_5CO_2$ (50)	247
C_{28}	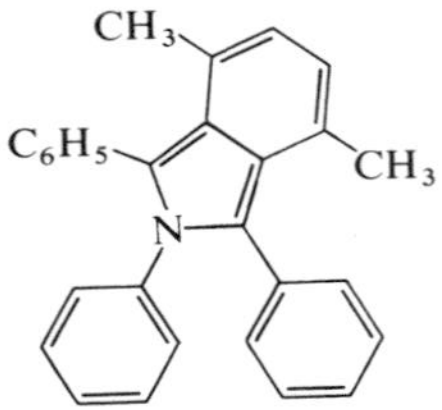 CH_3, C_6H_5, CH_3, N	CH_2Cl_2, O_2	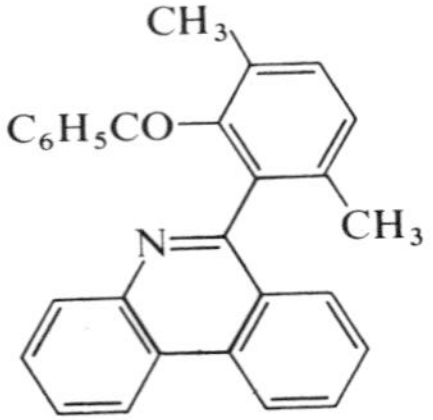 CH_3, C_6H_5CO, CH_3, N (42)	297

	Reactant	Conditions	Product (yield %)	Ref.
	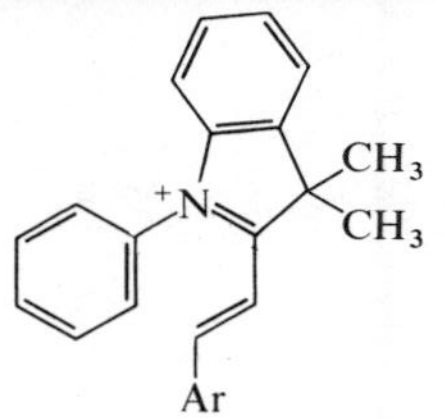 (I)		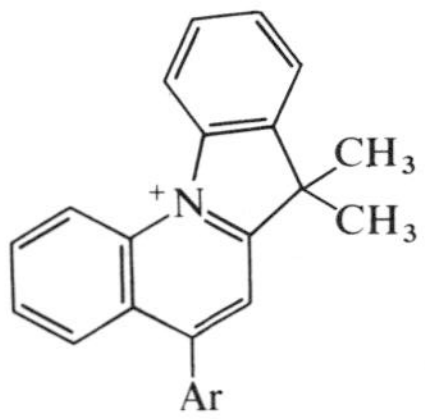 (II)	
	I, Ar = 1-Naphthyl	MeOH, air	II, Ar = 1-Naphthyl (very good)	628b
	I, Ar = 2-Naphthyl	”	II, Ar = 2-Naphthyl (very good)	628b
	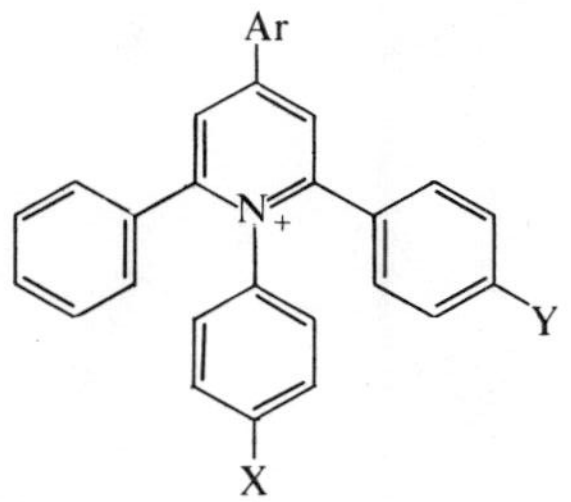 (I)		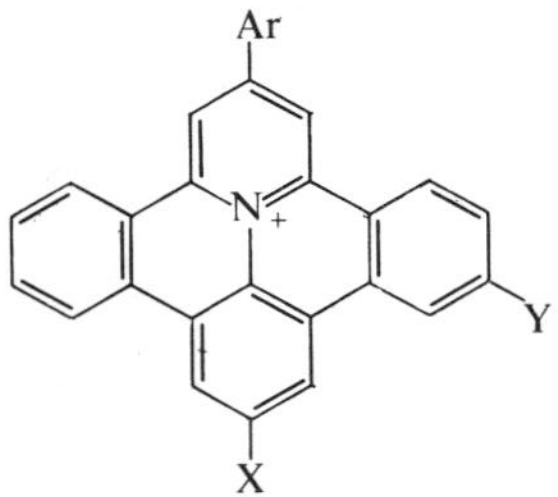 (II)	
C_{29}	I, Ar = Phenyl, X = Y = H	MeOH, O_2	II, Ar = Phenyl, X = Y = H (85)	627,628a
C_{30}	I, Ar = Phenyl, X = CH_3, Y = H	”	II, Ar = Phenyl, X = CH_3, Y = H (36)	627,628a
	I, Ar = Phenyl, X = H, Y = CH_3	”	II, Ar = Phenyl, X = H, Y = CH_3 (37)	627,628a
	I, Ar = *p*-Methylphenyl, X = Y = H	”	II, Ar = *p*-Methylphenyl, X = Y = H (61)	627,628a
	I, Ar = Phenyl, X = CO_2^-, Y = H	”	II, Ar = Phenyl, X = CO_2^-, Y = H (10)	627
	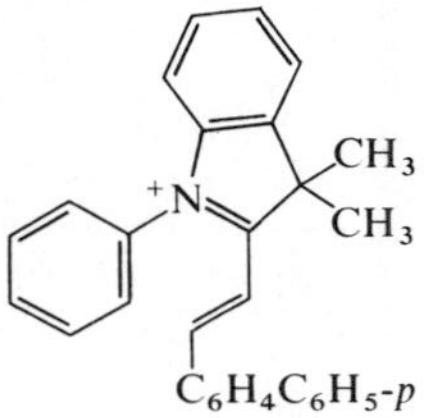 $C_6H_4C_6H_5$-*p*	MeOH, air	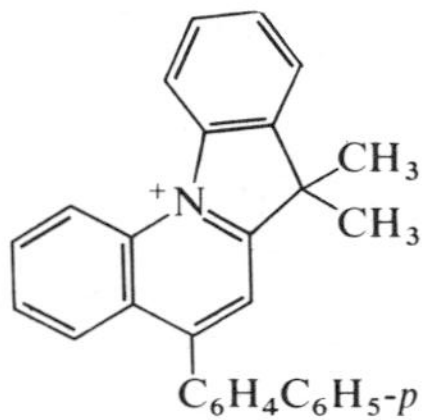 $C_6H_4C_6H_5$-*p* (very good)	628b
	$(C_6H_5)_2\overset{+}{N}{=}\overset{-}{B}[C_6H_2(CH_3)_3\text{-}2,4,6]_2$	C_6H_{12}, I_2, N_2	1,3,4-Trimethyl-10-(2,4,6-trimethylphenyl)-9-phenyl-**M** (48)	321b

TABLE V. NITROGEN HETEROCYCLIC SYSTEMS (*Continued*)[a]

A. Heteroaromatic Systems with One Ring Nitrogen (*Continued*)

	Starting Material	Reaction Conditions	Products and Yields (%)	Refs.
C_{31}	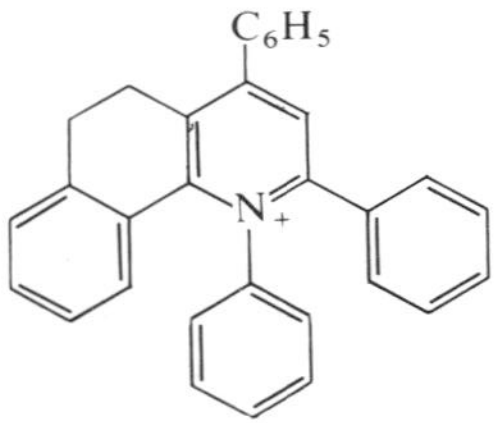 [structure: 4-C_6H_5 pyridinium]	MeOH, O_2	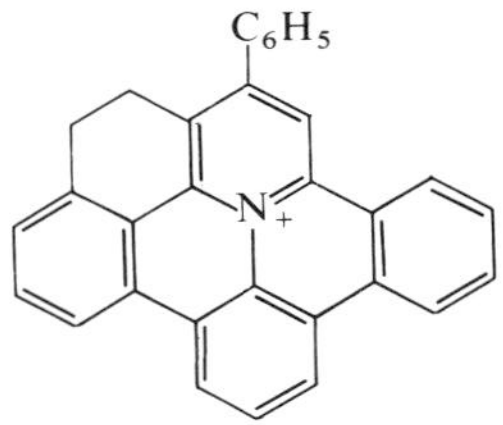 [structure: 4-C_6H_5 pyridinium] (18)	627
	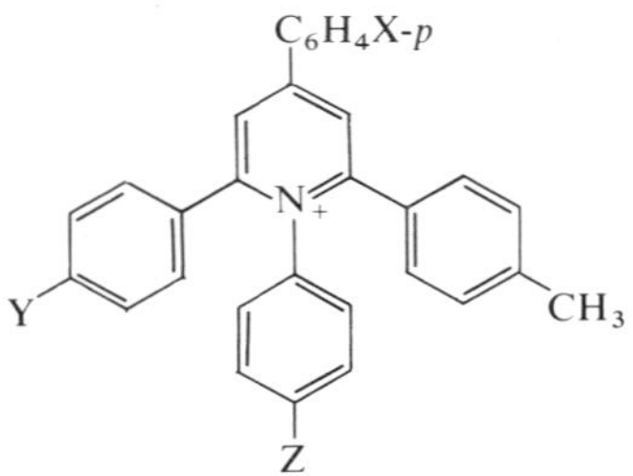 [structure: 4-C_6H_4X-*p* pyridinium, Y, Z, CH_3] (I)		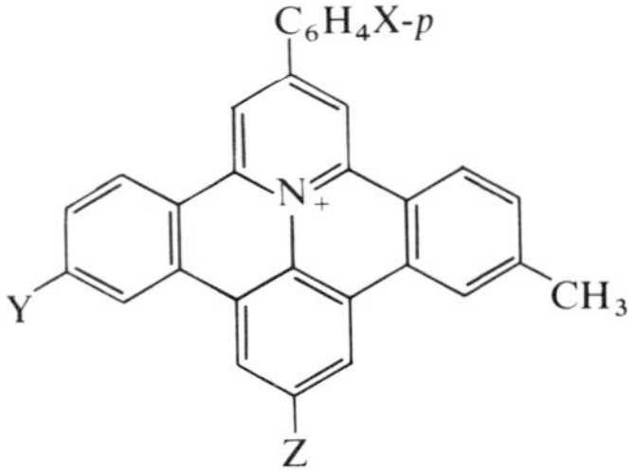 [structure: C_6H_4X-*p*, Y, Z, CH_3] (II)	
	I, X = Y = H, Z = CO_2^-	"	II, X = Y = H, Z = CO_2^- (18)	627
C_{32}	I, X = Y = CH_3, Z = H	"	II, X = Y = CH_3, Z = H (12)	627,628a
C_{33}	I, X = Y = CH_3, Z = CO_2^-	"	II, X = Y = CH_3, Z = CO_2^- (29)	627
	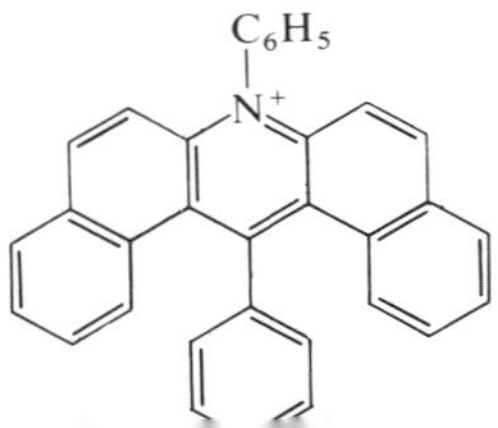 [structure: N-C_6H_5 pyridinium]	HOAc, air	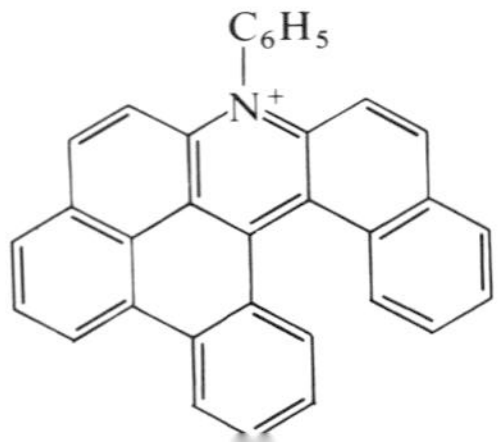 [structure: N-C_6H_5 pyridinium] (40)	340a

	Reactant	Conditions	Product (yield %)	Ref.
	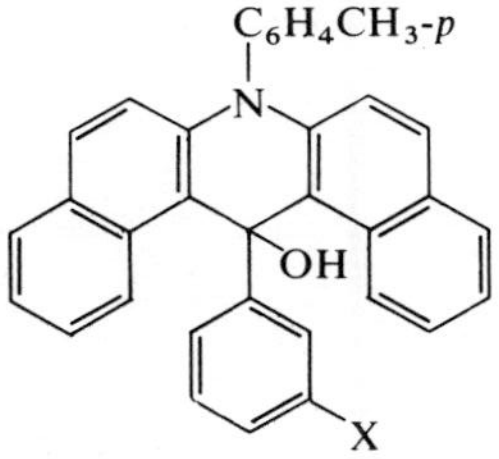 (I) (N-$C_6H_4CH_3$-p, OH, 3-X-phenyl)		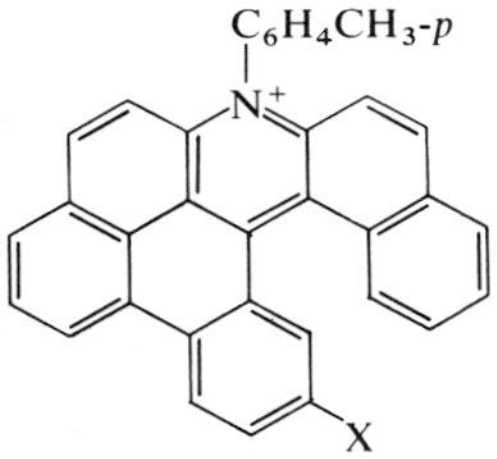 (II) (N^+-$C_6H_4CH_3$-p, X)	
C_{34}	I, X = H	”	II, X = H (—)	340a
C_{35}	I, X = CH_3O	”	II, X = CH_3O (38)	340a
C_{36}	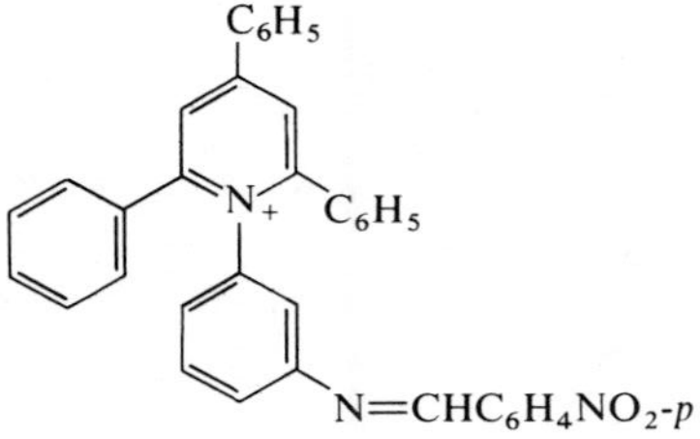 2,6-diphenyl-4-C_6H_5-pyridinium, N^+-aryl with $N{=}CHC_6H_4NO_2$-p	CH_2Cl_2, argon	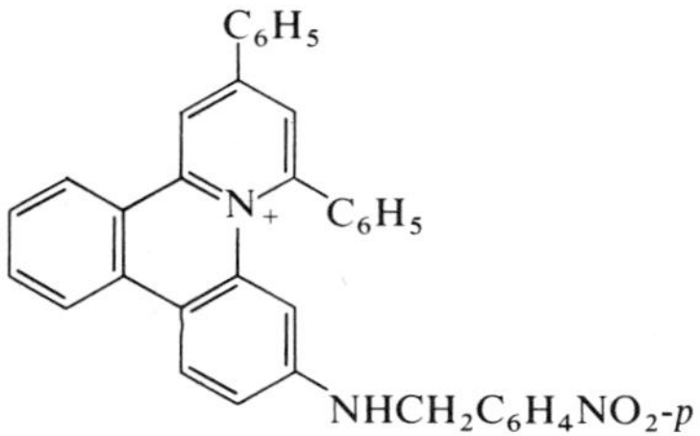 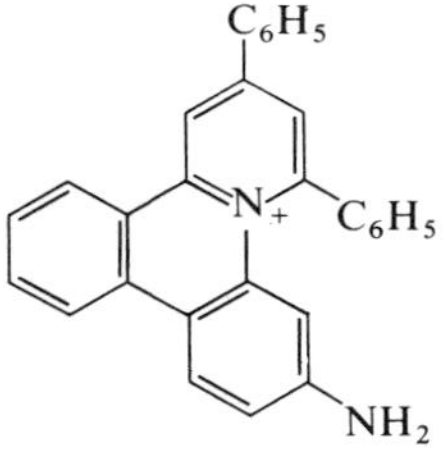 (C_6H_5, N^+, C_6H_5, $NHCH_2C_6H_4NO_2$-p) (35), (C_6H_5, N^+, C_6H_5, NH_2) (30)	629

TABLE V. NITROGEN HETEROCYCLIC SYSTEMS (*Continued*)[a]

B. Heteroaromatic Systems with Two or More Ring Nitrogens

	Starting Material	Reaction Conditions	Products and Yields (%)	Refs.
C_{10}	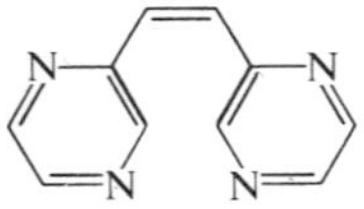	C_6H_6, O_2	1,4,5,8-Tetraaza-**L** (82)	630
C_{11}	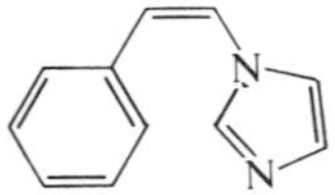	MeOH, I_2	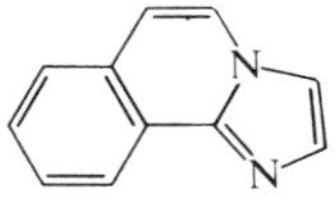 (12)	278
	3-Phenylazopyridine	Concd H_2SO_4	2,9,10-Triaza-**L** (19), 4,9,10-triaza-**L** (3)	316
	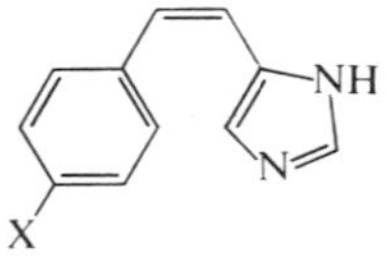 (I)		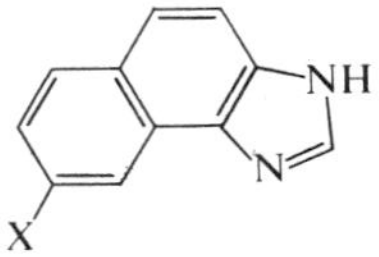 (II)	
	I, X = H	Abs EtOH, I_2	II, X = H (50)	631
	I, X = Cl	"	II, X = Cl (50)	631
C_{12}	I, X = CH_3	"	II, X = CH_3 (50)	631
	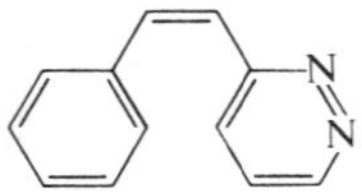	H_2SO_4	1,2-Diaza-**L** (20)	246a
	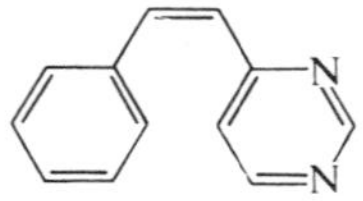	C_6H_{12}, air	1,3-Diaza-**L** (10)	246a

	"	" (9)	632
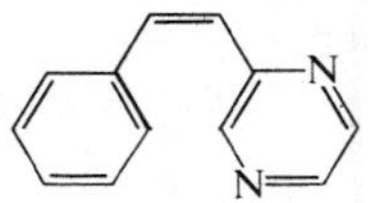	"	1,4-Diaza-**L** (22)	246a
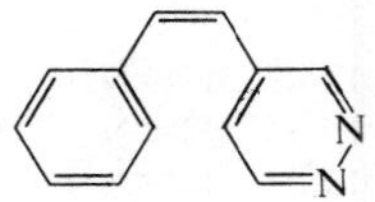	H_2SO_4	2,3-Diaza-**L** (59)	246a
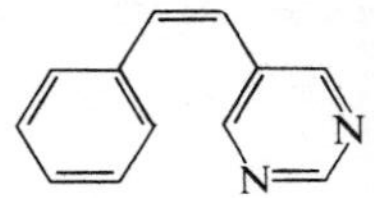	C_6H_{12}, air	2,4-Diaza-**L** (37)	246a
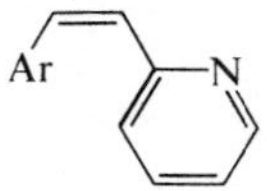 Ar = 2-Pyridyl	C_6H_6, air	1,8-Diaza-**L** (17)	249
Ar = 3-Pyridyl	"	1,7-Diaza-**L** (15)	249
Ar = 4-Pyridyl	"	1,6-Diaza-**L** (23)	249
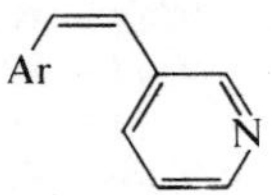 Ar = 3-Pyridyl	"	2,5-Diaza-**L** (20), 2,7-diaza-**L** (12), 4,5-diaza-**L** (1)	249
Ar = 4-Pyridyl	"	2,6-Diaza-**L** (40)	249
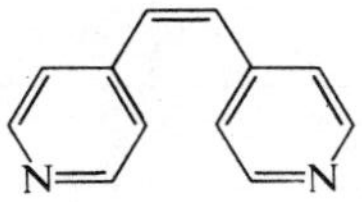	"	3,6-Diaza-**L** (47)	249
ArN=CHAr′			

TABLE V. NITROGEN HETEROCYCLIC SYSTEMS (*Continued*)[a]

B. Heteroaromatic Systems with Two or More Ring Nitrogens (*Continued*)

Starting Material	Reaction Conditions	Products and Yields (%)	Refs.
Ar = 2-Pyridyl, Ar′ = Phenyl	99% H_2SO_4	1,10-Diaza-**L** (—)	288
Ar = 3-Pyridyl, Ar′ = Phenyl	"	2,10-Diaza-**L** (—), 4,10-diaza-**L** (—)	288
Ar = 4-Pyridyl, Ar′ = Phenyl	"	3,10-Diaza-**L** (—)	288
Ar = Phenyl, Ar′ = 2-Pyridyl	"	1,9-Diaza-**L** (—)	288
Ar = Phenyl, Ar′ = 3-Pyridyl	"	2,9-Diaza-**L** (—), 4,9-diaza-**L** (—)	288
Ar = Phenyl, Ar′ = 4-Pyridyl	"	3,9-Diaza-**L** (—)	288
$C_6H_5N{=}NC_6H_5$	$ClCH_2CH_2Cl$, $SnCl_4$	9,10-Diaza-**L** (60)	29
	$ClCH_2CH_2Cl$, $AlCl_3$	" (57)	29,633
	$ClCH_2CH_2Cl$, $FeCl_3$	" (45)	29
	71% H_2SO_4, iron alum	" (55)	634
	22 *N* H_2SO_4	" (48)	308
	24 *N* H_2SO_4, EtOH	" (42)	307
	Concd H_2SO_4	" (20–40)	634
	HOAc, H_2SO_4	" (—)	634
	HOAc, CF_3CO_2H	" (—)	634
	HOAc, $FeCl_3$, degassed	" (37)	635
	HOAc, $FeCl_3$	" (—)	634
$C_6H_5N{=}NC_6H_4NH_2$-*m*	98% H_2SO_4	2-Amino-9,10-diaza-**L** (47)	310
$C_6H_5N{=}NC_6H_4Cl$-*o*	22 *N* H_2SO_4	1-Chloro-9,10-diaza-**L** (37), 9,10-diaza-**L** (12)	313
$C_6H_5N{=}NC_6H_4Cl$-*m*	"	2-Chloro-9,10-diaza-**L** (35), 4-chloro-9,10-diaza-**L** (11)	313
$C_6H_5N{=}NC_6H_4Cl$-*p*	"	3-Chloro-9,10-diaza-**L** (53)	313
	$ClCH_2CH_2Cl$, $SnCl_4$	" (49)	29
	$ClCH_2CH_2Cl$, $AlCl_3$	" (40)	29
	$ClCH_2CH_2Cl$, $FeCl_3$	" (45)	29
$C_6H_5N{=}NC_6H_4I$-*o*	22 *N* H_2SO_4	1-Iodo-9,10-diaza-**L** (29), 9,10-diaza-**L** (5)	313
$C_6H_5N{=}NC_6H_4I$-*m*	"	2-Iodo-9,10-diaza-**L** (25), 4-iodo-9,10-diaza-**L** (10)	313
$C_6H_5N{=}NC_6H_4I$-*p*	"	3-Iodo-9,10-diaza-**L** (34)	313
$C_6H_5N{=}NC_6H_4NO_2$-*m*	98% H_2SO_4	2-Nitro-9,10-diaza-**L** (90)	310

	Reactant	Conditions	Product (yield %)	Ref.
	$C_6H_5N{=}NC_6H_4NO_2$-*p*	”	3-Nitro-9,10-diaza-**L** (30)	310
C_{13}	$C_6H_5N{=}NC_6H_4CH_3$-*o*	$ClCH_2CH_2Cl$, $SnCl_4$	1-Methyl-9,10-diaza-**L** (58)	29
		$ClCH_2CH_2Cl$, $AlCl_3$	” (51)	29
		$ClCH_2CH_2Cl$, $FeCl_3$	” (50)	29
		22 *N* H_2SO_4	” (23), 9,10-diaza-**L** (11)	308
	$C_6H_5N{=}NC_6H_4CH_3$-*m*	”	2-Methyl-9,10-diaza-**L** (27), 4-methyl-9,10-diaza-**L** (13)	308
	$C_6H_5N{=}NC_6H_4CH_3$-*p*	”	3-Methyl-9,10-diaza-**L** (50)	308
	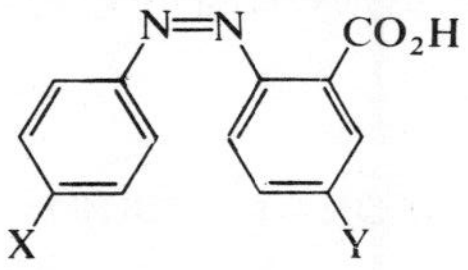 N=N, CO_2H, X, Y			
	X = Cl, Y = H	98 % H_2SO_4	1-Carboxy-6-chloro-9,10-diaza-**L** (75)	311
	X = H, Y = Cl	”	1-Carboxy-3-chloro-9,10-diaza-**L** (60)	311
	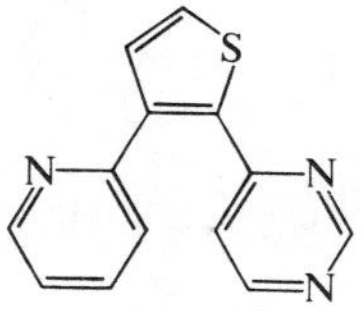 S, N, N, N	Et_2O, air	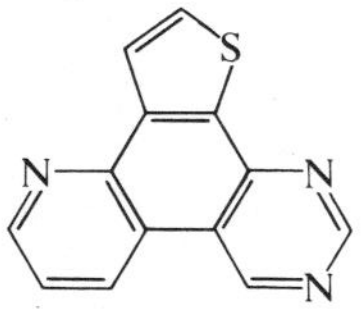 S, N, N, N (32)	274
	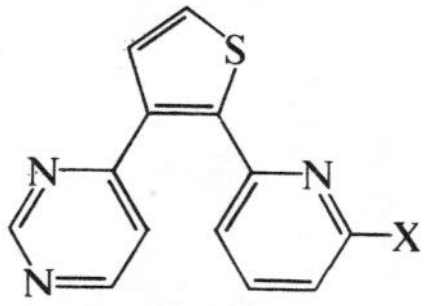 S, N, N, N, X (I)		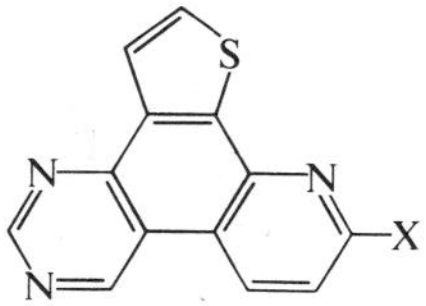 S, N, N, N, X (II)	
	I, X = H	”	II, X = H (34)	274
	I, X = Cl	”	II, X = Cl (22)	274
	$C_6H_5N{=}NC_6H_4CO_2H$-*o*	98 % H_2SO_4	1-Carboxy-9,10-diaza-**L** (75)	311
		22 *N* H_2SO_4	” (35), 9,10-diaza-**L** (6)	313
		$ClCH_2CH_2Cl$, air	” (10)	309

TABLE V. NITROGEN HETEROCYCLIC SYSTEMS (*Continued*)[a]

B. Heteroaromatic Systems with Two or More Ring Nitrogens (*Continued*)

	Starting Material	Reaction Conditions	Products and Yields (%)	Refs.
	$C_6H_5N{=}NC_6H_4CO_2H$-*m*	22 *N* H_2SO_4	2-Carboxy-9,10-diaza-**L** (14), 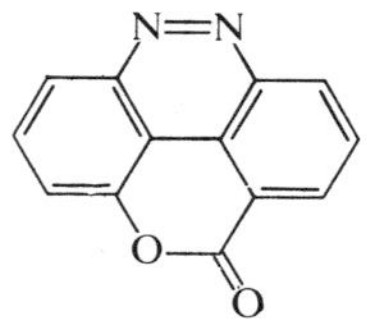	313
			(10)	
	$C_6H_5N{=}NC_6H_4CO_2H$-*p*	22 *N* H_2SO_4	3-Carboxy-9,10-diaza-**L** (44)	313
	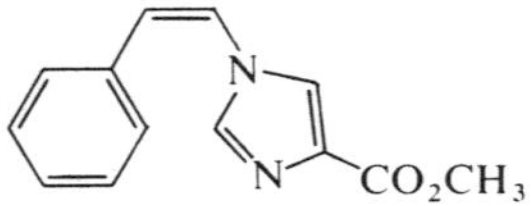 CO_2CH_3	MeOH, I_2	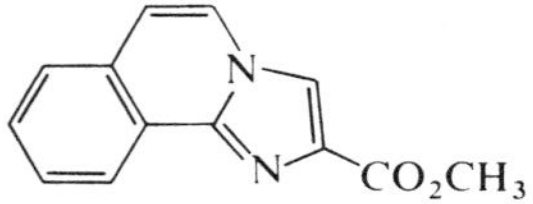 CO_2CH_3 (44)	278
	$C_6H_5N{=}NC_6H_4SO_2CH_3$-*o*	98% H_2SO_4	1-CH_3SO_2-9,10-diaza-**L** (69)	636
		$ClCH_2CH_2Cl$, $AlCl_3$	" (59)	636
C_{14}	*o*-$CH_3C_6H_4N{=}NC_6H_4CH_3$-*o*	$ClCH_2CH_2Cl$, $SnCl_4$	1,8-Dimethyl-9,10-diaza-**L** (47)	29
		$ClCH_2CH_2Cl$, $AlCl_3$	" (40)	29
		$ClCH_2CH_2Cl$, $FeCl_3$	" (43)	29
		22 *N* H_2SO_4	" (10), 1-methyl-9,10-diaza-**L** (19)	308
	m-$CH_3C_6H_4N{=}NC_6H_4CH_3$-*m*	"	2,5-Dimethyl-9,10-diaza-**L** (7), 2,7-dimethyl-9,10-diaza-**L** (16), 4,5-dimethyl-9,10-diaza-**L** (3)	308
	p-$CH_3C_6H_4N{=}NC_6H_4CH_3$-*p*	"	3,6-Dimethyl-9,10-diaza-**L** (57)	308
		$ClCH_2CH_2Cl$, $SnCl_4$	" (57)	29
		$ClCH_2CH_2Cl$, $AlCl_3$	" (56)	29
		$ClCH_2CH_2Cl$, $FeCl_3$	" (51)	29
	$C_6H_5N{=}NC_6H_4N(O)(CH_3)_2$-*p*	H_2SO_4, EtOH	3-Dimethylamino-9,10-diaza-**L** (12)	312
	$C_6H_5N{=}NC_6H_4COCH_3$-*m*	98% H_2SO_4	2-Acetyl-9,10-diaza-**L** (71)	310

$C_6H_5N{=}NC_6H_4COCH_3$-*p*	”	3-Acetyl-9,10-diaza-**L** (66)	310
$C_6H_5N{=}NC_6H_4CO_2CH_3$-*o*	”	1-Carbomethoxy-9,10-diaza-**L** (—)	309
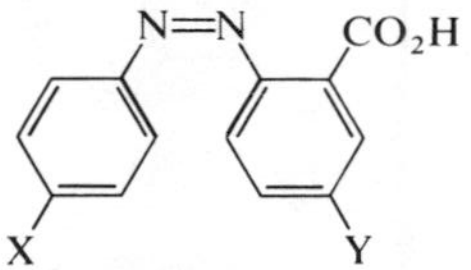 [structure: azobenzene with CO_2H, X, Y substituents]			
X = CH_3, Y = H	”	1-Carboxy-6-methyl-9,10-diaza-**L** (70)	311
X = H, Y = CH_3	”	1-Carboxy-3-methyl-9,10-diaza-**L** (63)	311
o-$CH_3C_6H_4N{=}NC_6H_4SO_2CH_3$-*o*	”	8-Methyl-1-CH_3SO_2-9,10-diaza-**L** (65)	636
	$ClCH_2CH_2Cl$, $AlCl_3$	” (58)	636
p-$CH_3C_6H_4N{=}NC_6H_4SO_2CH_3$-*o*	98% H_2SO_4	6-Methyl-1-CH_3SO_2-9,10-diaza-**L** (61)	636
	$ClCH_2CH_2Cl$, $AlCl_3$	” (61)	636
o-$HO_2CC_6H_4N{=}NC_6H_4CO_2H$-*o*	98% H_2SO_4	1,8-Dicarboxy-9,10-diaza-**L** (47), 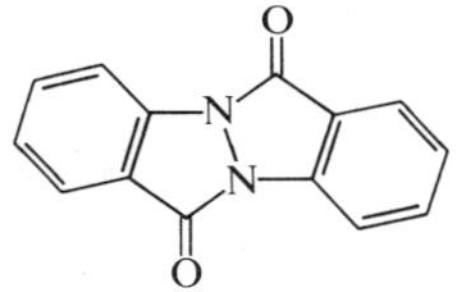 [structure] (46)	317
	$ClCH_2CH_2Cl$, $AlCl_3$	1,8-Dicarboxy-9,10-diaza-**L** (68)	317
	$ClCH_2CH_2Cl$, I_2	” (10)	317
	$ClCH_2CH_2Cl$, air	” (13), 1-carboxy-9,10-diaza-**L** (6)	309
m-$HO_2CC_6H_4N{=}NC_6H_4CO_2H$-*m*	98% H_2SO_4	2,7-Dicarboxy-9,10-diaza-**L** (62)	317
	$ClCH_2CH_2Cl$, $AlCl_3$	” (63)	317
p-$HO_2CC_6H_4N{=}NC_6H_4CO_2H$-*p*	98% H_2SO_4	3,6-Dicarboxy-9,10-diaza-**L** (65)	317
	$ClCH_2CH_2Cl$, $AlCl_3$	” (61)	317
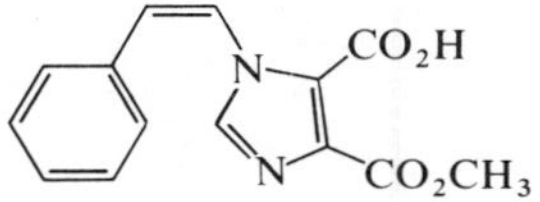 [structure: styryl imidazole with CO_2H, CO_2CH_3]	MeOH, I_2	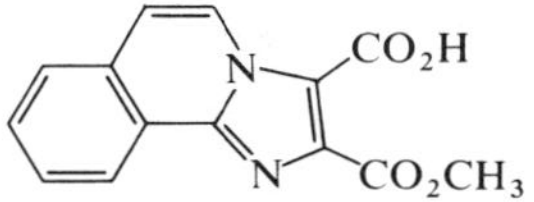 [structure with CO_2H, CO_2CH_3] (55)	278

TABLE V. NITROGEN HETEROCYCLIC SYSTEMS (*Continued*)[a]

B. Heteroaromatic Systems with Two or More Ring Nitrogens (*Continued*)

Starting Material	Reaction Conditions	Products and Yields (%)	Refs.
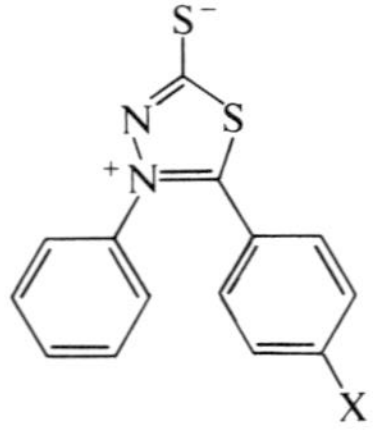 (I)		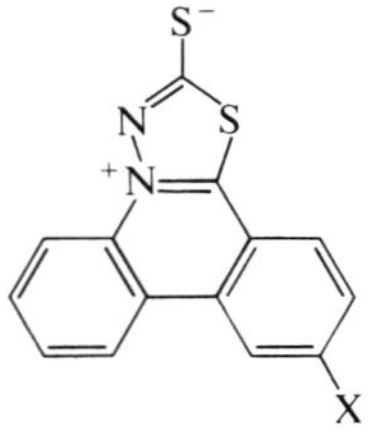 (II)	
I, X = H	MeCN, air	II, X = H (30)	275
I, X = Cl	"	II, X = Cl (—)	275
C_{15}	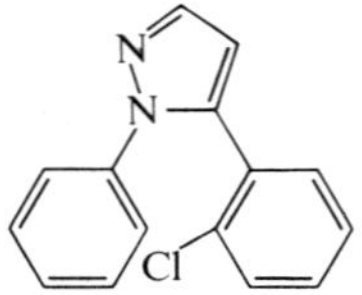 C_6H_{12}	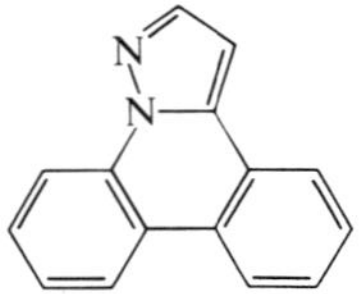 (92)	276
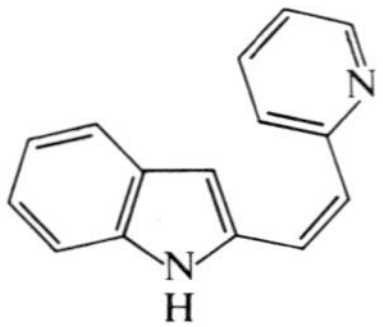	EtOH, I_2	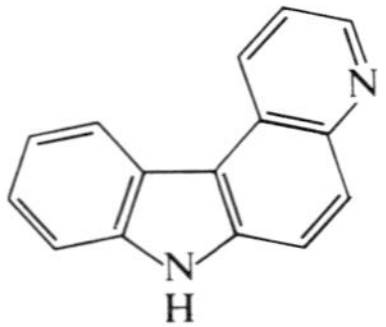 (48–56)	268,637

Substrate	Conditions	Product (yield %)	Ref.
	”	(54), (32)	268
	”	(85)	268
	”	(19), (23)	269
	MeOH, I_2	(64)	278
	”	(53)	278

TABLE V. NITROGEN HETEROCYCLIC SYSTEMS (*Continued*)[a]

B. Heteroaromatic Systems with Two or More Ring Nitrogens (Continued)

Starting Material	Reaction Conditions	Products and Yields (%)	Refs.
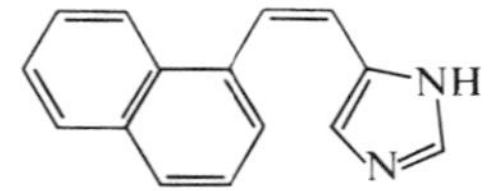 [structure]	Abs EtOH, I_2	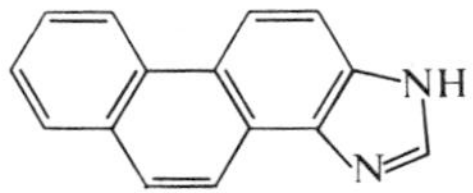 [structure] (50)	631
$C_6H_5N{=}NC_6H_2(CH_3)_3$-2,4,6	20.5 *N* H_2SO_4	1,3-Dimethyl-9,10-diaza-**L** (20), 1,3,4-trimethyl-9,10-diaza-**L** (2)	313
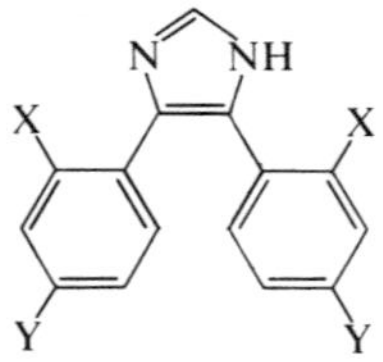 [structure] (I)		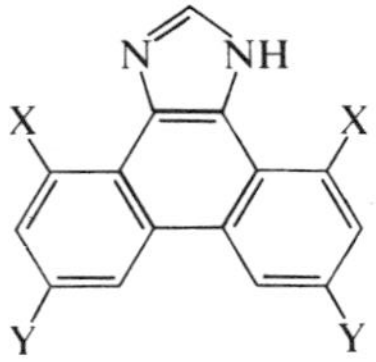 [structure] (II)	
I, X = Y = H	EtOH, I_2, air	II, X = Y = H (75)	271
	$CHCl_3$, I_2	" (59)	638
	C_6H_6, air	" (—)	160
	Solvent, N_2[e]	" (—)	639
I, X = Cl, Y = H	MeOH, I_2	II, X = Cl, Y = H (54)	638
I, X = H, Y = Cl	$CHCl_3$, I_2	II, X = H, Y = Cl (61)	638
$C_6H_5N{=}NC_6H_4CO_2C_2H_5$-*o*	98% H_2SO_4	1-Carbethoxy-9,10-diaza-**L** (—)	309
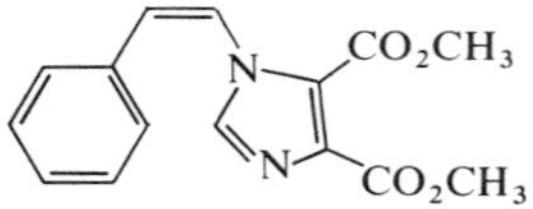 [structure]	MeOH, I_2	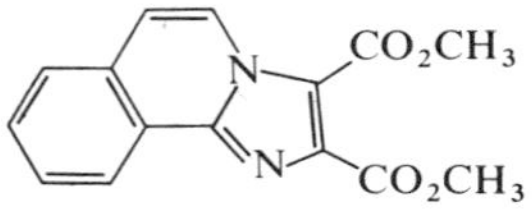 [structure] (44)	278

C_{16}	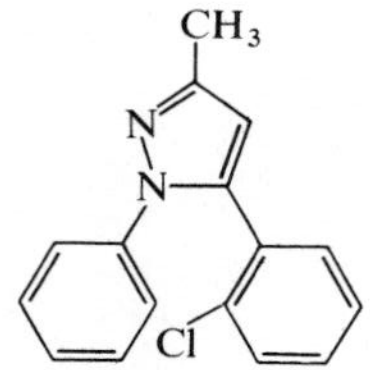	C_6H_{12}	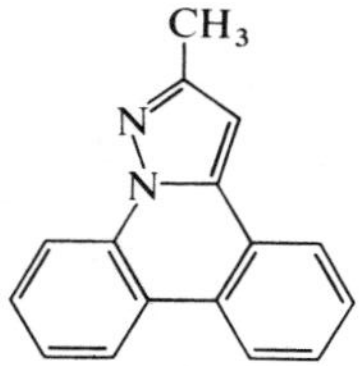 (94)	276
	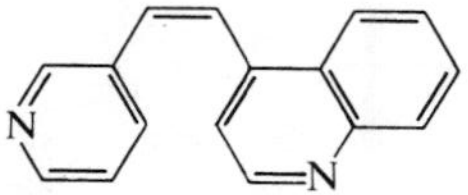	C_6H_{14}, air	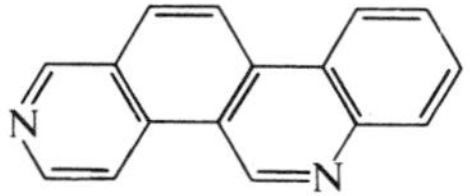 (10)	257
	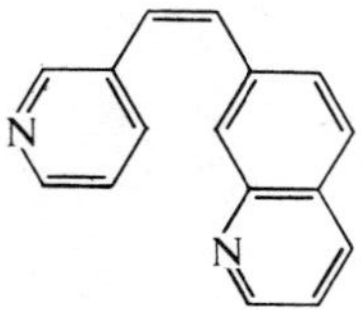	Solvent, air	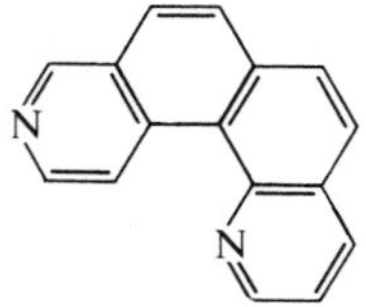 (40)	175a,250
	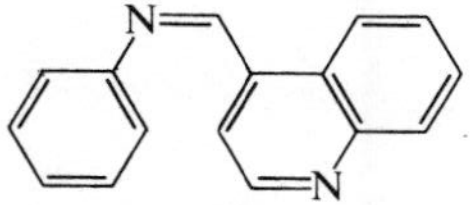	H_2SO_4, O_2	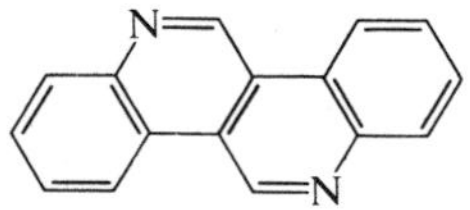 (53)	298
	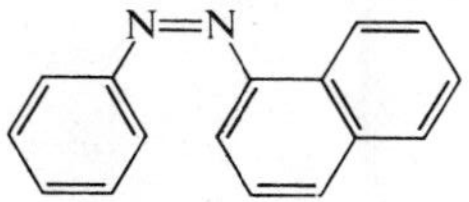	22 *N* H_2SO_4	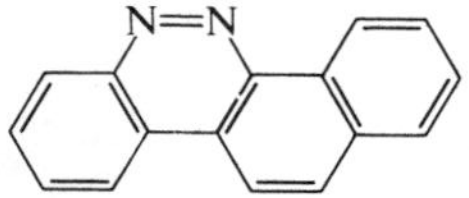 (42)	318
	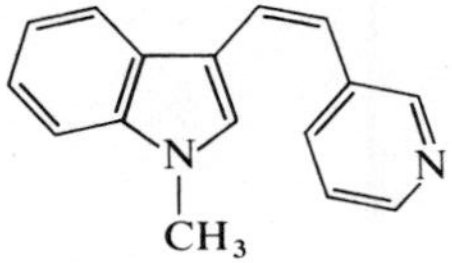	EtOH, I_2	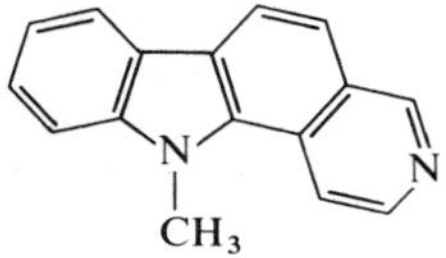 (17), 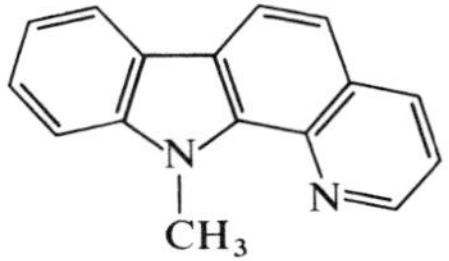 (20)	269

TABLE V. NITROGEN HETEROCYCLIC SYSTEMS (*Continued*)[a]

B. Heteroaromatic Systems with Two or More Ring Nitrogens (*Continued*)

Starting Material	Reaction Conditions	Products and Yields (%)	Refs.
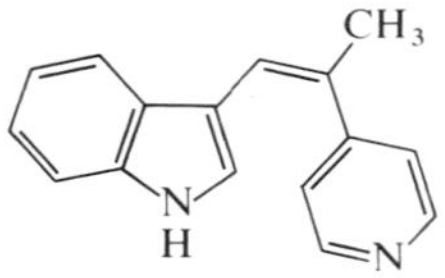	EtOH, I_2	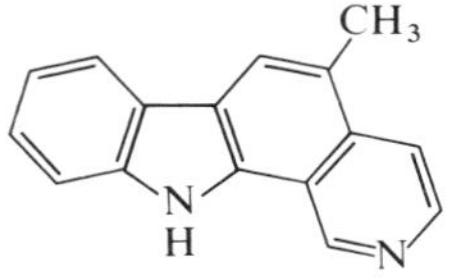 CH_3 (28)	269
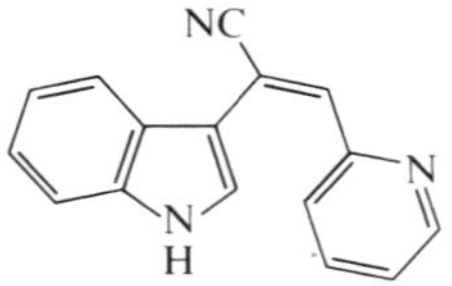 NC	EtOH, I_2, N_2	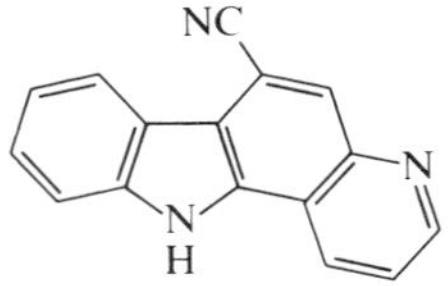 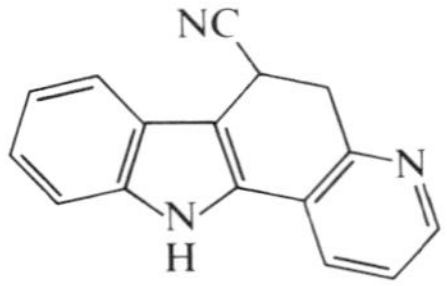 NC (25), NC (18)	623a
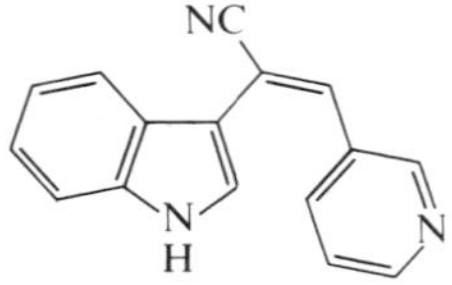 NC	EtOH, air	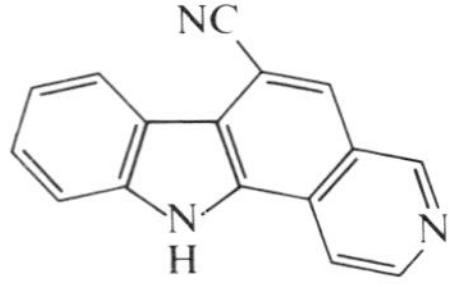 NC (80)	623a
	EtOH, I_2, N_2	" (50), 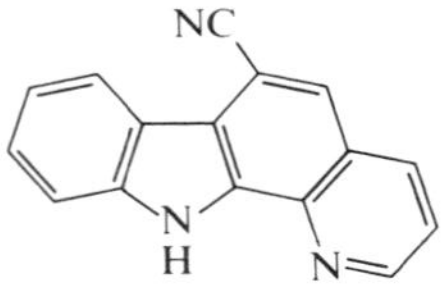 NC (21)	623a

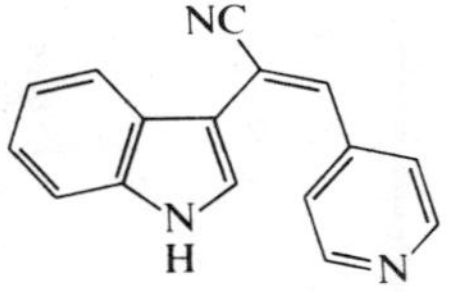	EtOH, I_2 (or $FeCl_3$)	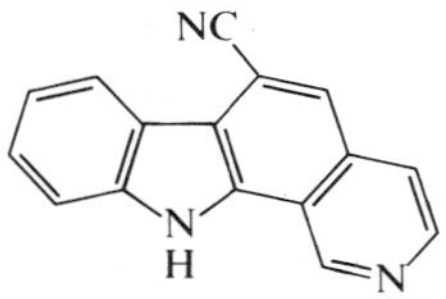 (—)	622
	EtOH, N_2	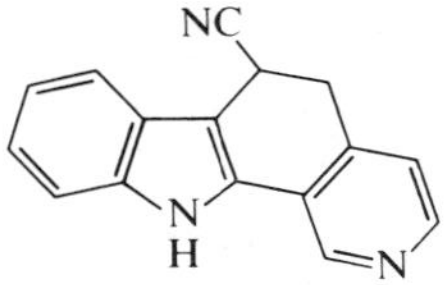 (—)	623a
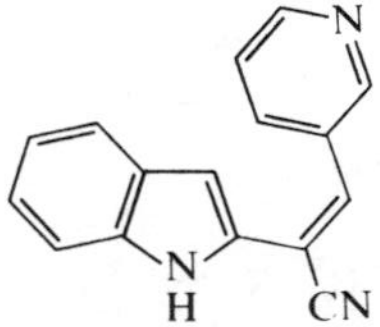	EtOH, I_2 (or $FeCl_3$)	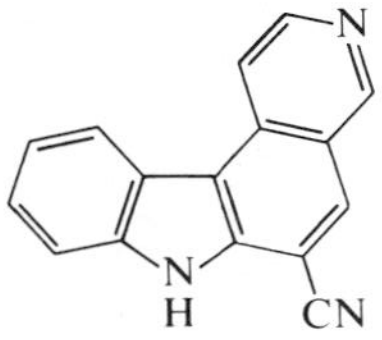 (30), 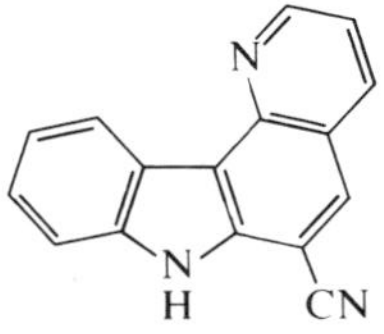 (5)	640
o-$CH_3O_2CC_6H_4N{=}NC_6H_4CO_2CH_3$-*o*	98% H_2SO_4	1,8-Dicarbomethoxy-9,10-diaza-**L** (35), 1-carbomethoxy-8-carboxy-9,10-diaza-**L** (23)	29
	"	1,8-Dicarbomethoxy-9,10-diaza-**L** (—)	309
m-$CH_3O_2CC_6H_4N{=}NC_6H_4CO_2CH_3$-*m*	"	2,7-Dicarbomethoxy-9,10-diaza-**L** (61)	29
p-$CH_3O_2CC_6H_4N{=}NC_6H_4CO_2CH_3$-*p*	"	3,6-Dicarbomethoxy-9,10-diaza-**L** (62)	29
C_{17} 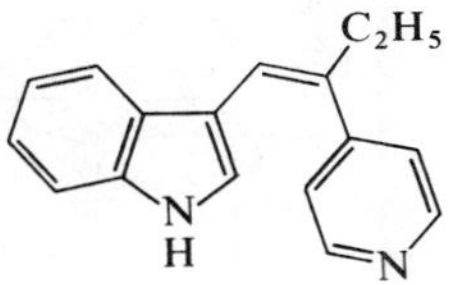	EtOH, I_2	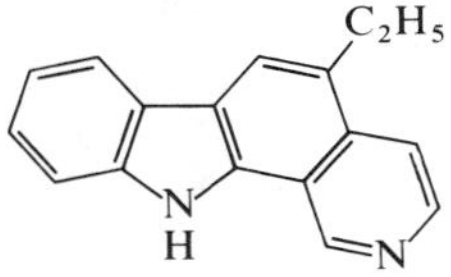 (48)	269
	EtOH, I_2, air	" (18)	641
$Ar\overset{+}{N}H{=}\overset{-}{B}(C_6H_5)_2$ Ar = 2-Pyridyl	C_6H_{12}, I_2, N_2	8-Aza-10-phenyl-**M** (15)	642

TABLE V. NITROGEN HETEROCYCLIC SYSTEMS (*Continued*)[a]

B. Heteroaromatic Systems with Two or More Ring Nitrogens (*Continued*)

	Starting Material	Reaction Conditions	Products and Yields (%)	Refs.
	CO_2CH_3 N H N	EtOH, I_2 (or $FeCl_3$)	CO_2CH_3 N H N (—)	622
C_{18}	$C_6H_5N{=}NC_6H_4C_6H_5$-*m*	22 *N* H_2SO_4, EtOH	2-Phenyl-9,10-diaza-**L** (18), 4-phenyl-9,10-diaza-**L** (9)	315
	$C_6H_5N{=}NC_6H_4C_6H_5$-*p*	"	3-Phenyl-9,10-diaza-**L** (43)	318
	$(C_6H_5)_2\overset{+}{N}{=}NC_6H_5$	MeCN, 2,6-lutidine	C_6H_5 $^+N{=}N$ (50)	643
	CH_3 CH_3 N N CH_3	MeOH, I_2	CH_3 CH_3 N N CH_3 (27)	278
	$C_6H_5N{=}N$–C_6H_4–$N{=}NC_6H_5$	98% H_2SO_4	3-Amino-4-(*p*-aminophenyl)-9,10-diaza-**L** (33), 3-(*p*-aminoanilino)-9,10-diaza-**L** (20)	315
	$N{=}N$ $N{=}NC_6H_5$	"	" (44), " (32)	315

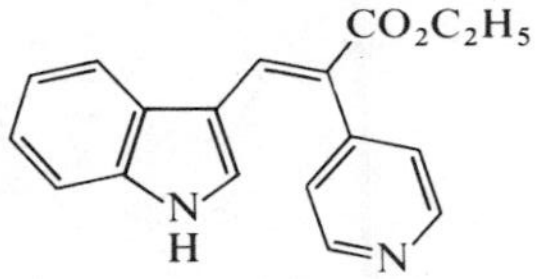	EtOH, I_2	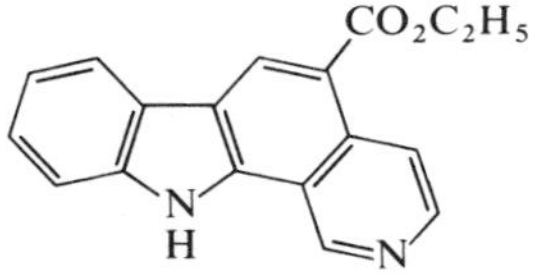 (79)	269
o-$C_2H_5O_2CC_6H_4N{=}NC_6H_4CO_2C_2H_5$-$o$	98% H_2SO_4	1,8-Dicarbethoxy-9,10-diaza-**L** (43), 1-carbethoxy-8-carboxy-9,10-diaza-**L** (18)	29
	”	1,8-Dicarbethoxy-9,10-diaza-**L** (—)	309
m-$C_2H_5O_2CC_6H_4N{=}NC_6H_4CO_2C_2H_5$-$m$	”	2,7-Dicarbethoxy-9,10-diaza-**L** (60)	29
p-$C_2H_5O_2CC_6H_4N{=}NC_6H_4CO_2C_2H_5$-$p$	”	3,6-Dicarbethoxy-9,10-diaza-**L** (60)	29
C_{19} $C_6H_5N{=}N{-}C_6H_4{-}N{=}CHC_6H_5$	 ”	3-Amino-9,10-diaza-**L** (97)	310
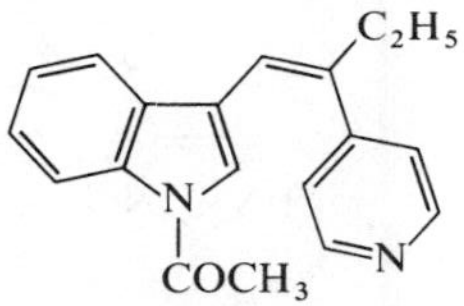	EtOH, I_2, air	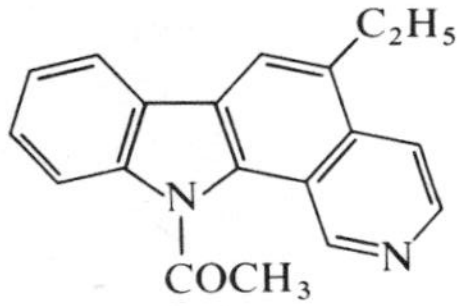 (64)	641
C_{20}	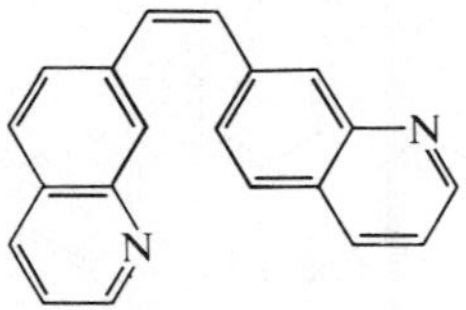 Solvent, air	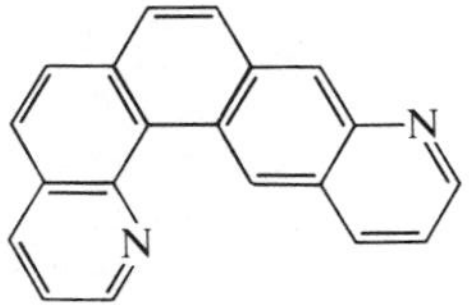 (20)	250
	”	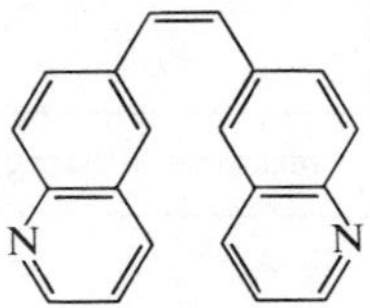 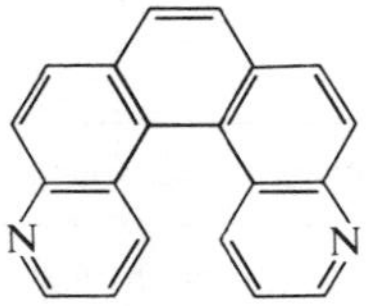 (15)	250

TABLE V. NITROGEN HETEROCYCLIC SYSTEMS (*Continued*)[a]

B. Heteroaromatic Systems with Two or More Ring Nitrogens (*Continued*)

	Starting Material	Reaction Conditions	Products and Yields (%)	Refs.
	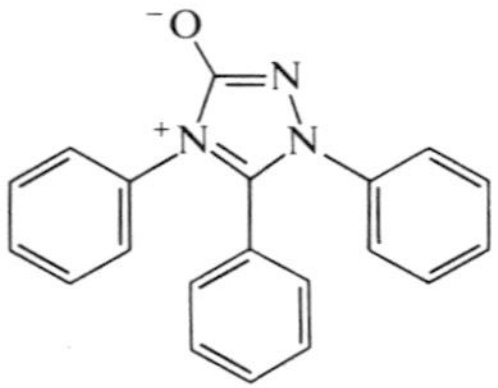 ^{-}O, N, ^{+}N, N (structure)	MeOH, air	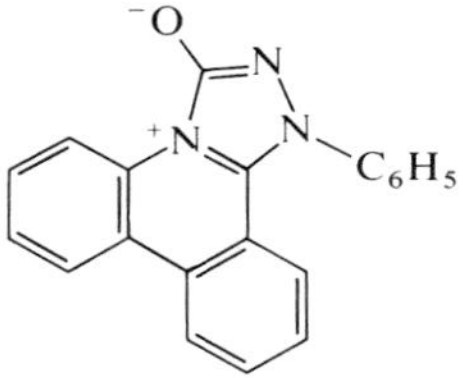 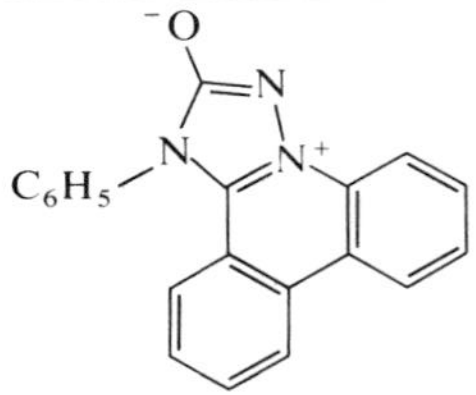 ^{-}O, N, ^{+}N, N, C_6H_5 (structure) or ^{-}O, N, N^{+}, N, C_6H_5 (structure) (83)	644
C_{21}	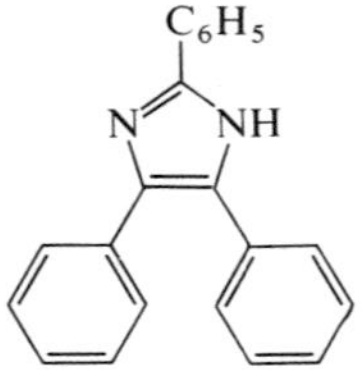 C_6H_5, N, NH (structure)	Solvent, N_2[e]	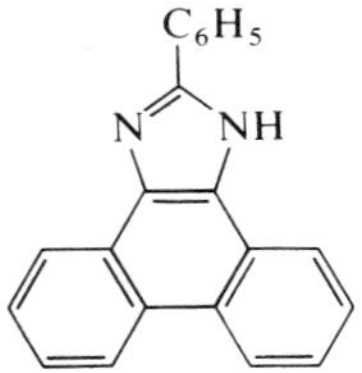 C_6H_5, N, NH (structure) (—)	639
	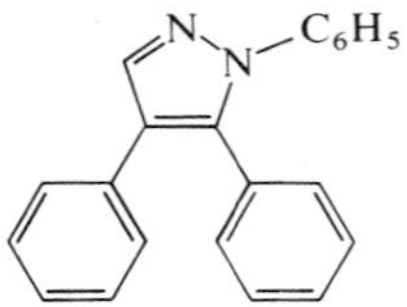 N, N, C_6H_5 (structure)	Dioxane, air	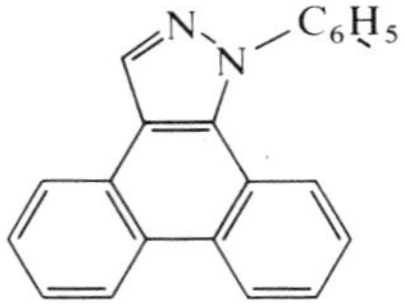 N, N, C_6H_5 (structure) (67)	272
	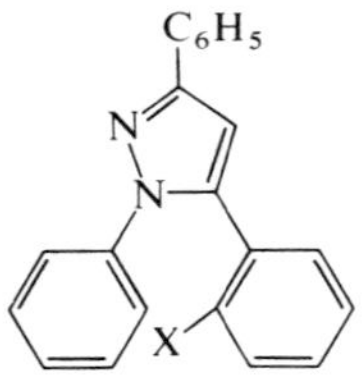 C_6H_5, N, N, X (structure)		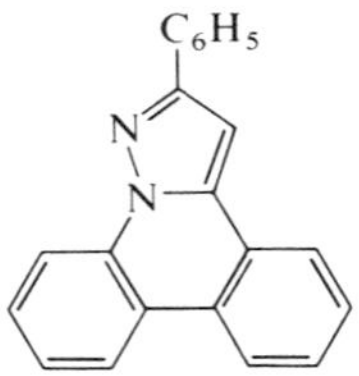 C_6H_5, N, N (structure)	

Substrate	Conditions	Product (yield %)	Ref.
X = Br	MeOH, N_2	" (84)	37a
	C_6H_{12}, N_2	" (81)	276
	C_6H_6, Ph_2CO, N_2	" (77)	37a
X = Cl	MeOH, N_2	" (80)	37a
	MeOH, air	" (61)	37a
	MeOH, Et_3N, Et_2NPh, N_2	" (56)	37a
	C_6H_{12}, N_2	" (—)	276
	C_6H_{12}, Ph_2CO, N_2	" (70)	37a
	C_6H_6, Ph_2CO, N_2	" (66)	37a
X = I	MeOH, N_2	" (83)	37a
	C_6H_6, Ph_2CO, N_2	" (69)	37a
C_{22} X = CH_3O	MeOH, N_2	" (52)	37a
C_{23} X = CH_3CO	"	" (61)	37a
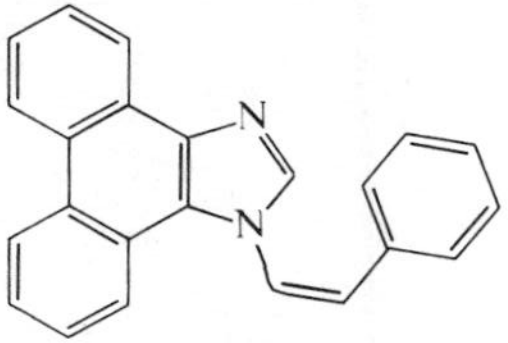	MeOH, I_2	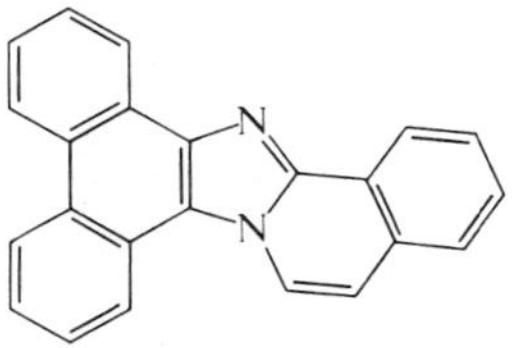 (75)	279
	"	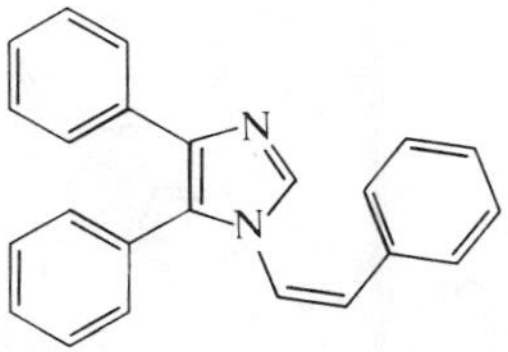 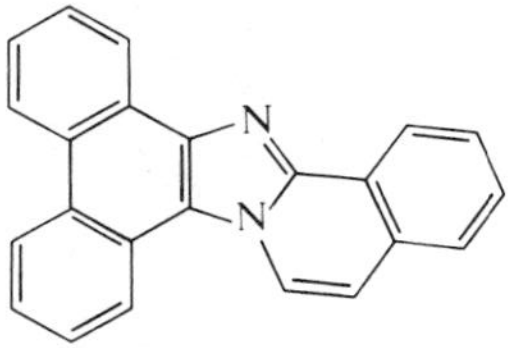 (35)	279
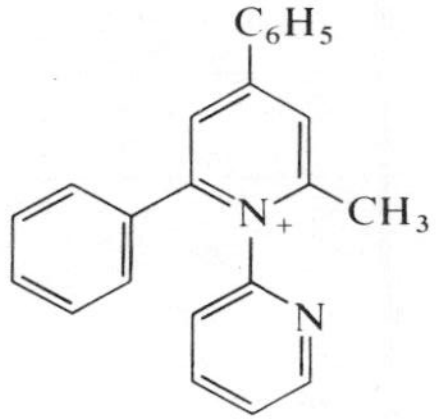 C_6H_5, CH_3, N^+, N	MeOH, O_2	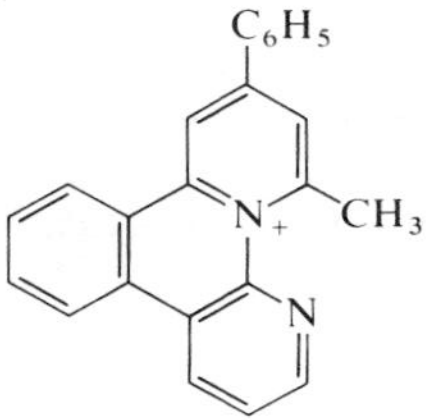 C_6H_5, CH_3, N^+, N (38)	627,628a

TABLE V. NITROGEN HETEROCYCLIC SYSTEMS (*Continued*)[a]

B. Heteroaromatic Systems with Two or More Ring Nitrogens (Continued)

	Starting Material	Reaction Conditions	Products and Yields (%)	Refs.
	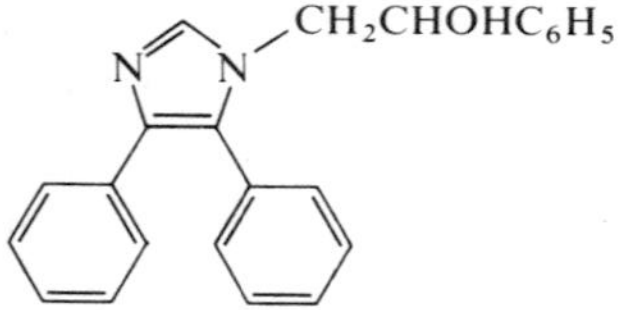 $CH_2CHOHC_6H_5$	MeOH, I_2	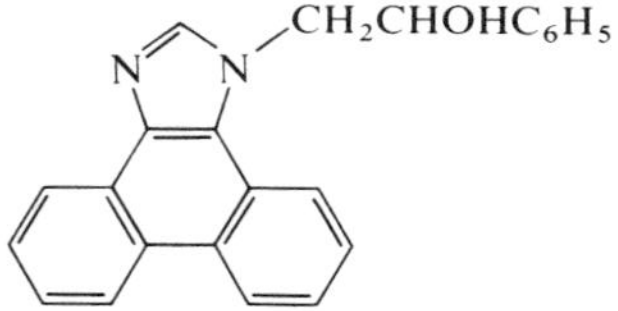 $CH_2CHOHC_6H_5$ (70)	279
	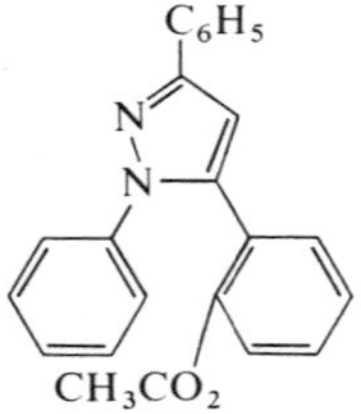 C_6H_5, CH_3CO_2	MeOH, deoxygenated	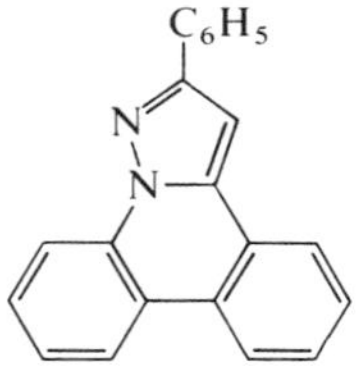 C_6H_5 (—)	277
C_{24}	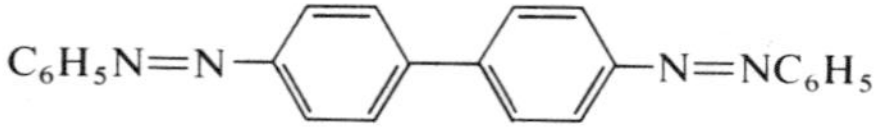 $C_6H_5N{=}N{-}C_6H_4{-}C_6H_4{-}N{=}NC_6H_5$	22 *N* H_2SO_4	3-(*p*-Aminophenyl)-9,10-diaza-**L** (21), 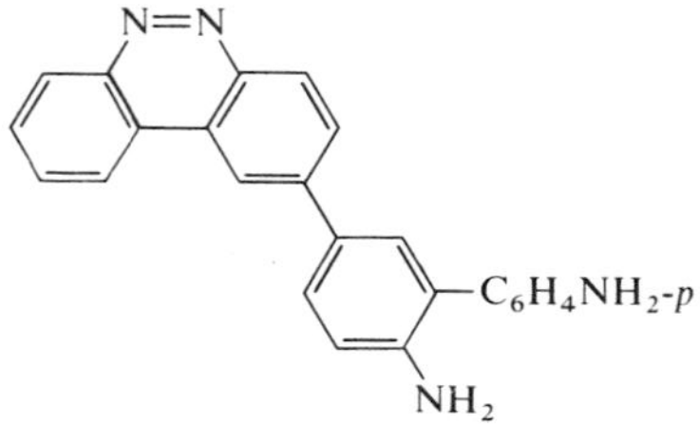 N=N, $C_6H_4NH_2$-*p*, NH_2 (20)	315,318

	Starting material	Conditions	Product (yield, %)	Ref.
C_{25} 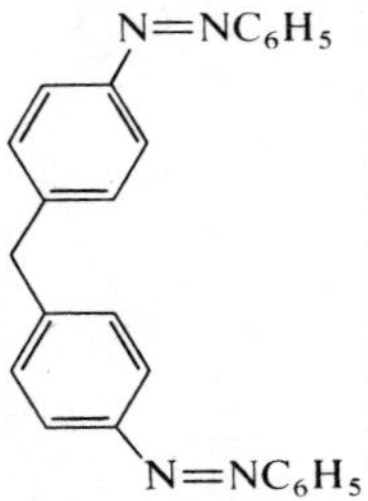		98% H_2SO_4	(64)	645
C_{26} 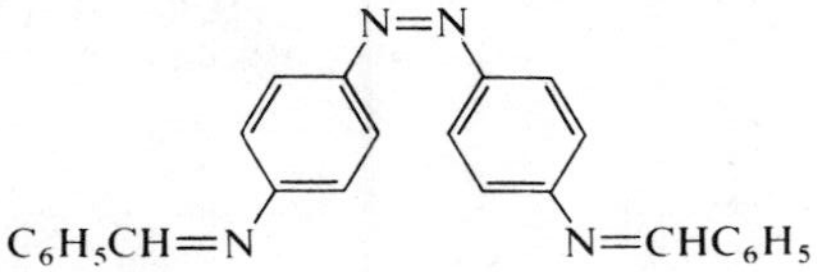		1. 98% H_2SO_4 2. H_2O	3,6-Diamino-9,10-diaza-**L** (85)	310
C_{27} 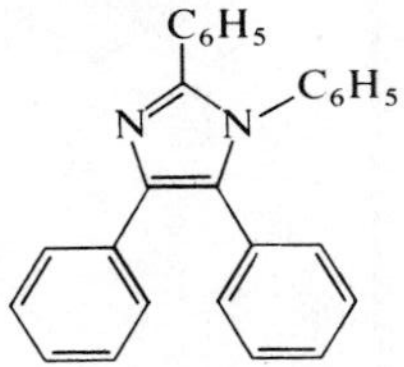		EtOH, I_2, air	(90)	271
C_{28} 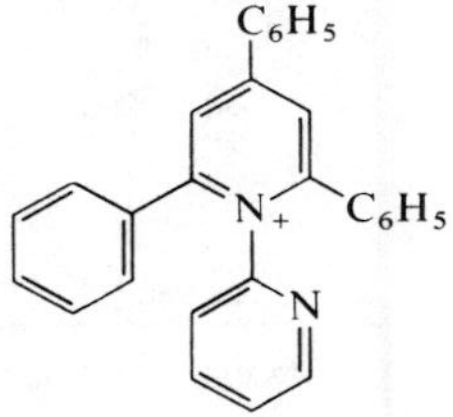		MeOH, O_2	(21)	627,628a

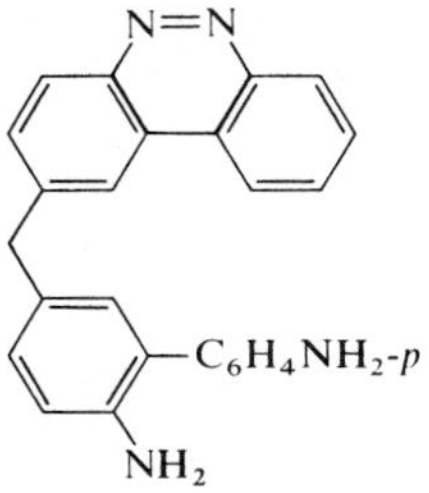

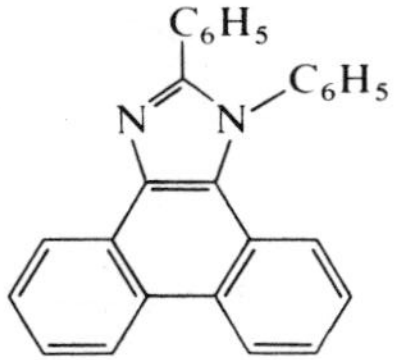

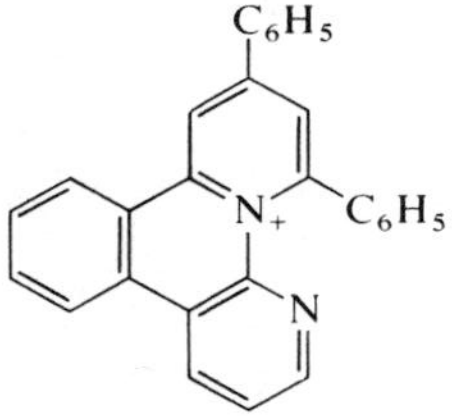

TABLE V. NITROGEN HETEROCYCLIC SYSTEMS (*Continued*)[a]

B. Heteroaromatic Systems with Two or More Ring Nitrogens (*Continued*)

	Starting Material	Reaction Conditions	Products and Yields (%)	Refs.
	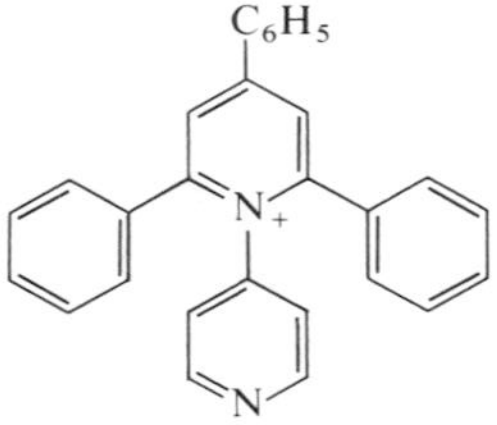 C_6H_5	MeOH, O_2	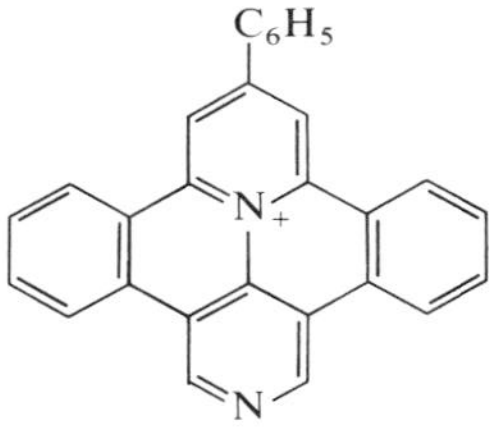 C_6H_5 (91)	627,628a
C_{29}	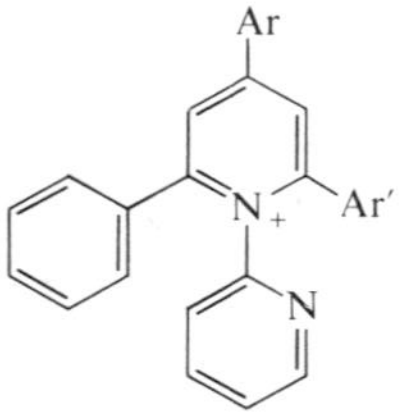 Ar, Ar′ (I)		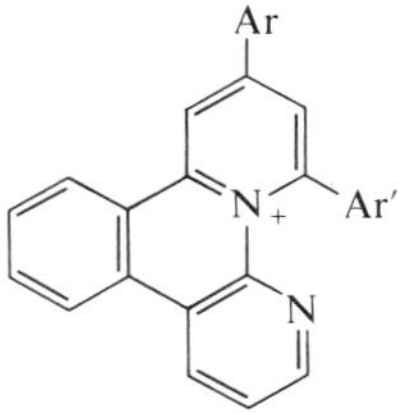 Ar, Ar′ (II)	
	I, Ar = Phenyl, Ar′ = *p*-Tolyl	"	II, Ar = Phenyl, Ar′ = *p*-Tolyl (22)	627,628a
	I, Ar = *p*-Tolyl, Ar′ = Phenyl	"	II, Ar = *p*-Tolyl, Ar′ = Phenyl (27)	627,628a
	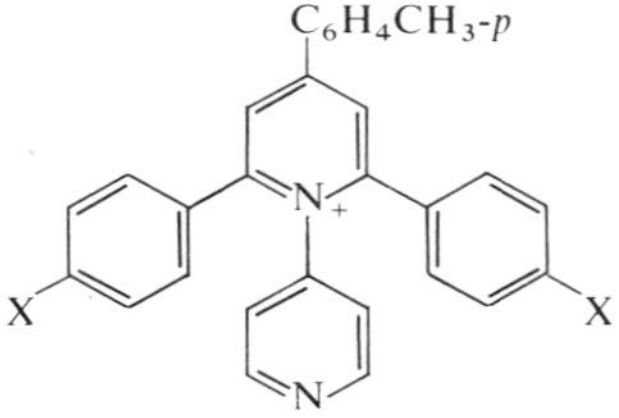 $C_6H_4CH_3$-*p* (I)		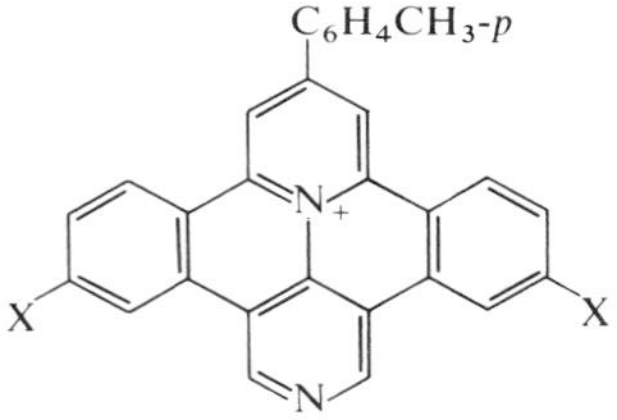 $C_6H_4CH_3$-*p* (II)	
	I, X = H	"	II, X = H (60)	627,628a
C_{31}	I, X = CH_3	"	II, X = CH_3 (8)	627,628a

	Reactant	Conditions	Product(s) and Yield(s) (%)	Refs.
C_{13}	**N** (2,3-Diphenyltetrazolium cation)	H_2O, HNO_3	**O** (Benzo[*c*]tetrazolo[2,3-*a*]cinnolin-4-ium cation) (69)	280
C_{14}	5-Methyl-**N**	H_2O, EtOH, HNO_3, N_2	2-Methyl-**O** (42)	280
C_{16}	5-Carbethoxy-**N**	H_2O, EtOH, HNO_3	2-Carbethoxy-**O** (30)	646
C_{17}	5-*tert*-Butyl-**N**	H_2O, EtOH, HNO_3, N_2	2-*tert*-Butyl-**O** (50)	647
C_{18}	3′-Aza-5-phenyl-**N**	H_2O, EtOH, HNO_3	6(or 8)-Aza-2-phenyl-**O** (52)	282
		"	8-Aza-2-phenyl-**O** (—)	316
	4′-Aza-5-phenyl-**N**	"	7-Aza-2-phenyl-**O** (26)	282
C_{19}	5-Phenyl-**N**	H_2O, EtOH, HNO_3, N_2	2-Phenyl-**O** (80)	280
		H_2O, NH_3, N_2[e]	" (—)	648a
		Abs EtOH, air	" (85)	648b
	5-*tert*-Butyl-4′,4″-dimethyl-**N**	H_2O, EtOH, HNO_3, N_2	2-*tert*-Butyl-7,10-dimethyl-**O** (30)	283
	4′-Aza-4″-methyl-5-phenyl-**N**	H_2O, HNO_3	7-Aza-10-methyl-2-phenyl-**O** (18)	282
	3′-Chloro-5-phenyl-**N**	"	6(or 8)-Chloro-2-phenyl-**O** (85)	280
	4′-Chloro-5-phenyl-**N**	H_2O, EtOH, HNO_3	7-Chloro-2-phenyl-**O** (70)	280
	3′,4′-Dichloro-5-phenyl-**N**	"	6,7(or 7,8)-Dichloro-2-phenyl-**O** (40)	280
	3′,3″-Dichloro-5-phenyl-**N**	"	6,11(or 6,9)-Dichloro-2-phenyl-**O** (35)	280
	5-*tert*-Butyl-4′,4″-dicyano-**N**	H_2O, EtOH, HNO_3, N_2	2-*tert*-Butyl-7,10-dicyano-**O** (27), 3,6-dicyano-9,10-diaza-**L** (33)	283
C_{20}	3′-Methoxy-5-phenyl-**N**	H_2O, EtOH, HNO_3	6(or 8)-Methoxy-2-phenyl-**O** (40)	280
	4′-Carboxy-5-phenyl-**N**	H_2O, HNO_3	7-Carboxy-2-phenyl-**O** (65)	280
C_{21}	5-*tert*-Butyl-4′,4″-dicarbethoxy-**N**	H_2O, EtOH, HNO_3, N_2	2-*tert*-Butyl-7,10-dicarbethoxy-**O** (40), 3,6-dicarbethoxy-9,10-diaza-**L** (12)	283

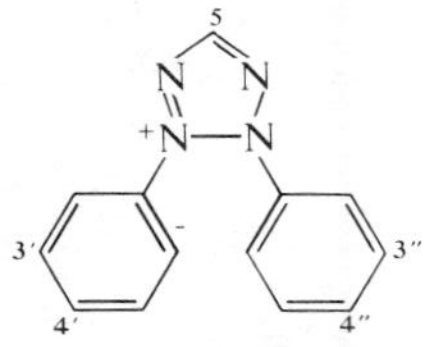

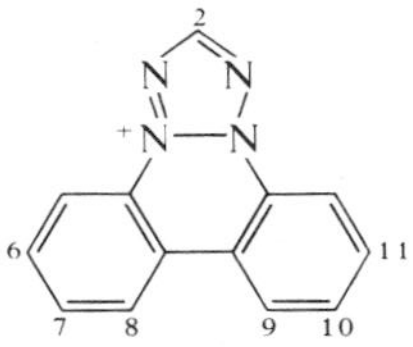

TABLE V. Nitrogen Heterocyclic Systems (*Continued*)[a]

C. *Tetrazolium Cations* (*Continued*)

	Starting Material	Reaction Conditions	Products and Yields (%)	Refs.
C_{22}	4′,4″-Dimethyl-5-(*p*-tolyl)-**N**	EtOH, air	7,10-Dimethyl-2-(*p*-tolyl)-**O** (—)	280
	4′-Carbethoxy-5-phenyl-**N**	H_2O, HNO_3	7-Carbethoxy-2-phenyl-**O** (83)	280
	4′,4″,5-Tricarbethoxy-**N**	H_2O, EtOH, HNO_3	2,7,10-Tricarbethoxy-**O** (23)	646
C_{28}	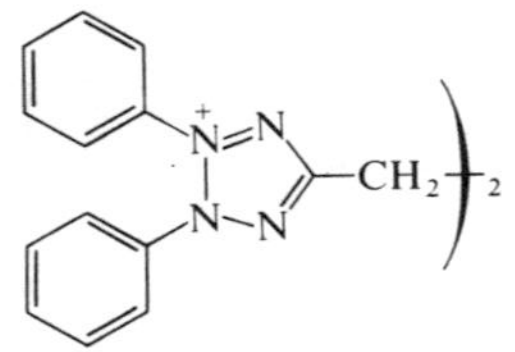	H_2O, EtOH, HNO_3	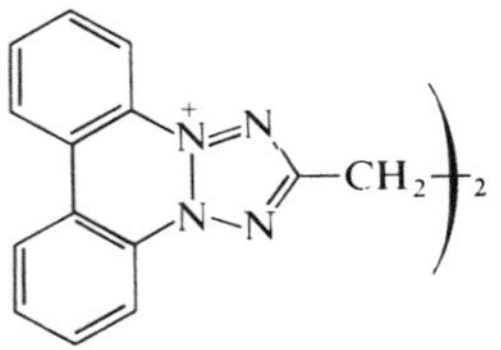 (73)	281
	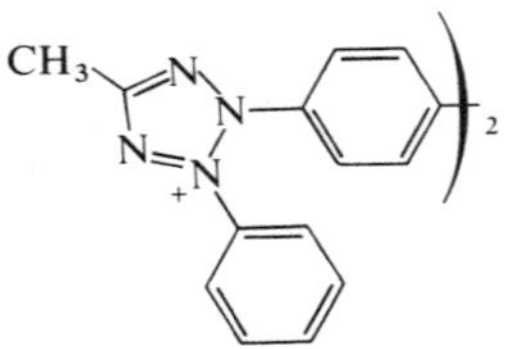	”	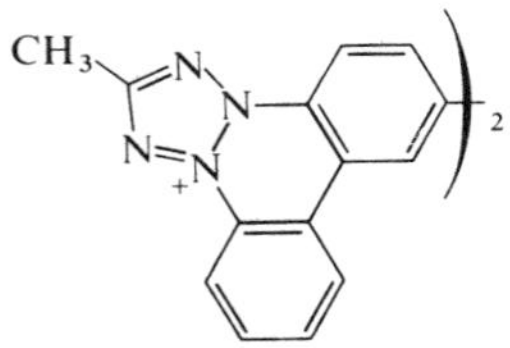 (18)	281
C_{32}	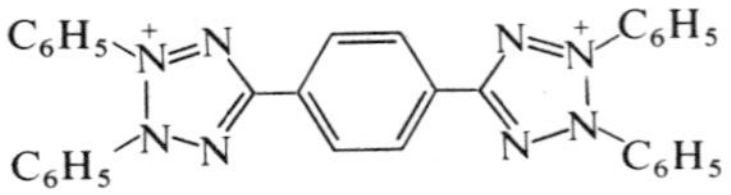	H_2O, HNO_3	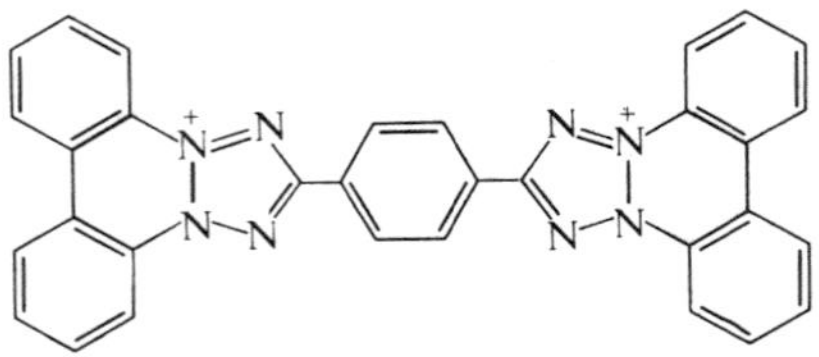 (60)	281

C_{38}	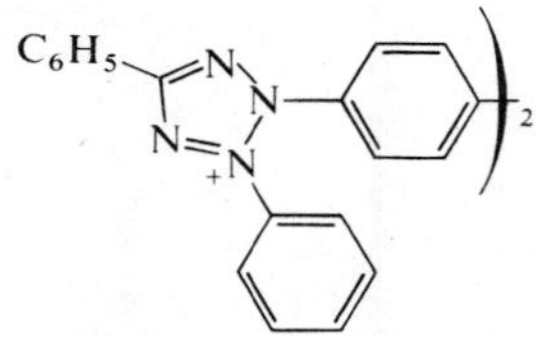	H_2O, EtOH, HNO_3	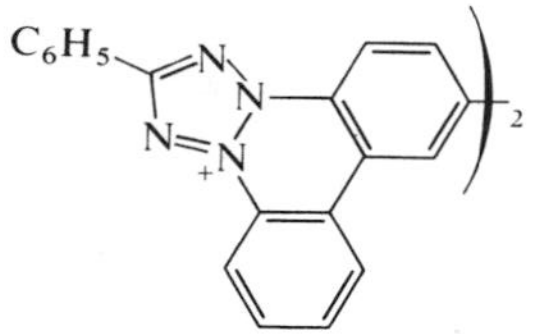 (40)	281
C_{39}	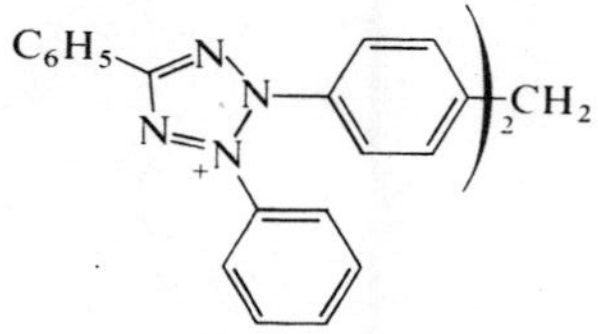	"	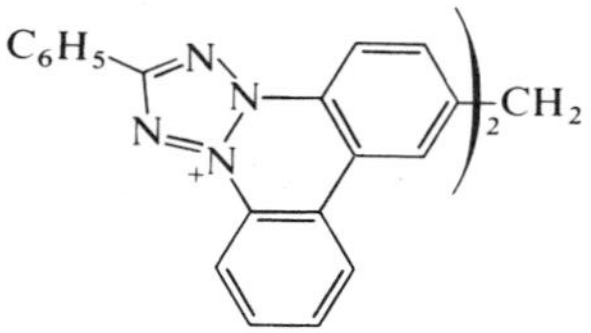 (44)	281

D. Miscellaneous

C_{13}	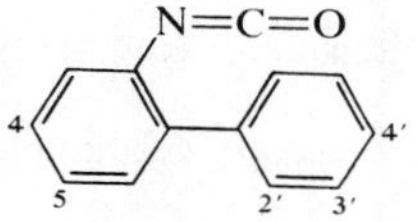 (2-Biphenylyl isocyanate)	Me_2CO, N_2	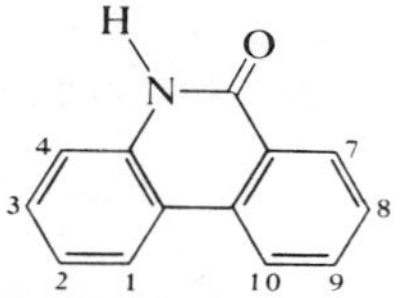 (70) **P** [6(5*H*)-Phenanthridinone]	301
		Et_2O, N_2	**P** (10), 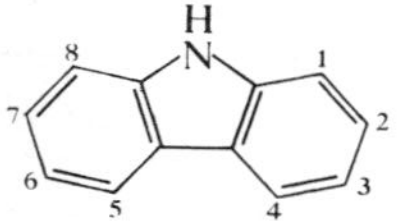 (13–15) **Q** (9*H*-Carbazole)	301

TABLE V. NITROGEN HETEROCYCLIC SYSTEMS (*Continued*)[a]

D. Miscellaneous (*Continued*)

	Starting Material	Reaction Conditions	Products and Yields (%)	Refs.
C_{14}	2′-Methoxy-2-biphenylyl isocyanate	Me_2CO, N_2	10-Methoxy-**P** (10)	301
		Et_2O, N_2	" (9), 4-methoxy-**Q** (15)	301
	3′-Methoxy-2-biphenylyl isocyanate	Me_2CO, N_2	9-Methoxy-**P** (15)	301
		Et_2O, N_2	" (14), 3-methoxy-**Q** (16), 1-methoxy-**Q** (16)	301
	4′-Methoxy-2-biphenylyl isocyanate	Me_2CO, N_2	8-Methoxy-**P** (91)	301
		Et_2O, N_2	" (29), 2-methoxy-**Q** (20)	301
	4-Methoxy-2-biphenylyl isocyanate	Me_2CO, N_2	3-Methoxy-**P** (100)	301
		Et_2O, N_2	" (55), 2-methoxy-**Q** (15)	301
	5-Methoxy-2-biphenylyl isocyanate	Me_2CO, N_2	2-Methoxy-**P** (91)	301
		Et_2O, N_2	" (20), 3-methoxy-**Q** (17)	301
C_{15}	$C_6H_5C(N{=}C{=}O){=}CHC_6H_5$	C_6H_6	3-Phenyl-1(2H)-isoquinolinone (structure: C_6H_5, N–H, C=O) (30)	649a
	2-[$N{=}C(CH_3)_2$]-biphenyl bearing X and Y (I)		6,6-Dimethyl-5,6-dihydrophenanthridine (N–H, CH_3, CH_3) bearing X and Y (II)	
	I, X = Y = H	Me_2CO, N_2	II, X = Y = H (66)	301

	Reactant	Conditions	Product (yield, %)	Refs.
C_{16}	I, $X = CH_3O$, $Y = H$	”	II, $X = CH_3O$, $Y = H$ (72)	301
	I, $X = H$, $Y = CH_3O$	”	II, $X = H$, $Y = CH_3O$ (74)	301
C_{20}	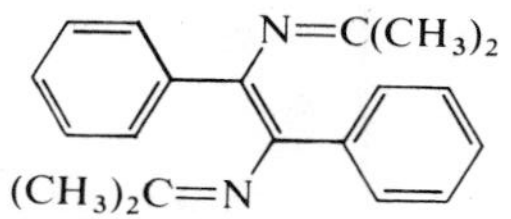 $N{=}C(CH_3)_2$; $(CH_3)_2C{=}N$	MeOH[f]	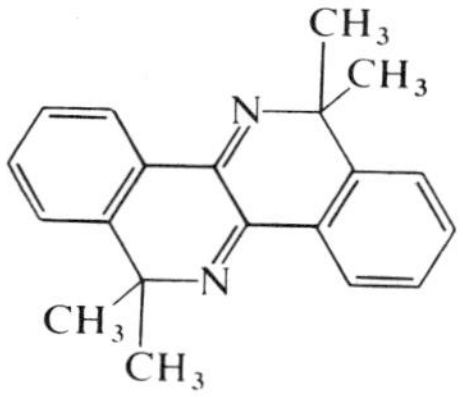 CH_3, CH_3, N, N, CH_3, CH_3 (53)	649b,302

[a] The following symbols are used as abbreviations in Table V: **L** = phenanthrene; **M** = 10,9-borazarophenanthrene; **N** = 2,3-diphenyltetrazolium cation; **O** = benzo[*c*]tetrazolo[2,3-*a*]cinnolin-4-ium cation; **P** = 6(5*H*)-phenanthridinone; and **Q** = 9*H*-carbazole. The numbering schemes for these systems are shown for **L** under C_{13} and for **M** under C_{18} in part A, for **N** and **O** under C_{13} in part C, and for **P** and **Q** under C_{13} in part D. The nomenclature for aza- and diazaphenanthrenes is as follows: 1-aza-**L** is benzo[*f*]quinoline; 2-aza-**L** is benz[*f*]isoquinoline; 3-aza-**L** is benz[*h*]isoquinoline; 4-aza-**L** is benzo[*h*]quinoline; 9-aza-**L** is phenanthridine; 1,2-diaza-**L** is benzo[*f*]cinnoline; 1,3-diaza-**L** is benzo[*f*]quinazoline; 1,4-diaza-**L** is benzo[*f*]quinoxaline; 1,5-diaza-**L** is 1,7-phenanthroline; 1,6-diaza-**L** is 2,7-phenanthroline; 1,7-diaza-**L** is 3,7-phenanthroline; 1,8-diaza-**L** is 4,7-phenanthroline; 1,9-diaza-**L** is benzo[*f*][1,7]naphthyridine; 1,10-diaza-**L** is benzo[*c*][1,8]naphthyridine; 2,3-diaza-**L** is benzo[*f*]phthalazine; 2,4-diaza-**L** is benzo[*h*]quinazoline; 2,5-diaza-**L** is 1,8-phenanthroline; 2,6-diaza-**L** is 2,8-phenanthroline; 2,7-diaza-**L** is 3,8-phenanthroline; 2,9-diaza-**L** is benzo[*c*][2,7]naphthyridine; 2,10-diaza-**L** is benzo[*c*][1,7]naphthyridine; 3,4-diaza-**L** is benzo[*h*]cinnoline; 3,5-diaza-**L** is 1,9-phenanthroline; 3,6-diaza-**L** is 2,9-phenanthroline; 3,9-diaza-**L** is benzo[*c*][2,6]naphthyridine; 3,10-diaza-**L** is benzo[*c*][1,6]naphthyridine; 4,5-diaza-**L** is 1,10-phenanthroline; 4,9-diaza-**L** is benzo[*h*][1,6]naphthyridine; 4,10-diaza-**L** is benzo[*c*][1,5]naphthyridine; and 9,10-diaza-**L** is benzo[*c*]cinnoline. 10,9-Borazarophenanthrene (**M**) is 5,6-dihydrodibenz[*c,e*][1,2]azaborine.

[b] Subsequent workers[618] report being unable to duplicate the results presented in Ref. 617.

[c] It is not established whether the product has the indicated structure or the alternative structure resulting from ring closure at the carbon atom adjacent to the ring nitrogen atom.

[d] A different structure for the product was proposed earlier.[623a]

[e] The identity of the oxidant in this reaction is obscure. One possibility is residual dissolved oxygen.

[f] The indicated product is formed by air-oxidation during work-up.

TABLE VI. OXYGEN AND SULFUR HETEROAROMATIC SYSTEMS FROM FURANS, THIOPHENES, AND PYRYLIUM IONS

	Starting Material	Reaction Conditions	Products and Yields (%)	Refs.
C_{10}	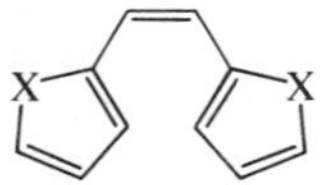 (I)		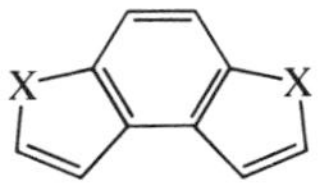 (II)	
	I, X = Y = O	EtOH, $CuCl_2$, I_2	II, X = Y = O (24)	322a
		C_6H_{12}, air	" (5)	332
	I, X = O, Y = S	C_6H_6, I_2, degassed	II, X = O, Y = S (42)	322a
		C_6H_{12}, air	" (22)	332
	I, X = Y = S	C_6H_6, I_2, air	II, X = Y = S (90)	322a
		C_6H_{12}, air	" (30)	332
	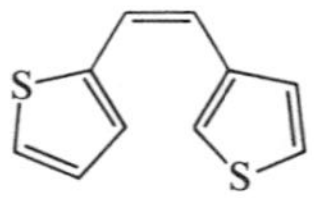	C_6H_6, I_2, degassed	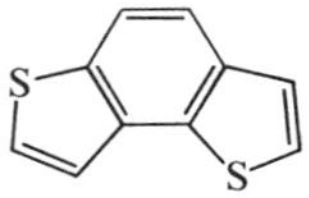 (47)	322a
	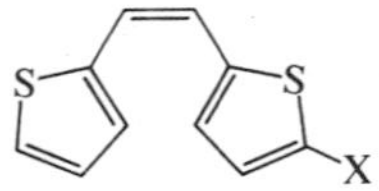 (I)		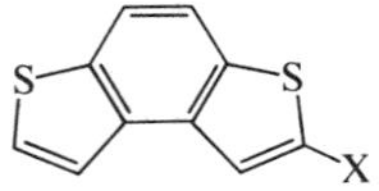 (II)	
	I, X = Br	C_6H_{12}, air	II, X = H (50)	650a
	I, X = Cl	C_6H_{12}, I_2	II, X = Cl (59)	650a
C_{11}	I, X = CH_3	C_6H_6, I_2	II, X = CH_3 (71)	327
		C_6H_6, I_2, air	" (68)	322a
		Solvent, air	" (—)	650a
	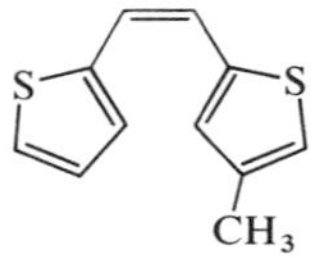 CH_3	C_6H_6, I_2, air	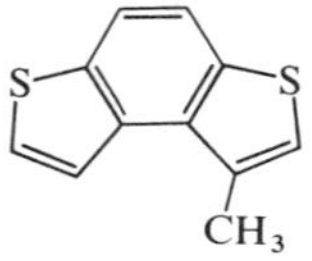 CH_3 (69)	322a

	Starting material	Conditions	Product (yield %)	Ref.
	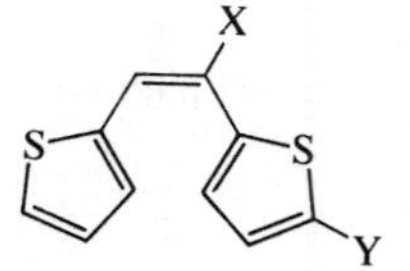 (I)		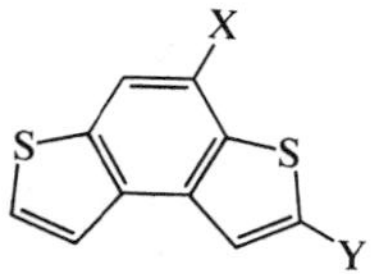 (II)	
	I, X = CH_3, Y = H	”	II, X = CH_3, Y = H (50)	322a
	I, X = CO_2H, Y = H	EtOH, I_2, air	II, X = CO_2H, Y = H (57)	650b
	I, X = CO_2H, Y = Cl	”	II, X = CO_2H, Y = Cl (15)	650b
C_{12}	2-Styrylfuran	C_6H_{12}, I_2 (or air)	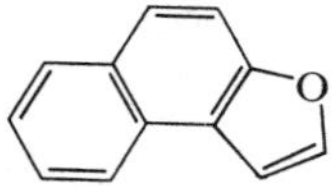 (9)	332
	2-Styrylthiophene	C_6H_{14}, I_2	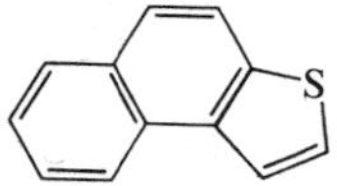 (100)	324
	X = Y = H	C_6H_{12}, I_2	” (40–50)	323
	X = H, Y = I	C_6H_{12}	” (50–60)	323
	X = I, Y = H	”	” (50–60)	323
	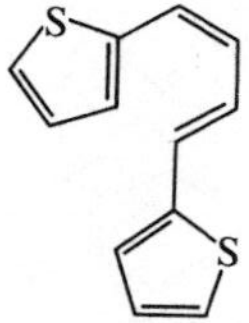	C_6H_6, I_2	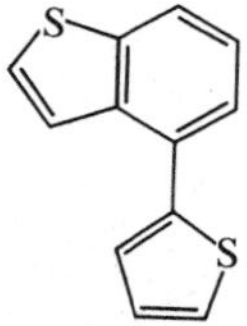 (49)	330

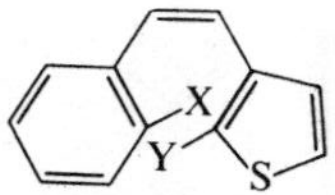

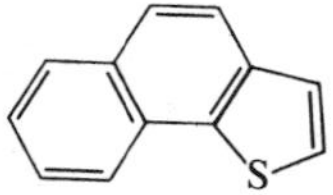

TABLE VI. OXYGEN AND SULFUR HETEROAROMATIC SYSTEMS FROM FURANS, THIOPHENES, AND PYRYLIUM IONS (*Continued*)

	Starting Material	Reaction Conditions	Products and Yields (%)	Refs.
	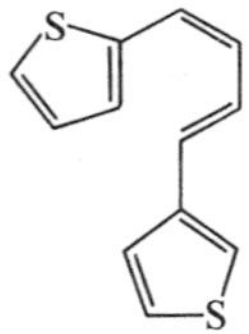 [structure]	C_6H_6, I_2	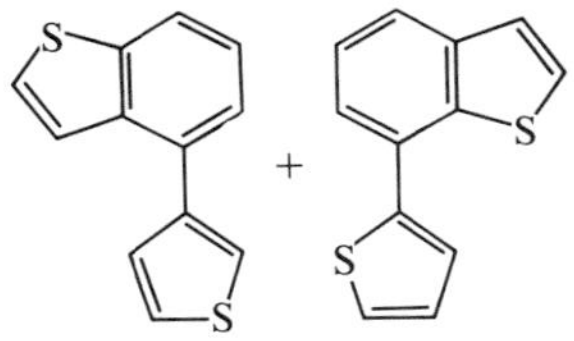 [structure] + [structure] (22 total)	330
	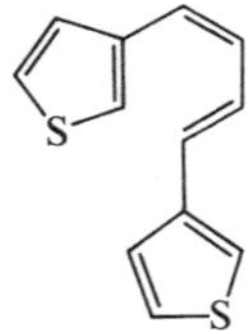 [structure]	"	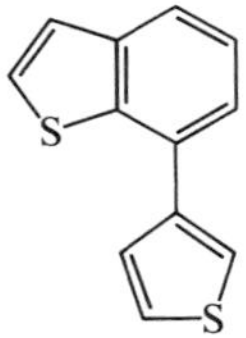 [structure] (62)	330
	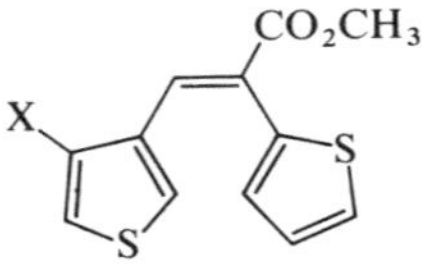 [structure: X, CO_2CH_3]		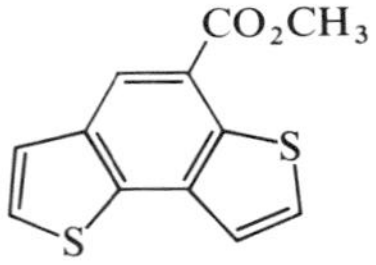 [structure: CO_2CH_3]	
	X = H	C_6H_{12}, I_2	" (40–50)	323
	X = I	C_6H_{12}	" (50–60)	323
C_{13}	2-(*o*-Methylstyryl)thiophene	C_6H_{14}, I_2	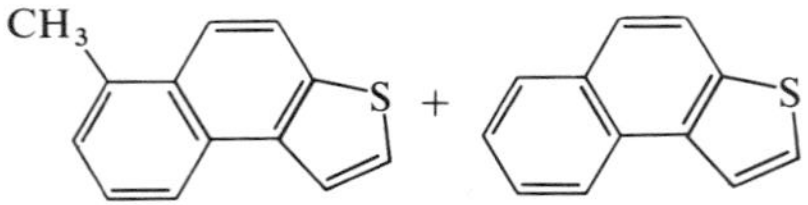 [structure: CH_3] + [structure] (4:1)	324
	2-(*m*-Methylstyryl)thiophene	"	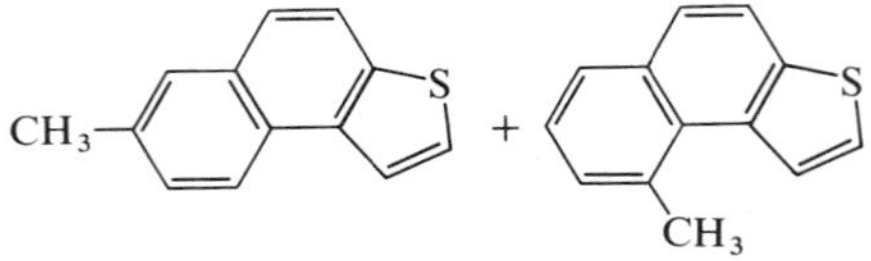 [structure: CH_3] + [structure: CH_3] (1:1)	324

2-(*p*-Methylstyryl)thiophene	”	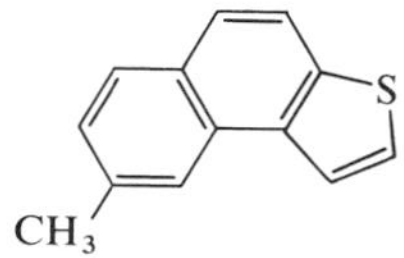 (high)	324
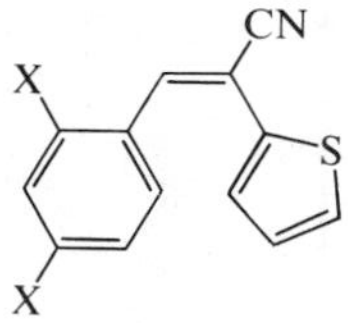 (I)		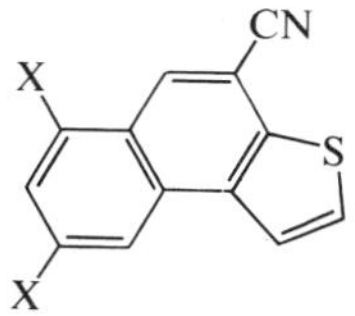 (II)	
I, X = H	EtOH, $CuBr_2$, I_2, air	II, X = H (60)	651
I, X = Cl	”	II, X = Cl (20)	651
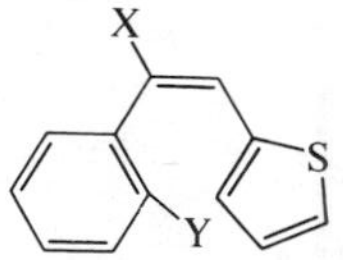 (I)		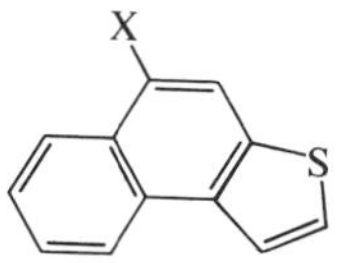 (II)	
I, X = CN, Y = H	”	II, X = CN (12)	652
I, X = CO_2H, Y = H	EtOH, I_2, air	II, X = CO_2H (10)	652
I, X = CO_2H, Y = I	*t*-BuOH, $Na_2S_2O_3$	” (21)	652
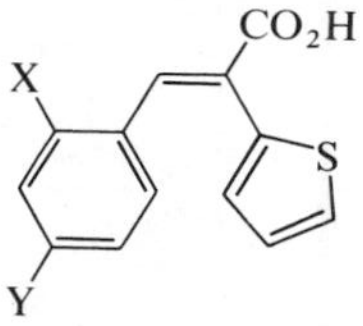 (I)		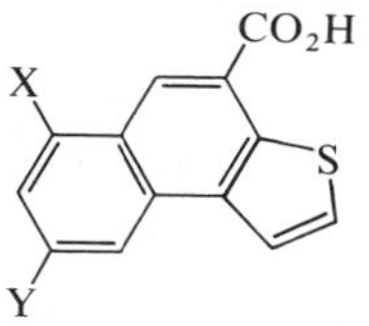 (II)	
I, X = Y = H	EtOH, I_2, air	II, X = Y = H (80)	651
I, X = H, Y = Br	”	II, X = H, Y = Br (15)	651
I, X = Y = Cl	”	II, X = Y = Cl (15)	651

TABLE VI. OXYGEN AND SULFUR HETEROAROMATIC SYSTEMS FROM FURANS, THIOPHENES, AND PYRYLIUM IONS (*Continued*)

	Starting Material	Reaction Conditions	Products and Yields (%)	Refs.
C_{14}	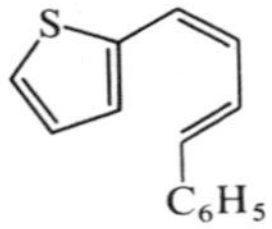 C_6H_5	C_6H_6, I_2	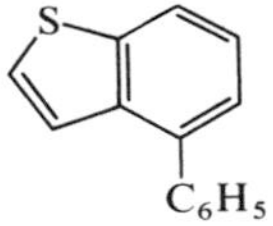 C_6H_5 (15)	330
	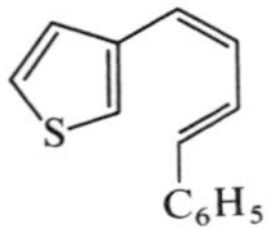 C_6H_5	"	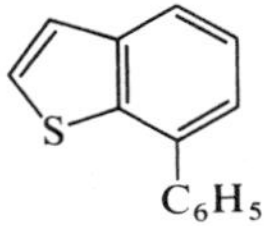 C_6H_5 (39)	330
	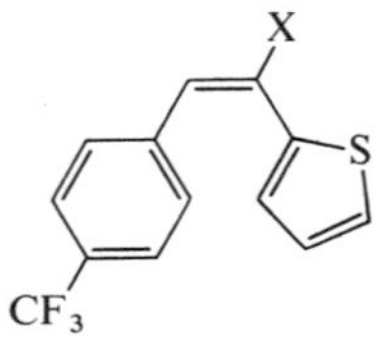 X (I) CF_3		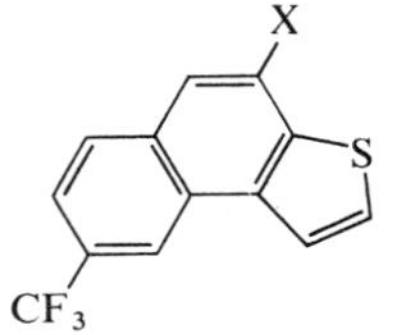 X (II) CF_3	
	I, X = CN	EtOH, $CuBr_2$, I_2, air	II, X = CN (51)	651
	I, X = CO_2H	EtOH, I_2, air	II, X = CO_2H (85)	651
	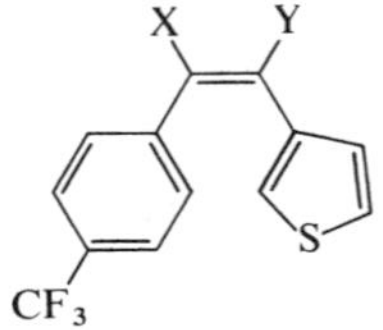 X Y (I) CF_3		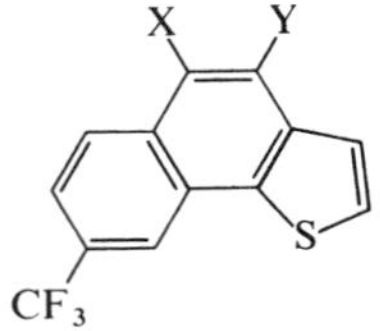 X Y (II) CF_3	
	I, X = CO_2H, Y = H	"	II, X = CO_2H, Y = H (30)	652
	I, X = H, Y = CO_2H	"	II, X = H, Y = CO_2H (63)	552
	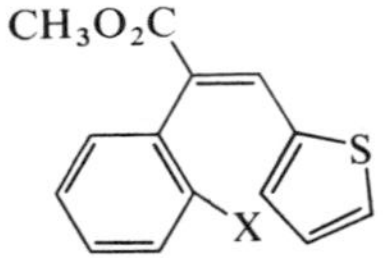 CH_3O_2C X		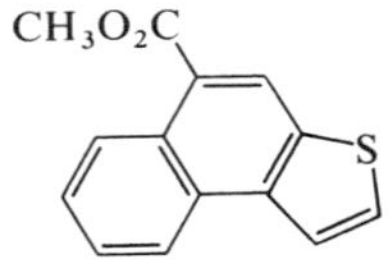 CH_3O_2C	

	Reactant	Conditions	Product (yield %)	Ref.
	X = H	C_6H_{12}, I_2	" (40–50)	323
	X = I	C_6H_{12}	" (50–60)	323
	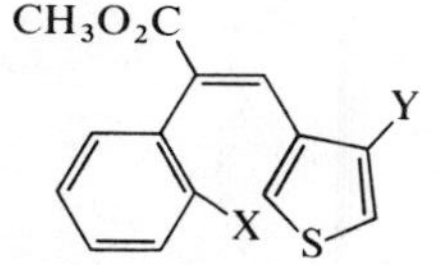 CH_3O_2C		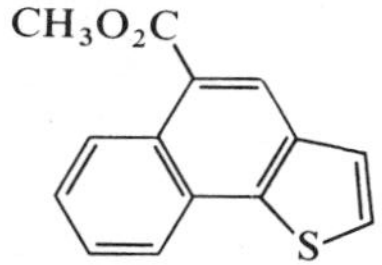 CH_3O_2C	
	X = Y = H	C_6H_{12}, I_2	" (40–50)	323
	X = I, Y = H	C_6H_{12}	" (50–60)	323
	X = H, Y = I	"	" (50–60)	323
	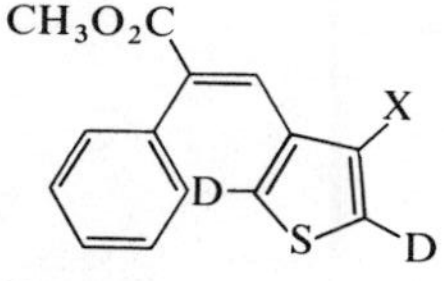 CH_3O_2C		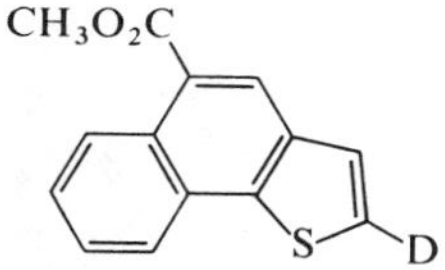 CH_3O_2C	
	X = H	C_6H_{12}, I_2	" (40–50)	323
	X = I	C_6H_{12}	" (50–60)	323
	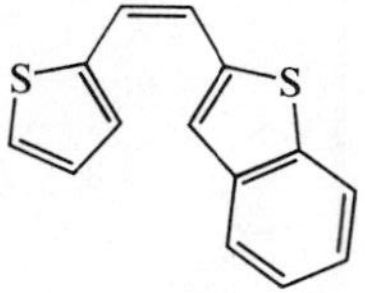	C_6H_6, I_2	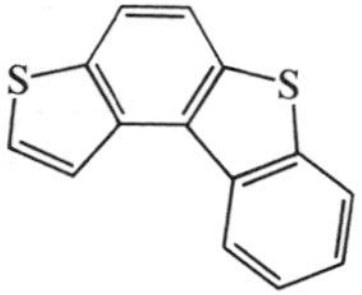 (73)	326
C_{16}	2,3-Diphenylfuran	C_6H_6, air	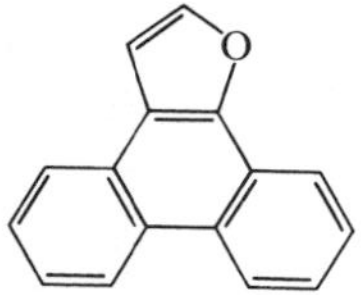 (40)	333

TABLE VI. Oxygen and Sulfur Heteroaromatic Systems from Furans, Thiophenes, and Pyrylium Ions (*Continued*)

Starting Material	Reaction Conditions	Products and Yields (%)	Refs.
2,3-Diphenylthiophene	Et_2O, air	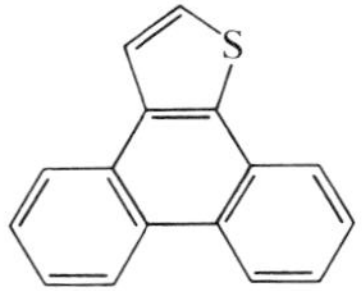 (70)	331
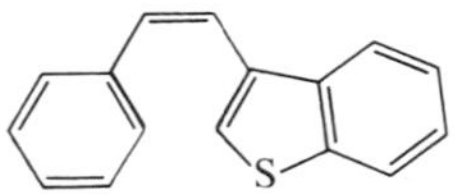 S	C_6H_{14}, I_2	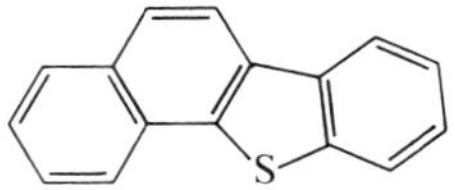 S (—)	324
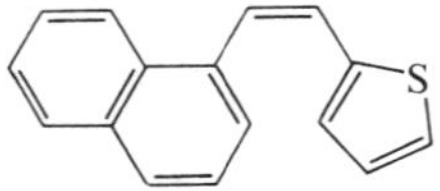 S	C_6H_{12}, I_2, air	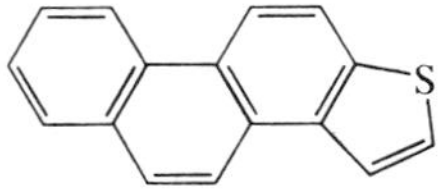 S (88)	653
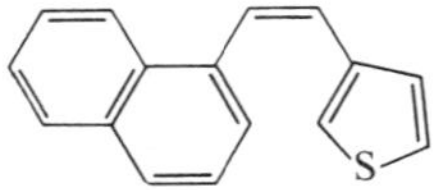 S	"	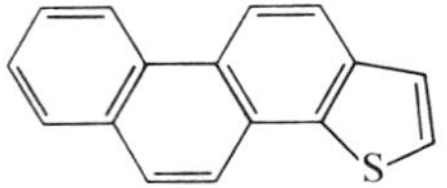 S (32)	653
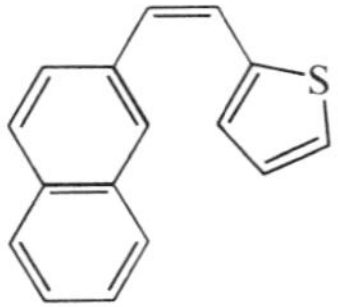 S	"	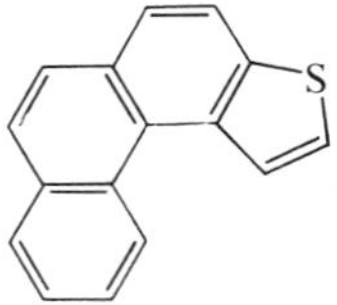 S (75)	653
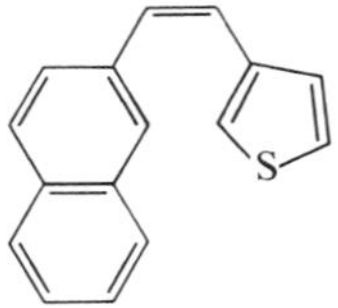 S	"	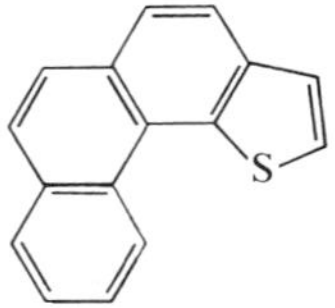 S (17)	653

	Reactant	Conditions	Product	Ref.
C_{17}	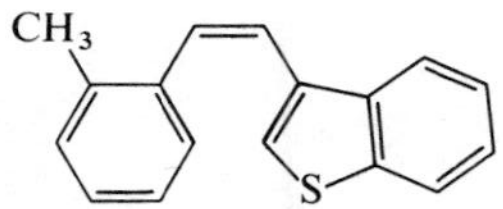 CH3 (structure)	C_6H_{14}, I_2	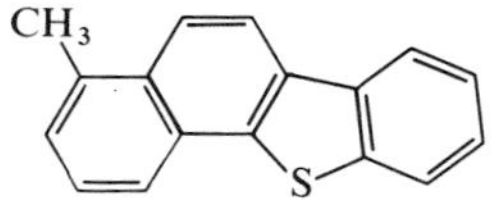 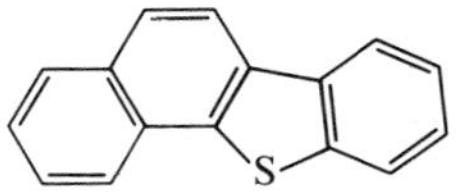 CH3 (structure) + (structure) (1:1)	324
	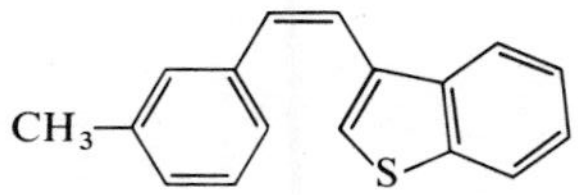 CH3 (structure)	”	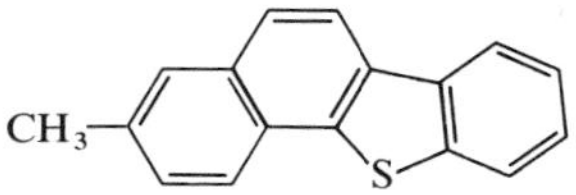 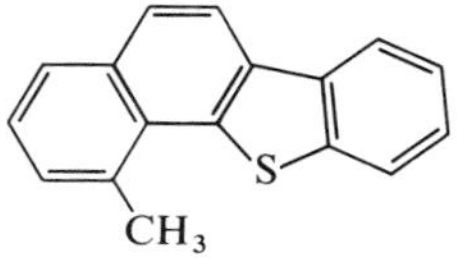 CH3 (structure) + CH3 (structure) (1:1)	324
	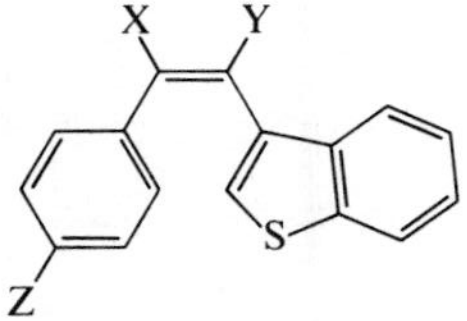 (I)		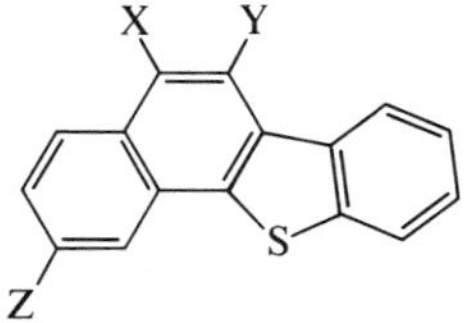 (II)	
	I, X = CH_3, Y = Z = H	”	II, X = CH_3, Y = Z = H (good)	324
	I, Y = CH_3, X = Z = H	”	II, Y = CH_3, X = Z = H (good)	324
	I, Z = CH_3, X = Y = H	”	II, Z = CH_3, X = Y = H (good)	324
	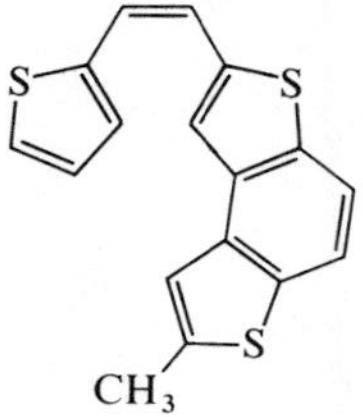 CH3 (structure)	C_6H_6, I_2	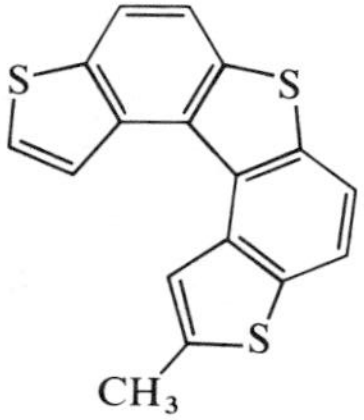 CH3 (structure) (73)	327

TABLE VI. OXYGEN AND SULFUR HETEROAROMATIC SYSTEMS FROM FURANS, THIOPHENES, AND PYRYLIUM IONS (*Continued*)

	Starting Material	Reaction Conditions	Products and Yields (%)	Refs.
C_{18}	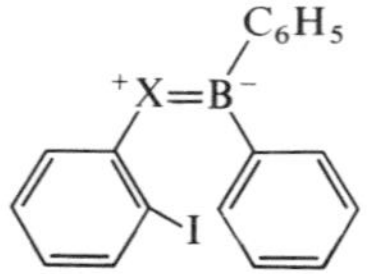 $^{+}X{=}B^{-}(C_6H_5)$ (I)		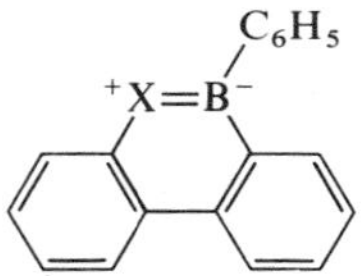 $^{+}X{=}B^{-}(C_6H_5)$ (II)	
	I, X = O	C_6H_{12}, N_2	II, X = O (31)	623b
	I, X = S	”	II, X = S (50)	623b
	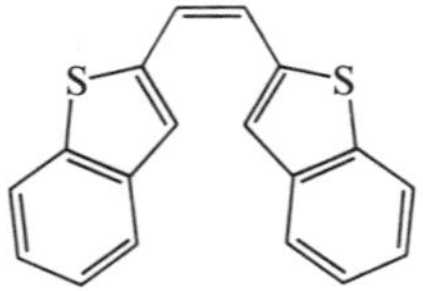	C_6H_6, I_2	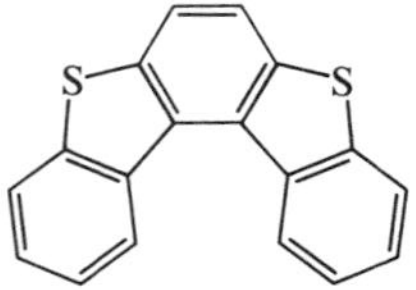 (57)	326
	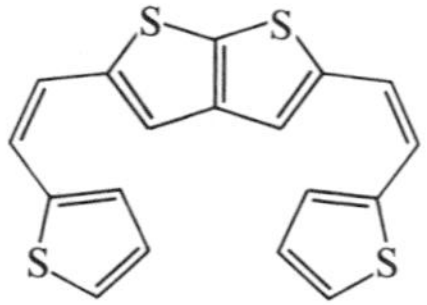	”	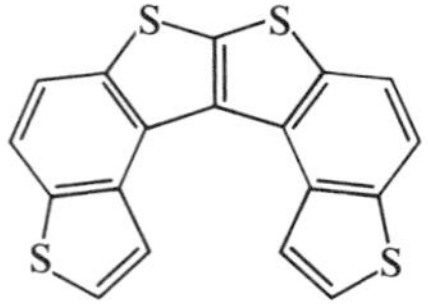 (51)	328
	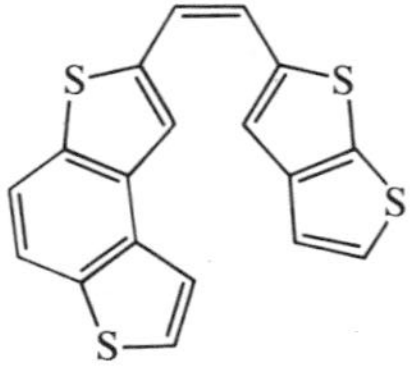	”	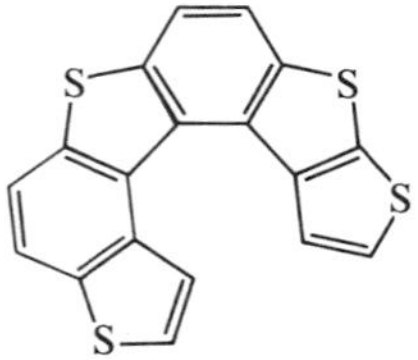 (69)	328

	Reactant	Conditions	Product(s) (Yield, %)		Refs.
C_{19}	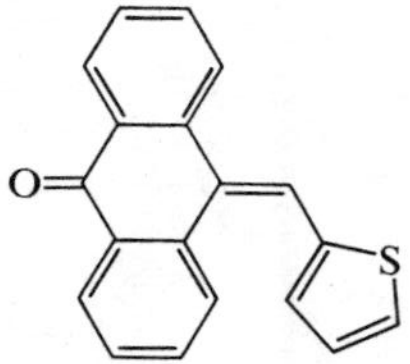	C_6H_6 or C_6H_{12}, I_2	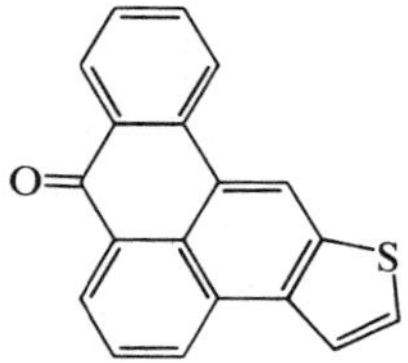	(20–90)	58
C_{20}	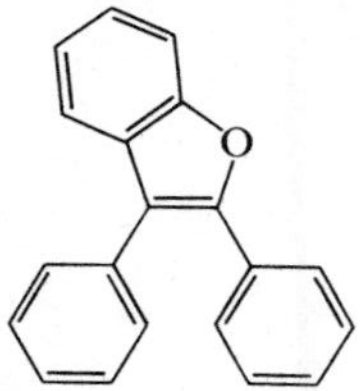	Et_2O, MeOH, EtOH, $CHCl_3$, t-$BuNH_2$, $(n$-$Pr)_2NH$, or Et_3N; air	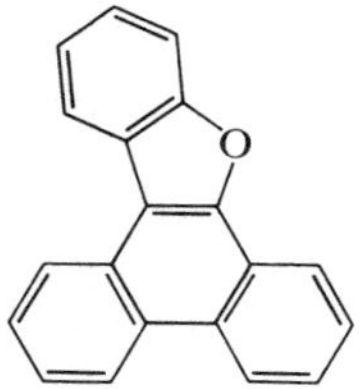	(52)	335
		n-$PrNH_2$ or n-$PrND_2$	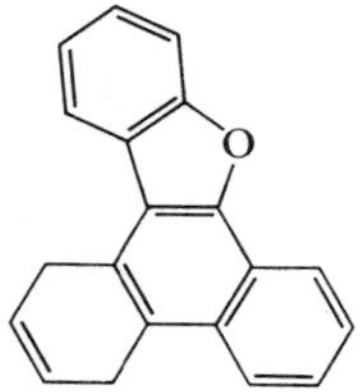	(65)	335,337
	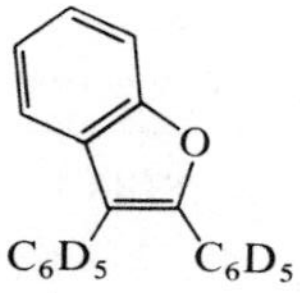 C_6D_5, C_6D_5	n-$PrNH_2$	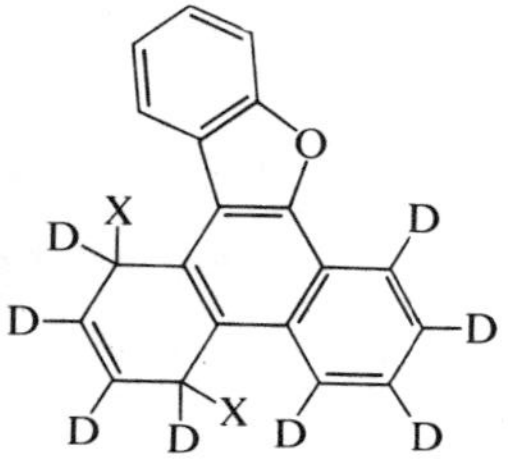 (63% X = H, 37% X = D)	(—)	336

TABLE VI. OXYGEN AND SULFUR HETEROAROMATIC SYSTEMS FROM FURANS, THIOPHENES, AND PYRYLIUM IONS (*Continued*)

Starting Material	Reaction Conditions	Products and Yields (%)	Refs.
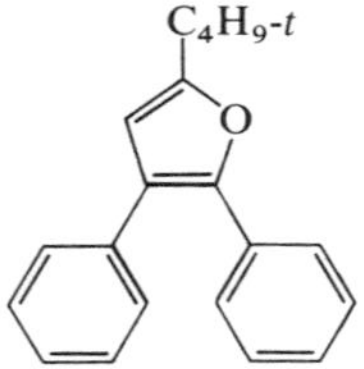 C_4H_9-*t*	EtOH, air	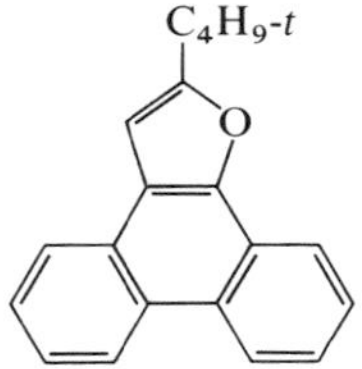 C_4H_9-*t* (50)	335
	n-$PrNH_2$, air	" (15), 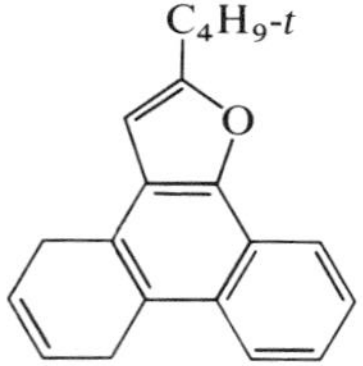 C_4H_9-*t* (35)	335
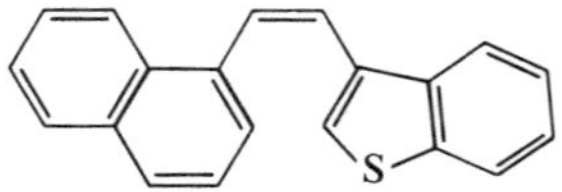	C_6H_{12}, I_2	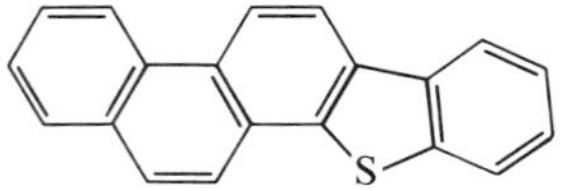 (35–38)	329
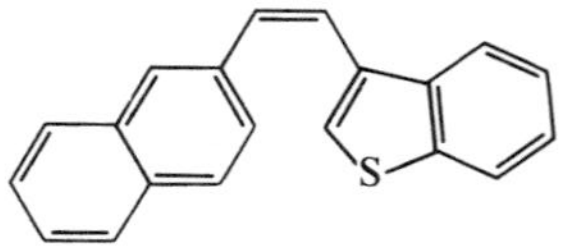	"	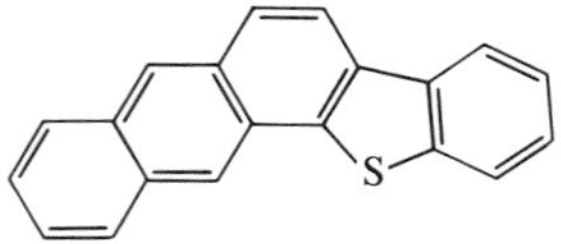 (35–38)	329

	Reactant	Conditions	Product (Yield %)	Ref.
	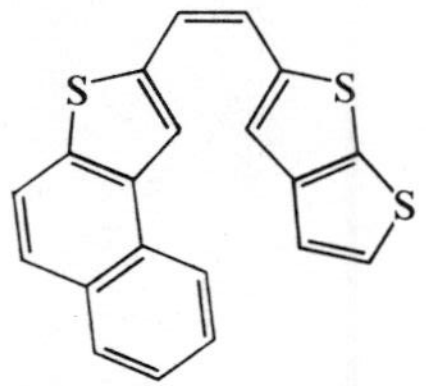	C_6H_6, I_2	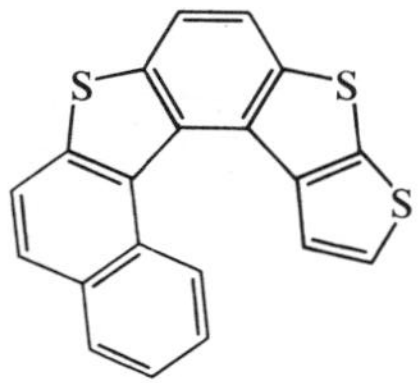 (60)	328
	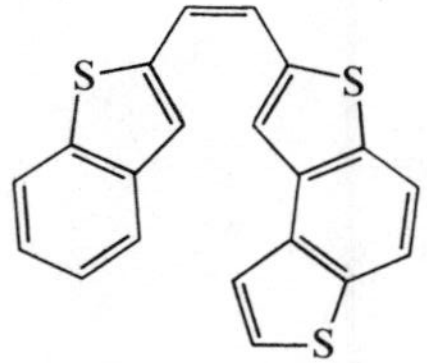	”	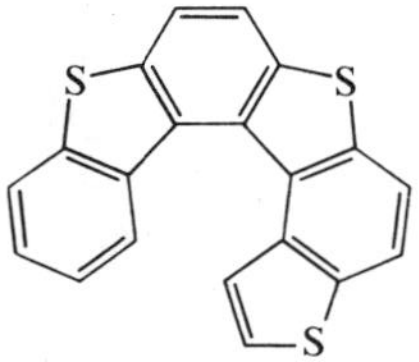 (40)	326
C_{21}	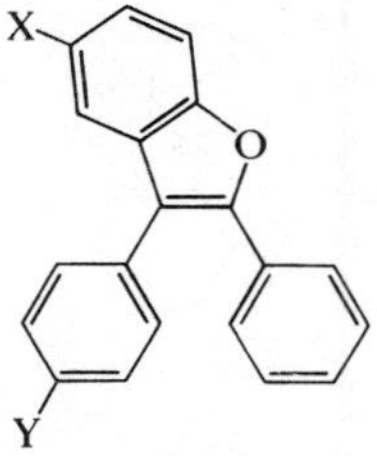 (I)		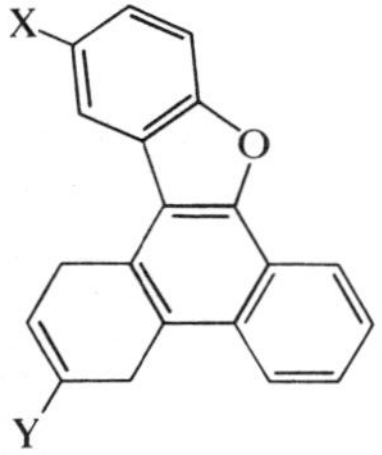 (II)	
	I, X = CH_3, Y = H	*n*-$PrNH_2$	II, X = CH_3, Y = H (—)	335
	I, X = H, Y = CH_3O	”	II, X = H, Y = CH_3O (—)	336
	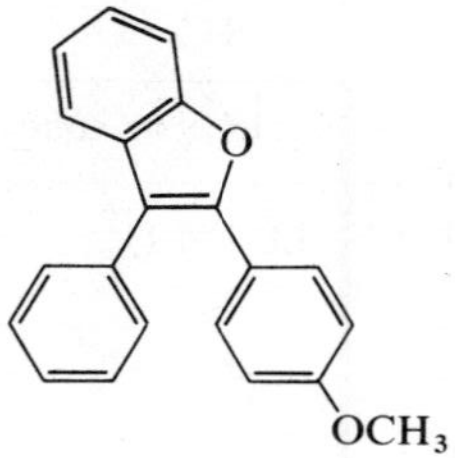 OCH_3	”	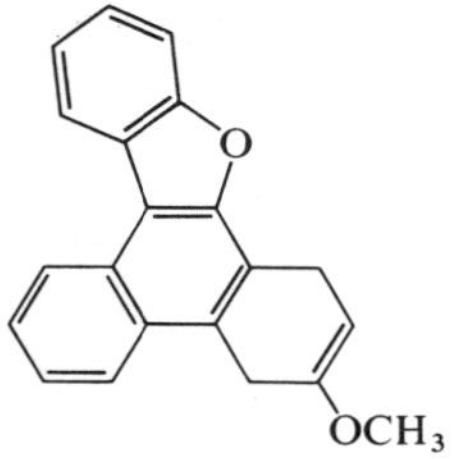 OCH_3 (—)	335,336

TABLE VI. OXYGEN AND SULFUR HETEROAROMATIC SYSTEMS FROM FURANS, THIOPHENES, AND PYRYLIUM IONS (*Continued*)

	Starting Material	Reaction Conditions	Products and Yields (%)	Refs.
C_{22}	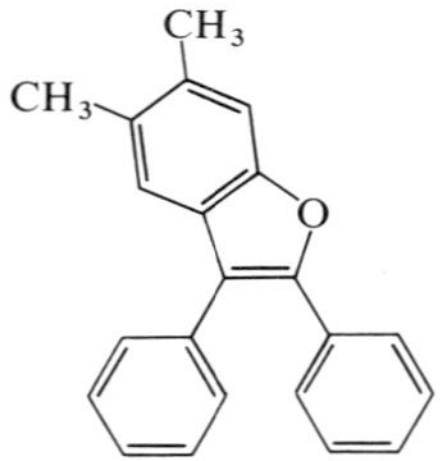 CH_3, CH_3	n-$PrNH_2$	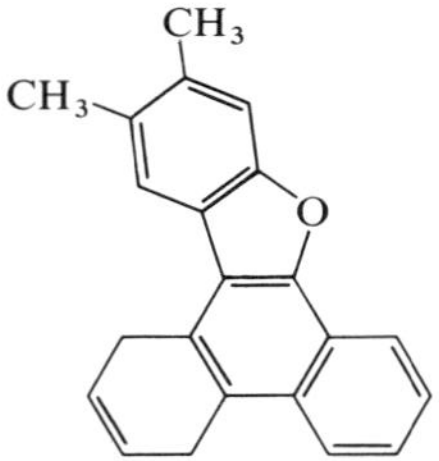 CH_3, CH_3 (—)	335
	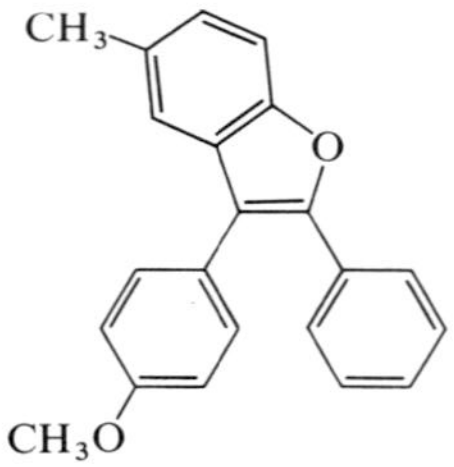 CH_3, CH_3O	EtOH, air	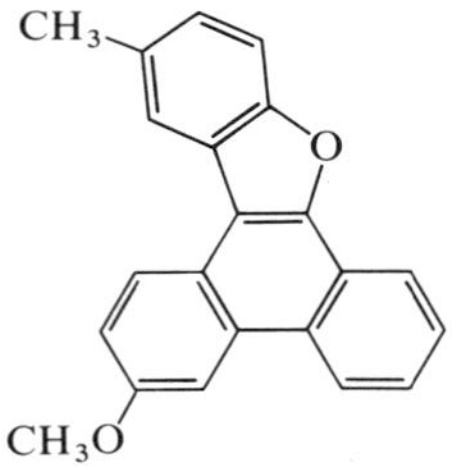 CH_3, CH_3O (45)	335
		n-$PrNH_2$	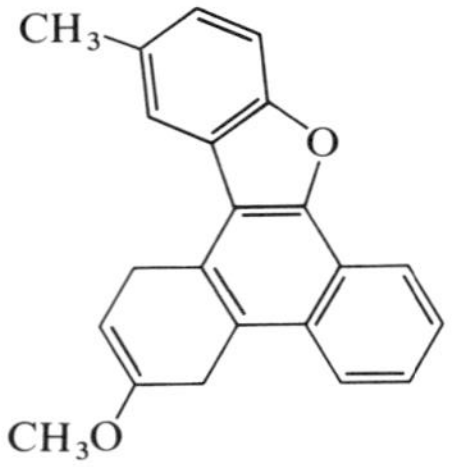 CH_3, CH_3O (—)	335

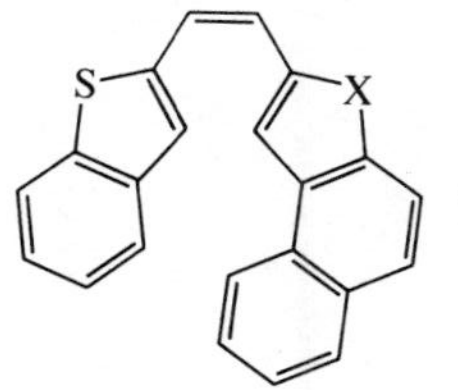

Reactant	Conditions	Product (Yield, %)	Refs.
(I)		(II)	
I, X = O	C_6H_6, I_2	II, X = O (40)	326
I, X = S	"	II, X = S (73)	326
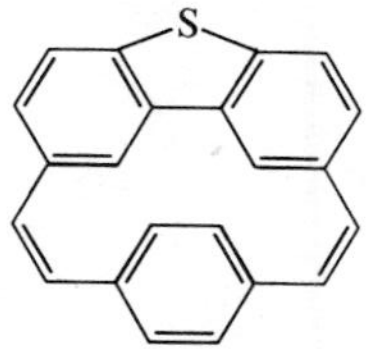	1. C_6D_6, degassed 2. O_2	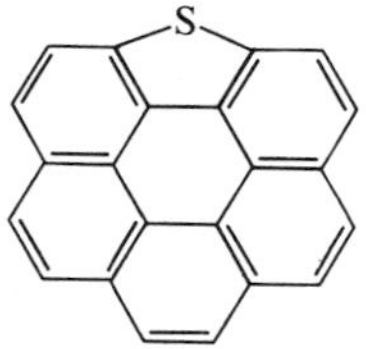 (—)	654
	THF-d_8, O_2	" (100)	492
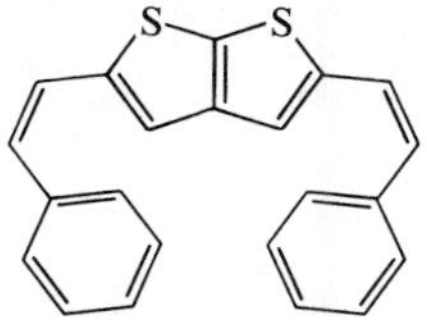	C_6H_6, I_2	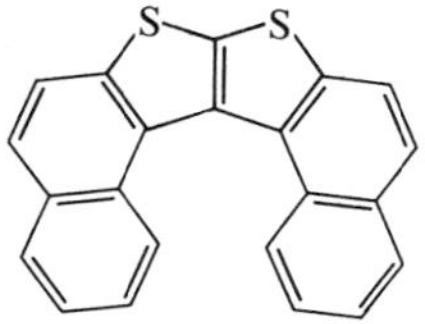 (50)	328
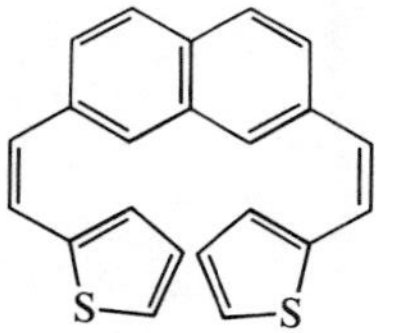	"	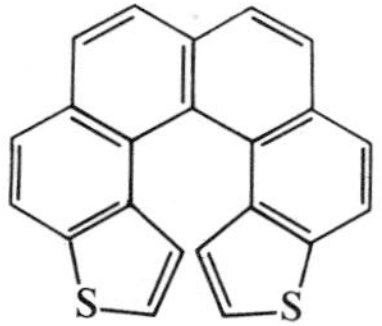 (50)	328

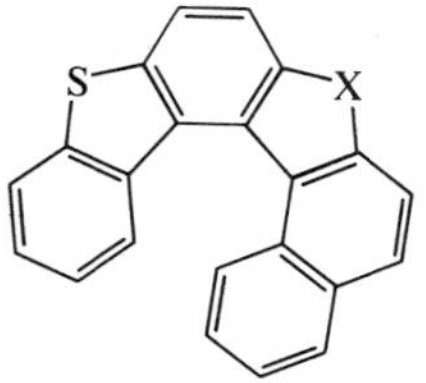

TABLE VI. OXYGEN AND SULFUR HETEROAROMATIC SYSTEMS FROM FURANS, THIOPHENES, AND PYRYLIUM IONS (*Continued*)

	Starting Material	Reaction Conditions	Products and Yields (%)	Refs.
	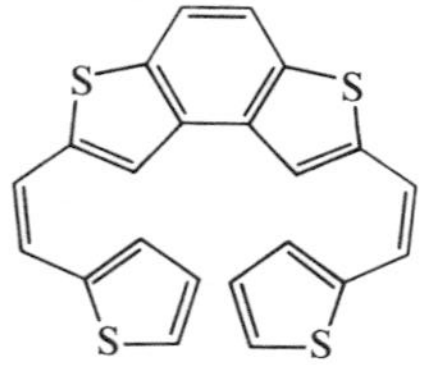	C_6H_6, I_2	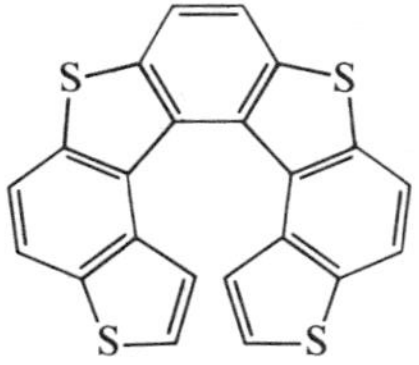 (26)	326
C_{23}	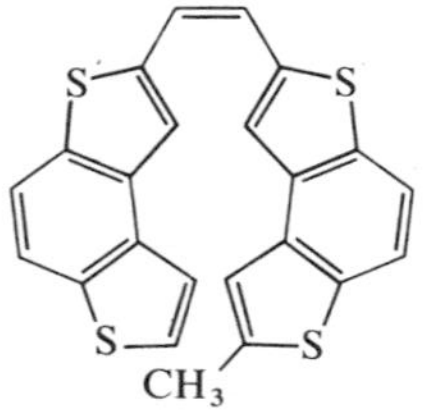 CH_3	"	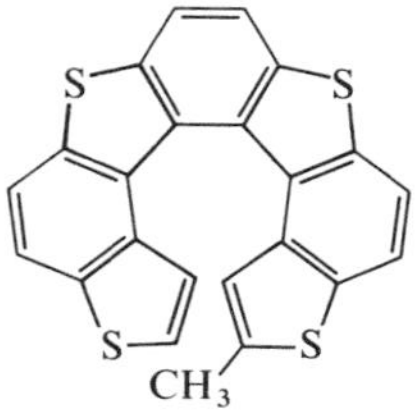 CH_3 (72)	327
C_{24}	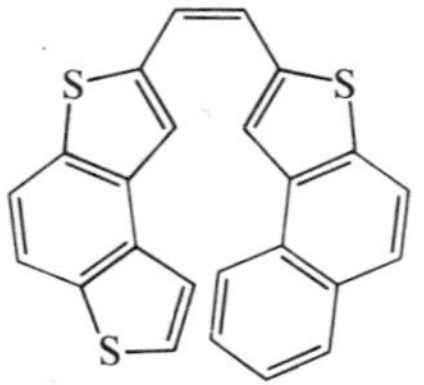	"	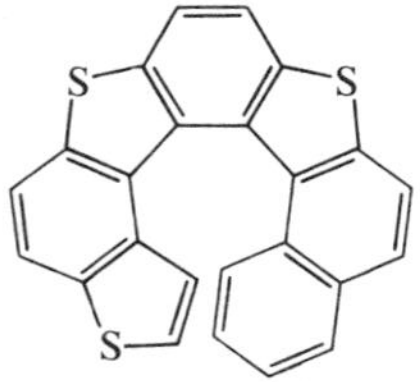 (49)	326
C_{26}	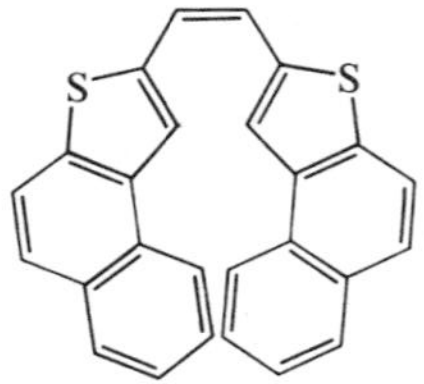	"	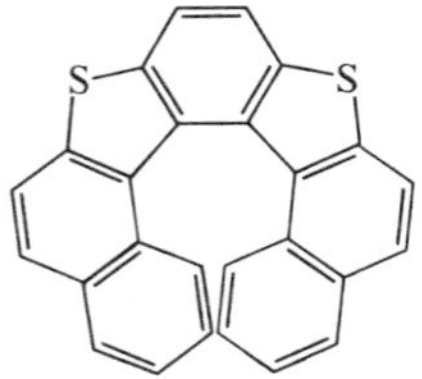 (66)	326

	Reactant	Conditions	Product (Yield, %)	Refs.
C_{27}		H_2O, air	(50)	340a
	(I)		(II)	
	I, Ar = Phenyl	HOAc, O_2	II, X = Y = Z = H (50)	338
	I, Ar = *o*-Chlorophenyl	HOAc, O_2 (or N_2)	" (91)	338
	I, Ar = *p*-Chlorophenyl	HOAc, O_2	II, X = Cl, Y = Z = H (—)	338
	I, Ar = *m*-Nitrophenyl	Boiling HOAc[a]	II, Y = NO_2, X = Z = H (50–60)	340a
C_{28}	I, Ar = *o*-Methylphenyl	HOAc, O_2	II, Z = CH_3, X = Y = H (85)	338
	I, Ar = *p*-Methylphenyl	Boiling HOAc[a]	II, X = CH_3, Y = Z = H (33)	338
	I, Ar = *m*-Methoxyphenyl	" [a]	II, Y = CH_3O, X = Z = H (50)	340a
C_{29}	I, Ar = *p*-Acetoxyphenyl	Ac_2O, air	II, X = CH_3CO_2, Y = Z = H (—)	340b
		C_6H_6, I_2	(18)	327

TABLE VI. OXYGEN AND SULFUR HETEROAROMATIC SYSTEMS FROM FURANS, THIOPHENES, AND PYRYLIUM IONS (*Continued*)

	Starting Material	Reaction Conditions	Products and Yields (%)	Refs.
C_{31}	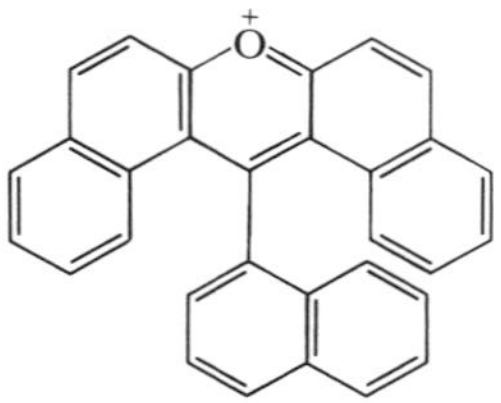	HOAc, O_2	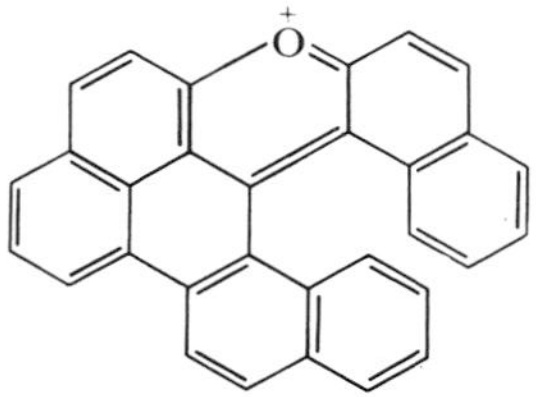 (65)	338
C_{32}	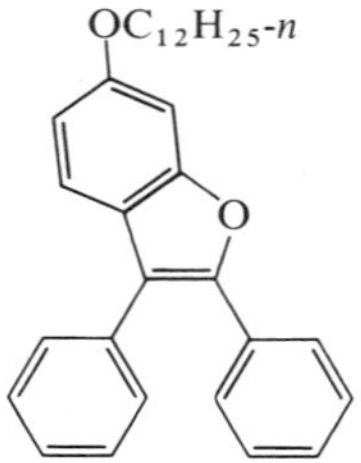 $OC_{12}H_{25}$-*n*	C_6H_{14}, air[b]	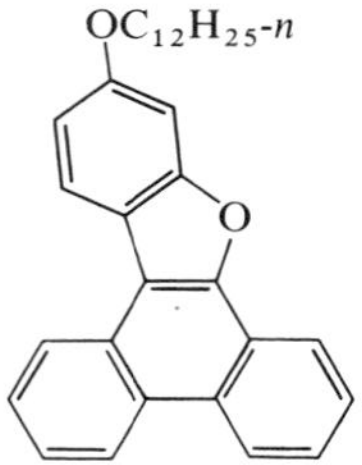 $OC_{12}H_{25}$-*n* (—)	655
	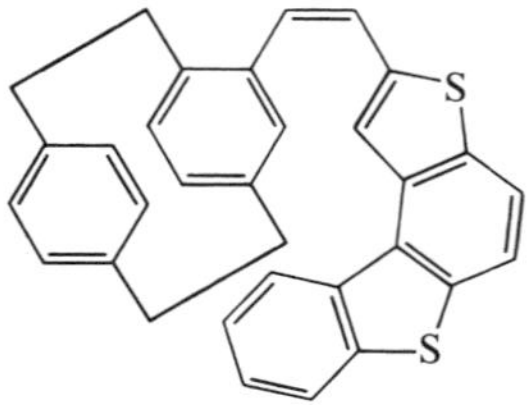 S S	Solvent, air	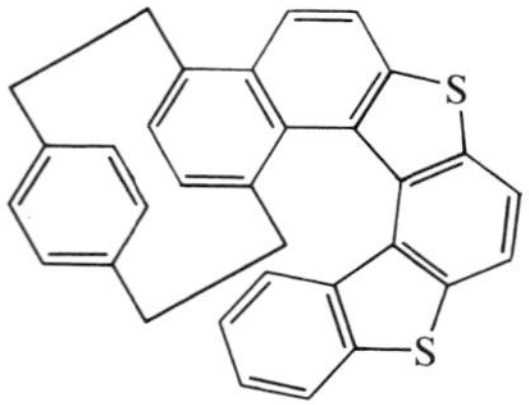 S S (—)	176

	Starting material	Conditions	Product (yield)	Ref.
C_{34}	OH; $O_2CC_6H_5$	HOAc, $HClO_4$, air	$O_2CC_6H_5$ (38)	340b
C_{35}	S; S; S; S; S; CH_3	C_6H_6, I_2	S; S; S; S; S; CH_3 (22)	327
C_{38}	S; S; S; S	"	S; S; S; S (14)	326

[a] The identity of the oxidant in this reaction is obscure. One possibility is residual dissolved oxygen.
[b] The starting material was generated *in situ* by irradiation of 2-benzyloxy-4-(*n*-dodecyloxy)benzophenone.

TABLE VII. LACTAMS AND OTHER HETEROCYCLIC SYSTEMS FROM AMIDES[a]

	Starting Material	Reaction Conditions	Products and Yields (%)	Refs.
C_8	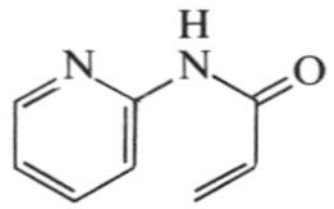	C_6H_6, HOAc, argon	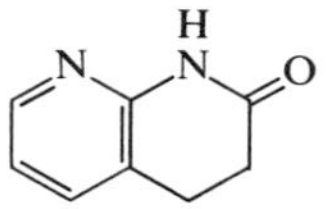 (17)	656
	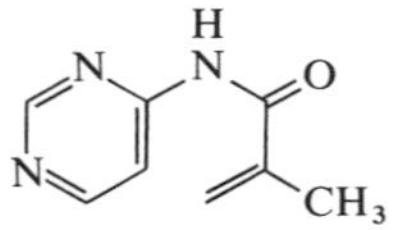	"	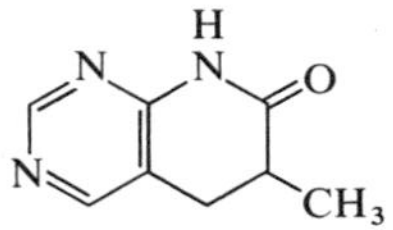 (24)	656
	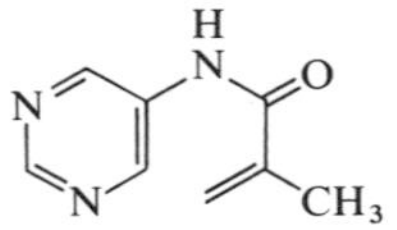	"	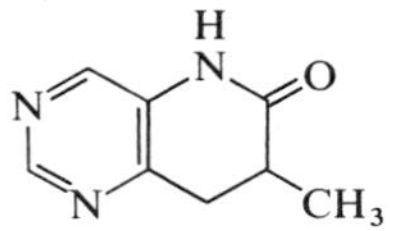 (14)	656
	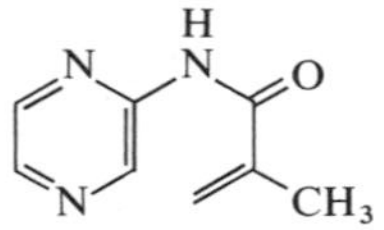	"	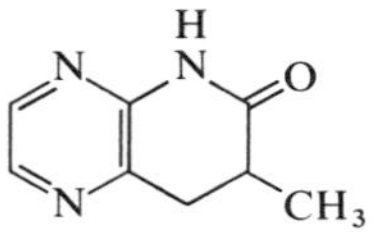 (19)	656
	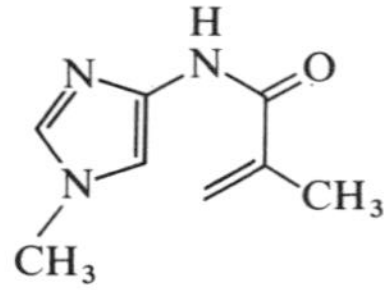	"	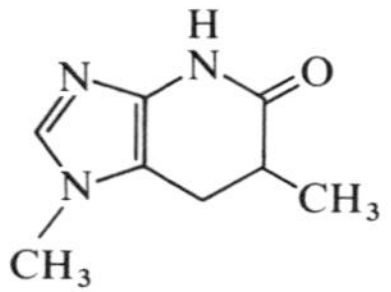 (6)	656
C_9	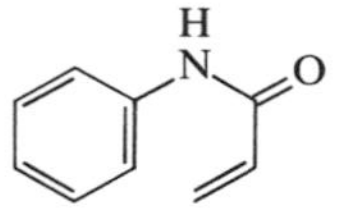	C_6H_6, N_2	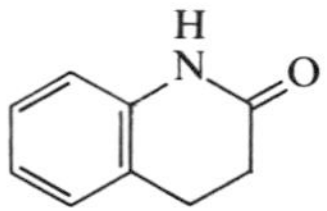 (4)	657a

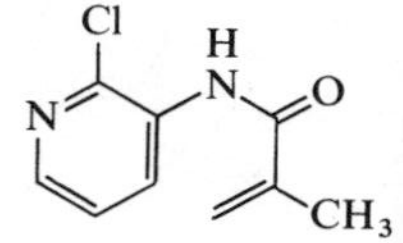	C_6H_6, HOAc, argon	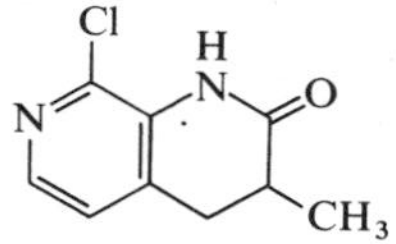 (25)	656
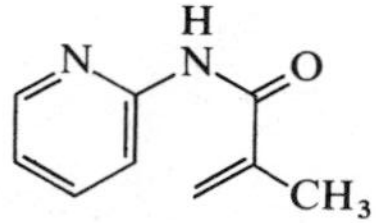	"	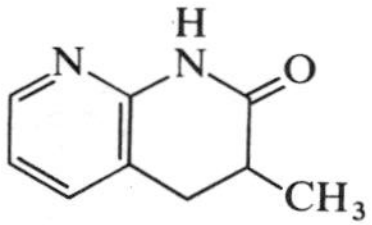 (78)	656
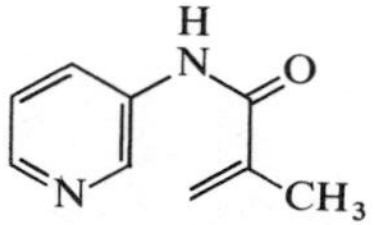	"	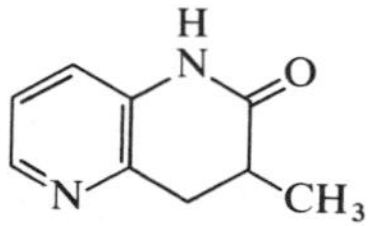 (53), 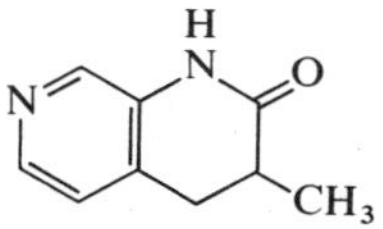 (22)	656
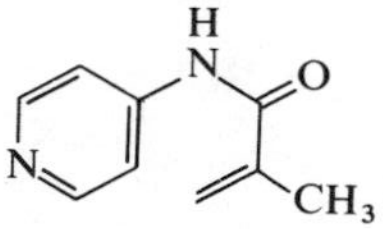	"	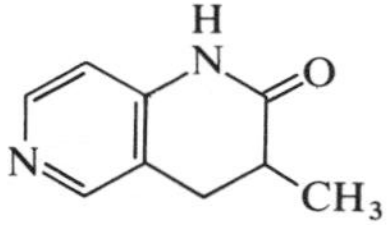 (72)	656
C_{10} 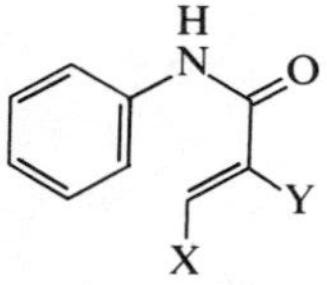 (I)		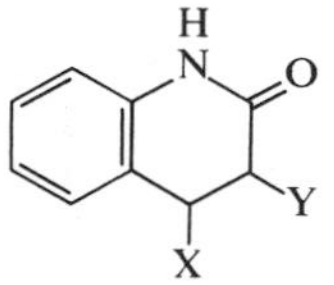 (II)	
I, X = CH_3, Y = H	Et_2O, HOAc	II, X = CH_3, Y = H (25)	372
I, X = H, Y = CH_3	C_6H_6, N_2	II, X = H, Y = CH_3 (61)	657a
	Et_2O, HOAc	" (50)	372
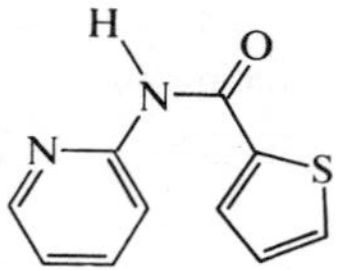	C_6H_6, EtOH, air	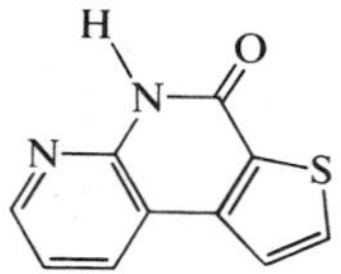 (27)	657b

TABLE VII. LACTAMS AND OTHER HETEROCYCLIC SYSTEMS FROM AMIDES (*Continued*)[a]

Starting Material	Reaction Conditions	Products and Yields (%)	Refs.
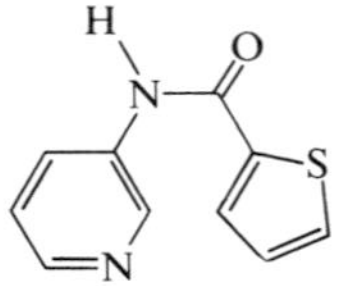	C_6H_6, EtOH, air	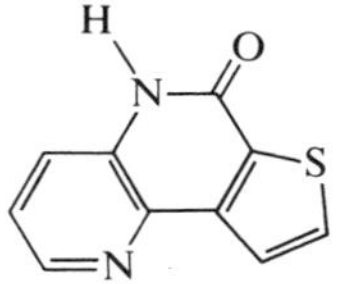 or 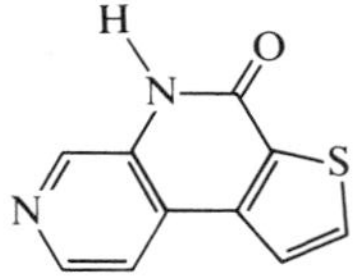 (62)	657b
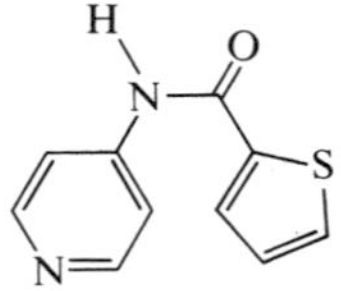	"	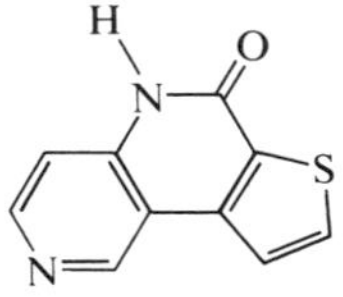 (45)	657b
C_{11} 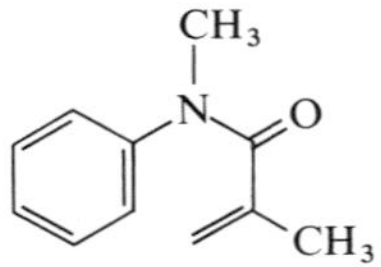	C_6H_6, N_2	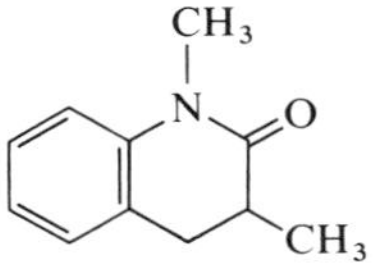 (57)	657a
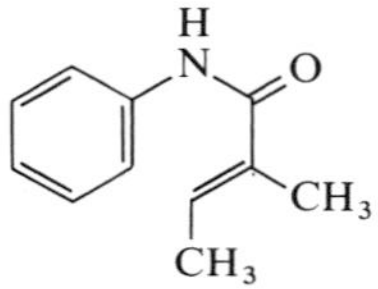	Et_2O, HOAc	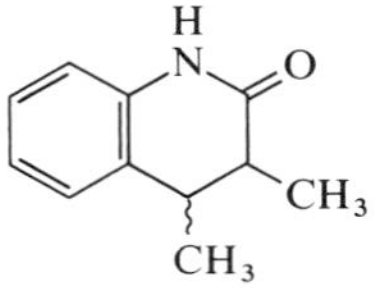 (58)	372
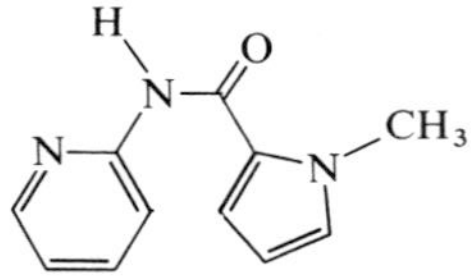	C_6H_6, EtOH, air	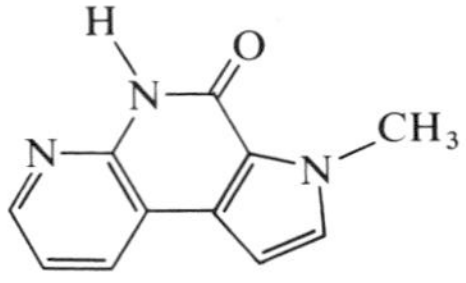 (10)	657b

Reactant	Conditions	Product (yield %)	Ref.
	"	(38)	657b
	"	(35)	657b
	"	(25)	342
	C_6H_6	(33)	368
	C_6H_6, EtOH, air	(50)	342
	Solvent	(56)	658a

TABLE VII. LACTAMS AND OTHER HETEROCYCLIC SYSTEMS FROM AMIDES (*Continued*)[a]

Starting Material	Reaction Conditions	Products and Yields (%)	Refs.
C_{12} 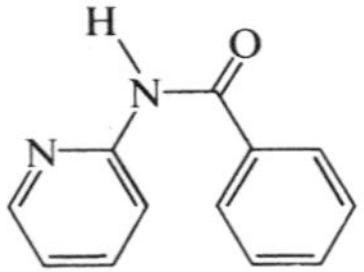 H N O N	C_6H_6, air	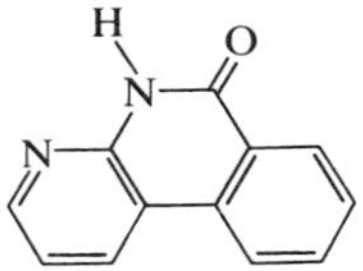 H N O N (29)	341
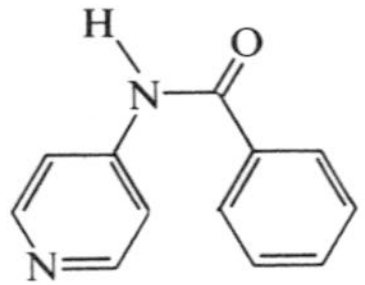 H N O N	EtOH, air	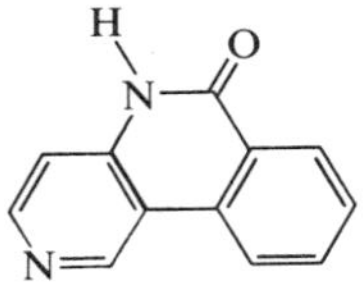 H N O N (28)	341
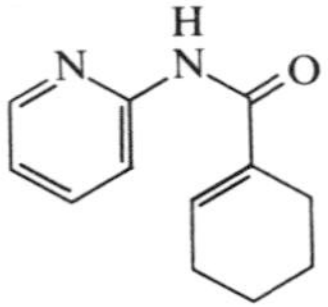 N H N O	MeOH, I_2	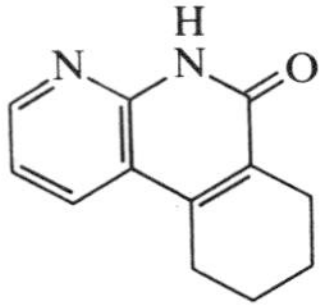 N H N O (10)	369
	MeOH, C_6H_6, or Et_2O	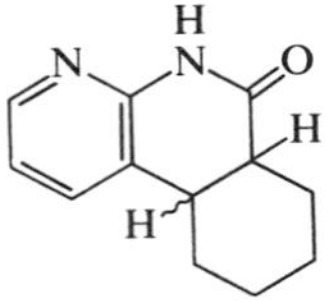 N H N O H H (38–41)	369
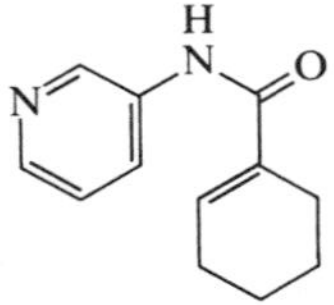 N H N O	Et_2O	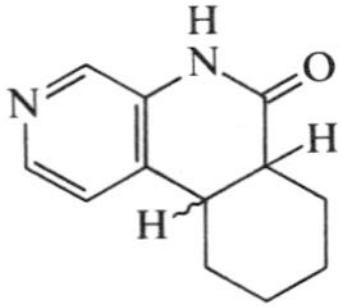 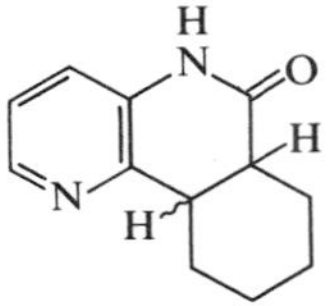 N H N O H H (18), H N O H N H (11)	369

Reactant	Solvent/Conditions	Product(s) (Yield %)	Refs.
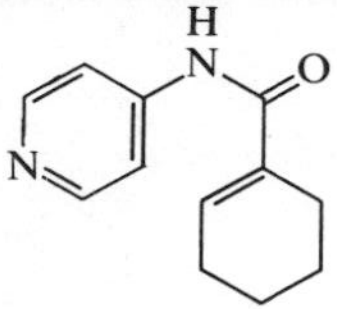	MeOH, I_2	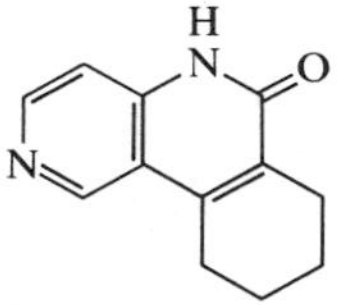 (78)	369
	MeOH, C_6H_6, or Et_2O	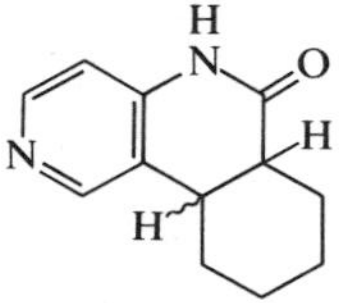 (73–88)	369
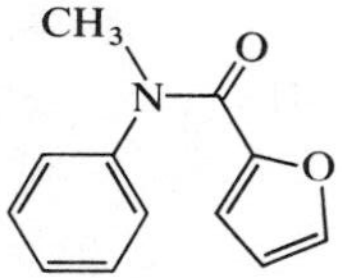 CH_3, N, O, O	C_6H_6 or C_6H_6–MeOH, I_2	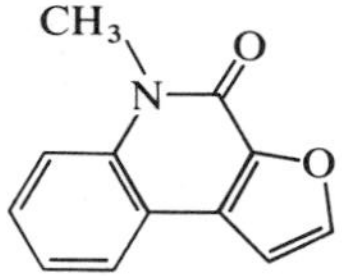 CH_3, N, O, O (29)	658b
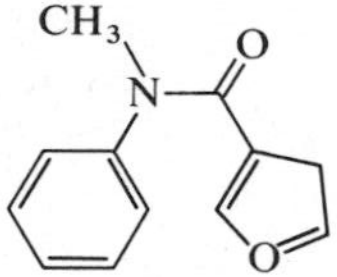 CH_3, N, O, O	"	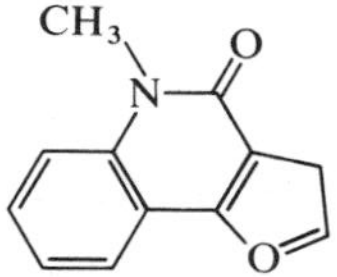 CH_3, N, O, O (5)	658b
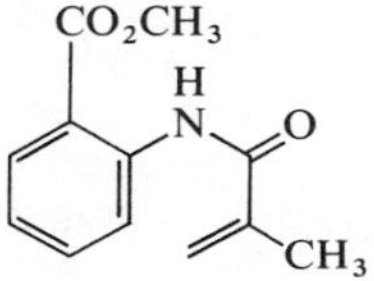 CO_2CH_3, H, N, O, CH_3	C_6H_6	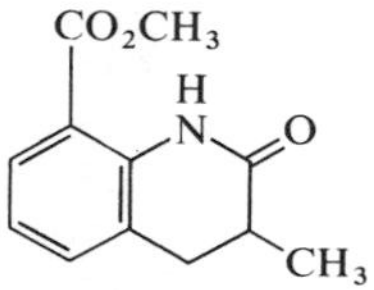 CO_2CH_3, H, N, O, CH_3 (61)	368
C_{13} 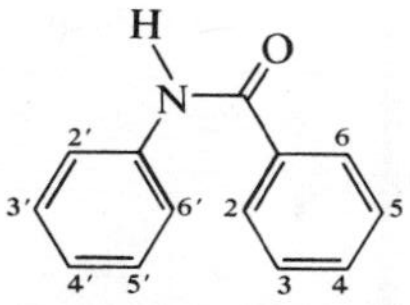 **R** (*N*-Phenylbenzamide)	C_6H_6, I_2	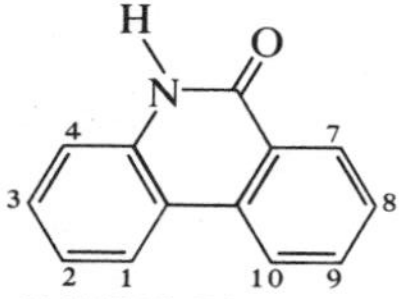 **P** (6(5*H*)-Phenanthridinone) (20)	77

TABLE VII. LACTAMS AND OTHER HETEROCYCLIC SYSTEMS FROM AMIDES (*Continued*)[a]

Starting Material	Reaction Conditions	Products and Yields (%)	Refs.
2-Bromo-**R**	C_6H_6, MeOH	" (8)	344
2′-Bromo-3-hydroxy-**R**	"	8-Hydroxy-**P** (58)	344
2-Bromo-3′-hydroxy-**R**	"	3-Hydroxy-**P** (22)	344
2-Chloro-**R**	C_6H_{12}, N_2	**P** (71)	343
2,4′-Dichloro-**R**	"	2-Chloro-**P** (67)	343
2′-Iodo-**R**	C_6H_6	**P** (48)	77
2-Iodo-**R**	C_6H_6, MeOH	" (57)	344
	C_6H_6	" (9)	77
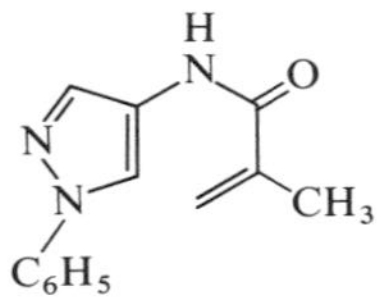	C_6H_6, HOAc, argon	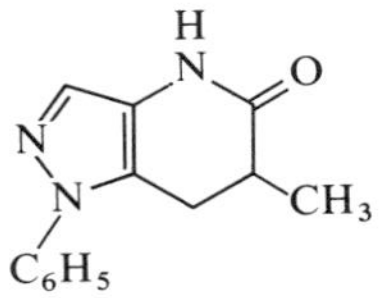 (3)	656
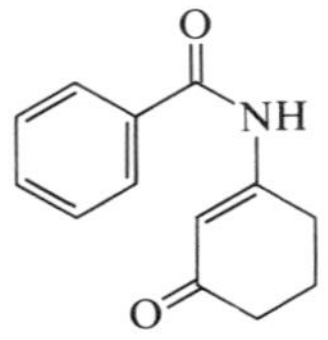	C_6H_6, air	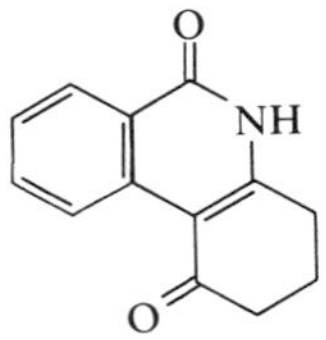 (30)	79
	MeOH, air	" (8)	79
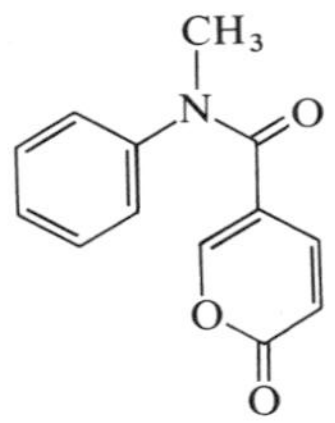	C_6H_6 or C_6H_6–MeOH, I_2	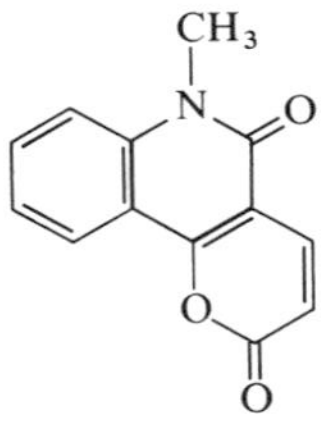 (30)	658b

	Reactant	Solvent	Product (yield)	Ref.
		C_6H_6 or C_6H_6–MeOH	(33)	658b
		”	(5)	658b
		”	(25)	658b
		”	(45)	658b
C_{14}		MeOH	(15)	79,659

TABLE VII. LACTAMS AND OTHER HETEROCYCLIC SYSTEMS FROM AMIDES (*Continued*)[a]

Starting Material	Reaction Conditions	Products and Yields (%)	Refs.
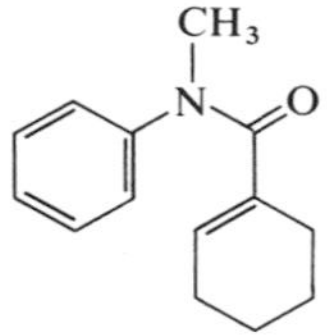 [structure]	MeOH, I_2	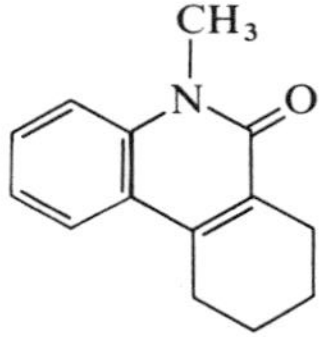 [structure] (30)	366
	Et_2O or C_6H_6 or MeOH	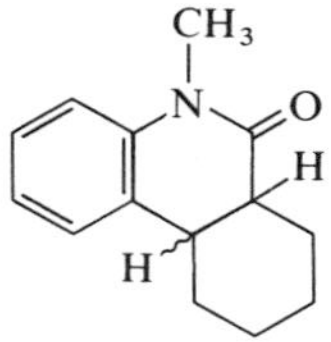 [structure] (30–59)[b]	366
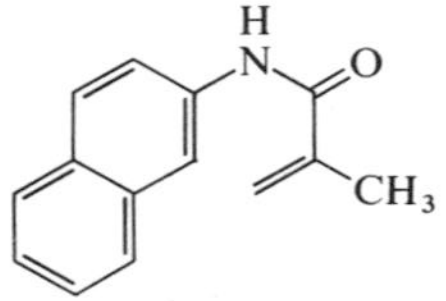 [structure]	C_6H_6	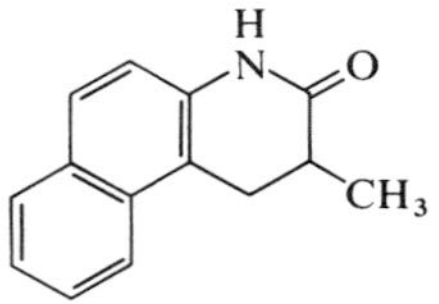 [structure] (59)	371
2′-Bromo-3,4-methylenedioxy-**R**	C_6H_6, MeOH	8,9-Methylenedioxy-**P** (20)	660a
2-Chloro-4′-methyl-**R**	C_6H_{12}, N_2	2-Methyl-**P** (67)	343
2-Chloro-4′-trifluoromethyl-**R**	"	2-Trifluoromethyl-**P** (74)	343
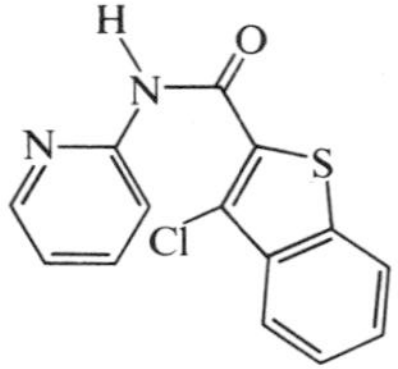 [structure]	EtOH, Et_3N, N_2	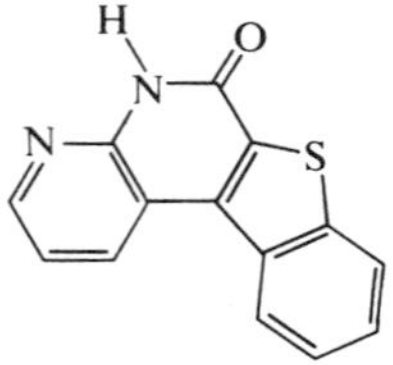 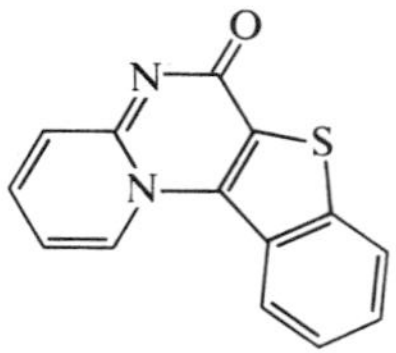 [structure] (48), [structure] (15)	660b

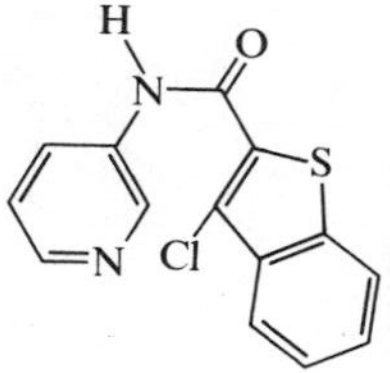

	Reactant	Conditions	Product(s) and Yield(s) (%)	Refs.
		”	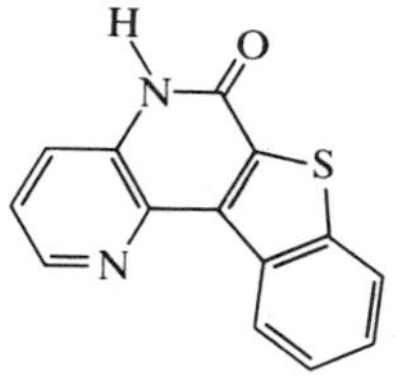 (48), 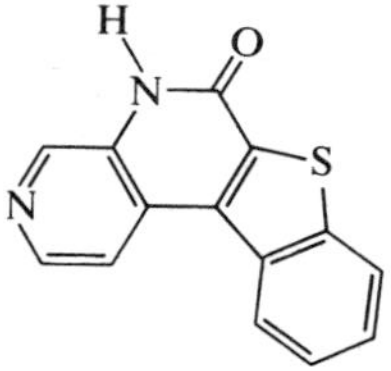 (34)	660b
	X = Cl	”	” (91)	660b
	X = H	Solvent, air	” (6)	660b
	2-Chloro-4′-methoxy-**R**	C_6H_{12}, N_2	2-Methoxy-**P** (23)	343
	2-Iodo-*N*-methyl-**R**	C_6H_6	*N*-Methyl-**P** (30)	345
	2-Methoxy-**R**	C_6H_6, EtOH	**P** (8)	78
	3,4-Methylenedioxy-**R**	C_6H_6, MeOH, I_2	8,9-Methylenedioxy-**P** (15–20)	660a
C_{15} 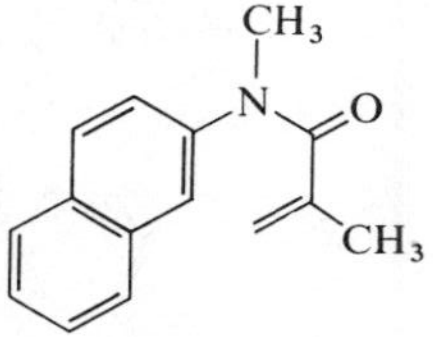		C_6H_6	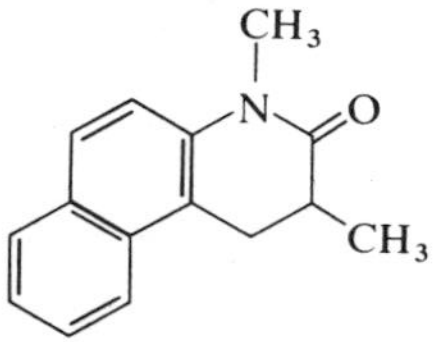 (50)	371
	2-Bromo-*N*-methyl-4,5-methylenedioxy-**R**	C_6H_6, MeOH	*N*-Methyl-8,9-methylenedioxy-**P** (22)	660a
		C_6H_6, Et_3N	” (45)	294b

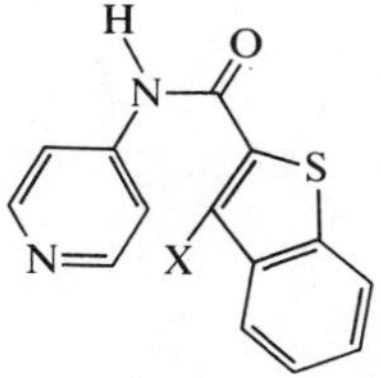

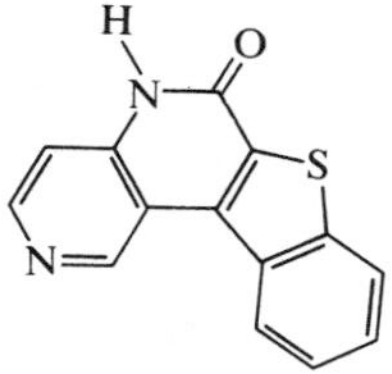

TABLE VII. LACTAMS AND OTHER HETEROCYCLIC SYSTEMS FROM AMIDES (*Continued*)[a]

Starting Material	Reaction Conditions	Products and Yields (%)	Refs.
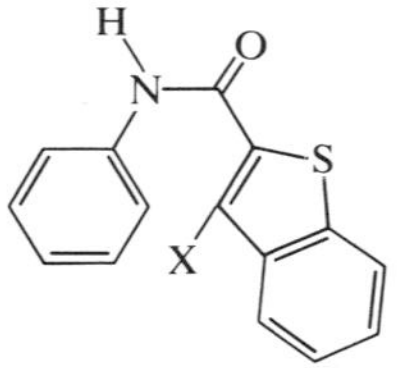 H, N, O, S, X		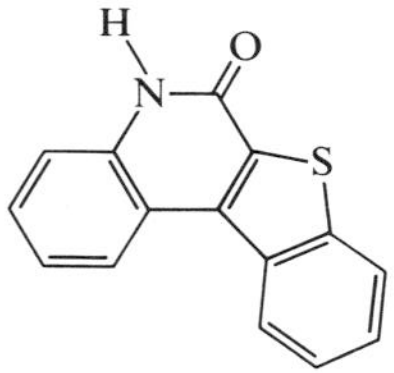 H, N, O, S	
X = Cl	C_6H_6, EtOH, air	" (100)	660b
X = H	"	" (45)	660b
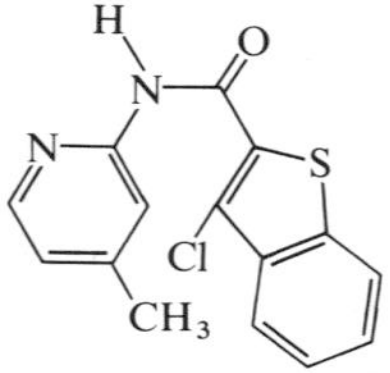 H, N, O, N, S, Cl, CH_3	EtOH, Et_3N, N_2	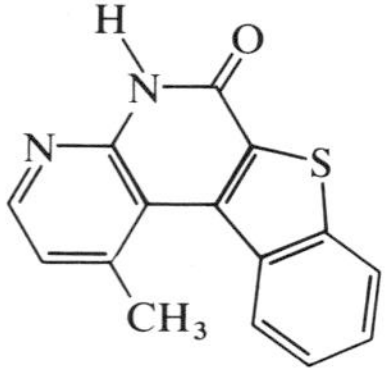 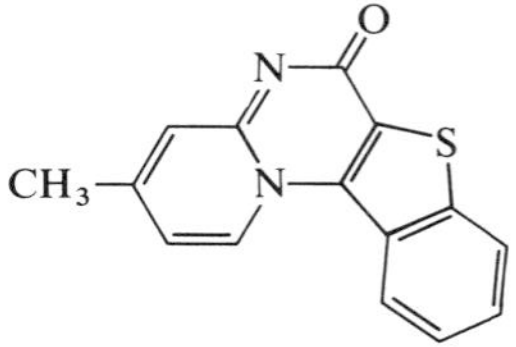 H, N, O, N, S, CH_3 (50), CH_3, N, O, N, S (21)	660b
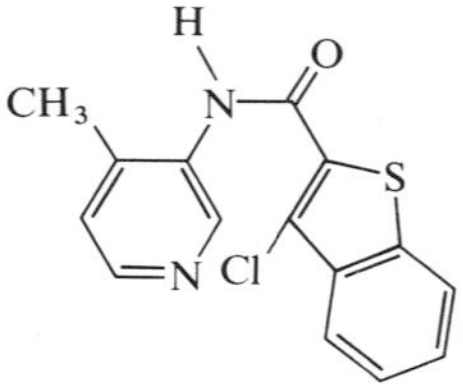 H, CH_3, N, O, S, N, Cl	"	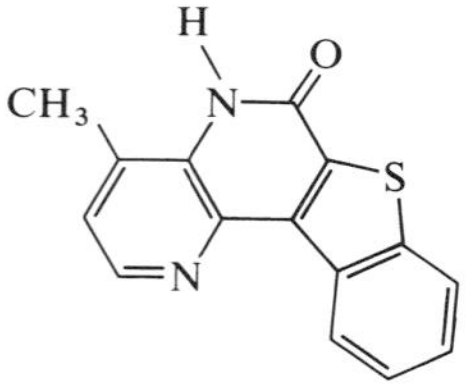 H, CH_3, N, O, S, N (80)	660b
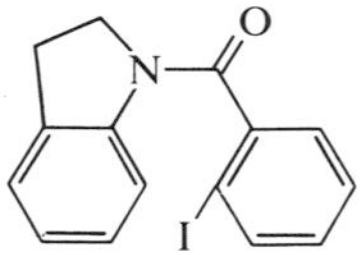 O, N, I	C_6H_6, H_2O, $Na_2S_2O_3$[c]	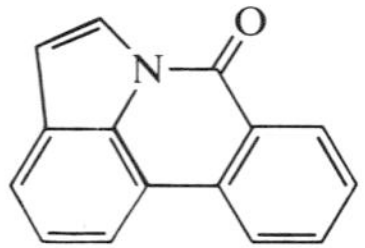 O, N (15)	662

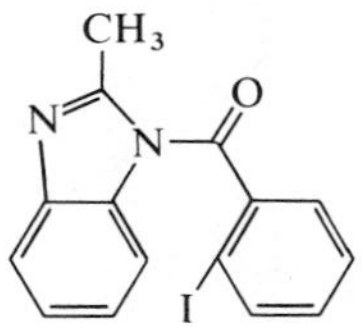	C_6H_6	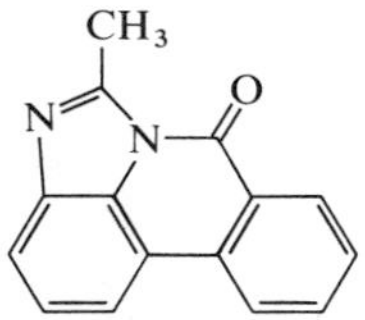 (19)	663a
2-Iodo-2′-methoxy-*N*-methyl-**R**	”	4-Methoxy-*N*-methyl-**P** (4)	661
2-Iodo-3′-methoxy-*N*-methyl-**R**	”	1-Methoxy-*N*-methyl-**P** (26), 3-methoxy-*N*-methyl-**P** (24)	661
2-Iodo-4′-methoxy-*N*-methyl-**R**	”	2-Methoxy-*N*-methyl-**P** + 3-methoxy-*N*-methyl-**P** (48 total)	661
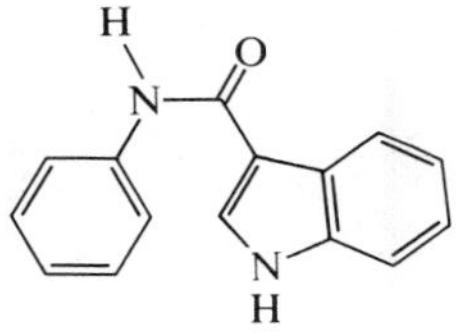	Dioxane, Me_2CO	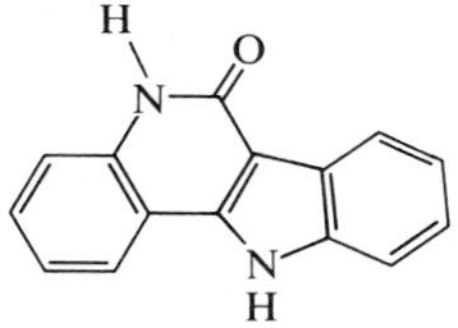 (15)	664
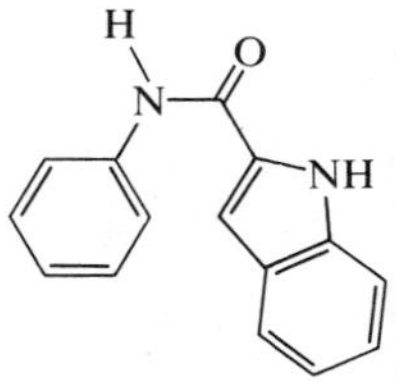	C_6H_6, EtOH, air	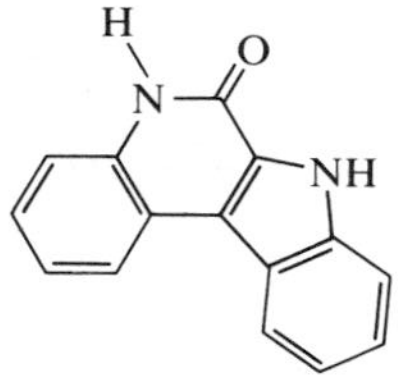 (77)	342,663b
	Dioxane, Me_2CO	” (15)	664
	C_6H_6, EtOH, N_2	” (22), 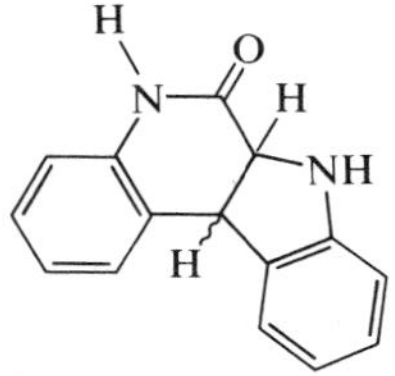 (*trans*, 29; *cis*, 25)	663b

TABLE VII. Lactams and Other Heterocyclic Systems from Amides (*Continued*)[a]

Starting Material	Reaction Conditions	Products and Yields (%)	Refs.
	MeCN, D_2O, N_2	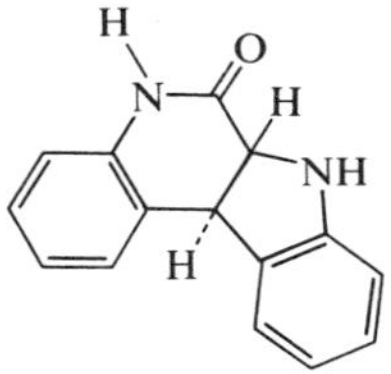 (17), 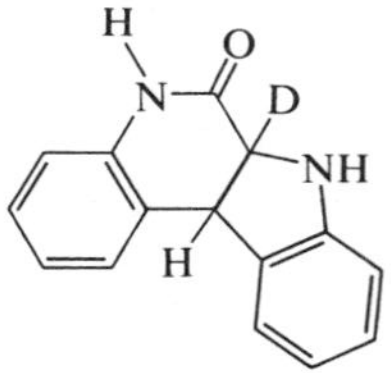 (11)	663b
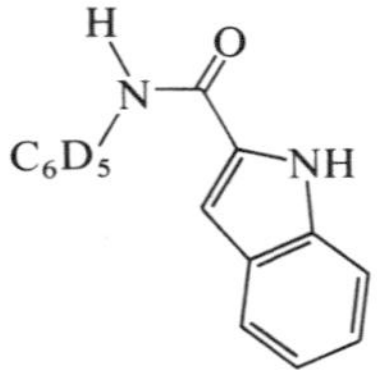	C_6H_6, EtOH, N_2	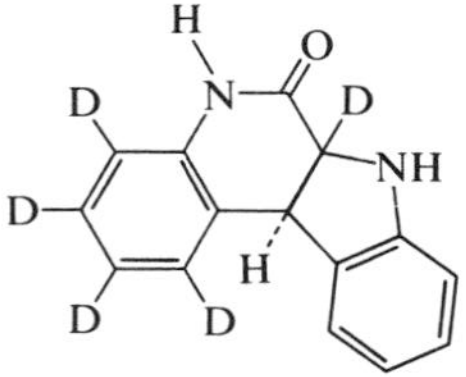 (10), 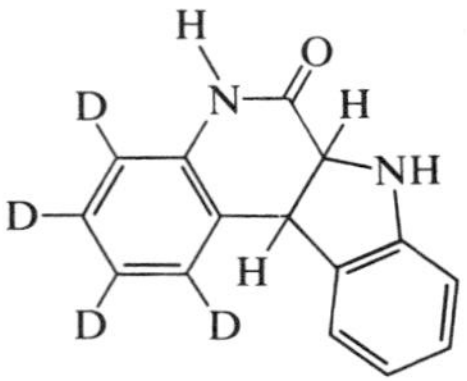 (15)	663b
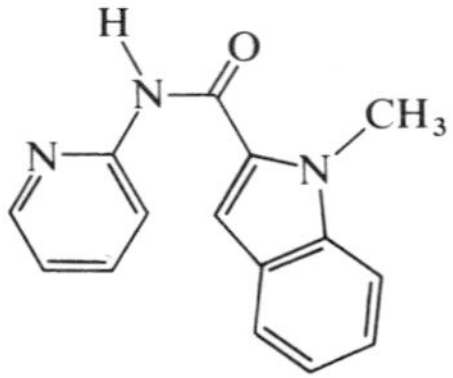	C_6H_6, EtOH, air	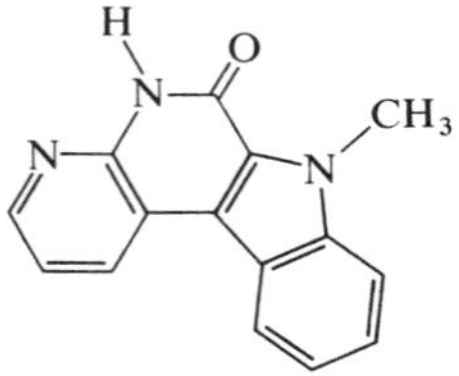 (8)	657b

Reactant	Conditions	Product(s) (Yield, %)	Refs.
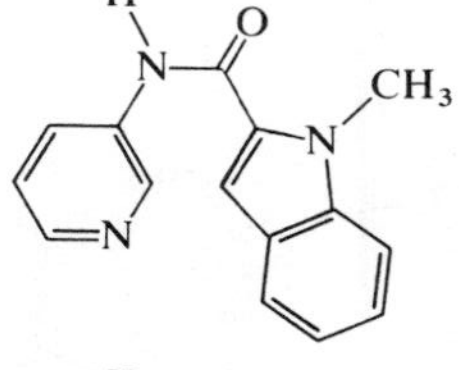	"	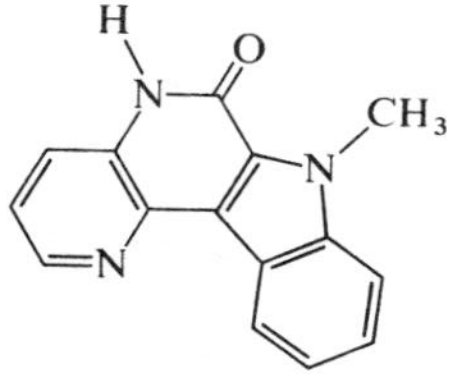 or 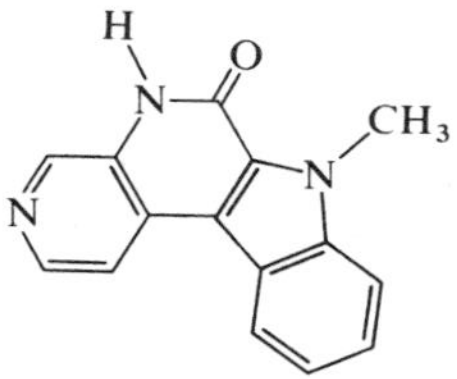 (60)	657b
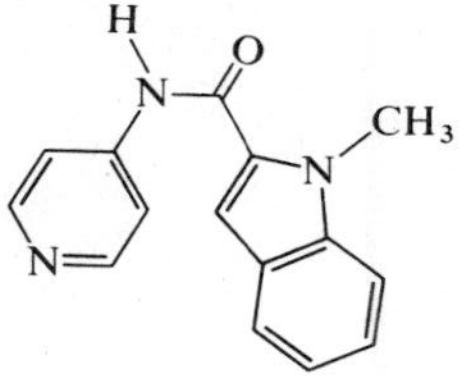	"	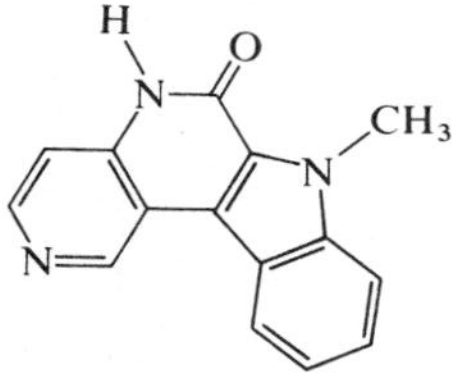 (58)	657b
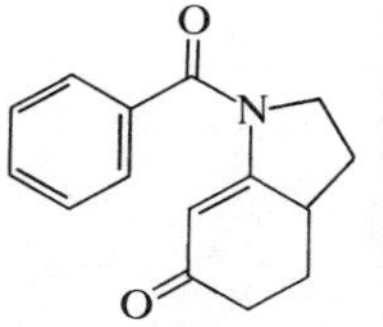	CH_2Cl_2, air	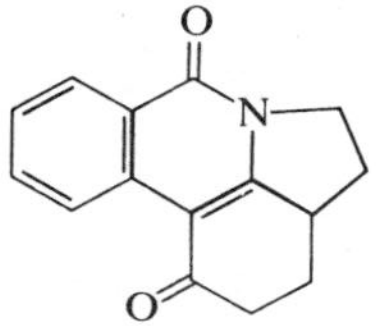 (69)	665,666
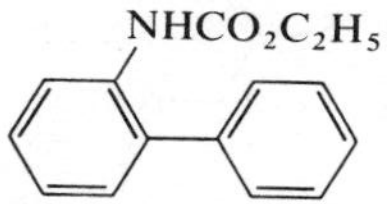 $NHCO_2C_2H_5$	MeOH	**P** (85)	376
2′-Carboxy-*N*-methyl-**R**	HOAc	*N*-Methyl-**P** (22)	368
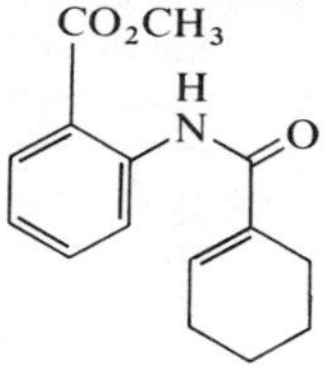 CO_2CH_3	Et_2O, C_6H_6, or MeOH	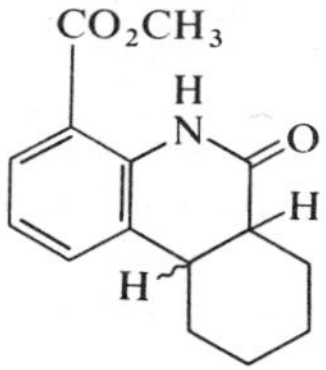 CO_2CH_3 (60)	368

TABLE VII. LACTAMS AND OTHER HETEROCYCLIC SYSTEMS FROM AMIDES (*Continued*)[a]

Starting Material	Reaction Conditions	Products and Yields (%)	Refs.
2,3-Dimethoxy-**R**	C_6H_6, EtOH	10-Methoxy-**P** (80)	78
2,4-Dimethoxy-**R**	"	9-Methoxy-**P** (50)	78
2,5-Dimethoxy-**R**	"	8-Methoxy-**P** (37)	78
2,6-Dimethoxy-**R**	"	7-Methoxy-**P** (45)	78
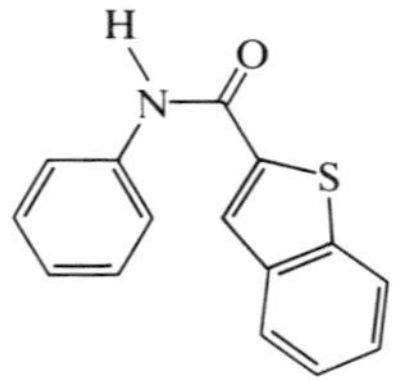 [structure: N-phenylbenzo[b]thiophene-2-carboxamide]	C_6H_6, EtOH, O_2	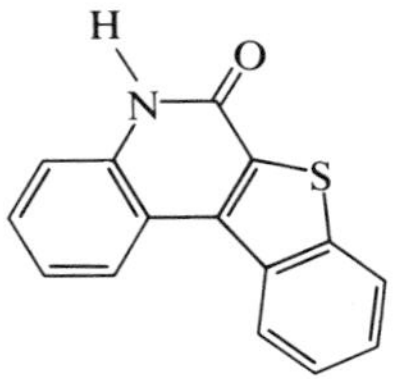 [structure] (45)	658a
		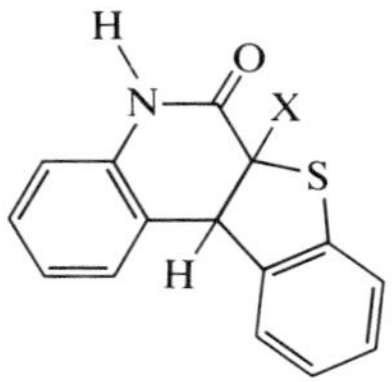 [structure] (I)	
	Solvent, N_2	I, X = H (50)	658a
	MeCN, D_2O, degassed	I, X = D (76)	658a
C_{16} 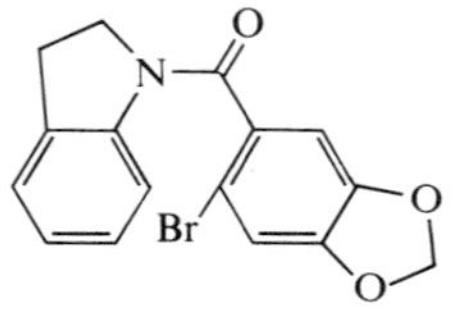 [structure]	*i*-PrOH	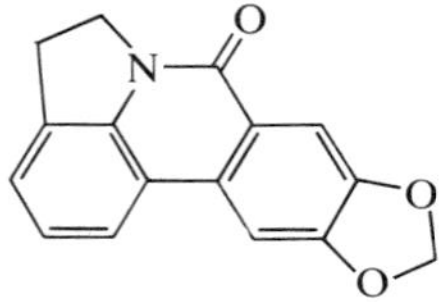 [structure] (67)	346
	C_6H_6, Et_3N	" (60)	294b

Reactant	Conditions	Product(s) (yield %)	Ref.
2-Bromo-4,5-dimethoxy-*N*-methyl-**R**	"	8,9-Dimethoxy-*N*-methyl-**P** (50)	294b
4′-Carbethoxy-2-chloro-**R**	C_6H_{12}, N_2	2-Carbethoxy-**P** (70)	343
2′-Carbomethoxy-2-chloro-5-methoxy-**R**	EtOH, C_6H_6	4-Carbomethoxy-8-methoxy-**P** (—)	667a
N-Ethyl-2-iodo-4′-methoxy-**R**	C_6H_6	*N*-Ethyl-2-methoxy-**P** (28), *N*-ethyl-3-methoxy-**P** (22)	661
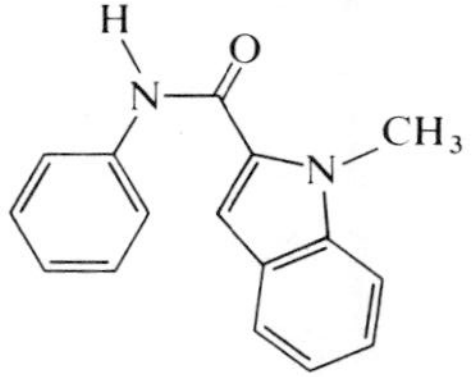 [structure: H, N, O, CH_3]	C_6H_6, EtOH, air	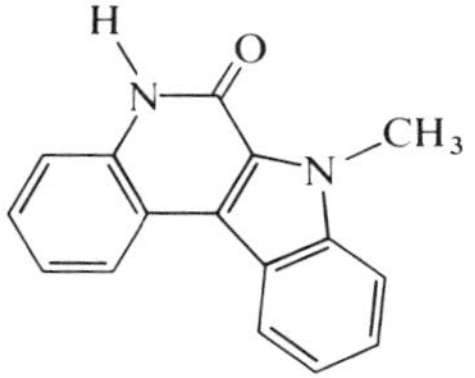 [structure: H, N, O, CH_3] (71)	663b
	C_6H_6, EtOH, N_2	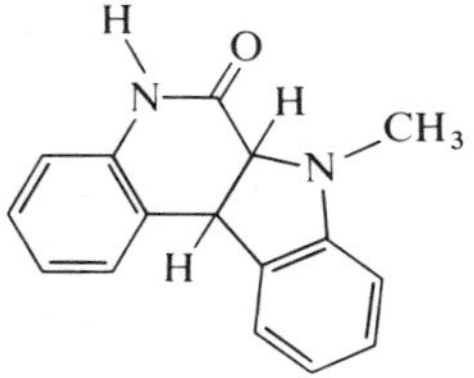 [structure: H, N, O, H, CH_3, H] (58)	663b
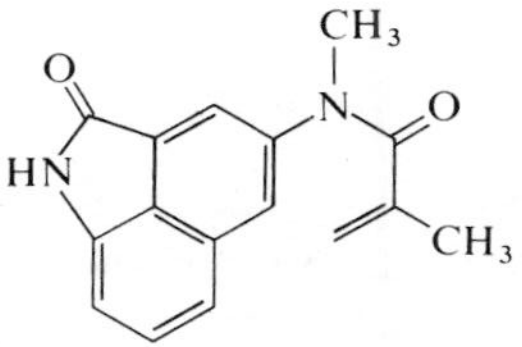 [structure: HN, O, CH_3, N, O, CH_3]	C_6H_6	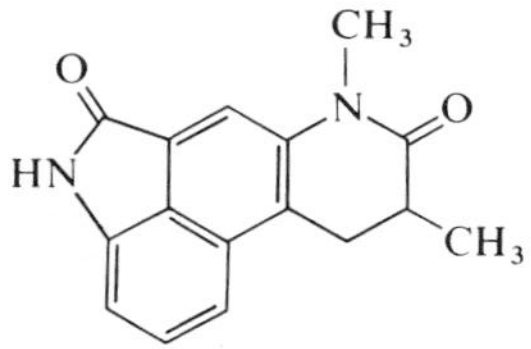 [structure: HN, O, CH_3, N, O, CH_3] (42)	371
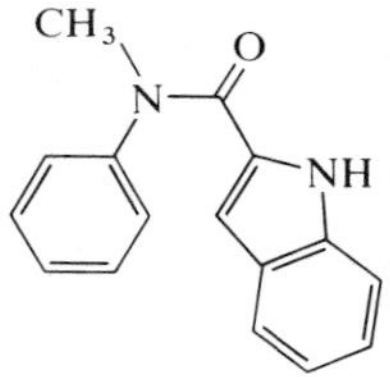 [structure: CH_3, N, O, NH]	C_6H_6, EtOH, air	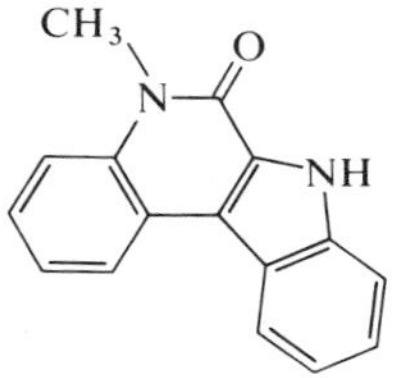 [structure: CH_3, N, O, NH] (46)	663b

TABLE VII. LACTAMS AND OTHER HETEROCYCLIC SYSTEMS FROM AMIDES (*Continued*)[a]

Starting Material	Reaction Conditions	Products and Yields (%)	Refs.
	C_6H_6, EtOH, N_2	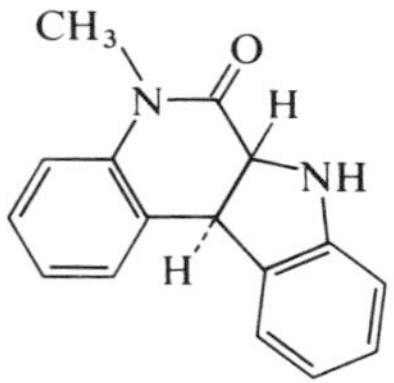 (26)	663b
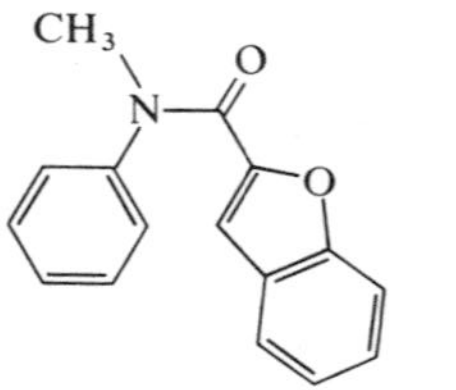	C_6H_6, N_2	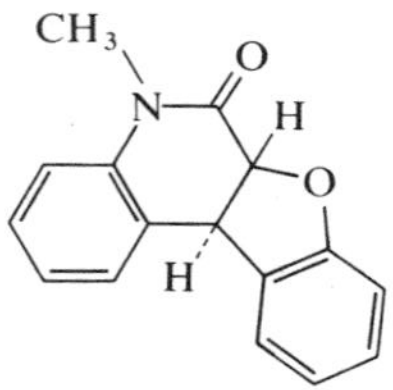 (50)	667b
	MeCN, N_2	" (66)	667b
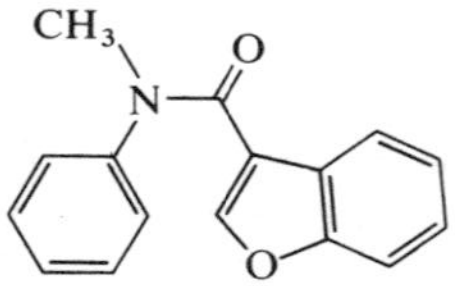	C_6H_6, N_2	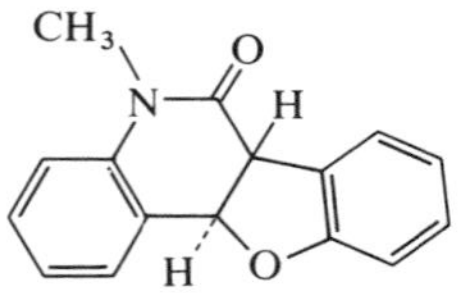 (7)	667b
	MeCN, N_2	" (22)	667b

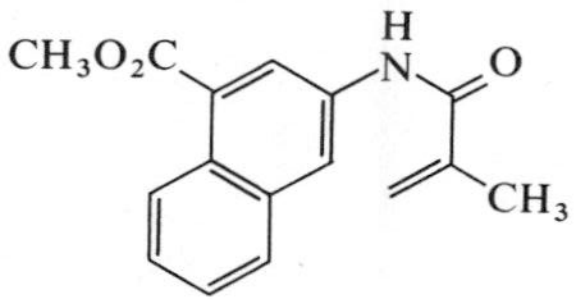	C_6H_6	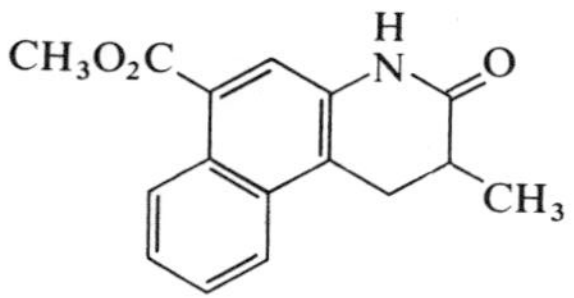 (43)	371
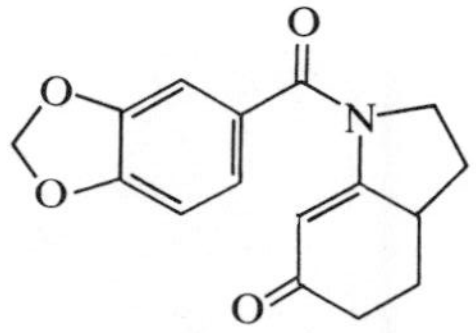	THF, air	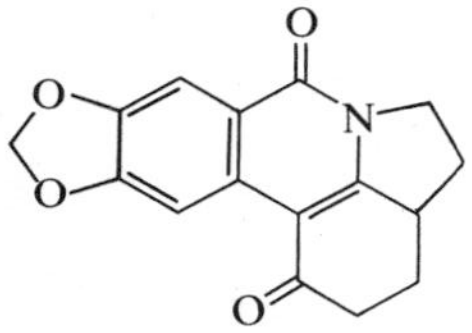 (70)	665,666
2′-Carbomethoxy-3-methoxy-**R**	EtOH, C_6H_6, air	4-Carbomethoxy-8-methoxy-**P** (—)	667
2,4,5-Trimethoxy-**R**	C_6H_6, EtOH	8,9-Dimethoxy-**P** (64)	78
	EtOAc, N_2	” (55)	80
2′-Acetoxy-3,4-methylenedioxy-**R**	C_6H_6, MeOH, I_2	4-Acetoxy-8,9-methylenedioxy-**P** (1)	660a
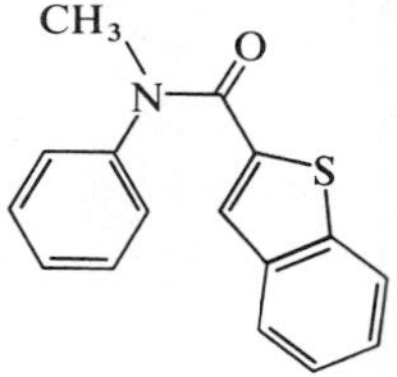	Solvent, O_2	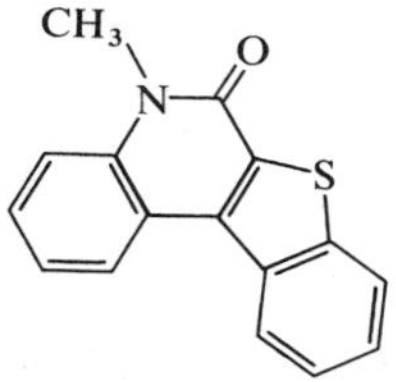 (—)	658a
	MeCN, degassed	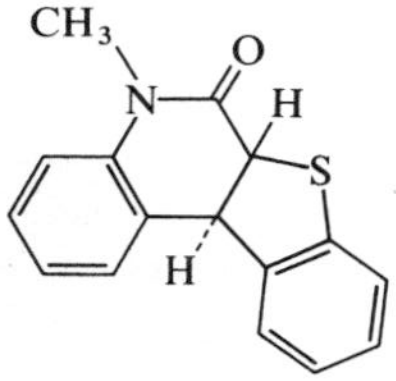 (52)	658a

TABLE VII. LACTAMS AND OTHER HETEROCYCLIC SYSTEMS FROM AMIDES (*Continued*)[a]

Starting Material	Reaction Conditions	Products and Yields (%)	Refs.
	MeCN, D_2O, degassed	" (28), 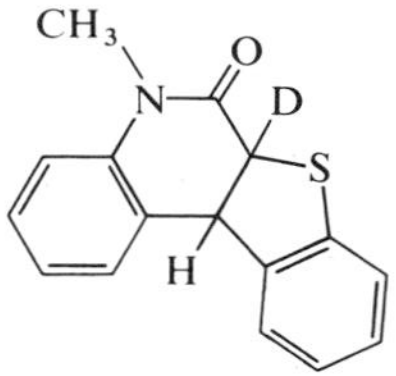 (43)	658a
C_{17} 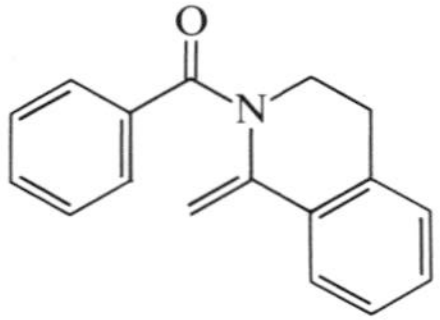	MeOH, O_2	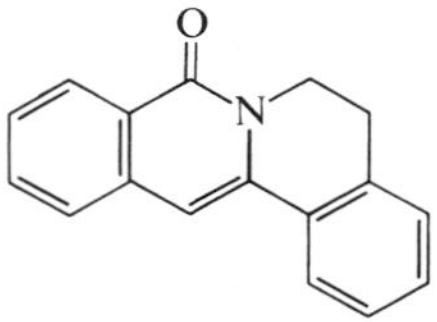 (good)	364
	MeOH	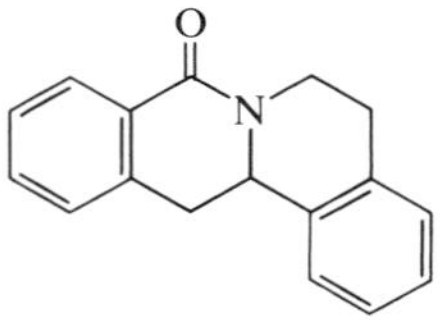 (70)	350,364
	"	" (40), 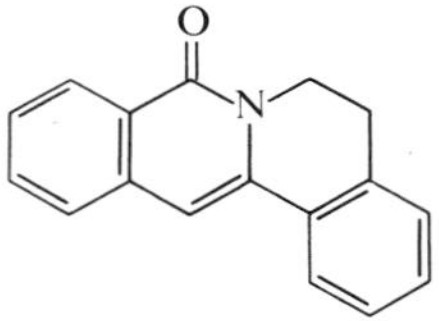 (20)[d]	363

Reactant	Conditions	Product(s) (Yield %)	Refs.
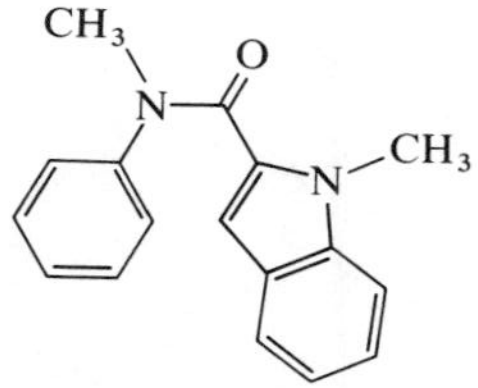 [structure]	C_6H_6, EtOH, air	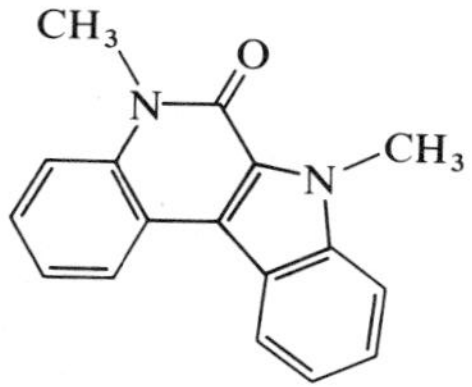 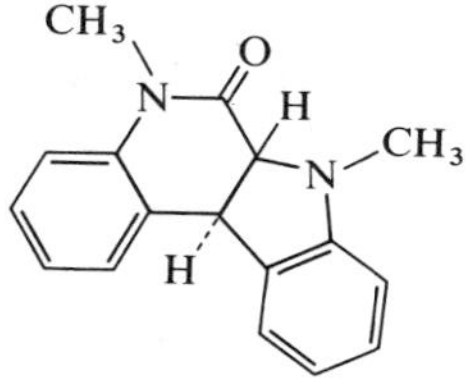 [structure] (35), [structure] (20)	663b
	C_6H_6, EtOH, N_2	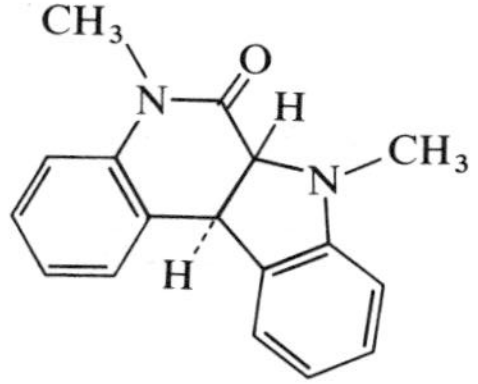 [structure] (58)	663b
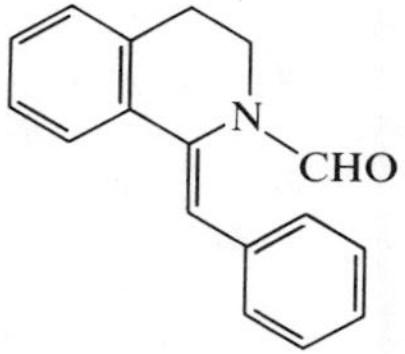 [structure]	Dioxane, *t*-BuOH, 47% HI, N_2	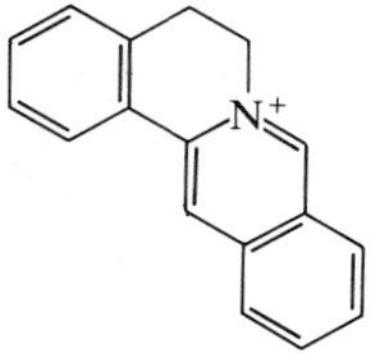 [structure] (45)	373
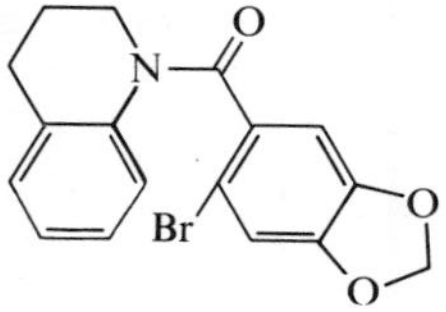 [structure]	C_6H_6, Et_3N	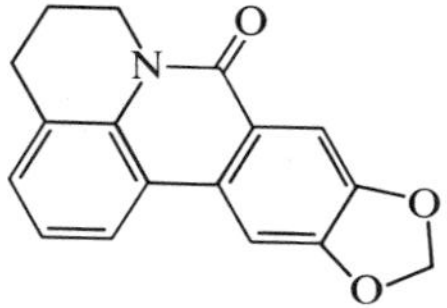 [structure] (62)	294b

TABLE VII. LACTAMS AND OTHER HETEROCYCLIC SYSTEMS FROM AMIDES (*Continued*)[a]

	Starting Material	Reaction Conditions	Products and Yields (%)	Refs.
	[structure] Br, OCH_3, OCH_3	C_6H_6, Et_3N	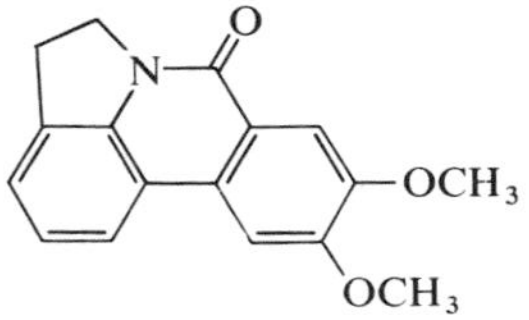 [structure] OCH_3, OCH_3 (75)	294b
	[structure] $CH_2C_6H_5$, N, O, Cl, CH_3	MeOH, air	[structure] $CH_2C_6H_5$, N, O, CH_3 (15)	368
	[structure] CH_3O_2C, CH_3, N, O, CH_3	C_6H_6	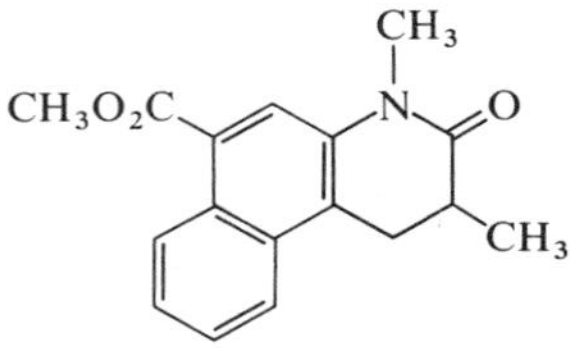 [structure] CH_3O_2C, CH_3, N, O, CH_3 (48)	371
	N-Methyl-2,4,5-trimethoxy-**R**	EtOAc, N_2	*N*-Methyl-8,9-dimethoxy-**P** (41)	80
	2′-Carbomethoxy-2,5-dimethoxy-**R**	EtOH, C_6H_6	4-Carbomethoxy-8-methoxy-**P** (16)	667a
C_{18} 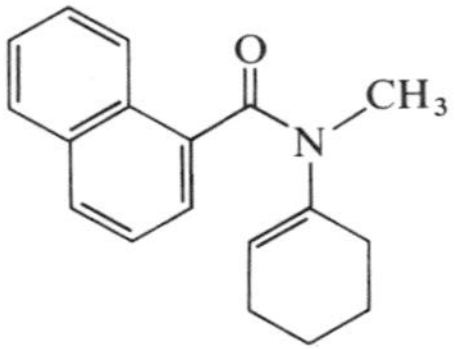	[structure] O, N, CH_3	MeOH, I_2	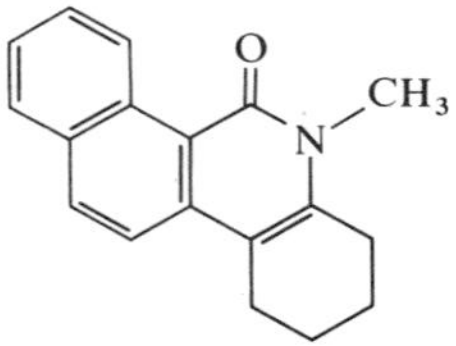 [structure] O, N, CH_3 (10)	668

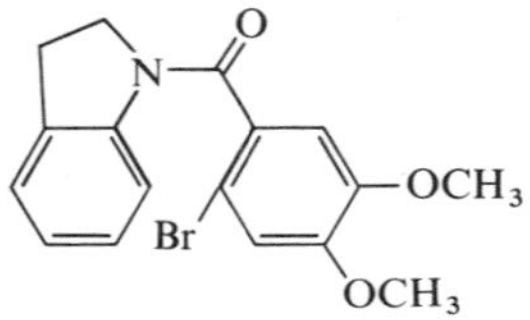

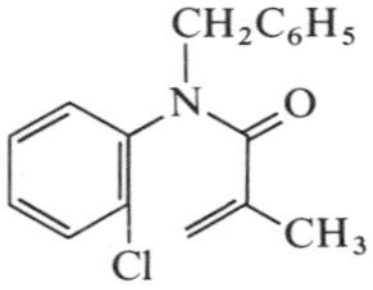

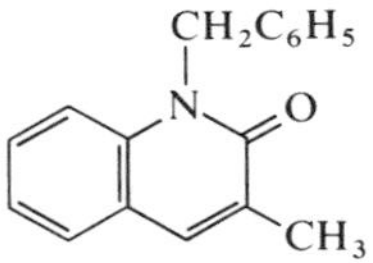

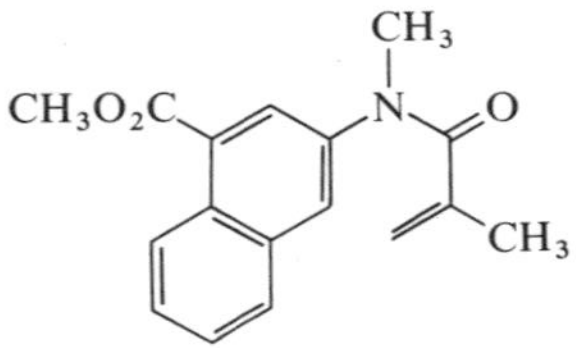

Reactant	Conditions	Product	Yield (%)	Ref.
	”		(21)	350,360
	MeOH		(51)	350,359,360
	Et_2O		(40–52)	361,362
	MeOH, I_2[e]		(27)	366
	MeOH		(*trans*, 17; *cis*, 26)[f]	366

TABLE VII. LACTAMS AND OTHER HETEROCYCLIC SYSTEMS FROM AMIDES (*Continued*)[a]

Starting Material	Reaction Conditions	Products and Yields (%)	Refs.
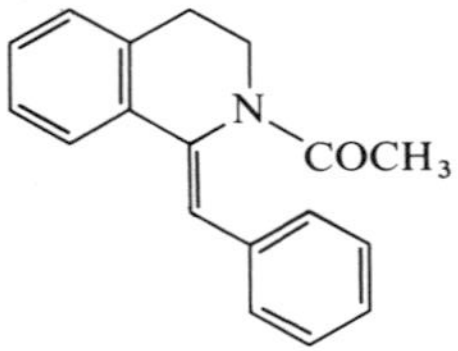 N-COCH$_3$	MeOH, I_2, N_2	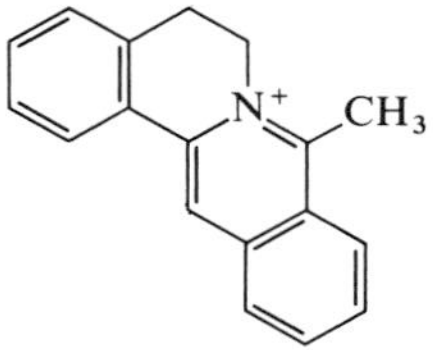 N$^+$ CH$_3$ (42)	374
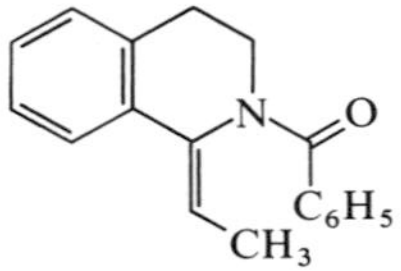 N, O, C$_6$H$_5$, CH$_3$	MeOH	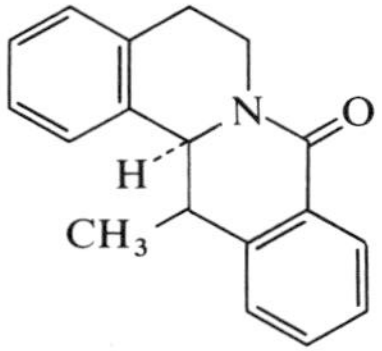 N, O, H, CH$_3$ (70)	350,364
	"	" (30)	358
N H, N, O, Br, N	MeOH, air	N H, N, O, Br, N (15),	669
		N H, N, O, N, Br (6)	

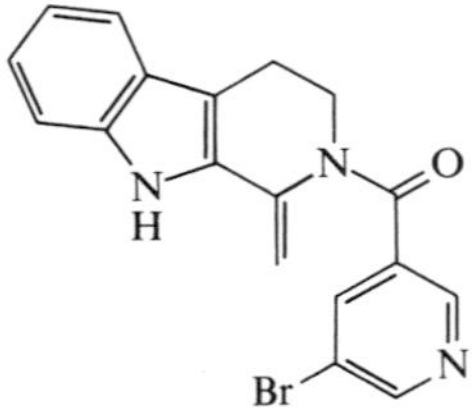

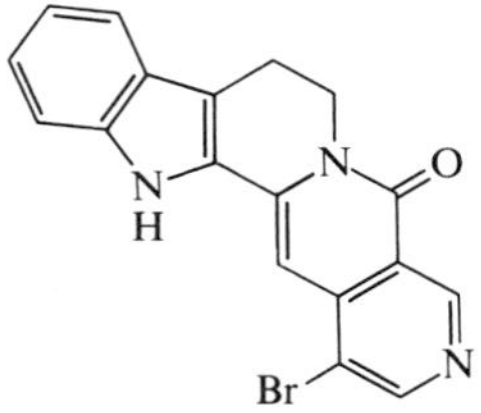

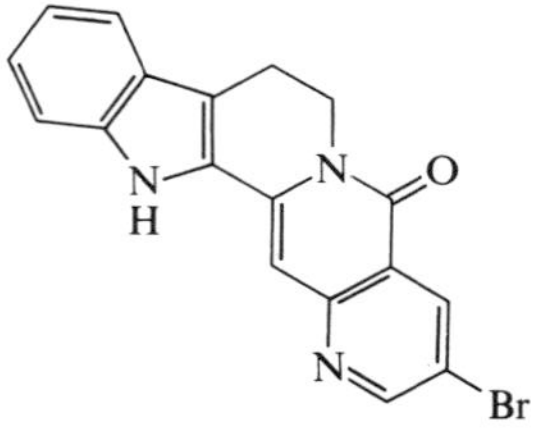

Reactant	Conditions	Product (Yield %)	Ref.
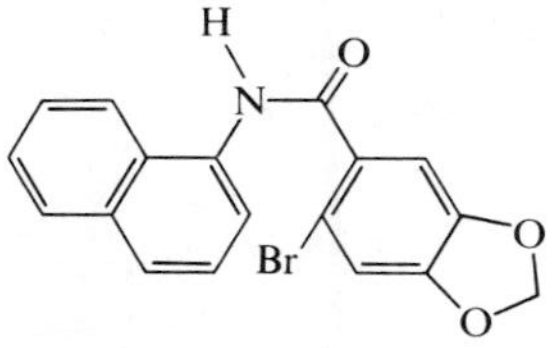 H, N, O, Br, O, O	C_6H_6, MeOH	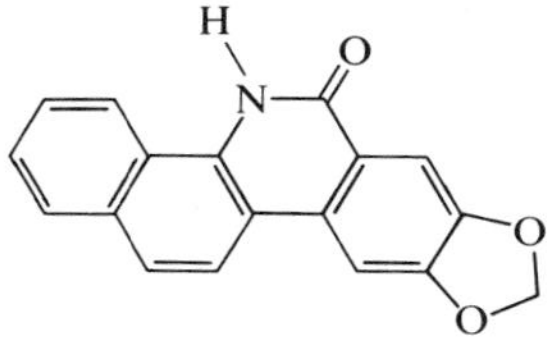 H, N, O, O, O (48)	299
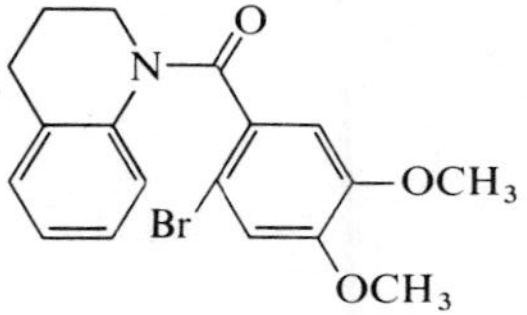 N, O, Br, OCH_3, OCH_3	C_6H_6, Et_3N	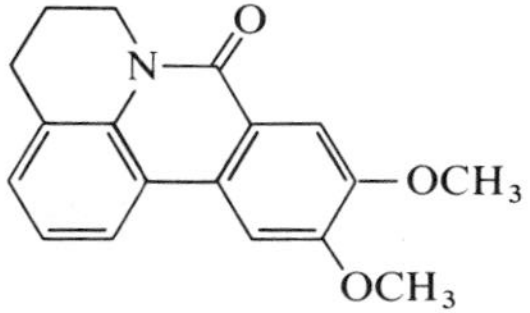 N, O, OCH_3, OCH_3 (81)	294b
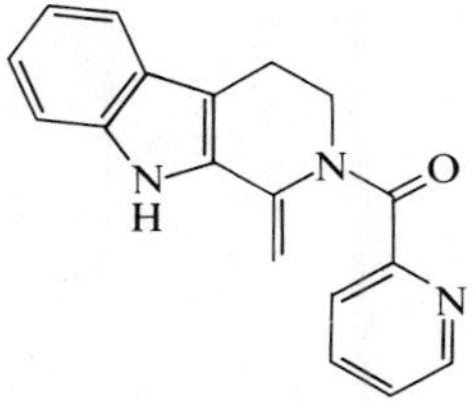 N, H, N, O, N	MeOH, air	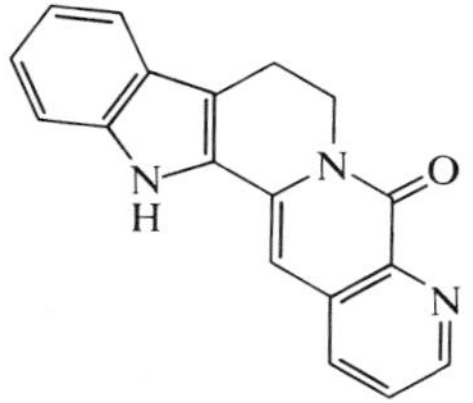 N, H, N, O, N (—)	669
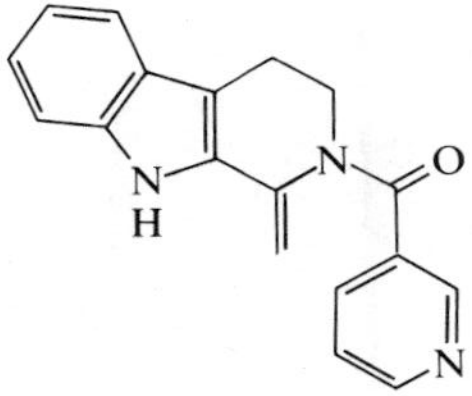 N, H, N, O, N	"	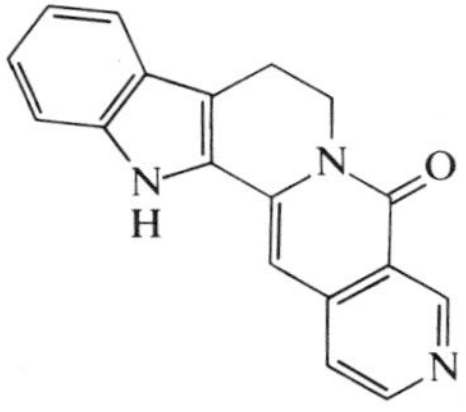 N, H, N, O, N (48)	670
	"	" (—)	669

TABLE VII. Lactams and Other Heterocyclic Systems from Amides (*Continued*)[a]

Starting Material	Reaction Conditions	Products and Yields (%)	Refs.
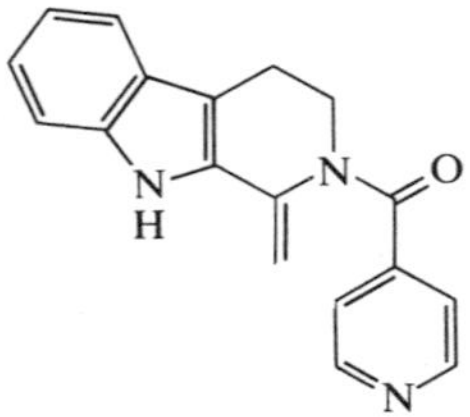	MeOH, air	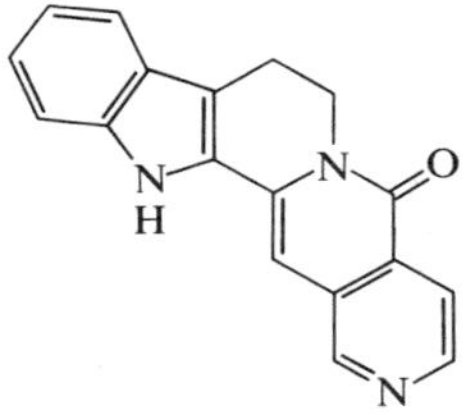 (—)	669
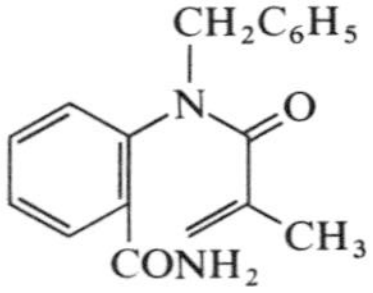	C_6H_6	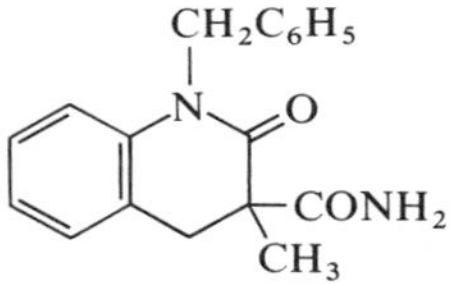 (54), 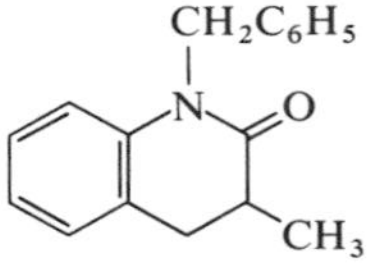 (11)	368
	MeOH	" (15), " (15)	368
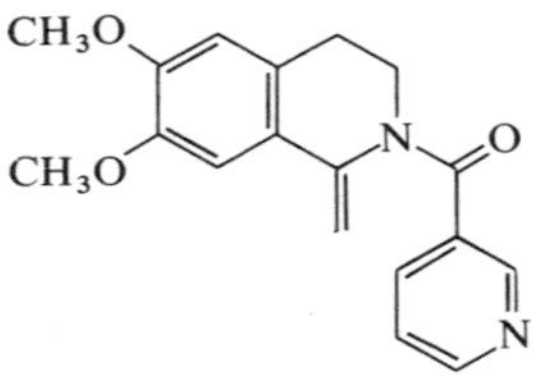	Dioxane, degassed[d]	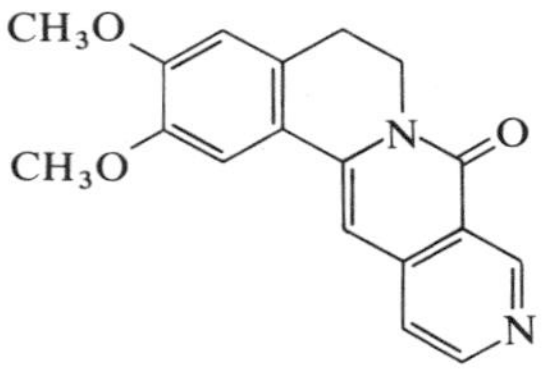 (14)	671
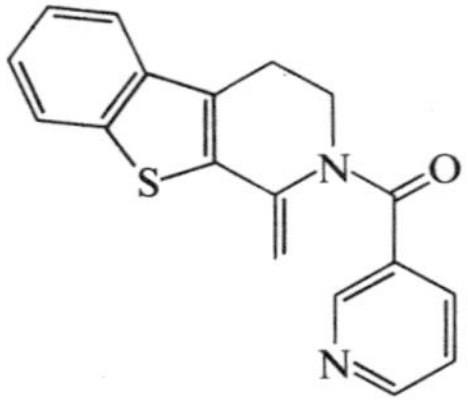	C_6H_6, air	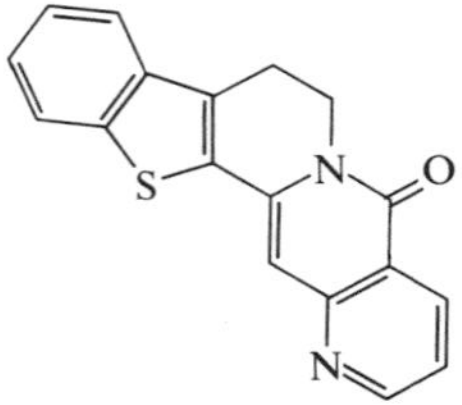 (45),	355

Reactant	Conditions	Product (Yield, %)	Ref.
		(24)	
CH_3O	MeOH	(50)	363
$CH_2C_6H_5$, CO_2H, CH_3	HOAc	$CH_2C_6H_5$, CH_3 (70)	368
	MeOH	" (41)	368
NH, CH_3O_2C	C_6H_6, air	NH, CH_3O_2C (40)	350

TABLE VII. LACTAMS AND OTHER HETEROCYCLIC SYSTEMS FROM AMIDES (*Continued*)[a]

Starting Material	Reaction Conditions	Products and Yields (%)	Refs.
C_{19} 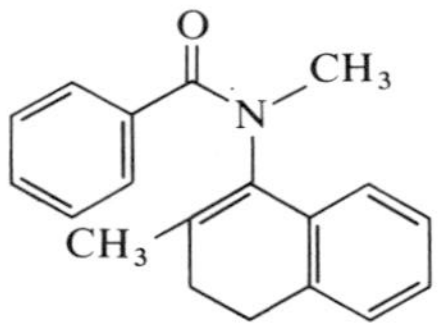	MeOH	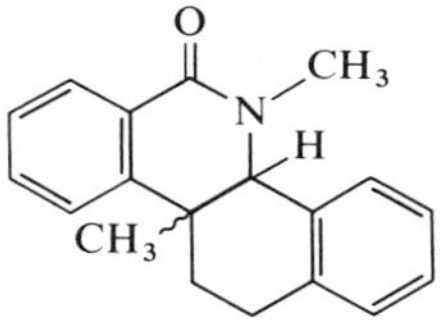 (8)	672,673
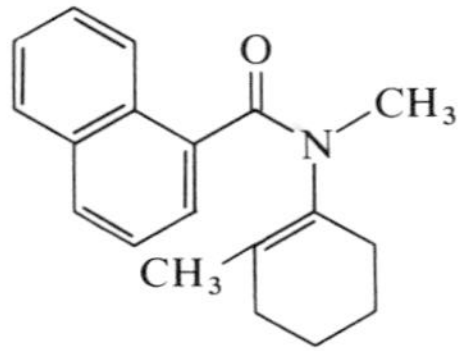	Et_2O	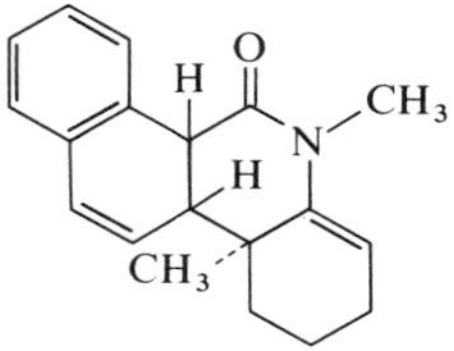 (44),	668
		(8)	
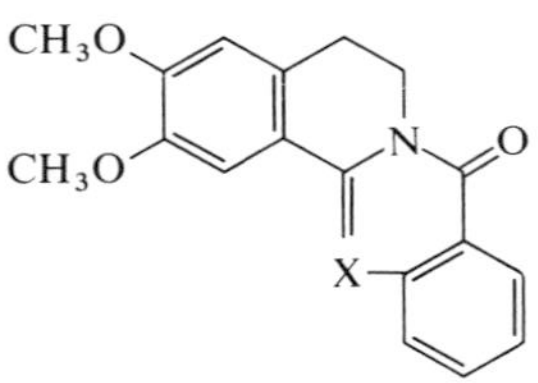 X = Br	*t*-BuOH, degassed	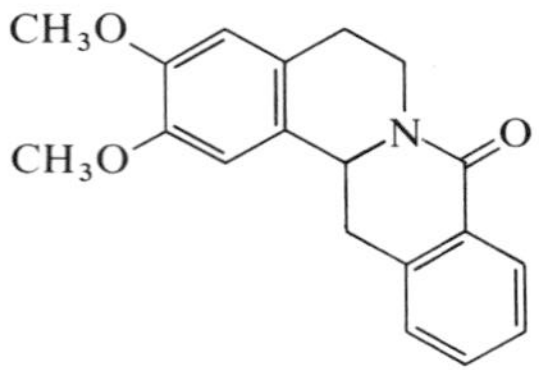 " (50)	80,356

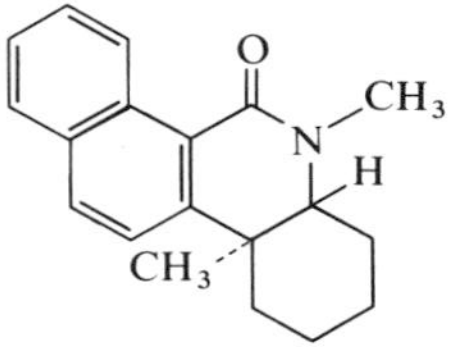

Reactant	Conditions	Product (Yield, %)	Ref.
X = Cl	”	” (50)	80,356
X = F	”	” (85)	80
X = NO_2	”	” (17)	80,356
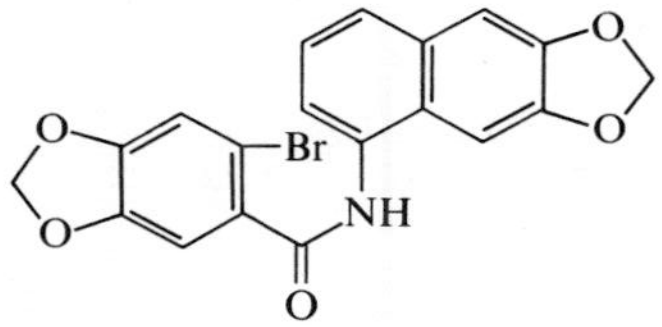	C_6H_6, MeOH	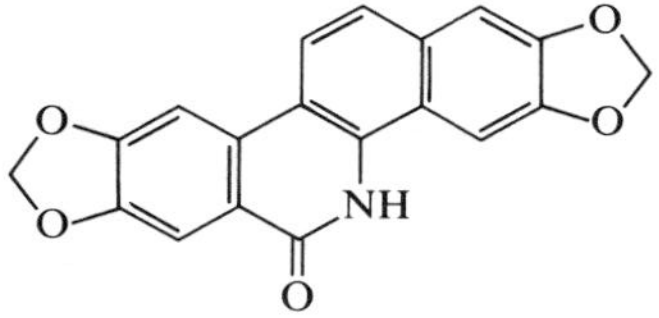 (70)	299
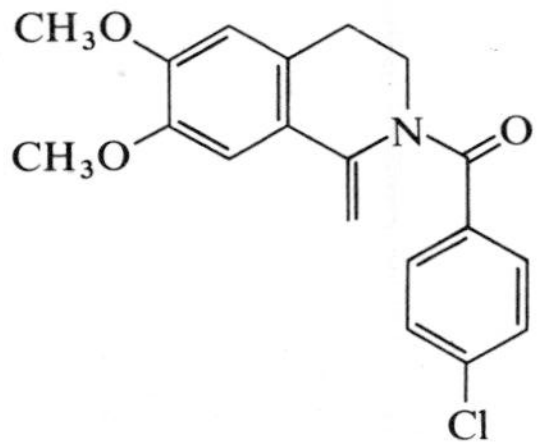	*t*-BuOH, degassed	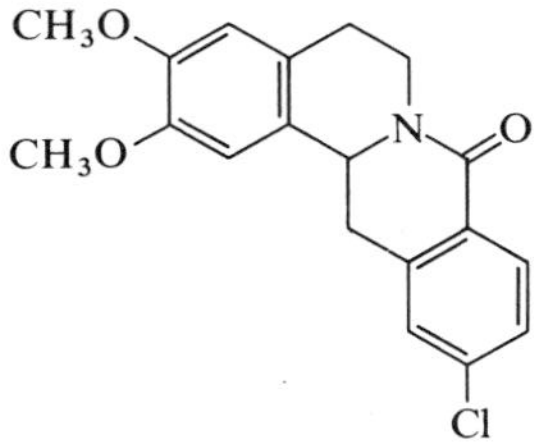 (76)	81
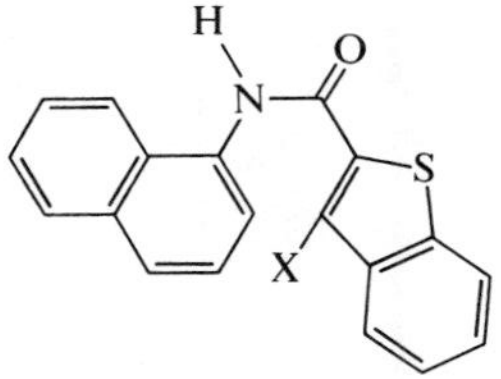 X = Cl	C_6H_6, EtOH, air	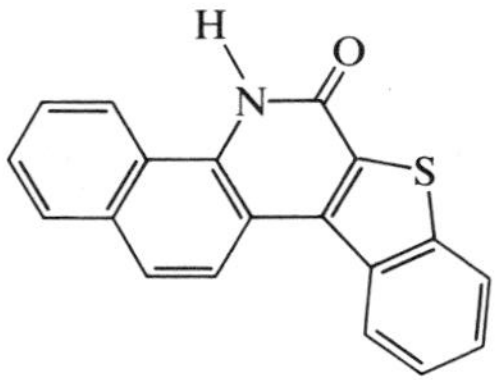 ” (74)	660b
X = H	”	” (30)	660b
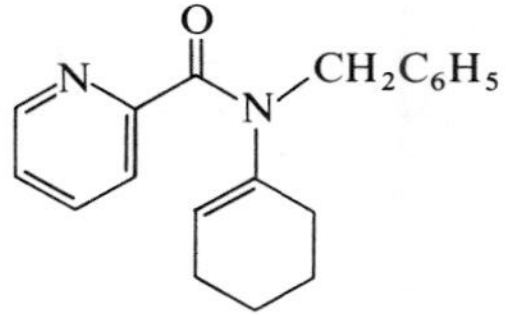	MeOH, I_2	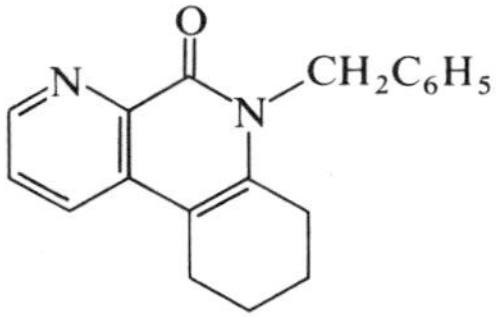 (1)	369

TABLE VII. LACTAMS AND OTHER HETEROCYCLIC SYSTEMS FROM AMIDES (*Continued*)[a]

Starting Material	Reaction Conditions	Products and Yields (%)	Refs.
	MeOH, air	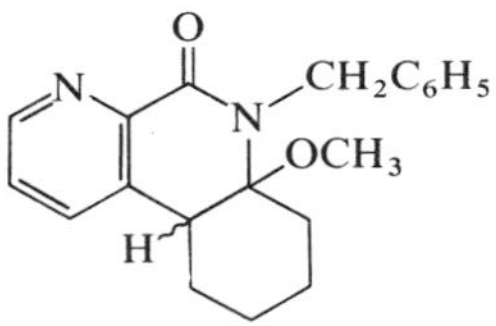 O, N, $CH_2C_6H_5$, OCH_3, H (1)	369
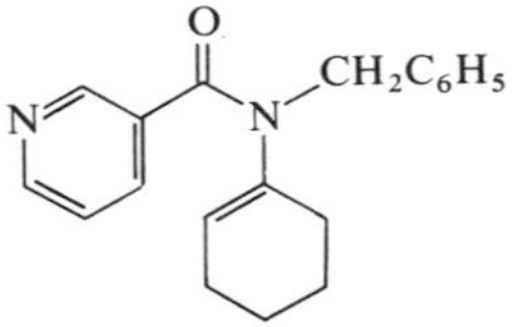 O, N, $CH_2C_6H_5$, N	MeOH, I_2	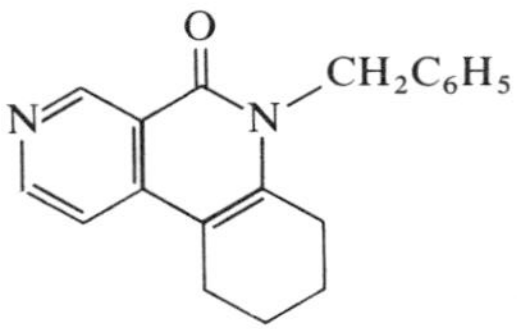 O, $CH_2C_6H_5$, N, N (4),	369
		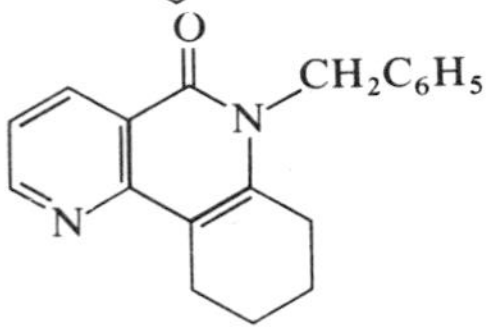 O, $CH_2C_6H_5$, N, N (24)	369
	MeOH, air	" (21), " (2)	369
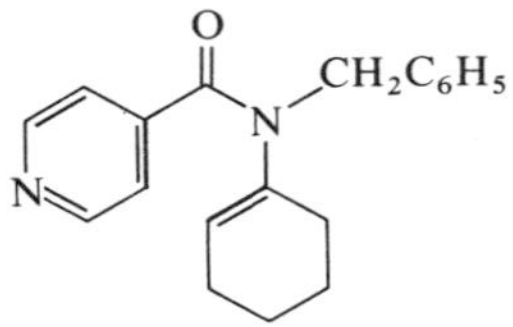 O, N, $CH_2C_6H_5$, N	MeOH, I_2	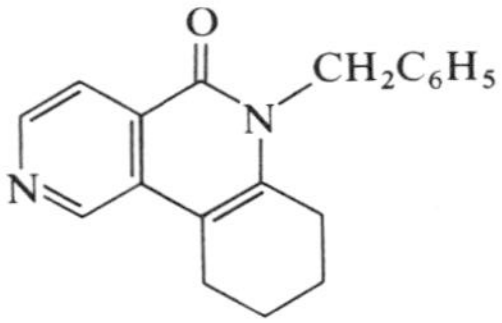 O, $CH_2C_6H_5$, N, N (16)	369

Reactant	Conditions	Product	Ref.
	MeOH, air	O, N, $CH_2C_6H_5$, OCH_3, N, H (9)	369
N, H, N, O	”	N, H, N, O (37)	353,354
N, H, N, O, N, CH_3	”	N, H, N, O, N, CH_3 (21),	353,354
		N, H, N, O, N, CH_3 (13)	

TABLE VII. LACTAMS AND OTHER HETEROCYCLIC SYSTEMS FROM AMIDES (*Continued*)[a]

Starting Material	Reaction Conditions	Products and Yields (%)	Refs.
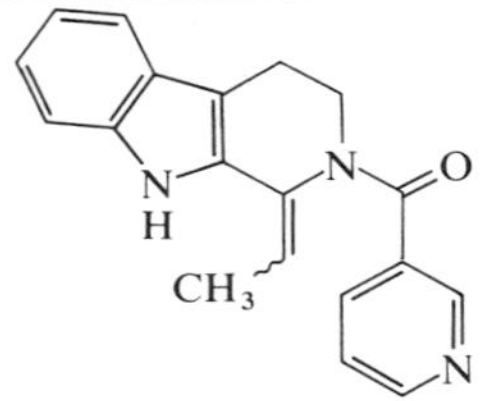	MeOH, air	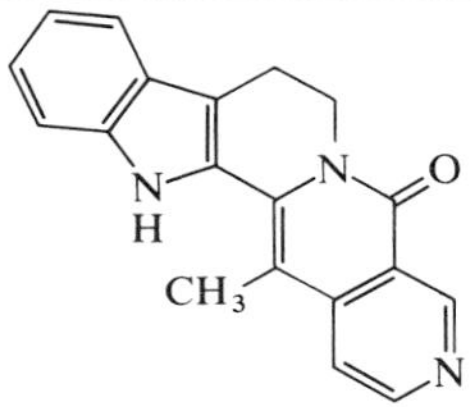 + 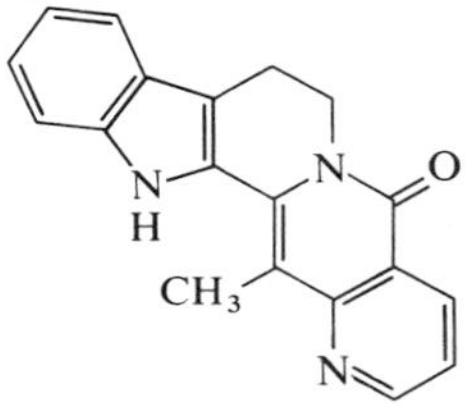 (7:3)	674
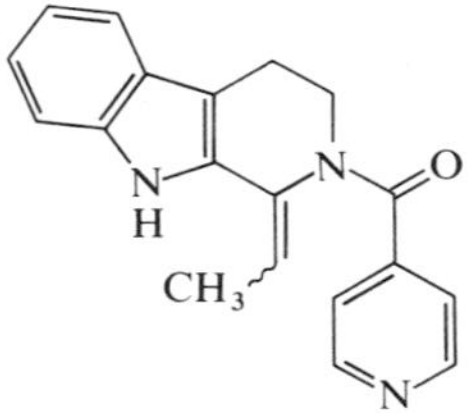	MeOH, degassed[d]	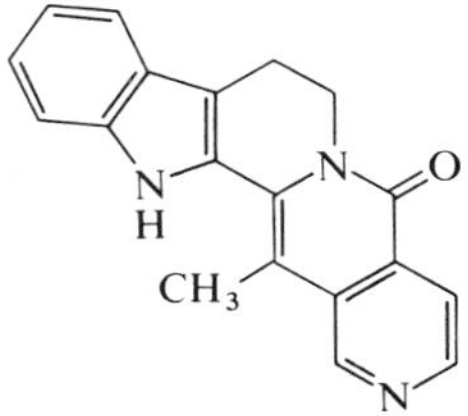 (—)	674
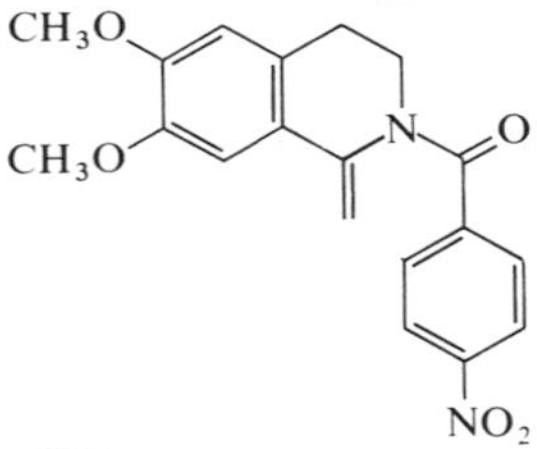	MeOH, air	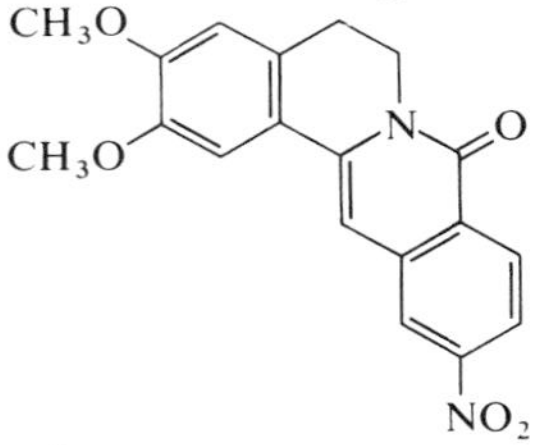 (8)	363
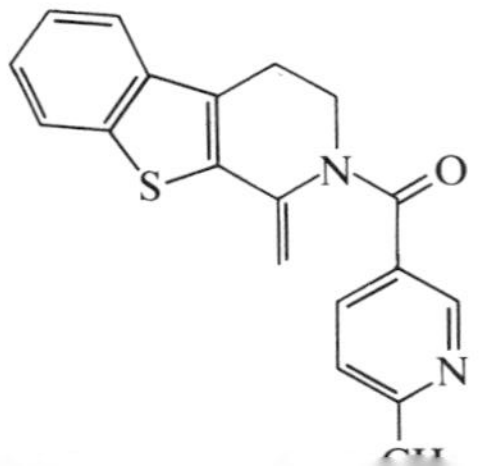	C_6H_6, air	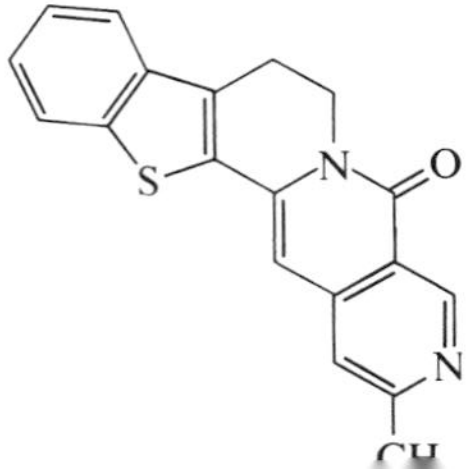 (25),	355

Reactant	Conditions	Product (yield, %)	Ref.
		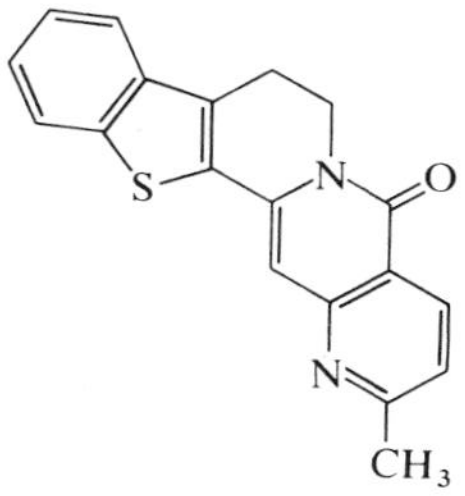 (24)	
$CH_2C_6H_5$, $COCH_3$, CH_3	MeOH	$CH_2C_6H_5$, $COCH_3$, CH_3 (25)	368
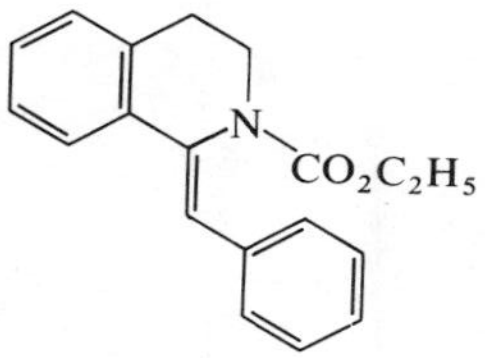 $CO_2C_2H_5$	MeOH, I_2	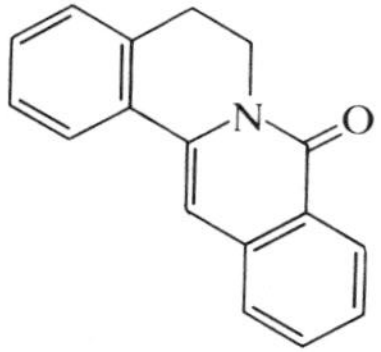 (10–21),	109,376
		$CO_2C_2H_5$ (65)	

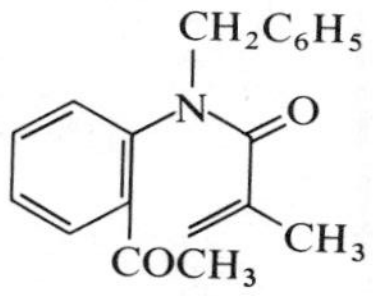

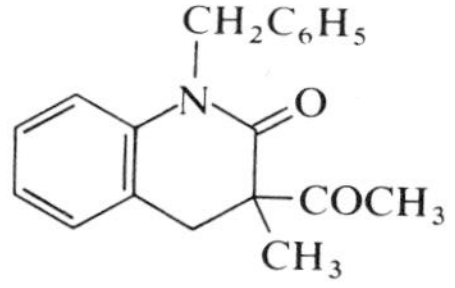

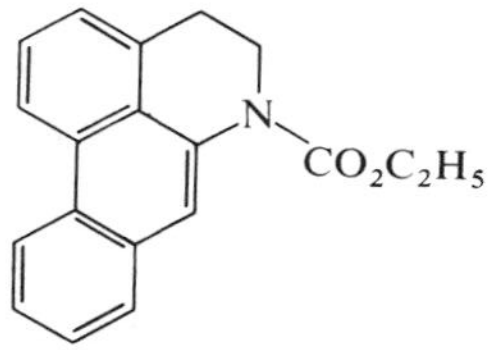

TABLE VII. LACTAMS AND OTHER HETEROCYCLIC SYSTEMS FROM AMIDES (*Continued*)[a]

Starting Material	Reaction Conditions	Products and Yields (%)	Refs.
CH_3, $C_6H_4OCH_3$-*o*	Et_2O	CH_3O, CH_3 (25)	358
$CH_2C_6H_5$, CO_2CH_3, CH_3	MeOH	$CH_2C_6H_5$, CO_2CH_3, CH_3 (40)	368
CH_3, CO_2H	HOAc	CH_3 (12),[d]	368
		CH_3, H, H (4)	

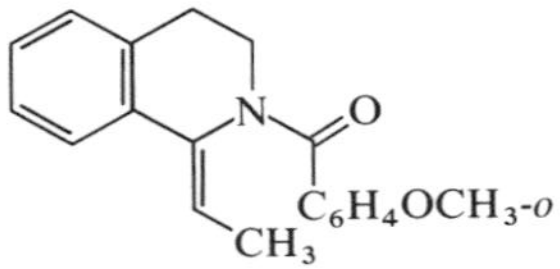

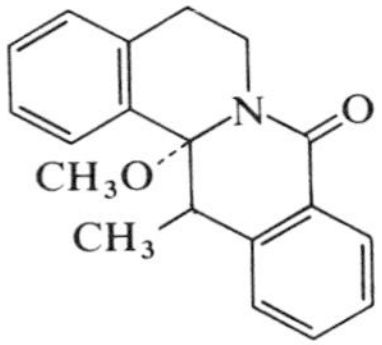

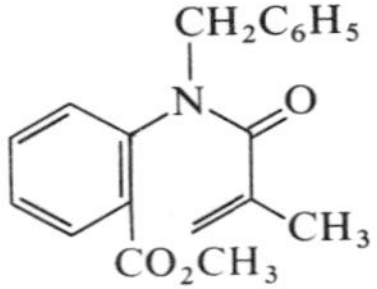

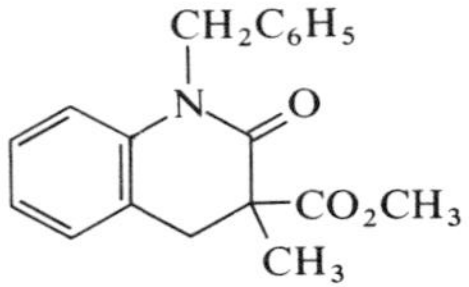

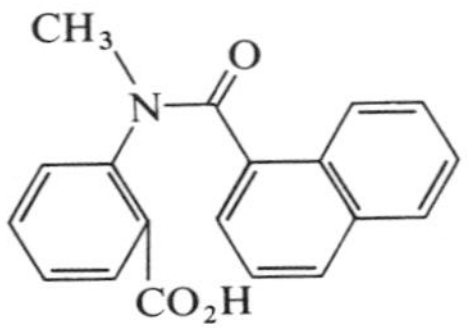

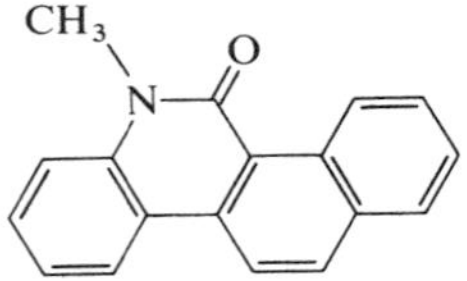

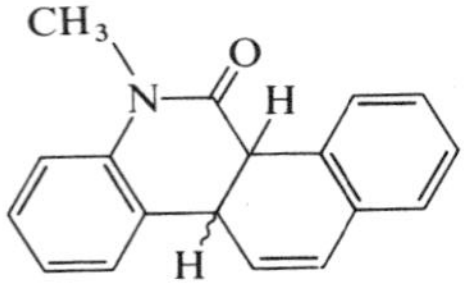

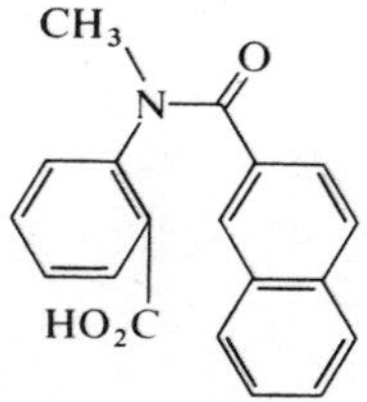

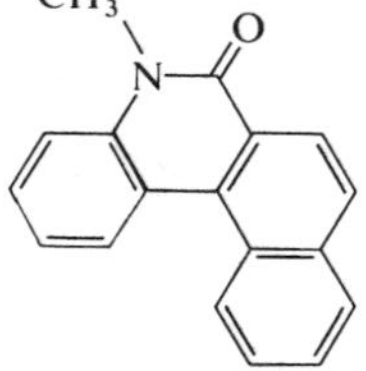

Reactant	Conditions	Product(s) and Yield(s) (%)	Refs.
CH_3, N, O, HO_2C	"	CH_3, N, O (2),[d] CH_3, N, O, H, H (16)	363
CH_3O, CH_3O, N, O	MeOH, air	CH_3O, CH_3O, N, O (24), CH_3O, CH_3O, N, O (18)	363
	t-BuOH, degassed	CH_3O, CH_3O, N, O (97)	81

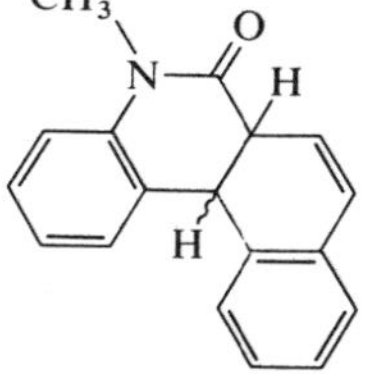

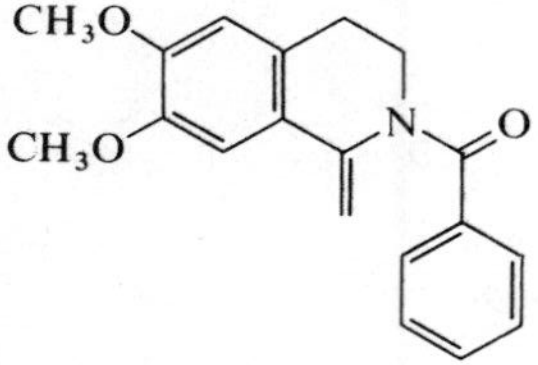

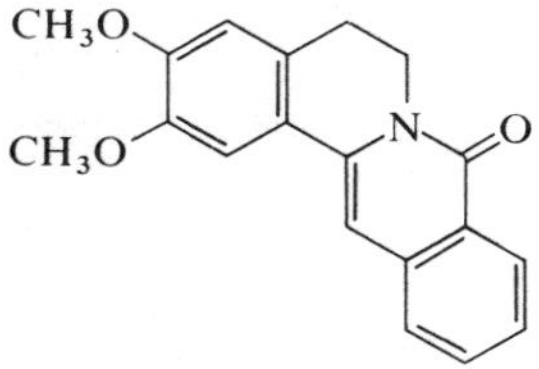

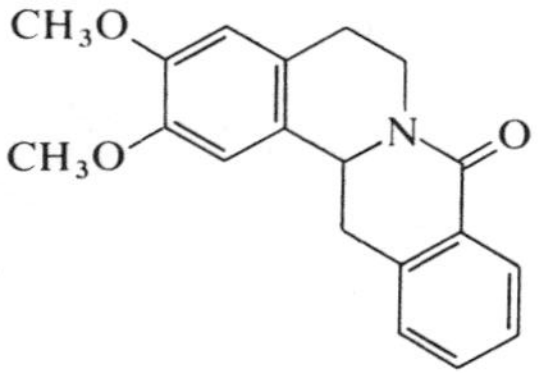

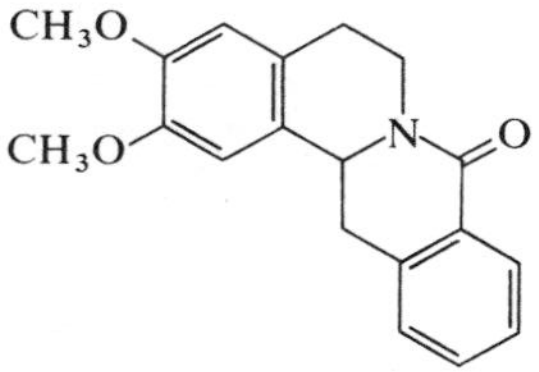

TABLE VII. LACTAMS AND OTHER HETEROCYCLIC SYSTEMS FROM AMIDES (*Continued*)[a]

	Starting Material	Reaction Conditions	Products and Yields (%)	Refs.
	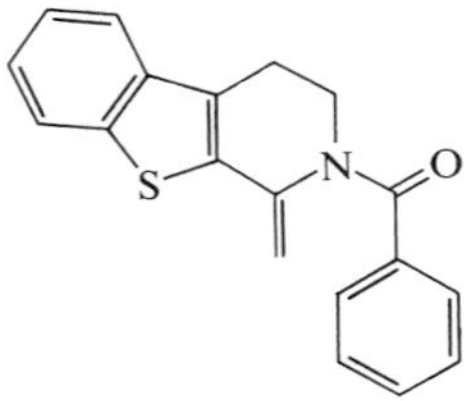 [structure]	C_6H_6, air	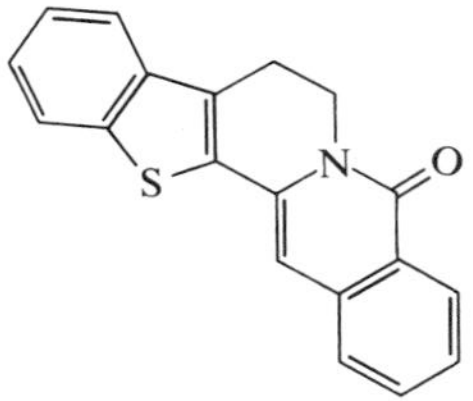 [structure] (72)	355
C_{20}	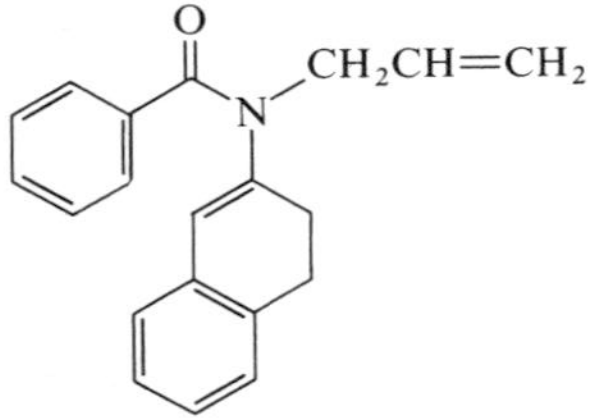 [structure] N-$CH_2CH{=}CH_2$	Et_2O	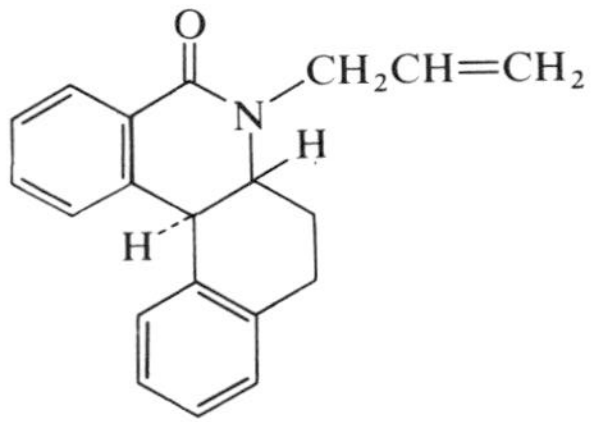 [structure] N-$CH_2CH{=}CH_2$, H, H (71)	361,362
	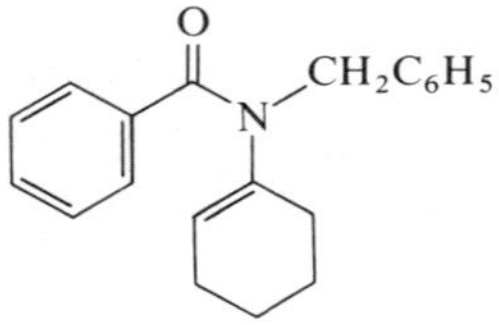 [structure] N-$CH_2C_6H_5$	MeOH, I_2	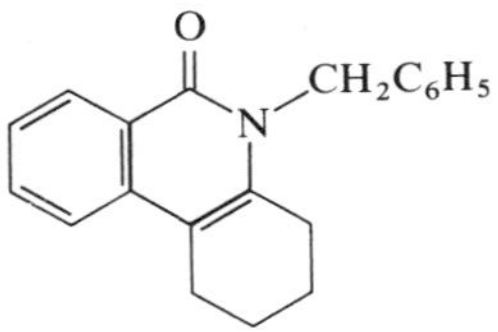 [structure] N-$CH_2C_6H_5$ (54)	79
		MeOH	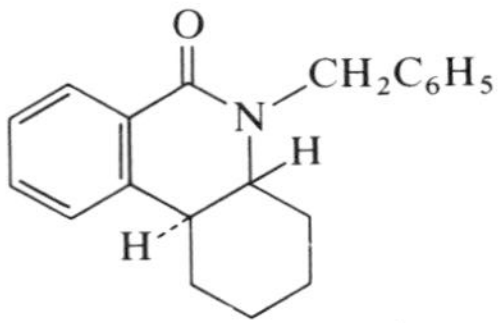 [structure] N-$CH_2C_6H_5$, H, H (35)	79,659

Reactant	Conditions	Product(s) and Yield(s) (%)	Refs.
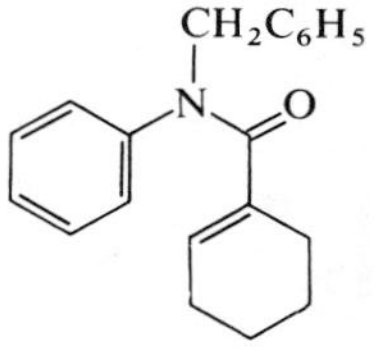	MeOH, I_2	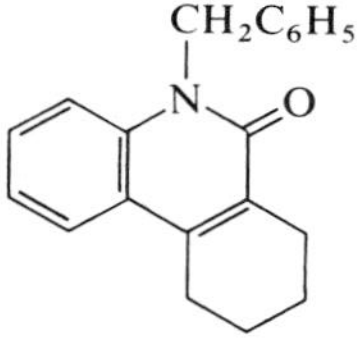 (35)	366
	Et_2O, Me_2SO, C_6H_6, or MeOH	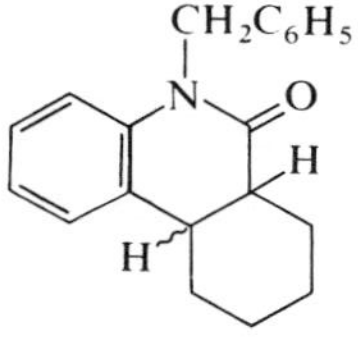 (48–67)[g]	366
	D_2O-saturated Et_2O	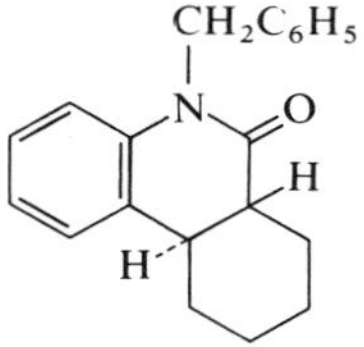 (53)	366
	MeOD	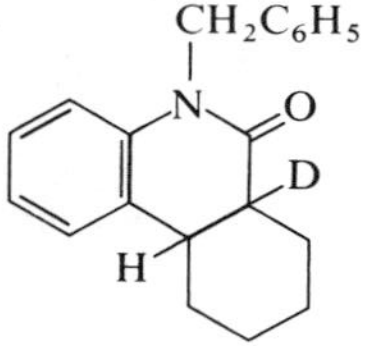 (36)	366
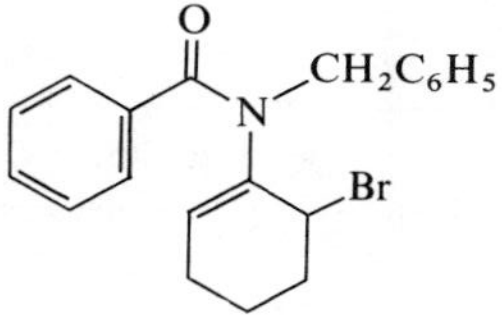	MeOH	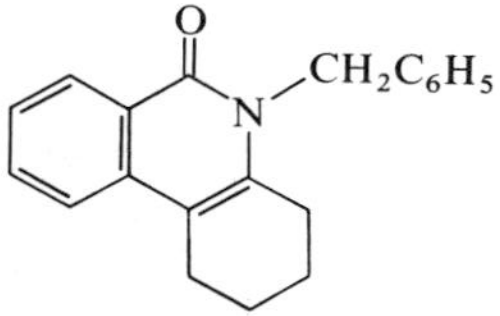 (5)	79

TABLE VII. Lactams and Other Heterocyclic Systems from Amides (*Continued*)[a]

Starting Material	Reaction Conditions	Products and Yields (%)	Refs.
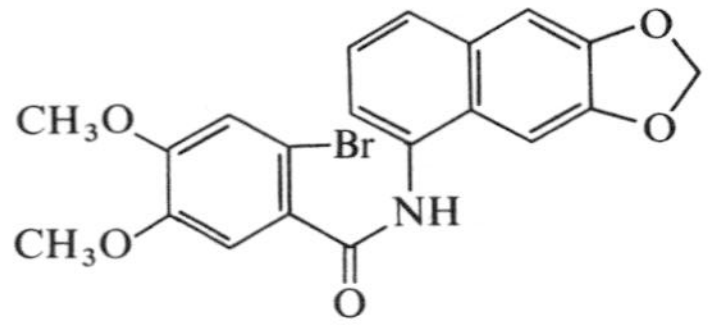 CH_3O, CH_3O, Br, NH, O, O, O	C_6H_6, MeOH	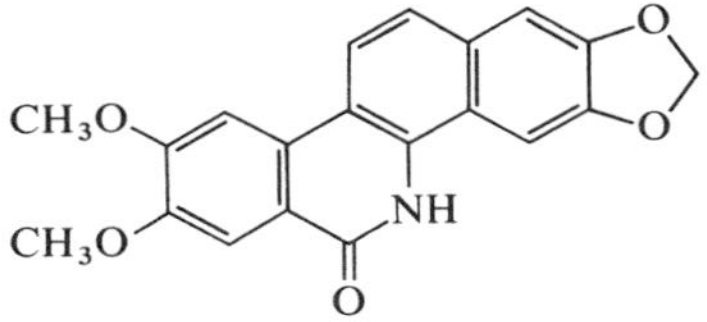 CH_3O, CH_3O, NH, O, O, O (70)	299
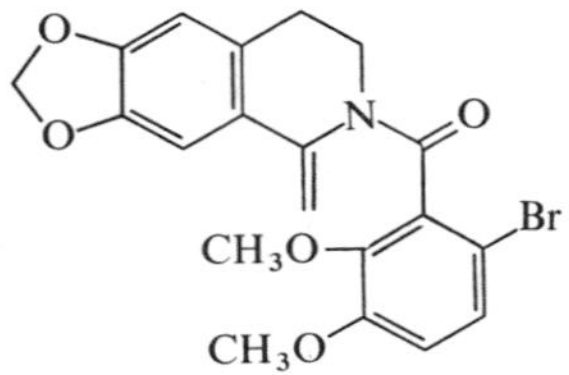 O, O, N, O, CH_3O, CH_3O, Br	C_6H_6	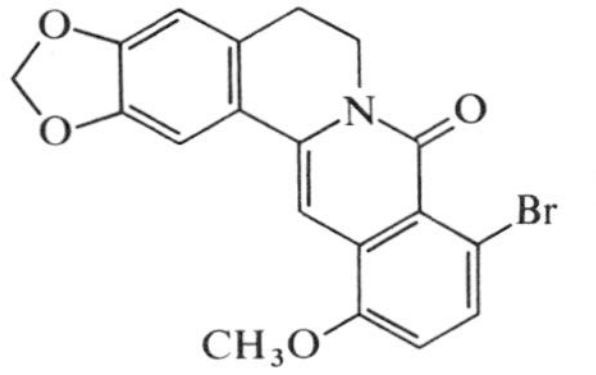 O, O, N, O, Br, CH_3O (72)	675
	MeOH	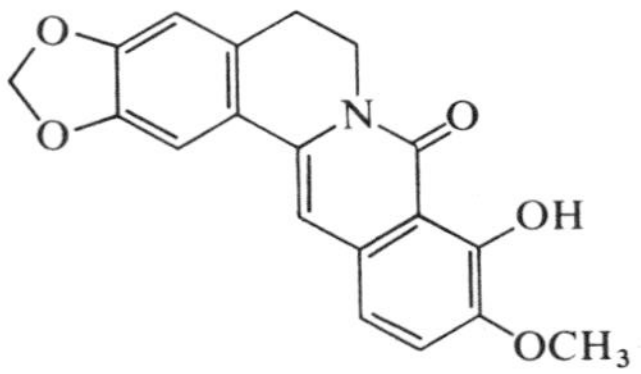 " + O, O, N, O, OH, OCH_3 (1:2)	675
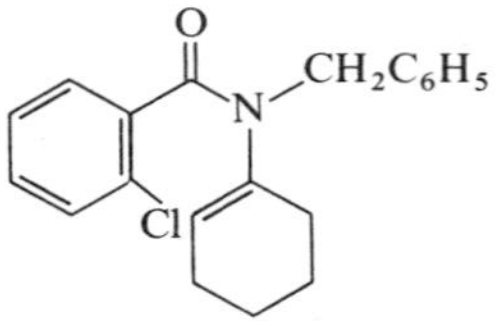 O, N, $CH_2C_6H_5$, Cl	MeOH, air	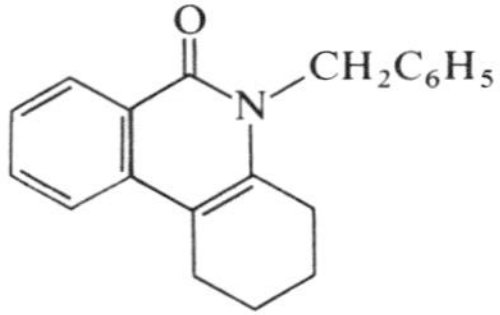 O, N, $CH_2C_6H_5$ (14),	103

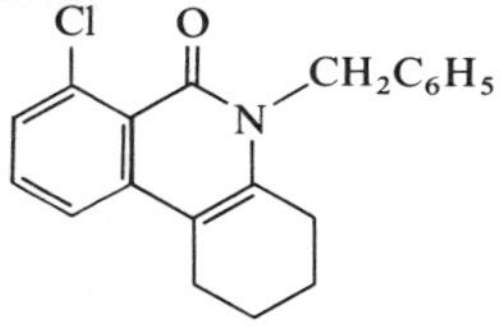

Reactant	Conditions	Product(s) (yield, %)	Refs.
		Cl, O, N, $CH_2C_6H_5$ (4)	
N, H, O, CH_3CHCl, N	MeOH	N, H, N, O, $CH_3CH(OCH_3)$, N (5)	354,676
NH_2, O, N, $CH_2C_6H_5$	Et_2O	NH_2, O, N, $CH_2C_6H_5$, H, H (71)	103
N, H, N, O, CH_3CO, N	MeOH, air	N, H, N, O, CH_3CO, N (30),	354,676
		N, H, N, O, N, $COCH_3$ (8)	

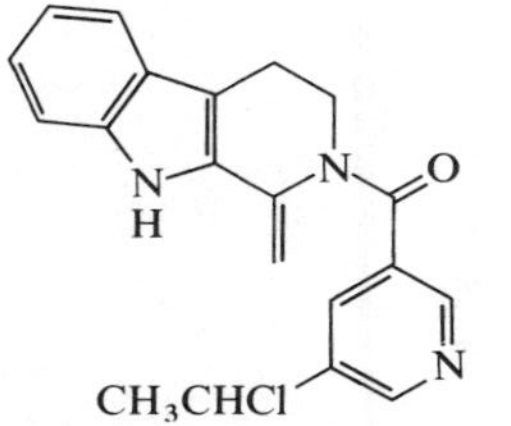

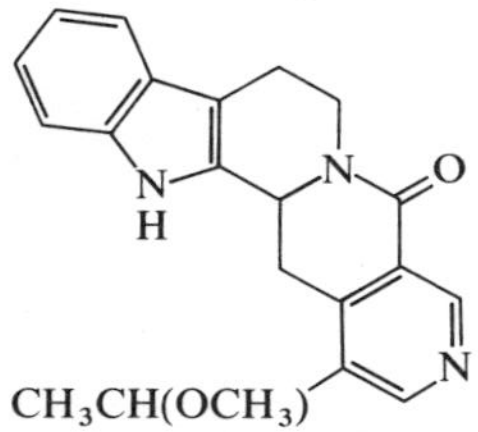

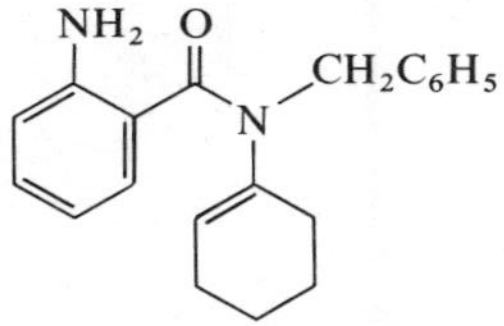

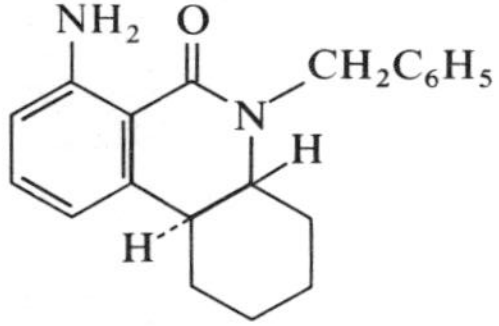

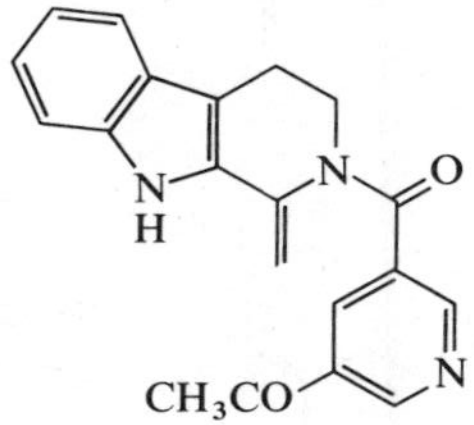

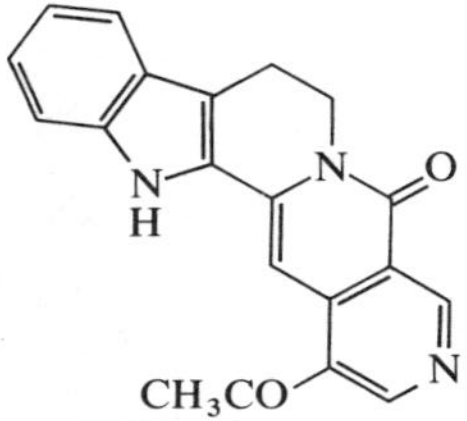

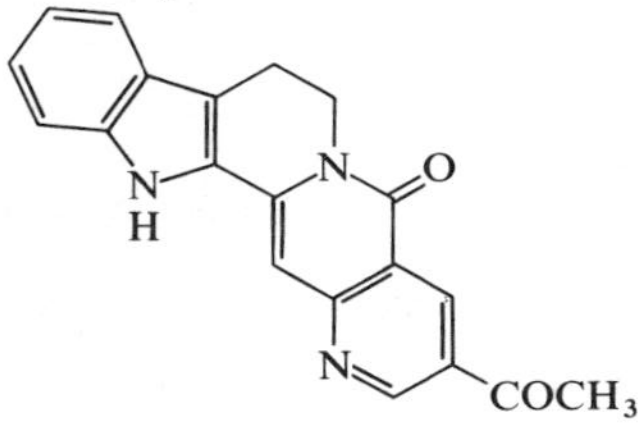

TABLE VII. Lactams and Other Heterocyclic Systems from Amides (*Continued*)[a]

Starting Material	Reaction Conditions	Products and Yields (%)	Refs.
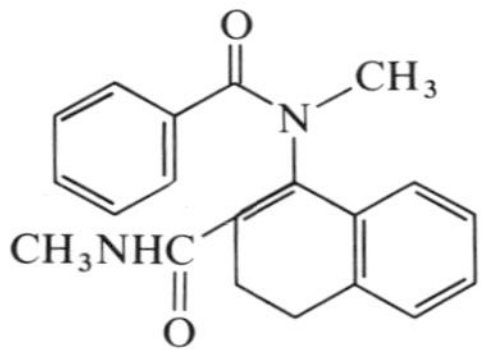 O, N, CH_3, CH_3NHC, O	MeOH	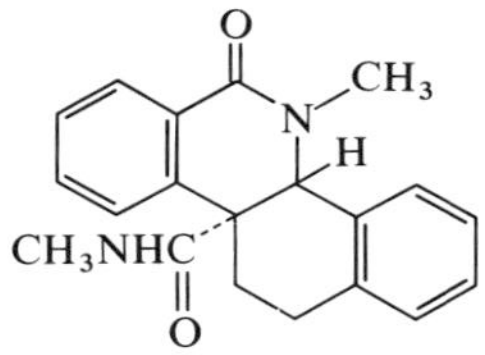 O, N, CH_3, H, CH_3NHC, O (77–78)	672,673
O, N, $CH_2C_6H_5$, O_2N	"	O, N, $CH_2C_6H_5$, O_2N (14)	103
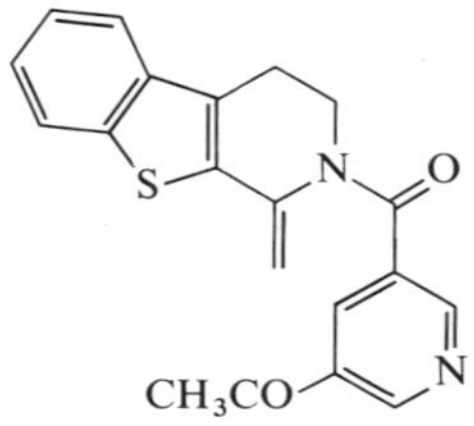 S, N, O, CH_3CO, N	C_6H_6, air	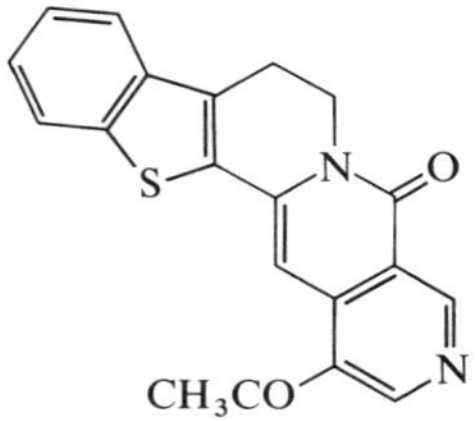 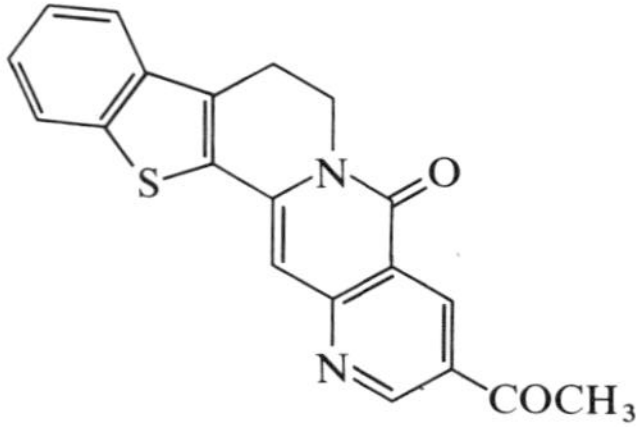 S, N, O, CH_3CO, N (34),	355
		S, N, O, N, $COCH_3$ (33)	

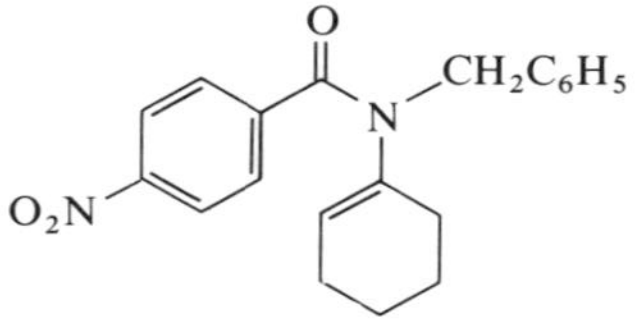

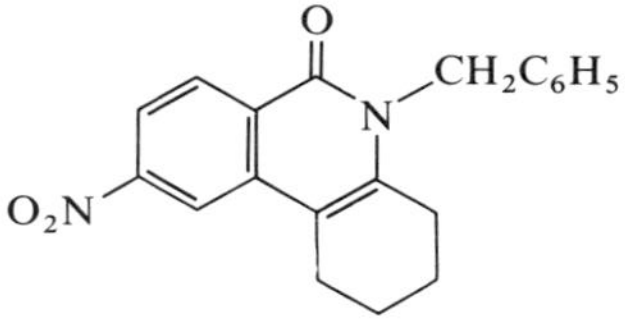

Reactant	Conditions	Product	(Yield, %)	Ref.
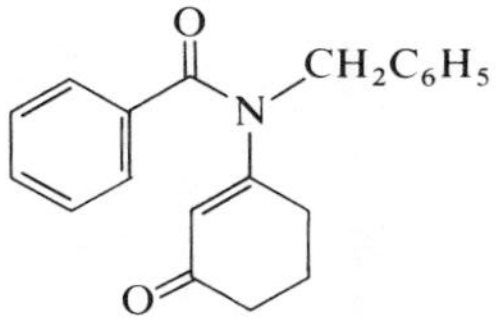 $CH_2C_6H_5$, N, O, O	MeOH, I_2	$CH_2C_6H_5$, N, O, O	(—)	79
	MeOH, air	”	(60)	79
CH_3CO_2, O, N, CH_3	MeOH	CH_3CO_2, O, N, CH_3, H, H	(33)	103
CH_3O_2C, O, N, CH_3	”	CH_3O_2C, O, N, CH_3, H, H	(24)	103
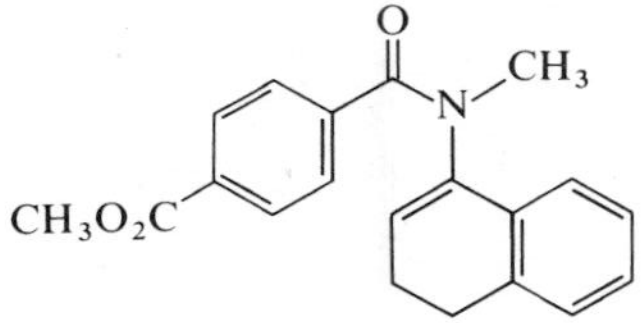 CH_3O_2C, O, N, CH_3	MeOH, I_2	CH_3O_2C, O, N, CH_3	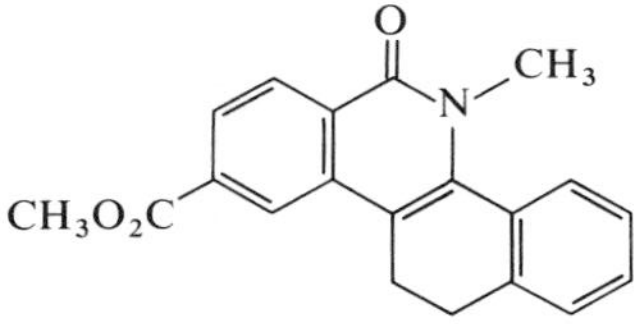 (23)	103
	C_6H_6, MeOH	CH_3O_2C, O, N, CH_3, H, H	(13)	103

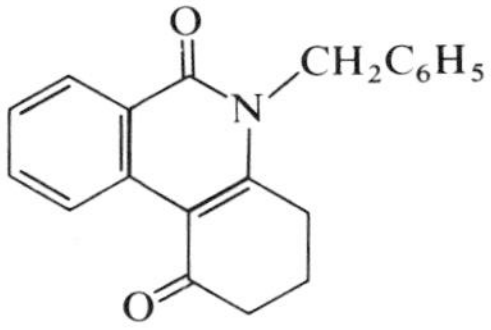

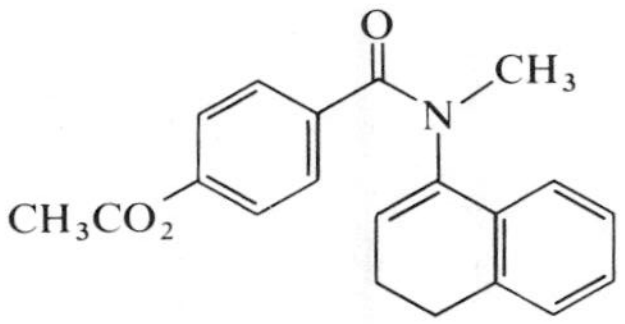

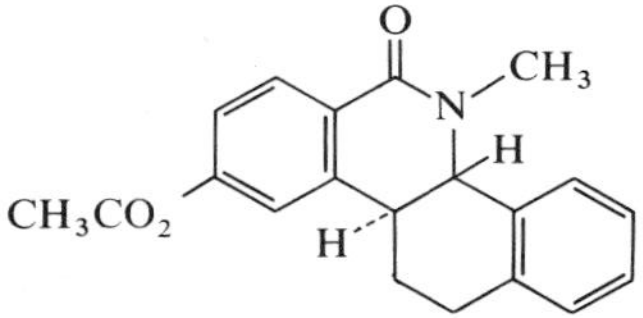

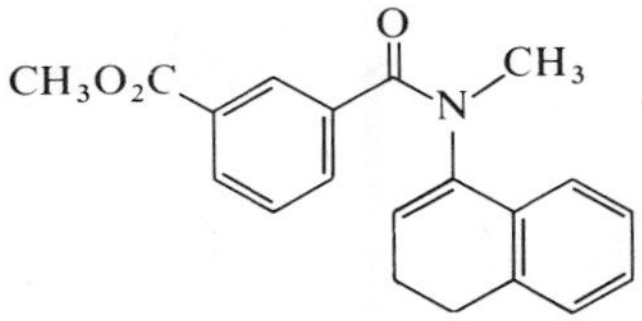

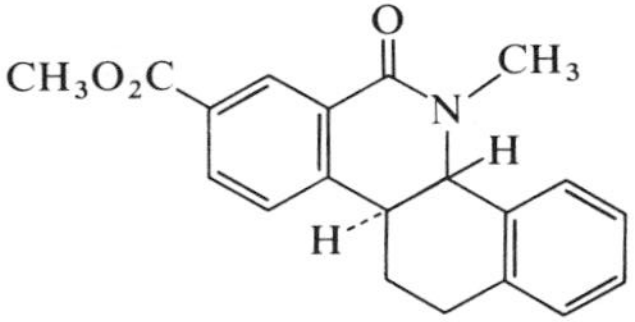

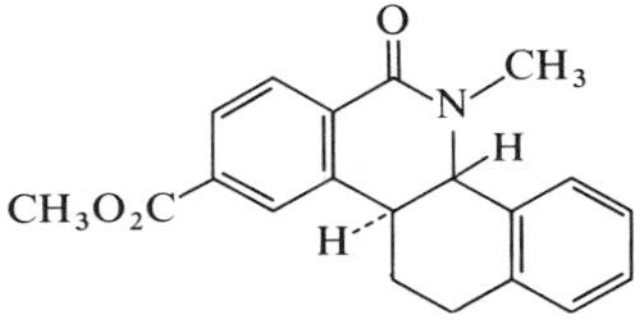

TABLE VII. LACTAMS AND OTHER HETEROCYCLIC SYSTEMS FROM AMIDES (*Continued*)[a]

Starting Material	Reaction Conditions	Products and Yields (%)	Refs.
	MeOH	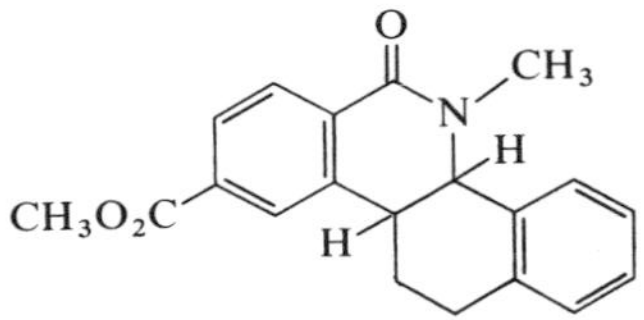 CH_3O_2C, H, H, N, CH_3, O (20)	103
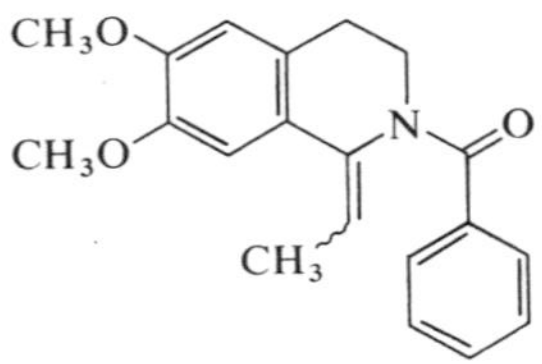 CH_3O, CH_3O, N, O, CH_3	MeOH, air	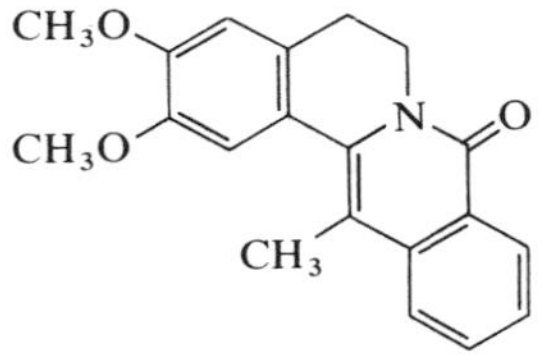 CH_3O, CH_3O, N, O, CH_3 (9)	358
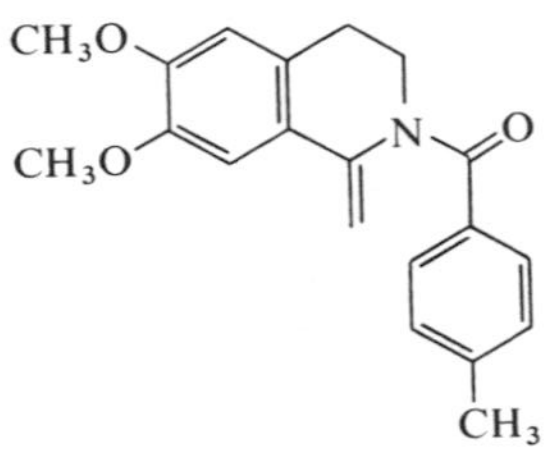 CH_3O, CH_3O, N, O, CH_3	*t*-BuOH, degassed	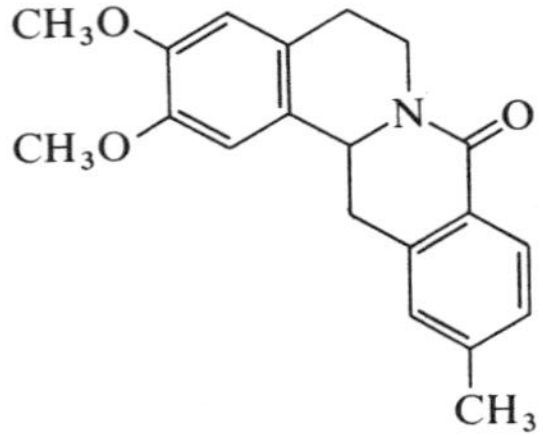 CH_3O, CH_3O, N, O, CH_3 (85)	81
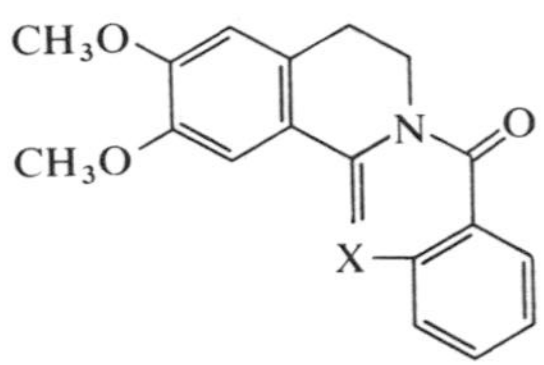 CH_3O, CH_3O, N, O, X		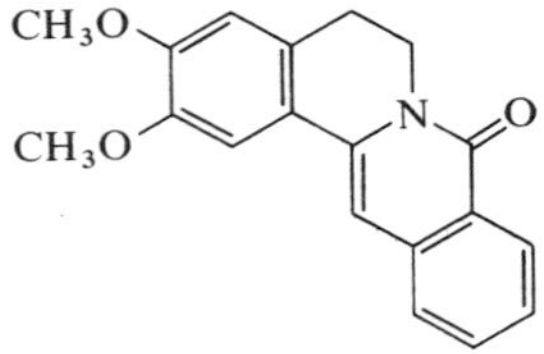 CH_3O, CH_3O, N, O	

$X = CH_3O$	MeOH	" (46)	363
$X = CH_3S$	Solvent	" (55)	80
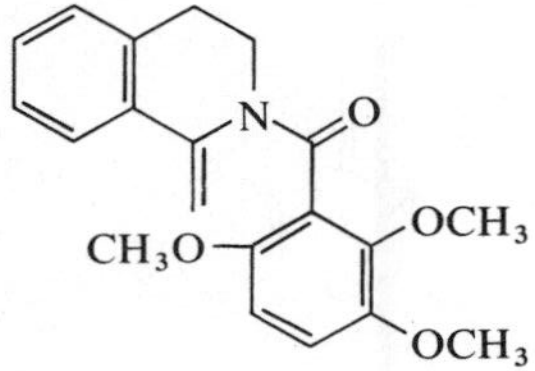	MeOH	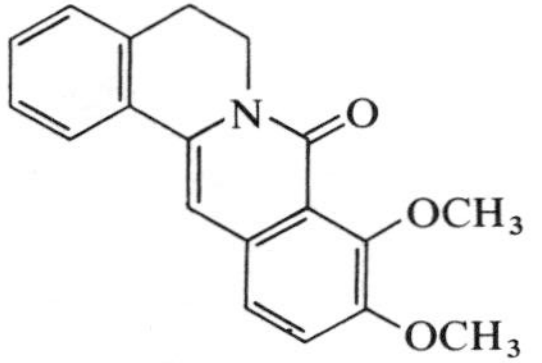 (25), 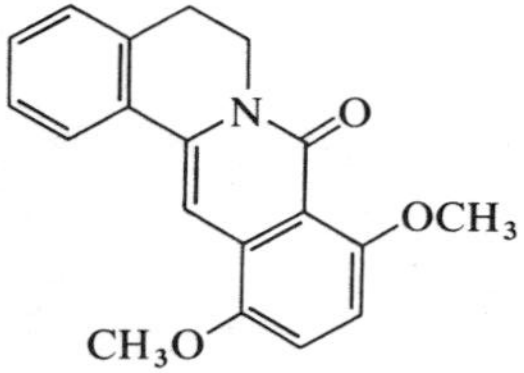 (10)	363
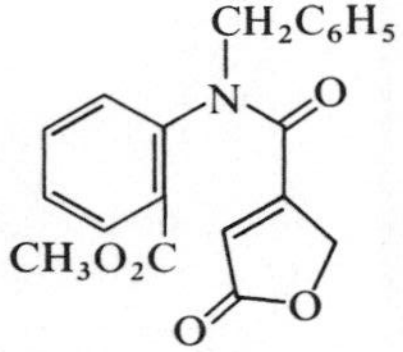	C_6H_6 or C_6H_6–MeOH	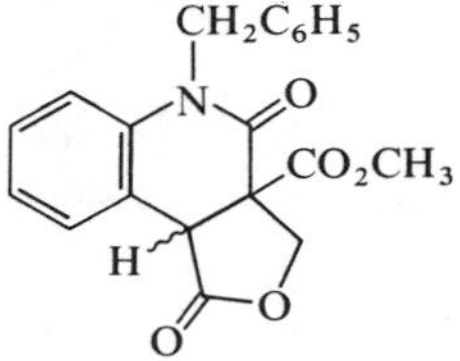 (28), 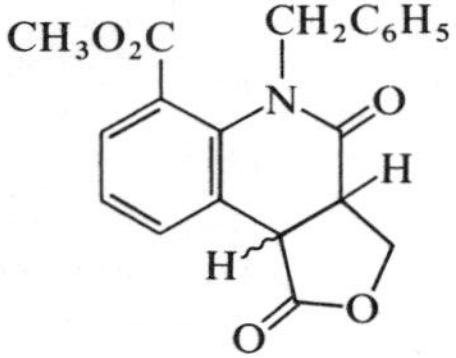 (22)	658b

TABLE VII. LACTAMS AND OTHER HETEROCYCLIC SYSTEMS FROM AMIDES (*Continued*)[a]

Starting Material	Reaction Conditions	Products and Yields (%)	Refs.
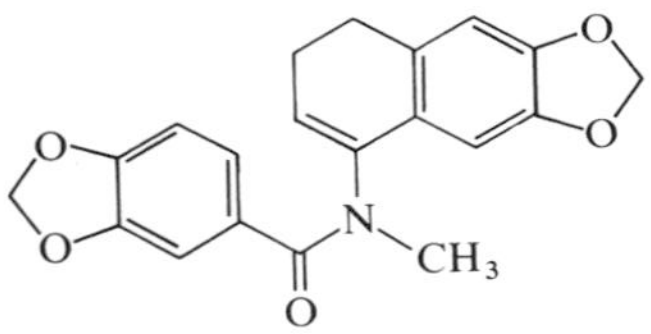 CH_3	MeOH, I_2	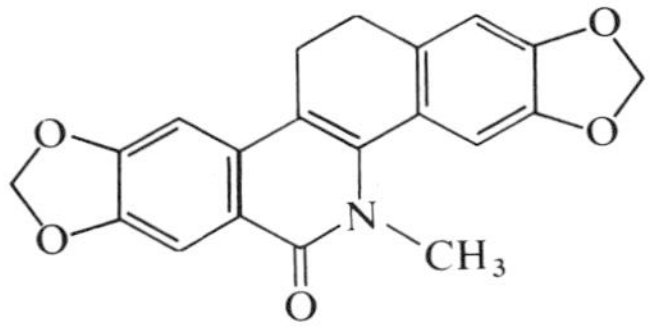 CH_3 (25),	350,365
		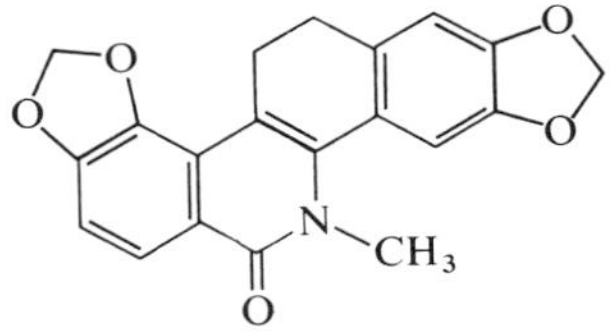 CH_3 (6)	
	Et_2O	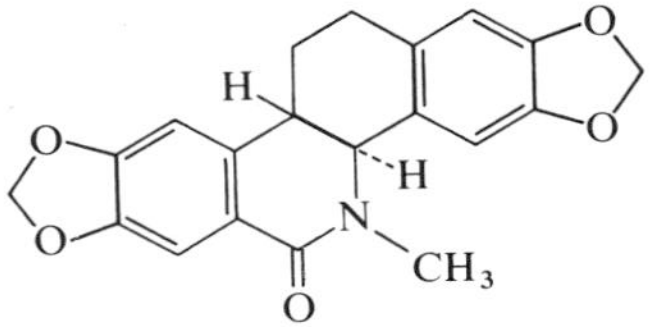 H H CH_3 (52)	350,365
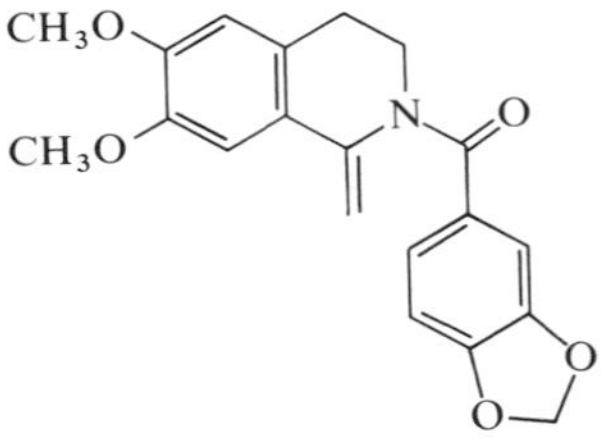 CH_3O CH_3O	*t*-BuOH, degassed	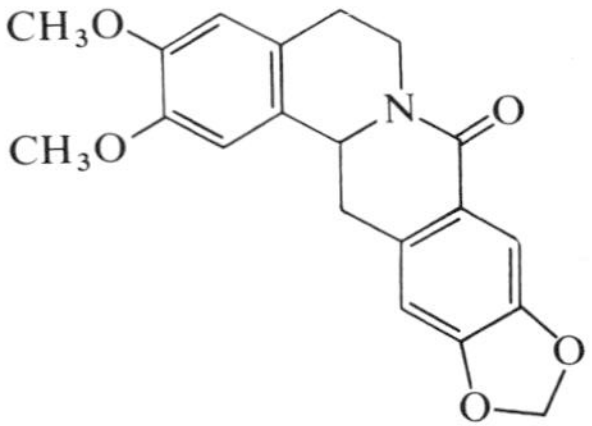 CH_3O CH_3O (75)	81

	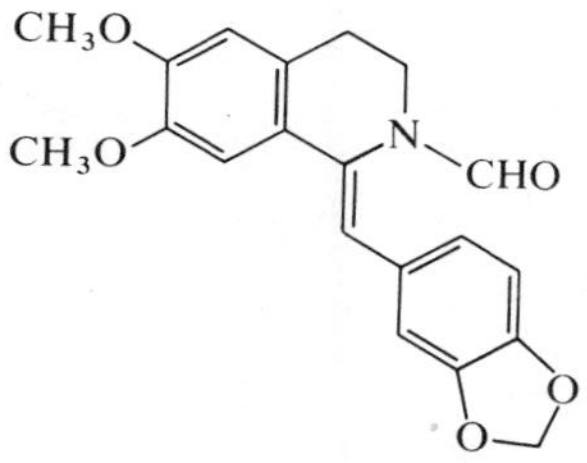	Dioxane, t-BuOH, 47% HI, N_2	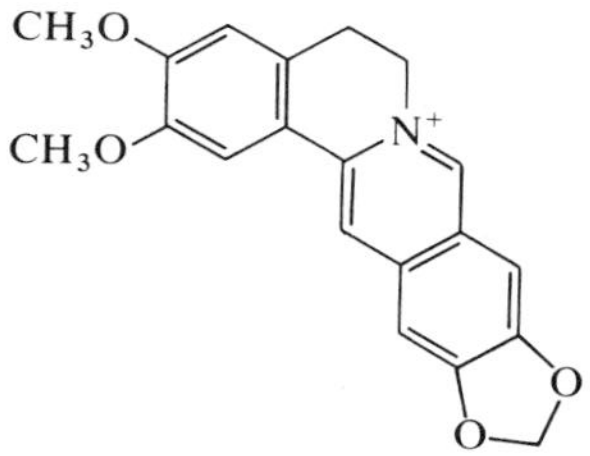 (100)	373
C_{21}	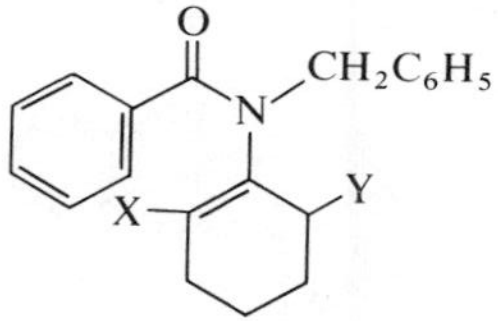 (I)		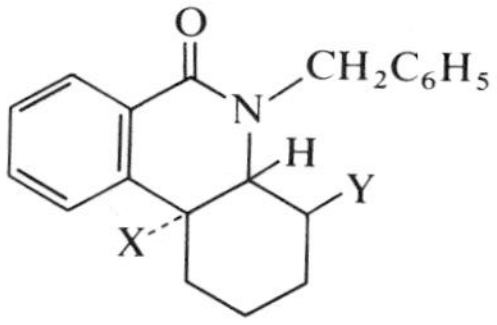 II	
	I, X = CH_3, Y = H	MeOH	II, X = CH_3, Y = H (55)	79
	I, X = H, Y = CH_3	"	II, X = H, Y = CH_3 (23)	79
	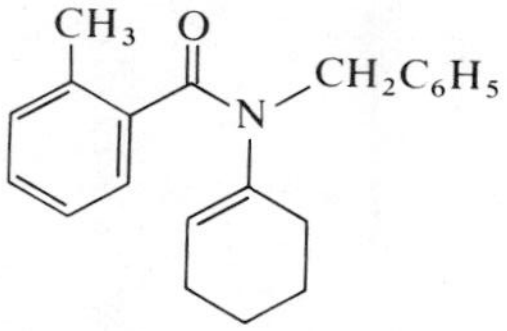	"	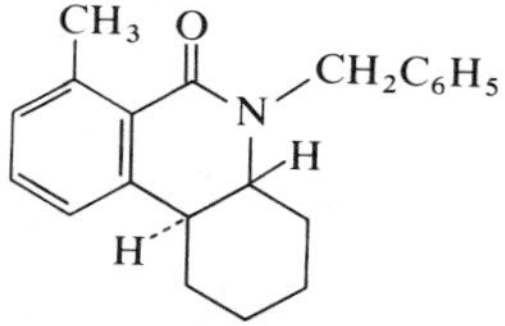 (22)	103
	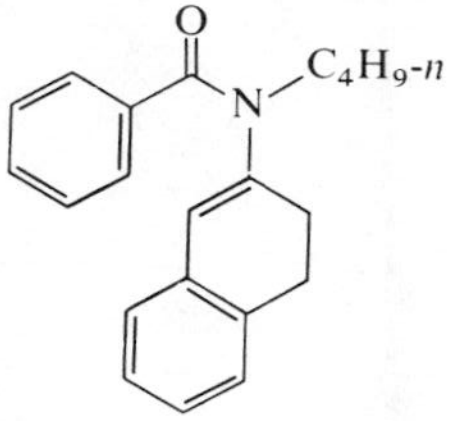	Et_2O	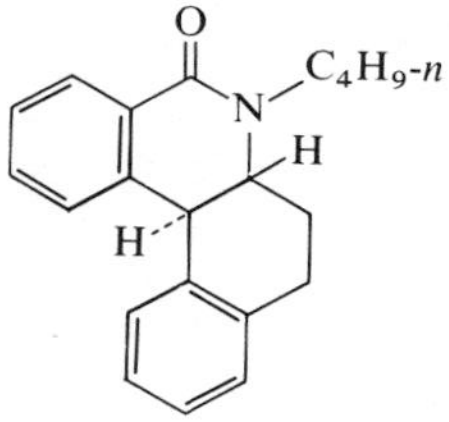 (63)	361,362

TABLE VII. LACTAMS AND OTHER HETEROCYCLIC SYSTEMS FROM AMIDES (*Continued*)[a]

Starting Material	Reaction Conditions	Products and Yields (%)	Refs.
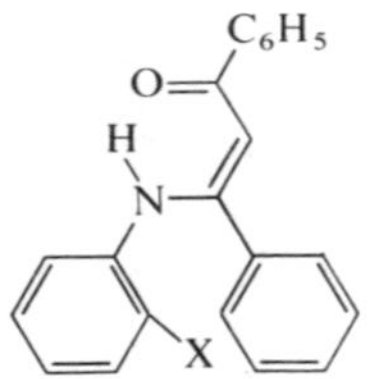 $X = Br$ or I	C_6H_{12}, N_2	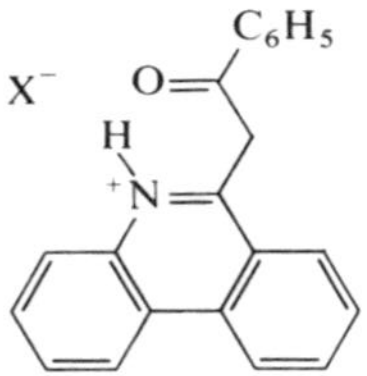 X^- (60)	347
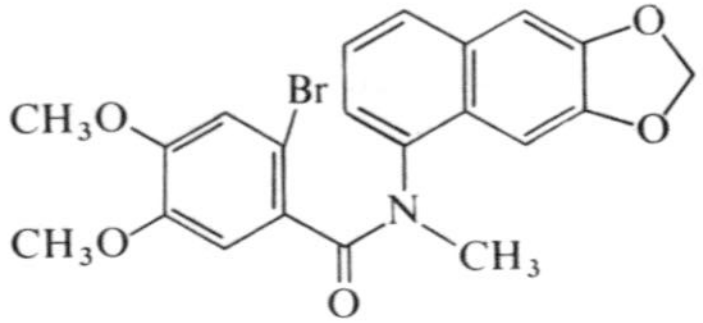 CH_3O, CH_3O, Br, N, CH_3, O	MeCN	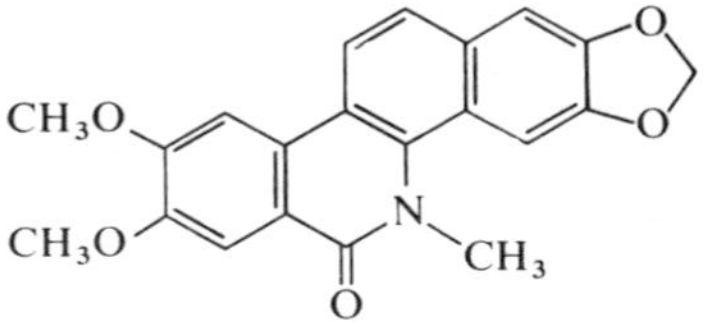 CH_3O, CH_3O, N, CH_3, O (35)	677a
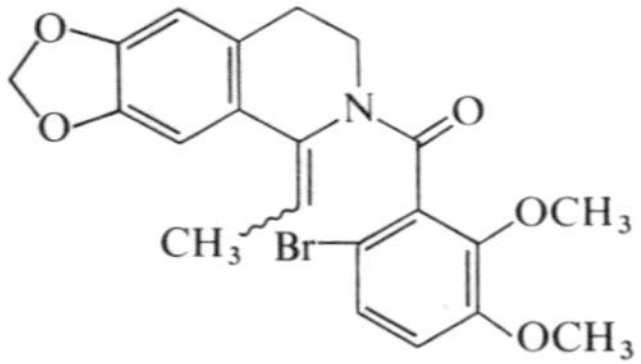 N, O, CH_3, Br, OCH_3, OCH_3	MeOH	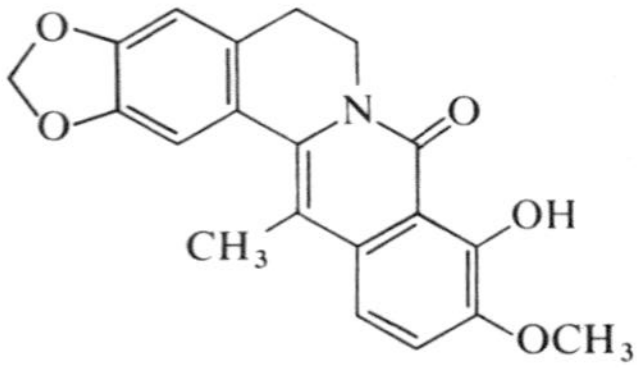 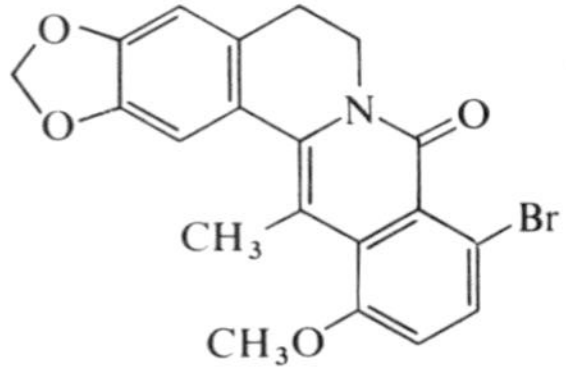 N, O, CH_3, OH, OCH_3 + N, O, CH_3, Br, CH_3O (3:2)	675

Reactant	Solvent	Product(s) (%)	Ref.
CH_3O, CH_3O, N, O, Br, OCH_3, OCH_3	C_6H_6	CH_3O, CH_3O, N, O, OCH_3, OCH_3 (80)	675
	„	„ (75)	677b
CH_3O, CH_3O, N, O, Br, CH_3O, CH_3O	„	CH_3O, CH_3O, N, O, Br, CH_3O (63)	675
	MeOH	CH_3O, CH_3O, N, O, OH, OCH_3 (65),	675
		CH_3O, CH_3O, N, O, CH_3O (20)	

TABLE VII. LACTAMS AND OTHER HETEROCYCLIC SYSTEMS FROM AMIDES (*Continued*)[a]

Starting Material	Reaction Conditions	Products and Yields (%)	Refs.
$CH_2C_6H_5$, N, O, CN	MeOH	$CH_2C_6H_5$, N, O, CN, H (40)	368
$CH_2C_6H_5$, N, O, CO, NH_2	Et_2O	$CH_2C_6H_5$, N, O, $CONH_2$, H (55),	368
		$CH_2C_6H_5$, N, O, H, H (14)	
	MeOH or HOAc	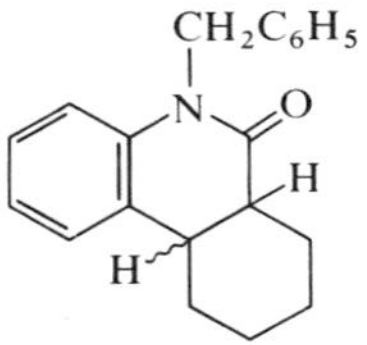 $CH_2C_6H_5$, N, O, H, H (70–80)	368

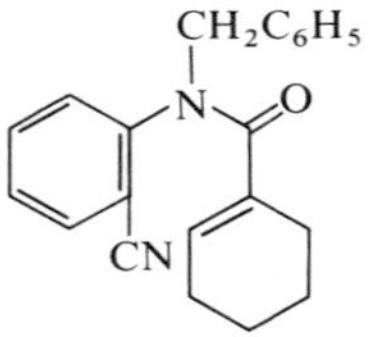

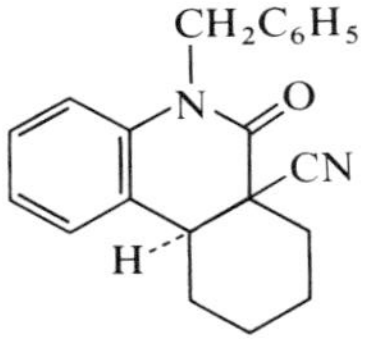

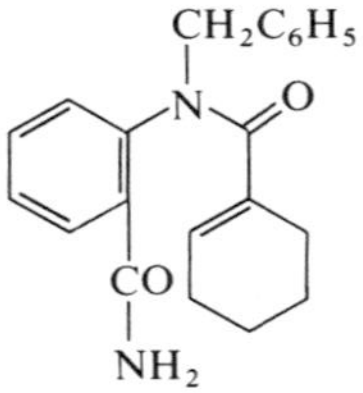

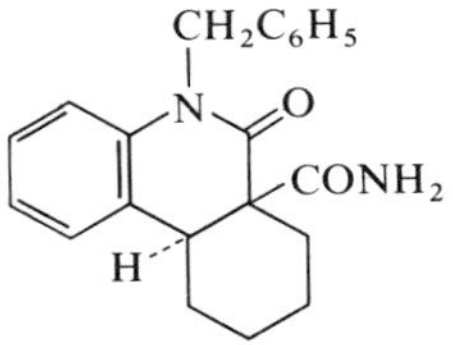

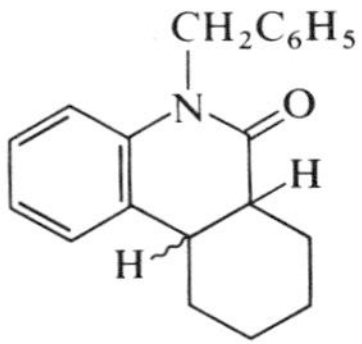

Reactant	Solvent	Product(s) (% yield)	Refs.
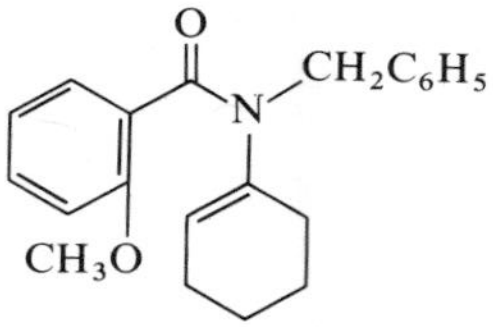 $CH_2C_6H_5$, CH_3O	Et_2O	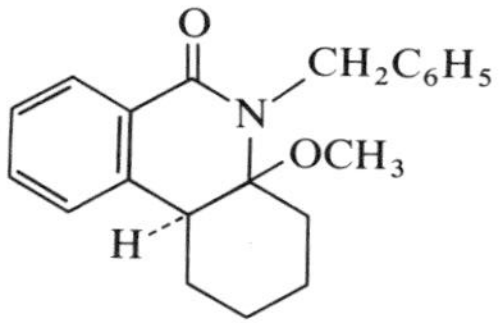 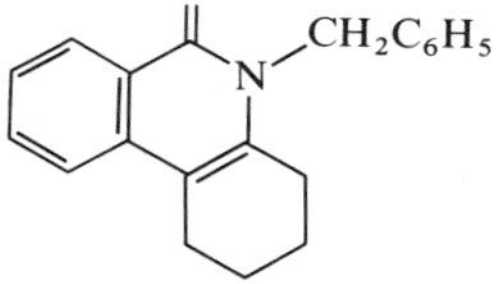 $CH_2C_6H_5$, OCH_3, H (30), $CH_2C_6H_5$ (18)	103,367
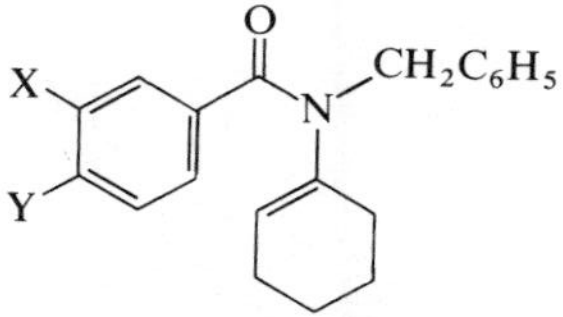 X, Y, $CH_2C_6H_5$ (I)		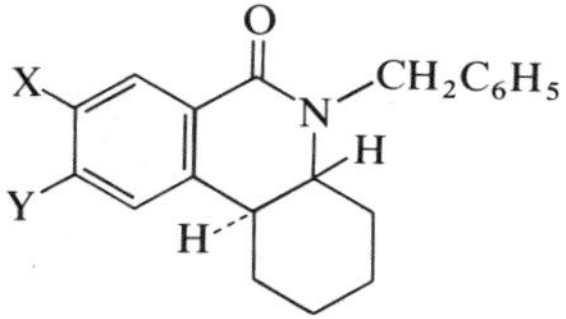 X, Y, $CH_2C_6H_5$, H, H (II)	
I, X = CH_3O, Y = H	MeOH	II, X = CH_3O, Y = H (36)	103
I, X = H, Y = CH_3O	"	II, X = H, Y = CH_3O (29)	103
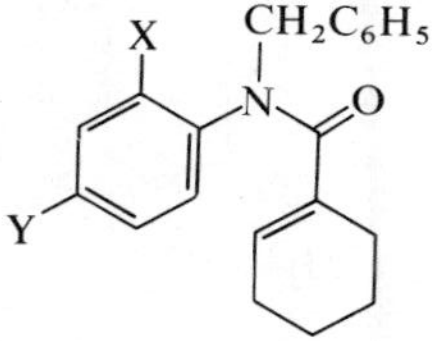 X, $CH_2C_6H_5$, Y (I)		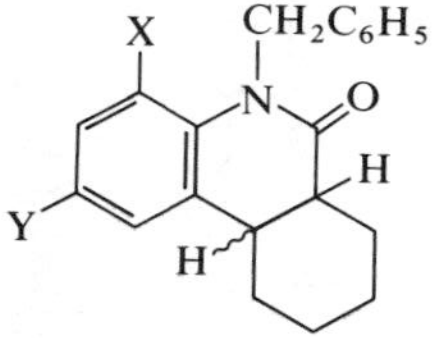 X, $CH_2C_6H_5$, Y, H, H (II)	
I, X = CH_3O, Y = H	Et_2O, C_6H_6, or MeOH	II, X = CH_3O, Y = H (52–65)	368
I, X = H, Y = CH_3O	"	II, X = H, Y = CH_3O (60–65)	368

TABLE VII. LACTAMS AND OTHER HETEROCYCLIC SYSTEMS FROM AMIDES (*Continued*)[a]

Starting Material	Reaction Conditions	Products and Yields (%)	Refs.
$CH_2C_6H_5$, N, O, HO_2C	MeOH	$CH_2C_6H_5$, N, O, H, H (53)	368
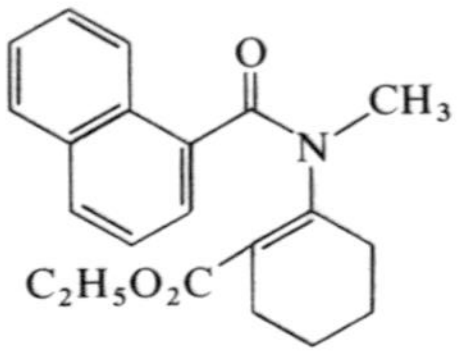 O, N, CH_3, $C_2H_5O_2C$	MeOH, I_2	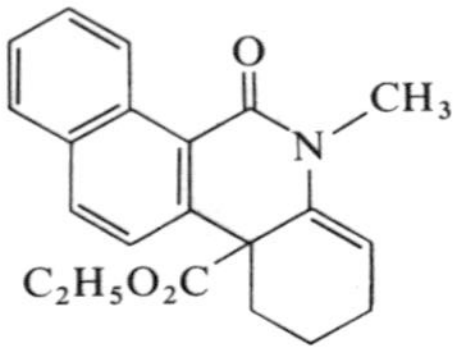 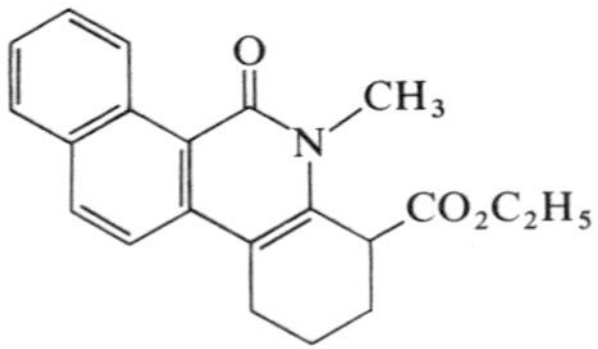 O, N, CH_3, $C_2H_5O_2C$ (58),	668
		O, N, CH_3, $CO_2C_2H_5$ (17)	
	Et_2O	H, O, N, CH_3, H, H, $CO_2C_2H_5$ (8)	668

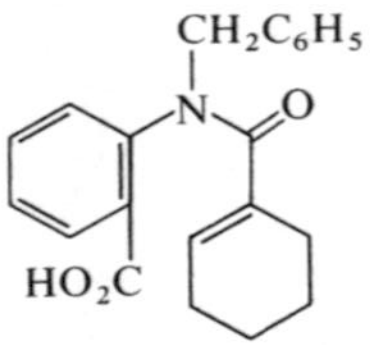

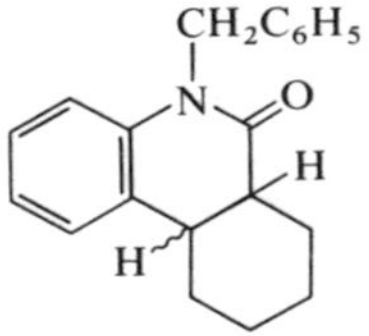

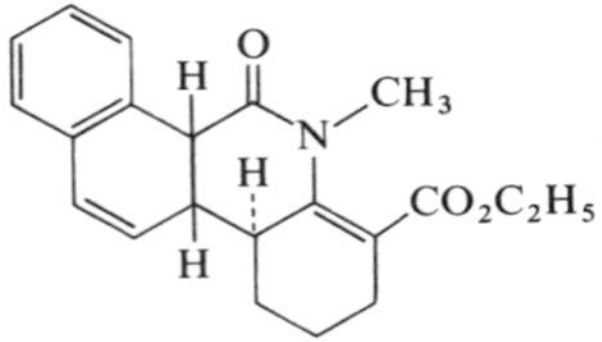

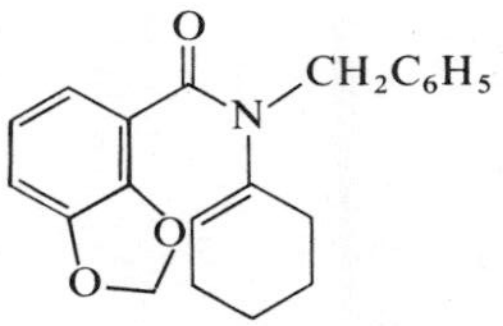

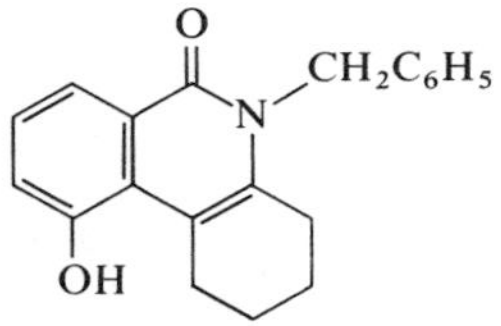

O, N, $CH_2C_6H_5$, O, O	"	O, N, $CH_2C_6H_5$, OH (14)	103
O, O, O, N, $CH_2C_6H_5$	MeOH	O, O, O, N, $CH_2C_6H_5$, H, H (*trans*, 13; *cis*, 15)	659,678a
CH_3O, CH_3O, N, O, C_6H_5, CH_3, CH_3	Dioxane, degassed	CH_3O, CH_3O, N, O, CH_3, CH_3 (43)	30a
CH_3O, CH_3O, N, O, C_6D_5, CH_3, CH_3	"	CH_3O, CH_3O, N, O, D, CH_3, CH_3, D, D, D, D (38)	30a

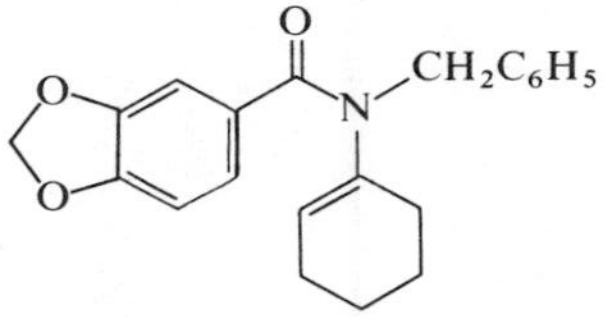

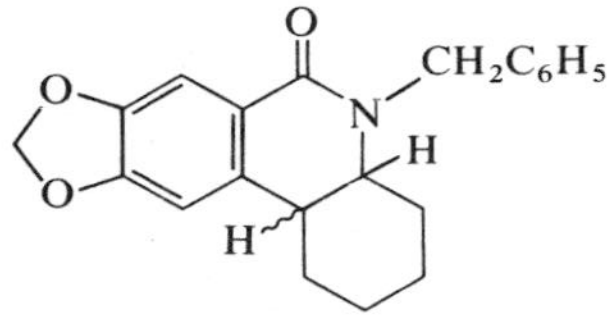

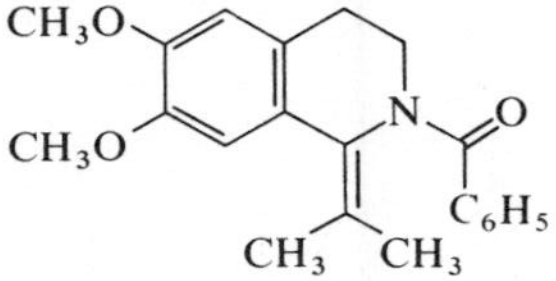

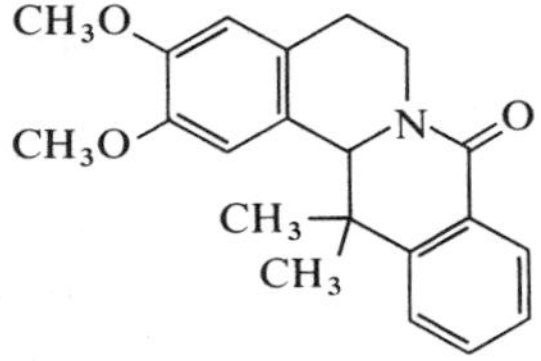

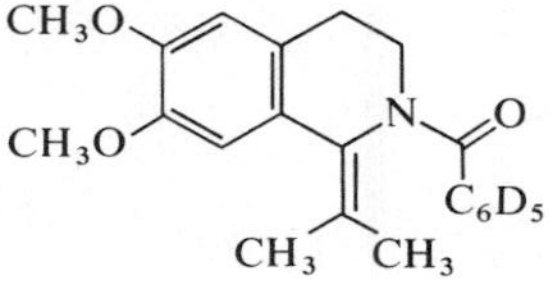

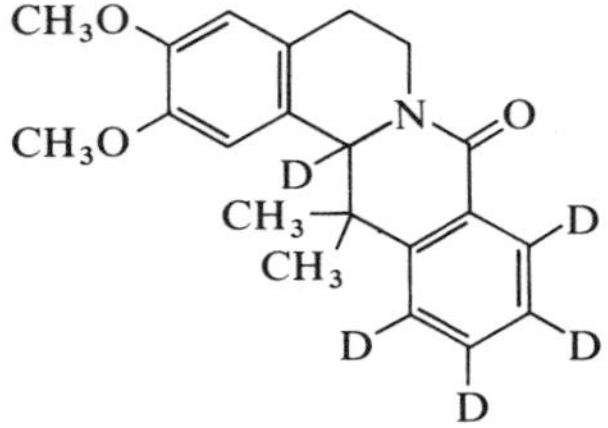

TABLE VII. LACTAMS AND OTHER HETEROCYCLIC SYSTEMS FROM AMIDES (*Continued*)[a]

Starting Material	Reaction Conditions	Products and Yields (%)	Refs.
CH_3O_2C, N–CH_3, OCH_3	MeOH	CH_3O_2C, N–CH_3, H, H, OCH_3 (5)	103
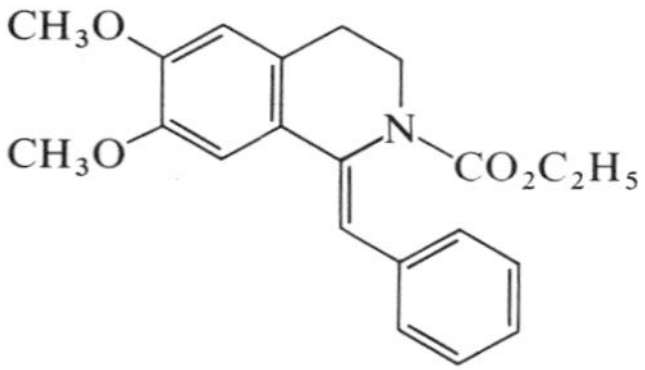 CH_3O, CH_3O, N–$CO_2C_2H_5$	Abs EtOH, N_2	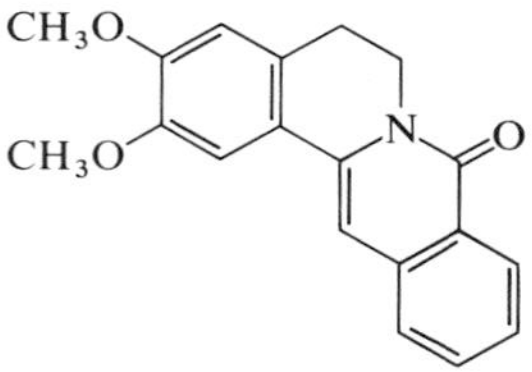 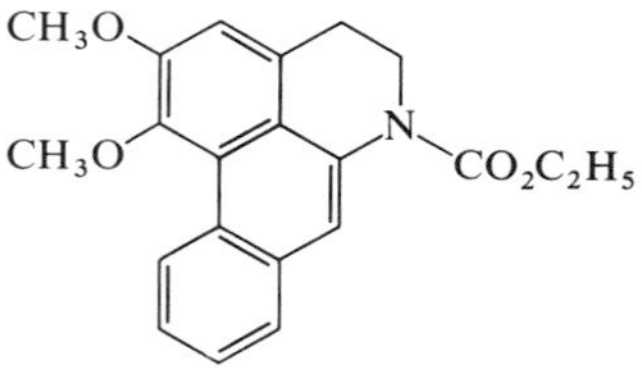 CH_3O, CH_3O, N, O (55),	80
		CH_3O, CH_3O, N–$CO_2C_2H_5$ (20)[d]	
CH_3O, CH_3O, N, O, $C_6H_4OCH_3$-*o*, CH_3	MeOH	CH_3O, CH_3O, N, O, CH_3 (35)	358

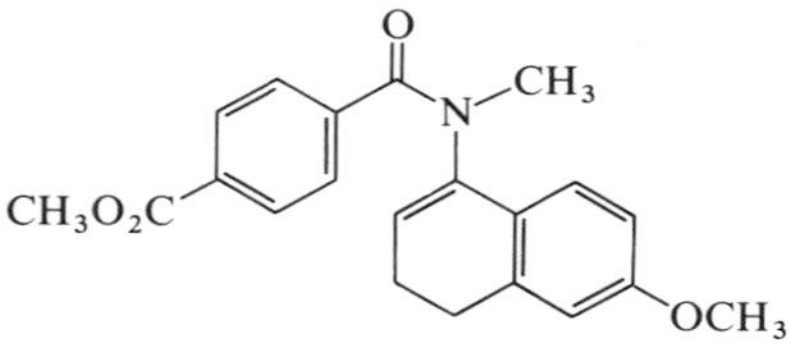

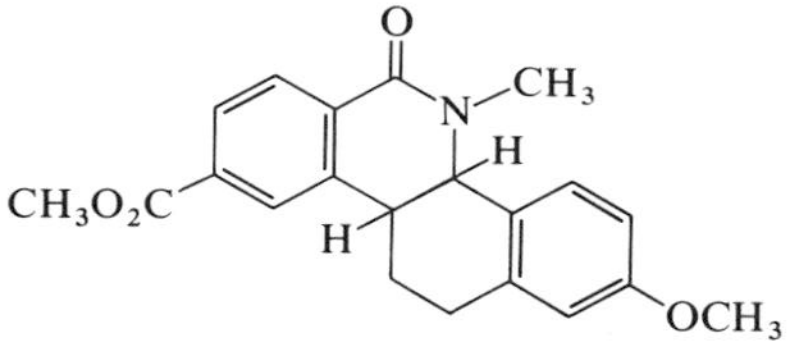

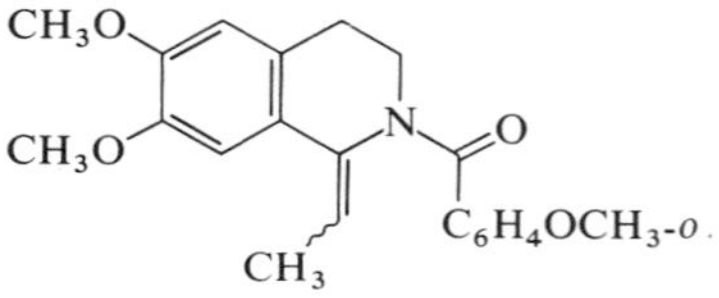

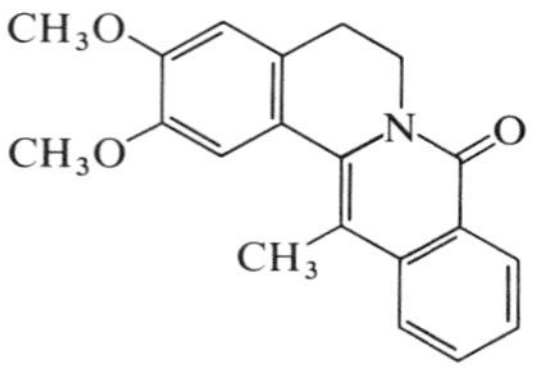

Starting material	Conditions	Product (yield, %)	Ref.
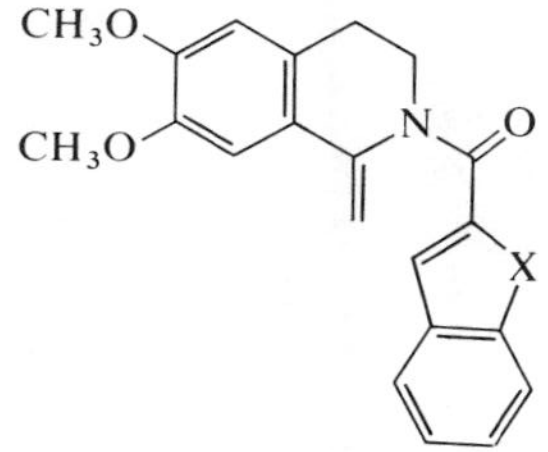 (I)		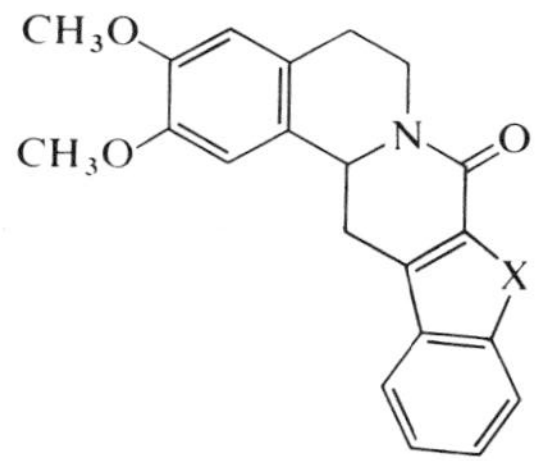 (II)	
I, X = O	C_6H_6, argon	II, X = O (53)	671
I, X = S	Dioxane, degassed	II, X = S (22)	671
	C_6H_6, O_2, N_2	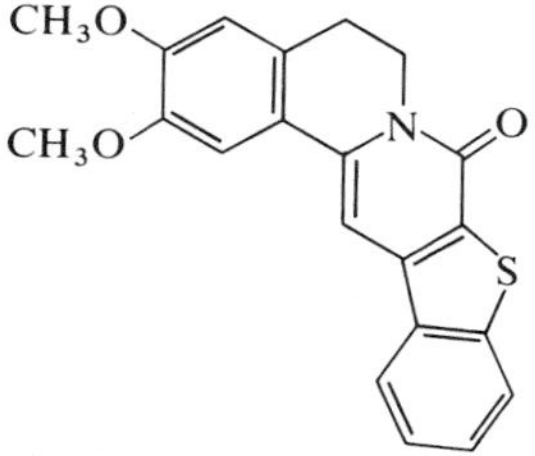 (33)	671
2′-Benzoyloxy-3,4-methylenedioxy-**R**	C_6H_6, MeOH, I_2	4-Benzoyloxy-8,9-methylenedioxy-**P** (4)	660a
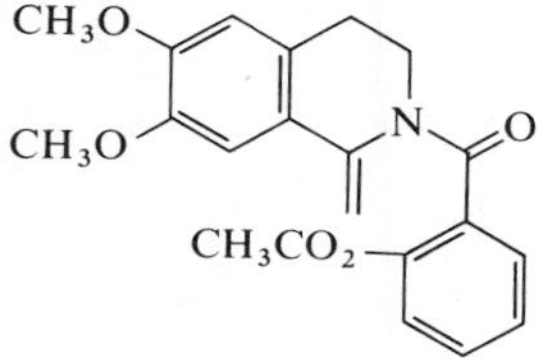	*t*-BuOH, degassed	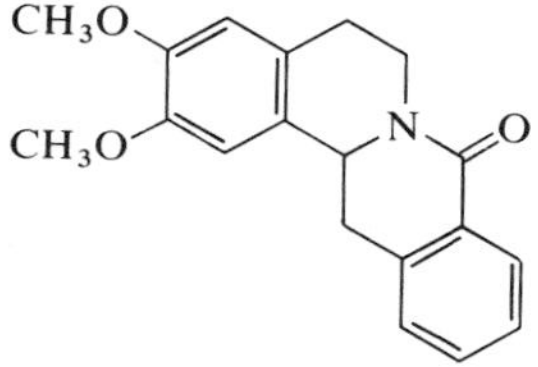 (80)	80,356

TABLE VII. LACTAMS AND OTHER HETEROCYCLIC SYSTEMS FROM AMIDES (*Continued*)[a]

Starting Material	Reaction Conditions	Products and Yields (%)	Refs.
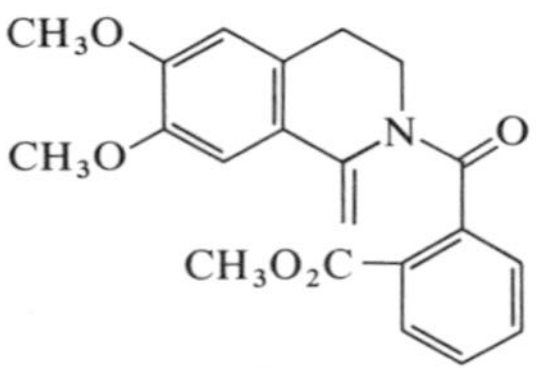 CH_3O, CH_3O, CH_3O_2C (structure)	MeOH, air	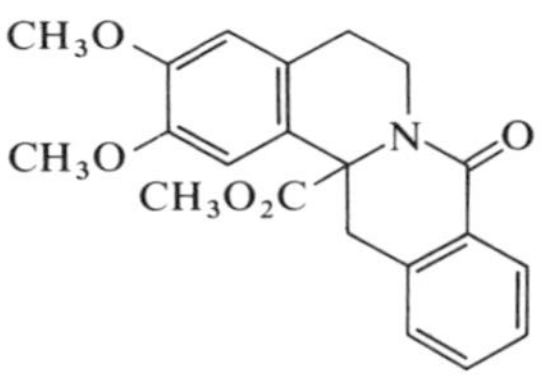 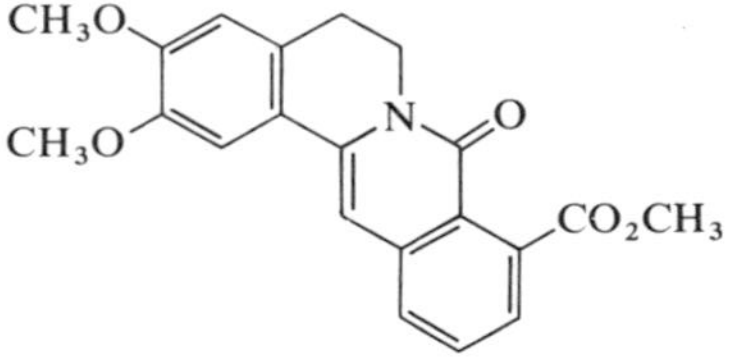 CH_3O, CH_3O, CH_3O_2C (structure) (8), CH_3O, CH_3O, CO_2CH_3 (structure) (5)	363
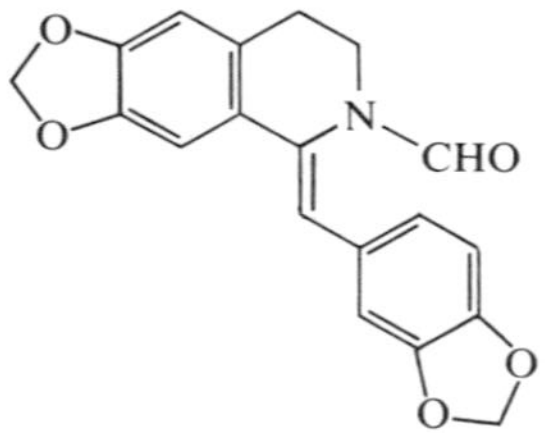 N–CHO (structure)	Dioxane, *t*-BuOH, 47% HI, N_2	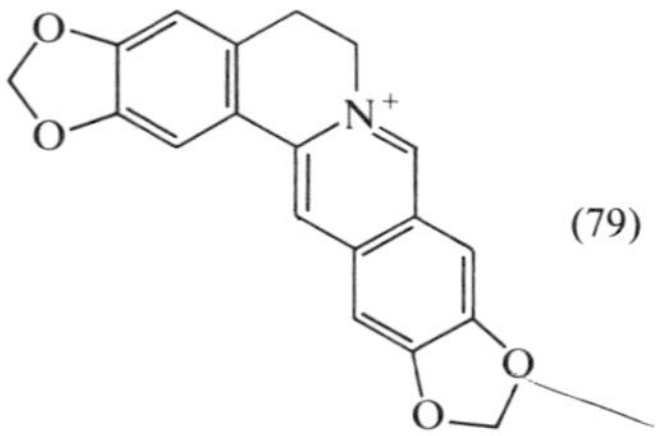 N^+ (structure) (79)	373
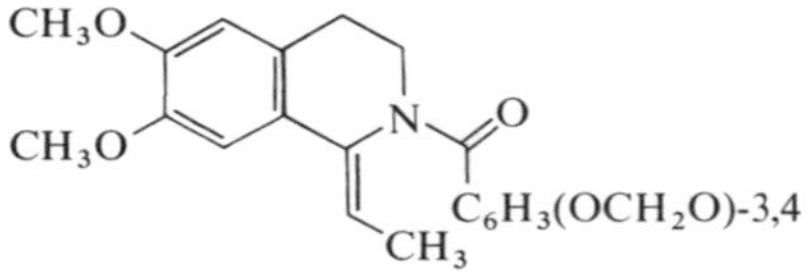 CH_3O, CH_3O, $C_6H_3(OCH_2O)$-3,4, CH_3 (structure)	Dioxane, degassed	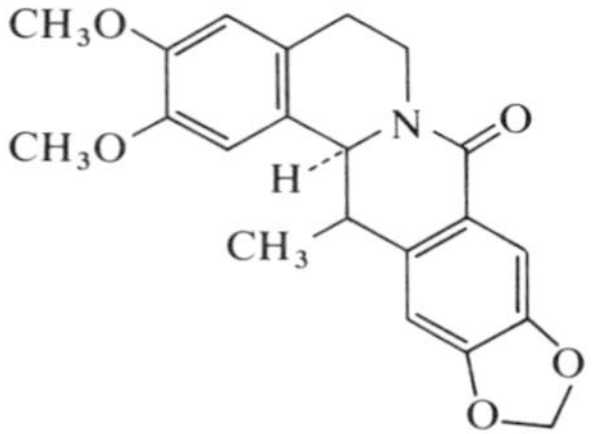 CH_3O, CH_3O, H, CH_3 (structure) (98)	30a

Reactant	Conditions	Product(s) and Yield(s) (%)	Refs.
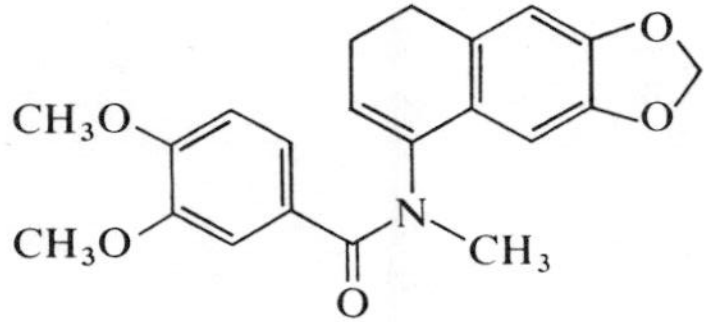 [structure]	Solvent	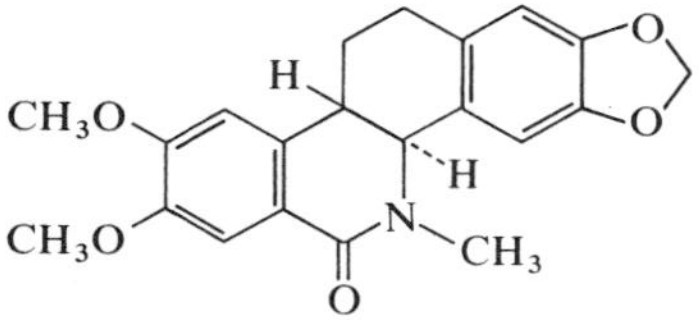 [structure] (53)	350,365
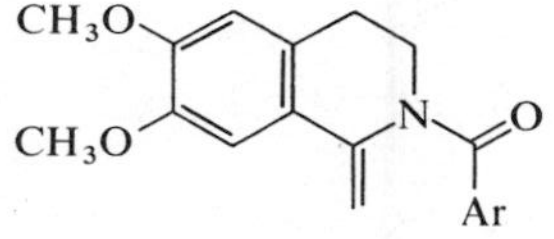 (I)		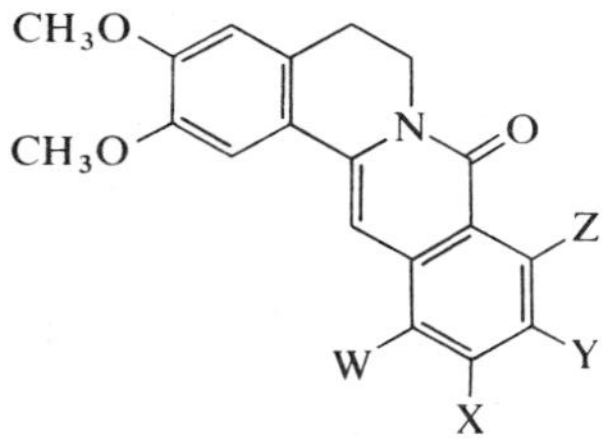 (II)	
I, Ar = 2,3-Dimethoxyphenyl	C_6H_6, N_2	II, W = CH_3O, X = Y = Z = H (85)	80,356
I, Ar = 2,4-Dimethoxyphenyl	EtOAc, argon	II, X = CH_3O, W = Y = Z = H (35)	373
I, Ar = 2,6-Dimethoxyphenyl	MeOH	II, Z = CH_3O, W = X = Y = H (56)	363
I, Ar = 3,4-Dimethoxyphenyl	MeOH, air	II, X = Y = CH_3O, W = Z = H (40), II, W = X = CH_3O, Y = Z = H (5)	363,364
	t-BuOH, degassed	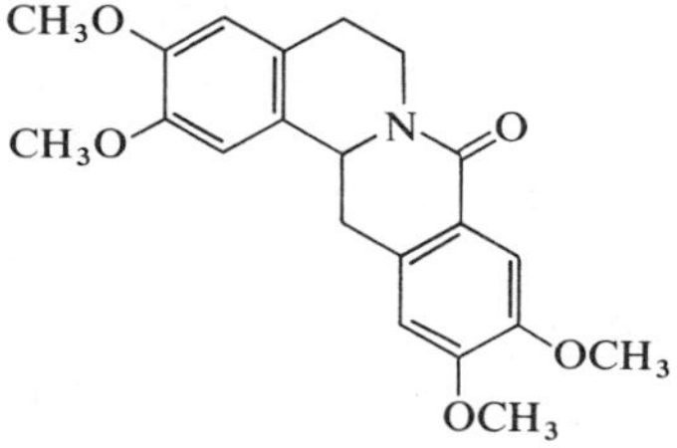 [structure] (94)	81

TABLE VII. LACTAMS AND OTHER HETEROCYCLIC SYSTEMS FROM AMIDES (*Continued*)[a]

Starting Material	Reaction Conditions	Products and Yields (%)	Refs.
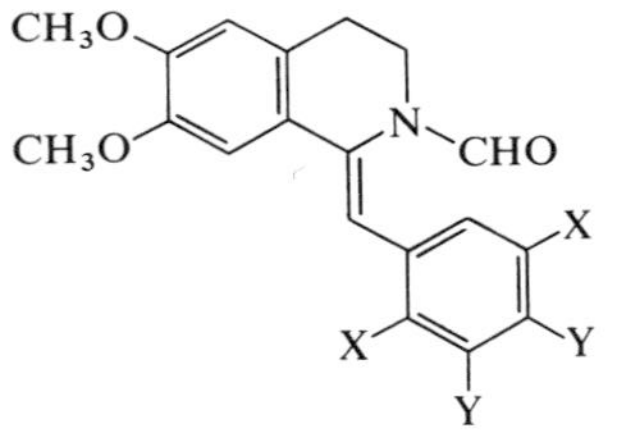 (I)		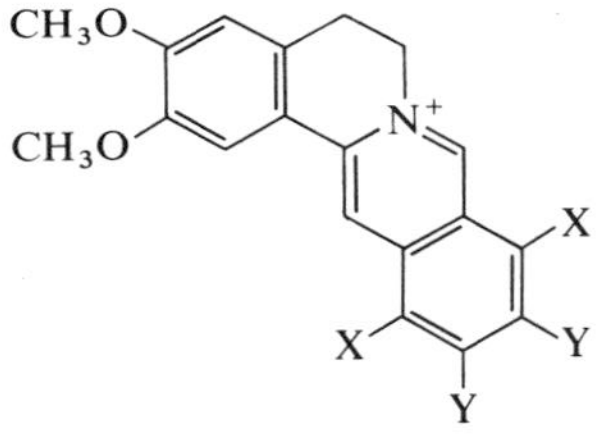 (II)	
I, X = CH_3O, Y = H	Dioxane, *t*-BuOH, 47% HI, N_2	II, X = CH_3O, Y = H (98)	373
I, X = H, Y = CH_3O	"	II, X = H, Y = CH_3O (75)	373
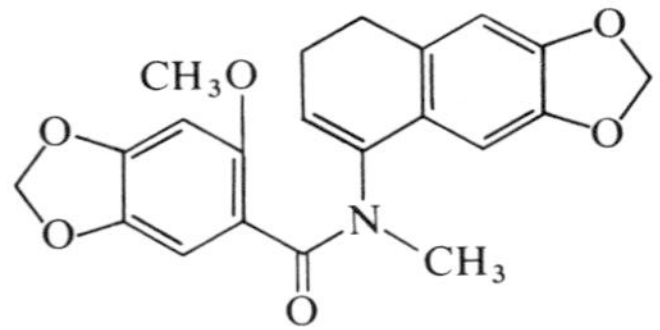	Et_2O	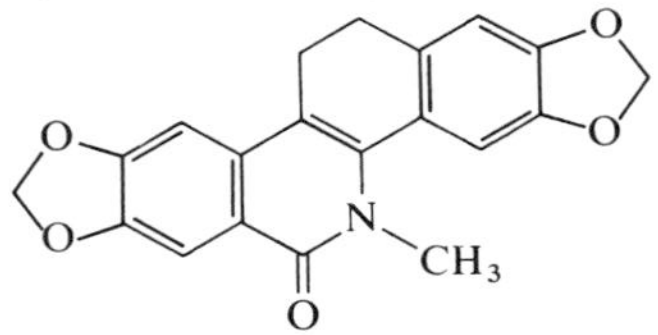 (41–45)	350,365
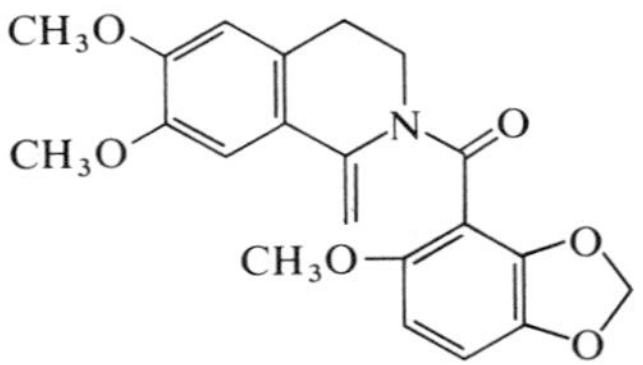	MeOH	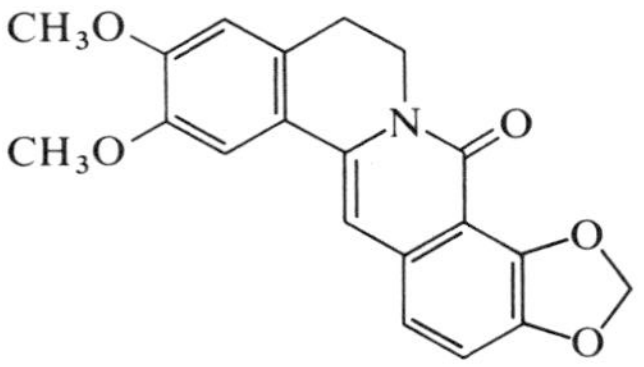 (44), 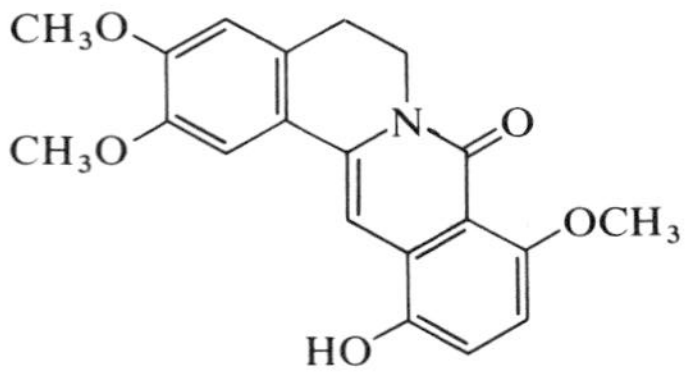 (21)	357,363

Carbon	Reactant	Conditions	Product (Yield, %)	Refs.
		Dioxane, t-BuOH, 47% HI, N_2	(55)	373
C_{22}		MeOH	(60)	675
		C_6H_6	" + (1:1)	675
		C_6H_6, MeOH, Et_3N	(36)	678b

TABLE VII. LACTAMS AND OTHER HETEROCYCLIC SYSTEMS FROM AMIDES (*Continued*)[a]

Starting Material	Reaction Conditions	Products and Yields (%)	Refs.
$CH_2C_6H_5$, N, O, CO, CH_3	MeOH	$CH_2C_6H_5$, N, O, $COCH_3$, H (45)	368
O_2CCH_3, O, N, $CH_2C_6H_5$	Et_2O	O_2CCH_3, O, N, $CH_2C_6H_5$, H, H (13),	103
		O, N, $CH_2C_6H_5$ (7)	
$CH_2C_6H_5$, N, O, CH_3O_2C	MeOH	$CH_2C_6H_5$, N, O, CO_2CH_3, H (71)	368

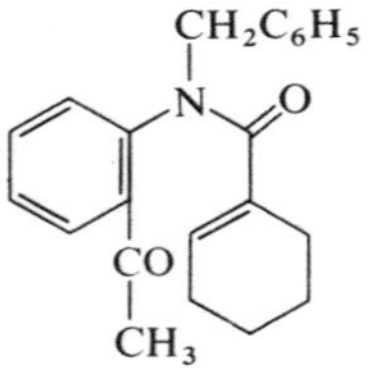

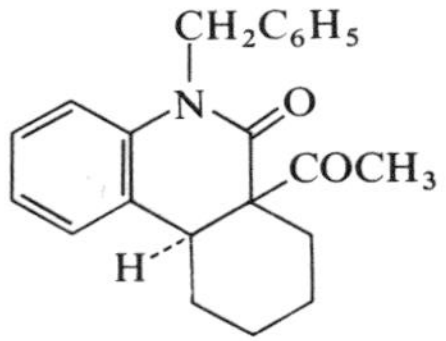

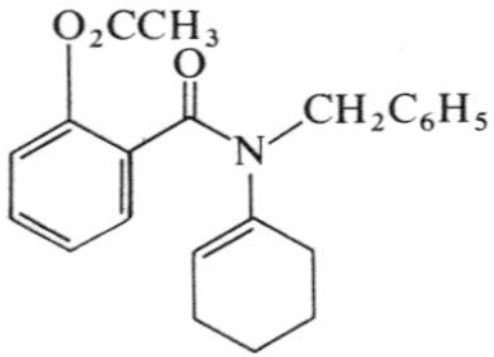

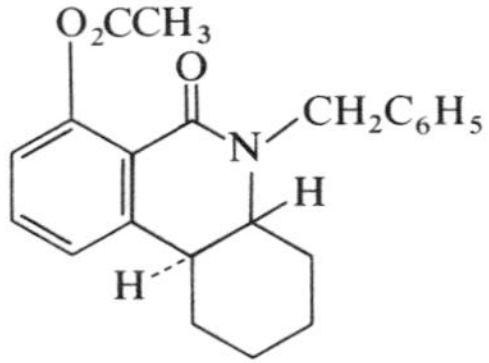

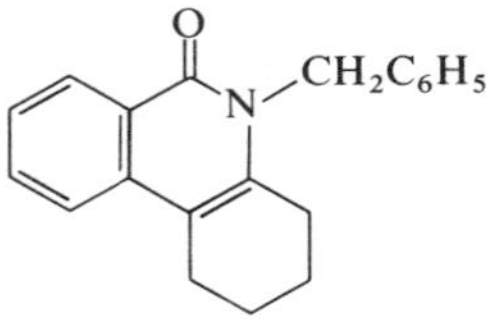

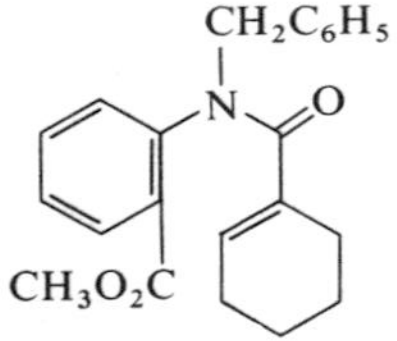

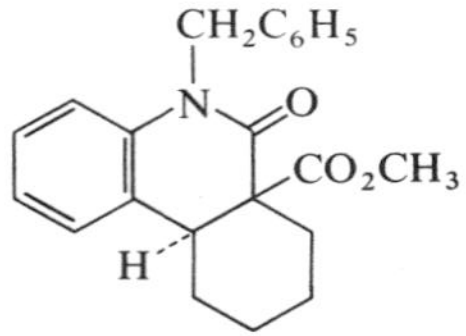

Reactant	Conditions	Product(s) and Yield(s) (%)	Refs.
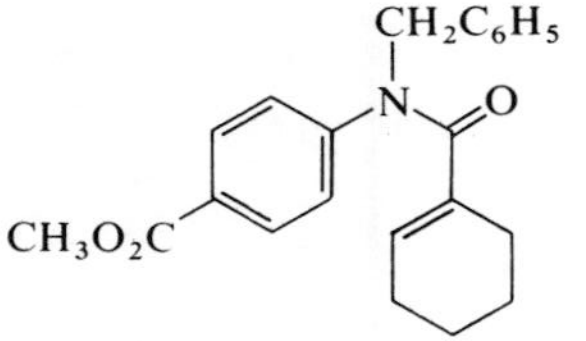 CH_3O_2C, $CH_2C_6H_5$, N, O	Et_2O, C_6H_6, or MeOH	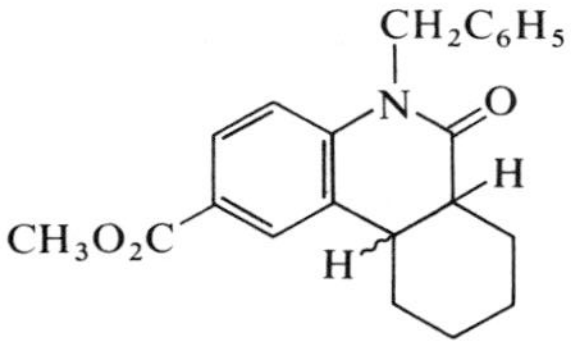 $CH_2C_6H_5$, N, O, H, CH_3O_2C, H (50–60)	368
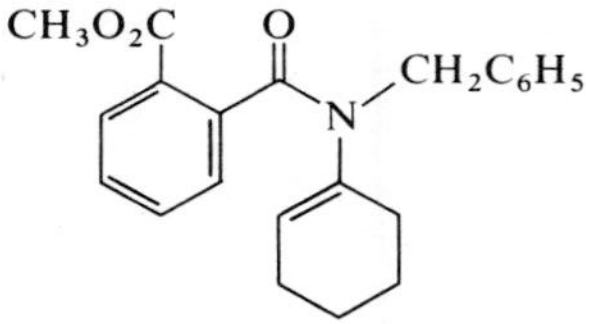 CH_3O_2C, O, N, $CH_2C_6H_5$	C_6H_6, air 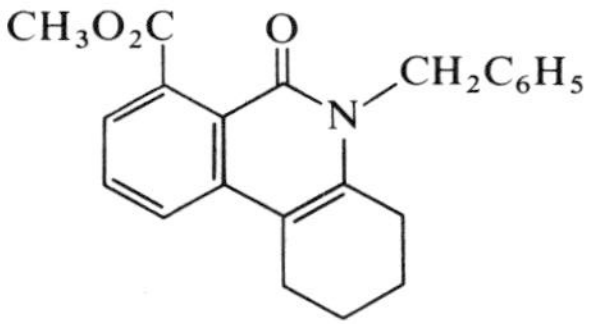	CH_3O_2C, O, N, $CH_2C_6H_5$ (3),	103
		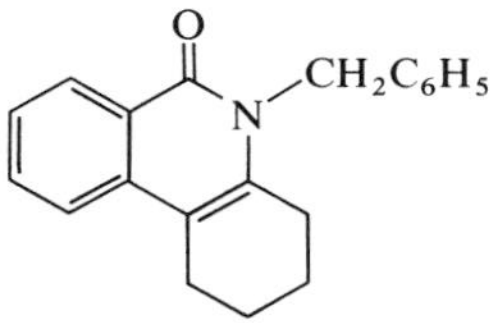 O, N, $CH_2C_6H_5$ (2)	
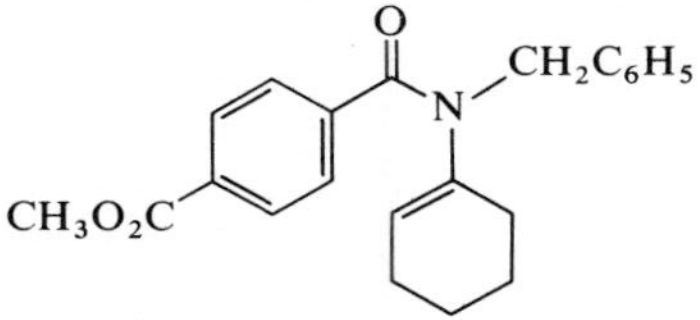 O, N, $CH_2C_6H_5$, CH_3O_2C	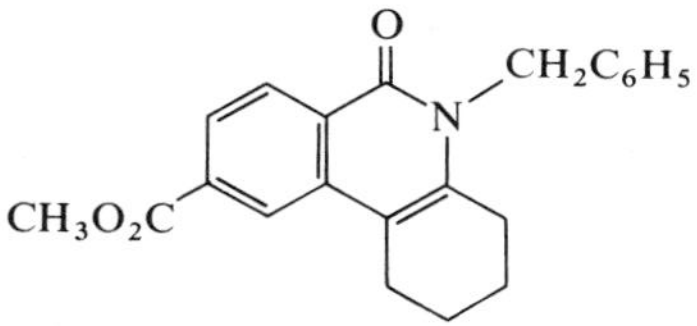 MeOH, air	O, N, $CH_2C_6H_5$, CH_3O_2C (15)	103
	MeOH, N_2 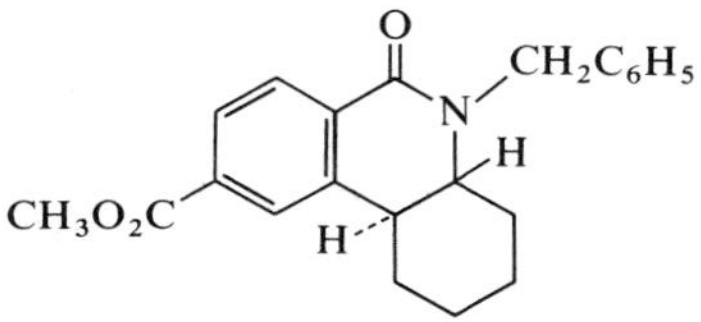	O, N, $CH_2C_6H_5$, H, CH_3O_2C, H (10)	103

TABLE VII. LACTAMS AND OTHER HETEROCYCLIC SYSTEMS FROM AMIDES (*Continued*)[a]

Starting Material	Reaction Conditions	Products and Yields (%)	Refs.
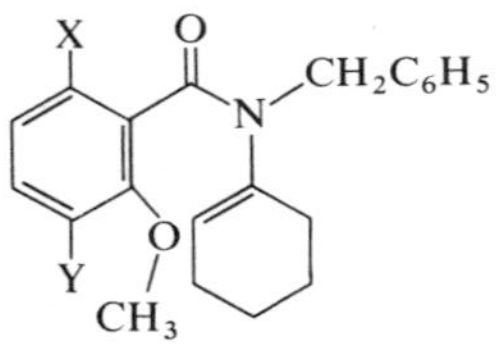 (I)		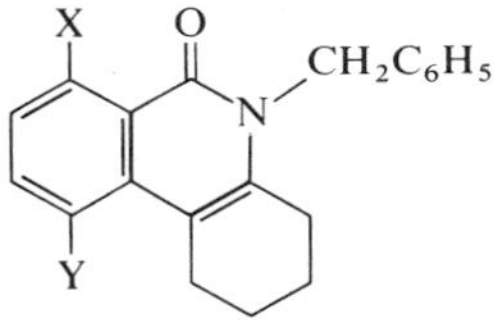 (II)	
I, X = CH_3O, Y = H	Et_2O	II, X = CH_3O, Y = H (10)	103
I, X = H, Y = CH_3O	MeOH	II, X = H, Y = CH_3O (15)	103
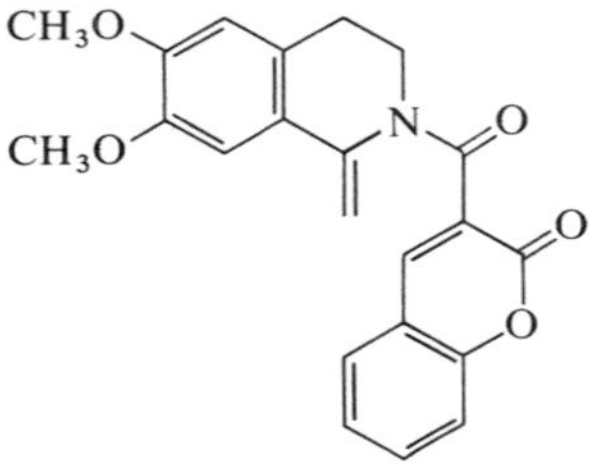	EtOAc, argon	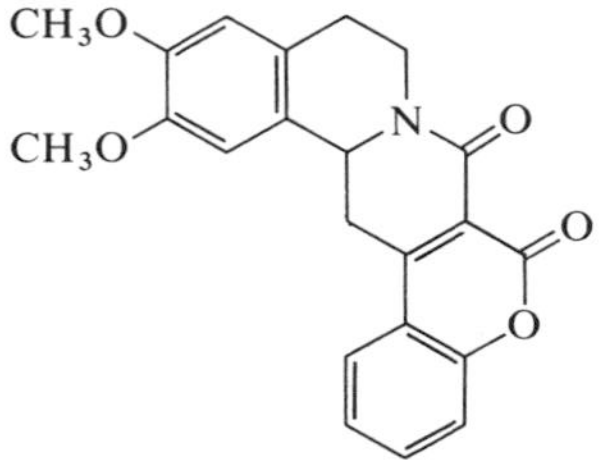 (65)	671
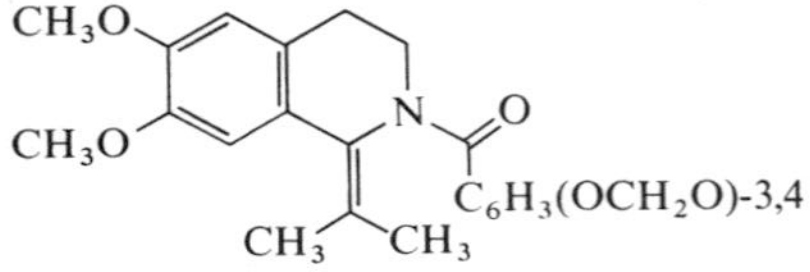 $C_6H_3(OCH_2O)$-3,4	PhMe, argon	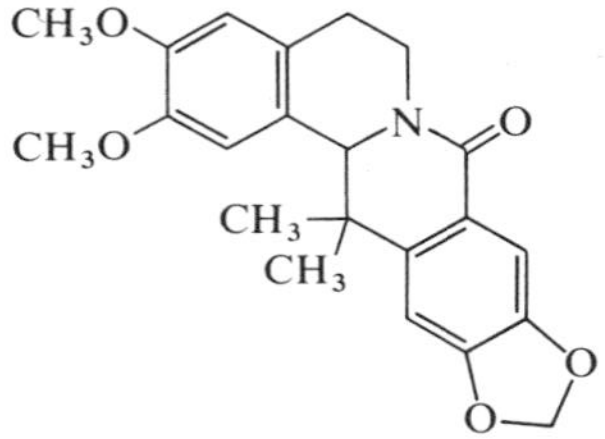 (97)	30a

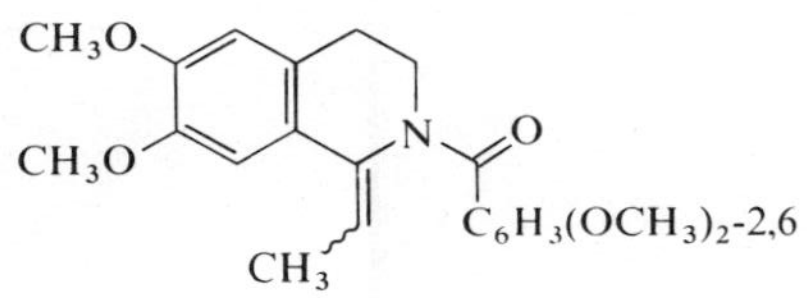

Reactant	Conditions	Product (Yield, %)		Ref.
CH_3O, CH_3O, N, O, $C_6H_3(OCH_3)_2$-2,6, CH_3	MeOH	CH_3O, CH_3O, N, O, CH_3, OCH_3	(34)	358
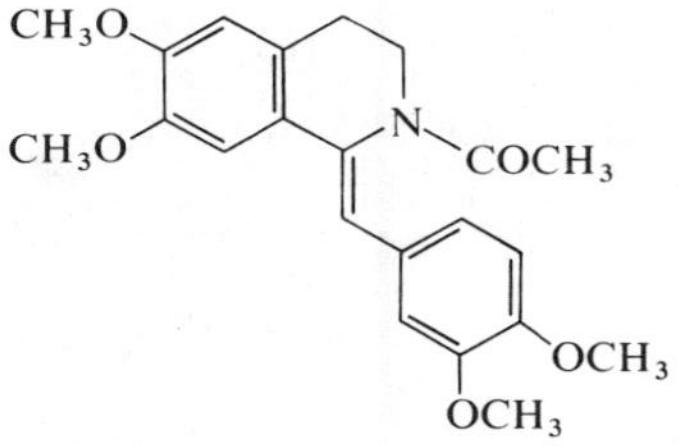 CH_3O, CH_3O, N, $COCH_3$, OCH_3, OCH_3	MeOH, THF, I_2, HI	CH_3O, CH_3O, N^+, CH_3, OCH_3, OCH_3	(75)	374
CH_3O, CH_3O, N, O, OCH_3, O_2CCH_3	EtOAc, air	CH_3O, CH_3O, N, O, OCH_3, O_2CCH_3	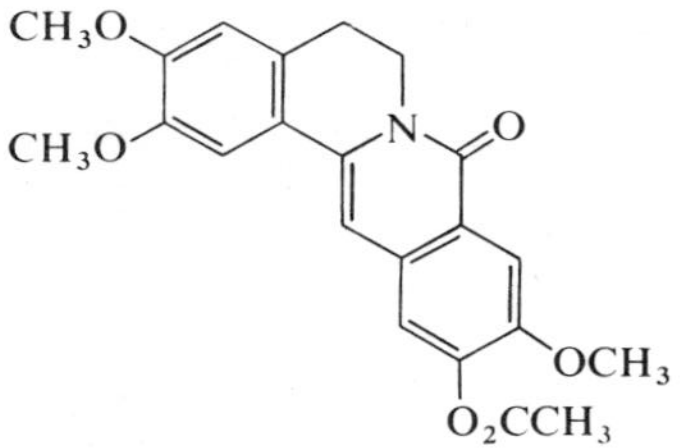 (62)	81
	t-BuOH, degassed	CH_3O, CH_3O, N, O, OCH_3, O_2CCH_3	(44)	81

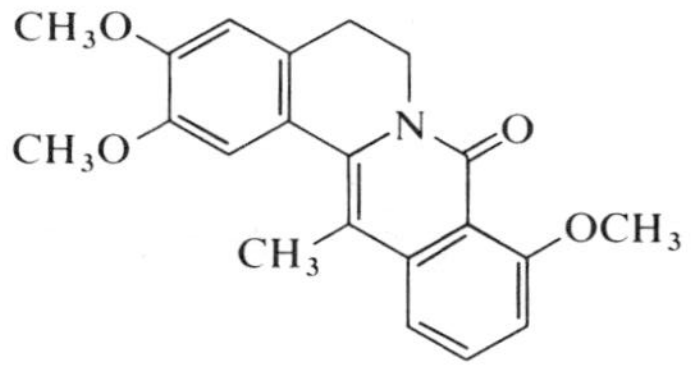

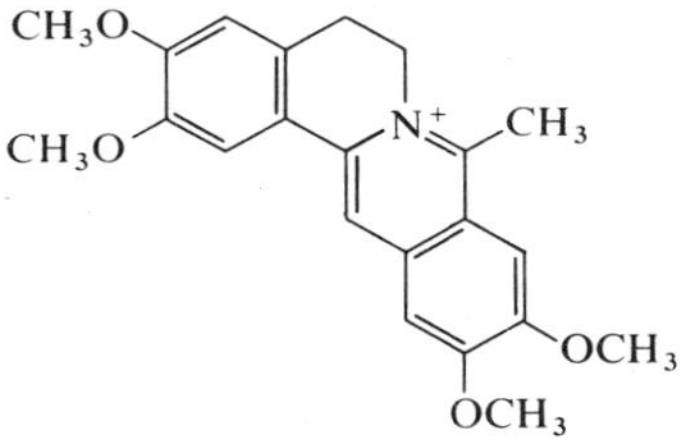

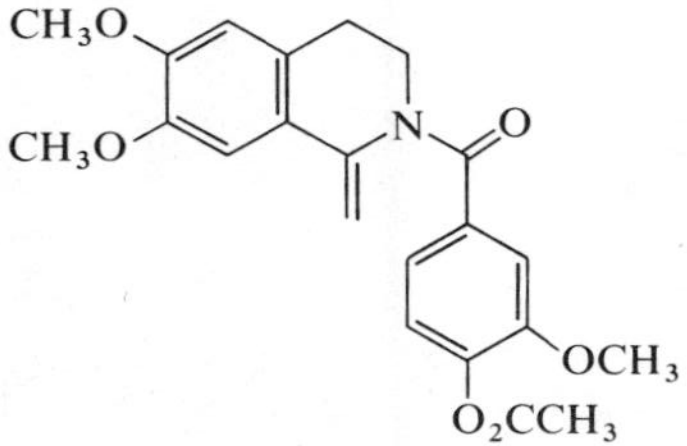

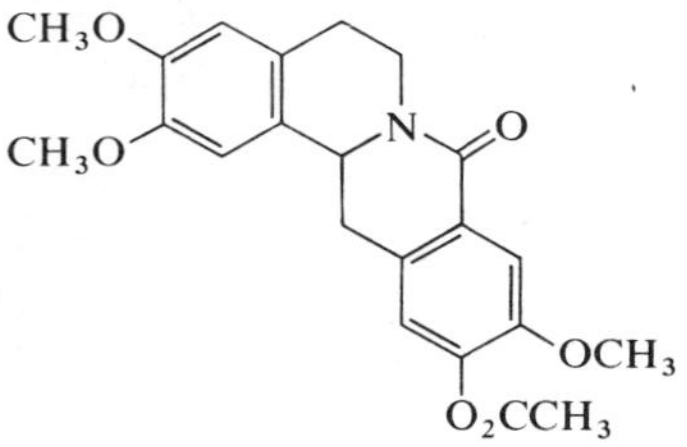

TABLE VII. LACTAMS AND OTHER HETEROCYCLIC SYSTEMS FROM AMIDES (*Continued*)[a]

Starting Material	Reaction Conditions	Products and Yields (%)	Refs.
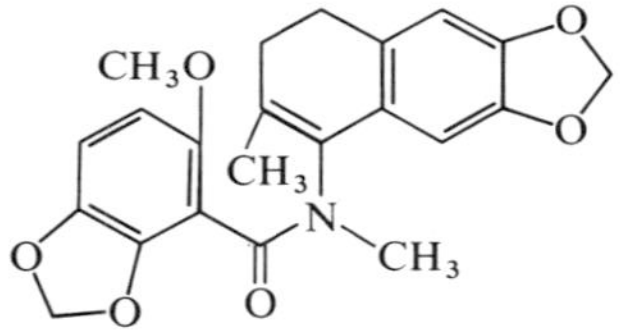	MeOH	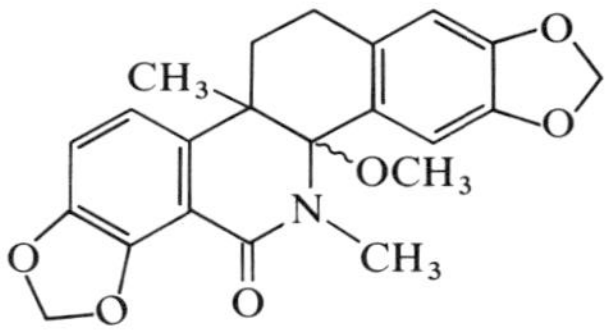 (20), 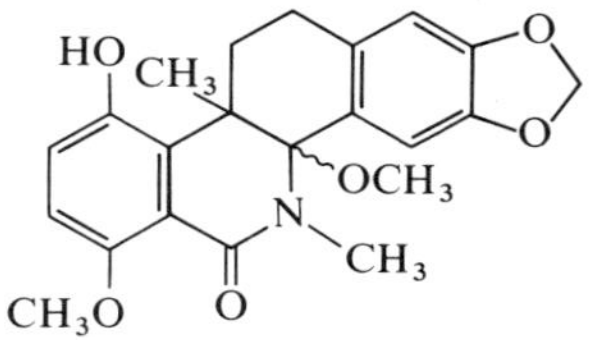 (10)	679,680
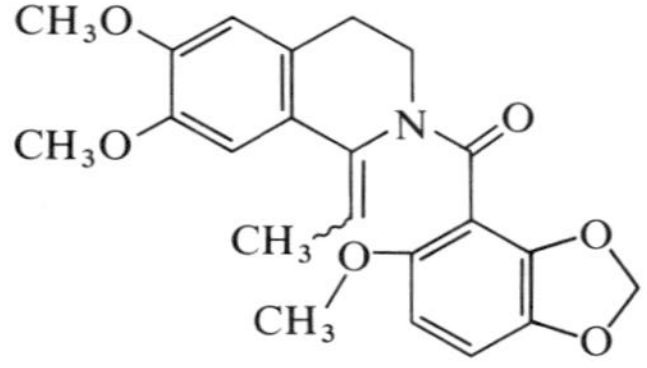	"	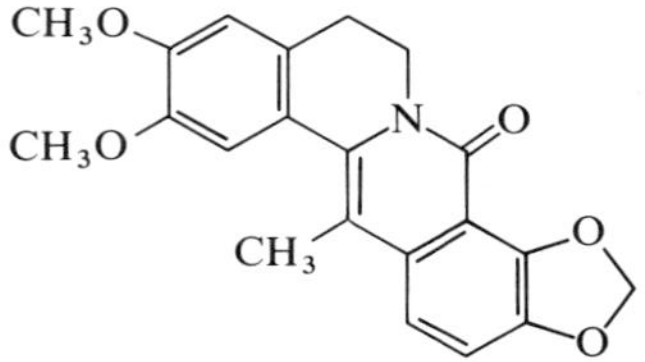 (41), 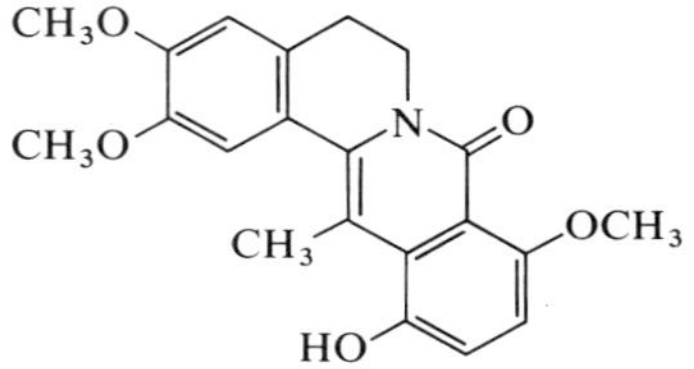 (29)	357,358

	Et_2O		(50)	365
	t-BuOH, O_2 (or I_2)		(—)	356
	t-BuOH, degassed		(90)	81,356
	"		(85)	80,356

TABLE VII. LACTAMS AND OTHER HETEROCYCLIC SYSTEMS FROM AMIDES (*Continued*)[a]

Starting Material	Reaction Conditions	Products and Yields (%)	Refs.
CH_3O, CH_3O, N, O, CH_3O, OCH_3, OCH_3	MeOH	CH_3O, CH_3O, N, O, OCH_3, OCH_3 (20),	350,363
		CH_3O, CH_3O, N, O, OCH_3, CH_3O (8)	
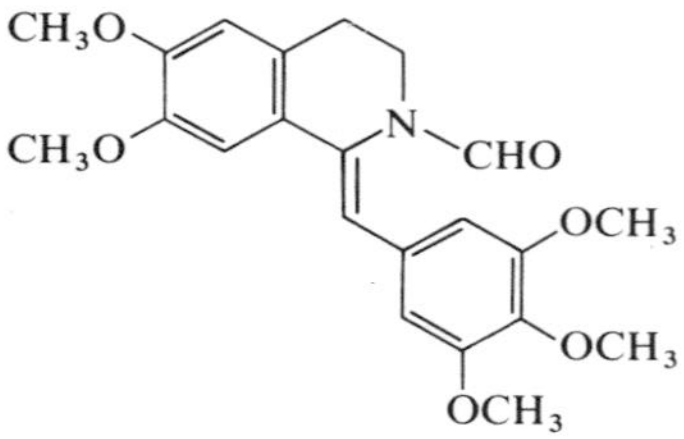 CH_3O, CH_3O, N, CHO, OCH_3, OCH_3, OCH_3	Dioxane, *t*-BuOH, 47% HI, N_2	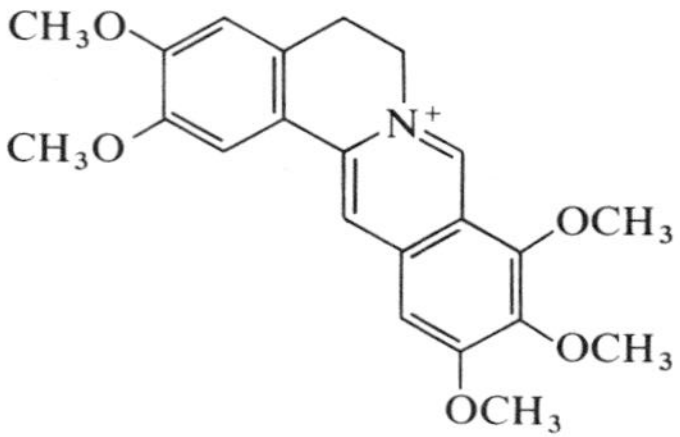 CH_3O, CH_3O, N^+, OCH_3, OCH_3, OCH_3 (75)	373

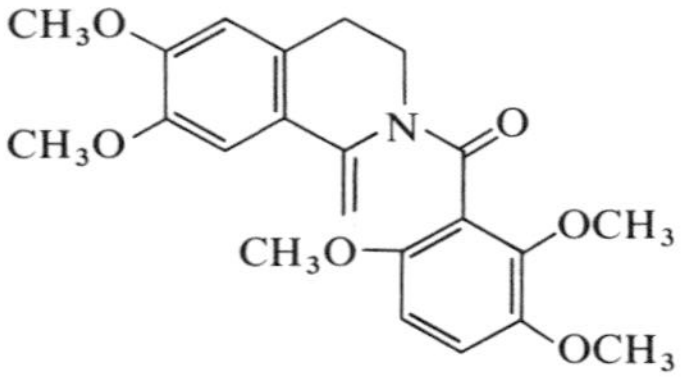

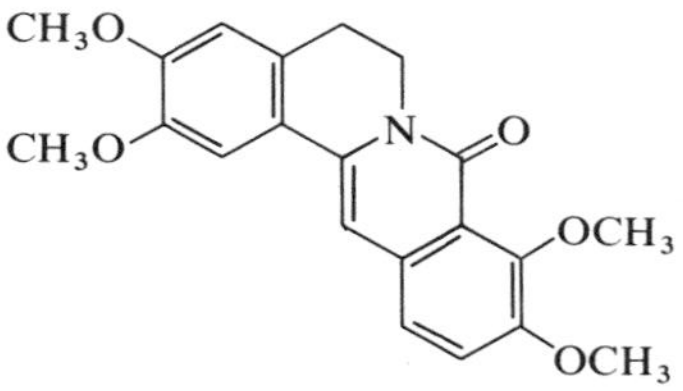

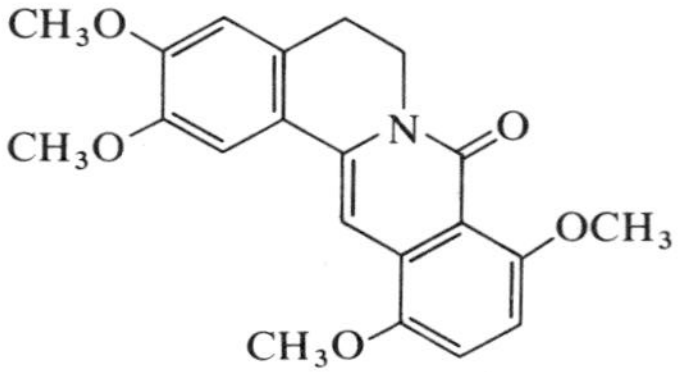

		MeOH		(38)	681,682
C_{23}		Solvent, air		$(-)^{h}$	109
		MeOH		(40)	79
		EtOAc, N_2		(69)	80

TABLE VII. LACTAMS AND OTHER HETEROCYCLIC SYSTEMS FROM AMIDES (*Continued*)[a]

Starting Material	Reaction Conditions	Products and Yields (%)	Refs.
C_{24} 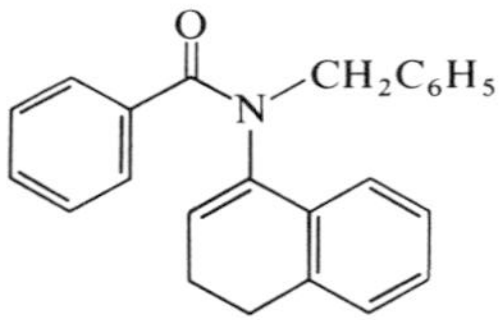	MeOH, I_2	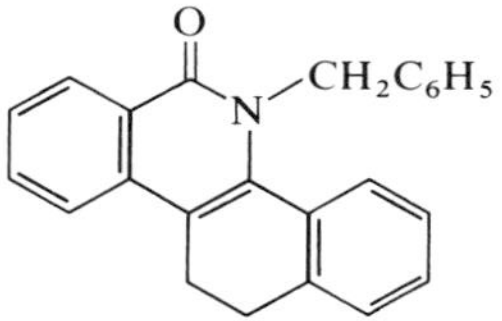 (good)	350
	MeOH	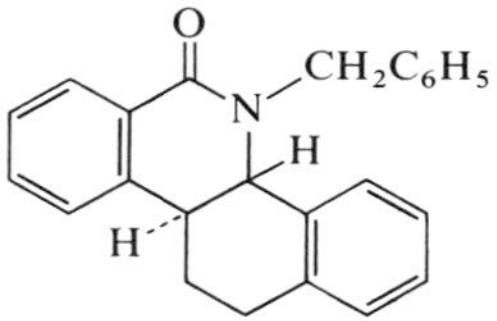 (55)	350,359,360
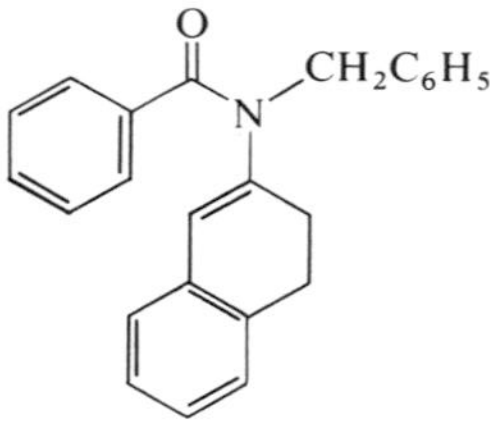	Et_2O	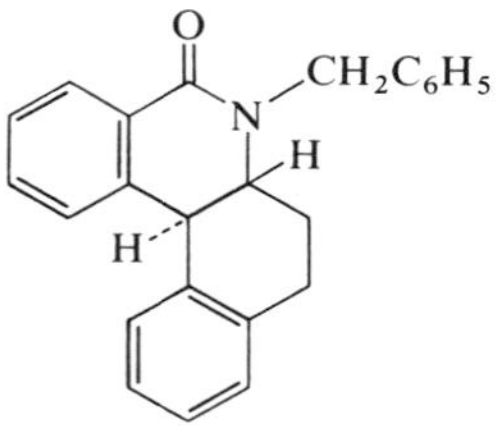 (47)	361,362
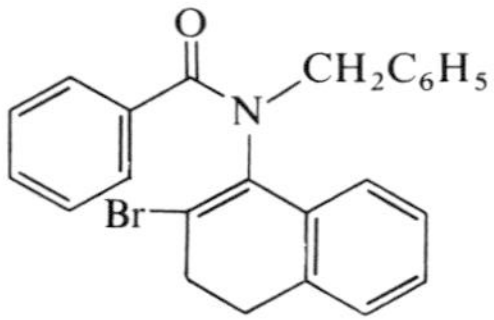	MeOH	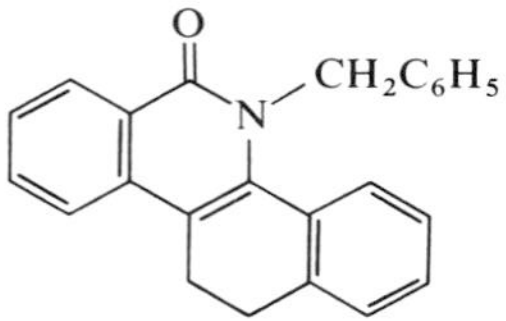 (28)	360

Reactant	Conditions	Product (Yield, %)		Refs.
CH_3O; O; N; $CH_2C_6H_5$	”	O; N; $CH_2C_6H_5$; H; CH_3O; H	(30)	103
CH_3O; CH_3O; N; O; C_6H_5	*t*-BuOH, degassed	CH_3O; CH_3O; N; O; C_6H_5	(76)	81
CH_3O; CH_3O; N; O; C_6H_5	EtOH, $Cu(OAc)_2$, I_2	CH_3O; CH_3O; N; O; C_6H_5	(50)	375
O; O; O; N; $CH_2C_6H_5$; $CH_2{=}CHCH_2$	MeOH	O; O; O; N; $CH_2C_6H_5$; H; $CH_2{=}CHCH_2$	(20)	678a,683

TABLE VII. Lactams and Other Heterocyclic Systems from Amides (*Continued*)[a]

	Starting Material	Reaction Conditions	Products and Yields (%)	Refs.
C_{25} 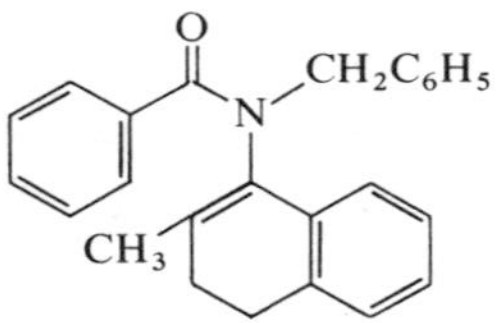	O, N, $CH_2C_6H_5$, CH_3	MeOH	O, N, $CH_2C_6H_5$, H, CH_3 (20)	360
	O, N, $CH_2C_6H_5$, NC	"	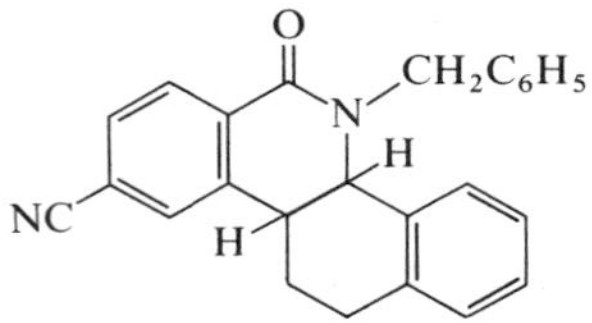 O, N, $CH_2C_6H_5$, H, NC, H (14)	103
C_{27} 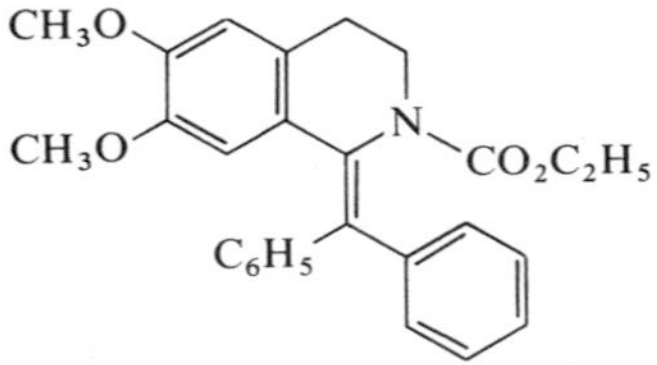	CH_3O, CH_3O, N, $CO_2C_2H_5$, C_6H_5	EtOH	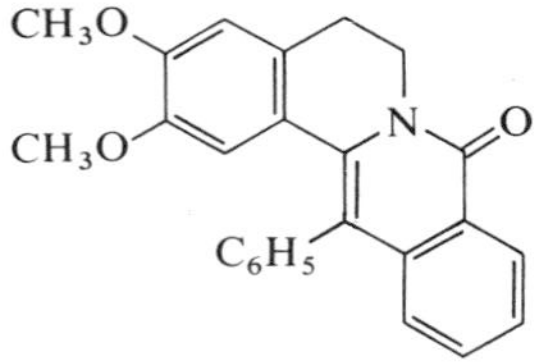 CH_3O, CH_3O, N, O, C_6H_5 (60)	375
C_{28} 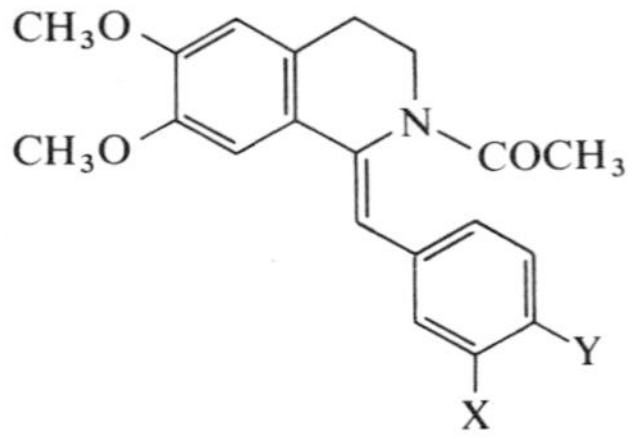	CH_3O, CH_3O, N, $COCH_3$, Y, X (I)		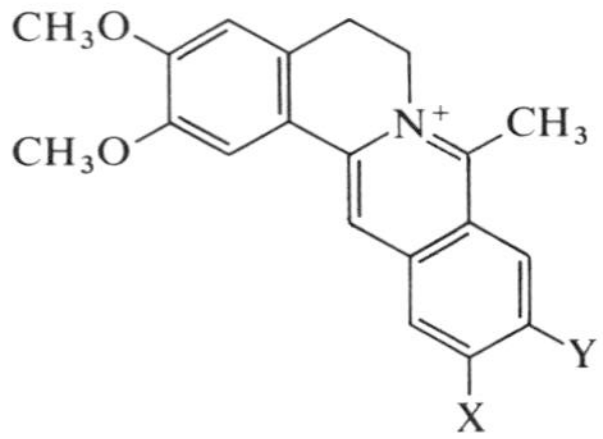 CH_3O, CH_3O, N^+, CH_3, Y, X (II)	

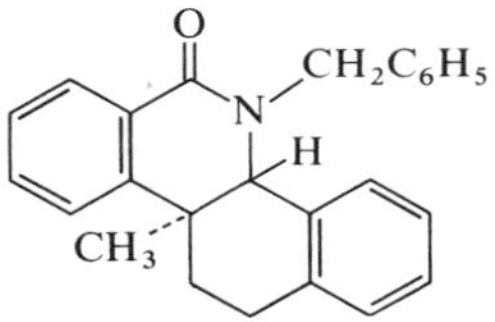

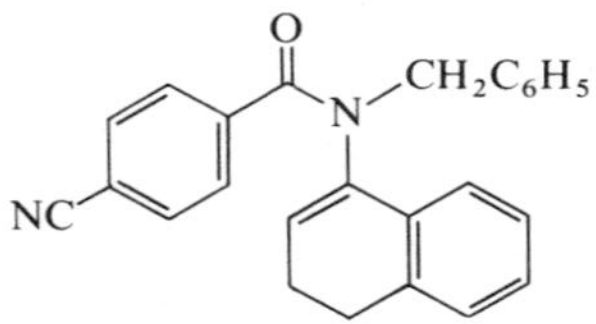

	Reactant	Conditions	Product(s) (Yield, %)	Refs.
	I, X = $C_6H_5CH_2O$, Y = CH_3O	MeOH, dioxane, HI, N_2	II, X = $C_6H_5CH_2O$, Y = CH_3O (56)	684
	I, X = CH_3O, Y = $C_6H_5CH_2O$	"	II, X = CH_3O, Y = $C_6H_5CH_2O$ (56)	684
C_{30}	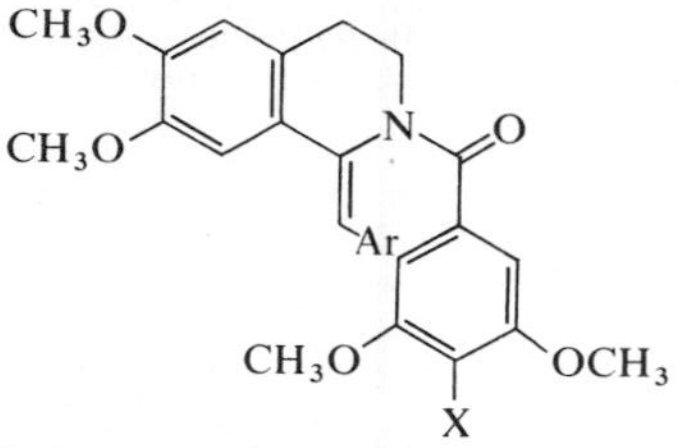 (I)		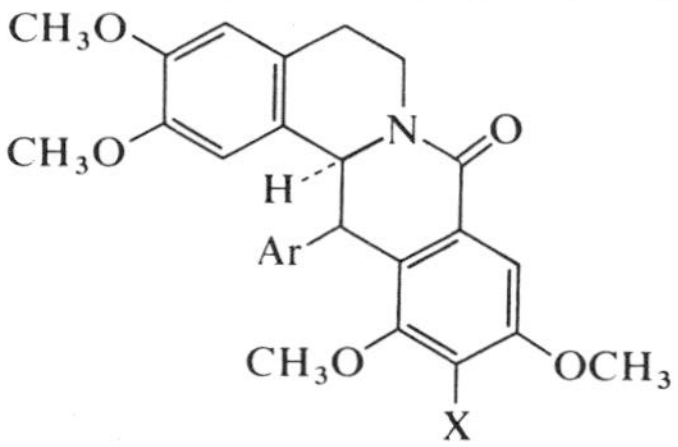 (II)	
	I, Ar = 3,4-Dimethoxyphenyl, X = CH_3O	Dioxane, degassed	II, Ar = 3,4-Dimethoxyphenyl, X = CH_3O (88)	30a
C_{31}	I, Ar = 3,4-Dimethoxyphenyl, X = CH_3CO_2	Dioxane, argon	II, Ar = 3,4-Dimethoxyphenyl, X = CH_3CO_2 (84)	30a

[a] The following symbols are used as abbreviations in Table VII: **P** = 6(5*H*)-phenanthridinone and **R** = *N*-phenylbenzamide. The numbering schemes for **P** and **R** are shown under the first entry for C_{13}.

[b] The *trans*:*cis* product ratio was 4.0 in Et_2O, 0.9 in C_6H_6, and 0.3 in MeOH.

[c] The pathway for dehydrogenation of the five-membered ring is uncertain.

[d] The identity of the oxidant in this reaction is obscure. One possibility is residual dissolved oxygen.

[e] The origin of the hydroxy group in the product is obscure.

[f] The *trans*:*cis* product ratio was 6.3 in Et_2O, 1.5 in C_6H_6, and 0.7 in MeOH.

[g] The *trans*:*cis* product ratio was 15.6 in Et_2O, 4.0 in Me_2SO, 1.6 in C_6H_6, and 0.4 in MeOH.

[h] The originally assigned structure was corrected in Ref. 375.

TABLE VIII. Photocyclizations Producing Five-Membered Rings

A. C_5 Ring Systems from 1-Vinylnaphthalenes and Related Molecules

	Starting Material	Reaction Conditions	Products and Yields (%)	Refs.
C_{13}	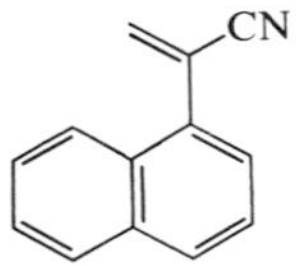 CN	C_6H_{12}, $H_2N(CH_2)_3NH_2$	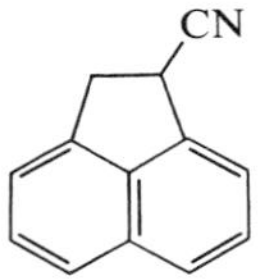 CN (95)	377
C_{16}	1-(*o*-Chlorophenyl)naphthalene	C_6H_6, N_2	Fluoranthene (72)	381
	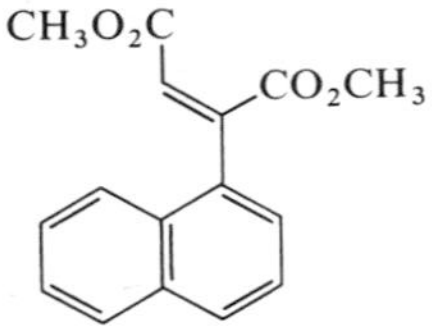 CH_3O_2C CO_2CH_3	MeOH, N_2	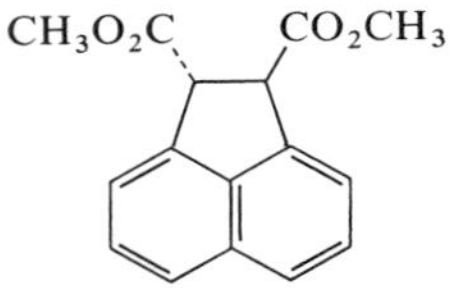 CH_3O_2C CO_2CH_3 (74)	378
	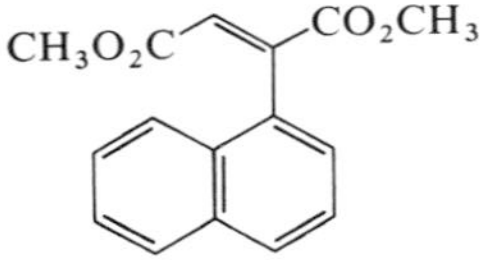 CH_3O_2C CO_2CH_3	"	" (59)	378
C_{17}	1-(*o*-Chlorophenyl)-5-methoxynaphthalene	MeOH	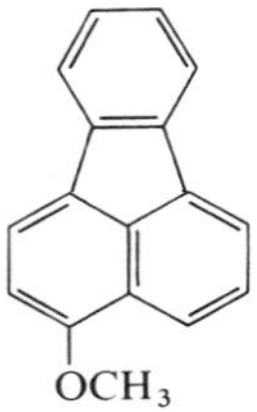 OCH_3 (56)	381

Substrate	Conditions	Product(s) and Yield(s) (%)	Refs.
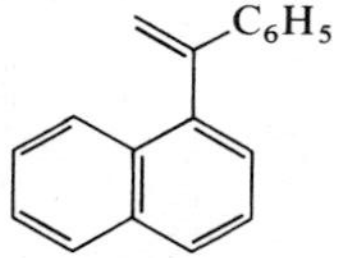 C_{18} — C_6H_5	C_6H_{12}, I_2, O_2	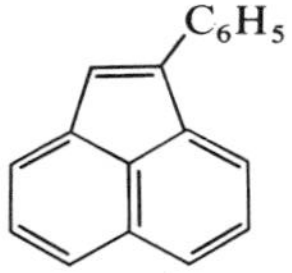 C_6H_5 (70)	73
	$CDCl_3$, I_2, O_2	" (—)	73
	MeCN, *n*-$PrNH_2$	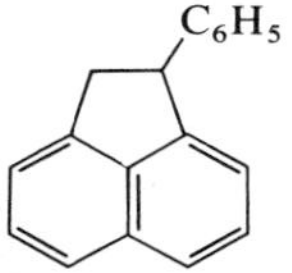 C_6H_5 (95)	377
	C_6H_{12}, *n*-$PrNH_2$	" (90)	377
	C_6H_{12}, Et_3N	" (80)	377
C_{19} — N, CO_2CH_3 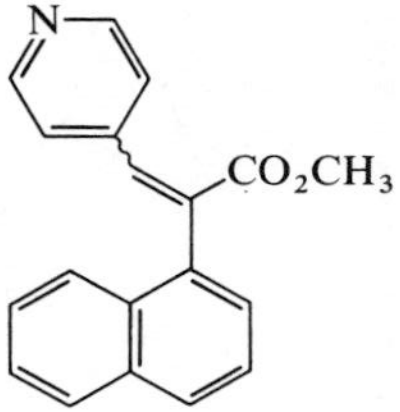 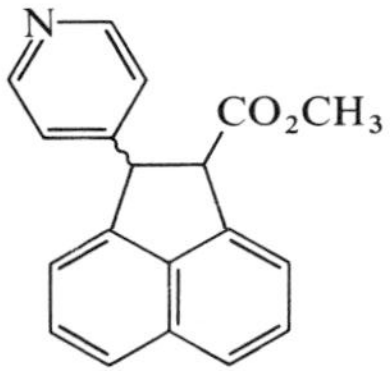 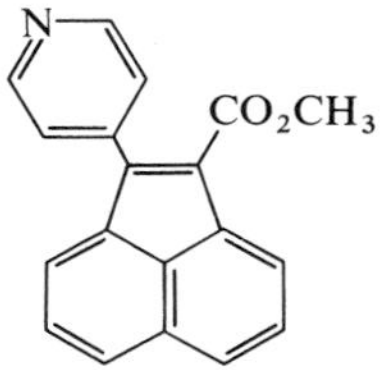	MeOH, air	N, CO_2CH_3 (56), N, CO_2CH_3 (19)	379
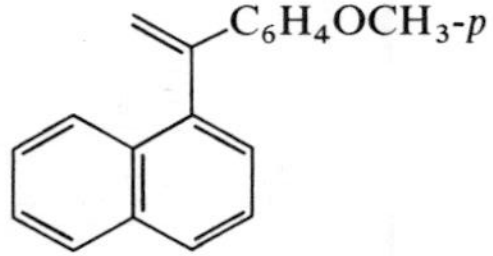 $C_6H_4OCH_3$-*p*	C_6H_{12}, I_2, O_2	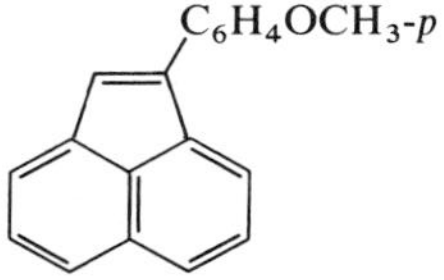 $C_6H_4OCH_3$-*p* (70)	73
C_6H_5, OCH_3	"	C_6H_5, OCH_3 (70)	73

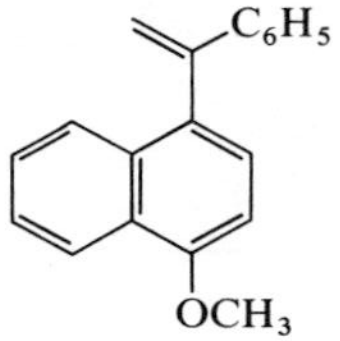

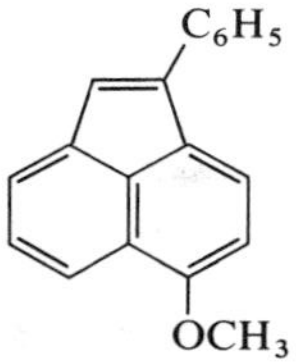

TABLE VIII. PHOTOCYCLIZATIONS PRODUCING FIVE-MEMBERED RINGS (*Continued*)

A. C_5 Ring Systems from 1-*Vinylnaphthalenes and Related Molecules* (*Continued*)

Starting Material	Reaction Conditions	Products and Yields (%)	Refs.
	C_6H_{12}, Et_3N	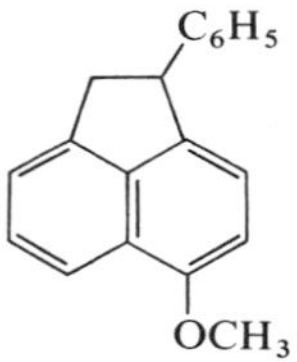 C_6H_5, OCH_3 (70)	377
C_{20} 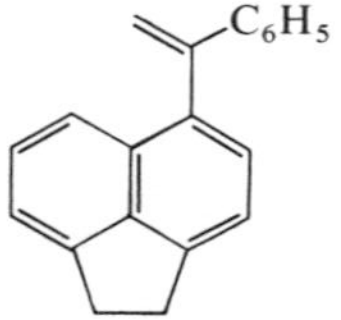 C_6H_5	"	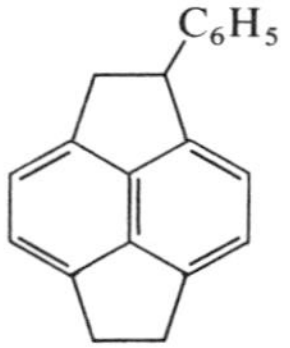 C_6H_5 (85)	377
C_{22} 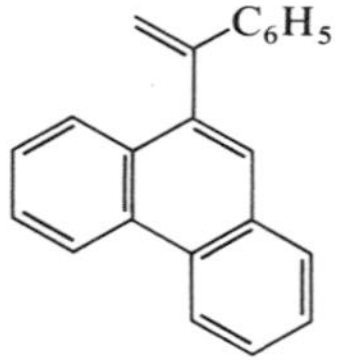 C_6H_5	C_6H_6, I_2	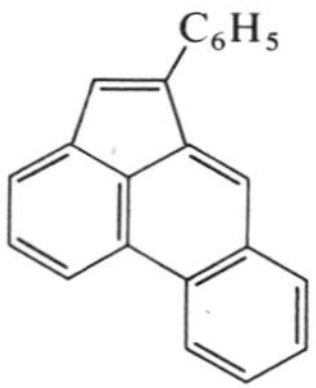 C_6H_5 (1)	122
C_{24} 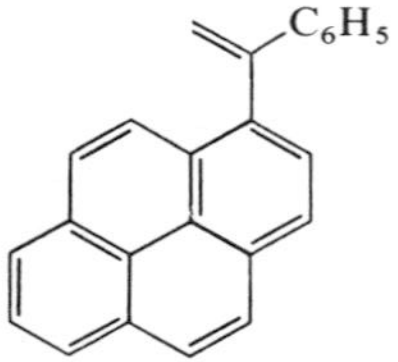 C_6H_5	C_6H_{12}, $H_2N(CH_2)_3NH_2$	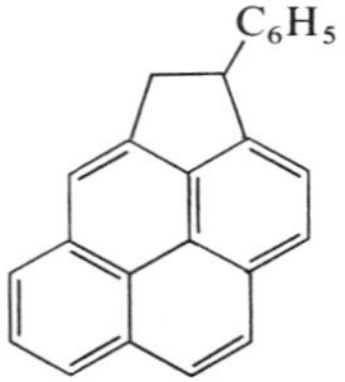 C_6H_5 (60)	377

	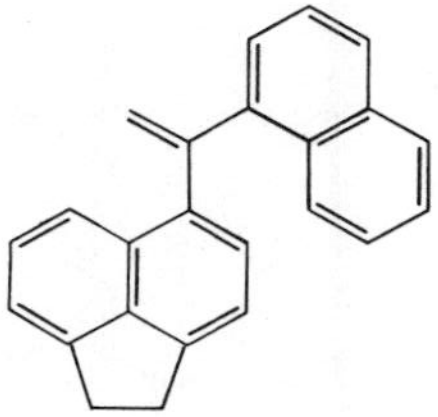	”	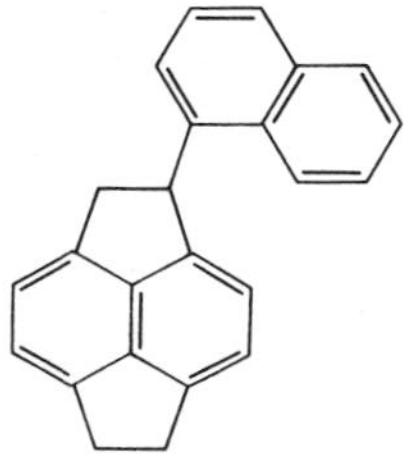 (85)	377
B. C_4N Ring Systems from Diarylamines and Aryl Vinyl Amines				
C_{10}	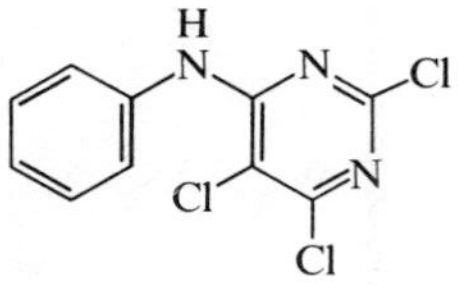	THF	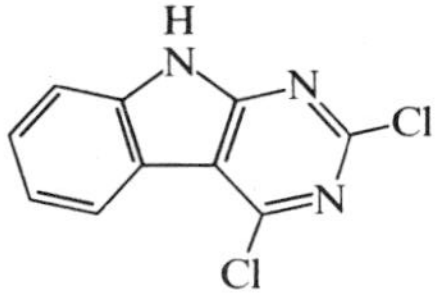 (17)	388b
	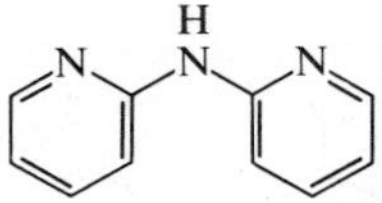	THF, N_2	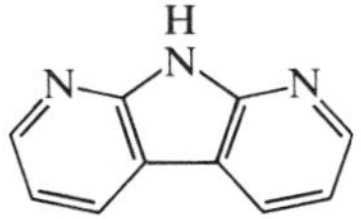 (53)	382
C_{11}	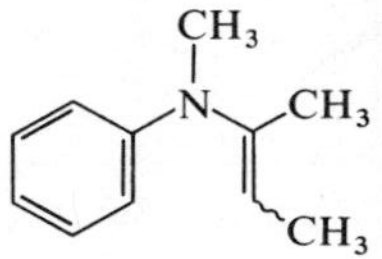	C_6H_6, O_2	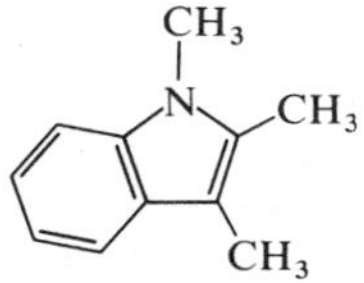 (20), *N*-methylaniline (60)	685
		C_6H_6, N_2	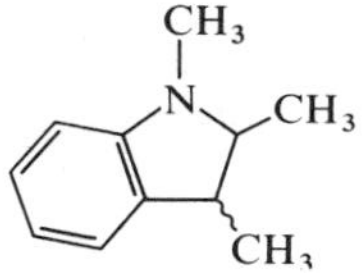 (*trans*, 11; *cis*, 9)	685

TABLE VIII. PHOTOCYCLIZATIONS PRODUCING FIVE-MEMBERED RINGS (*Continued*)

B. C_4N Ring Systems from Diarylamines and Aryl Vinyl Amines (*Continued*)

Starting Material	Reaction Conditions	Products and Yields (%)	Refs.
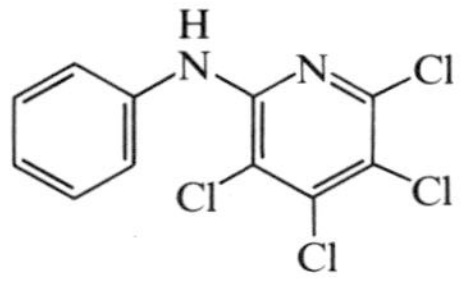	THF	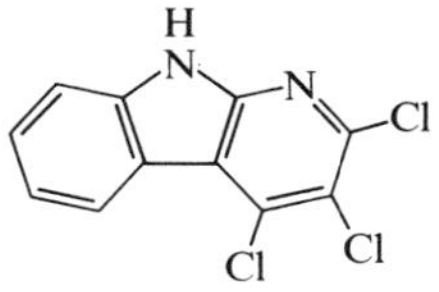 (12)	388a,b
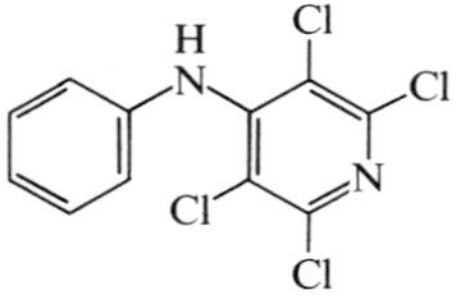	EtOH	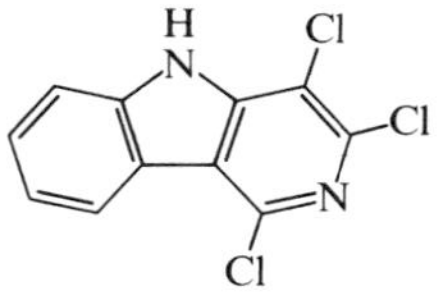 (79)	388a,b
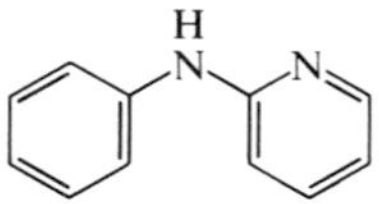	C_6H_{12} or THF, N_2[a]	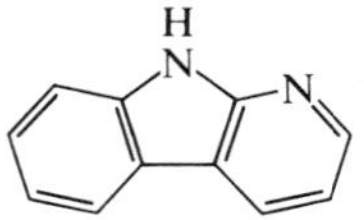 (81)	382
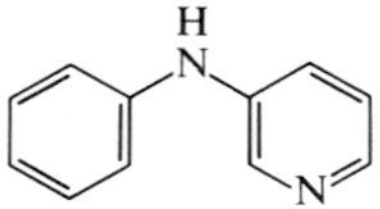	THF, N_2[a]	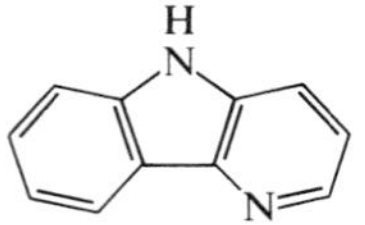 (46), 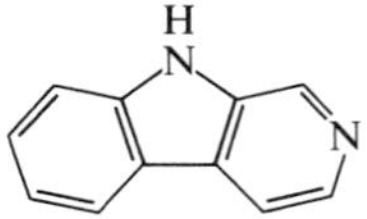 (24)	382
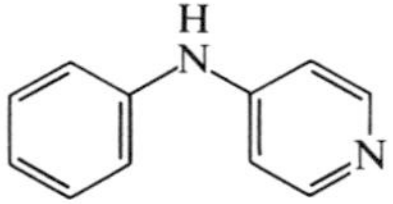	"	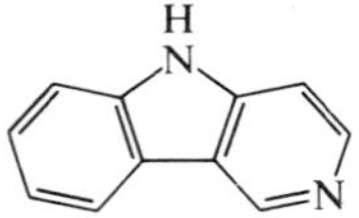 (70)	382

	Reactant	Conditions	Product (yield %)	Ref.
	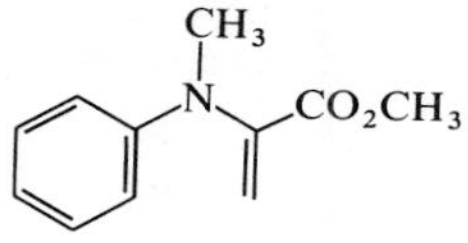	C_6H_6, argon	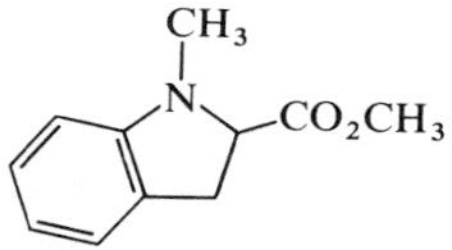 (87)	686
C_{12}	Diphenylamine	THF, N_2[a]	Carbazole (62)	382
	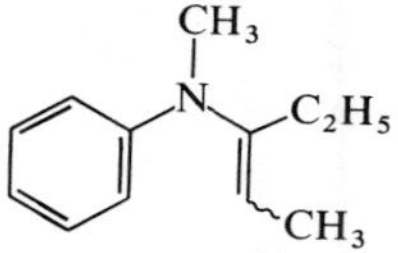	Et_2O, argon	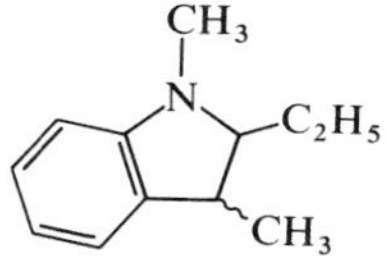 (43)	19,391a
	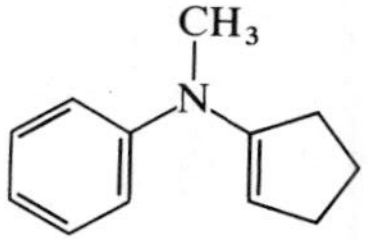	”	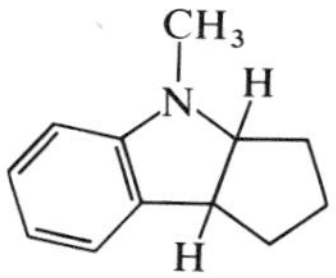 (52)	19,391a
		”	” (30)	391b
	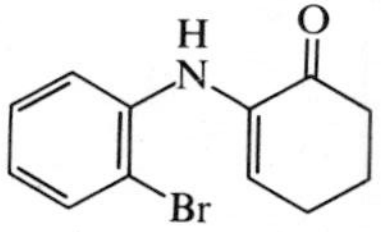	Dioxane, MeCN, Et_3N	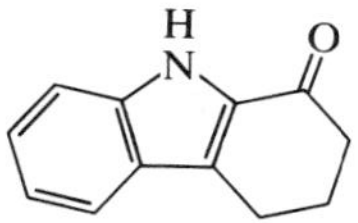 (80)	400
	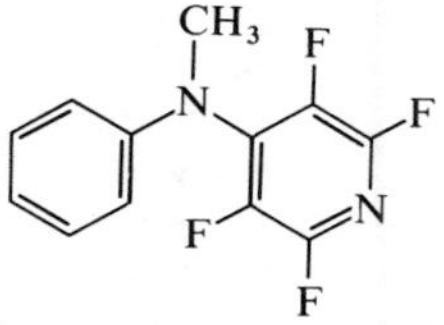	C_6H_{14}, Et_3N, argon	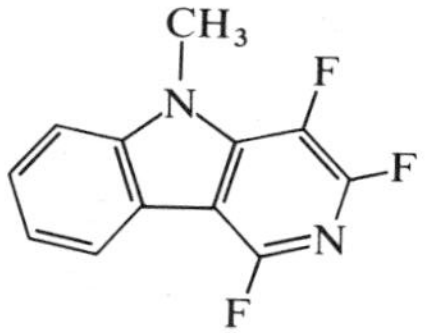 (92)	402
		C_6H_{14}, n-$BuNH_2$, argon	” (65)	402

TABLE VIII. PHOTOCYCLIZATIONS PRODUCING FIVE-MEMBERED RINGS (*Continued*)

B. C_4N Ring Systems from Diarylamines and Aryl Vinyl Amines (*Continued*)

	Starting Material	Reaction Conditions	Products and Yields (%)	Refs.
	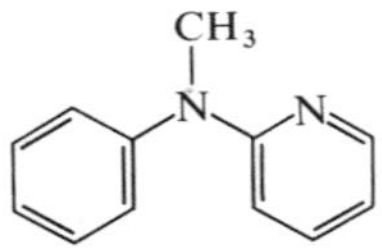	THF, N_2[b]	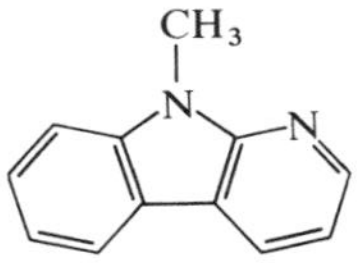 (74)	382
	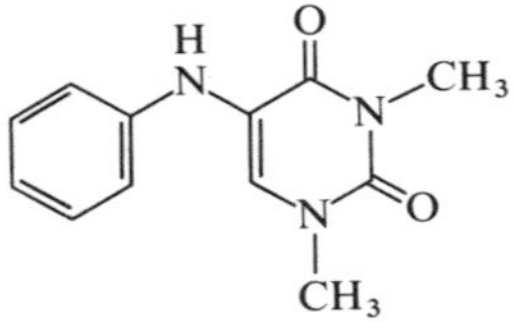	Me_2CO, air	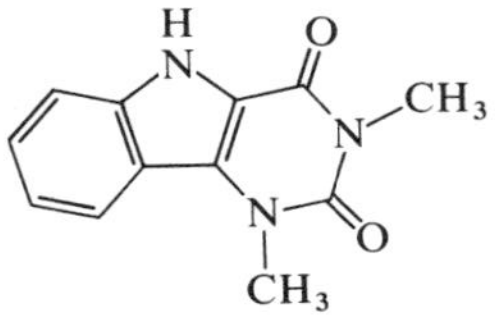 (16)	398
	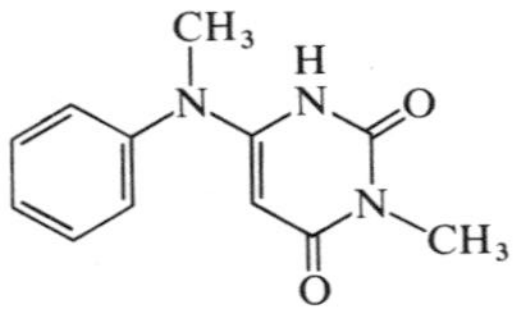	"	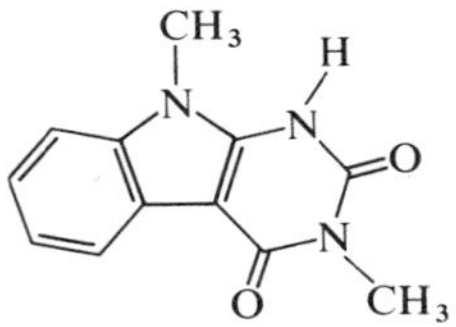 (31)	398
	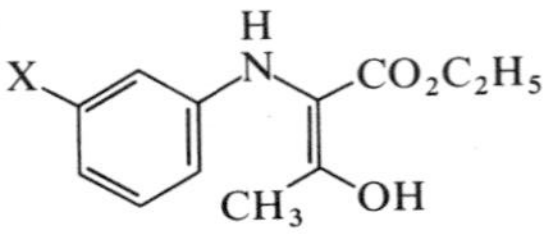 (I)		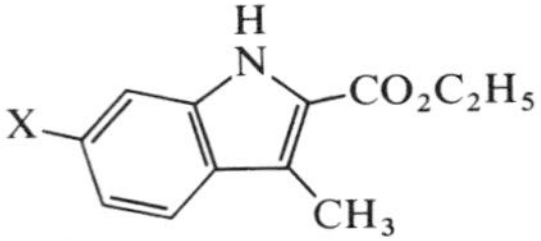 (II)	
	I, X = H	C_6H_6, MeOH, HOAc, degassed	II, X = H (96)	396
	I, X = Br	"	II, X = Br (50)	396
C_{13}	*N*-Methyldiphenylamine	C_6H_{14}; air	*N*-Methylcarbazole (70)	385
	$C_6H_5NHC_6H_4CH_3$-*o*	Pet. ether, I_2, air	1-Methylcarbazole (60–70)	384

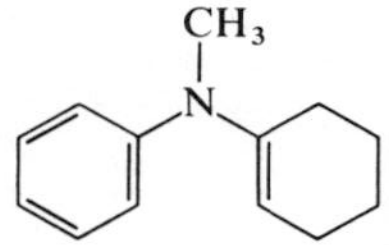 [structure]	Et_2O, argon	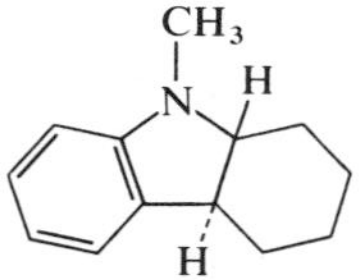 [structure] (55)	19,86,391a
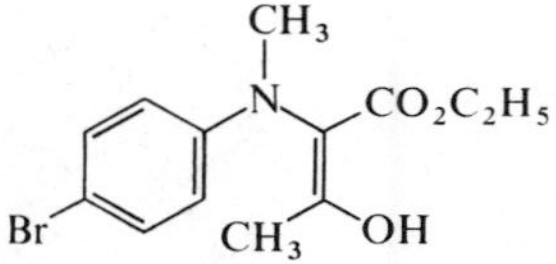 [structure]	C_6H_6, MeOH, HOAc, argon	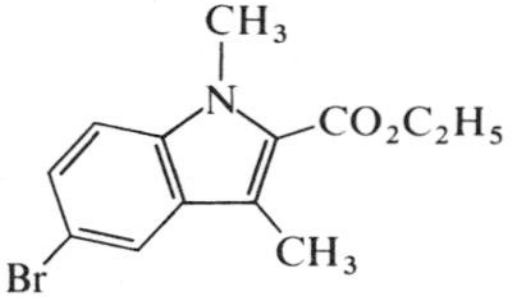 [structure] (84)	395
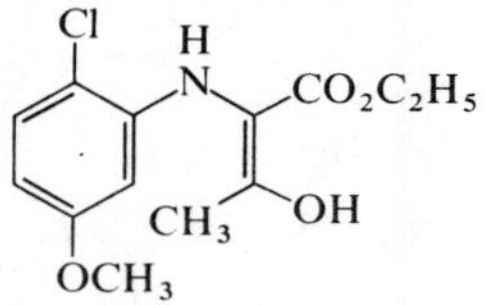 [structure]	C_6H_6, MeOH, HOAc, degassed	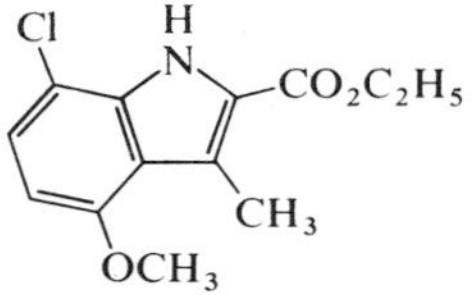 [structure] (95)	396
$C_6H_5N(CH_3)C_6H_4F$-*o*	$C_{6}H_{14}$, argon	*N*-Methylcarbazole (37)	402
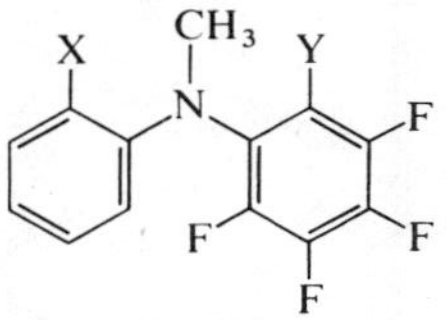 [structure] (I)		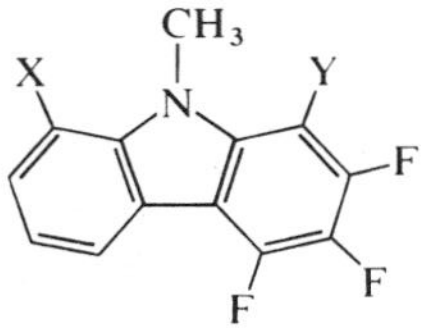 [structure] (II)	
I, X = Y = H	$C_{6}H_{14}$, argon	II, X = Y = H (66)	402
	$C_{6}H_{14}$, air	" (52)	402
I, X = F, Y = H	$C_{6}H_{14}$, argon	II, X = F, Y = H (36)	402
I, X = H, Y = F	$C_{6}H_{14}$, Et_3N, argon	II, X = H, Y = F (74)	402
	$C_{6}H_{14}$, argon	" (64)	402
	$C_{6}H_{14}$, air	" (52)	402
$C_6H_5NHC_6H_4CN$-*o*	MeCN, EtOH, degassed	Carbazole (87)	401

TABLE VIII. PHOTOCYCLIZATIONS PRODUCING FIVE-MEMBERED RINGS (*Continued*)

B. C_4N Ring Systems from Diarylamines and Aryl Vinyl Amines (*Continued*)

Starting Material	Reaction Conditions	Products and Yields (%)	Refs.
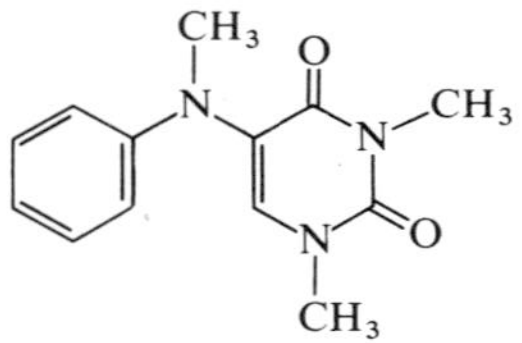	Me_2CO, air	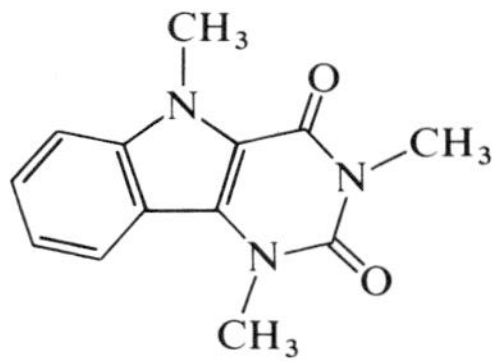 (31)	398
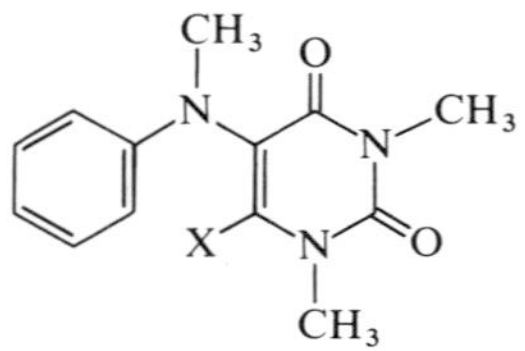 (I)		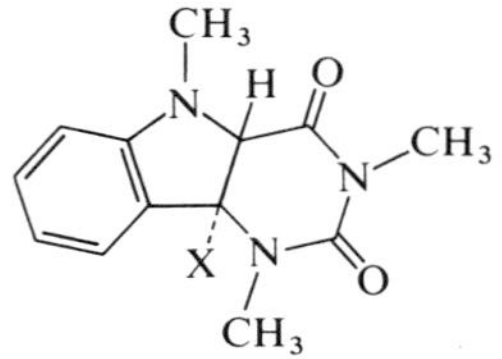 (II)	
I, X = H	*i*-PrOH, N_2	II, X = H (87)	398
I, X = D	"	II, X = D (86)	398
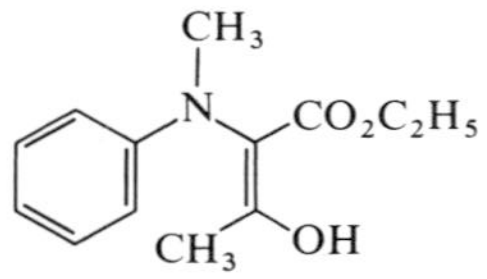	C_5H_{12}, Na_2CO_3, argon	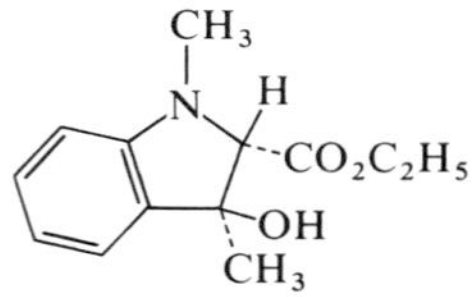 (100)	394,395
	C_6H_6, MeOH, HOAc, argon	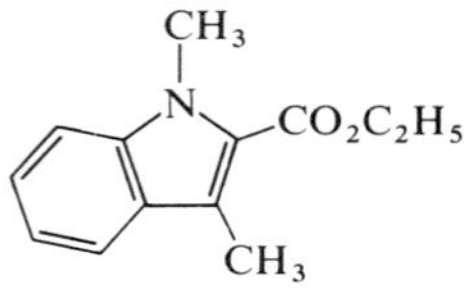 (92)	395

	Reactant	Conditions	Product(s) (Yield, %)	Refs.
		EtOH, N_2	" (—)	687
	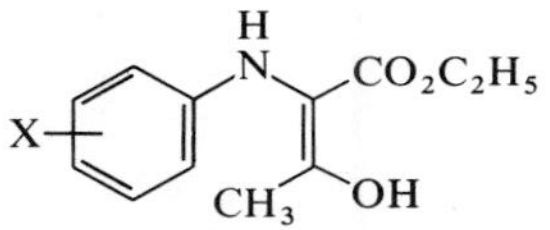 (I)		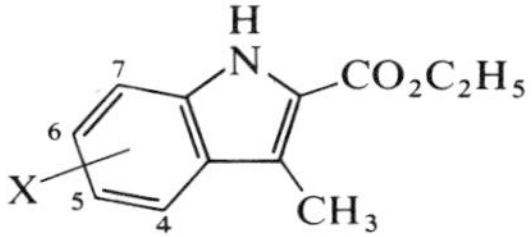 (II)	
	I, X = o-CH_3O	C_6H_6, MeOH, HOAc, degassed	II, X = 7-CH_3O (94)	396
	I, X = m-CH_3O	"	II, X = 6-CH_3O (47); II, X = 4-CH_3O (47)	396
	I, X = p-CH_3O	"	II, X = 5-CH_3O (89)	396
C_{14}	$C_6H_5NHC_6H_3(CH_3)_2$-2,5	THF, air	1,4-Dimethylcarbazole (98)	688
	o-$CH_3C_6H_4NHC_6H_4CH_3$-o	Pet. ether, I_2, air	1,8-Dimethylcarbazole (—)	384
	p-$CH_3C_6H_4NHC_6H_4CH_3$-p	Pet. ether, air	3,6-Dimethylcarbazole (—)	689
	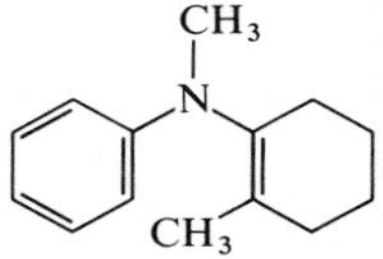	Et_2O, argon	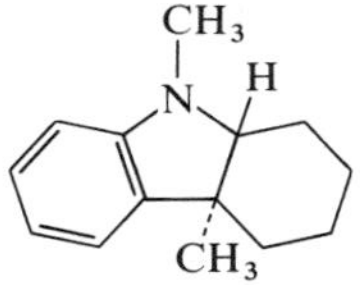 (70)	19,391a
		"	" (64)	391b
	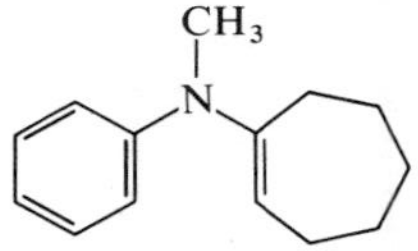	"	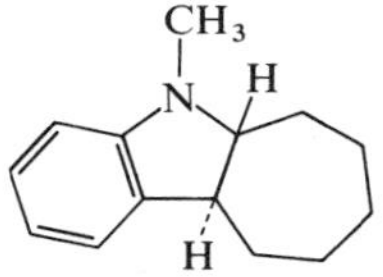 (66)	19,391a
	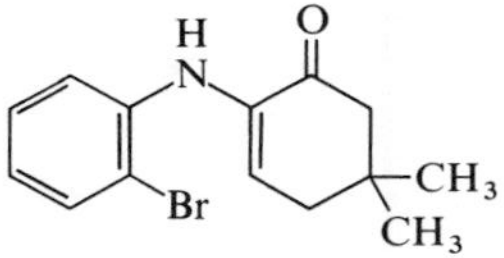	Dioxane, MeCN, Et_3N	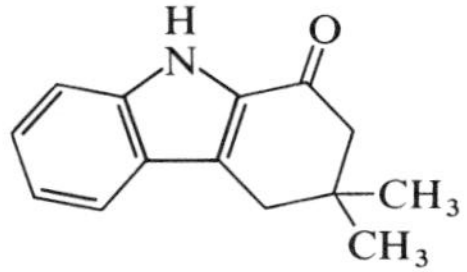 (86)	400

TABLE VIII. PHOTOCYCLIZATIONS PRODUCING FIVE-MEMBERED RINGS (*Continued*)

B. C_4N Ring Systems from Diarylamines and Aryl Vinyl Amines (*Continued*)

Starting Material	Reaction Conditions	Products and Yields (%)	Refs.
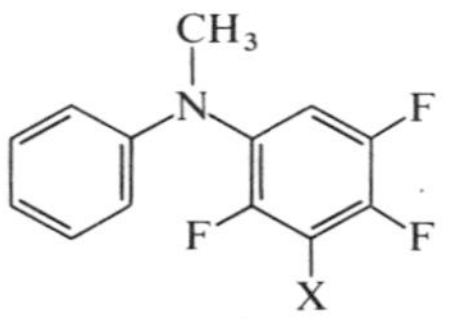 (I)		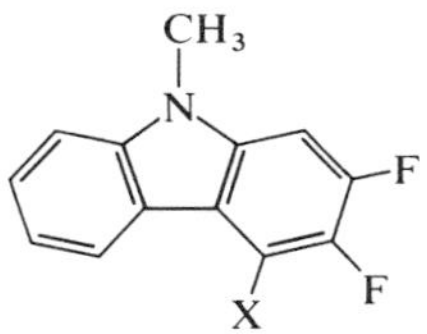 (II)	
I, X = CF_3	C_6H_{14}, argon	II, X = CF_3 (78)	402
I, X = CH_3O	"	II, X = CH_3O (47)	402
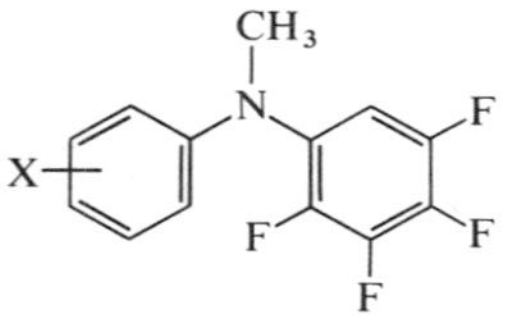 (I)		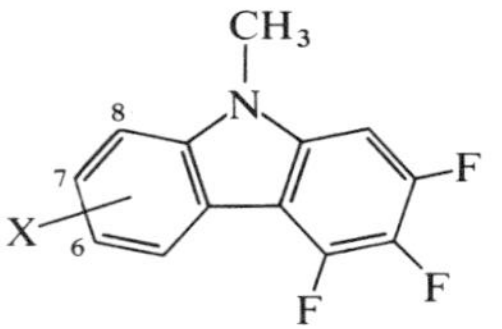 (II)	
I, X = *o*-CH_3O	"	II, X = 8-CH_3O (80)	402
I, X = *p*-CH_3O	"	II, X = 6-CH_3O (80)	402
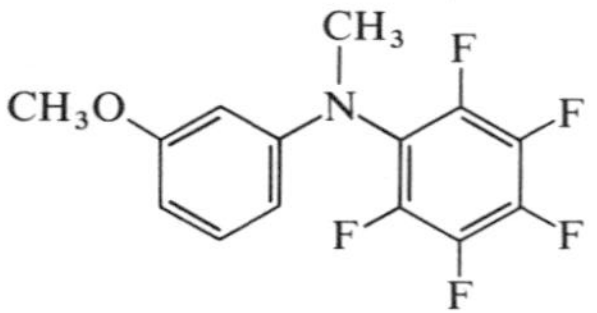 CH_3O	"	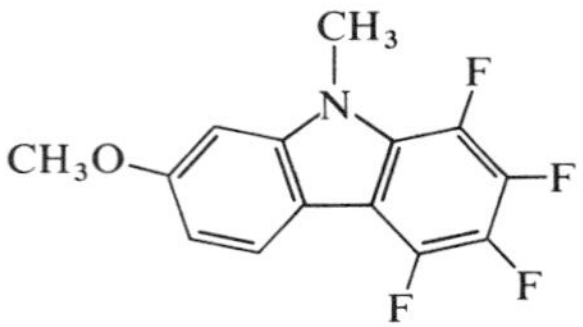 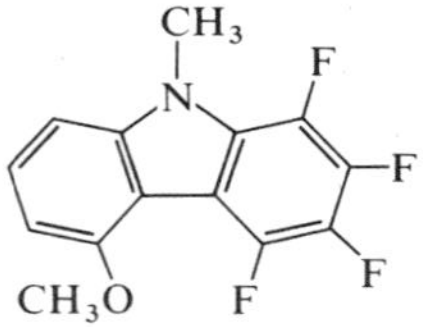 CH_3O + CH_3O (1:1, 80 total)	402

Reactant	Conditions	Product (yield, %)	Ref.
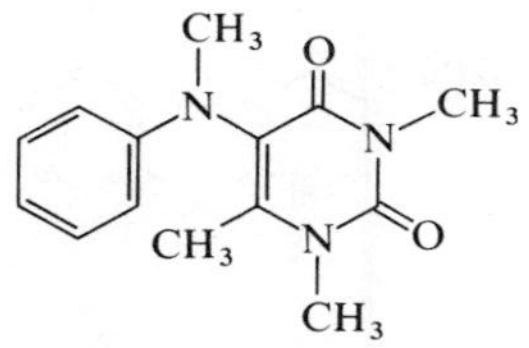	i-PrOH, N_2	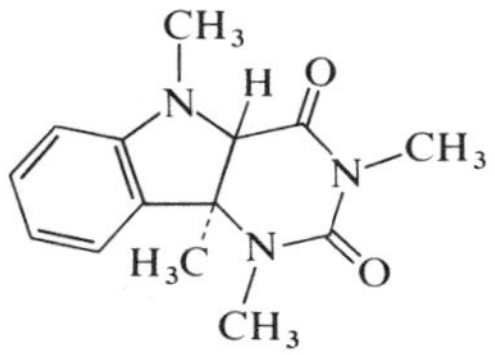 (56)	398
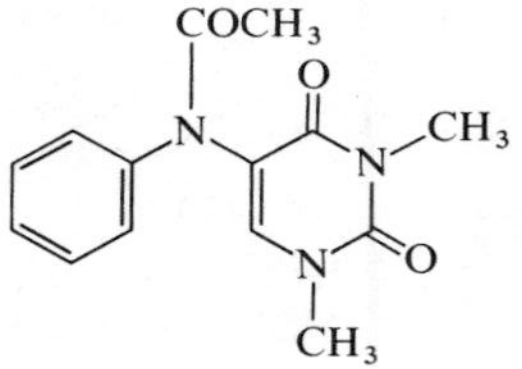	”	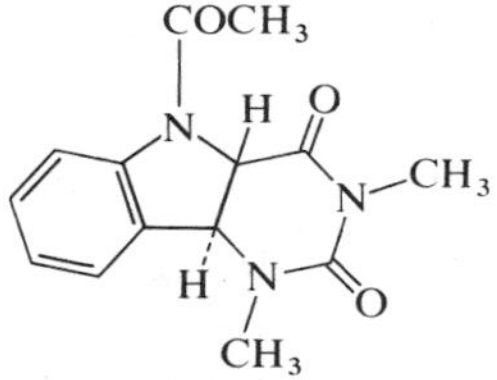 (91)	398
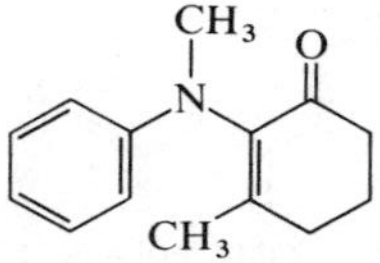	C_6H_6, degassed	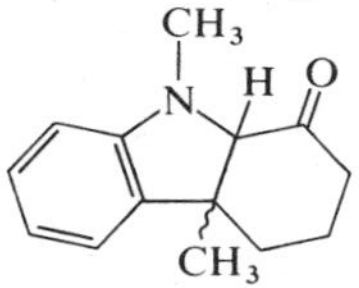 (*cis*, 90; *trans*, 7)	392
p-$CH_3C_6H_4NHC_6H_4OCH_3$-p	Pet. ether, I_2, air	3-Methoxy-6-methylcarbazole (—)	384
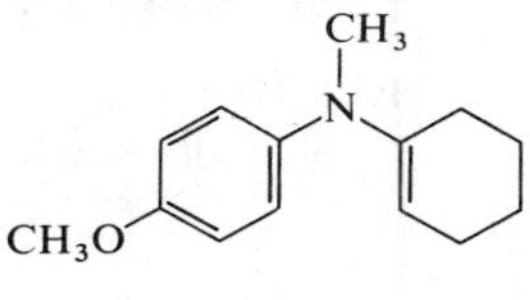	Et_2O, argon	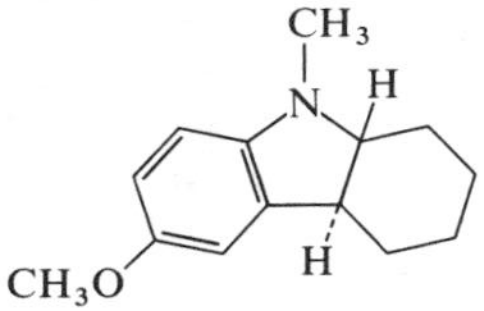 (12)	19,391a
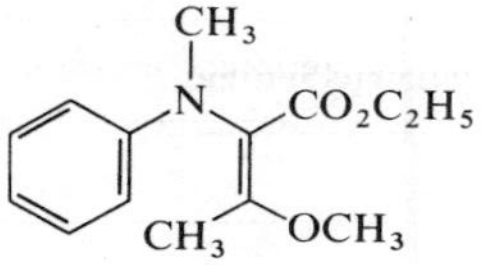	C_5H_{12}, Na_2CO_3	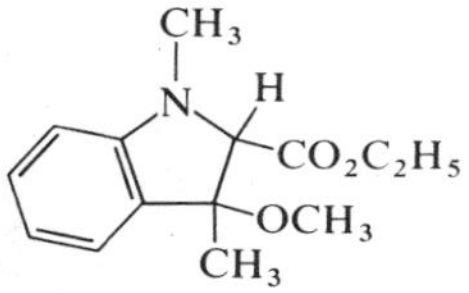 (100)	395

TABLE VIII. PHOTOCYCLIZATIONS PRODUCING FIVE-MEMBERED RINGS (*Continued*)

B. C_4N Ring Systems from Diarylamines and Aryl Vinyl Amines (*Continued*)

Starting Material	Reaction Conditions	Products and Yields (%)	Refs.
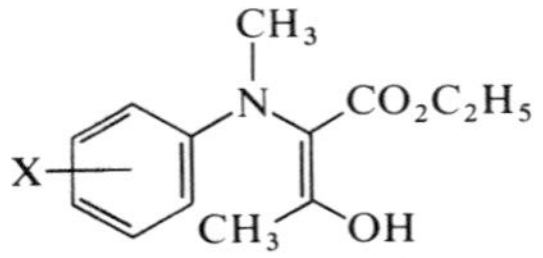 (I)		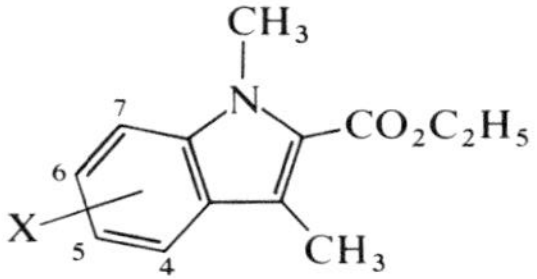 (II)	
I, X = *o*-CH_3O	C_6H_6, MeOH, HOAc, argon	II, X = 7-CH_3O (88)	395
I, X = *m*-CH_3O	"	II, X = 6-CH_3O (51); II, X = 4-CH_3O (42)	395
I, X = *p*-CH_3O	"	II, X = 5-CH_3O (86)	395
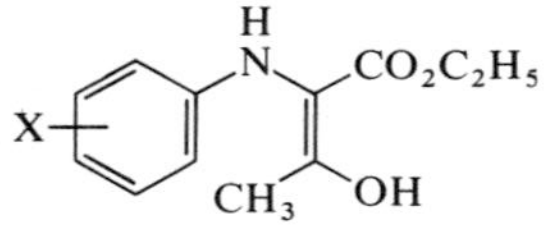 (I)		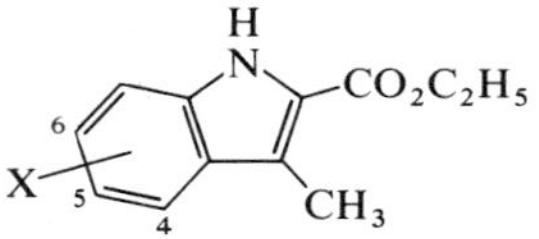 (II)	
I, X = *m*-CO_2CH_3	C_6H_6, MeOH, HOAc, degassed	II, X = 6-CO_2CH_3 (47); II, X = 4-CO_2CH_3 (37)	396
I, X = *p*-CO_2CH_3	"	II, X = 5-CO_2CH_3 (90)	396
C_{15} 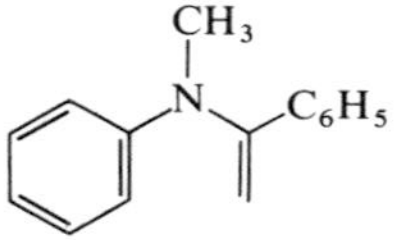	Et_2O, argon	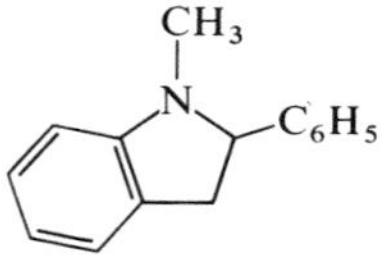 (73)	19,391a
	"	" (69–70)	391b
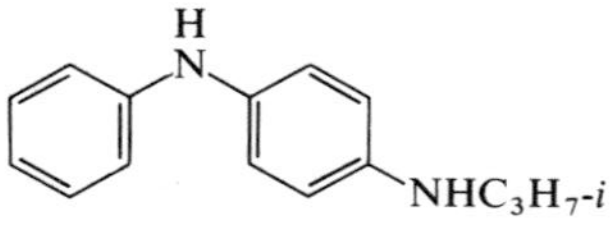	MeOH, H_2O, air	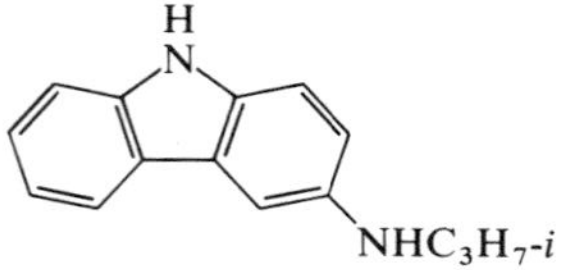 (—),	84

	Reactant	Conditions	Product (yield %)	Refs.
			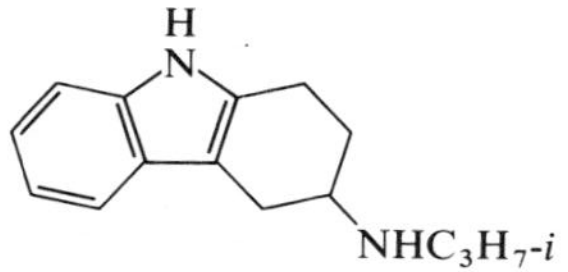 (—)	
	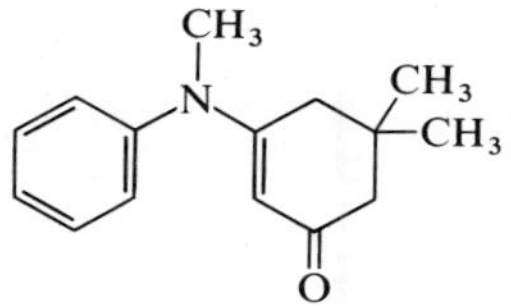	Et_2O, N_2[b]	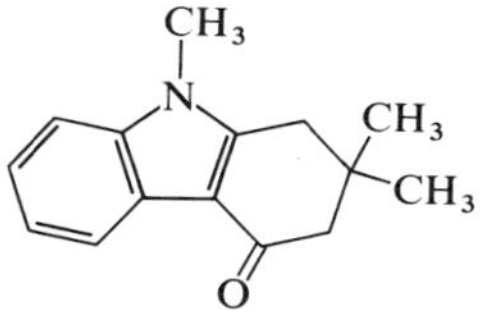 (38)	690,691
		C_6H_6[b]	” (—)	691
	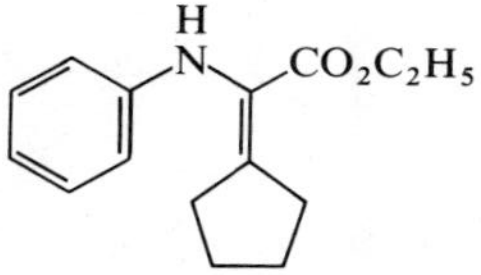	C_6H_6, argon	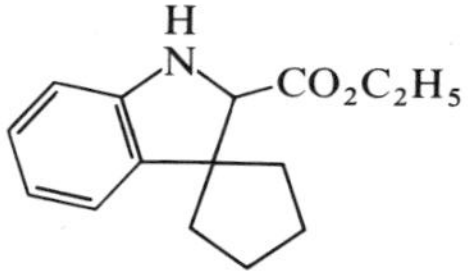 (70)	686
	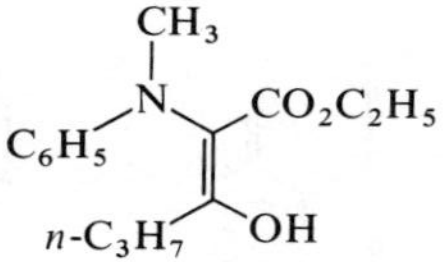	C_5H_{12}, Na_2CO_3, argon	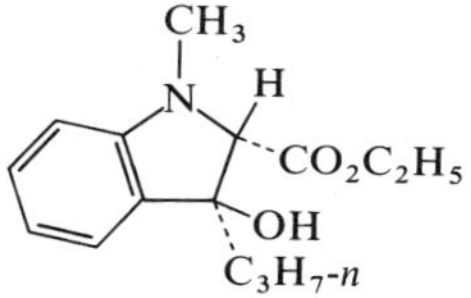 (> 76)	394
C_{16}	2,3-$(CH_3)_2C_6H_3NHC_6H_3(CH_3)_2$-2,3	Pet. ether, air	1,2,7,8-Tetramethylcarbazole (20), 1,2,5-trimethylcarbazole (18)	384,692
	2,4-$(CH_3)_2C_6H_3NHC_6H_3(CH_3)_2$-2,4	”	1,3,6,8-Tetramethylcarbazole (34)	384,692
	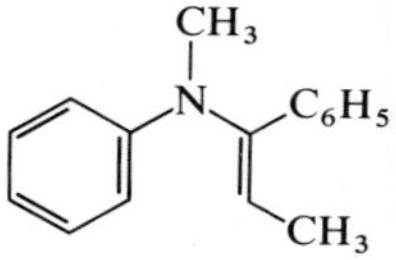	Et_2O or $CH_3C_6H_{11}$, N_2	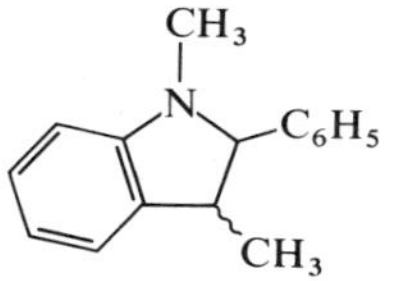 (—)	26c

TABLE VIII. PHOTOCYCLIZATIONS PRODUCING FIVE-MEMBERED RINGS (*Continued*)

B. C_4N Ring Systems from Diarylamines and Aryl Vinyl Amines (*Continued*)

Starting Material	Reaction Conditions	Products and Yields (%)	Refs.
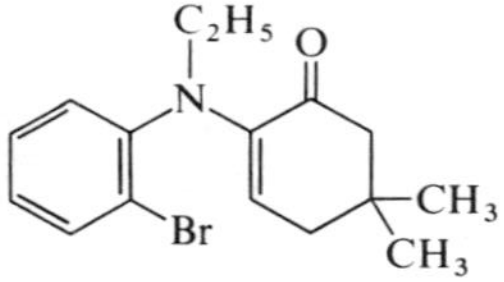	Dioxane, MeCN, Et_3N	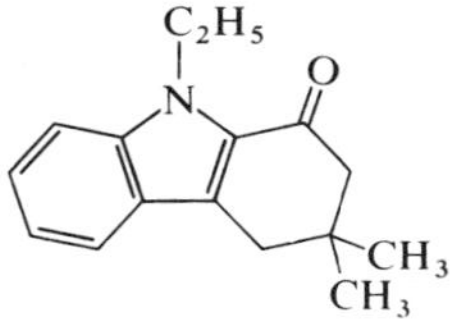 (64)	400
$C_6H_5N(C_2H_5)C_6H_4NHC_2H_5$-*p*	C_6H_{12}, degassed[b]	*N*-Ethyl-3-(ethylamino)carbazole (31)	693
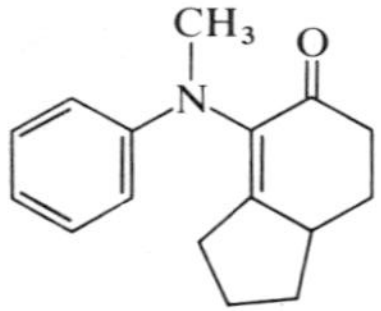	C_6H_6, degassed	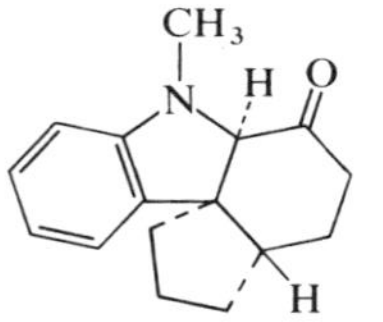 (71)	392
C_{17} 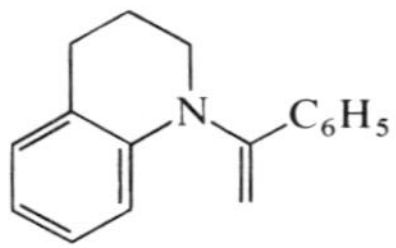	Et_2O, argon	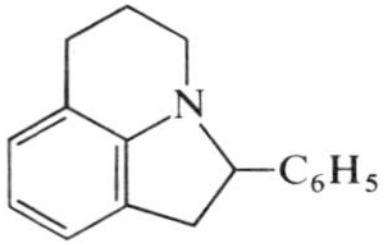 (70)	391b
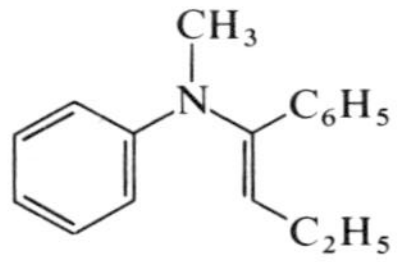	Et_2O or $CH_3C_6H_{11}$, N_2	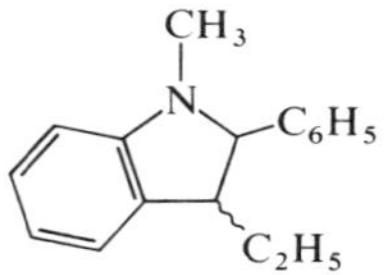 (—)	26c

	Reactant	Conditions	Product (% yield)	Refs.
	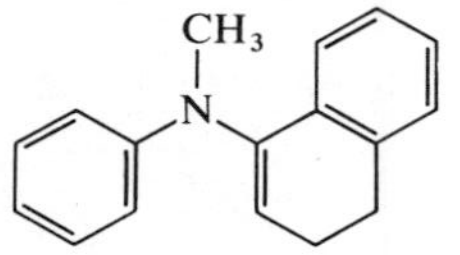	Et_2O, argon	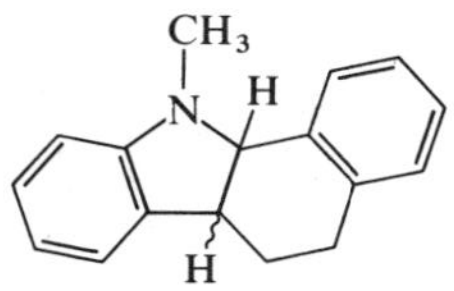 (*trans*, 36–40; *cis*, 19–20)	19,26c,391a,b
	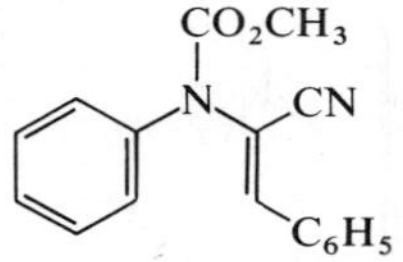	C_6H_6, argon	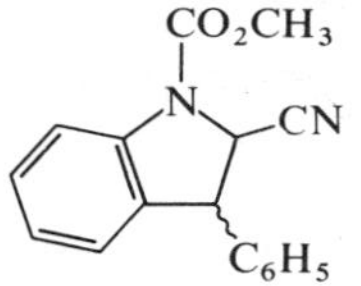 (64)	686
C_{18}	Triphenylamine	C_6H_{14}, air	*N*-Phenylcarbazole (65)	385
		Et_2O, air	" (38)	694
	$(p\text{-}FC_6H_4)_3N$	"	3,6-Difluoro-*N*-(*p*-fluorophenyl)carbazole (28)	694
	$(C_6H_5)_2NC_6HF_4$-2,3,4,5	C_6H_{14}, argon	2,3,4-Trifluoro-*N*-phenylcarbazole (53)	402
		C_6H_{14}, air	" (44)	402
	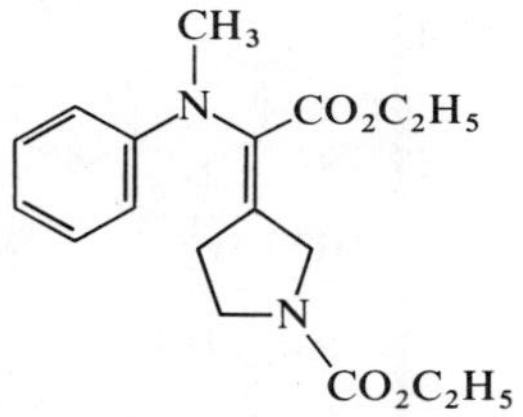	C_6H_6, argon	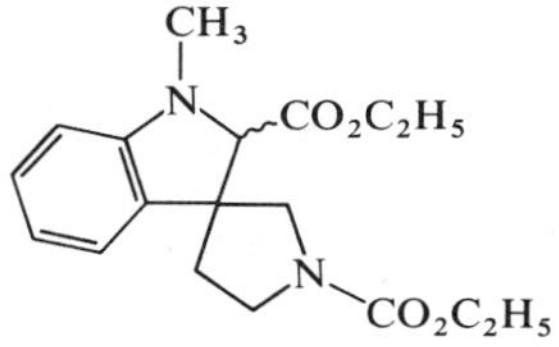 (87)	686
	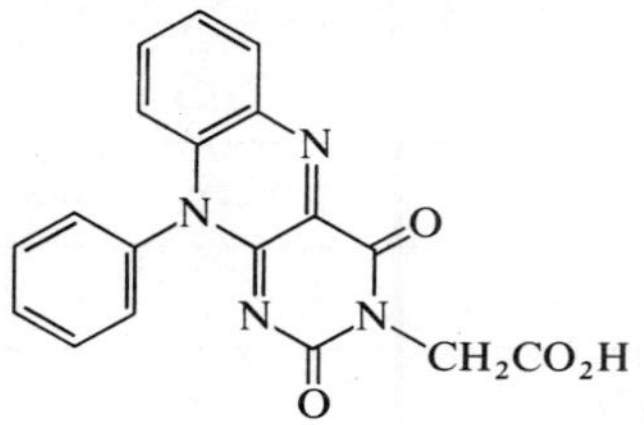	H_2O, acetate buffer, pH 4	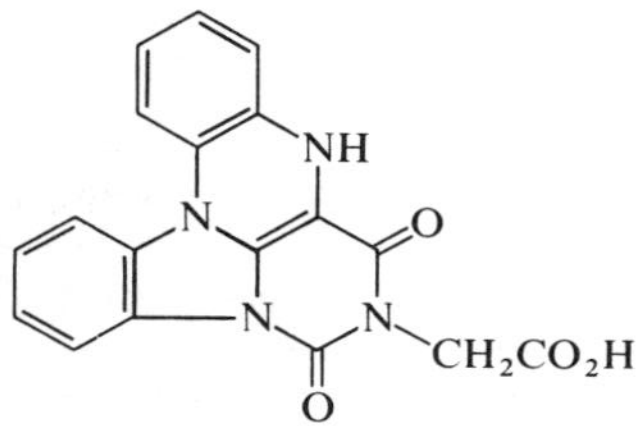 (45)	399

TABLE VIII. PHOTOCYCLIZATIONS PRODUCING FIVE-MEMBERED RINGS (*Continued*)

B. C_4N Ring Systems from Diarylamines and Aryl Vinyl Amines (*Continued*)

	Starting Material	Reaction Conditions	Products and Yields (%)	Refs.
	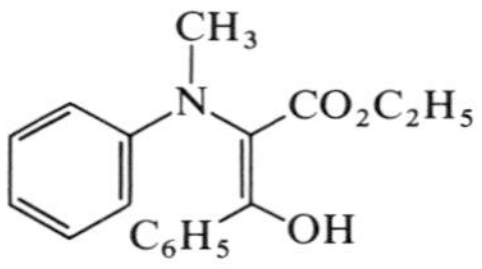 CH_3, N, $CO_2C_2H_5$, C_6H_5, OH	C_5H_{12}, Na_2CO_3, argon	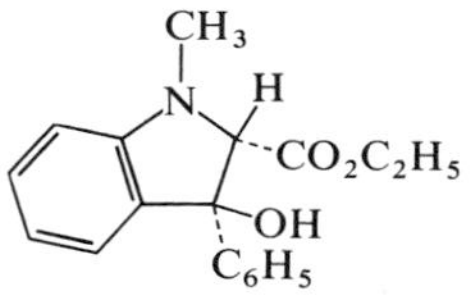 CH_3, N, H, $CO_2C_2H_5$, OH, C_6H_5 (>70)	394
C_{19}	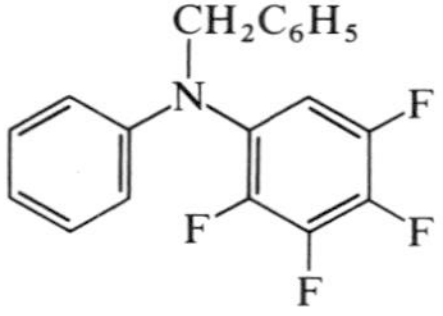 $CH_2C_6H_5$, N, F, F, F, F	C_6H_{14}, argon	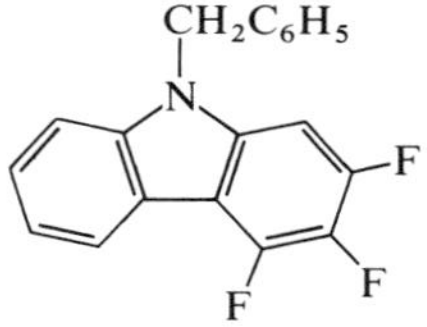 $CH_2C_6H_5$, N, F, F, F (65)	402
	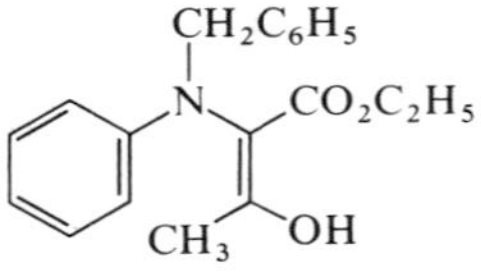 $CH_2C_6H_5$, N, $CO_2C_2H_5$, CH_3, OH	C_6H_6, MeOH, HOAc, degassed	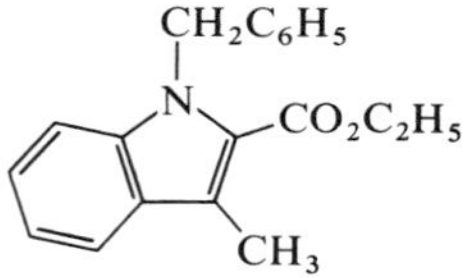 $CH_2C_6H_5$, N, $CO_2C_2H_5$, CH_3 (93)	396
	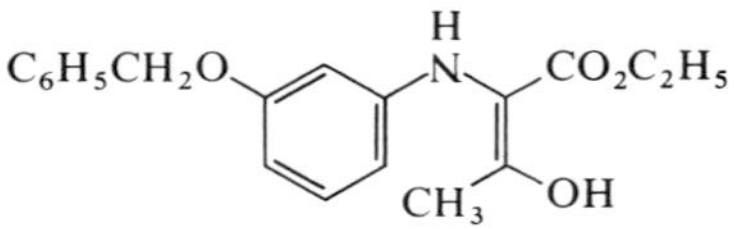 $C_6H_5CH_2O$, H, N, $CO_2C_2H_5$, CH_3, OH	"	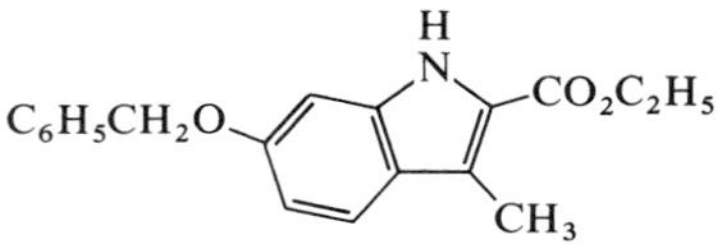 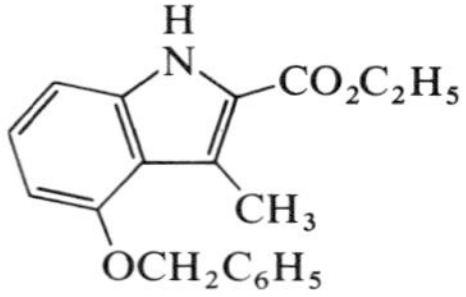 $C_6H_5CH_2O$, H, N, $CO_2C_2H_5$, CH_3 (48), H, N, $CO_2C_2H_5$, CH_3, $OCH_2C_6H_5$ (48)	396

	Reactant	Conditions	Product (yield %)	Ref.
C_{20}	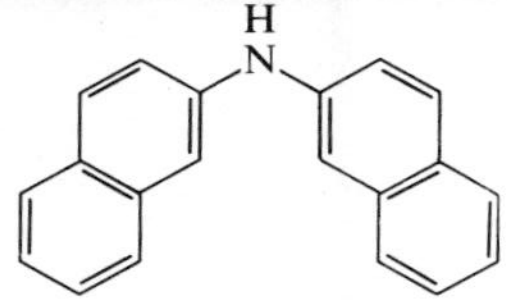 Di-2-naphthylamine (structure)	C_6H_{12}, air	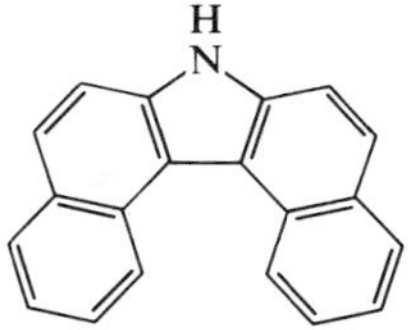 Dibenzocarbazole (structure) (18)	
		n-PrOH, N_2[a]	" (11)	
	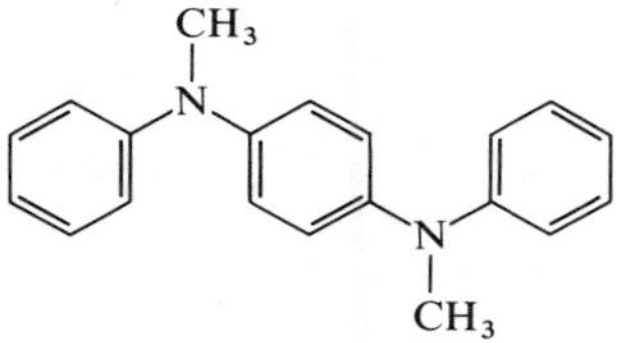 N,N'-Dimethyl-N,N'-diphenyl-p-phenylenediamine (structure)	Et_2O, air	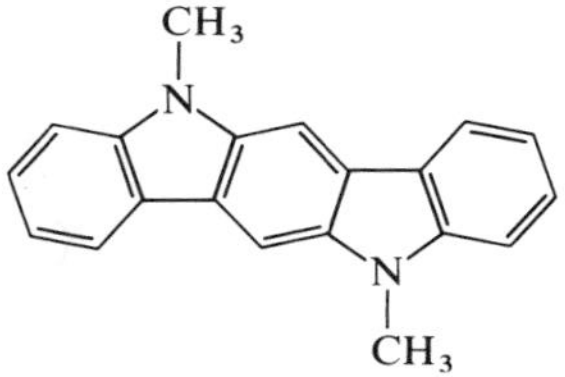 N,N'-Dimethylindolocarbazole (structure) (10)	694
C_{21}	$(p\text{-}CH_3OC_6H_4)_3N$	C_6H_{14} or Et_2O, air	3,6-Dimethoxy-*N*-(*p*-methoxyphenyl)carbazole (60)	694
	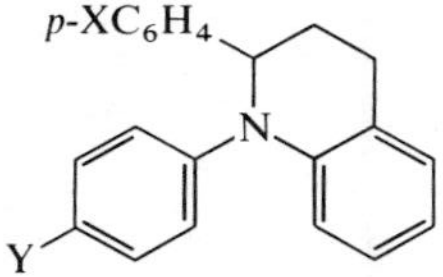 (I) $p\text{-}XC_6H_4$, Y (structure)		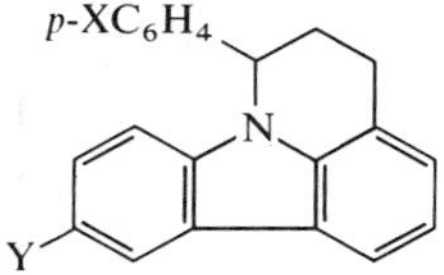 (II) $p\text{-}XC_6H_4$, Y (structure)	
	I, X = Y = H	EtOH, air	II, X = Y = H (70)	261
C_{22}	I, X = H, Y = CH_3O	"	II, X = H, Y = CH_3O (32–36)	261
C_{23}	I, X = Y = CH_3O	"	II, X = Y = CH_3O (84)	261
	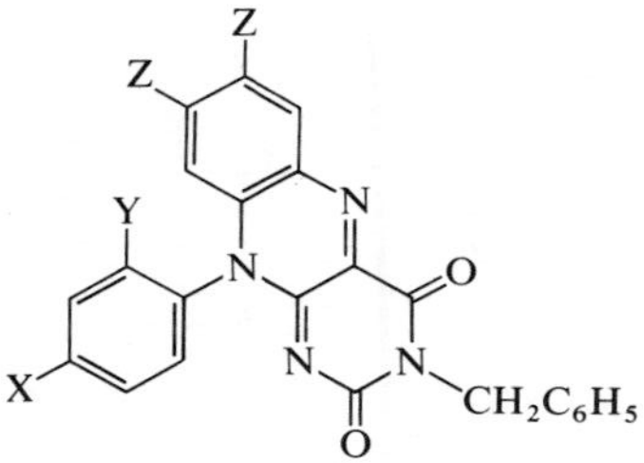 (I) X, Y, Z, N, O, $CH_2C_6H_5$ (structure)		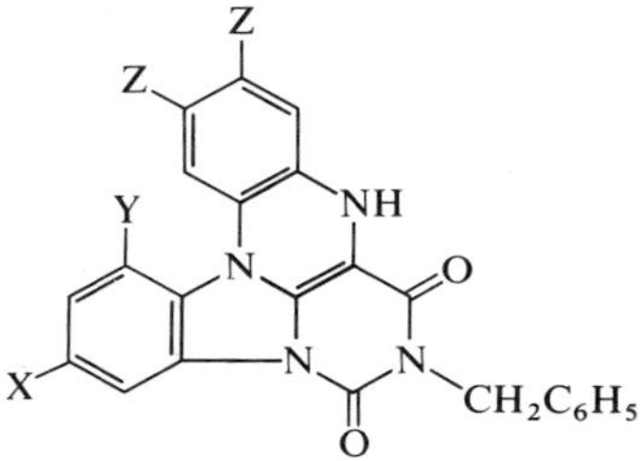 (II) X, Y, Z, NH, N, O, $CH_2C_6H_5$ (structure)	

TABLE VIII. PHOTOCYCLIZATIONS PRODUCING FIVE-MEMBERED RINGS (*Continued*)

	Starting Material	Reaction Conditions	Products and Yields (%)	Refs.
B. C_4N Ring Systems from Diarylamines and Aryl Vinyl Amines (Continued)				
	I, X = Y = Z = H	MeCN, HOAc, N_2	II, X = Y = Z = H (92)	399
	I, X = F, Y = Z = H	"	II, X = F, Y = Z = H (100)	399
C_{24}	I, Y = CH_3, X = Z = H	"	II, Y = CH_3, X = Z = H (100)	399
C_{25}	I, Z = CH_3, X = Y = H	"	II, Z = CH_3, X = Y = H (—)	399
C_{28}	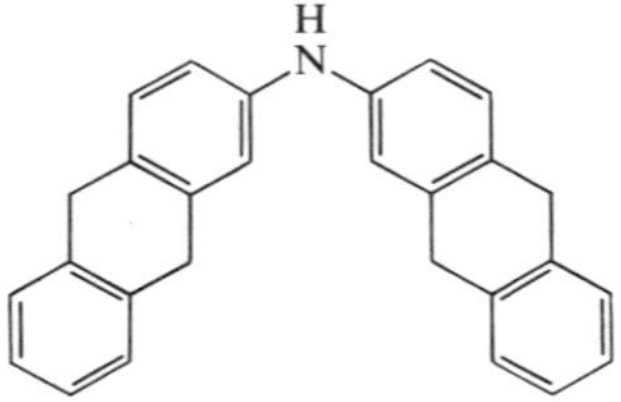 [structure] H N	*n*-PrOH, N_2 (or CO_2)[a]	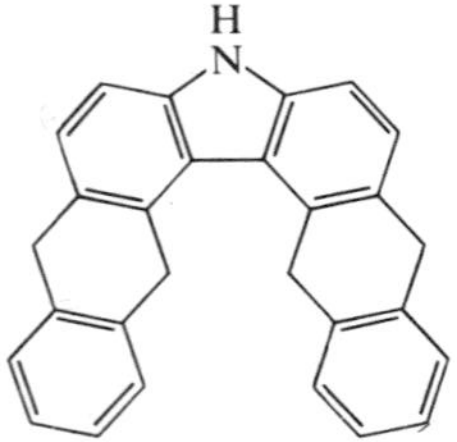 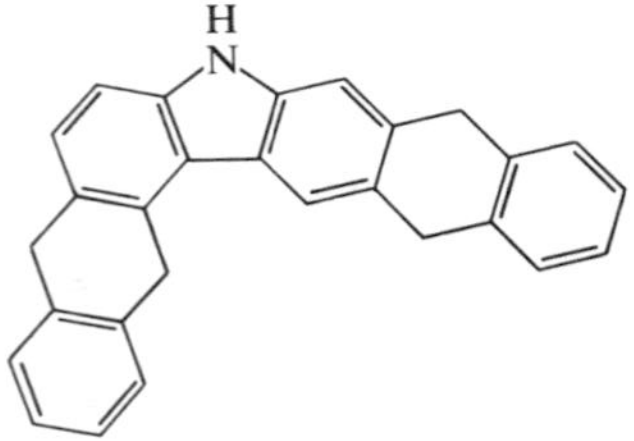 [structure] H N (21), [structure] H N (30)	695a

	Reactant	Conditions	Product (Yield %)	Ref.
C_{30}	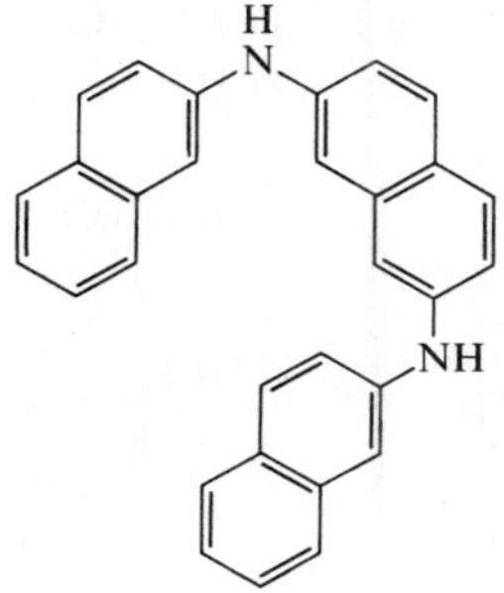	Xylene, I_2, CO_2	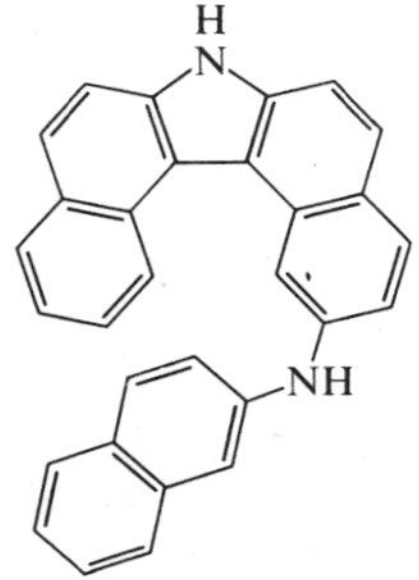 (83)	695b

C. C_4O Ring Systems from Diaryl Ethers and Aryl Vinyl Ethers

	Reactant	Conditions	Product (Yield %)	Ref.
C_{11}	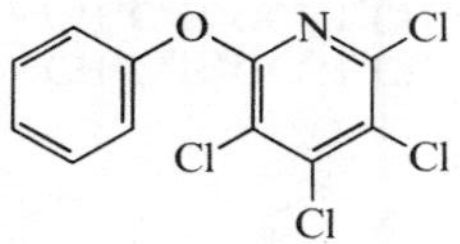	THF	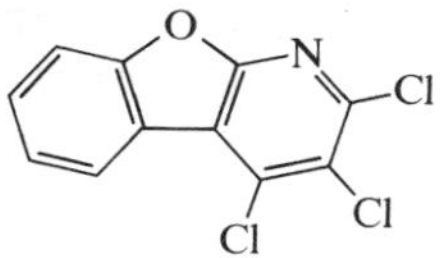 (38)	388a,b
	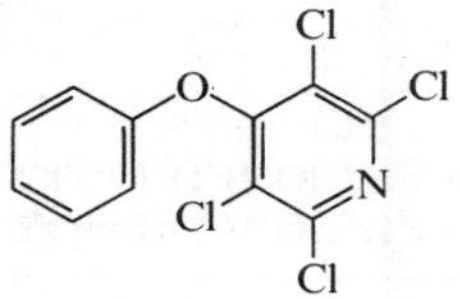	EtOH	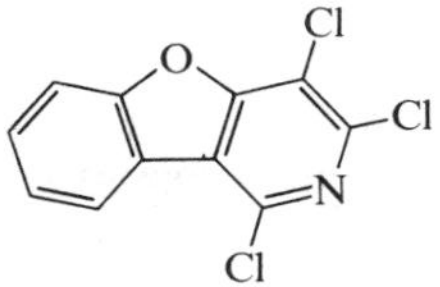 (54)	388a,b
C_{12}	Diphenyl ether	C_6H_{12}, I_2, N_2	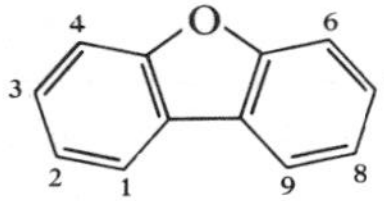 (Dibenzofuran) (55)	386
	4-$ClC_6H_4OC_6H_3Cl_2$-2,4	C_6H_{14}, N_2	2,8-Dichlorodibenzofuran (20)	696a
	2,4-$Cl_2C_6H_3OC_6H_3Cl_2$-2,4	"	2,4,8-Trichlorodibenzofuran (—)	696a

TABLE VIII. PHOTOCYCLIZATIONS PRODUCING FIVE-MEMBERED RINGS (*Continued*)

C. C_4O Ring Systems from Diaryl Ethers and Aryl Vinyl Ethers (*Continued*)

	Starting Material	Reaction Conditions	Products and Yields (%)	Refs.
	2,4,6-$Cl_3C_6H_2OC_6H_4Cl$-4	C_6H_{14}, EtOH, or MeOH; N_2	" (—)	696b
	2,4,5-$Cl_3C_6H_2OC_6H_4Cl$-4	"	2,8-Dichlorodibenzofuran (—)	696b
		C_6H_6 or MeCN	(1)	697
C_{14}	*o*-$CH_3C_6H_4OC_6H_4CH_3$-*o*	C_6H_{12}, I_2, N_2	4,6-Dimethyldibenzofuran (45)[c]	386a
	p-$CH_3C_6H_4OC_6H_4CH_3$-*o*	"	2,6-Dimethyldibenzofuran (45)[c]	386a,b
	2,5-$(CH_3)_2C_6H_3OC_6H_5$	"	1,4-Dimethyldibenzofuran (39)[c]	386b
	p-$CH_3C_6H_4OC_6H_4CH_3$-*p*	"	2,8-Dimethyldibenzofuran (45)[c]	386a,b
	m-$CH_3OC_6H_4OC_6H_4OCH_3$-*o*	EtOH, H_2O, N_2	1-Methoxydibenzofuran (1)[c]	390
C_{15}		C_6H_6, MeOH, HOAc, argon	(88)	35,698
	p-$CH_3C_6H_4O$ (with CH_3, CH_3O substituted benzene)	EtOH, H_2O, N_2	2,6-Dimethyldibenzofuran (2)[c]	390
	4-$CH_3C_6H_4OC_6H_3(OCH_3)_2$-2,4	"	8-Methoxy-2-methyldibenzofuran (1)[c]	390

	Reactant	Conditions	Product(s) (Yield, %)	Refs.
C_{16}	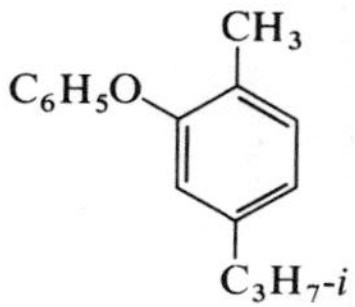 (structure: C_6H_5O, CH_3, C_3H_7-*i*)	C_6H_{12}, I_2, N_2	1-Isopropyl-4-methyldibenzofuran (25)[c]	386b
	2,5-$(CH_3)_2C_6H_3OC_6H_3(CH_3)_2$-2,5	"	1,4,6,9-Tetramethyldibenzofuran (35)[c]	386b
	2,3,4,5-$(CH_3)_4C_6HOC_6H_5$	"	1,2,3,4-Tetramethyldibenzofuran (40)[c]	386b
	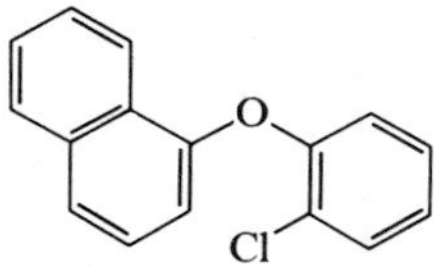 (structure: Cl)	C_6H_6	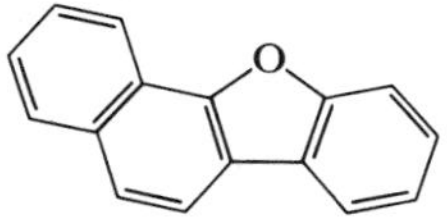 (structure) (45)	387
	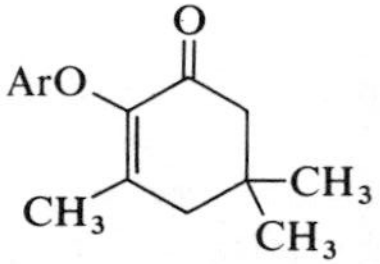 (structure: ArO, O, CH_3, CH_3, CH_3) (I)		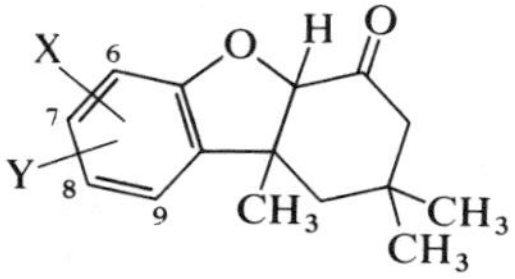 (structure: X, Y, 6, 7, 8, 9, O, H, O, CH_3, CH_3, CH_3) (II)	
	I, Ar = 2-Chloro-5-methylphenyl	C_6H_6, MeOH, argon	II, X = 6-Cl, Y = 9-CH_3 (83)	35
	I, Ar = 4-Chloro-3-methylphenyl	"	II, X = 8-Cl, Y = 9-CH_3 + II, X = 8-Cl, Y = 7-CH_3 (70:30)	35
	I, Ar = 3-Chloro-5-methoxyphenyl	"; or C_6H_6, MeOH, HOAc, argon	II, X = 7-Cl, Y = 9-CH_3O + II, X = 9-Cl, Y = 7-CH_3O (58:42 or 42:58)	35
	I, Ar = *o*-Methylphenyl	C_6H_6, MeOH, HOAc, argon	II, X = 6-CH_3, Y = H (80)	35
		C_6H_6, degassed	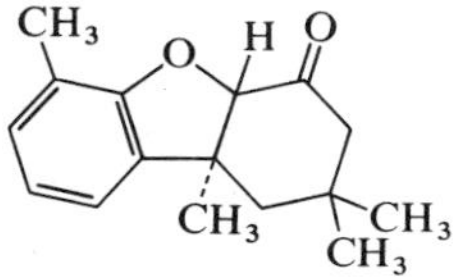 (structure: CH_3, O, H, O, CH_3, CH_3, CH_3) (30)	35
	I, Ar = *m*-Methylphenyl	C_6H_6, MeOH, HOAc, argon	II, X = 9-CH_3, Y = H + II, X = 7-CH_3, Y = H (75:25)	35

TABLE VIII. PHOTOCYCLIZATIONS PRODUCING FIVE-MEMBERED RINGS (*Continued*)

C. C_4O Ring Systems from Diaryl Ethers and Aryl Vinyl Ethers (*Continued*)

	Starting Material	Reaction Conditions	Products and Yields (%)	Refs.
	I, Ar = *p*-Methylphenyl	C_6H_6, MeOH, HOAc, argon	II, X = 8-CH_3, Y = H (80)	35
	I, Ar = *o*-Methoxyphenyl	"	II, X = 6-CH_3O, Y = H (30)	35
	I, Ar = *m*-Methoxyphenyl	"	II, X = 9-CH_3O, Y = H (90)	35
	I, Ar = *p*-Methoxyphenyl	"	II, X = 8-CH_3O, Y = H (32)	35
	I, Ar = *m*-Carboxyphenyl	C_6H_6, MeOH, argon	II, X = 9-CO_2H, Y = H + II, X = 7-CO_2H, Y = H (63:37)	35
C_{17}	I, Ar = 5-Cyano-2-methoxyphenyl	C_6H_6, MeOH, HOAc, argon	II, X = 9-CN, Y = 6-CH_3O (87)	35
	I, Ar = *p*-Acetylphenyl	"	II, X = 8-CH_3CO, Y = H (94)	35
	I, Ar = *m*-Carbomethoxyphenyl	C_6H_6, MeOH, argon	II, X = 9-CO_2CH_3, Y = H + II, X = 7-CO_2CH_3, Y = H (65:35)	35
	I, Ar = *p*-Carbomethoxyphenyl	"	II, X = 8-CO_2CH_3, Y = H (100)	35
	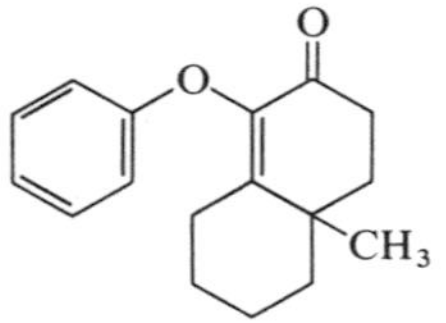 [structure: O, CH$_3$]	C_6H_6, *p*-TSA, degassed	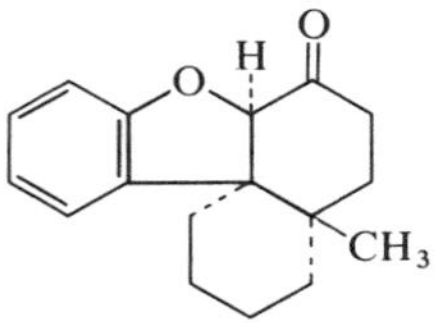 [structure: H, O, CH$_3$] (90)	699
		C_6H_6, *p*-TSA, argon	" (95)	35
C_{18}	2,3,4,5-$(CH_3)_4C_6HOC_6H_3(CH_3)_2$-2,5	C_6H_{12}, I_2, N_2	1,2,3,4,6,9-Hexamethyldibenzofuran (40)[c]	386b
	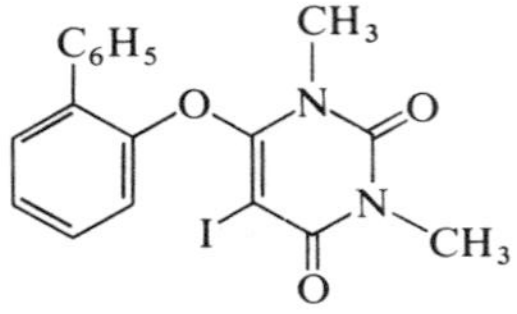 [structure: C_6H_5, O, CH_3, N, O, I, N, CH_3, O]	C_6H_6	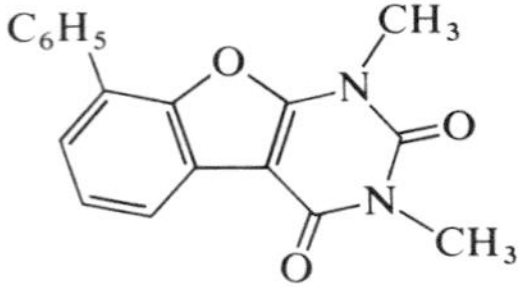 [structure: C_6H_5, O, CH_3, N, O, N, CH_3, O] (3),	700

			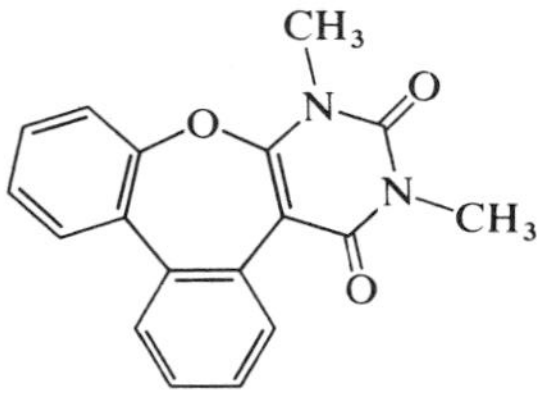	(21)	
	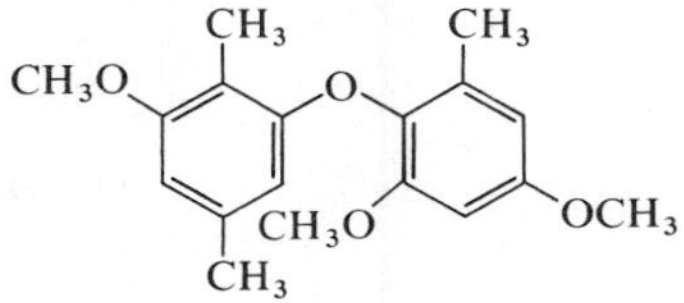	EtOH, H_2O, N_2	3,8-Dimethoxy-1,4,6-trimethyldibenzofuran	(35)[c]	389,390
	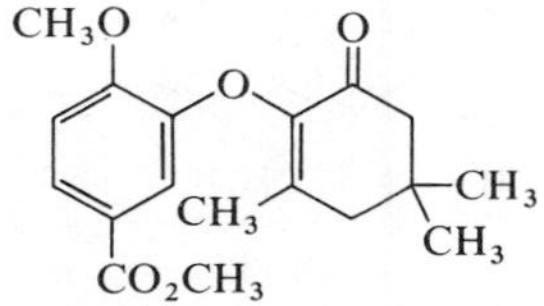	C_6H_6, MeOH, argon	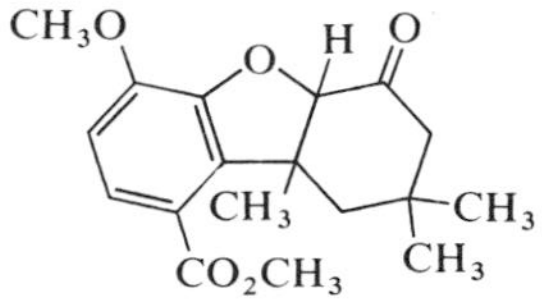	(96)	35
C_{19}	2,3,4,5-$(CH_3)_4C_6HOC_6H_2(CH_3)_3$-2,3,5	C_6H_{12}, I_2, N_2	1,2,3,4,6,7,9-Heptamethyldibenzofuran	(36)	386b
	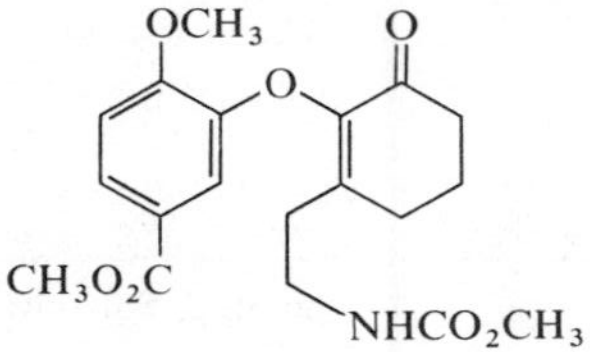	C_6H_6, MeOH, argon	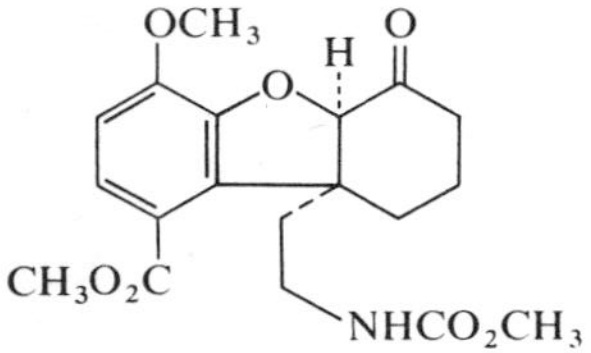	(86)	701
	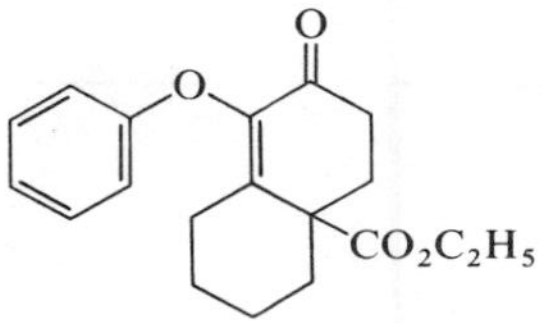	C_6H_6, Ph_2CO	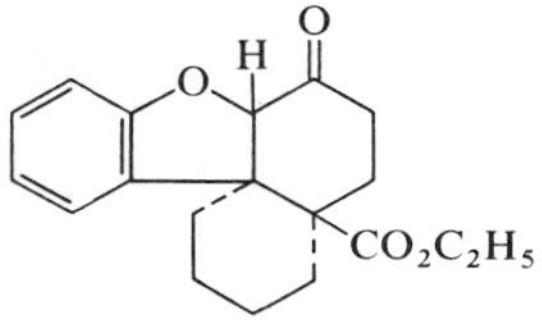	(67)	35,699

TABLE VIII. PHOTOCYCLIZATIONS PRODUCING FIVE-MEMBERED RINGS (*Continued*)

C. C_4O Ring Systems from Diaryl Ethers and Aryl Vinyl Ethers (*Continued*)

Starting Material	Reaction Conditions	Products and Yields (%)	Refs.
	C_6H_6, *p*-TSA	" (40)	35,699
	C_6H_6, MeOH, HOAc, argon	" (29), 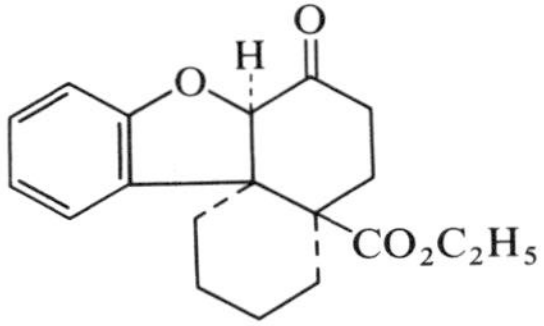 [structure] (35)	35,699
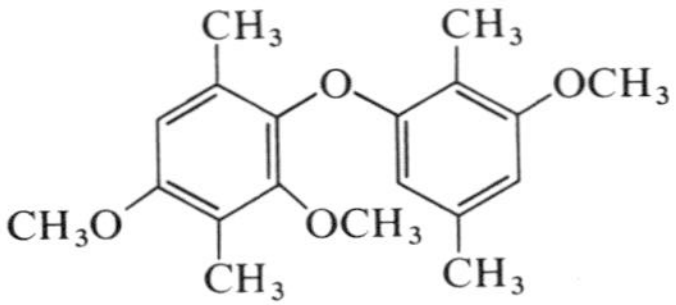 [structure: CH_3O, CH_3, OCH_3 substituted diaryl ether]	EtOH, H_2O, N_2	2,7-Dimethoxy-1,4,6,9-tetramethyldibenzofuran (36)	389,390
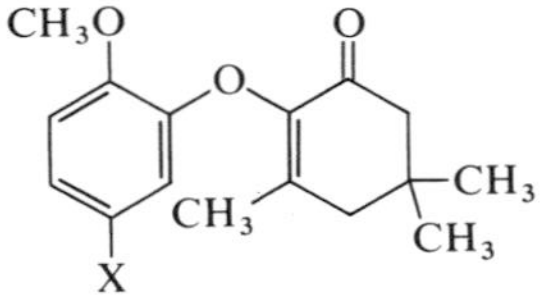 [structure] (I)		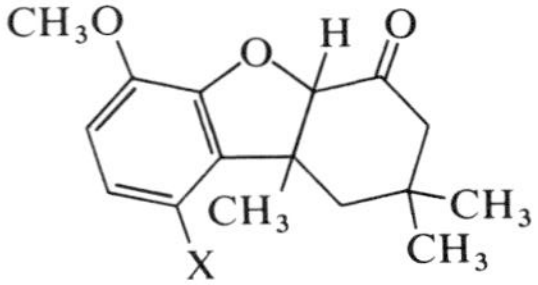 [structure] (II)	
I, X = $CH(OCH_3)_2$	Solvent	II, X = $CH(OCH_3)_2$ (—)	698
I, X 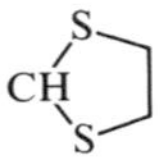 = CH(1,3-dithiolan-2-yl)	C_6H_6, MeOH, HOAc, argon	II, X 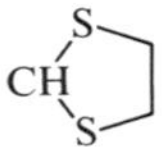 = CH(1,3-dithiolan-2-yl) (40)	35
C_{20} 3,5-(t-C_4H_9)$_2C_6H_3OC_6H_5$	C_6H_{12}, I_2, N_2	1,3-Di-*tert*-butyldibenzofuran (20)	386b

	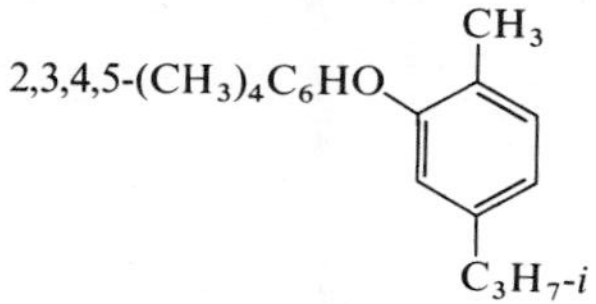 2,3,4,5-$(CH_3)_4C_6HO$–(ring with CH_3, C_3H_7-*i*)	"	9-Isopropyl-1,2,3,4,6-pentamethyldibenzofuran (5)	386b
	2,3,4,5-$(CH_3)_4C_6HOC_6H(CH_3)_4$-2,3,4,5	"	1,2,3,4,6,7,8,9-Octamethyldibenzofuran (34)	386b
	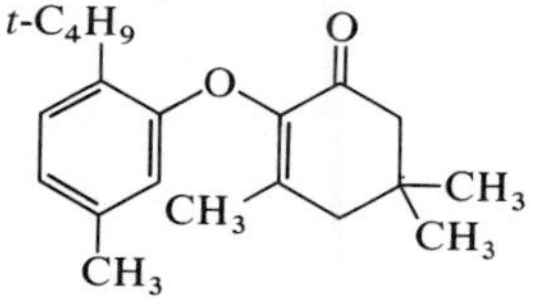 *t*-C_4H_9, O, O, CH_3, CH_3, CH_3, CH_3	C_6H_6, MeOH, argon	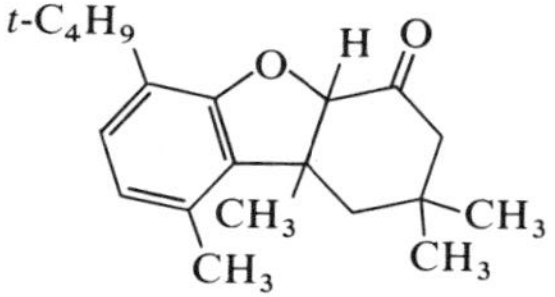 *t*-C_4H_9, O, H, O, CH_3, CH_3, CH_3, CH_3 (85)	35
	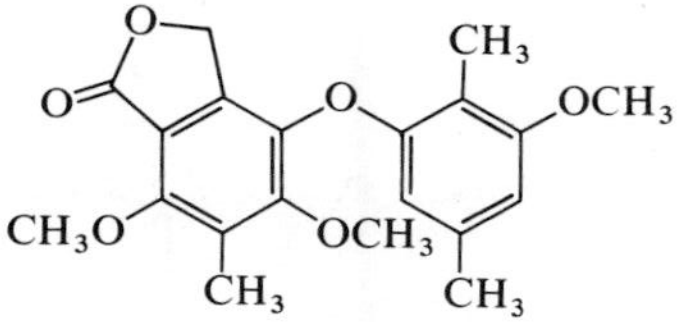 O, O, O, CH_3, OCH_3, CH_3O, OCH_3, CH_3, CH_3	EtOH, H_2O, N_2	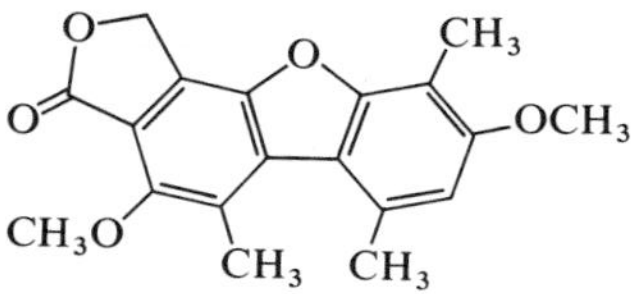 O, O, O, CH_3, OCH_3, CH_3O, CH_3, CH_3 (8–11)	389,390
C_{21}	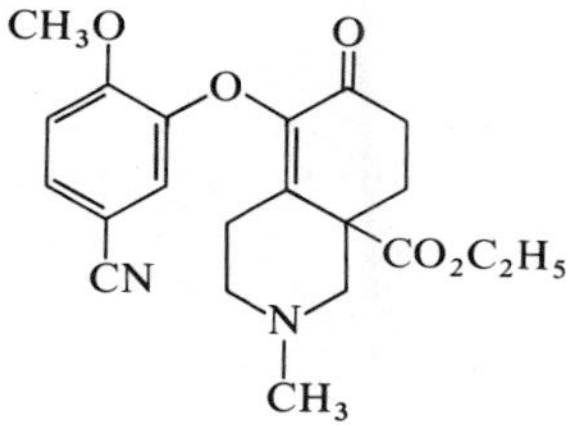 CH_3O, O, O, CN, $CO_2C_2H_5$, N, CH_3	C_6H_6, argon	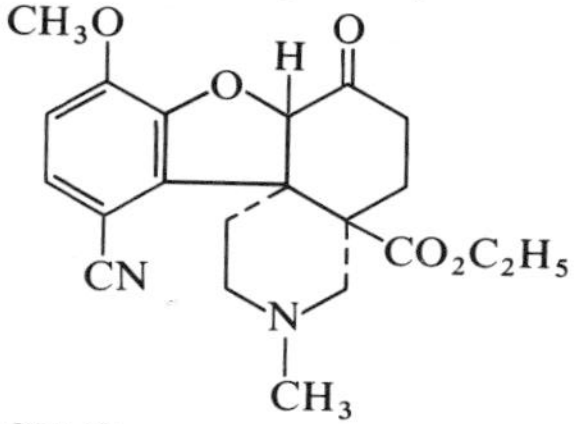 CH_3O, O, H, O, CN, $CO_2C_2H_5$, N, CH_3 (100)	35
		C_6H_6, MeOH, Na_2CO_3, argon	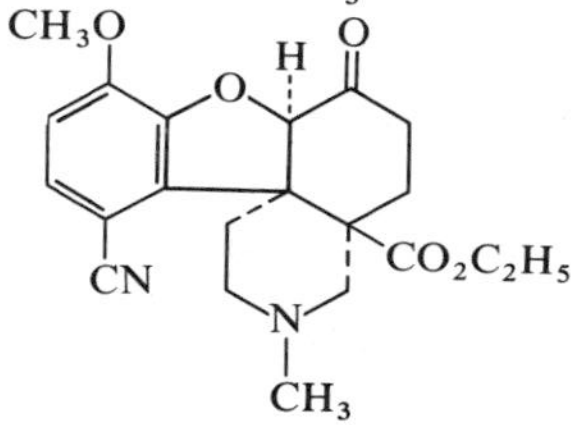 CH_3O, O, H, O, CN, $CO_2C_2H_5$, N, CH_3 (88)	35

TABLE VIII. PHOTOCYCLIZATIONS PRODUCING FIVE-MEMBERED RINGS (*Continued*)

C. C_4O Ring Systems from Diaryl Ethers and Aryl Vinyl Ethers (*Continued*)

	Starting Material	Reaction Conditions	Products and Yields (%)	Refs.
C_{28}	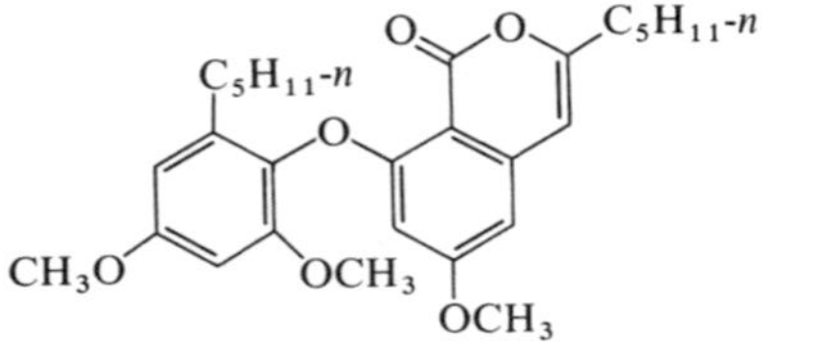 [structure: C_5H_{11}-n, CH_3O, OCH_3, OCH_3, O, C_5H_{11}-n]	EtOH, H_2O, N_2	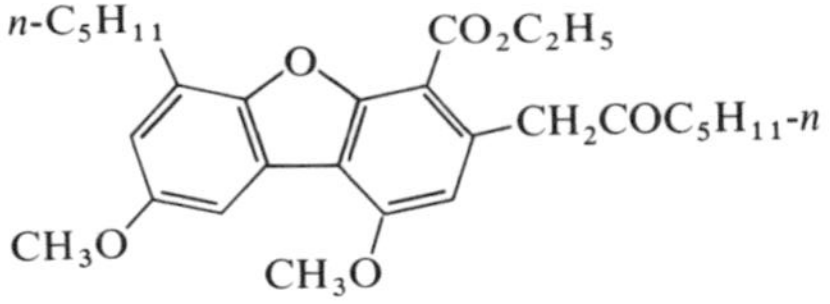 [structure: n-C_5H_{11}, $CO_2C_2H_5$, $CH_2COC_5H_{11}$-n, CH_3O, CH_3O] (2)	390
C_{29} 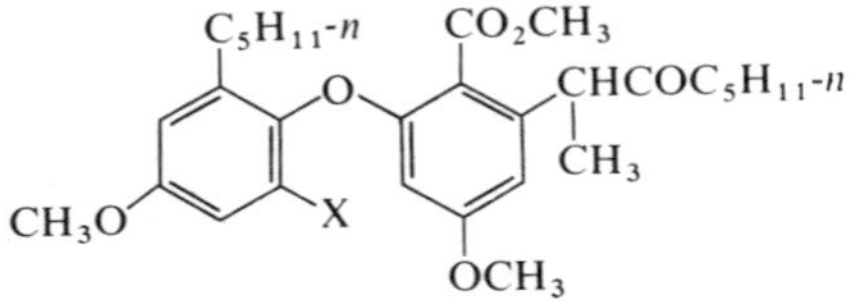 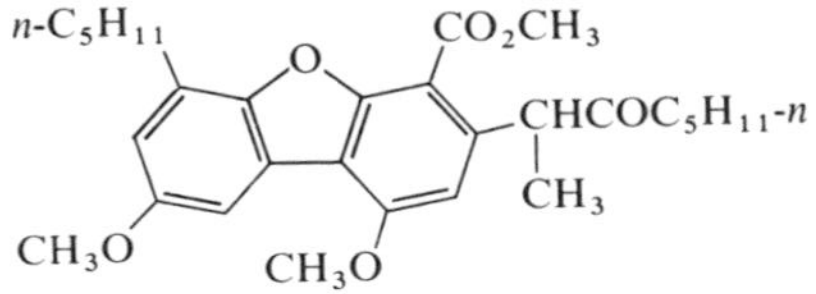	[structure: C_5H_{11}-n, CO_2CH_3, $CHCOC_5H_{11}$-n, CH_3, CH_3O, X, OCH_3]		[structure: n-C_5H_{11}, CO_2CH_3, $CHCOC_5H_{11}$-n, CH_3, CH_3O, CH_3O]	
	X = H	Solvent, air	" (—)	389
C_{30}	X = CH_3O	EtOH, H_2O, N_2	" (6–11)	389,390
C_{33} 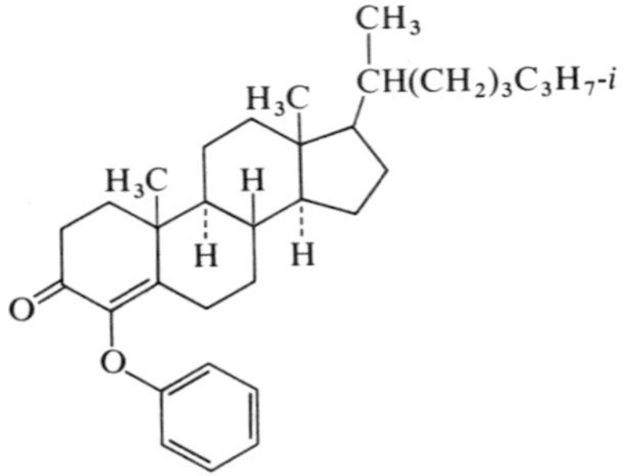 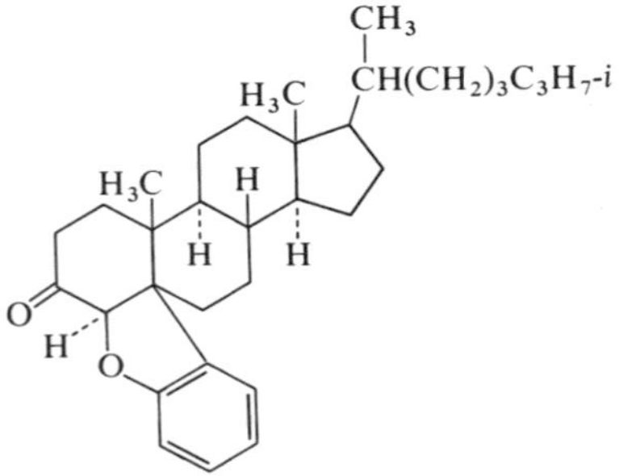	[structure: CH_3, $CH(CH_2)_3C_3H_7$-i, H_3C, H_3C, H, H, H, O, O]	Solvent	[structure: CH_3, $CH(CH_2)_3C_3H_7$-i, H_3C, H_3C, H, H, H, O, H, O] (high)	699

D. C_4S Ring Systems from Diaryl Sulfides and Aryl Vinyl Sulfides

	Starting Material	Reaction Conditions	Products and Yields (%)	Refs.
	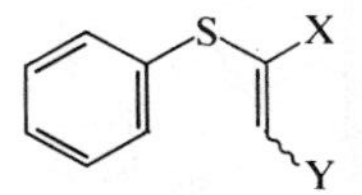 (I)		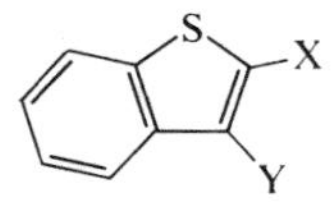 (II)	
C_8	I, X = Y = H	Et_2O, I_2, N_2	II, X = Y = H (10)	702
C_9	I, X = CH_3, Y = H	"	II, X = CH_3, Y = H (9)	702
	I, X = H, Y = CH_3	"	II, X = H, Y = CH_3 (3)	702
C_{10}	I, X = Y = CH_3	"	II, X = Y = CH_3 (5)	702
	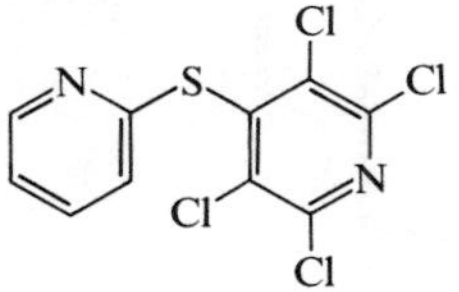	THF, EtOH	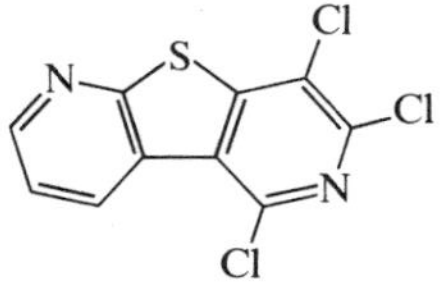 (13)	388b
	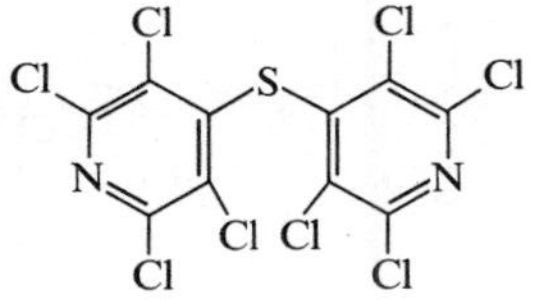	CCl_4	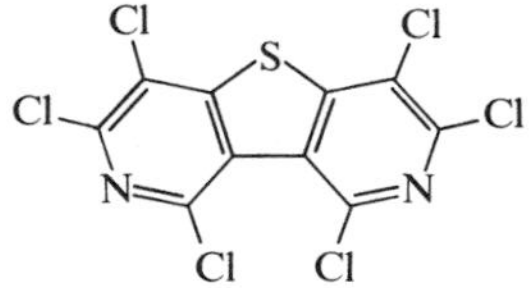 (44)	388b
	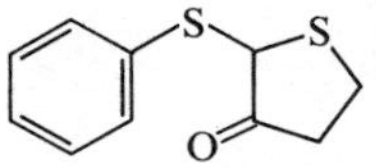	MeCN	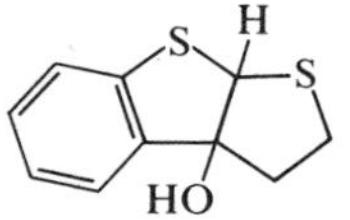 (43)	403
C_{11}	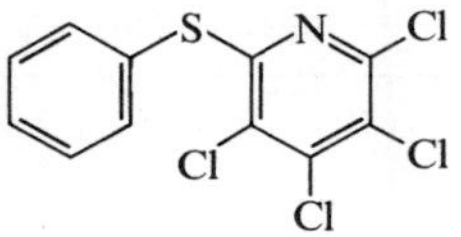	EtOH	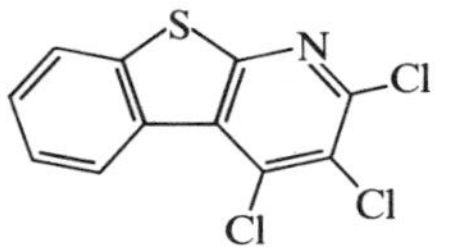 (87)	388a,b

TABLE VIII. PHOTOCYCLIZATIONS PRODUCING FIVE-MEMBERED RINGS (*Continued*)

D. C_4S Ring Systems from Diaryl Sulfides and Aryl Vinyl Sulfides (*Continued*)

	Starting Material	Reaction Conditions	Products and Yields (%)	Refs.
	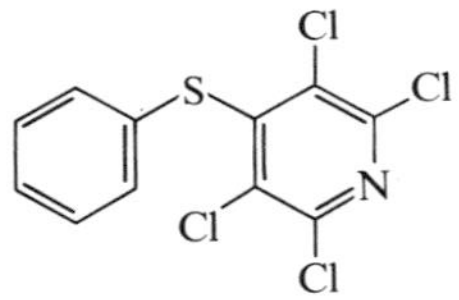 [structure]	EtOH	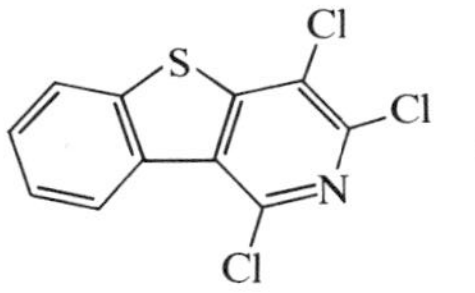 [structure] (68)	388a,b
	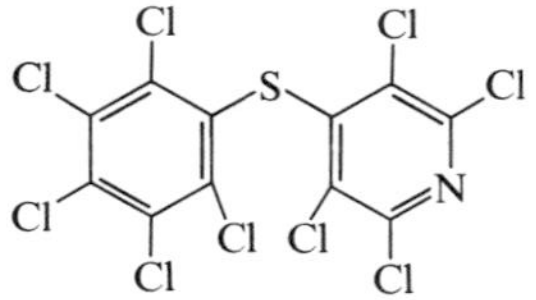 [structure]	CCl_4	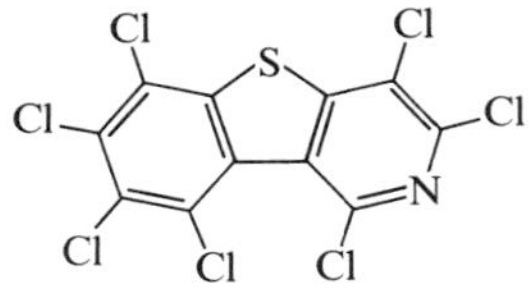 [structure] (36)	388b
	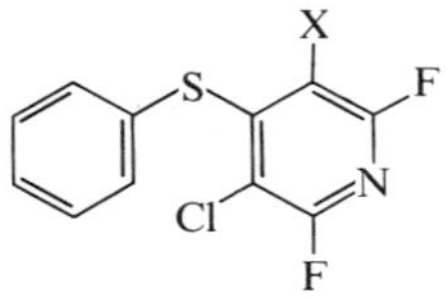 [structure] (I)		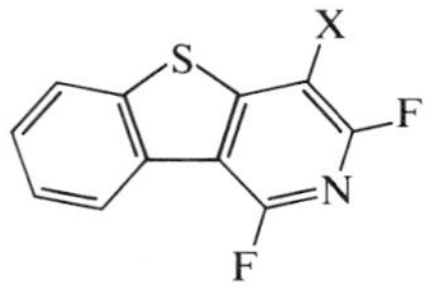 [structure] (II)	
	I, X = Cl	EtOH	II, X = Cl (83)	388a,b
	I, X = F	"	II, X = F (46–48)	388a,b
	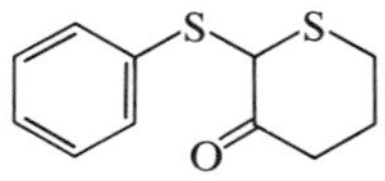 [structure]	MeCN	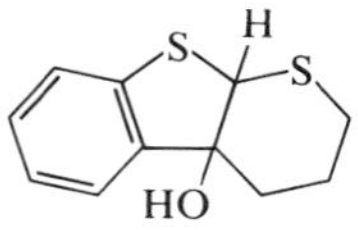 [structure] (43)	403
C_{12}	Diphenyl sulfide	C_6H_{12}, I_2, O_2	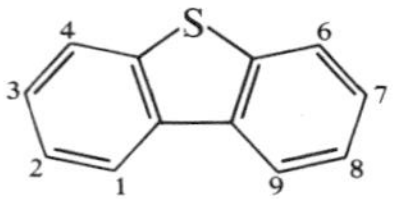 [structure] (54) (Dibenzothiophene)	386a

	Et_2O, I_2, N_2	" (3)	702
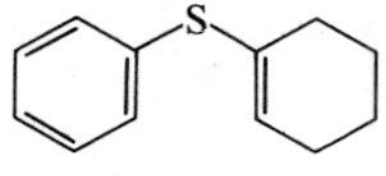	"	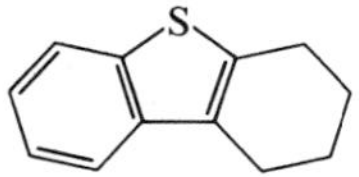 (6)	702
$C_6Cl_5SC_6Cl_5$	CCl_4	1,2,3,4,6,7,8,9-Octachlorodibenzothiophene (42)[d]	703
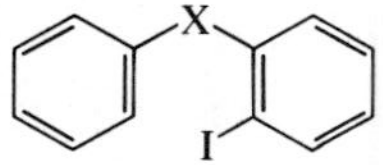 (I)		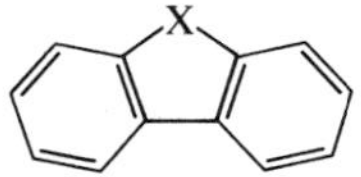 (II)	
I, X = S	C_6H_6	II, X = S (53)	704a
I, X = SO	"	II, X = SO (27)	704a
I, X = SO_2	"	II, X = SO_2 (3)	704a
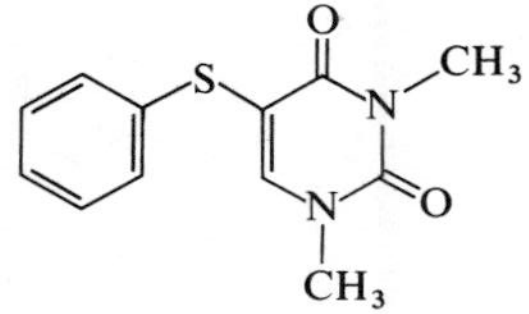	Me_2CO, air	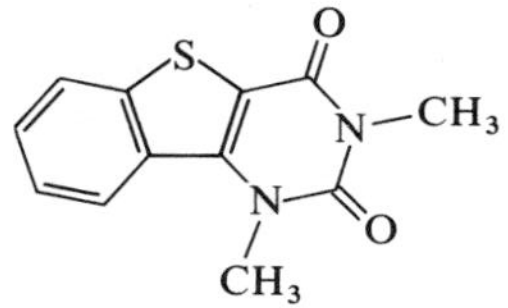 (9)	398
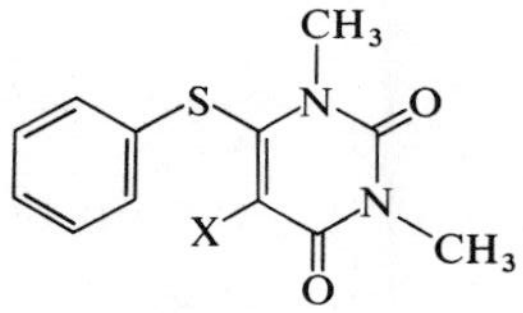 X = H	"	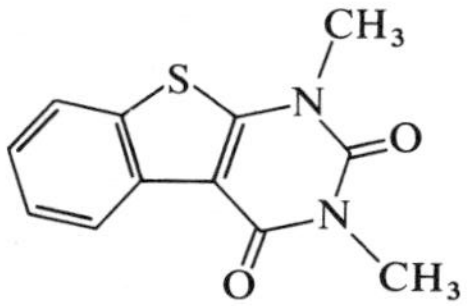 " (30)	398
X = I	MeCN	" (42)	697
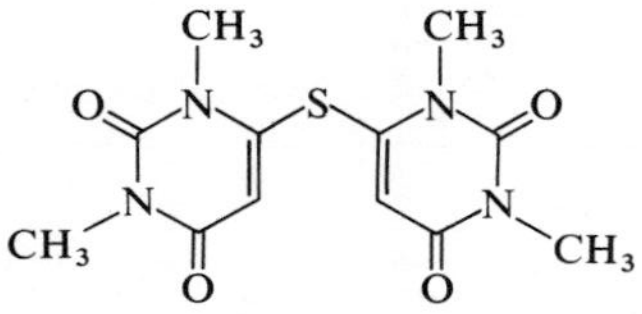	THF, N_2	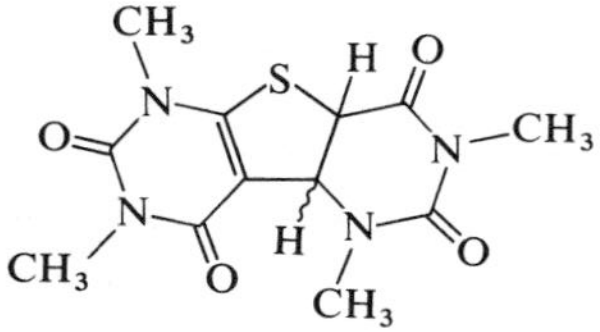 (9)	704b

TABLE VIII. PHOTOCYCLIZATIONS PRODUCING FIVE-MEMBERED RINGS (*Continued*)

D. C_4S Ring Systems from Diaryl Sulfides and Aryl Vinyl Sulfides (Continued)

	Starting Material	Reaction Conditions	Products and Yields (%)	Refs.
	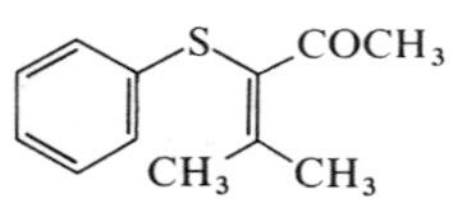 S, $COCH_3$, CH_3, CH_3	C_6H_6, MeOH, HOAc, argon	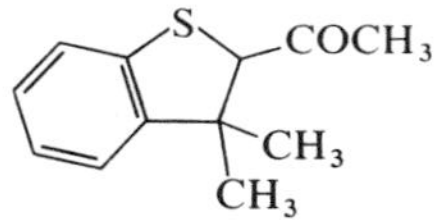 S, $COCH_3$, CH_3, CH_3 (84)	34
C_{14}	p-$CH_3C_6H_4SC_6H_4CH_3$-p	C_6H_{12}, I_2, O_2	2,8-Dimethyldibenzothiophene (43–47)[d]	386a
	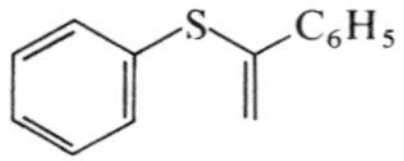 S, C_6H_5	Et_2O, I_2, N_2	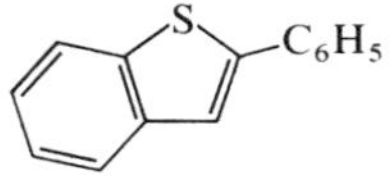 S, C_6H_5 (5), S, C_6H_5 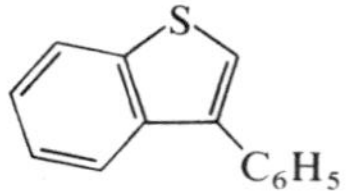 (4)	702
	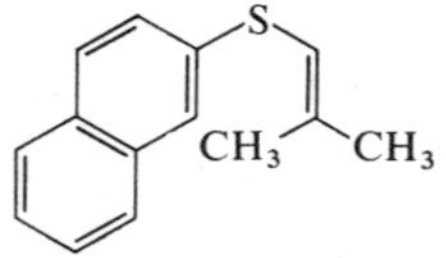 S, CH_3, CH_3	C_6H_6, degassed	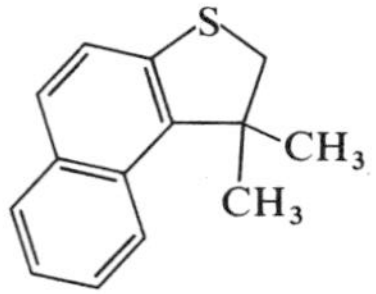 S, CH_3, CH_3 (—)	74
	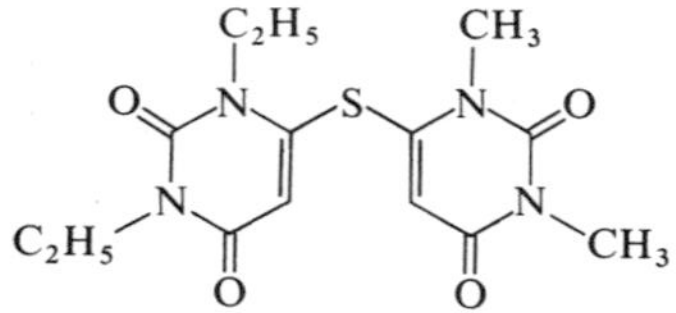 C_2H_5, CH_3, O, N, S, N, O, N, N, C_2H_5, CH_3, O, O	THF, N_2	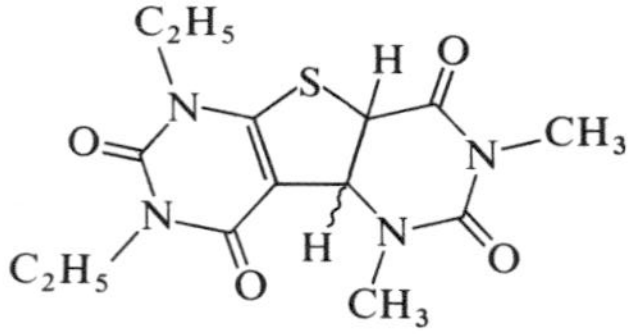 C_2H_5, N, S, H, O, O, N, CH_3, N, H, N, O, C_2H_5, O, CH_3 (7)	704b
	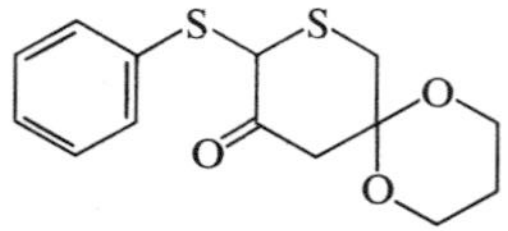 S, S, O, O, O	MeCN	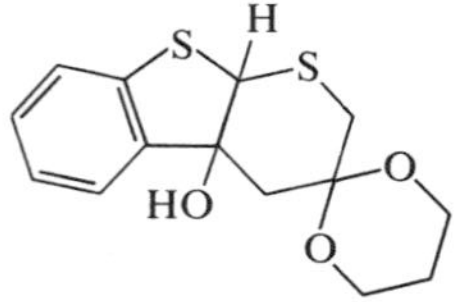 H, S, S, O, HO, O (53)	403

	Reactant	Conditions	Product(s) (yield %)	Ref.
C_{15}	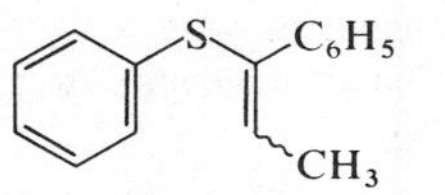 C_6H_5, CH_3	Et_2O, I_2, N_2	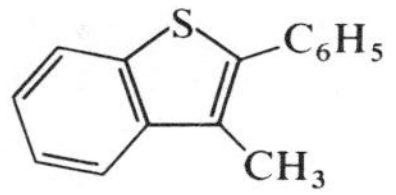 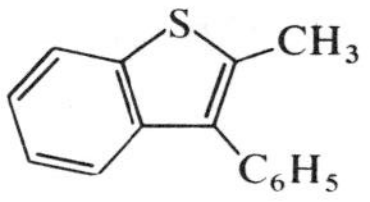 C_6H_5, CH_3 (8), CH_3, C_6H_5 (4)	702
	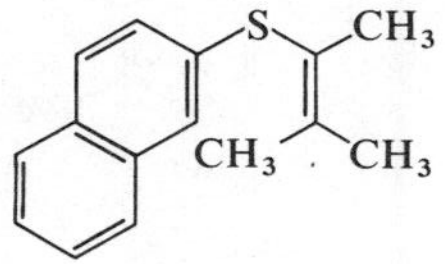 CH_3, CH_3, CH_3	C_6H_6, argon	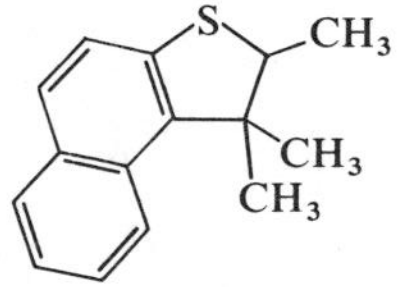 CH_3, CH_3, CH_3 (87–90)	74,705,706
	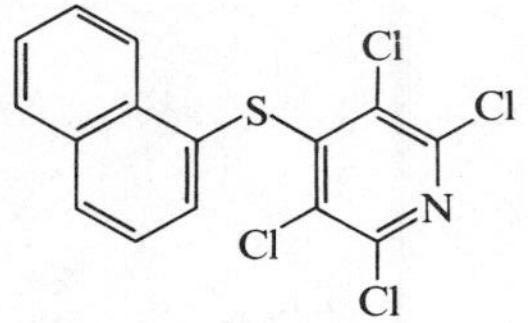 Cl, Cl, N, Cl, Cl	THF	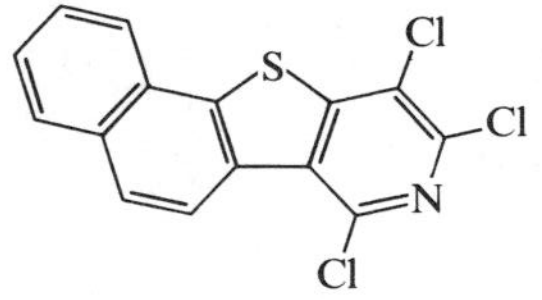 Cl, Cl, N, Cl (83)	388b
	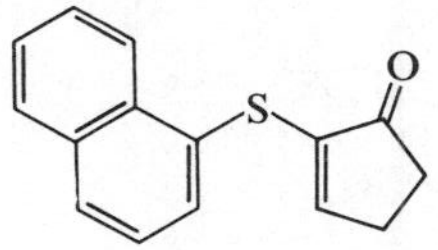 O	C_6H_6, MeOH, air	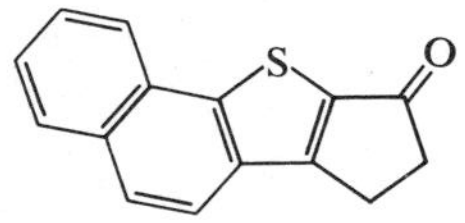 O (—)	34
		C_6H_6, MeOH, argon	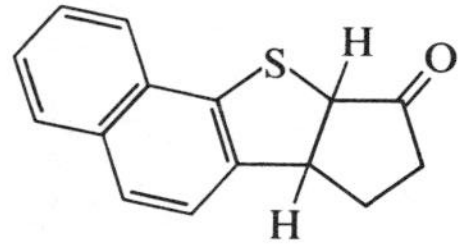 H, O, H (86)	34
	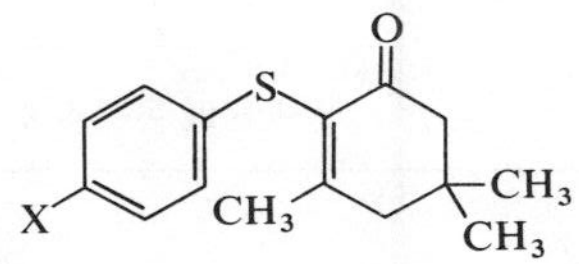 O, CH_3, CH_3, CH_3, X (I)		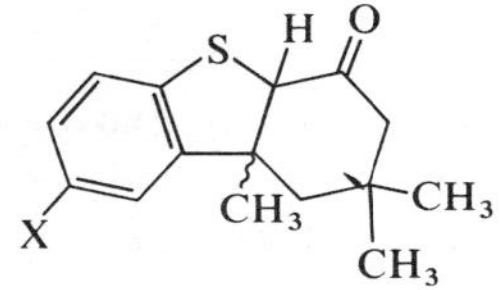 H, O, CH_3, CH_3, CH_3, X (II)	
	I, X = H	"	II, X = H (*cis*, 91)	34

TABLE VIII. PHOTOCYCLIZATIONS PRODUCING FIVE-MEMBERED RINGS (*Continued*)

D. C_4S Ring Systems from Diaryl Sulfides and Aryl Vinyl Sulfides (*Continued*)

	Starting Material	Reaction Conditions	Products and Yields (%)	Refs.
		C_6H_6, degassed	" (*trans*, 27; *cis*, 27)	34
	I, X = OH	C_6H_6, MeOH, argon	II, X = OH (*cis*, 83)	34
	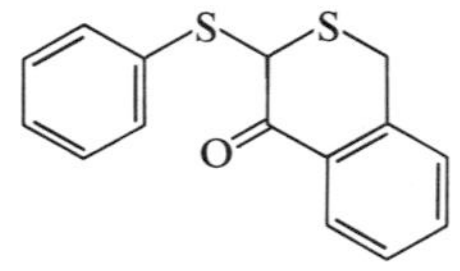 [structure]	MeCN	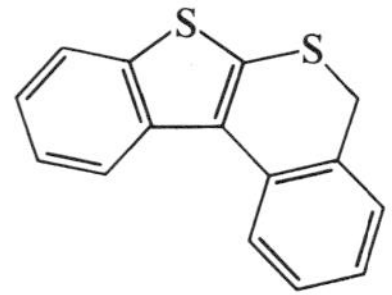 [structure] (30)	403
	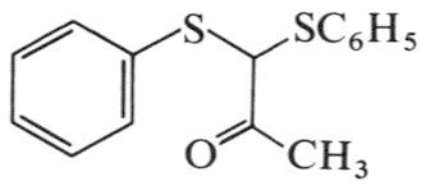 [structure] SC_6H_5, CH_3	"	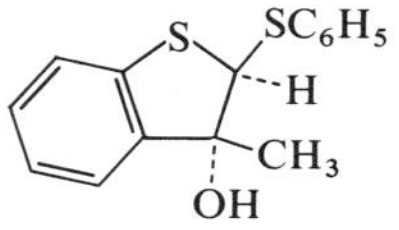 [structure] SC_6H_5, H, CH_3, OH (19)	403
C_{16}	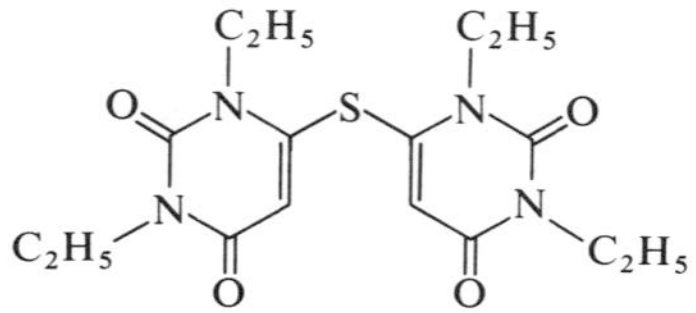 [structure]	THF, N_2[b]	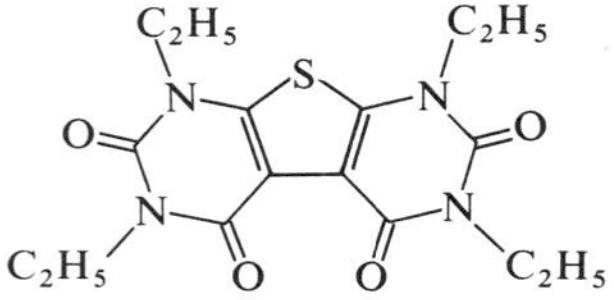 [structure] (13)	704b
	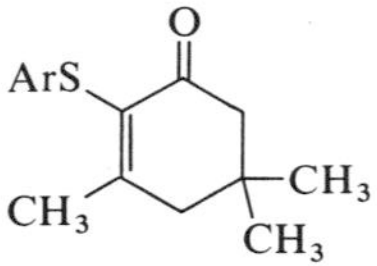 [structure] ArS, CH_3 (I)		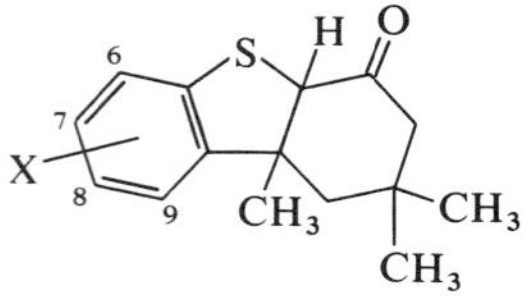 [structure] (II)	
	I, Ar = *o*-Methylphenyl	C_6H_6, MeOH, argon	II, X = 6-CH_3 (88)	34
	I, Ar = *m*-Methylphenyl	"	II, X = 9-CH_3 (64); II, X = 7-CH_3 (28)	34
	I, Ar = *p*-Methylphenyl	"	II, X = 8-CH_3 (84)	34

	Reactant	Conditions	Product (Yield %)		Ref.
	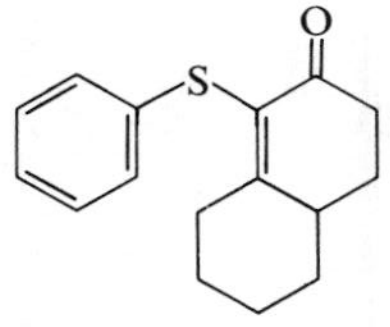 C_6H_5S-(octalone)	"	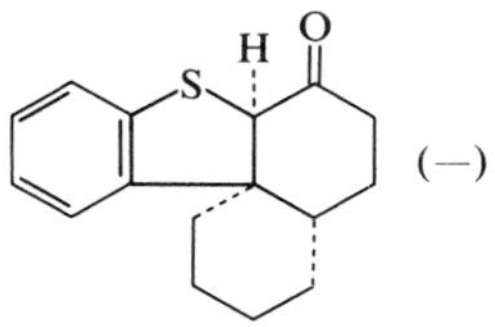	(—)	34
	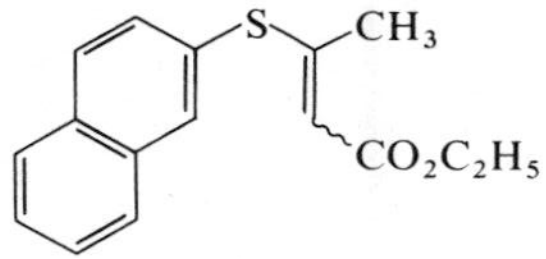 CH_3, $CO_2C_2H_5$	C_6H_6, $(p\text{-}Me_2NC_6H_4)_2CO$, argon	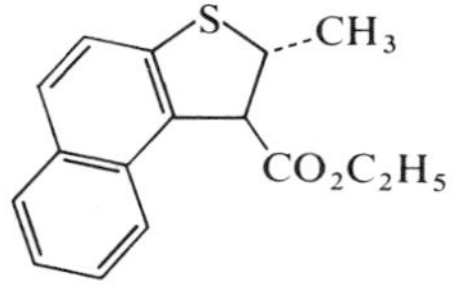 CH_3, $CO_2C_2H_5$	(83)	34
C_{17}	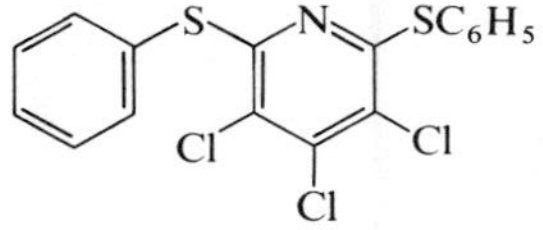 SC_6H_5, Cl, Cl, Cl	EtOH	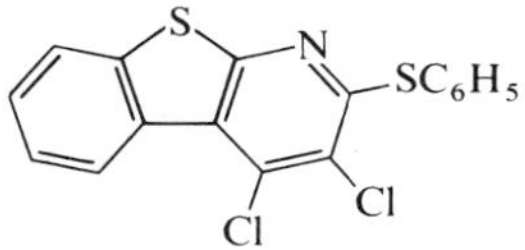 SC_6H_5, Cl, Cl	(26–40)	388a,b
	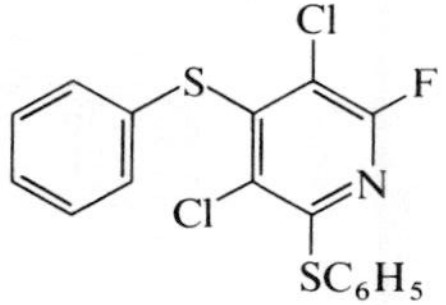 Cl, F, Cl, SC_6H_5	"	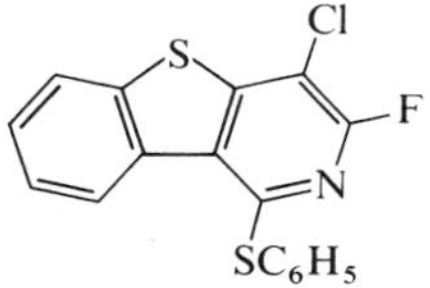 Cl, F, SC_6H_5	(66)	388b
	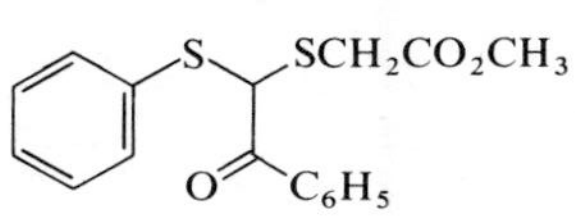 $SCH_2CO_2CH_3$, C_6H_5	MeCN	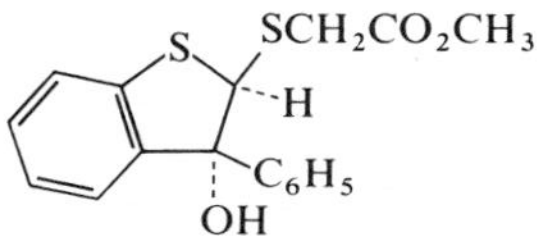 $SCH_2CO_2CH_3$, H, C_6H_5, OH	(36)	403
C_{18}	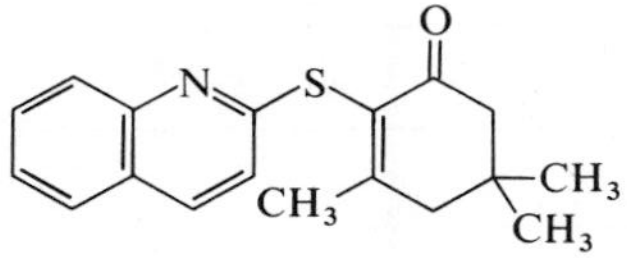 CH_3, CH_3, CH_3	C_6H_6, MeOH, argon	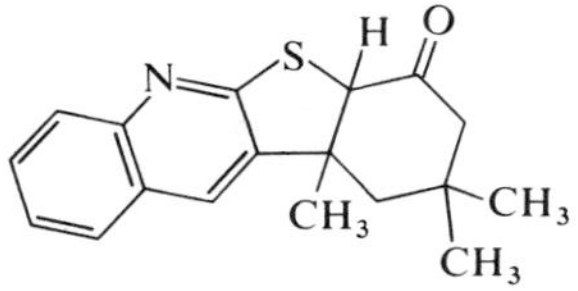 H, CH_3, CH_3, CH_3	(78)	34

TABLE VIII. PHOTOCYCLIZATIONS PRODUCING FIVE-MEMBERED RINGS (*Continued*)

D. C_4S Ring Systems from Diaryl Sulfides and Aryl Vinyl Sulfides (*Continued*)

	Starting Material	Reaction Conditions	Products and Yields (%)	Refs.
	N, CH_3, S, O, CH_3, CH_3, CH_3	C_6H_6, MeOH, argon	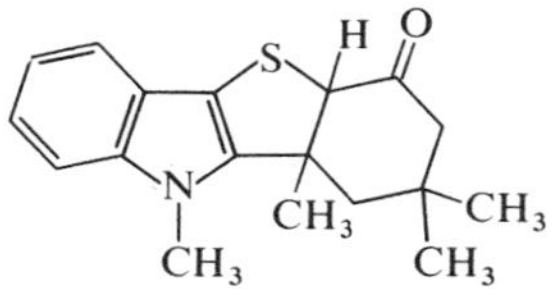 S, H, O, N, CH_3, CH_3, CH_3, CH_3 (63)	34
C_{19} 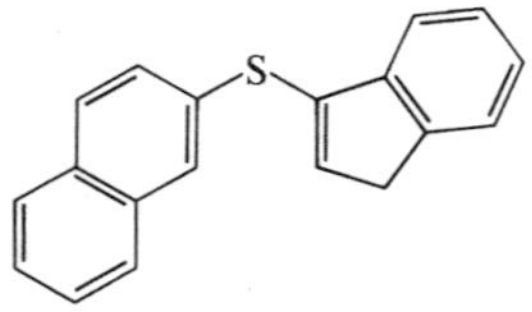	S	C_6H_6, degassed, uranyl glass filter	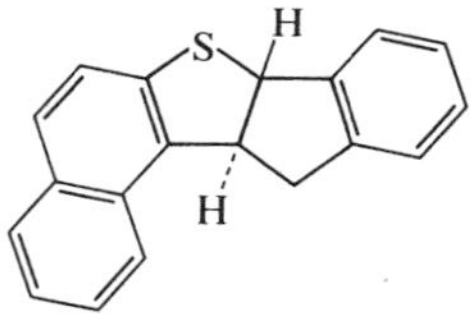 S, H, H (78)	74
		C_6H_6, degassed, Pyrex	Polymer (—)	74
		C_6H_6, argon, Pyrex, *N*-phenylmaleimide	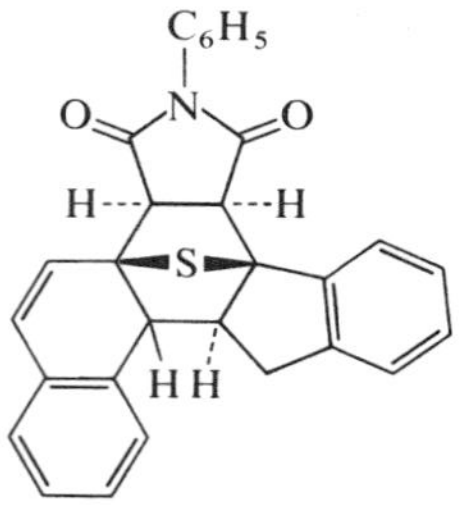 C_6H_5, N, O, O, H, H, S, H H (84)	74,393
	S, CH_3, C_6H_5	C_6H_6, degassed	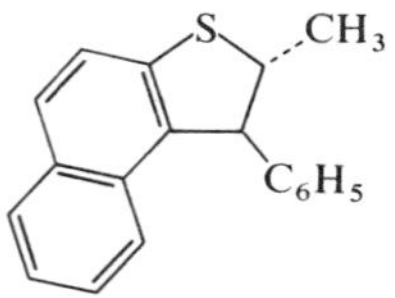 S, CH_3, C_6H_5 (50–59)	74,705

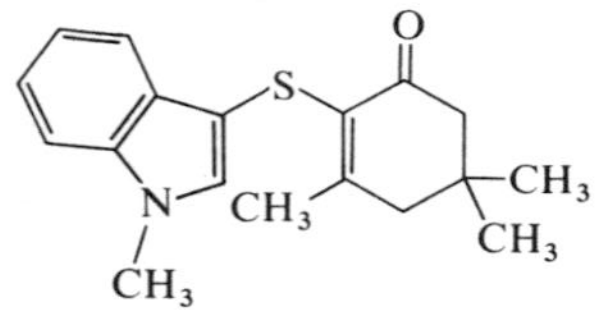

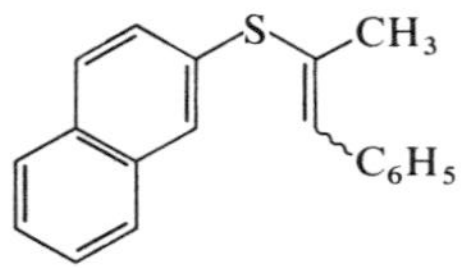

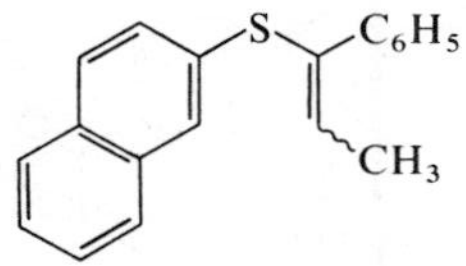

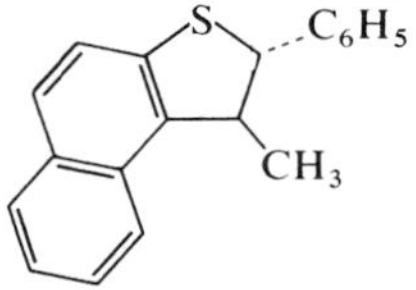

	Reactant	Conditions	Product(s) (yield %)	Ref.
	[structure: 2-naphthyl–S–C(C_6H_5)=CH–CH_3]	"	[structure: 2-C_6H_5-1-CH_3-dihydronaphtho[2,1-b]thiophene] (70–74),	74,705
			[structure: 2-CH_3-1-C_6H_5-dihydronaphtho[2,1-b]thiophene] (10–15)	
		C_6H_6, MeOD	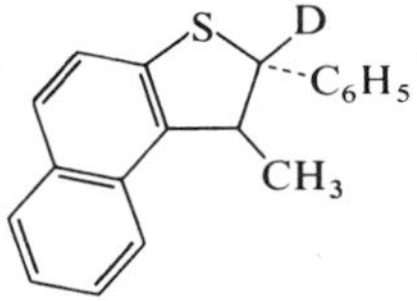 [structure: 2-D-2-C_6H_5-1-CH_3-dihydronaphtho[2,1-b]thiophene] (—)	74
	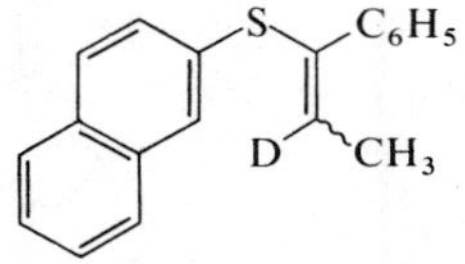 [structure: 2-naphthyl–S–C(C_6H_5)=C(D)–CH_3]	C_6H_6, degassed	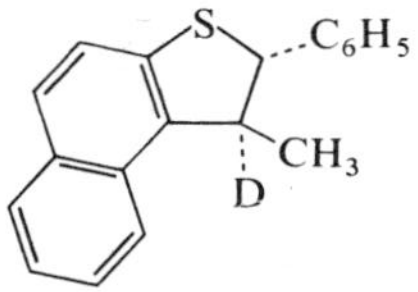 [structure: 2-C_6H_5-1-CH_3-1-D-dihydronaphtho[2,1-b]thiophene] (—)	74,705
	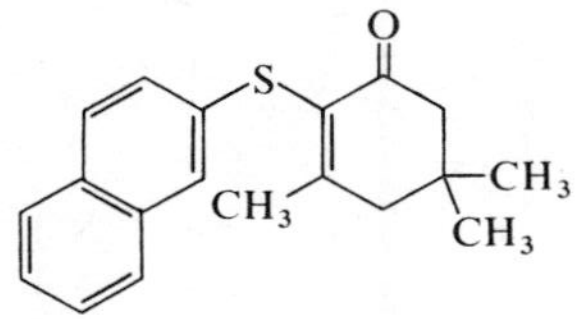 [structure: 2-(2-naphthylthio)-3,5,5-trimethylcyclohex-2-en-1-one]	C_6H_6, MeOH, argon	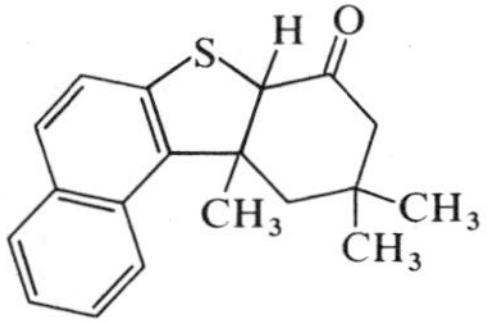 [structure: cyclized ketone, H, O, CH_3, CH_3, CH_3] (89)	34
C_{20}	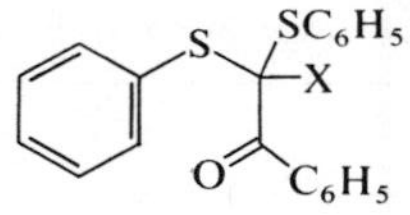 [structure: C_6H_5S–C(X)(SC_6H_5)–C(=O)C_6H_5] (I)		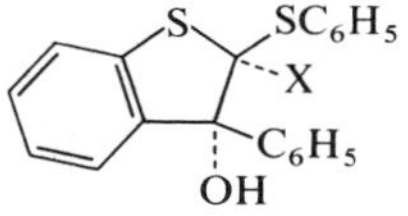 [structure: 2-SC_6H_5-2-X-3-C_6H_5-3-OH-dihydrobenzo[b]thiophene] (II)	

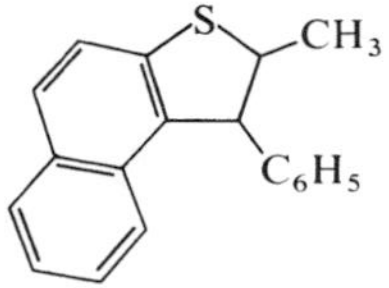

TABLE VIII. PHOTOCYCLIZATIONS PRODUCING FIVE-MEMBERED RINGS (*Continued*)

D. C_4S Ring Systems from Diaryl Sulfides and Aryl Vinyl Sulfides (*Continued*)

	Starting Material	Reaction Conditions	Products and Yields (%)	Refs.
	I, X = H	MeCN	II, X = H (71)	403
	I, X = D	"	II, X = D (45)	403
C_{21}	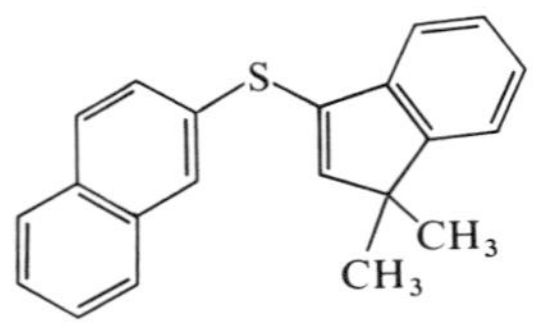 [structure: CH_3, CH_3, S]	C_6H_6, degassed	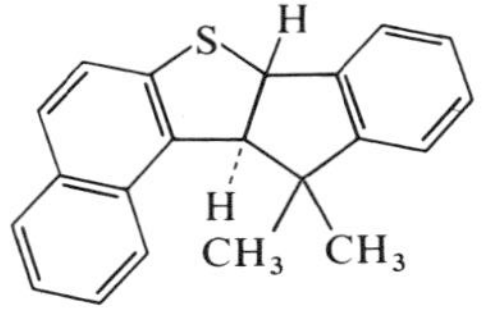 [structure: S, H, H, CH_3, CH_3] (50)	74
	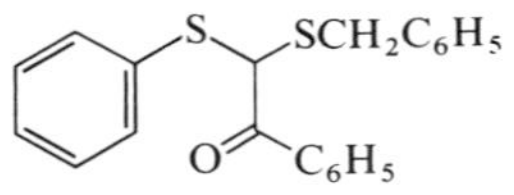 [structure: S, $SCH_2C_6H_5$, O, C_6H_5]	MeCN	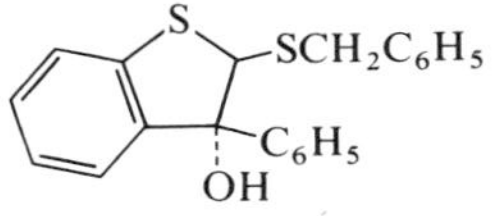 [structure: S, $SCH_2C_6H_5$, C_6H_5, OH] (46)	403
C_{23}	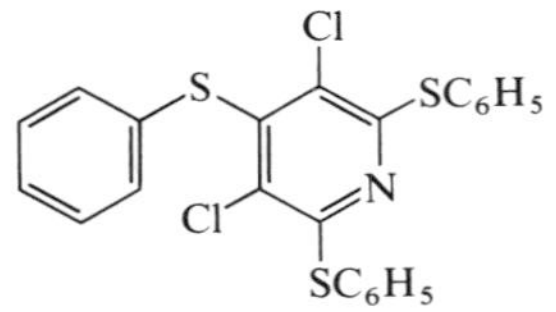 [structure: S, Cl, SC_6H_5, N, Cl, SC_6H_5]	EtOH	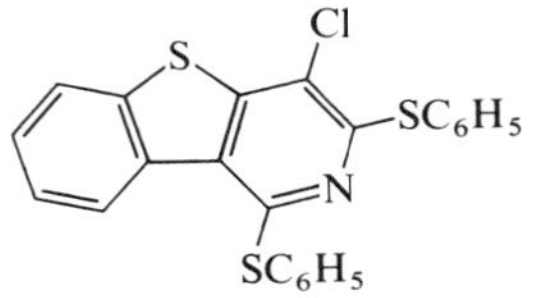 [structure: S, Cl, SC_6H_5, N, SC_6H_5] (30)	388a,b
C_{33}	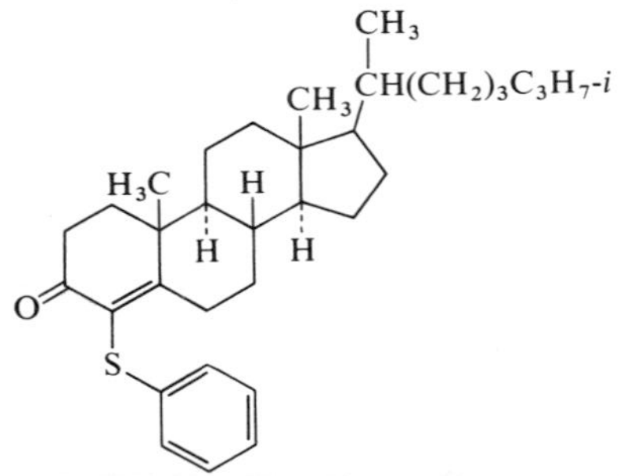 [structure: CH_3, $CH(CH_2)_3C_3H_7$-*i*, CH_3, H_3C, H, H, H, O, S]	C_6H_6, MeOH, argon	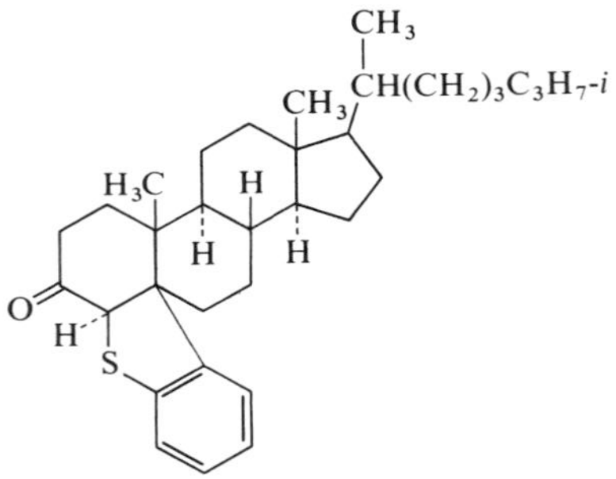 [structure: CH_3, $CH(CH_2)_3C_3H_7$-*i*, CH_3, H_3C, H, H, H, O, H, S] (90)	34

	Reactant	Conditions	Product(s) and Yield(s) (%)	Refs.
C_{37}	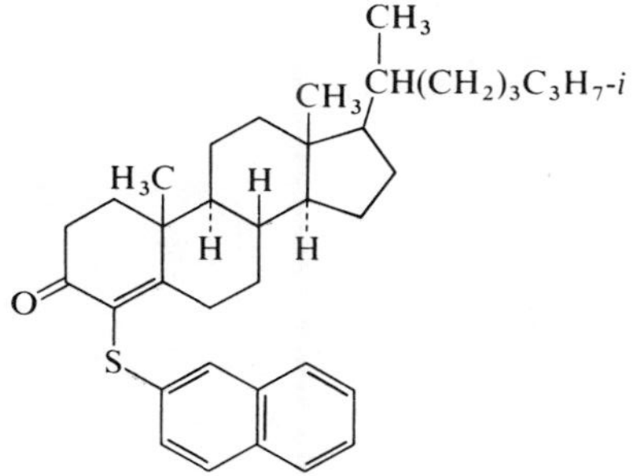 CH_3, $CH(CH_2)_3C_3H_7$-i, CH_3, H_3C, H, H, H, O, S (structure)	C_6H_6, MeOH, argon	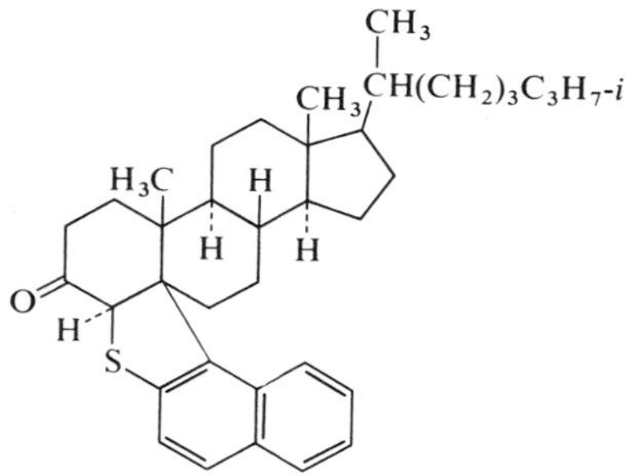 CH_3, $CH(CH_2)_3C_3H_7$-i, CH_3, H_3C, H, H, H, O, H, S (structure) (69)	34

E. Miscellaneous Ring Systems (C_3NS, C_4Se, and C_4S)

	Reactant	Conditions	Product(s) and Yield(s) (%)	Refs.
C_8	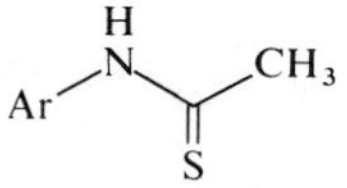 Ar–NH–C(=S)–CH_3 (I)		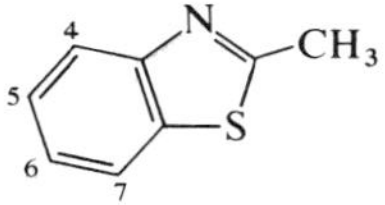 2-methylbenzothiazole, ring positions 4, 5, 6, 7 numbered (II)	
	I, Ar = 2,3-Dichlorophenyl	MeOH	7-Chloro-II (64)	707
	I, Ar = 2,4-Dichlorophenyl	"	6-Chloro-II (69)	707
	I, Ar = 2,4,6-Trichlorophenyl	"	4,6-Dichloro-II (94)	707
	I, Ar = 6-Bromo-2,4-dichlorophenyl	"	" (66)	707
C_9	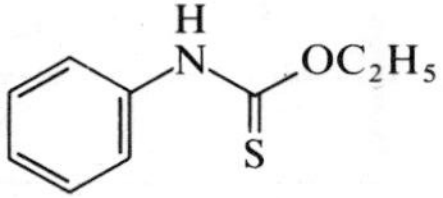 C_6H_5–NH–C(=S)–OC_2H_5	t-BuOH, dioxane, or C_6H_6; O_2	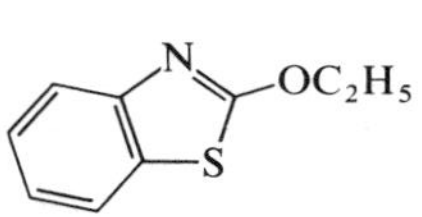 2-ethoxybenzothiazole (OC_2H_5) (2–3)	708
C_{11}	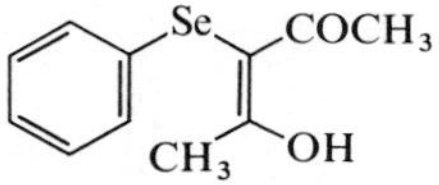 C_6H_5–Se–C($COCH_3$)=C(OH)CH_3	C_6H_6, p-TSA, argon	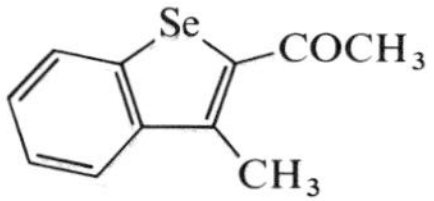 2-acetyl-3-methylbenzoselenophene ($COCH_3$, CH_3) (60)	397

TABLE VIII. PHOTOCYCLIZATIONS PRODUCING FIVE-MEMBERED RINGS (*Continued*)

E. Miscellaneous Ring Systems (C_3NS, C_4Se, and C_4S) (Continued)

Starting Material	Reaction Conditions	Products and Yields (%)	Refs.
	C_6D_6, HOAc	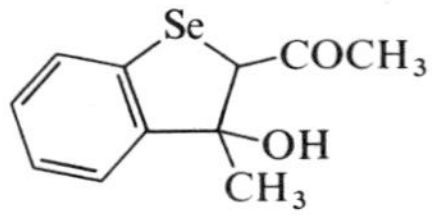 Se, $COCH_3$, OH, CH_3 (30)	397
C_{13} 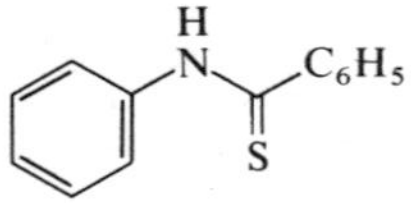 H, N, C_6H_5, S	C_6H_{14}, O_2	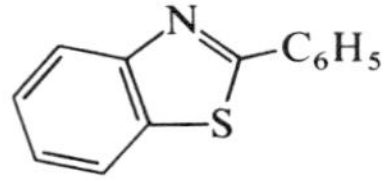 N, S, C_6H_5 (—)	405
C_{17} 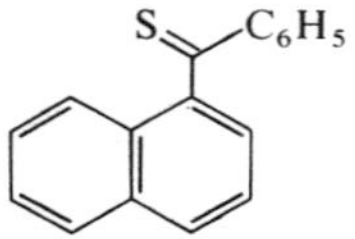 S, C_6H_5	C_6H_6, degassed	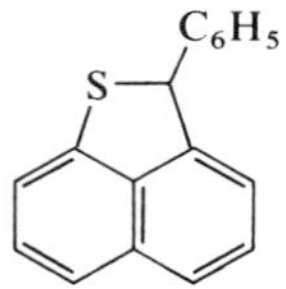 S, C_6H_5 (51)	404
	MeCN, MeOH, degassed	" (—)	404
	MeCN, H_2O, degassed	" (—)	404
	MeCN, D_2O, degassed	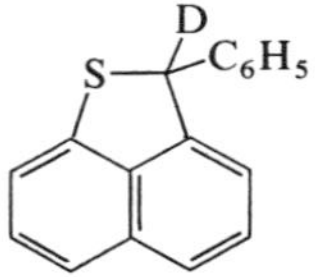 S, D, C_6H_5 (—)	404

	Reactant	Conditions	Product (yield)	Ref.
C_{23}	[structure: $S{=}C(C_6H_5)$ on pyrene]	C_6H_6, degassed	[structure, C_6H_5, S] (59)	404
C_{25}	[structure: $S{=}C(C_6H_5)$ on chrysene-type ring]	"	[structure, C_6H_5, S] (58)	404
C_{27}	[structure: $S{=}C(C_6H_5)$ on perylene]	"	[structure, C_6H_5, S] (60)	404

[a] It is uncertain whether the net loss of two hydrogens in this photocyclization occurs by elimination of H_2 or by oxidation by an adventitious oxidant such as residual dissolved oxygen.

[b] The identity of the oxidant in this reaction is obscure. One possibility is residual dissolved oxygen.

[c] The numbering scheme for dibenzofuran is indicated in a C_{12} entry in part C of this table.

[d] The numbering scheme for dibenzothiophene is indicated in a C_{12} entry in part D of this table.

REFERENCES

[1] F. R. Stermitz, *Org. Photochem.*, **1**, 247 (1967).

[2] M. Scholz, F. Dietz, and M. Mühlstädt, *Z. Chem.*, **7**, 329 (1967).

[3] A. Schönberg, *Preparative Organic Photochemistry*, Springer-Verlag, New York, 1968, pp. 127–145.

[4] E. V. Blackburn and C. J. Timmons, *Quart. Rev., Chem. Soc.*, **23**, 482 (1969).

[5] S. T. Reid, *Adv. Heterocycl. Chem.*, **11**, 87 (1970).

[6] (a) P. G. Sammes, *Quart. Rev., Chem. Soc.*, **24**, 37 (1970); (b) G. Kaupp, *Angew. Chem., Int. Ed. Engl.*, **19**, 243 (1980); (c) J. Grimshaw and A. P. de Silva, *Chem. Soc. Rev.*, **10**, 181 (1981).

[7] H. Stegemeyer, *Z. Naturforsch. Teil B*, **17**, 153 (1962).

[8] F. B. Mallory, C. S. Wood, J. T. Gordon, L. C. Lindquist, and M. L. Savitz, *J. Am. Chem. Soc.*, **84**, 4361 (1962).

[9] (a) W. M. Moore, D. D. Morgan, and F. R. Stermitz, *J. Am. Chem. Soc.*, **85**, 829 (1963); (b) R. Srinivasan and J. C. Powers, Jr., *J. Am. Chem. Soc.*, **85**, 1355 (1963); R. Srinivasan and J. C. Powers, Jr., *J. Chem. Phys.*, **39**, 580 (1963).

[10] F. B. Mallory, C. S. Wood, and J. T. Gordon, *J. Am. Chem. Soc.*, **86**, 3094 (1964).

[11] G. S. Hammond, J. Saltiel, A. A. Lamola, N. J. Turro, J. S. Bradshaw, D. O. Cowan, R. C. Counsell, V. Vogt, and C. Dalton, *J. Am. Chem. Soc.*, **86**, 3197 (1964).

[12] A. A. Lamola, G. S. Hammond, and F. B. Mallory, *Photochem. Photobiol.*, **4**, 259 (1965).

[13] G. M. Badger, R. J. Drewer, and G. E. Lewis, *Aust. J. Chem.*, **19**, 643 (1966).

[14] K. A. Muszkat and E. Fischer, *J. Chem. Soc. B*, **1967**, 662.

[15] J. Saltiel, *J. Am. Chem. Soc.*, **90**, 6394 (1968).

[16] A. Bylina and Z. R. Grabowski, *Trans. Faraday Soc.*, **65**, 458 (1969).

[17] T. D. Doyle, N. Filipescu, W. R. Benson, and D. Banes, *J. Am. Chem. Soc.*, **92**, 6371 (1970).

[18] (a) K. A. Muszkat and W. Schmidt, *Helv. Chim. Acta*, **54**, 1195 (1971); (b) K. A. Muszkat, S. Sharafi-Ozeri, G. Seger, and T. A. Pakkanen, *J. Chem. Soc., Perkin Trans. 2*, **1975**, 1515.

[19] O. L. Chapman, G. L. Eian, A. Bloom, and J. Clardy, *J. Am. Chem. Soc.*, **93**, 2918 (1971).

[20] T. Sato, S. Shimada, and K. Hata, *Bull. Chem. Soc. Jpn.*, **44**, 2484 (1971).

[21] F. B. Mallory and C. W. Mallory, *J. Am. Chem. Soc.*, **94**, 6041 (1972).

[22] A. Bromberg, K. A. Muszkat, and E. Fischer, *Isr. J. Chem.*, **10**, 765 (1972).

[23] T. Knittel-Wismonsky, G. Fischer, and E. Fischer, *Tetrahedron Lett.*, **1972**, 2853.

[24] Th. J. H. M. Cuppen and W. H. Laarhoven, *J. Am. Chem. Soc.*, **94**, 5914 (1972).

[25] M. P. Cava, P. Stern, and K. Wakisaka, *Tetrahedron*, **29**, 2245 (1973).

[26] (a) E. W. Förster, K. H. Grellmann, and H. Linschitz, *J. Am. Chem. Soc.*, **95**, 3108 (1973); (b) G. Fischer, E. Fischer, K. H. Grellmann, H. Linschitz, and A. Temizer, *J. Am. Chem. Soc.*, **96**, 6267 (1974); (c) T. Wolff and R. Waffenschmidt, *J. Am. Chem. Soc.*, **102**, 6098 (1980); (d) K. H. Grellmann, W. Kühnle, H. Weller, and T. Wolff, *J. Am. Chem. Soc.*, **103**, 6889 (1981).

[27] (a) T. Wismonski-Knittel, G. Fischer, and E. Fischer, *J. Chem. Soc., Perkin Trans. 2*, **1974**, 1930; (b) T. Wismonski-Knittel and E. Fischer, *J. Chem. Soc., Perkin Trans. 2*, **1979**, 449.

[28] R. G. F. Giles and M. V. Sargent, *J. Chem. Soc., Perkin Trans. 1*, **1974**, 2447.

[29] C. P. Joshua and V. N. Rajasekharan Pillai, *Tetrahedron*, **30**, 3333 (1974).

[30] (a) G. R. Lenz, *J. Org. Chem.*, **41**, 2201 (1976); (b) Th. Kindt, E.-P. Resewitz, Ch. Goedicke, and E. Lippert, *Z. Phys. Chem. (Frankfurt am Main)*, **101**, 1 (1976).

[31] J. Bendig, M. Beyermann, and D. Kreysig, *Tetrahedron Lett.*, **1977**, 3659.

[32] P. H. G. op het Veld and W. H. Laarhoven, *J. Am. Chem. Soc.*, **99**, 7221 (1977).

[33] A. Padwa, C. Doubleday, and A. Mazzu, *J. Org. Chem.*, **42**, 3271 (1977).

[34] A. G. Schultz, W. Y. Fu, R. D. Lucci, B. G. Kurr, K. M. Lo, and M. Boxer, *J. Am. Chem. Soc.*, **100**, 2140 (1978).

[35] A. G. Schultz, R. D. Lucci, W. Y. Fu, M. H. Berger, J. Erhardt, and W. K. Hagmann, *J. Am. Chem. Soc.*, **100**, 2150 (1978).

[36] (a) R. Koussini, R. Lapouyade, and P. Fornier de Violet, *J. Am. Chem. Soc.*, **100**, 6679 (1978); (b) P. Fornier de Violet, R. Bonneau, R. Lapouyade, R. Koussini, and W. R. Ware, *J. Am. Chem. Soc.*, **100**, 6683 (1978); (c) R. Lapouyade, R. Koussini, A. Nourmamode, and C. Courseille, *J. Chem. Soc., Chem. Commun.*, **1980**, 740.

[37] (a) J. Grimshaw and A. P. de Silva, *Can. J. Chem.*, **58**, 1880 (1980); (b) K. A. Muszkat, *Fortschr. Chem. Forsch.*, **88**, 89 (1980).

[38] (a) Ch. Goedicke and H. Stegemeyer, *Ber. Bunsenges. Phys. Chem.*, **73**, 782 (1969); (b) G. Ciamician and P. Silber, *Chem. Ber.*, **35**, 4128 (1902); (c) R. A. Caldwell, K. Mizuno, P. E. Hansen, L. P. Vo, M. Frentrup, and C. D. Ho, *J. Am. Chem. Soc.*, **103**, 7263 (1981).

[39] T. D. Doyle, W. R. Benson, and N. Filipescu, *J. Am. Chem. Soc.*, **98**, 3262 (1976).

[40] Ch. Goedicke and H. Stegemeyer, *Chem. Phys. Lett.*, **17**, 492 (1972).

[41] K. Ichimura and S. Watanabe, *Bull. Chem. Soc. Jpn.*, **49**, 2220 (1976).

[42] E. Fischer, *Mol. Photochem.*, **5**, 227 (1973).

[43] J. Saltiel, *Mol. Photochem.*, **5**, 231 (1973).

[44] (a) J. Saltiel, D. W.-L. Chang, and E. D. Megarity, *J. Am. Chem. Soc.*, **96**, 6521 (1974); (b) J. Saltiel, A. Marinari, D. W.-L. Chang, J. C. Mitchener, and E. D. Megarity, *J. Am. Chem. Soc.*, **101**, 2982 (1979).

[45] (a) F. Momicchioli, M. C. Bruni, I. Baraldi, and G. R. Corradini, *J. Chem. Soc., Faraday Trans, 2*, **1974**, 1325; (b) F. Momicchioli, G. R. Corradini, M. C. Bruni, and I. Baraldi, *J. Chem. Soc., Faraday Trans. 2*, **1975**, 215.

[46] H. Jungmann, H. Güsten, and D. Schulte-Frohlinde, *Chem. Ber.*, **101**, 2690 (1968).

[47] K. A. Muszkat, H. Kessel, and S. Sharafi-Ozeri, *Isr. J. Chem.*, **16**, 291 (1977).

[48] G. E. Lewis, *J. Org. Chem.*, **25**, 2193 (1960).

[49] H. H. Jaffe, S. J. Yeh, and R. W. Gardner, *J. Mol. Spectrosc.*, **2**, 120 (1958).

[50] H. Rau, *Ber. Bunsenges. Phys. Chem.*, **71**, 48 (1967).

[51] H. Mauser, D. J. Francis, and H.-J. Niemann, *Z. Phys. Chem. (Frankfurt am Main)*, **82**, 318 (1972).

[52] E. H. White and J. P. Anhalt, *Tetrahedron Lett.*, **1965**, 3937.

[53] C. D. DeBoer and R. H. Schlessinger, *J. Am. Chem. Soc.*, **90**, 803 (1968).

[54] G. Kaupp and M. Stark, *Chem. Ber.*, **111**, 3608 (1978).

[55] C. S. Wood and F. B. Mallory, *J. Org. Chem.*, **29**, 3373 (1964).

[56] T. E. Colman, Ph.D. Dissertation, Bryn Mawr College, 1970.

[57] P. L. Kumler and R. A. Dybas, *J. Org. Chem.*, **35**, 3825 (1970).

[58] M. Scholz, F. Dietz, and M. Mühlstädt, *Tetrahedron Lett.*, **1970**, 2835.

[59] F. B. Mallory and C. S. Wood, *Org. Synth.*, **45**, 91 (1965).

[60] D. J. Collins and J. J. Hobbs, *Aust. J. Chem.*, **20**, 1905 (1967).

[61] A. Bromberg and K. A. Muszkat, *J. Am. Chem. Soc.*, **91**, 2860 (1969).

[62] A. Bromberg, K. A. Muszkat, and A. Warshel, *J. Chem. Phys.*, **52**, 5952 (1970).

[63] S. W. Benson, *J. Chem. Educ.*, **42**, 502 (1965).

[64] E. V. Blackburn, C. E. Loader, and C. J. Timmons, *J. Chem. Soc. C*, **1968**, 1576.

[65] J. Bendig, M. Beyermann, D. Kreysig, R. Schöneich, and M. Siegmund, *J. Prakt. Chem.*, **320**, 798 (1978).

[66] (a) W. M. Horspool, *J. Chem. Soc. C*, **1971**, 400; (b) W. M. Horspool, *J. Chem. Soc., Chem. Commun.*, **1969**, 467.

[67] G. Rio and J.-C. Hardy, *Bull. Soc. Chim. Fr.*, **1970**, 3578.

[68] (a) T. Wismontski-Knittel and E. Fischer, *Mol. Photochem.*, **9**, 67 (1978); (b) J. H. Borkent, J. W. Diesveld, and W. H. Laarhoven, *Recl. Trav. Chim. Pays-Bas*, **100**, 114 (1981).

[69] M. V. Sargent and C. J. Timmons, *J. Chem. Soc. Suppl. 1*, **1964**, 5544.

[70] T. Wismontski-Knittel, K. A. Muszkat, and E. Fischer, *Mol. Photochem.*, **9**, 217 (1979).

[71] (a) K. Ichimura and S. Watanabe, *Bull. Chem. Soc. Jpn.*, **49**, 2224 (1976); (b) P. J. Darcy, R. J. Hart, and H. G. Heller, *J. Chem. Soc., Perkin Trans. 1*, **1978**, 571.

[72] (a) A. Buquet, A. Couture, and A. Lablache-Combier, *J. Org. Chem.*, **44**, 2300 (1979); (b) R. Lapouyade, A. Veyres, N. Hanafi, A. Couture, and A. Lablache-Combier, *J. Org. Chem.*, **47**, 1361 (1982).

[73] R. Lapouyade, R. Koussini, and J.-C. Rayez, *J. Chem. Soc., Chem. Commun.*, **1975**, 676.

[74] A. G. Schultz and M. B. DeTar, *J. Am. Chem. Soc.*, **98**, 3564 (1976).

[75] A. H. A. Tinnemans, W. H. Laarhoven, S. Sharafi-Ozeri, and K. A. Muszkat, *Recl. Trav. Chim. Pays-Bas*, **94**, 239 (1975).

[76] F. B. Mallory and C. W. Mallory, unpublished results.

[77] B. S. Thyagarajan, N. Kharasch, H. B. Lewis, and W. Wolf, *Chem. Commun.*, **1967**, 614.

[78] Y. Kanaoka and K. Itoh, *J. Chem. Soc., Chem. Commun.*, **1973**, 647.

[79] I. Ninomiya, T. Naito, and T. Kiguchi, *J. Chem. Soc., Perkin Trans. 1*, **1973**, 2257.

[80] G. Lenz, *J. Org. Chem.*, **39**, 2839 (1974).

[81] G. Lenz, *J. Org. Chem.*, **39**, 2846 (1974).

[82] (a) H. Linschitz and K. H. Grellmann, *J. Am. Chem. Soc.*, **86**, 303 (1964); (b) H. Shizuka, Y. Takayama, T. Morita, S. Matsumoto, and I. Tanaka, *J. Am. Chem. Soc.*, **93**, 5987 (1971).

[83] (a) C. A. Parker and W. J. Barnes, *Analyst (London)*, **82**, 606 (1957); (b) E. J. Bowen and J. H. D. Eland, *Proc. Chem. Soc., London*, **1963**, 202.

[84] D. Lopez, P. Boule, and J. Lemaire, *Nouv. J. Chim.*, **4**, 615 (1980).

[85] (a) K. H. Grellmann, W. Kühnle, and T. Wolff, *Z. Phys. Chem. (Frankfurt am Main)*, **101**, 295 (1976); (b) T. Wolff and R. Waffenschmidt, *J. Am. Chem. Soc.*, **102**, 6098 (1980).

[86] G. L. Eian and O. L. Chapman, *Org. Photochem. Synth.*, **1**, 72 (1971).

[87] (a) T. Wolff, *J. Am. Chem. Soc.*, **100**, 6157 (1978); (b) K. H. Grellmann, P. Hentzschel, T. Wismontski-Knittel, and E. Fischer, *J. Photochem.*, **11**, 197 (1979).

[88] G. J. Fonken, *Chem. Ind. (London)*, **1962**, 1327.

[89] E. V. Blackburn, C. E. Loader, and C. J. Timmons, *J. Chem. Soc. C*, **1970**, 163.

[90] (a) H. J. C. Jacobs and E. Havinga, *Adv. Photochem.*,**11**, 305 (1977); (b) G. M. Sanders, J. Pot, and E. Havinga, *Prog. Chem. Org. Nat. Prod.*, **27**, 131 (1969).

[91] A. Maercker, *Org. React.*, **14**, 270 (1965).

[92] P. A. Lowe, *Chem. Ind. (London)*, **1970**, 1070.

[93] E. J. Seus and C. V. Wilson, *J. Org. Chem.*, **26**, 5243 (1961).

[94] (a) W. S. Wadsworth, Jr., and W. D. Emmons, *J. Am. Chem. Soc.*, **83**, 1733 (1961); (b) W. S. Wadsworth, Jr., *Org. React.*, **25**, 73 (1977).

[95] J. Boutagy and R. Thomas, *Chem. Rev.*, **74**, 87 (1974).

[96] R. E. Buckles and K. Bremer, *Org. Synth. Coll. Vol. IV*, 777 (1963).

[97] R. E. Buckles and N. G. Wheeler, *Org. Synth. Coll. Vol. IV*, 857 (1963).

[98] C. S. Rondestvedt, Jr., *Org. React.*, **11**, 189 (1960).

[99] C. S. Rondestvedt, Jr., *Org. React.*, **24**, 225 (1976).

[100] A. E. Siegrist, *Helv. Chim. Acta*, **50**, 906 (1967).

[101] A. E. Siegrist and H. R. Meyer, *Helv. Chim. Acta*, **52**, 1282 (1969).

[102] A. E. Siegrist, P. Liechti, H. R. Meyer, and K. Weber, *Helv. Chim. Acta*, **52**, 2521 (1969).

[103] I. Ninomiya, T. Kiguchi, O. Yamamoto, and T. Naito, *J. Chem. Soc., Perkin Trans. 1*, **1979**, 1723.

[104] C. Manning and C. C. Leznoff, *Can. J. Chem.*, **53**, 805 (1975).

[105] J. Y. Wong, C. Manning, and C. C. Leznoff, *Angew. Chem., Int. Ed. Engl.*, **13**, 666 (1974).

[106] S. E. Potter and I. O. Sutherland, *J. Chem. Soc., Chem. Commun.*, **1973**, 520.

[107] S. C. Dickerman and I. Zimmerman, *J. Org. Chem.*, **39**, 3429 (1974).

[108] P. Hugelschofer, J. Kalvoda, and K. Schaffner, *Helv. Chim. Acta*, **43**, 1322 (1960).

[109] N. C. Yang, G. R. Lenz, and A. Shani, *Tetrahedron Lett.*, **1966**, 2941.

[110] M. P. Cava, M. J. Mitchell, S. C. Havlicek, A. Lindert, and R. J. Spangler, *J. Org. Chem.*, **35**, 175 (1970).

[111] F. E. Ziegler, Yale University, personal communication.

[112] H. Erdtman and A. Ronlan, *Acta Chem. Scand.*, **23**, 249 (1969).

[113] A. Stoessl, G. L. Rock, and M. H. Fisch, *Chem. Ind. (London)*, **1974**, 703.

[114] R. M. Letcher and K.-M. Wong, *J. Chem. Soc., Perkin Trans. 1*, **1977**, 178.

[115] P. H. Gore and F. S. Kamonah, *Synth. Commun.*, **1979**, 377.

[116] A. S. Kende and D. P. Curran, *J. Am. Chem. Soc.*, **101**, 1857 (1979).

[117] M. Joly, N. Defay, R. H. Martin, J. P. Declerq, G. Germain, B. Soubrier-Payen, and M. Van Meerssche, *Helv. Chim. Acta*. **60**, 537 (1977).

[118] M. S. Newman and H. M. Chung, *J. Org. Chem.*, **39**, 1036 (1974).

[119] E. Ghera, Y. Ben-David, and D. Becker, *Tetrahedron Lett.*, **1977**, 463.

[120] M. Mervič and E. Ghera, *J. Am. Chem. Soc.*, **99**, 7673 (1977).

[121] S. M. Kupchan and H. C. Wormser, *J. Org. Chem.*, **30**, 3792 (1965).

[122] W. H. Laarhoven and Th. J. H. M. Cuppen, *Recl. Trav. Chim. Pays-Bas*, **95**, 165 (1976).
[123] J. Blum, F. Grauer, and E. D. Bergmann, *Tetrahedron*, **25**, 3501 (1969).
[124] B. Thulin and O. Wennerström, *Acta Chem. Scand.*, **B30**, 369 (1976).
[125] S. M. Kupchan, J. L. Moniot, R. M. Kanojia, and J. B. O'Brien, *J. Org. Chem.*, **36**, 2413 (1971).
[126] G. S. Marx and E. D. Bergmann, *J. Org. Chem.*, **37**, 1807 (1972).
[127] B. Šket, M. Zupan, and A. Pollak, *Tetrahedron Lett.*, **1976**, 783.
[128] W. Schroth and H. Bahn, *Z. Chem.*, **17**, 56 (1977).
[129] T. Hiyama, S. Fujita, and H. Nozaki, *Bull. Chem. Soc. Jpn.*, **45**, 2797 (1972).
[130] K.-R. Stahlke, H.-G. Heine, and W. Hartmann, *Justus Liebigs Ann. Chem.*, **764**, 116 (1972).
[131] I. Lantos, *Tetrahedron Lett.*, **1978**, 2761.
[132] A. S. Dey and J. L. Neumeyer, *J. Med. Chem.*, **17**, 1095 (1974).
[133] G. H. Hakimelahi, C. B. Boyce, and H. S. Kasmai, *Helv. Chim. Acta*, **60**, 342 (1977).
[134] O. Tsuge, K. Oe, and Y. Ueyama, *Chem. Lett.*, **1976**, 425.
[135] I. Moritani, N. Toshima, S. Nakagawa, and M. Yakushiji, *Bull. Chem. Soc. Jpn.*, **40**, 2129 (1967).
[136] N. Toshima and I. Moritani, *Bull. Chem. Soc. Jpn.*, **40**, 1495 (1967).
[137] N. Ishibe, M. Sunami, and M. Odani, *J. Am. Chem. Soc.*, **95**, 463 (1973).
[138] R. Korenstein, K. A. Muszkat, and E. Fischer, *J. Chem. Soc., Perkin Trans. 2*, **1977**, 564.
[139] D. R. Maulding, *J. Org. Chem.*, **35**, 1221 (1970).
[140] H. Meyer, R. Bondy, and A. Eckert, *Monatsh. Chem.*, **33**, 1447 (1912).
[141] (a) R. Korenstein, K. A. Muszkat, and E. Fischer, *Helv. Chim. Acta*, **59**, 1826 (1976); (b) R. Korenstein, K. A. Muszkat, and E. Fischer, *J. Photochem.*, **5**, 447 (1976); (c) R. Korenstein, K. A. Muszkat, and E. Fischer, *J. Photochem.*, **5**, 345 (1976).
[142] A. Schönberg and K. Junghans, *Chem. Ber.*, **98**, 2539 (1965).
[143] R. Korenstein, K. A. Muszkat, M. A. Slifkin, and E. Fischer, *J. Chem. Soc., Perkin Trans. 2*, **1976**, 438.
[144] H. Brockmann and R. Mühlmann, *Chem. Ber.*, **82**, 348 (1949).
[145] E. D. Bergmann and H. J. E. Lowenthal, *Bull. Res. Counc. Isr.*, **3**, 72 (1953) [*C.A.*, **49**, 1686h (1955)].
[146] H. Brockmann, *Proc. Chem. Soc., London*, **1957**, 304.
[147] T. Bercovici, R. Korenstein, K. A. Muszkat, and E. Fischer, *Pure Appl. Chem.*, **24**, 531 (1970).
[148] R. S. Becker and C. E. Earhart, *J. Am. Chem. Soc.*, **92**, 5049 (1970).
[149] G. Körtum, *Ber. Bunsenges. Phys. Chem.*, **78**, 391 (1974).
[150] A. F. A. Ismail and Z. M. El-Shafei, *J. Chem. Soc.*, **1957**, 3393.
[151] (a) J. Nasielski, M. Jauquet, E. Vander Donckt, and A. Van Sinoy, *Tetrahedron Lett.*, **1969**, 4859; (b) E. Vander Donckt, P. Toussaint, C. Van Vooren, and A. Van Sinoy, *J. Chem. Soc., Faraday Trans. 1*, **72**, 2301 (1976); (c) G. P. DeGunst, *Recl. Trav. Chim. Pays-Bas*, **88**, 801 (1969).
[152] A. Schönberg and M. M. Sidky, *Chem. Ber.*, **107**, 1207 (1974).
[153] (a) H. Nimz, *Tappi*, **56**, 124 (1973); (b) H. K. Adam, M. A. Gay, and R. H. Moore, *J. Endocrinol.*, **84**, 35 (1980).
[154] (a) J. C. Russell, S. B. Costa, R. P. Seiders, and D. G. Whitten, *J. Am. Chem. Soc.*, **102**, 5678 (1980); (b) R. Arad-Yellin, S. Brunie, B. S. Green, M. Knossow, and G. Tsoucaris, *J. Am. Chem. Soc.*, **101**, 7529 (1979).
[155] (a) R. Göthe, C. A. Wachtmeister, B. Åkermark, P. Baeckström, B. Jansson, and S. Jensen, *Tetrahedron Lett.*, **1976**, 4501; (b) M. Maienthal, W. R. Benson, E. B. Sheinin, T. D. Doyle, and N. Filipescu, *J. Org. Chem.*, **43**, 972 (1978).
[156] (a) E. J. Levi and M. Orchin, *J. Org. Chem.*, **31**, 4302 (1966); (b) R. N. Nurmukhametov and G. I. Grishina, *Russ. J. Phys. Chem.* (*Engl. Transl.*), **43**, 1508 (1969); R. N. Nurmukhametov and G. I. Grishina, *Russ. J. Phys. Chem.* (*Engl. Transl.*), **43**, 1643 (1969).
[157] R. J. Hayward and C. C. Leznoff, *Tetrahedron*, **27**, 5115 (1971).
[158] W. Carruthers, *J. Chem. Soc. C*, **1967**, 1525.
[159] W. H. Laarhoven, Th. J. H. M. Cuppen, and R. J. F. Nivard, *Tetrahedron*, **26**, 4865 (1970).

[160] W. H. Laarhoven, Th. J. H. M. Cuppen, and R. J. F. Nivard, *Recl. Trav. Chim. Pays-Bas*, **87**, 687 (1968).

[161] E. V. Blackburn and C. J. Timmons, *J. Chem. Soc. C*, **1970**, 172.

[162] K. A. Muszkat, G. Seger, and S. Sharafi-Ozeri, *J. Chem. Soc., Faraday Trans. 2*, **1975**, 1529.

[163] (a) F. B. Mallory, C. W. Mallory, and E. J. Halpern, First Middle Atlantic Regional Meeting of the American Chemical Society, February 3, 1966, Philadelphia, Pa., Abstracts, p. 134; (b) D. D. Morgan, S. W. Horgan, and M. Orchin, *Tetrahedron Lett.*, **1972**, 1789.

[164] G. Snatzke and K. Kunde, *Chem. Ber.*, **106**, 1341 (1973).

[165] M. Scholz, M. Mühlstädt, and P. Dietz, *Tetrahedron Lett.*, **1967**, 665.

[166] W. H. Laarhoven and G. J. M. Brus, *J. Chem. Soc. B*, **1971**, 1433.

[167] H. Kagan, A. Moradpour, J. F. Nicoud, G. Balavoine, R. H. Martin, and J. P. Cosyn, *Tetrahedron Lett.*, **1971**, 2479.

[168] A. Moradpour, H. Kagan, M. Baes, G. Morren, and R. H. Martin, *Tetrahedron*, **31**, 2139 (1975).

[169] F. Dietz and M. Scholz, *Tetrahedron*, **24**, 6845 (1968).

[170] A. Moradpour, J. F. Nicoud, G. Balavoine, H. Kagan, and G. Tsoucaris, *J. Am. Chem. Soc.*, **93**, 2353 (1971).

[171] W. M. Ricker, Ph.D. Dissertation, Bryn Mawr College, 1974; F. B. Mallory, C. W. Mallory, and W. M. Ricker, *J. Am. Chem. Soc.*, **97**, 4770 (1975).

[172] W. H. Laarhoven and Th. J. H. M. Cuppen, *Recl. Trav. Chim. Pays-Bas*, **92**, 553 (1973).

[173] R. H. Martin, Ch. Eyndels, and N. Defay, *Tetrahedron Lett.*, **1972**, 2731.

[174] W. H. Laarhoven, Th. J. H. M. Cuppen, and R. J. F. Nivard, *Tetrahedron*, **30**, 3343 (1974).

[175] (a) R. H. Martin, *Angew. Chem., Int. Ed. Engl.*, **13**, 649 (1974); (b) T. J. Katz and J. Pesti, *J. Am. Chem. Soc.*, **104**, 346 (1982).

[176] J. Tribout, R. H. Martin, M. Doyle, and H. Wynberg, *Tetrahedron Lett.*, **1972**, 2839.

[177] (a) Y. Cochez, J. Jespers, V. Libert, K. Mislow, and R. H. Martin, *Bull. Soc. Chim. Belg.*, **84**, 1033 (1975); (b) J.-M. Vanest and R. H. Martin, *Recl. Trav. Chim. Pays-Bas*, **98**, 113 (1979).

[178] W. J. Bernstein, M. Calvin, and O. Buchardt, *J. Am. Chem. Soc.*, **94**, 494 (1972).

[179] (a) O. Buchardt, *Angew. Chem., Int. Ed. Engl.*, **13**, 179 (1974); (b) W. H. Laarhoven and T. J. H. M. Cuppen, *J. Chem. Soc., Perkin Trans. 2*, **1978**, 315; W. H. Laarhoven and T. J. H. M. Cuppen, *J. Chem. Soc., Chem. Commun.*, **1977**, 47.

[180] (a) M. Nakazaki, K. Yamamoto, and K. Fujiwara, *Chem. Lett.*, **1978**, 863; (b) M. Nakazaki, K. Yamamoto, K. Fujiwara, and M. Maeda, *J. Chem. Soc., Chem. Commun.*, **1979**, 1086.

[181] W. H. Laarhoven, T. J. H. M. Cuppen, and R. J. F. Nivard, *Tetrahedron*, **26**, 1069 (1970).

[182] D. D. Morgan, S. W. Horgan, and M. Orchin, *Tetrahedron Lett.*, **1970**, 4347.

[183] J. Blum and M. Zimmerman, *Tetrahedron*, **28**, 275 (1972).

[184] E. Mueller, H. Meier, and M. Sauerbier, *Chem. Ber.*, **103**, 1356 (1970).

[185] R. H. Martin, M.-J. Marchant, and M. Baes, *Helv. Chim. Acta*, **54**, 358 (1971).

[186] R. H. Martin and M. Baes, *Tetrahedron*, **31**, 2135 (1975).

[187] R. H. Martin, G. Morren, and J. J. Schurter, *Tetrahedron Lett.*, **1969**, 3683.

[188] M. Pomerantz, *J. Am. Chem. Soc.*, **89**, 694 (1967).

[189] J. Meinwald and P. H. Mazzocchi, *J. Am. Chem. Soc.*, **89**, 696 (1967).

[190] D. F. Tavares and W. H. Ploder, *Tetrahedron Lett.*, **1970**, 1567.

[191] M. Pomerantz and G. W. Gruber, *J. Am. Chem. Soc.*, **93**, 6615 (1971).

[192] (a) M. Sindler-Kulyk and W. H. Laarhoven, *J. Am. Chem. Soc.*, **100**, 3819 (1978); (b) P. M. op den Brouw and W. H. Laarhoven, *J. Org. Chem.*, **47**, 1546 (1982).

[193] M. Onda, K. Yonezawa, and K. Abe, *Chem. Pharm. Bull.*, **19**, 31 (1971).

[194] R. H. Mitchell and V. Boekelheide, *J. Am. Chem. Soc.*, **96**, 1547 (1974).

[195] R. H. Mitchell and R. J. Carruthers, *Can. J. Chem.*, **52**, 3054 (1974).

[196] P. J. Jessup and J. A. Reiss, *Aust. J. Chem.*, **29**, 173 (1976).

[197] P. J. Jessup and J. A. Reiss, *Aust. J. Chem.*, **30**, 851 (1977).

[198] J. R. Davy, M. N. Iskander, and J. A. Reiss, *Aust. J. Chem.*, **32**, 1067 (1979).

[199] D. N. Leach and J. A. Reiss, *Aust. J. Chem.*, **32**, 361 (1979).

[200] J. R. Davy and J. A. Reiss, *Aust. J. Chem.*, **29**, 163 (1976).

[201] T. Otsubo, R. Gray, and V. Boekelheide, *J. Am. Chem. Soc.*, **100**, 2449 (1978).
[202] F. Diederich and H. A. Staab, *Angew. Chem., Int. Ed. Engl.*, **17**, 372 (1978).
[203] B. Thulin and O. Wennerström, *Tetrahedron Lett.*, **1977**, 929.
[204] D. N. Leach and J. A. Reiss, *J. Org. Chem.*, **43**, 2484 (1978).
[205] A. S. Kende and D. P. Curran, *Tetrahedron Lett.*, **1978**, 3003.
[206] E. McDonald and R. T. Martin, *Tetrahedron Lett.*, **1978**, 4723.
[207] A. Bowd, D. A. Swann, and J. H. Turnbull, *J. Chem. Soc., Chem. Commun.*, **1975**, 797.
[208] P. H. G. op het Veld and W. H. Laarhoven, *J. Chem. Soc., Perkin Trans. 2*, **1978**, 915.
[209] S. C. Dickerman and I. Zimmerman, *J. Am. Chem. Soc.*, **86**, 5048 (1964).
[210] P. H. G. op het Veld and W. H. Laarhoven, *J. Chem. Soc., Perkin Trans. 2*, **1977**, 268.
[211] W. H. Laarhoven and Th. J. H. M. Cuppen, *J. Chem. Soc., Perkin Trans. 1*, **1972**, 2074.
[212] C. D. Tulloch and W. Kemp, *J. Chem. Soc. C*, **1971**, 2824.
[213] H. G. Heller and K. Salisbury, *J. Chem. Soc. C*, **1970**, 1997.
[214] C. C. Leznoff and R. J. Hayward, *Can. J. Chem.*, **50**, 528 (1972).
[215] F. G. Baddar, L. S. El-Assal, N. A. Doss, and A. W. Shehab, *J. Chem. Soc.*, **1959**, 1016.
[216] H. G. Heller, D. Auld, and K. Salisbury, *J. Chem. Soc. C*, **1967**, 682.
[217] R. J. Hart, H. G. Heller, and K. Salisbury, *Chem. Commun.*, **1968**, 1627.
[218] H. G. Heller and K. Salisbury, *J. Chem. Soc. C*, **1970**, 399.
[219] H. G. Heller and K. Salisbury, *J. Chem. Soc. C*, **1970**, 873.
[220] R. J. Hart and H. G. Heller, *J. Chem. Soc., Perkin Trans. 1*, **1972**, 1321.
[221] H. G. Heller and R. M. Megit, *J. Chem. Soc., Perkin Trans. 1*, **1974**, 923.
[222] R. Srinivasan, V. Y. Merritt, J. N. C. Hsu, P. H. G. op het Veld, and W. H. Laarhoven, *J. Org. Chem.*, **43**, 980 (1978).
[223] W. Carruthers, N. Evans, and R. Pooranamoorthy, *J. Chem. Soc., Perkin Trans. 1*, **1975**, 76.
[224] A. H. A. Tinnemans and W. H. Laarhoven, *J. Chem. Soc., Perkin Trans. 1*, **1976**, 1111.
[225] A. H. A. Tinnemans and W. H. Laarhoven, *J. Chem. Soc., Perkin Trans. 1*, **1976**, 1115.
[226] A. H. A. Tinnemans and W. H. Laarhoven, *Tetrahedron Lett.*, **1973**, 817.
[227] A. H. A. Tinnemans and W. H. Laarhoven, *Org. Photochem. Synth.*, **2**, 93 (1976).
[228] W. Wolf and N. Kharasch, *J. Org. Chem.*, **30**, 2493 (1965).
[229] R. J. F. M. van Arendonk and W. H. Laarhoven, *Tetrahedron Lett.*, **1977**, 2629.
[230] R. J. F. M. van Arendonk, W. H. Laarhoven, and P. A. J. Prick, *Recl. Trav. Chim. Pays-Bas*, **97**, 197 (1978).
[231] A. H. A. Tinnemans and W. H. Laarhoven, *J. Am. Chem. Soc.*, **96**, 4617 (1974).
[232] A. H. A. Tinnemans and W. H. Laarhoven, *J. Am. Chem. Soc.*, **96**, 4611 (1974).
[233] T. Sato, Y. Goto, and K. Hata, *Bull. Chem. Soc. Jpn.*, **40**, 1994 (1967).
[234] N. Kharasch, T. G. Alston, H. B. Lewis, and W. Wolf, *Chem. Commun.*, **1965**, 242.
[235] T. Sato, S. Shimada, and K. Hata, *Bull. Chem. Soc. Jpn.*, **42**, 766 (1969).
[236] R. J. Hayward and C. C. Leznoff, *Tetrahedron*, **27**, 2085 (1971).
[237] R. J. Hayward, A. C. Hopkinson, and C. C. Leznoff, *Tetrahedron*, **28**, 439 (1972).
[238] N. C. Yang, L. C. Lin, A. Shani, and S. S. Yang, *J. Org. Chem.*, **34**, 1845 (1969).
[239] M. Tada, H. Saiki, K. Miura, and H. Shinozaki, *Bull. Chem. Soc. Jpn.*, **49**, 1666 (1976).
[240] C. E. Loader and C. J. Timmons, *J. Chem. Soc. C*, **1966**, 1078.
[241] P. L. Kumler and R. A. Dybas, *J. Org. Chem.*, **35**, 125 (1970).
[242] G. Galiazzo, P. Bortolus, and G. Cauzzo, *Tetrahedron Lett.*, **1966**, 3717.
[243] P. Bortolus, G. Cauzzo, V. Mazzucato, and G. Galiazzo, *Z. Phys. Chem.* (*Frankfurt am Main*), **63**, 29 (1969).
[244] K. A. Muszkat and S. Sharafi-Ozeri, *Chem. Phys. Lett.*, **20**, 397 (1973).
[245] C. E. Loader, M. V. Sargent, and C. J. Timmons, *Chem. Commun.*, **1965**, 127.
[246] (a) H.-H. Perkampus and Th. Bluhm, *Tetrahedron*, **28**, 2099 (1972); (b) H. Fehn and H.-H. Perkampus, *Tetrahedron*, **34**, 1971 (1978).
[247] R. E. Doolittle and C. K. Bradsher, *J. Org. Chem.*, **31**, 2616 (1966).
[248] P. S. Mariano, E. Krochmal, Jr., and A. Leone, *J. Org. Chem.*, **42**, 1122 (1977).
[249] H.-H. Perkampus and G. Kassebeer, *Justus Liebigs Ann. Chem.*, **696**, 1 (1966).
[250] K. A. Muszkat and S. Sharafi-Ozeri. *Chem. Phys. Lett.*, **42**, 99 (1976).

[251] Y. J. Lee, D. G. Whitten, and L. Pedersen, *J. Am. Chem. Soc.*, **93**, 6330 (1971).

[252] D. G. Whitten and Y. J. Lee, *J. Am. Chem. Soc.*, **94**, 9142 (1972).

[253] H.-H. Perkampus, Th. Bluhm, and J. V. Knop, *Spectrochim. Acta*, **28A**, 2163 (1972).

[254] G. Bartocci, P. Bortolus, and U. Mazzucato, *J. Chem. Phys.*, **77**, 605 (1973).

[255] A. R. Gregory, W. Siebrand, and D. F. Williams, *J. Am. Chem. Soc.*, **101**, 1903 (1979).

[256] G. Orlandi, G. Poggi, and G. Marconi, *J. Chem. Soc., Faraday Trans. 2*, **76**, 598 (1980).

[257] C. E. Loader and C. J. Timmons, *J. Chem. Soc. C*, **1968**, 330.

[258] C. E. Loader and C. J. Timmons, *J. Chem. Soc. C*, **1967**, 1457.

[259] F. Andreani, R. Andrisano, G. Salvadori, and M. Tramontini, *J. Chem. Soc., Perkin Trans. 1*, **1977**, 1737.

[260] R. H. Martin and M. Deblecker, *Tetrahedron Lett.*, **1969**, 3597.

[261] C. Mortelmans and G. van Binst, *Tetrahedron*, **34**, 363 (1978).

[262] G. van Binst, R. B. Baert, and R. Salsmans, *Synth. Commun.*, **3**, 59 (1973).

[263] S. F. Dyke and M. Sainsbury, *Tetrahedron*, **23**, 3161 (1967).

[264] S. F. Dyke, M. Sainsbury, and B. J. Moon, *Tetrahedron*, **24**, 1467 (1968).

[265] V. Šmula, R. H. F. Manske, and R. Rodrigo, *Can. J. Chem.*, **50**, 1544 (1972).

[266] J. Glinka, *Rocz. Chem.*, **51**, 577 (1977) [*C.A.*, **87**, 102140j (1977)].

[267] R. Huisgen, *Justus Liebigs Ann. Chem.*, **564**, 16 (1949).

[268] O. de Silva and V. Snieckus, *Synthesis*, **1971**, 254.

[269] S. O. de Silva and V. Snieckus, *Can. J. Chem.*, **52**, 1294 (1974).

[270] C. A. Mudry and A. R. Frasca, *Tetrahedron*, **30**, 2983 (1974).

[271] J. L. Cooper and H. H. Wasserman, *J. Chem. Soc., Chem. Commun.*, **1969**, 200.

[272] J. Grimshaw and D. Mannus, *J. Chem. Soc., Perkin Trans. 1*, **1977**, 2096.

[273] (a) M. Kojima and M. Maeda, *Tetrahedron Lett.*, **1969**, 2379; (b) M. Maeda and M. Kojima, *J. Chem. Soc., Perkin Trans. 1*, **1977**, 239.

[274] A. Mitschker, U. Brandl, and T. Kauffmann, *Tetrahedron Lett.*, **1974**, 2343.

[275] R. M. Moriarty, J. M. Kleigman, and R. B. Desai, *Chem. Commun.*, **1967**, 1255.

[276] W. J. Begley, J. Grimshaw, and J. Trocha-Grimshaw, *J. Chem. Soc., Perkin 1*, **1974**, 2633.

[277] J. Grimshaw and A. P. de Silva, *J. Chem. Soc., Chem. Commun.*, **1979**, 193.

[278] G. Cooper and W. J. Irwin, *J. Chem. Soc., Perkin Trans. 1*, **1976**, 75.

[279] G. Cooper and W. J. Irwin, *J. Chem. Soc., Perkin Trans. 1*, **1976**, 2038.

[280] D. Jerchel and H. Fischer, *Justus Liebigs Ann. Chem.*, **590**, 216 (1954).

[281] D. Jerchel and H. Fischer, *Chem. Ber.*, **88**, 1595 (1955).

[282] D. Jerchel and H. Fischer, *Chem. Ber.*, **89**, 563 (1956).

[283] F. A. Neugebauer, *Tetrahedron*, **26**, 4843 (1970).

[284] E. A. Rick, Ph.D. Dissertation, Yale University, 1959.

[285] F. B. Mallory and C. S. Wood, *Tetrahedron Lett.*, **1965**, 2643.

[286] G. M. Badger, C. P. Joshua, and G. E. Lewis, *Tetrahedron Lett.*, **1964**, 3711.

[287] A. V. El'tsov, O. P. Studzinskii, and N. V. Ogol'tsova, *Zh. Org. Khim.*, **6**, 405 (1970); *J. Org. Chem. USSR* (*Engl. Transl.*), **6**, 399 (1970) [*C.A.*, **72**, 111266x, 1970].

[288] H.-H. Perkampus and B. Behjati, *J. Heterocycl. Chem.*, **11**, 511 (1974).

[289] K. Geibel, B. Staudinger, K. H. Grellmann, and H. Wendt, *J. Photochem.*, **3**, 241 (1974).

[290] K. H. Grellmann and E. Tauer, *J. Am. Chem. Soc.*, **95**, 3104 (1973).

[291] K. Maeda, and E. Fischer, *Isr. J. Chem.*, **16**, 294 (1977).

[292] D. G. Anderson and G. Wettermark, *J. Am. Chem. Soc.*, **87**, 1433 (1965).

[293] G. Wettermark and E. Wallström, *Acta Chem. Scand.*, **22**, 675 (1968).

[294] (a) T. Onaka, Y. Kanda, and M. Natsume, *Tetrahedron Lett.*, **1974**, 1179; (b) R. K.-Y. Zee-Cheng, S.-J. Yan, and C. C. Cheng, *J. Med. Chem.*, **21**, 199 (1978).

[295] M. P. Cava and R. H. Schlessinger, *Tetrahedron Lett.*, **1964**, 2109.

[296] F. B. Mallory and B. S. Alvord, unpublished results.

[297] M. Ahmed, L. J. Kricka, and J. M. Vernon, *J. Chem. Soc., Perkin Trans. 1*, **1975**, 71.

[298] V. M. Clark and A. Cox, *Tetrahedron*, **22**, 3421 (1966).

[299] S. V. Kessar, G. Singh, and P. Balakrishnan, *Tetrahedron Lett.*, **1974**, 2269.

[300] S. Prabhakar, A. M. Lobo, and M. R. Tavares, *J. Chem. Soc., Chem. Commun.*, **1978**, 884.

[301] J. S. Swenton, T. J. Ikeler, and G. L. Smyser, *J. Org. Chem.*, **38**, 1157 (1973).
[302] A. Padwa, J. Smolanoff, and S. I. Wetmore, Jr., *J. Org. Chem.*, **38**, 1333 (1973).
[303] P. J. Collin, J. S. Shannon, H. Silberman, S. Sternhell, and G. Sugowdz, *Tetrahedron*, **24**, 3069 (1968).
[304] M. Scholz, H. Herzschuh, and M. Mühlstädt, *Tetrahedron Lett.*, **1968**, 3685.
[305] A. Padwa, *Chem. Rev.*, **77**, 37 (1977).
[306] A. C. Pratt, *Chem. Soc. Rev.*, **6**, 63 (1977).
[307] G. E. Lewis, *Tetrahedron Lett.*, **1960**, 12.
[308] G. M. Badger, R. J. Drewer, and G. E. Lewis, *Aust. J. Chem.*, **16**, 1042 (1963).
[309] C. P. Joshua and V. N. R. Pillai, *Tetrahedron Lett.*, **1973**, 3559.
[310] G. M. Badger, C. P. Joshua, and G. E. Lewis, *Aust. J. Chem.*, **18**, 1639 (1965).
[311] C. P. Joshua and G. E. Lewis, *Aust. J. Chem.*, **20**, 929 (1967).
[312] G. E. Lewis and J. A. Reiss, *Aust. J. Chem.*, **20**, 1451 (1967).
[313] G. M. Badger, R. J. Drewer, and G. E. Lewis, *Aust. J. Chem.*, **17**, 1036 (1964).
[314] A. Marchetti and D. R. Kearns, *J. Am. Chem. Soc.*, **89**, 5335 (1967).
[315] N. C. Jamieson and G. E. Lewis, *Aust. J. Chem.*, **20**, 321 (1967).
[316] J. W. Barton and R. B. Walker, *Tetrahedron Lett.*, **1975**, 569.
[317] C. P. Joshua and V. N. R. Pillai, *Indian J. Chem.*, **12**, 60 (1974).
[318] G. M. Badger, N. C. Jamieson, and G. E. Lewis, *Aust. J. Chem.*, **18**, 190 (1965).
[319] H. Rau and O. Schuster, *Angew. Chem., Int. Ed. Engl.*, **15**, 114 (1976).
[320] J. Griffiths, *Chem. Soc. Rev.*, **1**, 481 (1972).
[321] (a) P. J. Grisdale, M. E. Glogowski, and J. L. R. Williams, *J. Org. Chem.*, **36**, 3821 (1971); (b) M. E. Glogowski, P. J. Grisdale, J. L. R. Williams, and T. H. Regan, *J. Organomet. Chem.*, **54**, 51 (1973); (c) M. E. Glogowski, P. J. Grisdale, J. L. R. Williams, and L. Costa, *J. Organomet. Chem.*, **74**, 175 (1974).
[322] (a) R. M. Kellogg, M. B. Groen, and H. Wynberg, *J. Org. Chem.*, **32**, 3093 (1967); (b) H. Güsten, L. Klasinc, and O. Volkert, *Z. Naturforsch. Teil B*, **24**, 12 (1969).
[323] G. De Luca, G. Martelli, P. Spagnolo, and M. Tiecco, *J. Chem. Soc. C*, **1970**, 2504.
[324] W. Carruthers and H. N. M. Stewart, *J. Chem. Soc.*, **1965**, 6221.
[325] H. Wynberg, *Acc. Chem. Res.*, **4**, 65 (1971).
[326] M. B. Groen, H. Schadenberg, and H. Wynberg, *J. Org. Chem.*, **36**, 2797 (1971).
[327] P. G. Lehman and H. Wynberg, *Aust. J. Chem.*, **27**, 315 (1974).
[328] J. H. Dopper, D. Oudman, and H. Wynberg, *J. Am. Chem. Soc.*, **95**, 3692 (1973).
[329] A. Croisy, P. Jacquignon, and F. Perin, *J. Chem. Soc., Chem. Commun.*, **1975**, 106.
[330] C. C. Leznoff, W. Lilie, and C. Manning, *Can. J. Chem.*, **52**, 132 (1974).
[331] H. Wynberg, H. van Driel, R. M. Kellogg, and J. Buter, *J. Am. Chem. Soc.*, **89**, 3487 (1967).
[332] C. E. Loader and C. J. Timmons, *J. Chem. Soc. C*, **1967**, 1677.
[333] A. Padwa and R. Hartman, *J. Am. Chem. Soc.*, **88**, 3759 (1966).
[334] D. T. Anderson and W. M. Horspool, *J. Chem. Soc., Perkin Trans. 1*, **1972**, 536.
[335] A. Couture, A. Lablache-Combier, and H. Ofenberg, *Tetrahedron*, **31**, 2023 (1975).
[336] A. Couture, A. Lablache-Combier, and H. Ofenberg, *Tetrahedron Lett.*, **1974**, 2497.
[337] A. Couture, A. Lablache-Combier, and H. Ofenberg, *Org. Photochem. Synth.*, **2**, 7 (1976).
[338] W. Dilthey and F. Quint, *Ber. Dtsch. Chem. Ges.*, **69**, 1575 (1936).
[339] E. Hertel and G. Sock, *Z. Phys. Chem. Abt. A*, **189**, 95 (1941).
[340] (a) W. Dilthey, F. Quint, and J. Heinen, *J. Prakt. Chem.*, **152**, 49 (1939); (b) W. Dilthey, F. Quint, and H. Stephan, *J. Prakt. Chem.*, **152**, 99 (1939).
[341] K. Itoh and Y. Kanaoka, *Chem. Pharm. Bull.*, **22**, 1431 (1974).
[342] Y. Kanaoka and K. Itoh, *Synthesis*, **1972**, 36.
[343] J. Grimshaw and A. P. de Silva, *J. Chem. Soc., Chem. Commun.*, **1980**, 302.
[344] S. Lalitha, S. Rajeswari, B. R. Pai, and H. Suguna, *Indian J. Chem.*, **15B**, 180 (1977).
[345] D. H. Hey, G. H. Jones, and M. J. Perkins, *J. Chem. Soc. C*, **1971**, 116.
[346] H. Hara, O. Hoshino, and B. Umezawa, *Tetrahedron Lett.*, **1972**, 5031.
[347] J. Grimshaw and A. P. de Silva, *J. Chem. Soc., Chem. Commun.*, **1980**, 301.
[348] (a) G. R. Lenz, *Synthesis*, **1978**, 489; (b) I. Ninomiya, *Heterocycles*, **14**, 1567 (1980).

[349] T. Kametani and K. Fukumoto, *Acc. Chem. Res.*, **5**, 212 (1972).
[350] I. Ninomiya, *Heterocycles*, **2**, 105 (1974).
[351] T. Kametani, K. Fukumoto, and F. Satoh, *Bioorg. Chem.*, **3**, 430 (1974).
[352] Y. Kondo, *Heterocycles*, **4**, 197 (1976).
[353] I. Ninomiya, H. Takasugi, and T. Naito, *J. Chem. Soc., Chem. Commun.*, **1973**, 732.
[354] I. Ninomiya, T. Naito, and H. Takasugi, *J. Chem. Soc., Perkin Trans. 1*, **1976**, 1865.
[355] A. Shafiee and A. Rashidbaigi, *J. Heterocycl. Chem.*, **14**, 1317 (1977).
[356] G. R. Lenz, *Tetrahedron Lett.*, **1973**, 1963.
[357] I. Ninomiya, H. Takasugi, and T. Naito, *Heterocycles*, **1**, 17 (1973).
[358] I. Ninomiya, T. Naito, and H. Takasugi, *J. Chem. Soc., Perkin Trans. 1*, **1975**, 1791.
[359] I. Ninomiya, T. Naito, and T. Mori, *Tetrahedron Lett.*, **1969**, 3643.
[360] I. Ninomiya, T. Naito, T. Kiguchi, and T. Mori, *J. Chem. Soc., Perkin Trans. 1*, **1973**, 1696.
[361] I. Ninomiya, T. Naito, and T. Mori, *Tetrahedron Lett.*, **1969**, 2259.
[362] I. Ninomiya, T. Naito, and T. Mori, *J. Chem. Soc., Perkin Trans. 1*, **1973**, 505.
[363] I. Ninomiya, T. Naito, and H. Takasugi, *J. Chem. Soc., Perkin Trans. 1*, **1975**, 1720.
[364] I. Ninomiya and T. Naito, *J. Chem. Soc., Chem. Commun.*, **1973**, 137.
[365] I. Ninomiya, T. Naito, H. Ishii, T. Ishida, M. Ueda, and K. Harada, *J. Chem. Soc., Perkin Trans. 1*, **1975**, 762.
[366] I. Ninomiya, S. Yamauchi, T. Kiguchi, A. Shinohara, and T. Naito, *J. Chem. Soc., Perkin Trans. 1*, **1974**, 1747.
[367] I. Ninomiya, T. Kiguchi, and T. Naito, *J. Chem. Soc., Chem. Commun.*, **1974**, 81.
[368] I. Ninomiya, T. Kiguchi, S. Yamauchi, and T. Naito, *J. Chem. Soc., Perkin Trans. 1*, **1980**, 197.
[369] I. Ninomiya, T. Kiguchi, S. Yamauchi, and T. Naito, *J. Chem. Soc., Perkin Trans. 1*, **1976**, 1861.
[370] M. Ogata and H. Matsumoto, *Chem. Pharm. Bull.*, **20**, 2264 (1972).
[371] I. Ninomiya, T. Kiguchi, and T. Naito, *Heterocycles*, **4**, 973 (1976).
[372] P. G. Cleveland and O. L. Chapman, *Chem. Commun.*, **1967**, 1064.
[373] G. R. Lenz, *J. Org. Chem.*, **42**, 1117 (1977).
[374] G. R. Lenz and N. C. Yang, *Chem. Commun.*, **1967**, 1136.
[375] M. P. Cava and S. C. Havlicek, *Tetrahedron Lett.*, **1967**, 2625.
[376] N. C. Yang, A. Shani, and G. R. Lenz, *J. Am. Chem. Soc.*, **88**, 5369 (1966).
[377] R. Lapouyade, R. Koussini, and H. Bouas-Laurent, *J. Am. Chem. Soc.*, **99**, 7374 (1977).
[378] E. Grovenstein, Jr., T. C. Campbell, and T. Shibata, *J. Org. Chem.*, **34**, 2418 (1969).
[379] W. C. Fleming, W. W. Lee, and D. W. Henry, *J. Org. Chem.*, **38**, 4404 (1973).
[380] W. A. Henderson, Jr., and A. Zweig, *J. Am. Chem. Soc.*, **89**, 6778 (1967).
[381] W. A. Henderson, Jr., R. Lopresti, and A. Zweig, *J. Am. Chem. Soc.*, **91**, 6049 (1969).
[382] V. M. Clark, A. Cox, and E. J. Herbert, *J. Chem. Soc. C*, **1968**, 831.
[383] F. Fratev, O. E. Polansky, and M. Zander, *Z. Naturforsch. Teil A*, **31**, 987 (1976).
[384] W. Carruthers, *Chem. Commun.*, **1966**, 272.
[385] K. H. Grellmann, G. M. Sherman, and H. Linschitz, *J. Am. Chem. Soc.*, **85**, 1881 (1963).
[386] (a) K.-P. Zeller and H. Petersen, *Synthesis*, **1975**, 532; (b) K.-P. Zeller and S. Berger, *J. Chem. Soc., Perkin Trans. 2*, **1977**, 54.
[387] W. A. Henderson, Jr., and A. Zweig, *Tetrahedron Lett.*, **1969**, 625.
[388] (a) J. Bratt and H. Suschitzky, *J. Chem. Soc., Chem. Commun.*, **1972**, 949; (b) J. Bratt, B. Iddon, A. G. Mack, H. Suschitzky, J. A. Taylor, and B. J. Wakefield, *J. Chem. Soc., Perkin Trans. 1*, **1980**, 648.
[389] J. A. Elix, D. P. H. Murphy, and M. V. Sargent, *Synth. Commun.*, **2**, 427 (1972).
[390] J. A. Elix and D. P. Murphy, *Aust. J. Chem.*, **28**, 1559 (1975).
[391] (a) O. L. Chapman and G. L. Eian, *J. Am. Chem. Soc.*, **90**, 5329 (1968); (b) J. F. Nicoud and H. B. Kagan, *Isr. J. Chem.*, **15**, 78 (1976/77).
[392] A. G. Schultz and I-C. Chiu, *J. Chem. Soc., Chem. Commun.*, **1978**, 29.
[393] A. G. Schultz and M. B. DeTar, *Org. Photochem. Synth.*, **2**, 47 (1976).
[394] A. G. Schultz and W. K. Hagmann, *J. Org. Chem.*, **43**, 4231 (1978).

[395] A. G. Schultz and W. K. Hagmann, *J. Chem. Soc., Chem. Commun.*, **1976**, 726.
[396] A. G. Schultz and W. K. Hagmann, *J. Org. Chem.*, **43**, 3391 (1978).
[397] A. G. Schultz, *J. Org. Chem.*, **40**, 3466 (1975).
[398] S. Senda, K. Hirota, and M. Takahashi, *J. Chem. Soc., Perkin Trans. 1*, **1975**, 503.
[399] W.-R. Knappe, *Chem. Ber.*, **107**, 1614 (1974).
[400] H. Iida, Y. Yuasa, and C. Kibayashi, *J. Org. Chem.*, **44**, 1236 (1979).
[401] J. P. Ferris and F. R. Antonucci, *J. Am. Chem. Soc.*, **96**, 2010 (1974).
[402] E. P. Fokin, T. N. Gerasimova, T. V. Fomenko, and N. V. Semikolenova, *J. Org. Chem. USSR* (*Engl. Transl.*), **14**, 772 (1978).
[403] T. Sasaki, K. Hayakawa, and S. Nishida, *Tetrahedron Lett.*, **1980**, 3903.
[404] A. Cox, D. R. Kemp, R. Lapouyade, P. de Mayo, J. Joussot-Dubien, and R. Bonneau, *Can. J. Chem.*, **53**, 2386 (1975).
[405] K. H. Grellmann and E. Tauer, *Tetrahedron Lett.*, **1967**, 1909.
[406] W. H. Laarhoven and P. G. F. Boumans, *Recl. Trav. Chim. Pays-Bas*, **94**, 114 (1975).
[407] F. B. Mallory, C. W. Mallory, and W. M. Ricker, Tenth Middle Atlantic Regional Meeting of the American Chemical Society, February 25, 1976, Philadelphia, Pa., Abstracts, p. 78; *see also* ref. 535.
[408] M. E. Glogowski and J. L. R. Williams, *J. Organomet. Chem.*, **195**, 123 (1980).
[409] S. D. Cohen, M. V. Mijovic, and G. A. Newman, *Chem. Commun.*, **1968**, 722.
[410] A. J. Floyd, S. F. Dyke, and S. E. Ward, *Chem. Rev.*, **76**, 509 (1976).
[411] R. Srinivasan, *Org. Photochem. Synth.*, **1**, 1 (1971).
[412] C. W. Mallory and F. B. Mallory, *Org. Photochem. Synth.*, **1**, 55 (1971).
[413] R. A. Dybas, *Org. Photochem. Synth.*, **1**, 25 (1971).
[414] C. O. Parker and P. E. Spoerri, *Nature* (*London*), **166**, 603 (1950).
[415] (a) T. Sato, Y. Goto, T. Tohyama, S. Hayashi, and K. Hata, *Bull. Chem. Soc. Jpn.*, **40**, 2975 (1967); (b) H. Moesta, *Faraday Discuss. Chem. Soc.*, **58**, 244 (1975).
[416] W. H. Laarhoven and Th. J. H. M. Cuppen, *Tetrahedron Lett.*, **1966**, 5003.
[417] G. Montaudo, *Gazz. Chim. Ital.*, **94**, 127 (1964).
[418] R. B. Buckles, *J. Am. Chem. Soc.*, **77**, 1040 (1955).
[419] J. A. Moore and T. Isaacs, *Tetrahedron Lett.*, **1973**, 5033.
[420] W. Templeton, *J. Chem. Soc., Chem. Commun.*, **1970**, 1412.
[421] T. D. Roberts, *J. Chem. Soc., Chem. Commun.*, **1971**, 362.
[422] A. M. Belcher, Ph.D. Dissertation, Bryn Mawr College, 1981.
[423] J. L. Charlton, B. Kostyk, and H. K. Lai, *Can. J. Chem.*, **52**, 3577 (1974).
[424] P. E. Hansen, O. K. Poulsen, and A. Berg, *Org. Mag. Reson.*, **12**, 43 (1979).
[425] J. C. Miller, K. U. Breakstone, J. S. Meek, and S. J. Strickler, *J. Am. Chem. Soc.*, **99**, 1142 (1977).
[426] J. W. Raniseski, Ph.D. Dissertation, Bryn Mawr College, 1967.
[427] T. L. Newirth, Honors Paper, Bryn Mawr College, 1967.
[428] S. M. Kupchan and H. C. Wormser, *Tetrahedron Lett.*, **1965**, 359.
[429] H. Güsten and L. Klasinc, *Tetrahedron*, **24**, 5499 (1968).
[430] R. E. Harmon, M. Mazharuddin, and S. K. Gupta, *J. Chem. Soc., Perkin Trans. 1*, **1973**, 1160.
[431] M. M. J. LeGuen, Y. E.-D. Shafig, and R. Taylor, *J. Chem. Soc., Perkin Trans. 2*, **1979**, 803.
[432] H. V. Ansell, P. J. Sheppard, C. F. Simpson, M. A. Stroud, and R. Taylor, *J. Chem. Soc., Perkin Trans. 2*, **1979**, 381.
[433] R. H. Martin and J. J. Schurter, *Tetrahedron*, **28**, 1749 (1972).
[434] O. L. Chapman and W. R. Adams, *J. Am. Chem. Soc.*, **89**, 4243 (1967).
[435] R. G. F. Giles and M. V. Sargent, *J. Chem. Soc., Chem. Commun.*, **1974**, 215.
[436] C. S. Barnes, D. J. Collins, J. J. Hobbs, P. I. Mortimer, and W. H. F. Sasse, *Aust. J. Chem.*, **20**, 699 (1967).
[437] G. R. Krow, K. M. Damodaran, E. Michener, R. Wolf, and J. Guare, *J. Org. Chem.*, **43**, 3950 (1978).
[438] M. B. Rubin and S. Welner, *J. Org. Chem.*, **45**, 1847 (1980).

[439] W. H. Laarhoven, W. H. M. Peters, and A. H. A. Tinnemans, *Tetrahedron*, **34**, 769 (1978).

[440] U. Meissner, B. Meissner, and H. A. Staab, *Angew. Chem., Int. Ed. Engl.*, **12**, 916 (1973).

[441] H. A. Staab, U. E. Meissner, and B. Meissner, *Chem. Ber.*, **109**, 3875 (1976).

[442] J. R. Davy, P. J. Jessup, and J. A. Reiss, *J. Chem. Educ.*, **52**, 747 (1975).

[443] J. T. Craig, B. Halton, and S.-F. Lo, *Aust. J. Chem.*, **28**, 913 (1975).

[444] M. Sindler-Kulyk and W. H. Laarhoven, *J. Am. Chem. Soc.*, **98**, 1052 (1976).

[445] H. Blaschke and V. Boekelheide, *J. Am. Chem. Soc.*, **89**, 2747 (1967).

[446] C. E. Ramey and V. Boekelheide, *J. Am. Chem. Soc.*, **92**, 3681 (1970).

[447] H. Blaschke, C. E. Ramey, I. Calder, and V. Boekelheide, *J. Am. Chem. Soc.*, **92**, 3675 (1970).

[448] R. Naef and E. Fischer, *Helv. Chim. Acta*, **57**, 2224 (1974).

[449] J. Pataki and R. Balick, *J. Med. Chem.*, **15**, 905 (1972).

[450] T.-S. Lin, R. E. Harmon, W. Pierantoni, and J. C. W. Su, *Org. Prep. Proced. Int.*, **6**, 185 (1974).

[451] K. L. Servis and K.-N. Fang, *J. Am. Chem. Soc.*, **90**, 6712 (1968).

[452] K. L. Servis and K.-N. Fang, *Tetrahedron Lett.*, **1968**, 967.

[453] R. A. Caldwell, N. I. Ghali, C.-K. Chien, D. DeMarco, and L. Smith, *J. Am. Chem. Soc.*, **100**, 2857 (1978).

[454] D. G. Coe, E. W. Garnish, M. M. Gale, and C. J. Timmons, *Chem. Ind.* (*London*), **1957**, 665.

[455] J. H. Boyer and R. Selvarajan, *J. Am. Chem. Soc.*, **91**, 6122 (1969).

[456] M. V. Sargent and C. J. Timmons, *J. Am. Chem. Soc.*, **85**, 2186 (1963).

[457] K. Ichimura and S. Watanabe, *Tetrahedron Lett.*, **1972**, 821.

[458] (a) G. Rio and D. Masure, *Bull. Soc. Chim. Fr.*, **1971**, 3232; (b) I. Ninomiya, I. Furutani, O. Yamamoto, T. Kiguchi, and T. Naito, *Heterocycles*, **9**, 853 (1978).

[459] R. Srinivasan and J. N. C. Hsu, *J. Am. Chem. Soc.*, **93**, 2816 (1971).

[460] A. Padwa and D. Dehm, *J. Am. Chem. Soc.*, **97**, 4779 (1975).

[461] R. M. Letcher and K.-M. Wong, *J. Chem. Soc., Perkin Trans. 1*, **1978**, 739.

[462] B. I. Rosen and W. P. Weber, *J. Org. Chem.*, **42**, 3463 (1977).

[463] C. W. Bird, *J. Chem. Soc., Chem. Commun.*, **1968**, 1537.

[464] A. R. Leed, S. D. Boettger, and B. Ganem, *J. Org. Chem.*, **45**, 1098 (1980).

[465] T. M. Cresp, R. G. F. Giles, M. V. Sargent, C. Brown, and D. O'N. Smith, *J. Chem. Soc., Perkin Trans. 1*, **1974**, 2435.

[466] K. L. Servis and F. R. Jerome, *J. Am. Chem. Soc.*, **93**, 1535 (1971).

[467] E. N. Marvell, J. K. Reed, W. Gänzler, and H. Tong, *J. Org. Chem.*, **42**, 3783 (1977).

[468] R. N. Warrener and J. B. Bremner, *Rev. Pure Appl. Chem.*, **16**, 117 (1966).

[469] D. J. Collins and J. J. Hobbs, *Chem. Ind.* (*London*), **1965**, 1725.

[470] M. V. Sargent and D. O'N. Smith, *J. Chem. Soc. C*, **1970**, 329.

[471] H.-D. Becker, *J. Org. Chem.*, **34**, 2026 (1969).

[472] P. Perkins and K. P. C. Vollhardt, *Angew. Chem., Int. Ed. Engl.*, **17**, 615 (1978).

[473] W. Carruthers and H. N. M. Stewart, *J. Chem. Soc. C*, **1967**, 556.

[474] D. W. Cameron and M. Mingin, *Aust. J. Chem.*, **30**, 859 (1977).

[475] S. S. Hixson, *J. Am. Chem. Soc.*, **97**, 1981 (1975).

[476] B. Chauncy and E. Gellert, *Aust. J. Chem.*, **22**, 993 (1969).

[477] (a) D. Banes, *J. Assoc. Off. Agric. Chem.*, **44**, 323 (1961); (b) H. W. Knoche and A. M. Gawienowski, *Endocrinology* (*Baltimore*), **71**, 514 (1962).

[478] T. D. Doyle, W. R. Benson, and N. Filipescu, *Photochem. Photobiol.*, **27**, 3 (1978).

[479] R. J. Hartle, U.S. Pat. 3,953,498 (1976) [*C.A.*, **85**, 32705e (1976)].

[480] A. Ronlán, O. Hammerich, and V. D. Parker, *J. Am. Chem. Soc.*, **95**, 7132 (1973).

[481] J. Reich, M. Báthory, I. Novák, and K. Szendrei, *Herba Hung.*, **9**, 43 (1970).

[482] G. Kaupp, *Angew. Chem., Int. Ed. Engl.*, **10**, 340 (1971).

[483] G. Kaupp, *Justus Liebigs Ann. Chem.*, **1973**, 844.

[484] M. Pailer, H. Berner, and S. Makleit, *Monatsh. Chem.*, **98**, 102 (1967).

[485] T. M. Cresp, R. G. F. Giles, and M. V. Sargent, *J. Chem. Soc., Chem. Commun.*, **1974**, 11.

[486] R. M. Letcher, *Phytochemistry*, **12**, 2789 (1973).

[487] R. M. Letcher and L. R. M. Nhamo, *J. Chem. Soc., Perkin Trans. 1*, **1972**, 2941.

[488] R. M. Letcher and K.-M. Wong, *J. Chem. Soc., Perkin Trans.* 1, **1979**, 2449.

[489] A. P. Bindra, J. A. Elix, and M. V. Sargent, *Aust. J. Chem.*, **24**, 1721 (1971).

[490] R. G. Harvey, P. P. Fu, and P. W. Rabideau, *J. Org. Chem.*, **41**, 3722 (1976).

[491] (a) F. B. Mallory, C. W. Mallory, and S. P. Varimbi, unpublished results; (b) L. Castedo, C. Saá, J. M. Saá, and R. Suau, *J. Org. Chem.*, **47**, 513 (1982); (c) L. Cleaver, S. Nimgirawath, E. Ritchie, and W. C. Taylor, *Aust. J. Chem.*, **29**, 2003 (1976); (d) I. Ninomiya and T. Naito, *Heterocycles*, **10**, 237 (1978).

[492] R. B. DuVernet, O. Wennerström, J. Lawson, T. Otsubo, and V. Boekelheide, *J. Am. Chem. Soc.*, **100**, 2457 (1978).

[493] R. B. Herbert and C. J. Moody, *J. Chem. Soc., Chem. Commun.*, **1970**, 121.

[494] G. Krow, K. M. Damodaran, E. Michener, S. I. Miller, and D. R. Dalton, *Synth. Commun.*, **6**, 261 (1976).

[495] M. P. Cava, S. C. Havlicek, A. Lindert, and R. J. Spangler, *Tetrahedron Lett.*, **1966**, 2937.

[496] G. Y. Moltrasio, R. M. Sotelo, and D. Giacopello, *J. Chem. Soc., Perkin Trans. 1*, **1973**, 349.

[497] N. Ishibe, S. Yutaka, J. Masui, and Y. Ishida, *J. Chem. Soc., Chem. Commun.*, **1975**, 241.

[498] N. Ishibe and S. Yutaka, *Tetrahedron*, **32**, 1331 (1976).

[499] R. M. Letcher, L. R. M. Nhamo, and I. T. Gumiro, *J. Chem. Soc., Perkin Trans. 1*, **1972**, 206.

[500] N. Sindler-Kulyk and W. H. Laarhoven, *Recl. Trav. Chim. Pays-Bas*, **98**, 452 (1979).

[501] M. P. Cava and S. S. Libsch, *J. Org. Chem.*, **39**, 577 (1974).

[502] G. Rio and J.-C. Hardy, *Bull. Soc. Chim. Fr.*, **1967**, 2642.

[503] E. D. Middlemas and L. D. Quin, *J. Am. Chem. Soc.*, **102**, 4838 (1980).

[504] T. Hashimoto, K. Hasegawa, H. Yamaguchi, M. Saito, and S. Ishimoto, *Phytochemistry*, **13**, 2849 (1974).

[505] A. Ch. Greiner, C. Spyckerelle, and P. Albrecht, *Tetrahedron*, **32**, 257 (1976).

[506] R. J. Olsen and R. E. Buckles, *J. Photochem.*, **10**, 215 (1979).

[507] G. W. Griffin, R. L. Smith, and A. Manmade, *J. Org. Chem.*, **41**, 338 (1976).

[508] (a) R. C. Petterson, A. H. Hebert, G. W. Griffin, I. Sarkar, O. P. Strausz, and J. Font, *J. Heterocycl. Chem.*, **10**, 879 (1973); (b) D. Billen, N. Boens, and F. C. DeSchryver, *J. Chem. Res., Synop.*, **1979**, 79.

[509] K. H. Grellmann, *Ber. Bunsenges. Phys. Chem.*, **73**, 827 (1969).

[510] H.-G. Heine, *Tetrahedron Lett.*, **1971**, 1473.

[511] (a) J. O. Stoffer and J. T. Bohanon, *J. Chem. Soc., Perkin Trans. 2*, **1978**, 692; (b) B. Halton, M. Kulig, M. A. Battiste, J. Perreten, D. M. Gibson, and G. W. Griffin, *J. Am. Chem. Soc.*, **93**, 2327 (1971).

[512] E. Clar and W. Müller, *Ber. Dtsch. Chem. Ges.*, **63**, 869 (1930).

[513] R. H. Mitchell, L. Mazuch, B. Shell, and P. R. West, *Can. J. Chem.*, **56**, 1246 (1978).

[514] N. K. Cuong, F. Fournier, and J.-J. Basselier, *Bull. Soc. Chim. Fr.*, **1974**, 2117.

[515] (a) H. Dürr, P. Herbst, P. Heitkämper, and H. Leismann, *Chem. Ber.*, **107**, 1835 (1974); (b) I. Moritani and N. Toshima, *Tetrahedron Lett.*, **1967**, 467.

[516] N. Toshima and I. Moritani, *Tetrahedron Lett.*, **1967**, 357.

[517] N. Ishibe, M. Odani, and M. Sunami, *J. Chem. Soc., Chem. Commun.*, **1971**, 1034.

[518] A. G. Brook, K. H. Pannell, and D. G. Anderson, *J. Am. Chem. Soc.*, **90**, 4374 (1968).

[519] R. S. Becker, R. O. Bost, J. Kolc, N. R. Bertoniere, R. L. Smith, and G. W. Griffin, *J. Am. Chem. Soc.*, **92**, 1302 (1970).

[520] M. Croisy-Delcey, Y. Ittah, and D. M. Jerina, *Tetrahedron Lett.*, **1979**, 2849.

[521] J. LeGuen and R. Taylor, *J. Chem. Soc., Perkin Trans. 2*, **1974**, 1274.

[522] D. L. Nagel, R. Kupper, K. Antonson, and L. Wallcave, *J. Org. Chem.*, **42**, 3626 (1977).

[523] W. H. Laarhoven and R. G. M. Veldhuis, *Tetrahedron*, **28**, 1811 (1972).

[524] R. H. Martin, M. Flammang-Barbieux, J. P. Cosyn, and M. Gelbcke, *Tetrahedron Lett.*, **1968**, 3507.

[525] L. Weiss, M. Loy, S. S. Hecht, and D. Hoffmann, *J. Labelled Compd. Radiopharm.*, **14**, 119 (1978).

[526] C. E. Browne, T. K. Dobbs, S. S. Hecht, and E. J. Eisenbraun, *J. Org. Chem.*, **43**, 1656 (1978).

[527] R. H. Martin and J. J. Schurter, *Tetrahedron Lett.*, **1969**, 3679.

[528] P. H. Gore and F. S. Kamonah, *Synthesis*, **1978**, 773.

[529] S. S. Hecht, M. Loy, R. Mazzarese, and D. Hoffmann, *J. Med. Chem.*, **21**, 38 (1978).

[530] J.-M. Vanest, M. Gorsane, V. Libert, J. Pecher, and R. H. Martin, *Chimia*, **29**, 343 (1975).

[531] R. H. Mitchell and R. J. Carruthers, *Tetrahedron Lett.*, **1975**, 4331.

[532] C. I. Lewis, J. Y. Chang, and A. W. Spears, *J. Org. Chem.*, **34**, 1176 (1969).

[533] W. Carruthers and H. N. M. Stewart, *J. Chem. Soc. C*, **1967**, 560.

[534] P. J. Jessup and J. A. Reiss, *Aust. J. Chem.*, **29**, 1267 (1976).

[535] F. B. Mallory and C. W. Mallory, *J. Org. Chem.*, **48**, 526 (1983).

[536] W. Carruthers, *Chem. Commun.*, **1966**, 548.

[537] P. J. Jessup and J. A. Reiss, *Aust. J. Chem.*, **30**, 843 (1977).

[538] W. H. Laarhoven and J. A. M. van Broekhoven, *Tetrahedron Lett.*, **1970**, 73.

[539] T. Wismontski-Knittel, M. Kaganowitch, G. Seger, and E. Fischer, *Recl. Trav. Chim. Pays-Bas.* **98**, 114 (1979).

[540] W. J. Bernstein, M. Calvin, and O. Buchardt, *Tetrahedron Lett.*, **1972**, 2195.

[541] W. J. Bernstein, M. Calvin, and O. Buchardt, *J. Am. Chem. Soc.*, **95**, 527 (1973).

[542] P. J. Jessup and J. A. Reiss, *Tetrahedron Lett.*, **1975**, 1453.

[543] W. H. Laarhoven and R. J. F. Nivard, *Tetrahedron*, **32**, 2445 (1976).

[544] R. H. Martin, Ch. Eyndels, and N. Defay, *Tetrahedron*, **30**, 3339 (1974).

[545] W. H. Laarhoven and Th. J. H. M. Cuppen, *Tetrahedron*, **30**, 1101 (1974).

[546] J. H. Borkent and W. H. Laarhoven, *Tetrahedron*, **34**, 2565 (1978).

[547] C. Jutz and H.-G. Löbering, *Angew. Chem., Int. Ed. Engl.*, **14**, 418 (1975).

[548] R. H. Martin and J. P. Cosyn, *Synth. Commun.*, **1**, 257 (1971).

[549] M. Flammang-Barbieux, J. Nasielski, and R. H. Martin, *Tetrahedron Lett.*, **1967**, 743.

[550] C. F. Wilcox, Jr., P. M. Lahti, J. R. Rocca, M. B. Halpern, and J. Meinwald, *Tetrahedron Lett.*, **1978**, 1893.

[551] W. H. Laarhoven and M. H. de Jong, *Recl. Trav. Chim. Pays-Bas*, **92**, 651 (1973).

[552] W. H. Laarhoven and N. P. J. Kuin, *Recl. Trav. Chim. Pays-Bas*, **94**, 105 (1975).

[553] Y. Cochez, R. H. Martin, and J. Jespers, *Isr. J. Chem.*, **15**, 29 (1976/77).

[554] W. H. Laarhoven and Th. J. H. M. Cuppen, *Tetrahedron Lett.*, **1971**, 163.

[555] B. Thulin and O. Wennerström, *Acta Chem. Scand.*, **B30**, 688 (1976).

[556] W. H. Laarhoven and R. G. M. Veldhuis, *Tetrahedron*, **28**, 1823 (1972).

[557] S. W. Horgan, D. D. Morgan, and M. Orchin, *J. Org. Chem.*, **38**, 3801 (1973).

[558] A. Padwa and A. Mazzu, *Tetrahedron Lett.*, **1974**, 4471.

[559] P. H. G. op het Veld, J. C. Langendam, and W. H. Laarhoven, *Tetrahedron Lett.*, **1975**, 231.

[560] C. C. Leznoff and R. J. Hayward, *Can. J. Chem.*, **48**, 1842 (1970).

[561] (a) K. Veeramani, K. Paramasivam, S. Ramakrishnasubramanian, and P. Shanmugam, *Synthesis*, **1978**, 855; (b) K. A. Kumar and G. Srimannarayana, *Indian J. Chem.*, **19B**, 615 (1980).

[562] W. Carruthers, N. Evans, and R. Pooranamoorthy, *J. Chem. Soc., Perkin Trans. 1*, **1973**, 44.

[563] (a) W. Carruthers, N. Evans, and D. Whitmarsh, *J. Chem. Soc., Chem. Commun.*, **1974**, 526; (b) B. Chittim and S. Safe, *Chemosphere*, **5**, 269 (1977).

[564] H. G. Heller and P. J. Strydom, *J. Chem. Soc., Chem. Commun.*, **1976**, 50.

[565] O. Crescente, H. G. Heller, and S. Oliver, *J. Chem. Soc., Perkin Trans. 1*, **1979**, 150.

[566] H. Stobbe, *Chem. Ber.*, **40**, 3372 (1907).

[567] B. Weinstein and D. N. Brattesani, *Chem. Ind. (London)*, **1967**, 1292.

[568] A. Santiago and R. S. Becker, *J. Am. Chem. Soc.*, **90**, 3654 (1968).

[569] H. G. Heller and M. Szewczyk, *J. Chem. Soc., Perkin Trans. 1*, **1974**, 1487.

[570] R. L. Funk and K. P. C. Vollhardt, *Angew. Chem., Int. Ed. Engl.*, **15**, 53 (1976).

[571] F. G. Baddar, L. S. El-Assal, and M. Gindy, *J. Chem. Soc.*, **1948**, 1270.

[572] M. Sindler-Kulyk and W. H. Laarhoven, *Recl. Trav. Chim. Pays-Bas*, **98**, 187 (1979).

[573] M. Onda, K. Yuasa, J. Okada, K. Kataoka, and K. Abe, *Chem. Pharm. Bull.*, **21**, 1333 (1973).

[574] M. Onda, Y. Harigaya, and T. Suzuki, *Heterocycles*, **4**, 1669 (1976).

[575] M. Onda, Y. Harigaya, and T. Suzuki, *Chem. Pharm. Bull.*, **25**, 2935 (1977).

[576] M. Onda and K. Kawakami, *Chem. Pharm. Bull.*, **20**, 1484 (1972).
[577] R. J. Hart, H. G. Heller, R. M. Megit, and M. Szewczyk, *J. Chem. Soc.*, *Perkin Trans. 1*, **1975**, 2227.
[578] G. Ege and K. Gilbert, *Chem. Ber.*, **112**, 3166 (1979).
[579] T. Bercovici, M. D. Cohen, E. Fischer, and D. Sinnreich, *J. Chem. Soc.*, *Perkin Trans. 2*, **1977**, 1.
[580] N. Campbell, P. S. Davison, and H. G. Heller, *J. Chem. Soc.*, **1963**, 993.
[581] J. S. Hastings and H. G. Heller, *J. Chem. Soc.*, *Perkin Trans. 1*, **1972**, 1839.
[582] G. Kaupp and H.-W. Grüter, *Angew. Chem.*, *Int. Ed. Engl.*, **17**, 52 (1978).
[583] N. Campbell and H. G. Heller, *J. Chem. Soc.*, **1965**, 5473.
[584] (a) F. Toda and Y. Todo, *Bull. Chem. Soc. Jpn.*, **50**, 3000 (1977); (b) F. Toda and Y. Todo, *Bull. Chem. Soc. Jpn.*, **49**, 2503 (1976).
[585] A. H. A. Tinnemans and W. H. Laarhoven, *Tetrahedron*, **35**, 1537 (1979).
[586] J. S. Hastings, H. G. Heller, and K. Salisbury, *J. Chem. Soc.*, *Perkin Trans. 1*, **1975**, 1995.
[587] J. S. Hastings, H. G. Heller, H. Tucker, and K. Smith, *J. Chem. Soc.*, *Perkin Trans. 1*, **1975**, 1545.
[588] H. G. Heller, D. Auld, and K. Salisbury, *J. Chem. Soc. C*, **1967**, 2457.
[589] H. G. Heller and K. Salisbury, *Tetrahedron Lett.*, **1968**, 2033.
[590] A. Schönberg, A. F. A. Ismail, and W. Asker, *J. Chem. Soc.*, **1946**, 442.
[591] N. Filipescu, E. Avram, and K. D. Welk, *J. Org. Chem.*, **42**, 507 (1977).
[592] W. Koch, T. Saito, and Z. Yoshida, *Tetrahedron*, **28**, 3191 (1972).
[593] G. Sauvage, *Ann. Chim.* (*Paris*), [12] **2**, 844 (1947).
[594] A. Eckert and R. Tomaschek, *Monatsh. Chem.*, **39**, 839 (1918).
[595] A. Eckert, *Ber. Dtsch. Chem. Ges.*, **58**, 322 (1925).
[596] R. W. Hardacre and A. G. Perkin, *J. Chem. Soc.*, **131**, 180 (1929).
[597] H. Brockmann, E. H. F. von Falkenhausen, R. Neeff, A. Dorlars, and G. Budde, *Chem. Ber.*, **84**, 865 (1951).
[598] A. Eckert and J. Hampel, *Chem. Ber.*, **60**, 1693 (1927).
[599] H. Brockmann, E. Lindemann, K.-H. Ritter, and F. Depke, *Chem. Ber.*, **83**, 583 (1950).
[600] H. Brockmann, R. Neeff, and E. Mühlmann, *Chem. Ber.*, **83**, 467 (1950).
[601] H. Brockmann and R. Randebrock, *Chem. Ber.*, **84**, 533 (1951).
[602] H. Brockmann, F. Pohl, K. Maier, and M. N. Haschad, *Justus Liebigs Ann. Chem.*, **553**, 1 (1942).
[603] H. Brockmann and A. Dorlars, *Chem. Ber.*, **85**, 1168 (1952).
[604] G. F. Attree and A. G. Perkin, *J. Chem. Soc.*, **1931**, 144.
[605] H. Brockmann and W. Sanne, *Naturwissenschaften*, **19**, 509 (1953).
[606] H. Brockmann and H. Eggers, *Chem. Ber.*, **91**, 81 (1958).
[607] G. Kortüm, W. Theilacker, H. Zeininger, and H. Elliehausen, *Chem. Ber.*, **86**, 294 (1953).
[608] J. W. E. Haller and A. G. Perkin, *J. Chem. Soc.*, **125**, 231 (1924).
[609] D. W. Cameron and P. E. Schütz, *J. Chem. Soc. C*, **1967**, 2121.
[610] H. Brockmann, F. Kluge, and H. Muxfeldt, *Chem. Ber.*, **90**, 2302 (1957).
[611] G. Sauvage, *C. R. Hebd. Seances Acad. Sci.*, **225**, 247 (1947).
[612] P. Bortolus, G. Cauzzo, and G. Galiazzo, *Tetrahedron Lett.*, **1966**, 239.
[613] R. E. Doolittle and C. K. Bradsher, *Chem. Ind.* (*London*), **1965**, 1631.
[614] G. Galiazzo, P. Bortolus, G. Cauzzo, and U. Mazzucato, *J. Heterocycl. Chem.*, **6**, 465 (1969).
[615] T. Matsuo and S. Mihara, *Bull. Chem. Soc. Jpn.*, **48**, 3660 (1975).
[616] N. R. Beller, D. C. Neckers, and E. P. Papadopoulos, *J. Org. Chem.*, **42**, 3514 (1977).
[617] S. Searles, Jr., and R. A. Clasen, *Tetrahedron Lett.*, **1965**, 1627.
[618] H. Ohta and K. Tokumaru, *Bull. Chem. Soc. Jpn.*, **48**, 1669 (1975); D. R. Arnold, V. Y. Abraitys, and D. McLeod, Jr., *Can. J. Chem.*, **49**, 923 (1971).
[619] (a) M. Kojima and M. Maeda, *J. Chem. Soc.*, *Chem. Commun.*, **1970**, 386; (b) W. Carruthers and N. Evans, *J. Chem. Soc.*, *Perkin Trans. 1*, **1974**, 421.
[620] A. Shafiee and A. Rashidbaigi, *J. Heterocycl. Chem.*, **13**, 141 (1976).
[621] V. N. R. Pillai and M. Ravindran, *Indian J. Chem.*, **15B**, 1043 (1977).

[622] C. Riche, A. Chiaroni, H. Doucerain, R. Besselièvre, and C. Thal, *Tetrahedron Lett.*, **1975**, 4567.

[623] (a) C. Dieng, C. Thal, H.-P. Husson, and P. Potier, *J. Heterocycl. Chem.*, **12**, 455 (1975); (b) P. J. Grisdale and J. L. R. Williams, *J. Org. Chem.*, **34**, 1675 (1969).

[624] G. Cauzzo, G. Galiazzo, P. Bortolus, and F. Coletta, *Photochem. Photobiol.*, **13**, 445 (1971).

[625] F. Andreani, R. Andrisano, G. Salvadori, and M. Tramontini, *J. Chem. Soc. C*, **1971**, 1008.

[626] J. Szmuszkovicz, *Org. Prep. Proced.*, **1**, 105 (1969).

[627] A. R. Katritzky, Z. Zakaria, and E. Lunt, *J. Chem. Soc., Perkin Trans. 1*, **1980**, 1879.

[628] (a) A. R. Katritzky, Z. Zakaria, E. Lunt, P. G. Jones, and O. Kennard, *J. Chem. Soc., Chem. Commun.*, **1979**, 268; (b) K. B. Soroka and J. A. Soroka, *Tetrahedron Lett.*, **1980**, 4631.

[629] G. E. Trukhan, Ya. R. Tymyanskii, Yu. P. Andreichikov, M. I. Knyazhanskii, and G. N. Dorofeenko, *Khim. Geterotsikl. Soedin.*, **1978**, 1226 [*C.A.*, **90**, 87203 (1979)].

[630] S. C. Shim and S. K. Lee, *Synthesis*, **1980**, 116.

[631] G. Lindgren, K.-E. Stensiö, and K. Wahlberg, *J. Heterocycl. Chem.*, **17**, 679 (1980).

[632] C. E. Loader and C. J. Timmons, *J. Chem. Soc. C*, **1967**, 1343.

[633] C. P. Joshua and V. N. Rajasekharan Pillai, *Tetrahedron Lett.*, **1972**, 2493.

[634] R. F. Evans, *J. Chem. Educ.*, **48**, 768 (1971).

[635] G. E. Lewis and R. J. Mayfield, *Aust. J. Chem.*, **19**, 1445 (1966).

[636] C. P. Joshua and V. N. R. Pillai, *Indian J. Chem.*, **13**, 1018 (1975).

[637] J. A. Eenkhoorn, S. O. de Silva, and V. Snieckus, *Can. J. Chem.*, **51**, 792 (1973).

[638] V. N. R. Pillai and E. Purushothaman, *Curr. Sci.*, **47**, 627 (1978).

[639] J. Hennessy and A. C. Testa, *J. Phys. Chem.*, **76**, 3362 (1972).

[640] H.-P. Husson, C. Thal, P. Potier, and E. Wenkert, *J. Org. Chem.*, **35**, 442 (1970).

[641] D. Cohylakis, G. J. Hignett, K. V. Lichman, and J. A. Joule, *J. Chem. Soc., Perkin Trans. 1*, **1974**, 1518.

[642] K.-D. Müller and K. Niedenzu, *Inorg. Chim. Acta*, **25**, L53 (1977).

[643] G. Cauquis and G. Reverdy, *Tetrahedron Lett.*, **1977**, 3267.

[644] H. Kato, T. Shiba, E. Kitajima, T. Kiyosawa, F. Yamada, and T. Nishiyama, *J. Chem. Soc., Perkin Trans. 1*, **1976**, 863.

[645] N. C. Jamieson and G. E. Lewis, *Aust. J. Chem.*, **20**, 2777 (1967).

[646] F. A. Neugebauer, *Chem. Ber.*, **102**, 1339 (1969).

[647] F. A. Neugebauer and H. Trischmann, *Justus Liebigs Ann. Chem.*, **706**, 107 (1967).

[648] (a) F. Weygand and I. Frank, *Z. Naturforsch. Teil B*, **3**, 377 (1948); (b) I. Hausser, D. Jerchel, and R. Kuhn, *Chem. Ber.*, **82**, 195 (1949).

[649] (a) J. H. Boyer and G. J. Mikol, *J. Heterocycl. Chem.*, **9**, 1325 (1972); (b) A. Padwa and S. I. Wetmore, Jr., *J. Chem. Soc., Chem. Commun.*, **1972**, 1116.

[650] (a) E. V. Blackburn, T. J. Cholerton, and C. J. Timmons, *J. Chem. Soc., Perkin Trans. 2*, **1972**, 101; (b) B. P. Das and D. W. Boykin, Jr., *J. Med. Chem.*, **16**, 413 (1973).

[651] B. P. Das, J. A. Campbell, F. B. Samples, R. A. Wallace, L. K. Whisenant, R. W. Woodard, and D. W. Boykin, Jr., *J. Med. Chem.*, **15**, 370 (1972).

[652] B. P. Das, R. T. Cunningham, D. W. Boykin, Jr., *J. Med. Chem.*, **16**, 1361 (1973).

[653] M. Iwao, M. L. Lee, and R. N. Castle, *J. Heterocycl. Chem.*, **17**, 1259 (1980).

[654] J. Lawson, R. DuVernet, and V. Boekelheide, *J. Am. Chem. Soc.*, **95**, 955 (1973).

[655] G. R. Lappin and J. S. Zannucci, *J. Org. Chem.*, **36**, 1808 (1971).

[656] M. Ogata and H. Matsumoto, *Chem. Pharm. Bull.*, **20**, 2264 (1972).

[657] (a) Y. Ogata, K. Takagi, and I. Ishino, *J. Org. Chem.*, **36**, 3975 (1971); (b) Y. Kanaoka, K. San-nohe, Y. Hatanaka, K. Itoh, M. Machida, and M. Terashima, *Heterocycles*, **6**, 29 (1977).

[658] (a) Y. Kanaoka, K. Itoh, Y. Hatanaka, J. L. Flippen, I. L. Karle, and B. Witkop, *J. Org. Chem.*, **40**, 3001 (1975); (b) I. Ninomiya, T. Kiguchi, and T. Naito, *Heterocycles*, **9**, 1023 (1978).

[659] I. Ninomiya, T. Naito, and T. Kiguchi, *Tetrahedron Lett.*, **1970**, 4451.

[660] (a) A. Mondon and K. Krohn, *Chem. Ber.*, **105**, 3726 (1972); (b) M. Terashima and K. Seki, *Heterocycles*, **8**, 421 (1977).

[661] D. H. Hey, G. H. Jones, and M. J. Perkins, *J. Chem. Soc., Perkin Trans. 1*, **1972**, 1150.

[662] W. Carruthers and N. Evans, *J. Chem. Soc., Perkin Trans. 1*, **1974**, 1523.

[663] (a) T. Nakano and A. Martin, *J. Heterocycl. Chem.*, **16**, 1235 (1979); (b) Y. Kanaoka, S. Nakao, and Y. Hatanaka, *Heterocycles*, **5**, 261 (1976).

[664] E. Winterfeldt and H. J. Altmann, *Angew. Chem., Int. Ed. Engl.*, **7**, 466 (1968).

[665] H. Iida, S. Aoyagi, and C. Kibayashi, *J. Chem. Soc., Chem. Commun.*, **1974**, 499.

[666] H. Iida, S. Aoyagi, and C. Kibayashi, *J. Chem. Soc., Perkin Trans. 1*, **1975**, 2502.

[667] (a) K. Ito, T. Naruchi, H. Tsuruta, and K. Komiriya, *Jpn. Kokai Tokkyo Koho*, **78**, 77,067 (1978) [*C.A.*, **89**, 163,430 (1978)]; (b) Y. Kanaoka and K. San-nohe, *Tetrahedron Lett.*, **1980**, 3893.

[668] I. Ninomiya, A. Shinohara, T. Kiguchi, and T. Naito, *J. Chem. Soc., Perkin Trans. 1*, **1976**, 1868.

[669] M. Sainsbury and N. L. Uttley, *J. Chem. Soc., Perkin Trans. 1*, **1976**, 2416.

[670] M. Sainsbury and B. Webb, *Phytochemistry*, **14**, 2691 (1975).

[671] G. R. Lenz, *J. Heterocycl. Chem.*, **16**, 433 (1979).

[672] I. Ninomiya, O. Yamamoto, and T. Naito, *Heterocycles*, **4**, 743 (1976).

[673] I. Ninomiya, O. Yamamoto, T. Kiguchi, and T. Naito, *J. Chem. Soc., Perkin Trans. 1*, **1980**, 203.

[674] M. Sainsbury and N. L. Uttley, *J. Chem. Soc., Perkin Trans. 1*, **1977**, 2109.

[675] T. Kametani, T. Sugai, Y. Shoji, T. Honda, F. Satoh, and K. Fukumoto, *J. Chem. Soc., Perkin Trans. 1*, **1977**, 1151.

[676] I. Ninomiya and T. Naito, *Heterocycles*, **2**, 607 (1974).

[677] (a) W. J. Begley and J. Grimshaw, *J. Chem. Soc., Perkin Trans. 1*, **1977**, 2324; (b) T. Kametani, T. Honda, T. Sugai, and K. Fukumoto, *Heterocycles*, **4**, 927 (1976).

[678] (a) I. Ninomiya, T. Naito, and T. Kiguchi, *J. Chem. Soc., Perkin Trans. 1*, **1973**, 2261; (b) I. Ninomiya, T. Naito, and H. Ishii, *Heterocycles*, **3**, 307 (1975).

[679] I. Ninomiya, O. Yamamoto, and T. Naito, *J. Chem. Soc., Chem. Commun.*, **1976**, 437.

[680] I. Ninomiya, O. Yamamoto, and T. Naito, *J. Chem. Soc., Perkin Trans. 1*, **1980**, 212.

[681] H. Ishii, K. Harada, T. Ishida, E. Ueda, K. Nakajima, I. Ninomiya, T. Naito, and T. Kiguchi, *Tetrahedron Lett.*, **1975**, 319.

[682] H. Ishii, E. Ueda, K. Nakajima, T. Ishida, T. Ishikawa, K. Harada, I. Ninomiya, T. Naito, and T. Kiguchi, *Chem. Pharm. Bull.*, **26**, 864 (1978).

[683] I. Ninomiya, T. Naito, and T. Kiguchi, *J. Chem. Soc., Chem. Commun.*, **1970**, 1669.

[684] T. Kametani, A. Ujiie, M. Ihara, K. Fukumoto, and S.-T. Lu, *J. Chem. Soc., Perkin Trans. 1*, **1976**, 1218.

[685] M. Riviere, N. Paillous, and A. Lattes, *Bull. Soc. Chim. Fr.*, **1974**, 1911.

[686] A. G. Schultz and C.-K. Sha, *Tetrahedron*, **36**, 1757 (1980).

[687] A. G. Schultz and V. Kane, *J. Chem. Educ.*, **56**, 555 (1979).

[688] R. B. Miller and T. Moock, *Tetrahedron Lett.*, **1980**, 3319.

[689] C. Wentrup and M. Gaugaz, *Helv. Chim. Acta*, **54**, 2108 (1971).

[690] K. Yamada, T. Konakahara, S. Ishihara, H. Kanamori, T. Itoh, K. Kimura, and H. Iida, *Tetrahedron Lett.*, **1972**, 2513.

[691] K. Yamada, T. Konakahara, and H. Iida, *Bull. Chem. Soc. Jpn.*, **46**, 2504 (1973).

[692] W. Carruthers, *J. Chem. Soc. C*, **1968**, 2244.

[693] J.-D. Cheng and H. J. Shine, *J. Org. Chem.*, **39**, 336 (1974).

[694] W. Lamm, W. Jugelt, and F. Pragst, *J. Prakt. Chem.*, **317**, 284 (1975).

[695] (a) M. Zander and W. H. Franke, *Chem. Ber.*, **99**, 2449 (1966); (b) M. Zander and W. H. Franke, *Chem. Ber.*, **105**, 3495 (1972).

[696] (a) A. Norström, K. Andersson, and C. Rappe, *Chemosphere*, **1**, 21 (1976); (b) A. Norström, K. Andersson, and C. Rappe, *Chemosphere*, **5**, 241 (1977).

[697] R. D. Youssefyeh and M. Weisz, *Tetrahedron Lett.*, **1973**, 4317.

[698] A. G. Schultz and R. D. Lucci, *J. Org. Chem.*, **40**, 1372 (1975).

[699] A. G. Schultz and W. Y. Fu, *J. Org. Chem.*, **41**, 1483 (1976).

[700] R. D. Youssefyeh and M. Weisz, *J. Am. Chem. Soc.*, **96**, 315 (1974).

[701] A. G. Schultz, Y. K. Yee, and M. H. Berger, *J. Am. Chem. Soc.*, **99**, 8065 (1977).

[702] S. H. Groen, R. M. Kellogg, J. Buter, and H. Wynberg, *J. Org. Chem.*, **33**, 2218 (1968).

[703] N. Kharasch and Z. S. Ariyan, *Chem. Ind. (London)*, **1965**, 302.

[704] (a) N. Kharasch and R. B. Langford, *Int. J. Sulfur Chem.*, **8**, 573 (1976); (b) T. Itoh, H. Ogura, and K. A. Watanabe, *Tetrahedron Lett.*, **1977**, 2595.

[705] A. G. Schultz and M. B. DeTar, *J. Am. Chem. Soc.*, **96**, 296 (1974).

[706] A. G. Schultz and M. B. DeTar, *Org. Photochem. Synth.*, **2**, 101 (1976).

[707] R. Paramasivam, R. Palaniappan, and V. T. Ramakrishnan, *J. Chem. Soc., Chem. Commun.* **1979**, 260.

[708] D. Belluš and K. Schaffner, *Helv. Chim. Acta*, **51**, 221 (1968).

CHAPTER 2

OLEFIN SYNTHESIS BY DEOXYGENATION OF VICINAL DIOLS

ERIC BLOCK

State University of New York at Albany, Albany, New York

CONTENTS

ACKNOWLEDGMENTS

It is a pleasure to acknowledge the assistance of Mrs. Marsha A. Lee of the E. I. du Pont de Nemours Experimental Station and Dr. Larry K. Revelle in searching the literature. I thank Dr. Roland Winter and Professor K. Barry Sharpless for providing unpublished data and Mrs. Virginia Dollar for her meticulous typing of the manuscript. Support during the preparation of this review from the National Science Foundation and the Petroleum Research Fund, administered by the American Chemical Society, is gratefully acknowledged.

INTRODUCTION

Since the discovery in 1963 that 1,3-dioxolane-2-thiones (cyclic thionocarbonates) undergo stereospecific fragmentation to olefins on heating with trivalent phosphorus compounds (the Corey–Winter reaction; Eq. 1),[1,2] a variety of approaches have been developed for the regio- and stereospecific (or stereoselective) deoxygenation of vicinal diols based on *syn* elimination from cyclic derivatives of type **1** (Eq. 2).

$$C_6H_5CH(OH)CH(OH)C_6H_5 \longrightarrow \text{cyclic thionocarbonate } (C=S) \xrightarrow{R_3P} C_6H_5CH{=}CHC_6H_5 + R_3P{=}S + CO_2$$

(92%)

$$C_6H_5CH(OH)CH(OH)C_6H_5 \longrightarrow \text{cyclic thionocarbonate } (C=S) \xrightarrow{R_3P} C_6H_5CH{=}CHC_6H_5 + R_3P{=}S + CO_2 \quad \text{(Eq. 1)}$$

(87%)

$$R_1R_3C(OH)\text{–}C(OH)R_2R_4 \longrightarrow \underset{\mathbf{1}}{\text{cyclic } O\text{–}X\text{–}O} \longrightarrow R_1R_3C{=}CR_2R_4 + \text{“}O\text{—}X\text{—}O\text{”} \quad \text{(Eq. 2)}$$

These methods have been particularly useful in synthesizing substrates with delicate structural features such as strained and twisted olefins, unsaturated carbohydrates, macrocyclic lactones, lipids, and polyenes, as well as in establishing the stereochemistry of diol functions in natural products.

This review deals primarily with deoxygenation of vicinal diols via cyclic species **1** with X = C, P, Ti, W, or S (suitably functionalized). Brief coverage is given to deoxygenation processes that do not involve cyclic intermediates, whereas routes proceeding via oxiranes or *vic*-dihalides as separately isolated intermediates are excluded from consideration. The stereochemical equilibration of vicinal diols and the conversion of vicinal hydroxy thiols and dithiols to olefins are also covered in this chapter. Brief reviews have appeared in the literature of the Corey–Winter reaction[3–5] and of the general topic of deoxygenation of vicinal diols.[6,7]

DESULFURIZATION OF 1,3-DIOXOLANE-2-THIONES (THIONOCARBONATES)

Mechanism

The 1,3-dioxolane-2-thione desulfurization olefin synthesis, often termed the Corey–Winter reaction, was based on the hypothesis that 2-carbena-1,3-dioxolanes, (**2**) "might be unstable relative to olefin and carbon dioxide" (Eq. 3).[1] This mechanistic hypothesis is apparently supported by pyrolysis studies involving norbornadienone (Eqs. 4 and 5)[8,9]and quadricyclanone ethylene ketals;[8] however, theoretical calculations and experimental studies of the decomposition of carbenes of type **2** suggest that this decomposition can lead

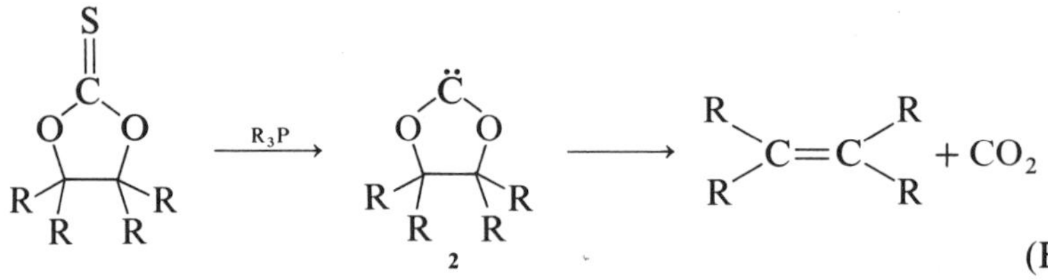

$\xrightarrow{R_3P}$ **2** $\longrightarrow$ $R_2C{=}CR_2 + CO_2$ (Eq. 3)

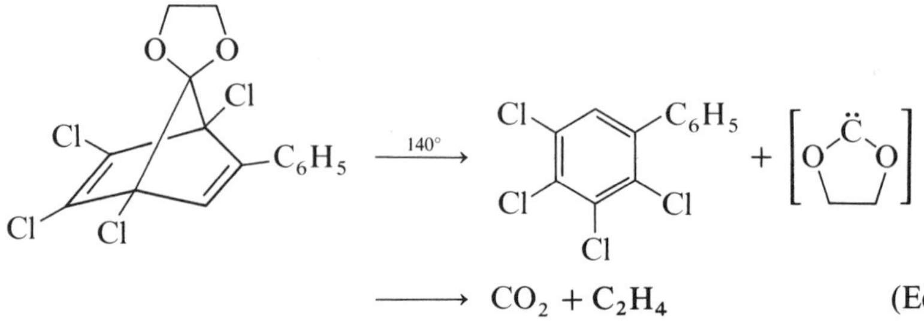

$\xrightarrow{140^\circ}$ $+$ $\longrightarrow CO_2 + C_2H_4$ (Eq. 4)

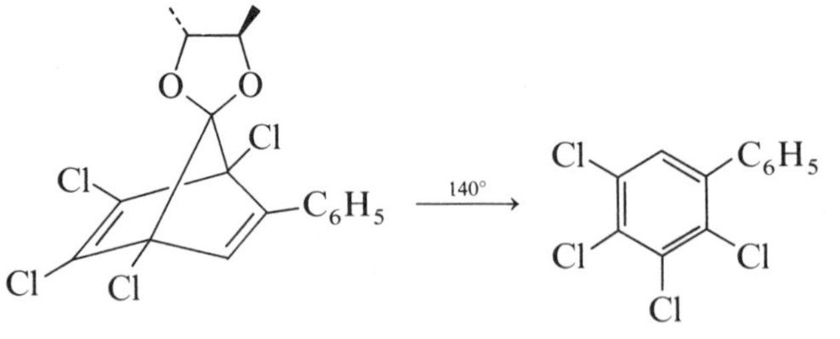

$\xrightarrow{140^\circ}$ $+$ *trans*-$CH_3CH{=}CHCH_3 + CO_2$ (Eq. 5)

to products strikingly different from those obtained under Corey–Winter conditions. The mechanism of the Corey–Winter reaction depicted in Eq. 3 has more recently been supplanted by another mechanism, shown below (Eq. 15), which incorporates carbenes **2**, ylides, and zwitterionic thionocarbonate–phosphite adducts. This newer proposal is presented following a discussion of the evidence against the involvement of the process of Eq. 3 in the Corey–Winter reaction.

Desulfurization of 4,4,5,5,-tetramethyl-1,3-dioxolane-2-thione with triethyl phosphite at 150° leads in 85% yield to 2,3-dimethyl-2-butene (Eq. 6).[1,10] In contrast, decomposition of tosylhydrazone salt **3** in refluxing tetraglyme (tetraethylene glycol dimethyl ether) solution or neat at 290° affords a whole array of products (Eq. 7).[11] An even more complex range of products is obtained from pyrolysis of tosylhydrazone salts derived from *erythro*- and *threo*-4-methylpentane-2,3-diol (**5**, Eq. 8, and **7**, Eq. 9, respectively).[12]

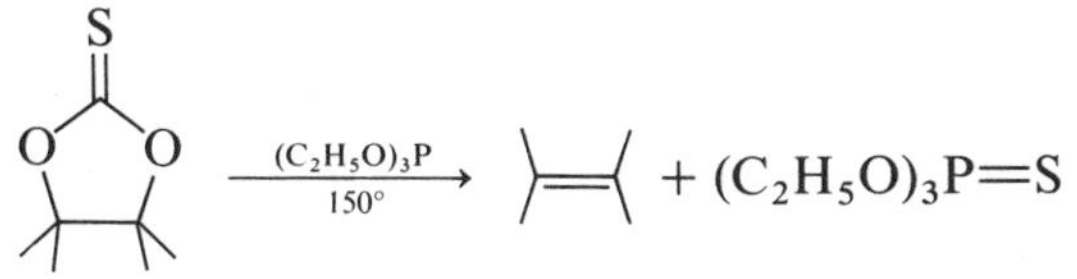

$\xrightarrow[150°]{(C_2H_5O)_3P}$ + $(C_2H_5O)_3P{=}S$ (Eq. 6)

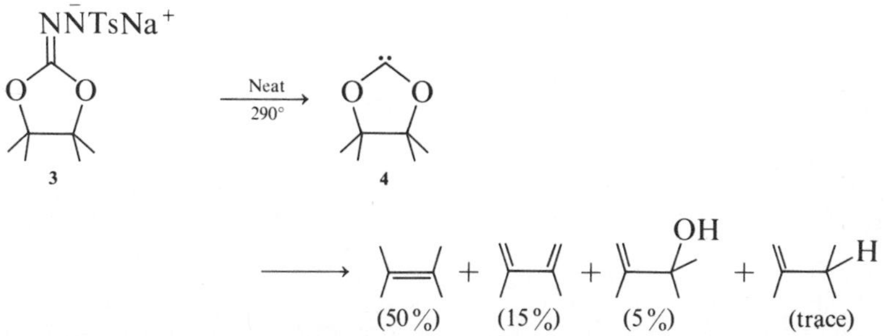

NN̄TsNa⁺ **3** $\xrightarrow[290°]{\text{Neat}}$ **4**

⟶ (50%) + (15%) + OH (5%) + H (trace) (Eq. 7)

Carbena-1,3-dioxolanes such as **4** (Eq. 7) are believed to be intermediates in the decomposition of tosylhydrazone salts **3**, **6**, and **8**. These carbenes are postulated to undergo homolytic cleavage of one C—O bond (Eq. 10), leading to diradical **9**; this is followed by loss of carbon dioxide (with partial loss of double-bond stereochemistry in the diradical intermediate), intramolecular hydrogen abstraction with subsequent loss of formic acid, or loss of carbon monoxide and subsequent hydrogen abstraction (leading to allylic alcohol), hydrogen shift (leading to ketone), or ring closure (leading to epoxide).

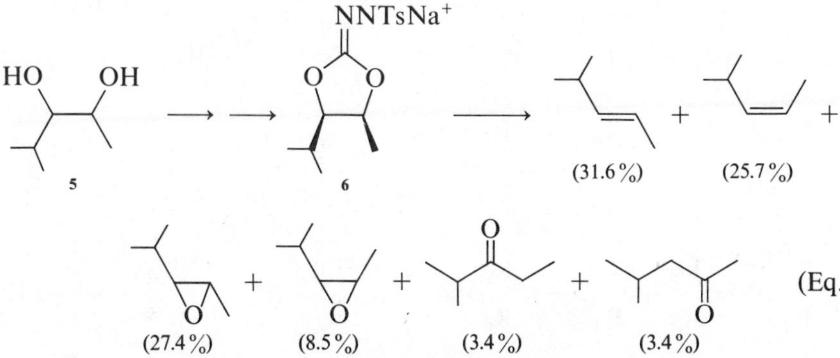

HO OH **5** ⟶ ⟶ NN̄TsNa⁺ **6** ⟶ (31.6%) + (25.7%) +

(27.4%) + (8.5%) + (3.4%) + (3.4%) (Eq. 8)

$NN^{-}TsNa^{+}$

HO OH

7 → → 8 → (72.0%) + (15.1%) +

+ (4.9%) + (4.9%) + O (1.5%) + O (1.5%) (Eq. 9)

R_1 R_3 + R_1 R_4

R_2 R_4 R_2 R_3

$-CO_2$

O

C·

O

R_1 R_3

R_2 R_4

9

$R_1 = R_3 = CH_3$

CH_3 O_2CH

R_2 R_4

$-HCO_2H$

R_2 R_4

$-CO$

R_1 O· R_3

R_2 R_4

$R_3 = CH_3$

R_1 OH

R_2 R_4

$R_2 = H$

R_1 O R_4 + R_1 O R_3 R_1 H R_3

R_2 R_3 R_2 R_4 O R_4 (Eq.10)

Ketones and epoxides are also produced in another reaction of 1,2-diols that is believed to involve carbena-1,3-dioxolanes, namely, the reaction of the diols with dichlorocarbene (formed by reaction of chloroform with base under phase-transfer catalysis conditions) (Eqs. 11, 12).[13] However, the transformations shown in Eqs. 11 and 12 may in fact involve mechanisms quite different from those shown in Eq. 10.[13]

$$\text{cis-cyclooctane-1,2-diol} \xrightarrow[\substack{NaOH \\ (C_2H_5)_3N^+CH_2C_6H_5Cl^-}]{CHCl_3,\ H_2O-CH_2Cl_2} \text{cyclooctene} + \text{cyclooctane}{>}CCl_2 + \text{cyclooctanone} \qquad \text{(Eq. 11)}$$

$$C_6H_5CH(OH)CH(OH)C_6H_5 \xrightarrow[\substack{NaOH \\ (C_2H_5)_3N^+CH_2C_6H_5\,Cl^-}]{CHCl_3,\ H_2O-CH_2Cl_2} \underset{(15\%)}{cis\text{-}C_6H_5CH{=}CHC_6H_5} + \underset{(1.5\%)}{trans\text{-}C_6H_5CH{=}CHC_6H_5} + \underset{(35\%)}{trans\text{-stilbene oxide}} \qquad \text{(Eq. 12)}$$

If rotation about the central C—C bond in diradical **9** is prevented, olefin formation through loss of carbon dioxide is favored (Eq. 13).[11]

$$\text{tricyclic dioxolane}{=}N\bar{N}TsNa^+ \xrightarrow[-CO_2]{300^\circ} \underset{(80\%)}{\text{bicyclo[3.3.0]oct-1(5)-ene}} \qquad \text{(Eq. 13)}$$

Theoretical calculations reveal that loss of carbon dioxide from carbena-1,3-dioxolanes in a concerted manner ($_{\sigma}2_s + {_{\sigma}2_s} + {_{\omega}2_s}$ cycloreversion) is about 9 kcal/mol *higher* in activation energy than homolytic cleavage of one C—O bond because the former process would lead to a nonlinear carbon dioxide (Eq. 14).[14]

$$\text{carbena-1,3-dioxolane } (R_4) \xrightarrow{E_a = 30.4\text{ kcal/mol}} [\text{concerted transition state}] \longrightarrow R_2C{=}CR_2 + CO_2$$

$$\text{carbena-1,3-dioxolane } (R_4) \xrightarrow{E_a = 21.6\text{ kcal/mol}} O{=}\dot{C}{-}O{-}CR_2{-}\dot{C}R_2 \longrightarrow R_2C{=}CR_2 + CO_2 \qquad \text{(Eq. 14)}$$

Equation 15 shows an alternative mechanism for the stereospecific desulfurization of 1,3-dioxolane-2-thiones (Eq. 1). Nucleophilic attack by phosphorus on sulfur thus affords 1,3-dipole **10**, which is converted into ylide **11** either by the cyclization–desulfurization pathway *a* or the α-elimination–carbene capture pathway *b*. It is assumed in pathway *b* that capture of 2-carbena-1,3-dioxolane **2** by trivalent phosphorus is more rapid than loss of

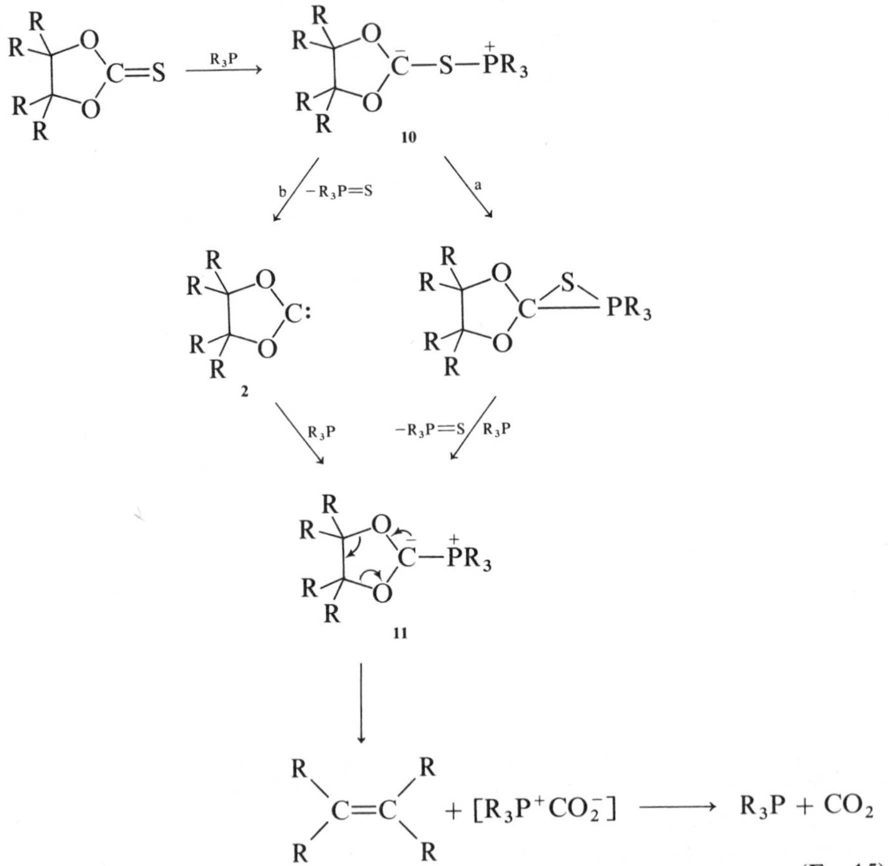

R_3P ; **10** ; b, $-R_3P{=}S$; a ; **2** ; R_3P ; $-R_3P{=}S$, R_3P ; **11**

$$\text{R}_2\text{C}{=}\text{CR}_2 + [\text{R}_3\text{P}^+\text{CO}_2^-] \longrightarrow \text{R}_3\text{P} + \text{CO}_2$$

(Eq. 15)

carbon dioxide from **2** according to Eq. 15. Concerted 1,3-dipolar cycloreversion[15] of ylide **11** would then yield stereospecifically the olefin and an unstable product that could correspond to the known phosphine–carbon disulfide adducts.[16] An alternative mechanism would involve direct stereospecific 1,3-dipolar cycloreversion of zwitterion **10** to afford the olefin and a species decomposing to carbon dioxide and phosphine sulfide (or thiophosphate). Since kinetic data indicate that decomposition of thionocarbonates is first order in phosphite concentration,[17] it is not possible to distinguish between these alternative mechanisms.

A consequence of the proposed mechanism is that the Corey–Winter reaction should prove useful for the synthesis of hindered olefins since attack of phosphorus takes place at the sulfur atom directly, a site located at a considerable distance from the crowded centers.

Support for the mechanism outlined in Eq. 15 is provided by the following observations on reactions involving *trithiocarbonates*:

1. Treatment of 1,3-dithiane-2-thione with excess trimethyl phosphite gives a 1:1 mixture of trimethyl thiophosphate and ylide **12** (which, unlike ylide **11**, cannot undergo cycloreversion) (Eq. 16).[18] Ylide **12** reacts with benzaldehyde affording ketene thioacetal **13**; on heating, ylide **12** is converted into a phosphonate derivative **14**.

S CH$_3$ S PO(OCH$_3$)$_2$ **14**

80° 24 hr

S =S $\xrightarrow{(CH_3O)_3P}$ S S $\bar{}$$\overset{+}{P}$(OCH$_3$)$_3$ **12** $\xrightarrow{C_6H_5CHO}$ S S =CHC$_6$H$_5$ **13** (90%)

(Eq. 16)

2. The formation of *trans*-cyclooctene by desulfurization of *trans*-1,2-cyclooctane trithiocarbonate with trimethyl phosphite, ordinarily a high-yield reaction,[2] is inhibited by the presence of excess benzaldehyde, and ketene thioacetal **15** is formed instead (Eq. 17).[18] Similarly, desulfurization of 1,3-benzodithiole-2-thione with triethyl phosphite in the presence of benzaldehyde gives ketene thioacetal **16**, presumably by the mechanism shown in Eq. 18.[19]
3. Desulfurization of *trans*-cyclohexane trithiocarbonate **17** with trialkyl phosphite affords coupling product **18**[2] that may arise as shown in Eq. 19.[19]

S S =S $\xrightarrow[Heat]{(CH_3O)_3P}$

$\xrightarrow[C_6H_5CHO, Heat]{(CH_3O)_3P}$ S S =CHC$_6$H$_5$ **15**

(Eq. 17)

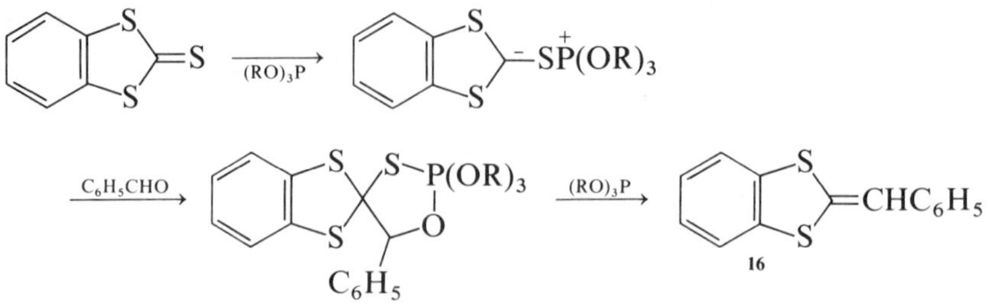

$$+ (RO)_3P{=}S + (RO)_3P{=}O \quad \text{(Eq. 18)}$$

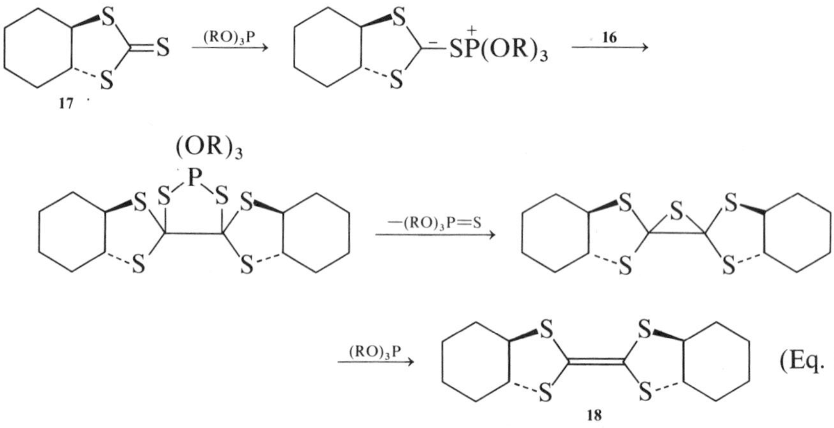

(Eq. 19)

4. In contrast to trialkyl phosphites, neither phosphorus trichloride nor trialkyl phosphines are effective in the desulfurization step in Eq. 15.[10,17]

Apparently, phosphorus trichloride is insufficiently nucleophilic to attack sulfur to give intermediate **10**. The formation of **10** should be favored with trialkyl phosphines compared to trialkylphosphites because the former are better nucleophiles. However, the conversion of **10** to ylide **11** by either path *a* or *b* (Eq. 15) may be less favorable with R = alkyl than with R = alkoxyl because of the reduced ability of phosphorus to accommodate positive charge in the former case.

Scope and Limitations

Preparation of Thionocarbonates. Thionocarbonates required for subsequent desulfurization can be prepared by a variety of routes, including reaction of diols with *N,N′*-thiocarbonyldiimidazole (TCDI) **19** in refluxing toluene, xylene, benzene, 2-butanone, tetrahydrofuran, pyridine, or acetone (e.g., Eq. 20).[20]

19

$\xrightarrow[\text{Reflux, 8 hr}]{CH_3COC_2H_5}$ (82%) (Eq. 20)

Alternatively, the desired thionocarbonates can be obtained by the reaction of tertiary dilithium or dipotassium 1,2-diolates (prepared from diol and excess *n*-butyllithium or potassium hydride, respectively) with TCDI in tetrahydrofuran;[21,22] by the reaction of diols with thiophosgene[23] in the presence of imidazole,[24] 4-dimethylaminopyridine,[25] or pyridine;[26] by the reaction of diols in sequence with base (e.g., *n*-butyllithium or potassium hydride), carbon disulfide, and either methyl iodide (e.g., Eq. 21)[27] or iodine;[28,29] and by cycloaddition of vinylene thionocarbonates to dienes (Eq. 22).[30,31]

The reagent TCDI is available commercially, or it can be synthesized.[32] The thiocarbonyl transfer reagent 1,1′-thiocarbonylbis(1,2,4-triazole)[33] is also useful in thionocarbonate synthesis. It should be recognized that secondary diols may undergo isomerization or rearrangement when strongly basic conditions are used in the preparation of thionocarbonates (see below).[34,35]

Applications. Desulfurization of 1,3-dioxolane-2-thiones can be used in conjunction with intermolecular and intramolecular carbon–carbon bond-forming processes to prepare unusual olefins. Intermolecular carbon–carbon bond-forming processes such as pinacol coupling and [2 + 2] cycloaddition provide the diols used in synthesis of pentasecododecahedrane isomer **20** (Eq. 23)[22] and spirohex-4-ene **21** (Eq. 24B),[36] respectively. Intramolecular carbon–carbon bond formation by acyloin condensation followed by reduction provides the diols utilized in a second synthesis of spirohex-4-ene **21** (Eq. 24), preparation of twistene **22** (Eq. 25),[37] and formation of unsaturated, deuterated propellane **23** (Eq. 26).[38,39]

$OCH_2C_6H_5$, $OCH_2C_6H_5$ $\xrightarrow{\text{1. } n\text{-}C_4H_9Li\text{, THF; 2. } CS_2\text{, heat; 3. } CH_3I\text{, heat}}$ $OCH_2C_6H_6$, $OCH_2C_6H_5$ (73%)

(Eq. 21)

(Eq. 22)

(60%)

$Mg(Hg)_x$, $(CH_3)_3SiCl$, $C_2H_5OH—H_2O$

OH OH

(46%)

1. KH; CS_2, CH_3I
2. $(C_2H_5O)_3P$, heat

(Eq. 23)

(36%)

20

A

CO_2CH_3

CO_2CH_3

Na, $(CH_3)_3SiCl$

$OSi(CH_3)_3$

$OSi(CH_3)_3$

H_2

$OSi(CH_3)_3$

$OSi(CH_3)_3$

H_2O, CH_3OH

OH

OH

B

$h\nu$, Sensitizer

NaOH

OH

OH

1. TCDI, $C_6H_5CH_3$, heat
2. $(n\text{-}C_4H_9O)_3P$, 120°

(63%) (Eq. 24)

21

HO OH

1. TCDI,
2. $(CH_3O)_3P$, heat

(40%) (Eq. 25)

(49%)

22

NaK, C_6H_6 → LiAlD$_4$ → 1. TCDI, $C_6H_5CH_3$, heat; 2. $(C_2H_5O)_3P$, 72 hr, heat

23

(Eq. 26)

Particularly reactive olefins such as tricyclo[3.3.3.0^{2,6}]undec-2(6)-ene **24** (Eq. 27),[21] *trans*-cycloheptene **25**[2] (Eq. 28), and $\Delta^{1,2}$-bicyclo[3.2.1]octene **26**[40] (Eq. 29) can be generated by desulfurization of thionocarbonates in the presence of a trapping agent such as diphenylisobenzofuran (DPIBF) to afford stable Diels–Alder adducts.

1. n-C_4H_9Li, THF, 0°; 2. TCDI; 3. $(C_2H_5O)_3P$, 36 hr, reflux

24

DPIBF →

(66%)

(Eq. 27)

$(CH_3O)_3P$, DPIBF

25

(42%)

(Eq. 28)

$$\xrightarrow[\text{2. } CS_2 \text{ 3. } CH_3I]{\text{1. } n\text{-}C_4H_9Li} (44\%) \xrightarrow[\text{DPIBF, 165°, 24 hr}]{(C_2H_5O)_3P} [\mathbf{26}] \longrightarrow (62\%) \quad \text{(Eq. 29)}$$

Alternatively, if the reactive olefin is volatile, a high-boiling liquid phosphite such as triphenyl, tri-*n*-butyl, or triisooctyl phosphite can be used for desulfurization, and the olefin can be continuously removed as formed with a carrier gas stream or by running the desulfurization under high-vacuum conditions. A rapid nitrogen stream is effective in the "external trapping" of *trans*-cycloheptene with diphenylisobenzofuran.[17]

Olefin inversion[41] is illustrated by the conversion of *cis*-cyclooctene to optically active *trans*-cyclooctene (Eq. 30).[17,42]

A useful method for assigning the configuration of dialkyl and trialkyl substituted olefins, diols, and epoxides takes advantage of the characteristic ^{1}H-nuclear magnetic resonance (NMR) chemical shift differences in stereoisomeric thionocarbonates together with their stereospecific formation and desulfurization to olefins.[43] A number of examples have been described of the use of the stereospecific 1,3-dioxolane-2-thione-desulfurization olefin synthesis to establish the stereochemistry of naturally occurring diols or diols derived from natural products.[44–46]

$$\xrightarrow{CH_3CO_3H} \xrightarrow[\text{2. KOH, } C_2H_5OH\text{, heat}]{\text{1. } CH_3CO_2Na\text{, } CH_3CO_2H\text{, heat}}$$

$$\xrightarrow[\text{2. TCDI, } C_6H_5CH_3\text{, heat} \quad \text{3. } (i\text{-}C_8H_{17}O)_3P\text{, 130°, 17 hr}]{\text{1. Resolve}} \quad \text{(Eq. 30)}$$

Optically active,
$>99\%$ isomeric purity

The Corey–Winter reaction is of considerable value in the preparation of unsaturated sugars as illustrated by Eq. 31[27] and Eq. 32[47] and of steroids,[48] terpenes,[49] lipids,[50–52] and erythronolides.[25]

1. $n\text{-}C_4H_9Li$, THF; 2. CS_2; 3. CH_3I — $OCH_2C_6H_5$, $OCH_2C_6H_5$ (73%)

$(CH_3O)_3P$, Heat, 70 hr — $OCH_2C_6H_5$, $OCH_2C_6H_5$ (73%)

(Eq. 31)

1. TCDI; 2. $(CH_3O)_3P$, heat — $C_6H_5CH_2O$, OCH_3 (65%)

(Eq. 32)

Modifications. In the original investigation of the desulfurization of 1,3-dioxolane-2-thiones a variety of desulfurizing agents were tried with *cis*-4,5-diphenyl-1,3-dioxolane-2-thione.[10] Although refluxing trimethyl phosphite (72 hours) gives exclusively *cis*-stilbene in 92% yield (see Eq. 1), deactivated Raney nickel in dioxane affords 68% *cis*- and 12% *trans*-stilbene, 1% sodium amalgam in dioxane yields 14% *cis*- and 9% *trans*-stilbene, silver amalgam gives 24% *trans*-stilbene free from *cis* isomer, and nickel tetracarbonyl in acetone at 25° (48 hours) affords 15% *cis*- and 4% *trans*-stilbene. Ultraviolet irradiation (254 nm) of *trans*-4,5-dimethyl-1,3-dioxolane-2-thione in tri-*n*-butyl phosphite (25°, 4 hours) gives in 78% yield a 3:1 mixture of *trans*- and *cis*-2-butenes, even with continuous removal of olefin (Eq. 33).[10] Raney nickel is used to desulfurize a thionocarbonate in an indole alkaloid synthesis (Eq. 34).[53,54]

$(n\text{-}C_4H_9O)_3P$, $h\nu$, 25°, 4 hr — 3:1 (78%)

(Eq. 33)

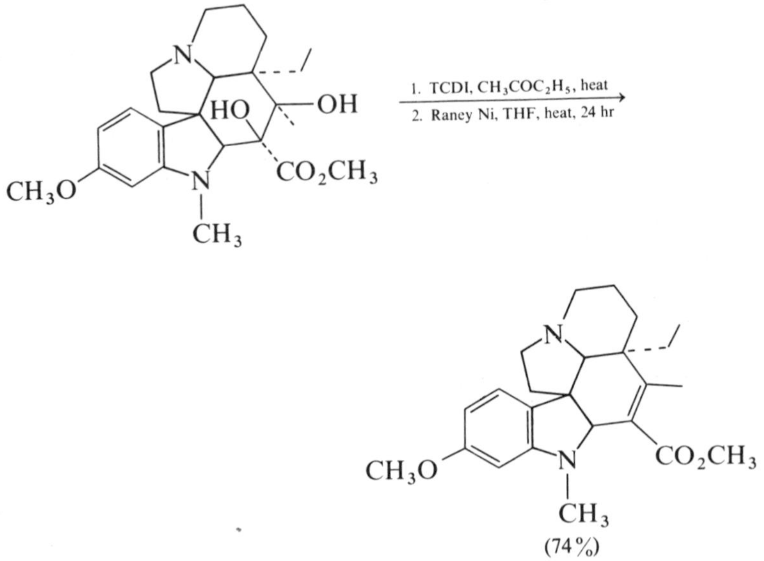

CH3O
N
CH3
HO
OH
CO2CH3
1. TCDI, CH3COC2H5, heat
2. Raney Ni, THF, heat, 24 hr
CH3O
N
CH3
CO2CH3
(74%)

(Eq. 34)

Bis(1,5-cyclooctadiene)nickel(0) is effective in the stereospecific desulfurization of the thionocarbonates from *erythro*- and *threo*-4-methylpentane-2,3-diol (Eq. 35)[55] and *exo*, *exo*-bicyclo[2.2.1]heptane-2,3-diol (Eq. 36).[55] However, the thionocarbonate from *trans*-1,2-cyclooctanediol gives only *cis*-cyclooctene (99% yield) under these same conditions, presumably as a result of isomerization of the *trans*-cyclooctene by nickel. These reactions are believed to involve the intermediacy of nickel-2-carbena-1,3-dioxolane complexes (Eq. 36).

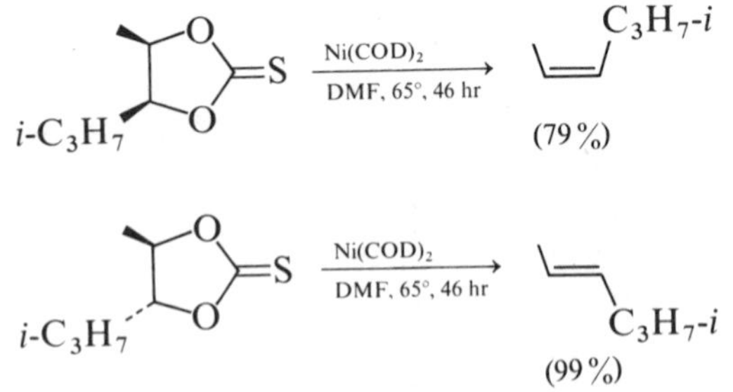

i-C3H7
Ni(COD)2
DMF, 65°, 46 hr
C3H7-i
(79%)
i-C3H7
Ni(COD)2
DMF, 65°, 46 hr
C3H7-i
(99%)

(Eq. 35)

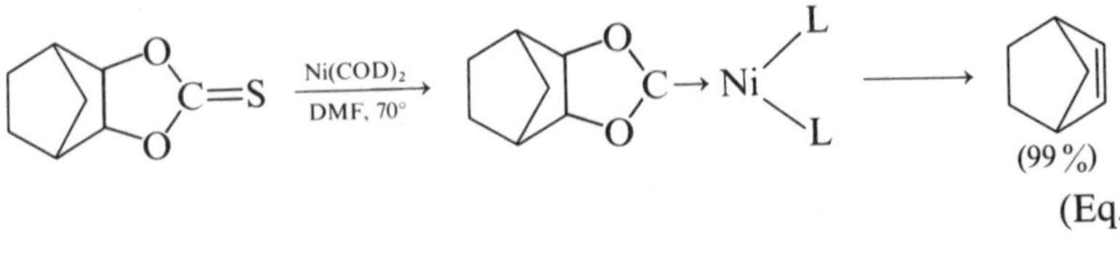

C=S
Ni(COD)2
DMF, 70°
C→ Ni
L
L
(99%)

(Eq. 36)

Iron pentacarbonyl can also be used to convert thionocarbonates to olefins (Eq. 37).[26,56–58] Here it appears that metal–carbene complexes are not precursors to olefins since these complexes can be isolated from the same reaction and are stable under the reaction conditions. Apparently, olefins are formed by another metal-promoted route, perhaps initiated by cleavage of one carbon–oxygen bond in a thionocarbonate–iron carbonyl complex. The thionocarbonate from *trans*-1,2-cyclooctanediol yields only *cis*-cyclooctene (35%) on

$\xrightarrow{Fe(CO)_5}$

(79%) + $Fe(CO)_4$ (18%) (Eq. 37)

desulfurization with iron pentacarbonyl. Nickel boride desulfurization of 1,3-dioxolane-2-thiones affords 1,3-dioxolanes rather than olefins.[23] Limited success is realized in the desulfurization of a thiocarbonate with tri-*n*-butylstannane.[59]

Whereas thionocarbonate **27** proves resistant to the action of boiling trimethyl phosphite even after 4 days, a two-step procedure involving treatment with 2-iodopropane followed by zinc dust is successful (Eq. 38).[60,61] This two-step desulfurization procedure is nonstereospecific (Eq. 39).[60,61] A modified version of this procedure, using methyl iodide instead of 2-iodopropane and either zinc dust or chromium(II) acetate in the second step, can be used in the preparation of an unsaturated sugar, although here the yield is lower than that obtained with trimethyl phosphite desulfurization (Eq. 40).[62]

27 $\xrightarrow[\text{Heat}]{i\text{-}C_3H_7I}$ I^- $\overset{+}{}SC_3H_7\text{-}i$

$\longrightarrow$ I $O_2CSC_3H_7\text{-}i$ $\xrightarrow[C_2H_5OH]{\text{Zn dust}}$ (Eq. 38)

S, O, O, C_6H_5, C_6H_5 — 1. CH_3I / 2. Zn or Mg(Hg) → C_6H_5, H, H, C_6H_5 (Eq. 39)

Both the iron pentacarbonyl and 2-iodopropane–zinc dust procedures are less effective than trivalent phosphorus reagents in the desulfurization in Eq. 41;[24] tris(diethylamino)phosphine, $[(C_2H_5)_2N]_3P$, gives better yields (30%) than trimethyl phosphite (12–20%). Other recommended alternatives to the use of refluxing trimethyl phosphite (bp 111°) or triethyl phosphite (bp 156°) as desulfurizing agents include dimethyl trimethylsilyl phosphite at 130–140°[63] and the highly thiophilic reagent 1,3-dimethyl-2-phenyl-1,3,2-diazaphospholidine at 40°.[25] (The latter reagent is available in one step in 70% yield from *sym*-dimethylethylenediamine and dichlorophenylphosphine.)

The Corey–Winter reaction can also be applied to the synthesis of alkynes (Eq. 42)[64] and to the conversion of derivatives of vicinal hydroxythiols (Eq. 43)[29] and dithiols to olefins (Eqs. 44[65] and 45).[2,17,42] (Note the use of the reagent 1,3-dibenzyl-2-methyl-1,3,2-diazaphospholidine, which effects desulfurization at 30°.) The latter sequence can also be accomplished by fragmentation of 1,3-dithiolanes (Eq. 46)[66] and their derived ylides (Eq. 47)[67] or carbenes (Eq. 48)[68] or by treatment of trithiocarbonates with dimethyl acetylenedicarboxylate (Eq. 49)[69] or *n*-butyllithium (Eq. 50).[70]

(Eq. 40)

Conditions	Yield
CH_3I, heat; Zn, C_2H_5OH	(48%)
CH_3I, heat; $Cr(OAc)_2$, C_2H_5OH, heat	(38%)
$(CH_3O)_3P$, reflux, 24 hr	(85%)

HO, OH — 1. $CSCl_2$, imidazole / 2. $[(C_2H_5)_2N]_3P$, heat → (Eq. 41)

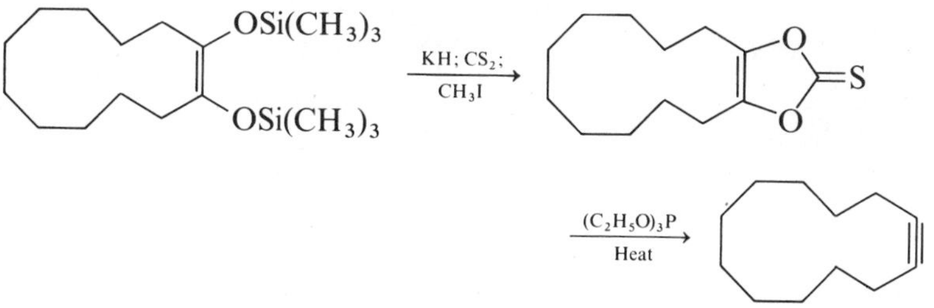

$OSi(CH_3)_3$ / $OSi(CH_3)_3$ $\xrightarrow[CH_3I]{KH;\ CS_2;}$ $\xrightarrow[Heat]{(C_2H_5O)_3P}$ (Eq. 42)

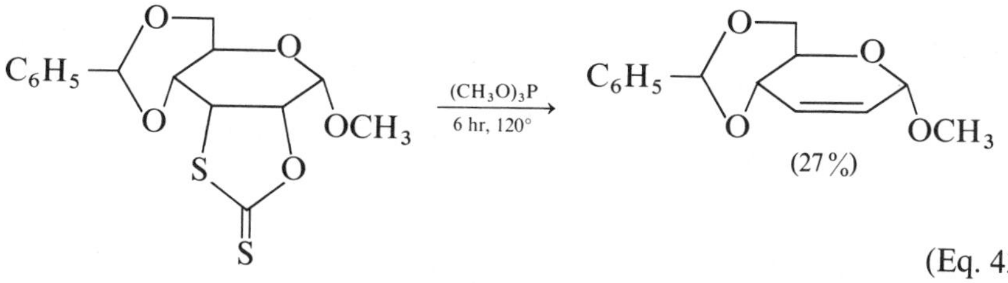

C_6H_5 … OCH_3 $\xrightarrow[6\ hr,\ 120^\circ]{(CH_3O)_3P}$ C_6H_5 … OCH_3 (27%)

(Eq. 43)

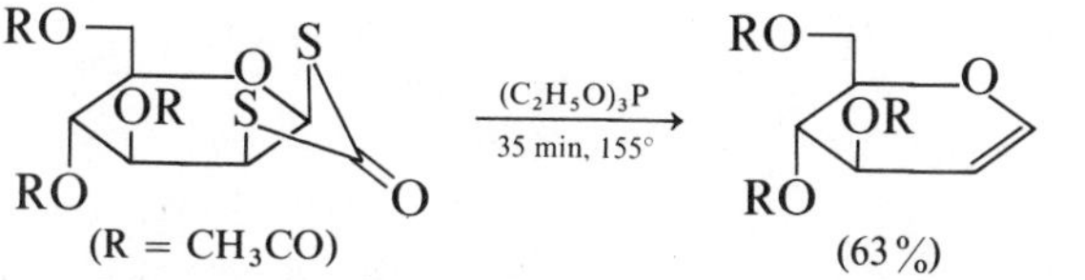

RO … OR … RO (R = CH_3CO) $\xrightarrow[35\ min,\ 155^\circ]{(C_2H_5O)_3P}$ RO … OR … RO (63%) (Eq. 44)

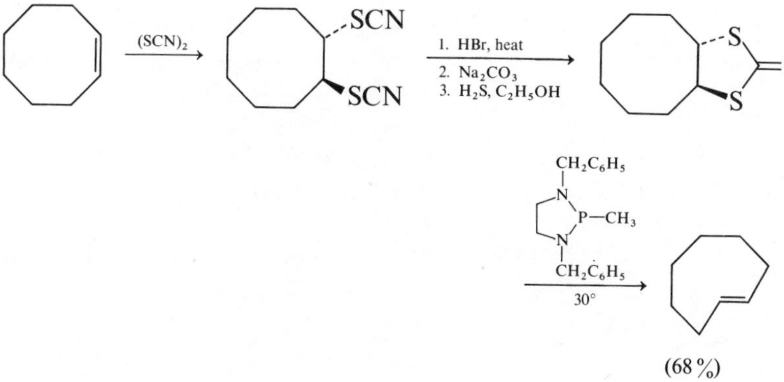

$\xrightarrow{(SCN)_2}$ SCN / SCN $\xrightarrow[2.\ Na_2CO_3\ \ 3.\ H_2S,\ C_2H_5OH]{1.\ HBr,\ heat}$ S / S / =S

$CH_2C_6H_5$, N, P—CH_3, N, $CH_2C_6H_5$ $\xrightarrow[30^\circ]{}$ (68%) (Eq. 45)

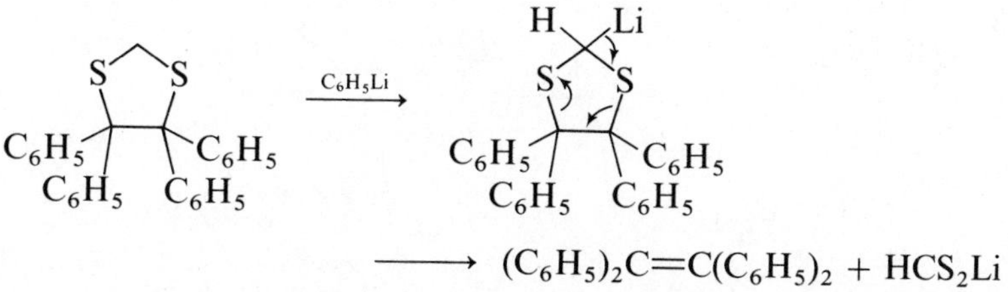

S S, C_6H_5 C_6H_5 C_6H_5 C_6H_5 $\xrightarrow{C_6H_5Li}$ H Li, S S, C_6H_5 C_6H_5 C_6H_5 C_6H_5

$\longrightarrow (C_6H_5)_2C{=}C(C_6H_5)_2 + HCS_2Li$ (Eq. 46)

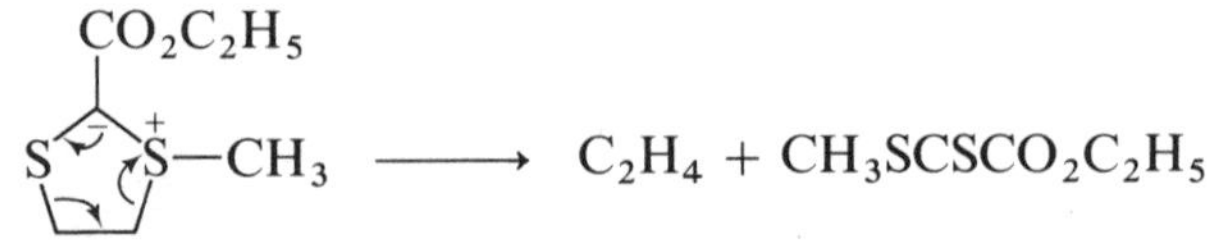
$CO_2C_2H_5$
S
S—CH_3
C_2H_4 + $CH_3SCSCO_2C_2H_5$

(Eq. 47)

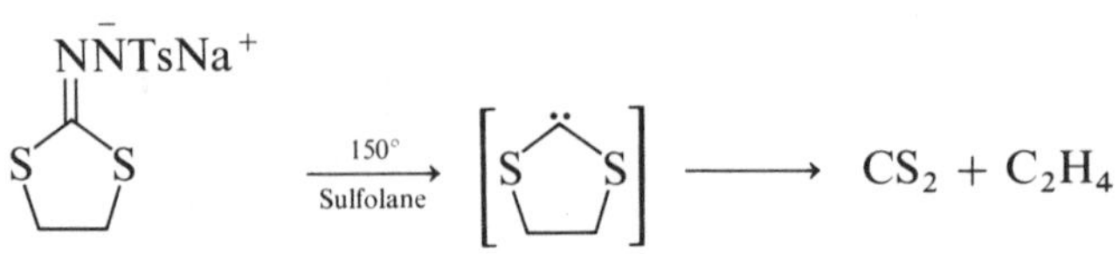
$N\bar{N}TsNa^+$
S
S
150°
Sulfolane
CS_2 + C_2H_4

(Eq. 48)

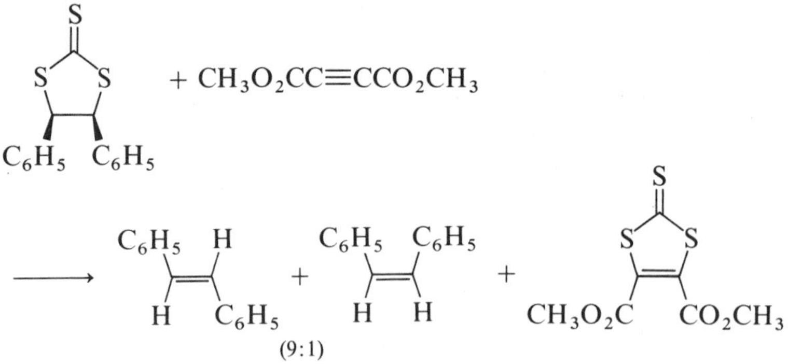
+ $CH_3O_2CC{\equiv}CCO_2CH_3$
C_6H_5
H
(9:1)
CH_3O_2C
CO_2CH_3

(Eq. 49)

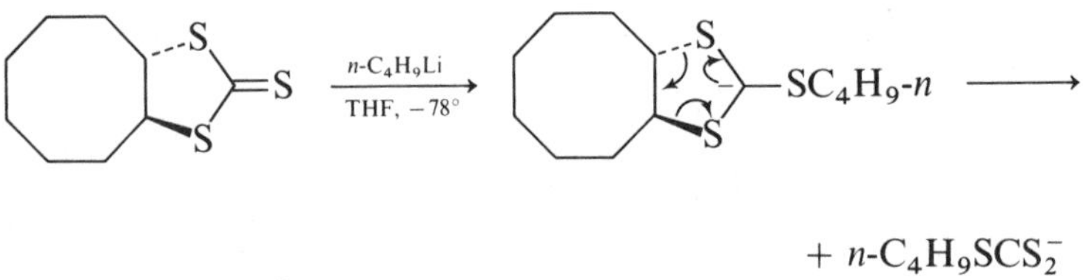
n-C_4H_9Li
THF, −78°
SC_4H_9-n
(39%)
+ n-$C_4H_9SCS_2^-$

(Eq. 50)

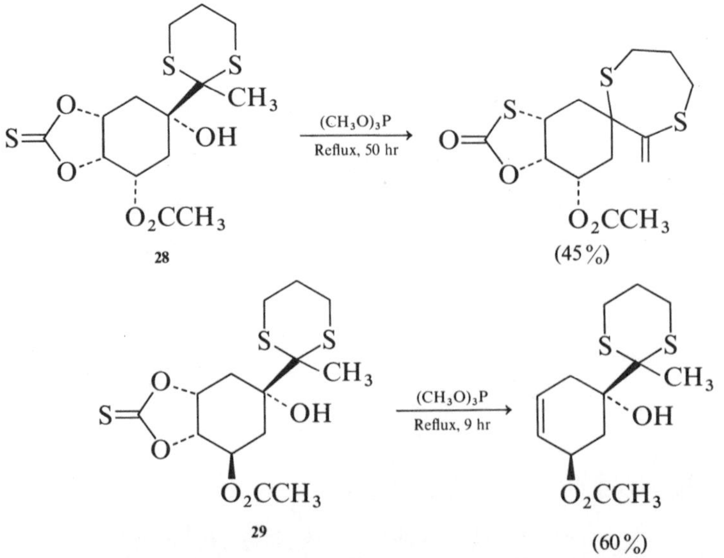
CH_3
OH
$(CH_3O)_3P$
Reflux, 50 hr
O_2CCH_3
28
(45%)
$(CH_3O)_3P$
Reflux, 9 hr
29
(60%)

(Eq. 51)

Side Reactions. One side reaction that may occur during attempts to desulfurize thionocarbonates is isomerization to the thermodynamically more stable monothiolcarbonates, sometimes termed the Schönberg reaction[71] (Eq. 51).[72] The reaction can be promoted by a nucleophilic catalyst[73,74] such as trimethyl phosphite, particularly when the approach of the phosphorus to sulfur is hindered as in **28**, but not **29**, by the *cis*-acetoxy group.[71] A second type of side reaction is orthoformate formation involving either an external (phosphite-derived?) nucleophile such as methanol (Eq. 52)[75] or an internal hydroxyl group (Eq. 53)[76] that presumably interacts with one of the zwitterionic intermediates in the desulfurization process.[19]

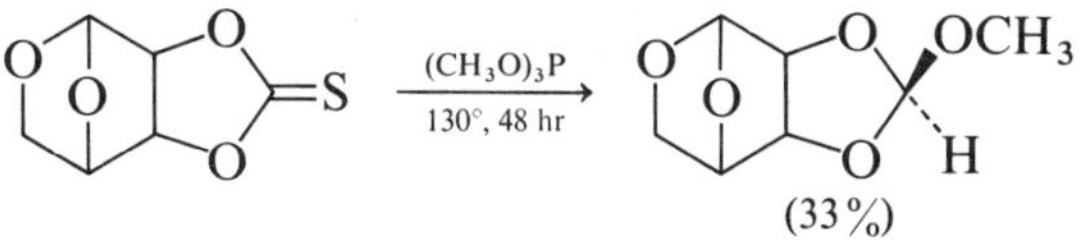

$(CH_3O)_3P$, 130°, 48 hr; OCH_3, H (33%) (Eq. 52)

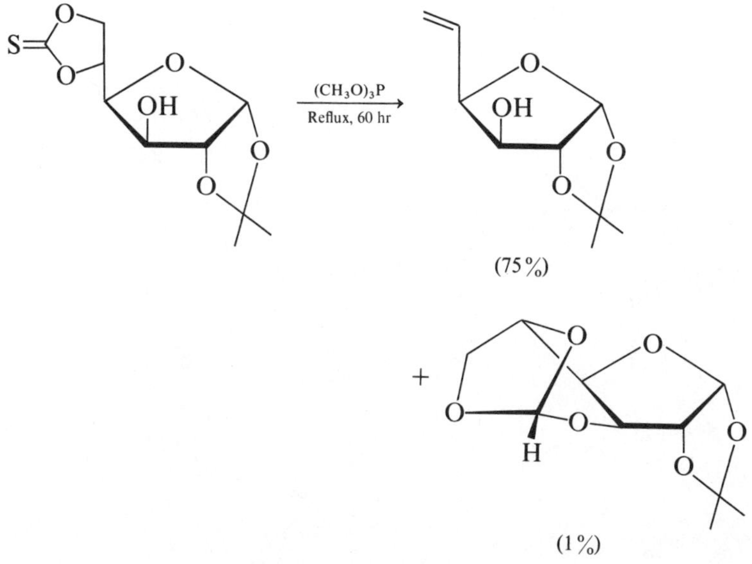

$(CH_3O)_3P$, Reflux, 60 hr; OH (75%) + H (1%) (Eq. 53)

A number of problems are encountered in the conversion of nucleoside **30** to the unsaturated nucleoside **31**, including intramolecular nucleophilic attack by oxygen during preparation of thionocarbonate **32** (or on exposure of **32** to base) and *N*-methylation by trimethyl phosphite, leading to **33** (Eq. 54).[77,78] The first problem can be minimized by preparing **32** from **30** at room temperature in tetrahydrofuran. The second problem can be avoided by using deactivated Raney nickel to desulfurize thionocarbonate **32**. Unfortunately, the yield in the latter reaction is low, and significant nucleophilic ring opening of the thionocarbonate occurs. Other examples of *N*-methylation during thionocarbonate desulfurization with trimethylphosphite have been reported.[79]

30

TCDI, THF, 25°, 5 days (90%)

TCDI, $C_6H_5CH_3$, Reflux, 0.5 hr (86%)

Base

32

Raney nickel

(26%)

(16%)

$(CH_3O)_3P$ Reflux, 65 hr (80%)

33

31

(Eq. 54)

The high temperature required for the thionocarbonate desulfurization step can result in isomerization or decomposition of the olefinic product as illustrated by Eq. 55,[80] Eq. 56,[81] Eq. 57,[57] Eq. 58,[82] and Eq. 59.[83] A number of reports indicate the failure of the Corey–Winter reaction without indicating the nature of the decomposition product.[84,85]

$(C_6H_5O)_3P$, 150°, 96 hr → (3 : 2) (Eq. 55)

C_6H_5 CO_2CH_3 — $(CH_3O)_3P$, 65 hr, reflux → C_6H_5 H / H CO_2CH_3 (84%) + C_6H_5 CO_2CH_3 / H H (4%)

(Eq. 56)

$(C_2H_5O)_3P$, 165° → (57%) + (43%)

(Eq. 57)

$(C_2H_5O)_3P$, 110°, 120 hr → (Eq. 58)

t-C_4H_9 — $(n\text{-}C_4H_9O)_3P$, 110°, 24 hr → t-C_4H_9 + C_4H_9-t

(Eq. 59)

DECOMPOSITION OF 2-ETHOXY-, 2-ACYLOXY-, AND 2-DIALKYLAMINO-1,3-DIOXOLANES

Mechanism

Treatment of vicinal diols with ethyl orthoformate at elevated temperatures (100–180°) followed by continued heating at somewhat higher temperatures

(160–220°) in the presence of a carboxylic acid leads to stereospecific olefin formation in high yield, sometimes referred to as Eastwood deoxygenation (Eq. 60).[86] Although the 2-ethoxy-1,3-dioxolane intermediate derived from *cis*-cyclohexane-1,2-diol decomposes cleanly to cyclohexene in 85% yield at 160°, the 2-ethoxy-1,3-dioxolane from *trans*-cyclohexane-1,2-diol is unscathed after 3 hours at 200–220°.[86]

HO OH H C6H5 H C6H5 → $(C_2H_5O)_3CH$, 110° → OC_2H_5 O O C_6H_5 C_6H_5

→ CH_3CO_2H → C_6H_5 C_6H_5 H H (96%) + CO_2 + C_2H_5OH (Eq. 60)

Decomposition of dioxolane **34** derived from pinacol is unaffected by the addition of benzoyl peroxide or hydroquinone, suggesting that free-radical intermediates are probably not involved in the olefin-forming reaction. The use of catalytic *p*-toluenesulfonic acid instead of a carboxylic acid leads to a different group of decomposition products from dioxolane **34** (Eq. 61).[86] 2-Acetoxy-4,4,5,5-tetramethyl-1,3-dioxolane (**35**) can be prepared from dioxolane **34** and decomposes at 130–140° to 2,3-dimethyl-2-butene in 93% yield (Eq. 61).[87]

HO OH

↓ 140° $(C_2H_5O)_3CH$

OC_2H_5 O O **34** (95%) → TsOH, 100–130° → (42%) + t-$C_4H_9COCH_3$ (55%) + $HCO_2C_2H_5$ (100%

b HCO_2H, $(CH_3CO)_2O$; *a* CH_3CO_2H, 160°, 5 hr, $-C_2H_5OH$

O_2CCH_3 O O **35** → 130–140°, $-CH_3CO_2H$ → $(CH_3)_2C{=}C(CH_3)_2$ + CO_2 (Eq. 61)

(67% via *a*)
(93% via *b*)

A detailed mechanism for the acid-catalyzed decomposition of dioxolane **34** is given in Scheme 1.[86] Although carbocation **36**, a key intermediate in this process, could undergo deprotonation to carbene **37**, the intermediacy of this carbene is unlikely for reasons previously developed (see Eqs. 7 and 14). A more probable course for carbocation **36** is capture of the conjugate base of the catalytic acid HX. When X^- is acetate, the product would be **35**, which has already been shown to undergo facile conversion to olefin and carbon dioxide (Eq. 61). Of the several decomposition pathways possible for acetoxydioxolane **35**, α-elimination path *a* is unlikely if carbene **37** is excluded as a likely intermediate.

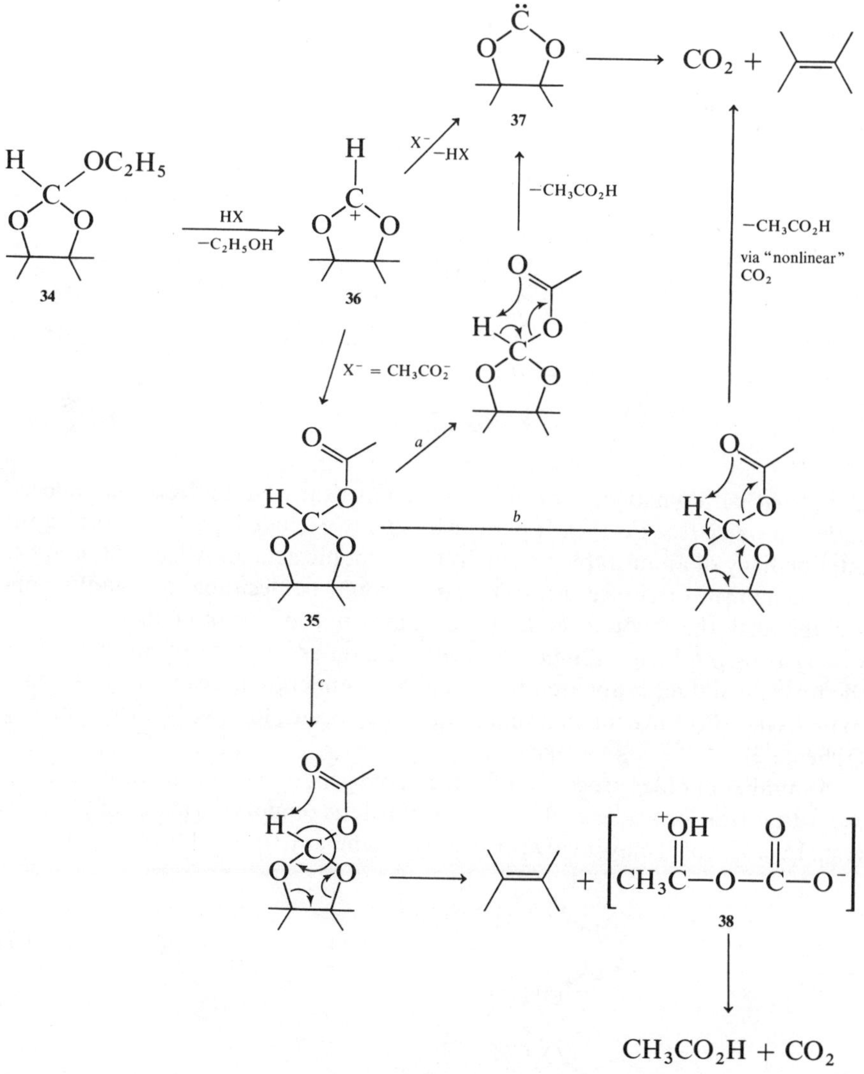

H, OC_2H_5; **34**; HX, $-C_2H_5OH$; **36**; X^-, $-HX$; **37**; CO_2 +; $-CH_3CO_2H$; $-CH_3CO_2H$ via "nonlinear" CO_2; $X^- = CH_3CO_2^-$; *a*; *b*; **35**; *c*; $[CH_3\overset{+OH}{\overset{\|}{C}}-O-\overset{O}{\overset{\|}{C}}-O^-]$ **38**; $CH_3CO_2H + CO_2$

Scheme 1

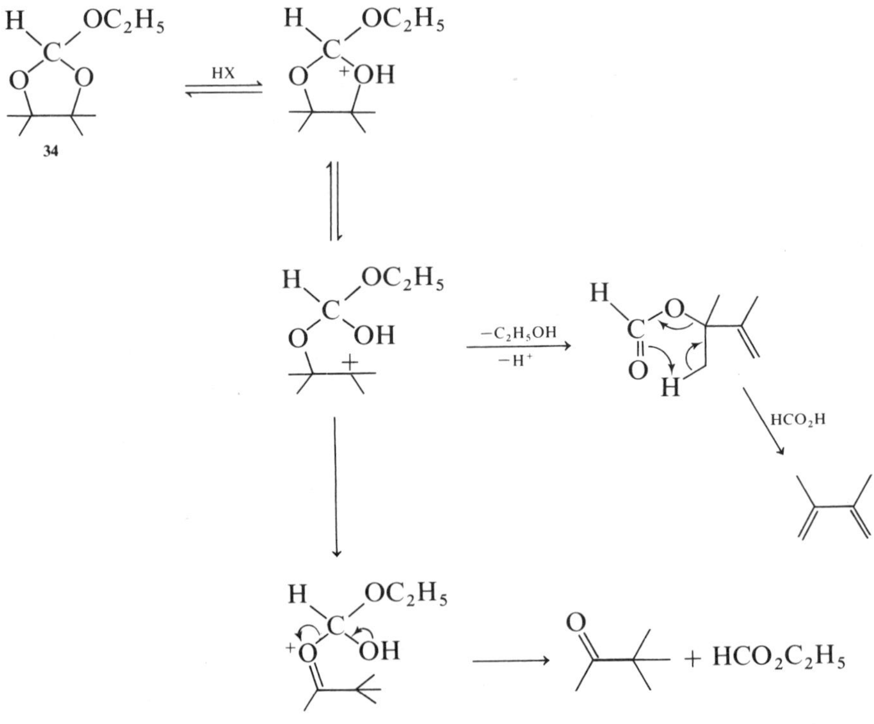

Scheme 2

Concerted path *b* may also be unlikely insofar as it would afford "nonlinear" carbon dioxide (see Eq. 14). If paths *a* and *b* are excluded, path *c* involving the intermediacy of an unstable mixed acetic–carbonic acid anhydride **38** becomes the mechanism of choice. More research would be desirable to establish unambiguously the favored mode of decomposition of compounds such as **35**. Catalysis by *p*-toluenesulfonic acid leads to a different series of products since the tosylate adduct is apparently incapable of undergoing reactions analogous to paths *a–c*. Protonation of a ring oxygen may occur, followed by ring opening (Scheme 2).

A number of older procedures for the deoxygenation of vicinal diols by heating with formic or oxalic acid[86,88] or by pyrolysis of monoformates of 1,2-diols (e.g., Eqs. 62, 63[86]) may involve mechanisms similar to those shown in Scheme 1.

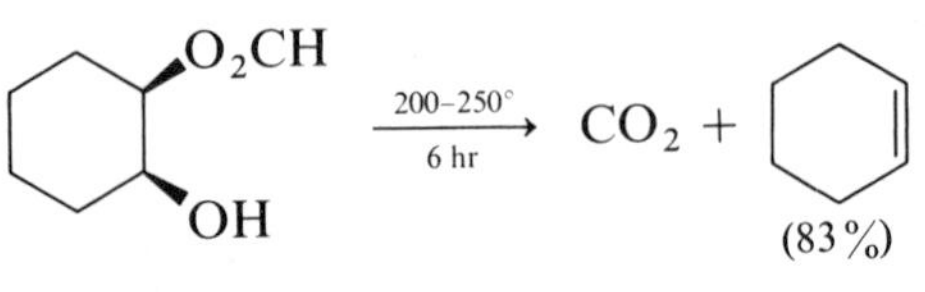

$\xrightarrow[\text{6 hr}]{200\text{–}250^\circ}$ CO_2 + (83%) (Eq. 62)

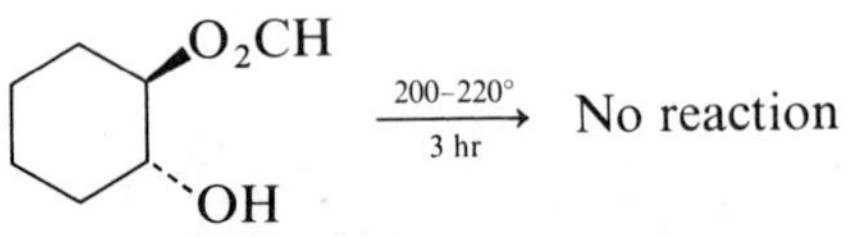

$\xrightarrow[\text{3 hr}]{200\text{–}220^\circ}$ No reaction (Eq. 63)

A reaction that is related to the acid-catalyzed decomposition of 2-ethoxy-1,3-dioxolanes is the conversion of 2-dialkylamino-1,3-dioxolanes into alkenes. This process requires acetic anhydride (Eq. 64)[89] or methyl iodide (Eq. 65)[90] as coreagent. Whereas the acetic-anhydride-promoted reaction may involve formation of 2-acetoxy-1,3-dioxolanes such as **35**, the reaction promoted by methyl iodide may involve ammonium ylides if 2-carbena-1,3-dioxolanes are excluded for reasons described above (Eq. 66).

$\xrightarrow{(CH_3O)_2CHN(CH_3)_2}$ $\xrightarrow[90°,\ 1.5\ hr]{(CH_3CO)_2O}$ (75%) $+ CO_2 + CH_3CO_2H + CH_3CON(CH_3)_2$ (Eq. 64)

$\xrightarrow[25°,\ 1\ hr]{CH_3I}$ $\xrightarrow[Reflux,\ 40\ min]{C_6H_5CH_3}$ $+ CO_2 + (CH_3)_4\overset{+}{N}\overset{-}{I}$ (Eq. 65)

$(CH_3)_3N$; Base ; $CO_2 + R_2C{=}CR_2$; $R_2C{=}CR_2 + [(CH_3)_3\overset{+}{N}CO_2^-] \rightarrow (CH_3)_3N + CO_2$ (Eq. 66)

Scope and Limitations

The 2-ethoxy-1,3-dioxolane/carboxylic acid olefin synthesis can be applied to stereospecific syntheses of fatty acids (Eq. 67),[91,92] terpenes (Eq. 68),[93] unsaturated sugars,[94] steroids (Eq. 69),[95] and the strained olefin *trans*-cyclooctene (Eq. 70).[96] The 2-dialkylamino-1,3-dioxolane–acetic anhydride procedure has been utilized in syntheses of juvenile hormone (Eq. 71),[97] prostaglandin precursors (Eq. 72),[98] terpenes (Eq. 73),[99] and unsaturated sugars,[100] among other applications, whereas the 2-dialkylamino-1,3-dioxolane–methyl iodide method has been used primarily in preparing unsaturated sugars (Eq. 74).[90]

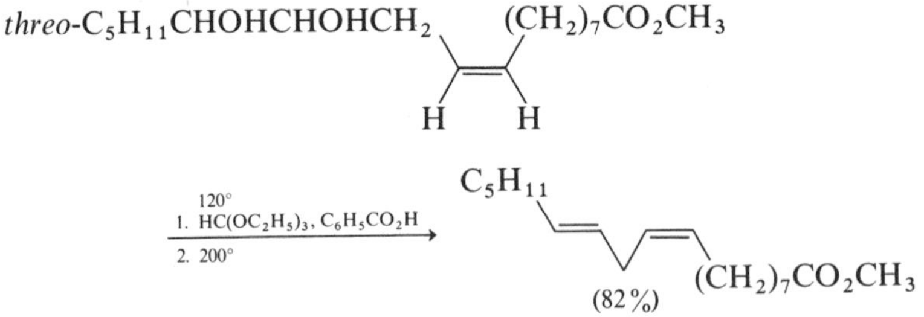

threo-$C_5H_{11}CHOHCHOHCH_2$ $(CH_2)_7CO_2CH_3$

H H

120°
1. $HC(OC_2H_5)_3$, $C_6H_5CO_2H$
2. 200°

C_5H_{11} $(CH_2)_7CO_2CH_3$ (82%) (Eq. 67)

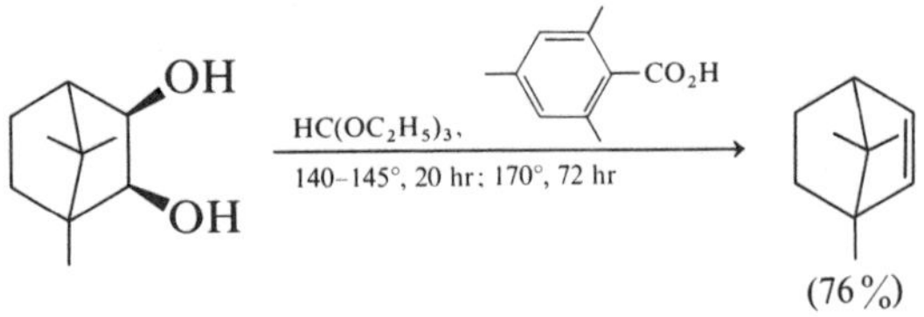

OH OH

$HC(OC_2H_5)_3$, CO_2H
140–145°, 20 hr; 170°, 72 hr

(76%) (Eq. 68)

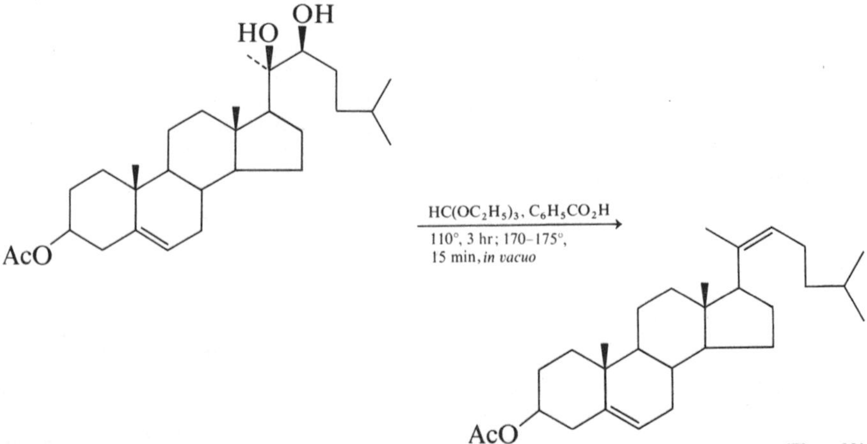

HO OH

AcO

$HC(OC_2H_5)_3$, $C_6H_5CO_2H$
110°, 3 hr; 170–175°, 15 min, *in vacuo*

AcO (48%) (Eq. 69)

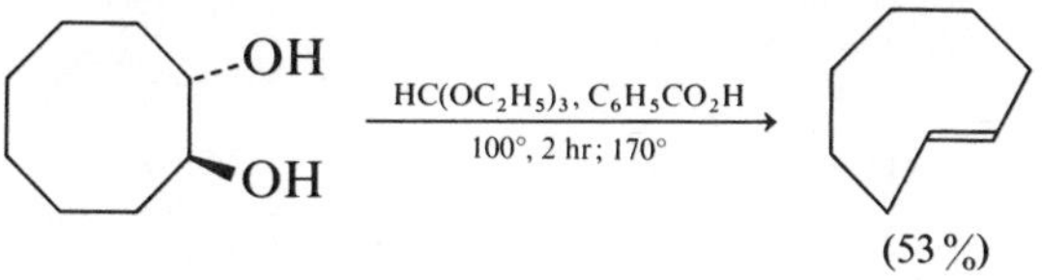

$HC(OC_2H_5)_3$, $C_6H_5CO_2H$; 100°, 2 hr; 170° (53%) (Eq. 70)

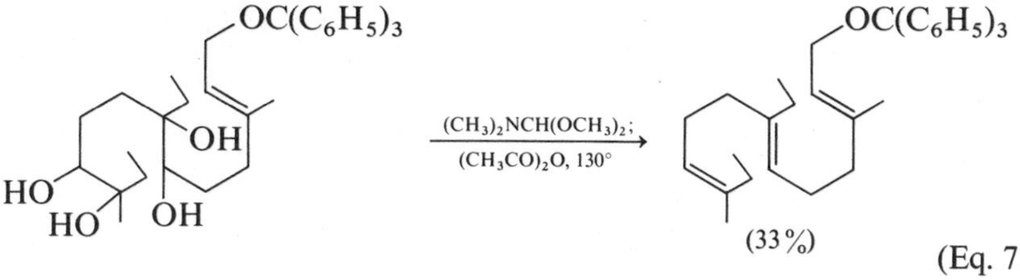

$(CH_3)_2NCH(OCH_3)_2$; $(CH_3CO)_2O$, 130° (33%) (Eq. 71)

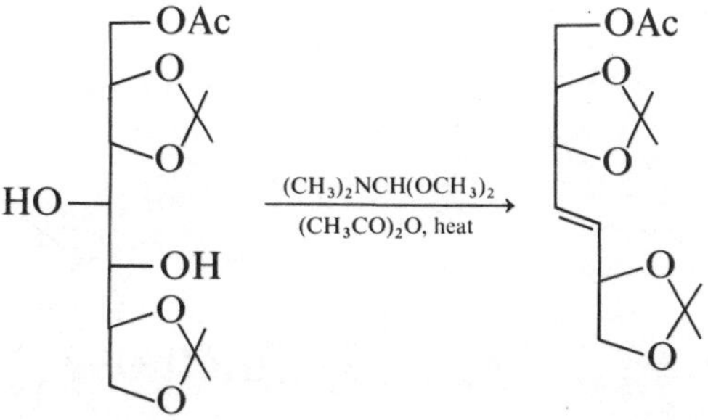

$(CH_3)_2NCH(OCH_3)_2$; $(CH_3CO)_2O$, heat (Eq. 72)

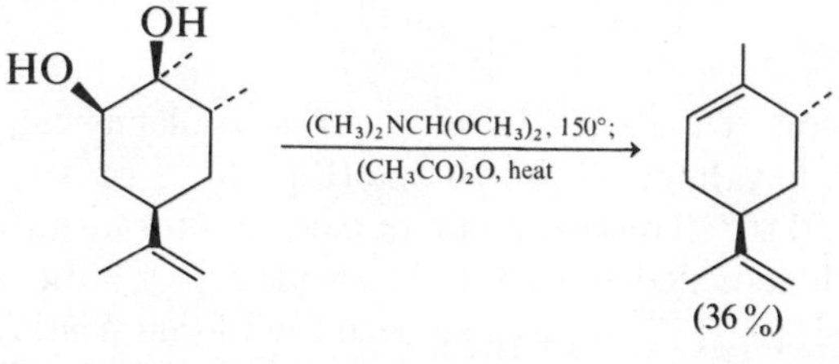

$(CH_3)_2NCH(OCH_3)_2$, 150°; $(CH_3CO)_2O$, heat (36%) (Eq. 73)

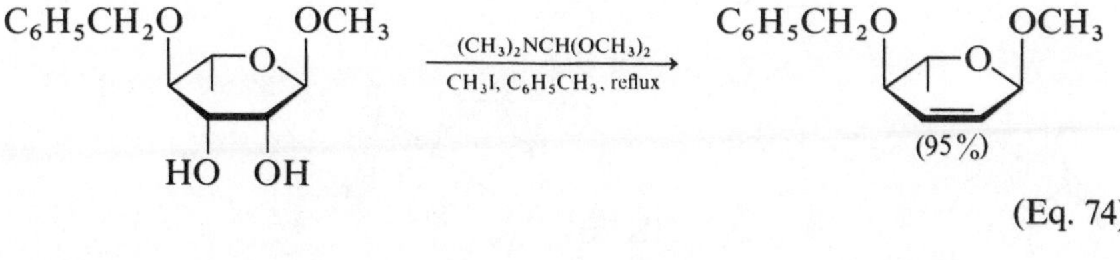

$(CH_3)_2NCH(OCH_3)_2$; CH_3I, $C_6H_5CH_3$, reflux (95%) (Eq. 74)

Acetic anhydride is favored in some reactions over methyl iodide in decomposing 2-dialkylamino-1,3-dioxolanes because the weakly nucleophilic acetate ion causes fewer side reactions than does the more nucleophilic iodide ion (Eq. 75).[101]

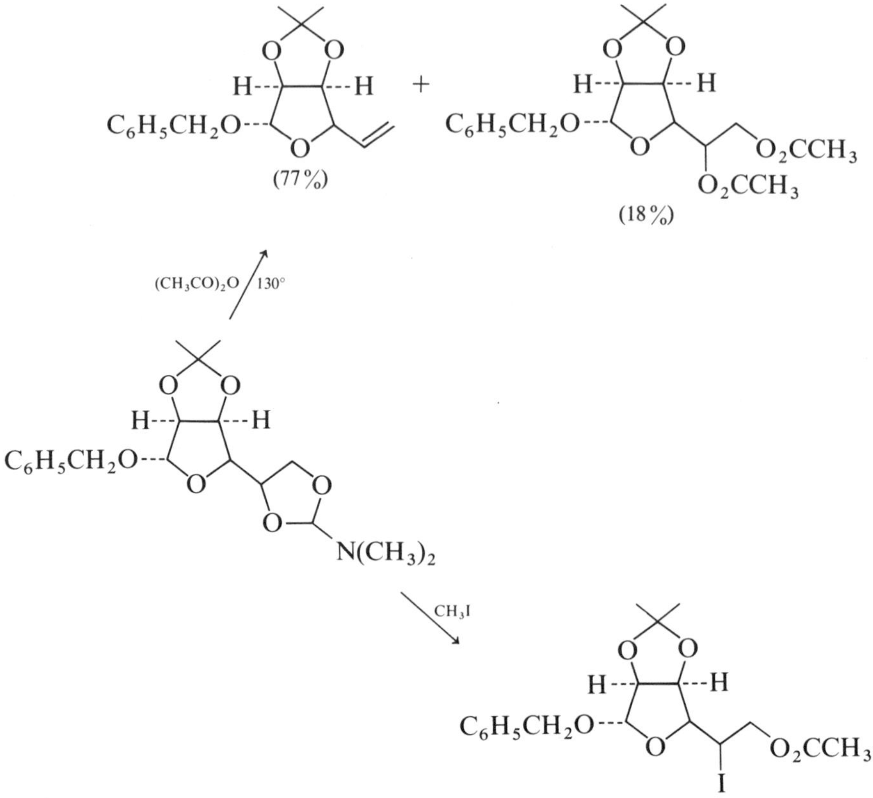

(Eq. 75)

The conversion of 2-alkoxy-1,3-dioxolanes to olefins can also be achieved by heating with hexafluoro-2-propanone (Eq. 76)[102] or with activated zinc in acetic acid (Eq. 77).[103] Treatment of monoacetates (or formates or other esters) of 1,2-diols with activated zinc (or in lower yields, with tin or aluminum) also affords olefins (Eq. 77).[103] It has also been found that heating of 2-ethoxy-1,3-dioxoles with catalytic amounts of acid yields alkynes (Eq. 78).[104]

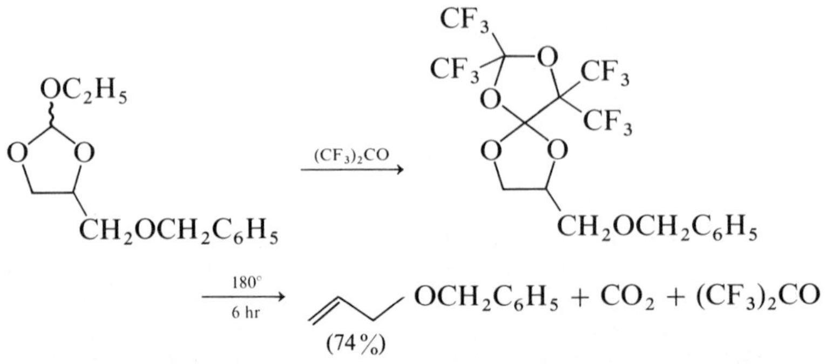

$\xrightarrow{180^\circ,\ 6\ hr}$ $CH_2{=}CHCH_2OCH_2C_6H_5 + CO_2 + (CF_3)_2CO$ (74%) (Eq. 76)

O_2CCH_3 OH

CH_3O OH OH

Zn, CH_3CO_2H (92%)

$(CH_3O)_3CCH_3$, TsOH

O OCH_3 O

Zn, CH_3CO_2H, reflux, 6 hr (93%)

(Eq. 77)

$$C_6H_5CHOHCOC_6H_5 \xrightarrow[\substack{CH_3CO_2H,\\130^\circ}]{(C_2H_5O)_3CH}$$

OC_2H_5 O O C_6H_5 C_6H_5

$$\xrightarrow[500^\circ]{n\text{-}C_6H_{13}CO_2H} C_6H_5C{\equiv}CC_6H_5 \quad \text{(Eq. 78)}$$

BASE-INDUCED FRAGMENTATION OF 2-PHENYL-1,3-DIOXOLANES

Mechanism

2-Phenyl-1,3-dioxolanes, -1,3-oxathiolanes, and -1,3-dithiolanes undergo stereospecific base-induced fragmentation to olefins and products derived from aryl carboxylates or thiocarboxylates (Eqs. 79,[105–108] 80,[66] 81,[15,109,110] and 82).[15] All of these reactions are believed to involve a concerted 1,3-anionic (or 1,3-dipolar) cycloreversion mechanism that, in orbital symmetry terms, can be described as a thermally allowed suprafacial $[_\sigma 2_s + {}_\sigma 2_s + {}_\omega 2_s]$ process (Eq. 83). This mechanism is favored when the anionic carbon, such as in **41**, is flanked by electronegative atoms such as oxygen or sulfur. This is because the negative charge is better stabilized on the triatomic anionic fragment—where it can be localized at the two end atoms—than on the pentaatomic cyclic anion, where it is confined to the central carbon atom.[15]

The best yield of *trans*-cyclooctene in Eq. 81 is realized when 2-phenyl-1,3-dioxolane (**40**) is treated with two equivalents of *n*-butyllithium in petroleum ether. No reaction occurs with potassium *tert*-butoxide in dimethyl sulfoxide, lithium diisopropylamide in tetrahydrofuran (THF), or potassium amide in liquid ammonia as base, whereas fair yields of *trans*-cyclooctene result with phenyllithium in refluxing THF or *n*-butyllithium in THF at 0°.[110] In the latter reaction a byproduct is *n*-butylcyclooctane, which presumably arises through

$$\mathbf{39} \xrightarrow{\text{RLi}} C_2H_4 + C_6H_5COR + R_2(C_6H_5)COH \quad \text{(Eq. 79)}$$

$$\xrightarrow{\text{NaH}} C_2H_4 + C_6H_5CS_2Na \quad \text{(Eq. 80)}$$

$$\xrightarrow[\text{H}_2\text{SO}_4,\ \text{C}_6\text{H}_5\text{CH}_3,\ \text{heat, 12 hr}]{\text{C}_6\text{H}_5\text{CHO}} \mathbf{40}\ (76\%) \xrightarrow[\text{Petroleum ether, 14 hr, 20°}]{n\text{-C}_4\text{H}_9\text{Li}}$$

$$\mathbf{41} \longrightarrow (73\%) + C_6H_5CO_2Li$$

$$C_6H_5CO_2Li + n\text{-}C_4H_9Li \longrightarrow C_6H_5COC_4H_9\text{-}n \quad \text{(Eq. 81)}$$

$$\mathbf{42} \xrightarrow[(\text{C}_2\text{H}_5)_2\text{O},\ 20°,\ 12\ \text{hr}]{2(\text{C}_2\text{H}_5)_2\text{NLi}} (45\%) + C_6H_5CON(C_2H_5)_2 \quad \text{(Eq. 82)}$$

addition of *n*-butyllithium to *trans*-cyclooctene. The reaction of *trans*-cyclooctene with *n*-butyllithium does not occur when tetrahydrofuran is replaced by petroleum ether as solvent. Addition of *n*-butyllithium to the strained *trans* double bond of *cis,trans*-1,5-cyclooctadiene is the dominant reaction even when petroleum ether is used as solvent in the deprotonation of dioxolane **43** (Eq. 84).[110]

A successful synthesis of *cis,trans*-1,5-cyclooctadiene is based on the deprotonation of oxathiolane **42** (more acidic than dioxolane **43**) by the less nucleophilic base lithium diethylamide (Eq. 82).[15] Deprotonation of dioxolanes **39**, **40**,

$$\longrightarrow \quad \text{(Eq. 83)}$$

$$\mathbf{43} \xrightarrow[\text{Petroleum ether, 14 hr, 20°}]{n\text{-}C_4H_9Li} \text{(cyclooctene)}C_4H_9\text{-}n + \text{(bicyclooctane)}C_4H_9\text{-}n + \text{(vinylcyclopentane)}C_4H_9\text{-}n \quad \text{(Eq. 84)}$$

and **43** with the hindered base neopentyllithium leads to the formation of both neopentyl phenyl ketone (**44**) and 2-benzoyl-2-phenyl-1,3-dioxolane (**45**) (Eq. 85).[110] Dioxolanes **45** are thought to arise through trapping of 2-lithio-2-phenyl-1,3-dioxolanes such as **46** (Eq. 85). The ratio of **45** to **44** varies from 65:35 with **43** to 30:70 with **40** to 8:92 with **39**, which has been interpreted as indicating that fragmentation occurs less readily the more strained the olefin being formed and that the transition state for fragmentation is reached early and has little olefin character.[110]

$$\text{2-phenyl-1,3-dioxolane} \xrightarrow{t\text{-}C_4H_9CH_2Li} \mathbf{46} \rightarrow RCH{=}CHR + C_6H_5CO_2Li \xrightarrow{t\text{-}C_4H_9CH_2Li} C_6H_5COCH_2C_4H_9\text{-}t\ (\mathbf{44}); \quad \mathbf{46} + C_6H_5CO_2Li \rightarrow \mathbf{45} \quad \text{(Eq. 85)}$$

Scope and Limitations

Base-induced fragmentation of 2-phenyl-1,3-dioxolanes has found only limited application in olefin synthesis, such as in the preparation of *trans*-thiacyclooct-4-ene (Eq. 86),[111] bicyclo[3.3.0]oct-1(5)-ene (Eq. 87),[112] and several substituted *trans*-cyclooctenes (Eq. 88).[113] The reaction fails with the 2-phenyl-1,3-dioxolane derived from either *cis*- or *trans*-1,2-cyclohexanediol and with tri- and tetrasubstituted dioxolanes such as those from 1-methyl-*trans*-1,2-cyclooctanediol and 2,3-dimethyl-2,3-butanediol (pinacol).[110] The reaction also fails when alternative modes of fragmentation are available (e.g., Eq. 89),[114] including in particular processes initiated by proton abstraction at C_4 (or C_5) of the dioxolane (Eqs. 90[115] and 91).[116]

The tendency for proton abstraction in 1,3-dioxolanes to occur at C_4 rather than C_2 may reflect the inability of oxygen to acidify adjacent carbon–hydrogen bonds.[117] In contrast, 2,4,5-triphenyl-1,3-oxathiolane undergoes efficient deprotonation at C_2 and subsequent fragmentation (Eq. 92).[118] The fact that 2-*p*-nitrophenyl-4,5-diphenyl-1,3-dioxolane undergoes C_2 deprotonation, giving stilbene rather than the C_4/C_5 deprotonation-initiated processes (Eq. 90),[119] suggests that olefin synthesis by base-induced fragmentation of 2-substituted 1,3-dioxolanes can be improved by the proper choice of a carbanion-stabilizing C_2 substituent.

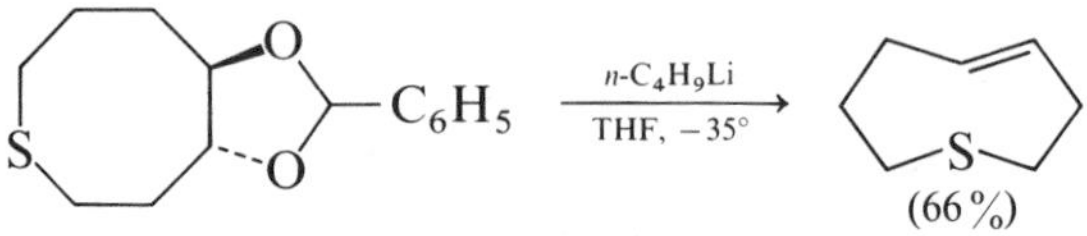

C_6H_5 ; n-C_4H_9Li, THF, −35° ; (66%) (Eq. 86)

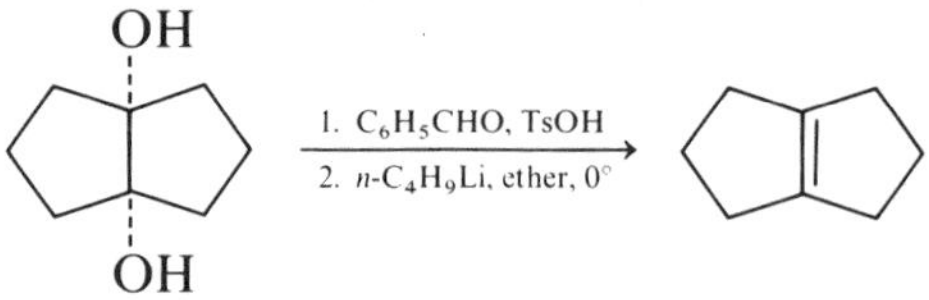

OH ; OH ; 1. C_6H_5CHO, TsOH ; 2. n-C_4H_9Li, ether, 0° (Eq. 87)

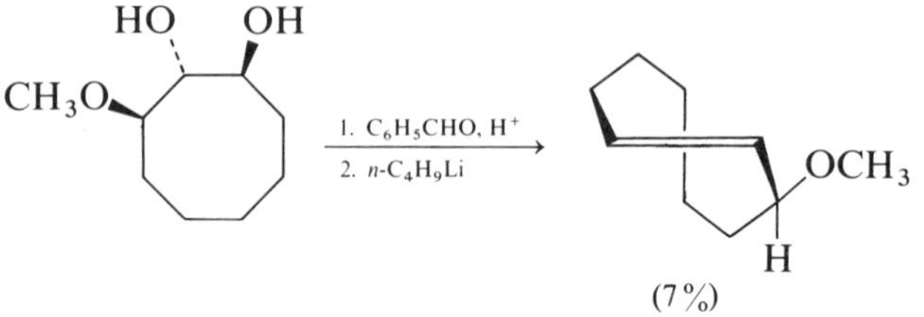

HO ; OH ; CH_3O ; 1. C_6H_5CHO, H^+ ; 2. n-C_4H_9Li ; OCH_3 ; H ; (7%) (Eq. 88)

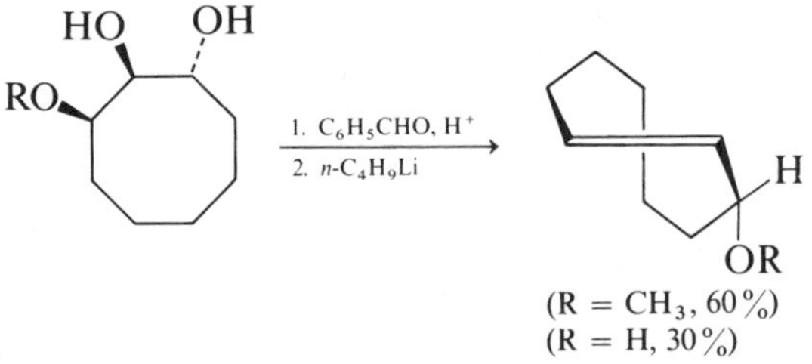

HO ; OH ; RO ; 1. C_6H_5CHO, H^+ ; 2. n-C_4H_9Li ; H ; OR ; (R = CH_3, 60%) ; (R = H, 30%) (Eq. 89)

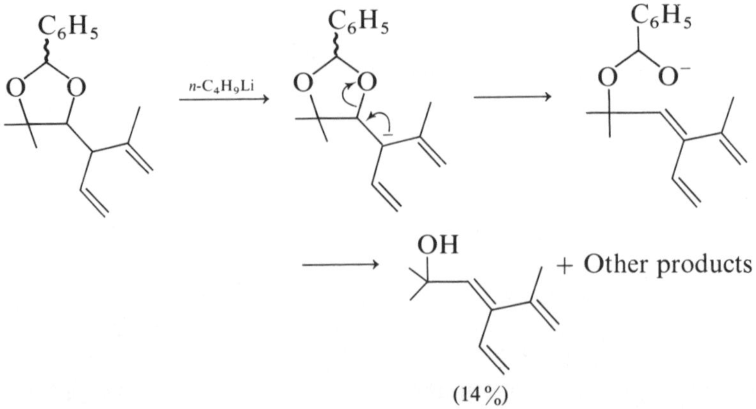

C_6H_5 ; O ; O ; n-C_4H_9Li ; C_6H_5 ; O ; O ; C_6H_5 ; O ; O^-

OH ; + Other products (Eq. 90)

(14%)

(9%)

$[C_6H_5CHO] + C_6H_5COCH_2C_6H_5$ (66%) (Eq. 91)

n-C_4H_9Li, 2 eq, −30°, THF

(69%)

(1%) (Eq. 92)

i-$C_3H_7(C_6H_{11})NLi$, 20°, ether, 48 hr

$+ C_6H_5COSLi$ (71%) (Eq. 93)

REDUCTION–ELIMINATION OF CYCLIC PHOSPHATES AND PHOSPHORAMIDATES

A stereoselective method has been developed for the conversion of vicinal diols to alkenes by reduction–elimination of cyclic phosphates (Eq. 94, Z = OC_2H_5) and phosphoramidates [Eq. 94, Z = $(CH_3)_2N$] with dissolving metals.[120] The mechanism of the reduction–elimination step is suggested to involve a two-step process. The predominant product results from *syn* elimination. Rotation about the carbon–carbon bond of the intermediate radical anion or dianion leads to the minor product by net *anti* elimination, with the *syn*:*anti* ratio depending on the relative timing of the first and second bond-breaking steps.[120] In addition to lithium in liquid ammonia, the following reducing agents are effective in stereoselectively converting phosphoramidates into olefins: sodium–naphthalene, sodium–xylene, titanium tetrachloride–magnesium amalgam, and titanium trichloride–potassium.[120]

This olefin synthesis has been used to prepare simple cyclic (Eq. 95) and acyclic (Eq. 96) olefins and [10.10]-betweenanenes, novel fused bicyclic *trans*-cycloalkenes (Eq. 97).[121] A related procedure involving magnesium metal reduction of an unsaturated cyclic thionophosphate affords acetylenes.[122]

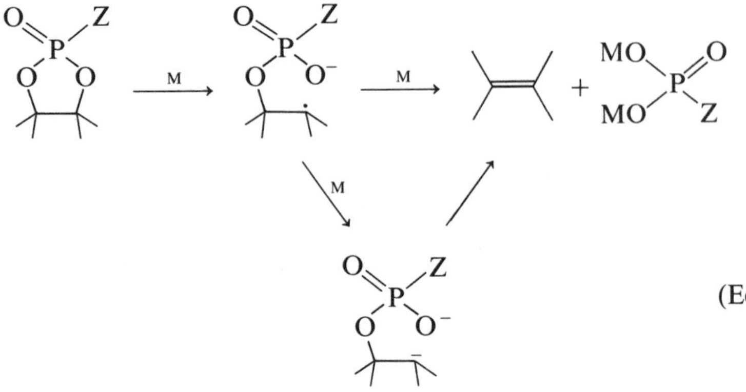

(Eq. 94)

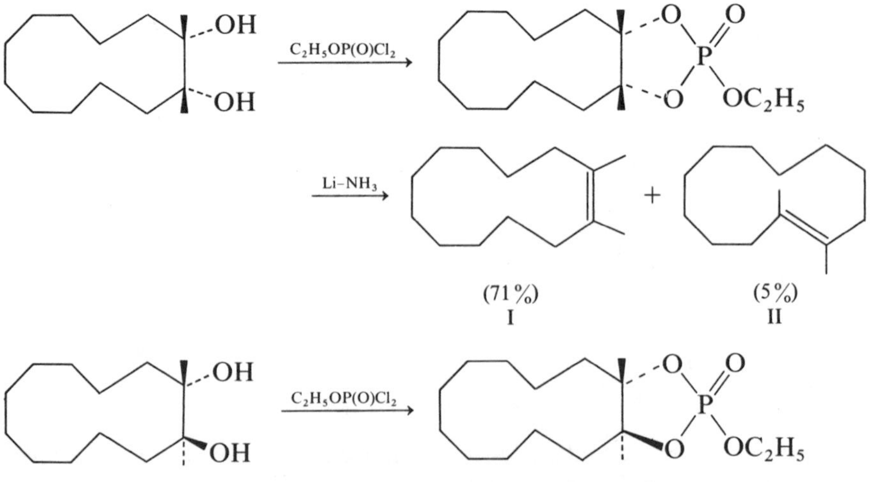

$C_2H_5OP(O)Cl_2$; Li-NH$_3$

(71%) I (5%) II

$C_2H_5OP(O)Cl_2$; Li-NH$_3$ → I (5%) + II (61%) (Eq. 95)

$\xrightarrow{(CH_3)_2NP(O)Cl_2}$ $\xrightarrow{Li\text{-}NH_3}$ (26%) I + (14%) II

$\xrightarrow{(CH_3)_2NP(O)Cl_2}$ $\xrightarrow{Li\text{-}NH_3}$ I (5%) + II (37%) (Eq. 96)

1. n-C_4H_9Li; 2. $C_2H_5OP(O)Cl_2$; 3. Li–NH_3 $\longrightarrow$ Several steps $\longrightarrow$ (Eq. 97)

TRANSITION-METAL-BASED DEOXYGENATION OF VICINAL DIOLS

Titanium Complexes

Mechanism. A one-step procedure for deoxygenation of vicinal diols to olefins involves the use of the system titanium trichloride–potassium metal, which is believed to produce active titanium [Ti(0)].[123] It is necessary to use excess potassium in order to first form the pinacol dianion. Both cyclic and acyclic diols are reduced in good yield. Only moderate stereoselectivity is realized, as shown by the results in Eq. 98, indicating that *concerted* loss of titanium dioxide from a cyclic titanium (II)–pinacol dianion complex (path *a*, Eq. 99) is not a viable mechanism.

$\xrightarrow[\text{THF, reflux, 16 hr}]{TiCl_3\text{–}K}$ I (30%) + II (45%)

$\xrightarrow[\text{THF, reflux, 16 hr}]{TiCl_3\text{–}K}$ I (7%) + II (73%) (Eq. 98)

47

(Eq. 99)

However, the fact that *different* mixtures of *cis*- and *trans*-5-decene are produced by deoxygenation of *meso*- and *d,l*-5,6-decanediols suggests that the reaction lies on the borderline of being concerted.[123] It has been separately established that both *cis*- and *trans*-5-decenes are stable to reaction conditions and that the corresponding *meso*- and *d,l*-diols do not interconvert prior to deoxygenation. The failure of *trans*-9,10-decalindiol to undergo deoxygenation rules out a mechanism in which the diol oxygens are bound to *different*, or at least noncontiguous, titanium atoms (path *c*, Eq. 99) and thus requires some type of cyclic diol–titanium complex. If five-membered intermediate **47** is involved in the rate-limiting steps of the deoxygenation reaction, it might be expected that 2-*exo*,3-*exo*-camphanediol would undergo deoxygenation considerably faster than 2-*exo*,3-*endo*-/2-*endo*,3-*exo*-camphanediol in view of the difficulty of forming a five-membered ring with the latter pair of diols. In fact, all the camphanediols are reduced at approximately the same rate; the deoxygenation is complete after 5 hours with each (Eq. 100). This result suggests that the deoxygenation of diols *does not* require the formation of a five-membered ring intermediate **47** but rather involves a heterogeneous process on the surface of an active titanium particle (path *d*, Eq. 99).[123]

Scope and Limitations. The titanium trichloride–potassium metal deoxygenation procedure has been used with a 2α,3β-dihydroxy steroid (Eq. 101) as well as with simpler diols (Eqs. 102 and 103).[123] Methyllithium or lithium aluminum hydride can be substituted for potassium metal.[124, 125]

An early example is known in which titanium trichloride alone effects deoxygenation of an activated diol (Eq. 104).[126] The mechanism of this particular reaction may be quite different from that involved in the other titanium-induced deoxygenations.

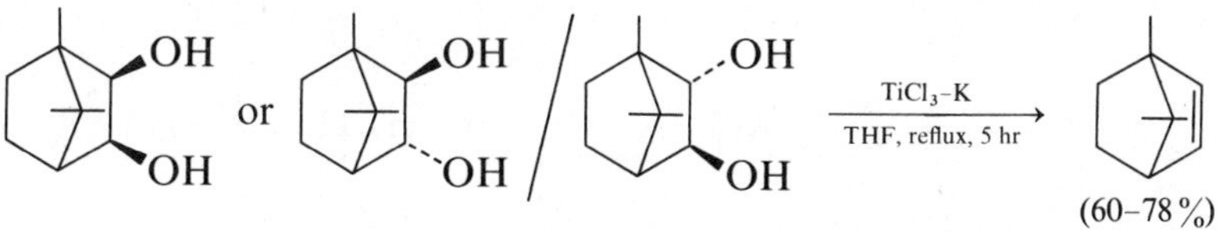

OH OH or OH OH / OH OH $\xrightarrow[\text{THF, reflux, 5 hr}]{TiCl_3\text{–K}}$ (60–78%)

(Eq. 100)

HO HO H $\xrightarrow{TiCl_3\text{–K}}$ H (80%)

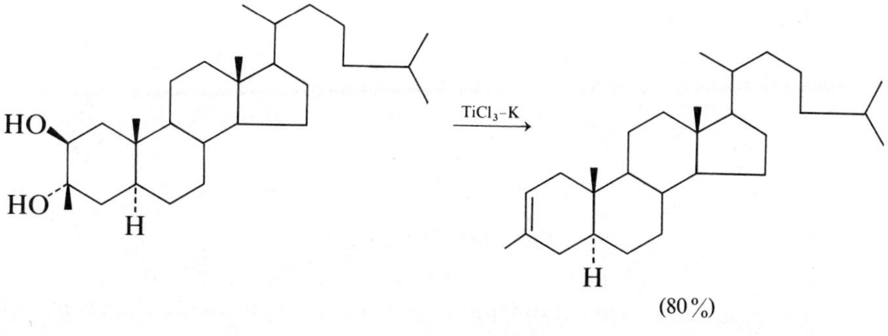

(Eq. 101)

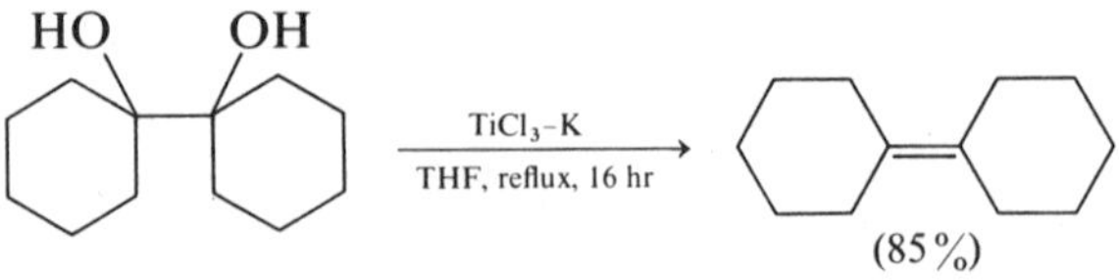

(Eq. 102)

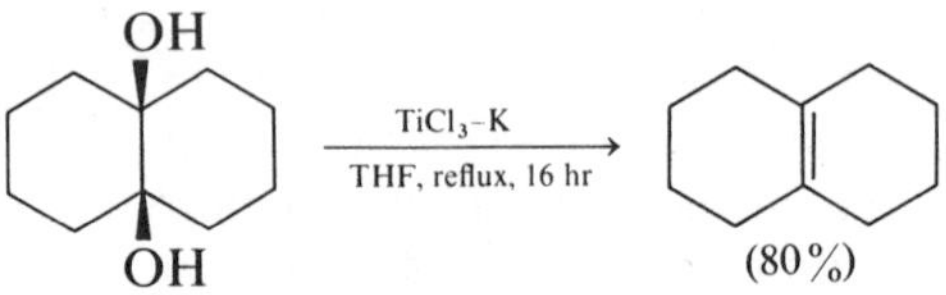

$\xrightarrow[\text{THF, reflux, 16 hr}]{\text{TiCl}_3\text{-K}}$ (85%)

(Eq. 103)

$\xrightarrow[\text{THF, reflux, 16 hr}]{\text{TiCl}_3\text{-K}}$ (80%)

The utility of the titanium(0) deoxygenation procedure is limited by the fact that titanium(0) reacts with many functional groups such as ketones, thioketones, acyloins, bromohydrins, epoxides, cyanohydrins, and allylic or benzylic alcohols.[123] Hydrogenolysis rather than olefin formation sometimes occurs, as in Eq. 105.[24] Benzylic 1,3-diols are cyclized to cyclopropanes (Eq. 106),[127] whereas 2-ene-1,4-diols afford dienes[128] with the titanium trichloride-lithium aluminum hydride reagent.

$$C_6H_5CH{=}CHCHOHCHOHCH{=}CHC_6H_5 \xrightarrow[C_2H_5OH,\ \text{reflux, 2hr}]{TiCl_3} C_6H_5(CH{=}CH)_3C_6H_5 \quad \text{(Eq. 104)}$$

(3%)

$\xrightarrow[\text{THF, reflux, 6 hr}]{\text{TiCl}_3\text{, LiAlH}_4}$ (Eq. 105)

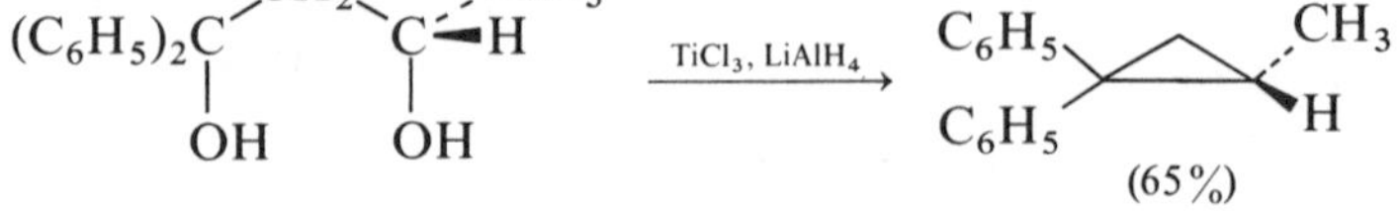

$(C_6H_5)_2C(OH)CH_2CH(OH)CH_3 \xrightarrow{TiCl_3,\ LiAlH_4}$ 1,1-$(C_6H_5)_2$-2-CH_3-cyclopropane (65%) (Eq. 106)

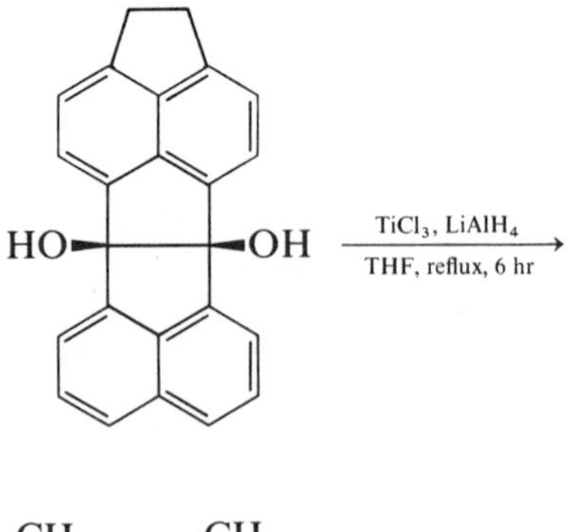

Tungsten Complexes

The reagent potassium hexachlorotungstate (K_2WCl_6) effects stereoselective deoxygenation of vicinal diols in a process that is formally the reverse of permanganate or osmium tetroxide *vic*-hydroxylation of olefins (Eq. 107).[129] In addition to the tetrahydrofuran-insoluble reagent potassium hexachlorotungstate, soluble tungsten reagents derived from treatment of tungsten hexachloride with alkyllithiums, lithium dispersion, or lithium iodide are also useful for deoxygenation of vicinal dialkoxides to olefins.[129] Applications of the deoxygenation are given in Eqs. 108 and 109[129] and Eq. 110.[130] Potassium hexachlorotungstate fails to cleanly deoxygenate the pinacol derived from cyclooctanone, giving rise to dienes in addition to the desired cyclooctylidene cyclooctane.[131] A mechanistic study indicates that the intermediacy of a cyclic metalloester (e.g., a 2-metallo-1,3-dioxolane) cannot be established unequivocally.[131] The bonding of the dialkoxide to the surface of a solid lattice or to a metallic dimer are possibilities.[131]

OH, OH —1. $2CH_3Li$, THF; 2. K_2WCl_6, reflux, 7 hr→ (O, O, Cl, Cl, Cl, Cl, W^{2-})

→ (69%) + (5%) + $WCl_4O_2^{2-}$ (Eq. 107)

OH, OH —1. $2CH_3Li$, THF; 2. K_2WCl_6, reflux, 9 hr→ (22%) + (50%)

(Eq. 108)

HO OH —$2CH_3Li$, THF; K_2WCl_6, reflux, 5 hr→ (74%)

(Eq. 109)

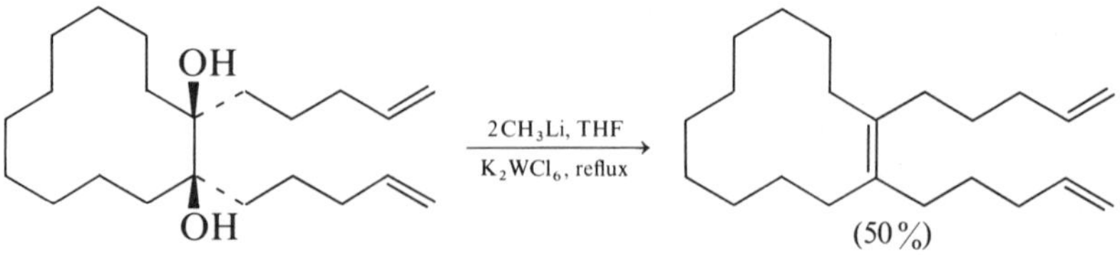

$$\xrightarrow[K_2WCl_6,\ \text{reflux}]{2CH_3Li,\ THF}$$

(50%)

(Eq. 110)

DEOXYGENATIONS INVOLVING CYCLIC SULFOXYLATES

A limited number of examples have been reported of the stereospecific deoxygenation of vicinal diols through loss of sulfur dioxide from cyclic sulfoxylate intermediates (Eq. 111).[132] A related example is thought to involve a tetracoordinate sulfur(IV) intermediate (Eq. 112).[133]

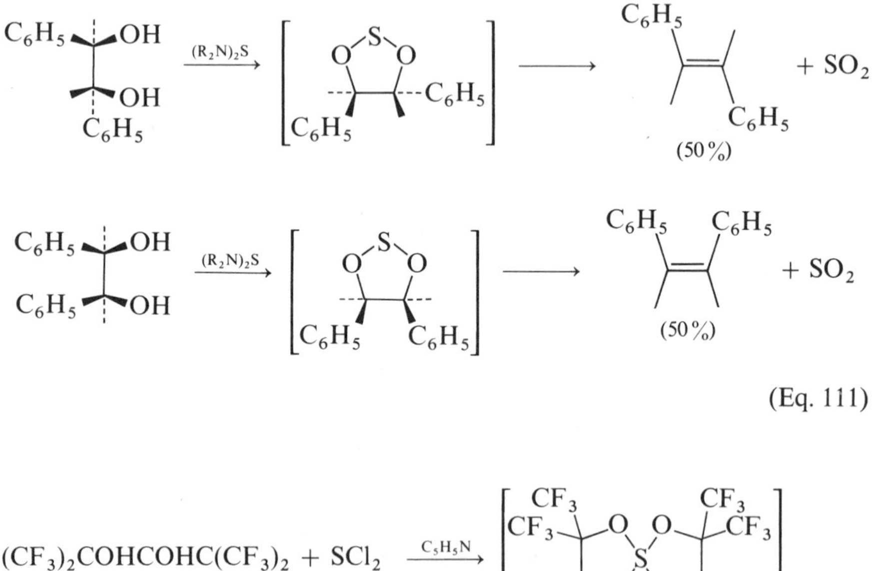

$$\xrightarrow{(R_2N)_2S} \longrightarrow C_6H_5CH{=}CHC_6H_5\ (50\%) + SO_2$$

$$\xrightarrow{(R_2N)_2S} \longrightarrow C_6H_5CH{=}CHC_6H_5\ (50\%) + SO_2$$

(Eq. 111)

$$(CF_3)_2COHCOHC(CF_3)_2 + SCl_2 \xrightarrow{C_5H_5N}$$

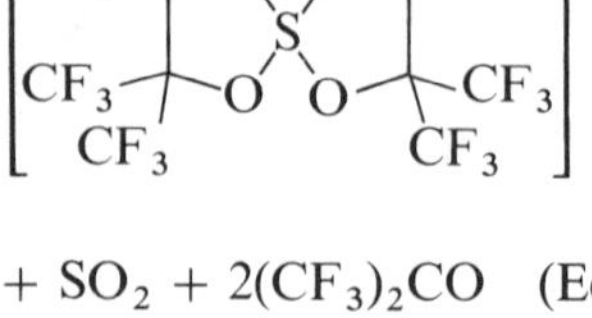

$$\longrightarrow (CF_3)_2C{=}C(CF_3)_2 + SO_2 + 2(CF_3)_2CO \quad \text{(Eq. 112)}$$

DEOXYGENATION OF VICINAL DIOLS VIA ACYCLIC INTERMEDIATES

While the stereospecific deoxygenation of vicinal diols via cyclic intermediates of type **1**

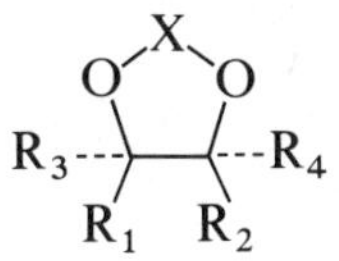

1

has found wide application in organic synthesis, older nonstereospecific procedures involving treatment of vicinal sulfonate esters with nucleophiles (e.g., Eq. 113[134]) (the so-called Tipson–Cohen reaction), are still used extensively, particularly in carbohydrate chemistry. The general lack of stereospecificity in these and related deoxygenations involving acyclic intermediates can actually be an advantage if the target molecule contains a double bond in a five-membered or six-membered ring and the precursor is a *trans* vicinal diol or polyol, for example, a sugar or a cyclitol.

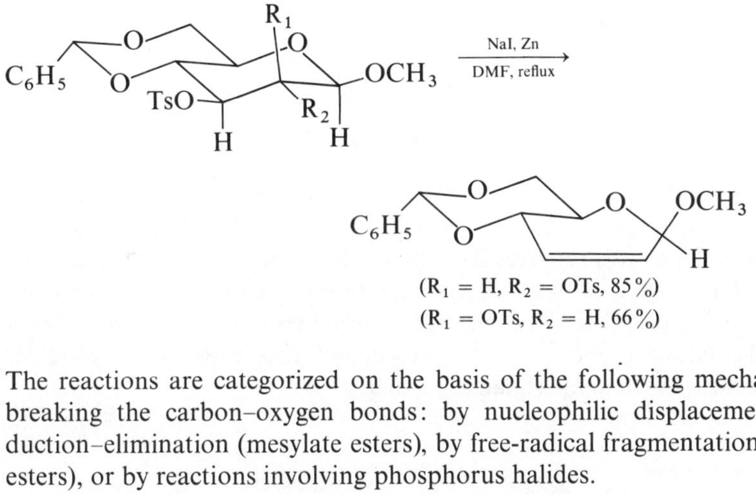

NaI, Zn / DMF, reflux

(R_1 = H, R_2 = OTs, 85%)
(R_1 = OTs, R_2 = H, 66%) (Eq. 113)

The reactions are categorized on the basis of the following mechanisms for breaking the carbon–oxygen bonds: by nucleophilic displacement, by reduction–elimination (mesylate esters), by free-radical fragmentation (xanthate esters), or by reactions involving phosphorus halides.

Nucleophilic Displacement Reactions

It was discovered in 1943 that 1,2,3,4-tetra-*O*-tosylerythritol reacts with sodium iodide in acetone to afford butadiene (Eq. 114).[135] This procedure for the elimination of vicinal disulfonyloxy groups has since been used for the preparation of unsaturated sugars,[136] cyclitols,[137] and steroids.[138] The mechanism is thought to involve nucleophilic displacement of a sulfonyloxy group by iodide ion, followed by elimination with *anti*periplanar geometry (Eq. 115).[134]

$$\text{OTs–CH}_2\text{–CH(OTs)–CH(OTs)–CH}_2\text{–OTs} \xrightarrow{4\,\text{NaI}} \text{CH}_2\text{=CH–CH=CH}_2 + 2\text{I}_2 + 4\,\text{NaOTs} \qquad \text{(Eq. 114)}$$

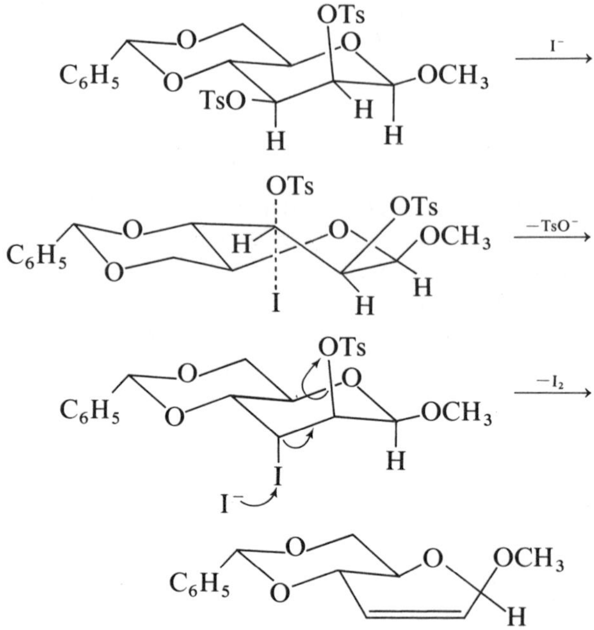

(Eq. 115)

A modification of this reaction (the Tipson–Cohen reaction) involves the use of a mixture of zinc dust and sodium iodide in boiling dimethylformamide (Eq. 113).[139] The zinc possibly removes the iodine, which could cause side reactions, and assists in the removal of the sulfonyloxy group, particularly in *cis* vicinal iodosulfonates derived from *trans* vicinal disulfonates as shown in **48.**[134] It is also possible that the iodine in a *cis* vicinal iodosulfonate is displaced by iodide, giving a precursor with *anti*periplanar geometry.

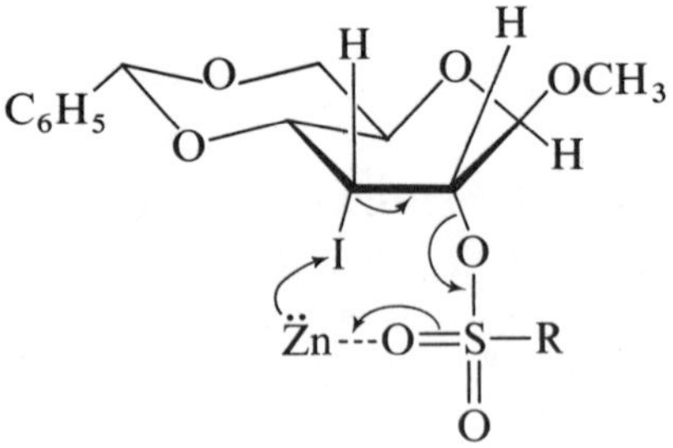

48

The Tipson–Cohen reaction is useful in the synthesis of unsaturated sugars,[139–144] nucleosides,[79] and antibiotics related to kanamycin B.[145,146] Variations of this procedure involve the use of zinc–copper couple and potassium iodide[141] and potassium ethyl xanthate or potassium selenocyanate rather than sodium iodide–zinc.[28,147,148]

Treatment of acyclic vicinal disulfonates with iodide ion can either be non-stereospecific (Eq. 116)[80] or stereospecific (Eq. 117),[139] depending on substrate and conditions. Side reactions include deacetylation or debenzoylation of vicinal dimesylates containing an adjacent ester grouping[141] and neighboring-group participation by azido[149] and *N*-tosylamino[145] functions adjacent to the sulfonate ester, leading to formation of nitrogen-containing heterocycles (Eq. 118).[149]

OH / OH — 3 : 2 *meso* : *d,l* — 1. TsCl, C_5H_5N; 2. NaI, $C_2H_5(OCH_2CH_2)_2OH$, 62 hr, 100° → (9%) + (76%)

(Eq. 116)

HO, OH — 1. MsCl, C_5H_5N; 2. NaI, Zn, DMF, reflux, 5 hr →

(Eq. 117)

MsO, MsO, N_3 — NaI, DMF, Zn, reflux, 1 hr → MsO, H, NH (68%)

(Eq. 118)

Vicinal diol monotosylates can also be converted to olefins under slightly modified Tipson–Cohen conditions (Eq. 119).[150] A one-pot diol deoxygenation procedure that should be mechanistically related to the Tipson–Cohen reaction involves treatment of *cis* and *trans* vicinal diols with the combination chlorotrimethylsilane–sodium iodide at room temperature for 1 hour or less.[151]

C_6H_5, HO, TsO, H, OCH_3 — Zn-Cu, NaI; DMF–DME, 130°, 3 hr → C_6H_5, OCH_3 (83%)

(Eq. 119)

Reductive Elimination of Vicinal Dimesylates

Treatment of vicinal dimesylates, but not ditosylates or di(*p*-bromobenzenesulfonates), with anion radicals such as sodium anthracene, sodium naphthalene, or sodium trimesitylborane gives good yields of olefins under mild conditions. The reaction is not stereospecific; the more stable alkene predominates (Eqs. 120, 121,[152] and 122[153]). The best yields are obtained when the anion radical solution is added to a chilled tetrahydrofuran solution of the dimesylate. The proposed mechanism for the reduction is shown in Eq. 123. A significant

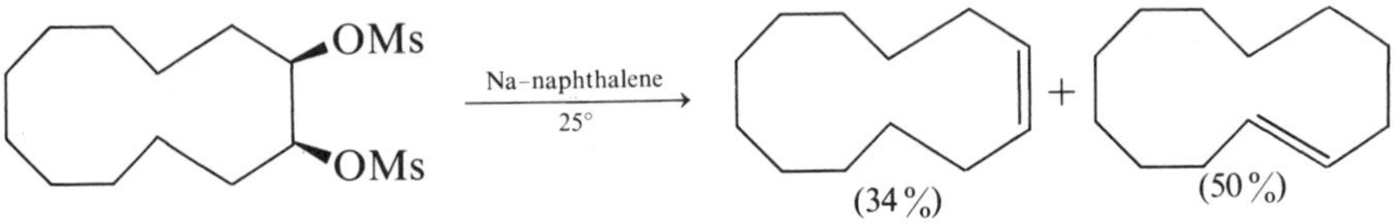

Na–naphthalene, 25° (34%) + (50%)

(Eq. 120)

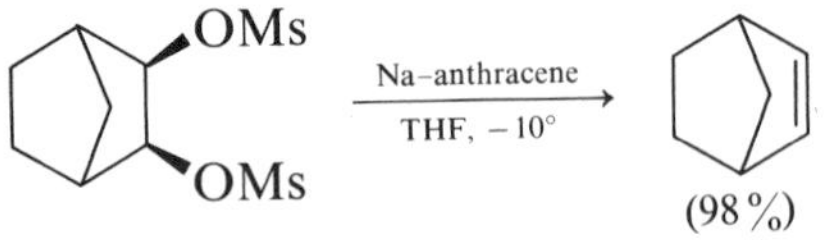

Na–anthracene, THF, −10° (98%)

(Eq. 121)

1. MsCl, C_5H_5N
2. Na–naphthalene, THF

(65%)

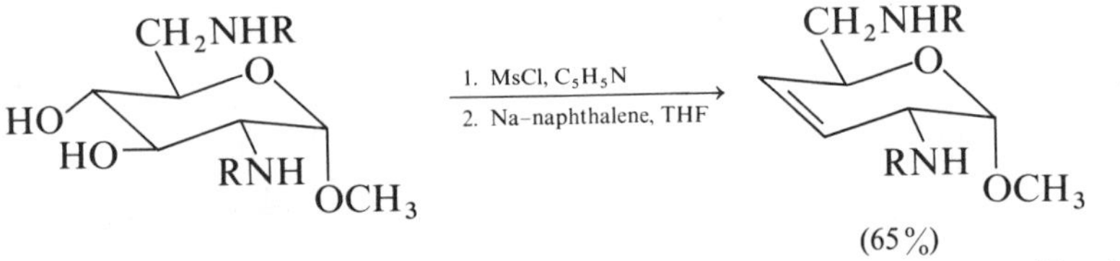

(Eq. 122)

limitation of this procedure is that a number of functional groups (e.g., carbonyl, nitro, cyano, halogen) undergo reduction with the radical anion reagents.

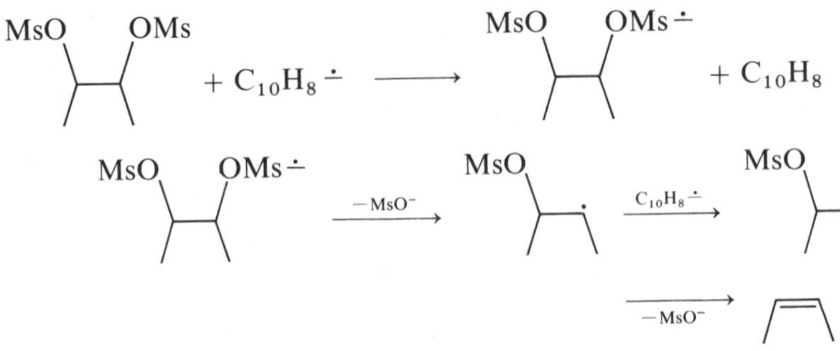

$+ C_{10}H_8^{\dot{-}} \longrightarrow$ $+ C_{10}H_8$

$\xrightarrow{-MsO^-}$ $\xrightarrow{C_{10}H_8^{\dot{-}}}$

$\xrightarrow{-MsO^-}$

(Eq. 123)

Decomposition of Vicinal Dixanthate Esters Induced by Tin Hydride

A radical fragmentation reaction is the basis for the tri-*n*-butylstannane-induced conversion of vicinal bisdithiocarbonates to olefins (Eq. 124).[154] The reaction occurs under neutral, relatively mild conditions (refluxing toluene) and gives predominantly or exclusively the more stable olefin isomer. Chromatography is usually necessary to remove tin compounds and minor byproducts; silica gel impregnated with silver nitrate is recommended for this purpose. The principal applications are in the synthesis of unsaturated sugars (Eqs. 125, 126,[154] and 127[155]).

$(n\text{-}C_4H_9)_3Sn\cdot$

$\longrightarrow (n\text{-}C_4H_9)_3SnSCOSCH_3 +$

$\longrightarrow$ + $CH_3S\cdot$ + COS (Eq. 124)

1. NaH; CS_2, CH_3I
2. $(n\text{-}C_4H_9)_3SnH$, $C_6H_5CH_3$; reflux

(59%) (Eq. 125)

$(n\text{-}C_4H_9)_3SnH$, $C_6H_5CH_3$, reflux, 60 hr

(62%)

(Eq. 126)

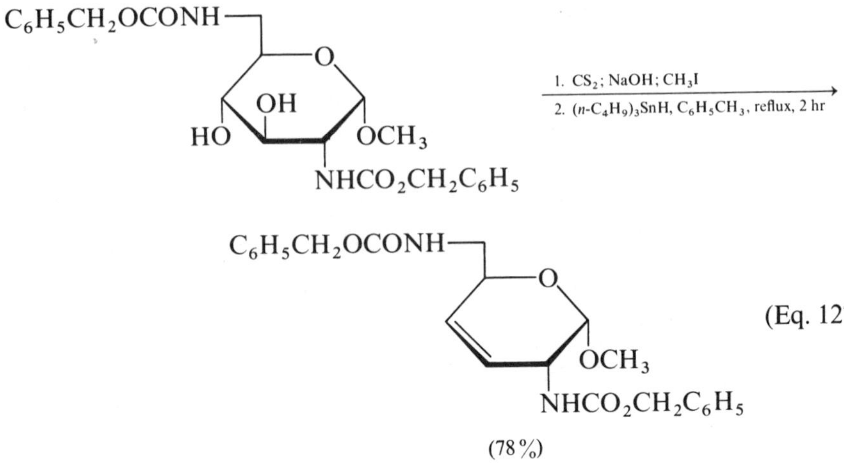

$C_6H_5CH_2OCONH$ — (pyranoside: HO, OH, OCH_3, $NHCO_2CH_2C_6H_5$) $\xrightarrow[\text{2. } (n\text{-}C_4H_9)_3SnH,\ C_6H_5CH_3,\ \text{reflux, 2 hr}]{\text{1. } CS_2;\ NaOH;\ CH_3I}$

$C_6H_5CH_2OCONH$ — (unsaturated pyranoside: OCH_3, $NHCO_2CH_2C_6H_5$) (Eq. 127)

(78%)

Procedures Involving Phosphorus Halides

It was discovered in 1928 that vicinal diols could be deoxygenated in a single step simply by stirring at room temperature with an ether solution of diphosphorus tetraiodide (Eq. 128).[126] The so-called Kuhn–Winterstein reaction has found only limited application in organic synthesis (Eqs. 129[156] and 130[24]) because the yields are very variable;[24] however, a number of recently developed and mechanistically related procedures involving combinations of phosphines and iodinating reagents have proved useful in one-step diol deoxygenations. These procedures involve refluxing the diol in toluene with a mixture of triphenylphosphine, imidazole, and either iodine,[157] triiodoimidazole,[158] or iodoform,[159] and can be used to prepare unsaturated sugars (Eq. 131).[157–159] A possible mechanism is given in Eq. 132.[157]

$$C_6H_5CH{=}CHCHOHCHOHCH{=}CHC_6H_5 \xrightarrow[(C_2H_5)_2O,\ 25^\circ,\ 30\ \text{min}]{P_2I_4} C_6H_5(CH{=}CH)_3C_6H_5 \quad \text{(Eq. 128)}$$

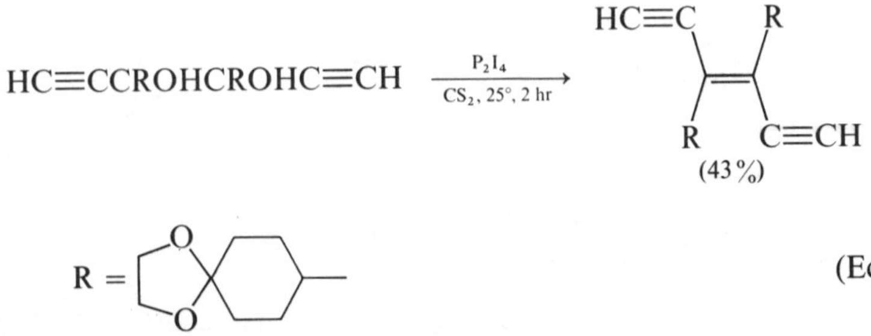

$$HC{\equiv}CCROHCROHC{\equiv}CH \xrightarrow[CS_2,\ 25^\circ,\ 2\ \text{hr}]{P_2I_4} (HC{\equiv}C)RC{=}CR(C{\equiv}CH) \quad (43\%)$$

R = (1,4-dioxaspiro[4.5]decan-8-yl) (Eq. 129)

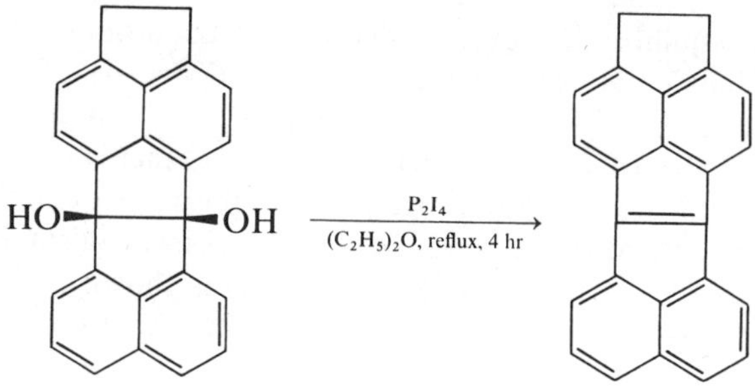

$$\xrightarrow[(C_2H_5)_2O,\ \text{reflux, 4 hr}]{P_2I_4} \quad (49\%) \qquad \text{(Eq. 130)}$$

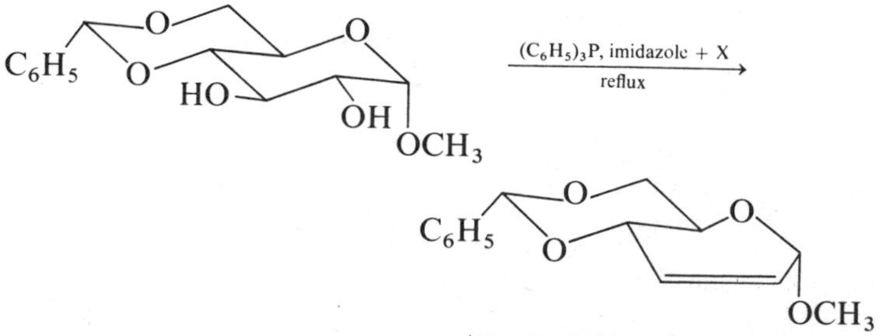

$$\xrightarrow[\text{reflux}]{(C_6H_5)_3P,\ \text{imidazole} + X} \qquad \text{(Eq. 131)}$$

($X = I_2$, 76%; X = triiodoimidazole, 87%; X = iodoform, 80%)

$$(C_6H_5)_3P + \text{“I source”} \longrightarrow (C_6H_5)_3PI^+I^- \xrightarrow{\text{imidazole}} (C_6H_5)_3P\text{—}\overset{+}{N}\text{(imidazolyl)}\ I^-$$

$$\xrightarrow{\text{RCHOHCHOHR}} \text{RCH}[\text{O}\overset{+}{P}(C_6H_5)_3]\text{CH}[\text{O}\overset{+}{P}(C_6H_5)_3]\text{R}$$

$$\xrightarrow{I^-} \text{RCHICH}[\text{O}\overset{+}{P}(C_6H_5)_3]\text{R} \xrightarrow{I^-} \text{RCH=CHR} + I_2 + (C_6H_5)_3P\text{=O}$$

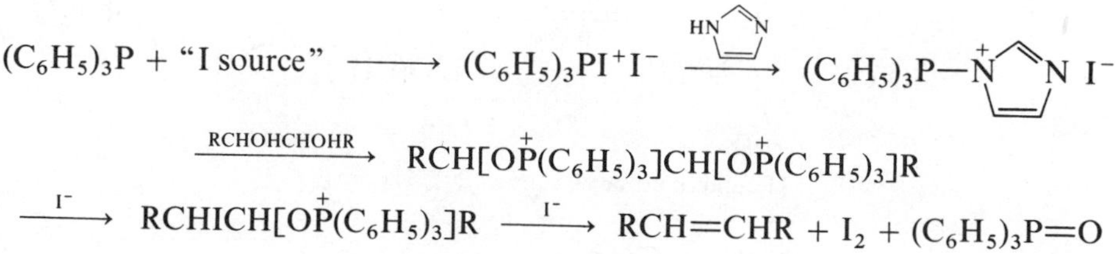

(Eq. 132)

STEREOCHEMICAL EQUILIBRATION OF VICINAL DIOLS

The development of stereospecific methods for the deoxygenation of vicinal diols requires that the starting diols retain their stereochemical integrity under all conditions used in the deoxygenation procedures. It is thus important to review what is known about the stereochemical equilibration and rearrangement of diols.

It has been reported that vicinal diols undergo skeletal rearrangement when treated with sodium methoxide in refluxing toluene (Eq. 133)[34] and that the dilithium derivative of 1,2-diols undergoes stereochemical equilibrium in diglyme (diethylene glycol dimethyl ether) at 155° (Eqs. 134[35,160] and 135[131]), or even in refluxing tetrahydrofuran (Eq. 136).[131] The dilithium derivatives of the ditertiary 1,2-diols, *cis*- and *trans*-1,2-dimethyl-1,2-cyclohexanediols and *cis*- and *trans*-1,2-dimethyl-1,2-cyclododecanediols failed to undergo base-induced stereochemical equilibration even under drastic conditions.[131,160] Mechanistic studies using deuterated *cis*- and *trans*- cyclododecane-1,2-diols indicated that diolate stereoisomerization was accompanied by considerable intermolecular deuterium transfer; note in particular the large increase in the percentage of d_1-labeling in the *trans*-1,2-diol together with the decrease in the percentage of d_2-labeling in the same compound (Eq. 137).[131] A mechanism that accommodates the results of all of the above experiments is shown in Eq. 138,[131,160] where reversible hydride transfer occurs to any carbonyl group present. The

OH, OH → ($NaOCH_3$, $C_6H_5CH_3$, reflux) → OH, CH_2OH (Eq. 133)

OH, OH ⇌ (n-C_4H_9Li / H_2O) OLi, OLi ⇌ (155°, 17 hr) OLi, OLi ⇌ (H_2O / n-C_4H_9Li) OH, OH

[Equilibrium: 97–98% *trans*, 2–3% *cis*]

(Eq. 134)

n-C_8H_{17}, H, OLi, $LiO(CH_2)_8$, OLi, H (*erythro*) ⇌ (155°, 3–37 hr / Diglyme) n-C_8H_{17}, H, OLi, $LiO(CH_2)_8$, H, OLi (*threo*) (Eq. 135)

[Equilibrium: ca. 80% *threo*, 20% *erythro*]

OLi, OLi ⇌ (2–4 days, THF, reflux) OLi, OLi

[Equilibrium: 18% *cis*, 82% *trans*] (Eq. 136)

H(D), OLi, OLi, H(D) (I) → (reflux, 2 hr / Diglyme) → H(D), OLi, OLi, H(D) (II (10% recovered)) + H(D), OLi, H(D), OLi (III (62% recovered))

d_2: I, 44.4%	II, 56.9%	III, 8.1%
d_1: I, 5.3%	II, 9.9%	III, 35.8%
d_0: I, 50.3%	II, 33.2%	III, 56.2%

(Eq. 137)

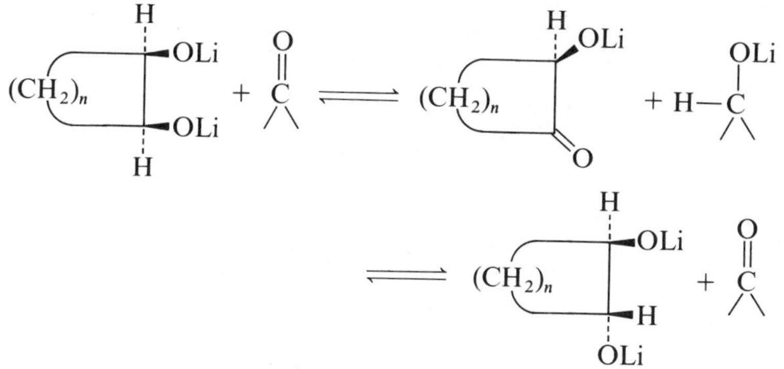

(Eq. 138)

results in Eq. 141 can be explained by skeletal rearrangement of the intermediate α-hydroxy ketone prior to reduction to diol.[160]

In the presence of potassium hexachlorotungstate, even dilithium derivatives of ditertiary 1,2-diols undergo stereochemical equilibration (Eq. 139).[131] Intermolecular hydride transfer is not observed in the tungstate-catalyzed isomerization: note the absence of significant deuterium exchange, after correcting for a small isotope effect, in the reaction shown in Eq. 140.[131] A tungsten–carbonyl complex is postulated to explain the results of the metal-catalyzed stereochemical equilibration of diolates (Eq. 141).[131] Photoisomerization of 1,2-diolates is also observed in the presence of tungsten reagents (Eq. 142).[131]

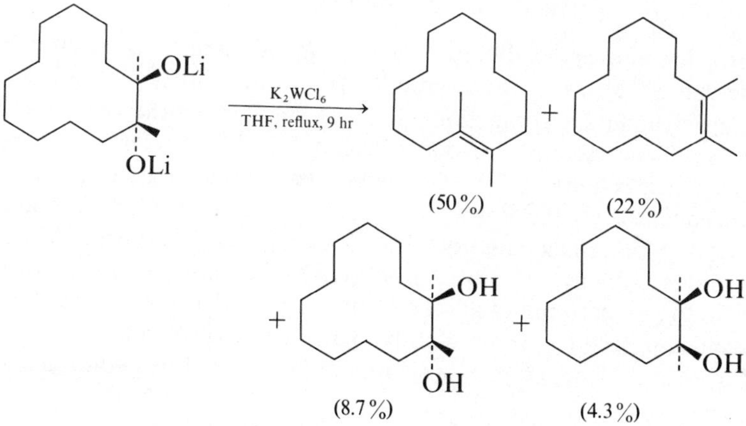

K_2WCl_6, THF, reflux, 9 hr

(50%) (22%) (8.7%) (4.3%)

(Eq. 139)

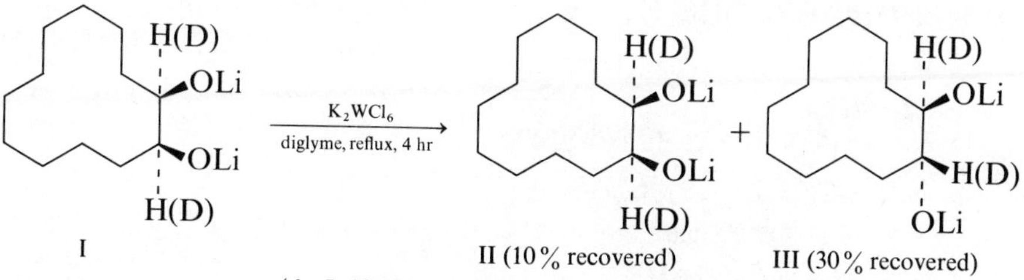

K_2WCl_6, diglyme, reflux, 4 hr

I

II (10% recovered)

III (30% recovered)

d_2: I, 44.4%	II, 51.6%	III, 50.8%
d_1: I, 5.3%	II, 6.3%	III, 6.9%
d_0: I, 50.3%	II, 42.1%	III, 42.3%

(Eq. 140)

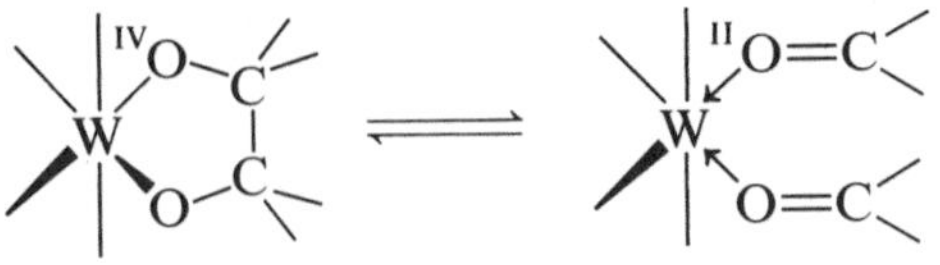

(Eq. 141)

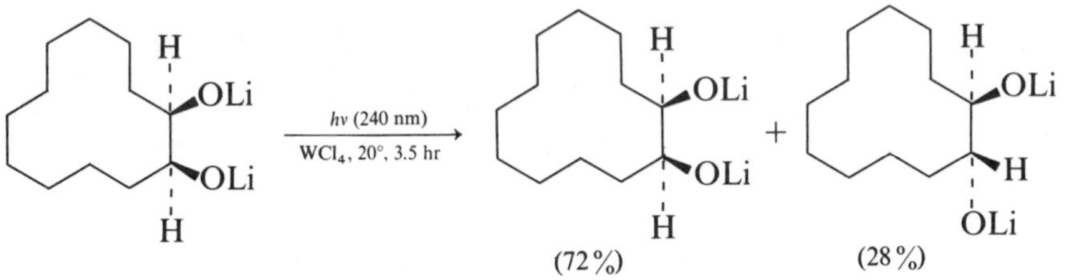

$$\text{CH}_3\text{CHOHC(CH}_3\text{)OHC}_2\text{H}_5 \xrightarrow{hv\ (240\ \text{nm}),\ \text{WCl}_4,\ 20°,\ 3.5\ \text{hr}}$$

(72%) (28%)

(Eq. 142)

The above findings on the stereochemical equilibration of vicinal diols indicate that both strongly basic conditions and oxygen or peroxides (which could lead to the formation of carbonyl compounds) should be avoided in attempting stereospecific deoxygenations of diols containing at least one secondary hydroxyl group.

COMPARISON OF PROCEDURES

The conversion of a vicinal diol into the corresponding olefin must often be done in the context of a multistep synthesis with a polyfunctional molecule, where delicate structural and functional features may restrict the use of available methods. Of the various methods considered in this chapter, only three show good stereoselectivity: (1) the Corey–Winter procedure (desulfurization of thionocarbonates), (2) the Eastwood procedure (acid-catalyzed decomposition of 2-ethoxy- or 2-dialkylamino-1,3-dioxolanes), and (3) the base-induced fragmentation of 2-phenyl-1,3-dioxolanes. Side-by-side comparison of these procedures in the preparation of acyclic olefins suggest that the second (Eastwood) and third procedures show slightly higher stereoselectivities than does the first (Eq. 143).[161] The use of *n*-butyllithium in the third procedure makes

$$\text{CH}_3\text{CHOHC(CH}_3\text{)OHC}_2\text{H}_5 \xrightarrow{a,\ b\ \text{or}\ c} cis\text{-CH}_3\text{CH=C(CH}_3\text{)C}_2\text{H}_5 + trans\text{-CH}_3\text{CH=C(CH}_3\text{)C}_2\text{H}_5$$

99%, *erythro*, 1% *threo*	*a*	95:1 *cis*:*trans*	*b*	99:1 *cis*:*trans*	*c*	99:1 *cis*:*trans*
1%, *erythro*, 99% *threo*	*a*	4:96 *cis*:*trans*	*c*	2:98 *cis*:*trans*		

a = 1. n-C_4H_9Li, CS_2, CH_3I
2. $(CH_3O)_3P$, heat

b = $(C_2H_5O)_3CH$, heat

c = 1. C_6H_5CHO, H^+
2. n-C_4H_9Li

(Eq. 143)

it inapplicable to many substrates. Furthermore, it has been shown that the third procedure fails with stilbenes and molecules possessing acidic protons on or adjacent to the 4 or 5 position of the 1,3-dioxolane ring.

The Eastwood procedure has the advantage of being a one-pot, one-reagent olefin synthesis. Electrophilic conditions (e.g., acetic anhydride or carboxylic acids on cyclic diol orthoesters or amides) and high temperature may reduce the usefulness of the Eastwood procedure, although the acid-sensitive *trans*-cyclooctene could be prepared in good yield by this method. Two diols that failed to give good yields of olefins under Eastwood conditions are **49**[162] and **50**.[84] Compound **49** also failed to give satisfactory yields of the triene on treatment

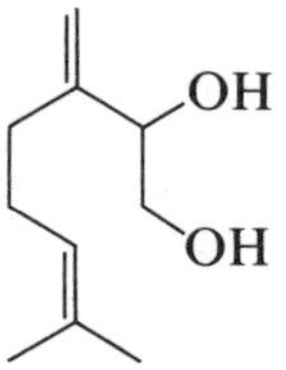

OH OH HO HO NCH_3 =O

49  **50**

of the 2-phenyl-1,3-dioxolane derivative with *n*-butyllithium. The most satisfactory procedure for deoxygenation of **49** was sequential treatment with cuprous bromide–phosphorus tribromide followed by zinc dust at 0° (58% yield of triene from this procedure).[162,163]

The Corey–Winter reaction requires prolonged heating at elevated temperatures with reagents (trialkylphosphites) that occasionally undergo undesired side reactions with the substrate. This deoxygenation procedure also fails in the preparation of bicyclo[2.2.1]hept-2-enes and related olefins from diols such as **50**[84] and thionocarbonates such as **51**,[93] **52**,[75] and **53**.[47] The diol corresponding

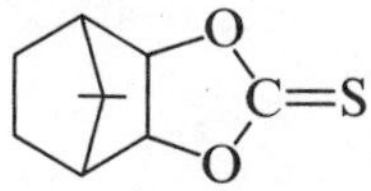

51 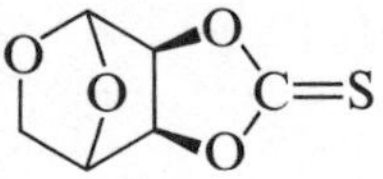 **52** 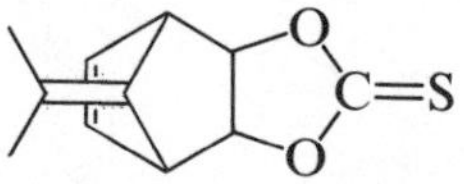 **53**

to **51** can be deoxygenated in good yield by the Eastwood procedure (ethyl orthoformate followed by heating at 176°, Eq. 68).[93] The failure of the Corey–Winter procedure with thionocarbonates **50–53** is surprising in view of the successful preparation by this method of even more highly strained olefins such as *trans*-cyclooctene and *trans*-cycloheptene.

Desulfurization of thionocarbonate **54** with triethylphosphite at 150° gave a better yield of olefin (18%) than did sodium-anthracene (in tetrahydrofuran) reduction–elimination of dimesylate **55** (4% olefin) or Tipson–Cohen reaction of **55** (sodium-iodide-activated zinc in hexamethylphosphoramide at 100°; 10% olefin).[164] A comparison of the Corey–Winter reaction of a potential dodecahedrane precursor (see Eq. 23)[22] with the tungsten hexachloride-*n*-butyllithium procedure[129] showed the former method to be clearly superior in this case. Both the Corey–Winter and the methyl iodide–2-dialkylamino-1,3-dioxolane procedures failed with diol **56** whereas the Eastwood method succeeded (67% olefin).[165] Deoxygenation of diol **57** by the Corey–Winter route succeeded, whereas the methyl iodide–2-dimethylamino-1,3-dioxolane method led only to the 4-*O*-formate ester.[63]

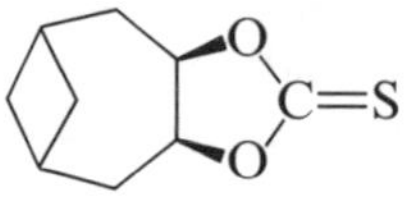

O, C=S, O

54

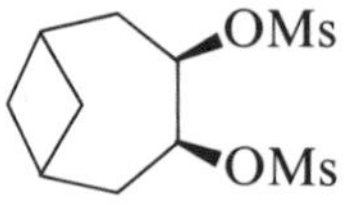

OMs, OMs

55

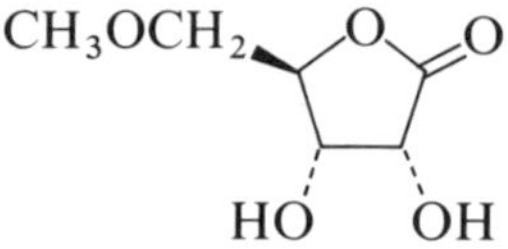

CH_3OCH_2, O, O, HO, OH

56

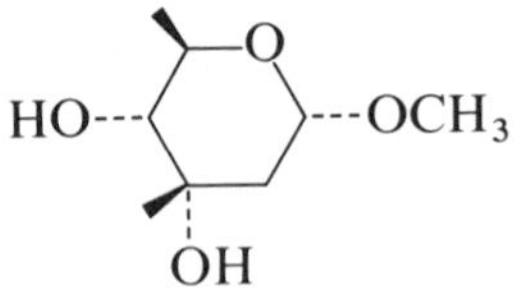

HO, O, OCH_3, OH

57

Occasionally special methods prove successful in the deoxygenation of unusual diols. Thus liquid hydrogen fluoride is superior to the Corey–Winter procedure in the deoxygenation of **58** and **59**.[24]

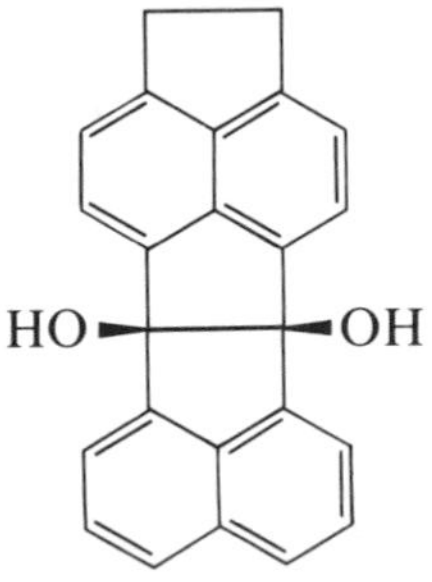

HO, OH

58

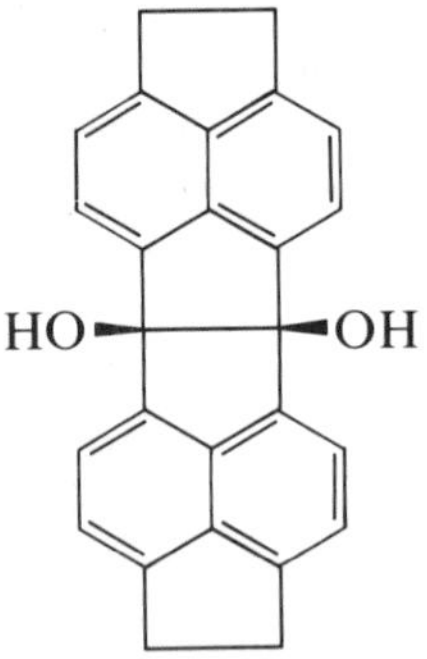

HO, OH

59

In transformations involving vicinal diols in six-membered rings, such as the preparation of unsaturated sugars, lack of stereospecificity can be a virtue, since attempted deoxygenation of the readily available *trans* vicinal diols by procedures that are stereospecific (e.g., the Corey–Winter and Eastwood methods) fails. The Tipson–Cohen reaction (sodium iodide–zinc–dimethylformamide on diol disulfonates) is often unsuccessful because of difficulty in the initial S_N2 displacement by iodide arising from steric and stereoelectronic factors.[154] Radical procedures, such as decomposition of vicinal dixanthate esters induced by tin hydride, offer advantages since they are not as susceptible to steric hindrance and involve mild, neutral reaction conditions.

EXPERIMENTAL PROCEDURES

Thiocarbonyldiimidazole (Reagent for Preparation of 1,3-Dioxolane-2-thiones).[32] In a dry 200-mL round-bottomed flask fitted with a reflux condenser carrying a drying tube was placed trimethylsilylimidazole (18.2 g, 0.13 mol) and 100 mL of anhydrous benzene (distilled from calcium hydride). The flask had a side arm closed off with a serum cap. The solution was stirred magnetically and cooled to 0°. Thiophosgene (7.5 g, 65 mmol) was introduced slowly from a syringe into the reaction flask through the serum-capped side arm. Reaction ensued immediately. After the addition was complete, the contents were stirred for 1 hour. The condenser was replaced with a vacuum take-off with a stream of dry nitrogen sweeping the apparatus. The reaction mixture was concentrated *in vacuo*, leaving thiocarbonyldiimidazole (ca. 100% yield) as a bright yellow powder (mp 105–106°) that can be used directly without additional purification.

5,6-Dideoxy-1,2-*O*-isopropylidene-α-D-xylohex-5-enofuranose (Preparation of an Unsaturated Sugar).[166] To a stirred solution of 1,2-*O*-isopropylidene-α-D-glucofuranose (13.5 g, 0.061 mol) in 250 mL of hot acetone was added 13.1 g (0.073 mol) of thiocarbonyldiimidazole. The mixture was heated at reflux for 1.5 hours under a slow stream of nitrogen. Activated charcoal (0.5 g) was added, and, after 15 minutes, the mixture was filtered. The filtrate was evaporated to give a light-brown solid. The solid was triturated with 60 mL of methanol, and the resulting suspension was filtered to give a white crystalline solid (12.1 g). Refrigeration of the filtrate gave an additional 2.4 g. Recrystallization of the product from 200 mL of methanol gave 11.2–13.5 g (0.043–0.052 mol, 70–80%) of pure 1,2-*O*-isopropylidene-α-D-glucofuranose 5,6-thionocarbonate (mp 180–185°), $[\alpha]_D^{20}$ − 18° (*c* 1% acetone).

A mixture of 5.0 g (0.019 mol) of the thionocarbonate and 20 mL of freshly distilled trimethyl phosphite was heated to reflux (bath temperature maintained at 150°) under nitrogen for 60 hours. The solution was cooled and poured into 250 mL of 1 *M* aqueous sodium hydroxide, giving a permanently basic, homogeneous solution. The solution was extracted with four 250-mL portions of chloroform, and the extracts were dried with anhydrous magnesium sulfate and evaporated to give 5,6-dideoxy-1,2-*O*-isopropylidene-α-D-xylohex-5-enofuranose as a colorless syrup that crystallized spontaneously, yield 2.78 g (0.015 mol, 78%). The chromatographed compound had mp 61–65°, $[\alpha]_D^{20}$ − 60° (*c* 2%, chloroform).

***trans*-3,4-Didehydro-3,4-dideoxy-1,2:5,6-di-*O*-isopropylidene-D-*threo*-hexitol (Preparation of an Unsaturated Sugar with 1,3-Dimethyl-2-phenyl-1,3,2-diazaphospholidine; Thionocarbonate Preparation with Thiophosgene).**[25] To a stirred solution of 262 mg (1.0 mmol) of 1,2:5,6-di-*O*-isopropylidene-D-mannitol[90] and 293 mg (2.4 mmol) of 4-dimethylaminopyridine in 4.0 mL of dry methylene chloride at 0° under argon was added 108 μL (162 mg, 1.2 mmol) of 85% thiophosgene in carbon tetrachloride. The contents were stirred for 1.0

hour at 0°. Silica gel (2.0 g, Merck) was added, and the mixture was allowed to warm to 25°. After removal of the methylene chloride *in vacuo*, the remaining solid was loaded onto a column of 6.0 g of silica gel and eluted with 50% ethyl acetate in *n*-hexane. Concentration *in vacuo* afforded the thionocarbonate (284 mg, 93%) as a colorless solid, mp 152–156°; TLC R_f 0.51 (50% ethyl acetate in *n*-hexane); IR ($CHCl_3$) 1320 cm^{-1} (C=S); ^1H NMR ($CDCl_3$) δ 1.34 (6*H*, s), 1.46 (6*H*, s), 4.15 (6*H*, m), 4.66 (2*H*, m).

A suspension of 164 mg (0.54 mmol) of the above thionocarbonate in 0.29 mL (310 mg, 1.6 mmol) of 1,3-dimethyl-2-phenyl-1,3,2-diazaphospholidine[167] was stirred under argon at 40° for 20 hours. After cooling to 25° the contents were directly chromatographed on a column of 20 g of silica gel (elution with 5% diethyl ether in methylene chloride) to afford 108 mg (88%) of pure *trans* olefin as a colorless solid, mp 80–81°; TLC R_f 0.55 (17% ether in methylene chloride); ^1H NMR ($CDCl_3$) δ 1.35 (5*H*, s), 1.39 (6*H*, s), 3.55 (2*H*, t, $J = 7.7$ Hz), 4.05 (2*H*, dd, $J = 8.0$ and 6.2 Hz), 4.45 (2*H*, m), 5.77 (2*H*, m); $[\alpha]_D^{20}$ + 56.7° (*c* 3.2%, chloroform).

(−)-*trans*-Cyclooctene (Preparation of a Volatile Olefin with Isolation Using a Carrier Gas Stream).[17] A mixture of thiocarbonyldiimidazole (3.82 g, 0.022 mol) and (+)-*trans*-1,2-cyclooctanediol (2.88 g, 0.020 mol) in 100 mL of toluene was heated at reflux for 1 hour. The reaction mixture was cooled, transferred to a separatory funnel using 30 mL of benzene as a rinse, and washed with two 20-mL portions of water, and then with two 20-mL portions of brine. The organic layer was dried, the solvents were removed, and the yellow solid residue was recrystallized from 30 mL of 1:2 chloroform-*n*-hexane (with cooling to −15°) to give colorless crystals of (+)-*trans*-4,5-hexamethylene-1,3-dioxolane-2-thione; yield 2.81 g (0.0167 mol, 76%), mp 141.4–142.2° (cf. mp 131.7–132.2° for the racemate); $[\alpha]_D^{21}$ + 17.2° (*c* 2.27% chloroform); IR (KBr) 2890, 2832, 1461, 1441, 1399, 1361, 1330, 1263, 1190, 1117, and 960 cm^{-1}; NMR δ 1.72 (12*H*, m) and 4.75 (2*H*, m).

Glassware for the conversion of the above thionocarbonate to *trans*-cyclooctene was washed with dilute ammonium hydroxide and then with distilled water, and was dried at 140° to minimize acid-catalyzed isomerization or polymerization of the product. A 250-mL flask equipped with a side arm was charged with the above thionocarbonate (2.00 g, 0.0108 mol) and 25 mL of triisooctyl phosphite. The reaction flask was attached to a train consisting of a drying-tube filled with Ascarite, an efficient, large-bore trap cooled in liquid nitrogen, and a soap-bubble flowmeter. The reaction mixture was heated at 130° while a rapid stream of predried nitrogen (600 mL/min) was passed through the solution to remove product as it was formed. After 17 hours, the contents of the trap were vacuum-transferred to a tared receiver, affording 1.014 g (0.0092 mol, 85%) of (−)-*trans*-cyclooctene; n_D^{23} 1.4743; $[\alpha]_{578}^{21}$ − 443°, $[\alpha]_D^{21}$ − 423° (*c* 0.650%, methylene chloride); NMR δ 0.70 (2*H*, m), 1.0–2.4 (10*H*, m), and 5.28 (2*H*, m). Gas chromatography showed the product to be contaminated with 1% *cis*-cyclooctene.

1,3-Diphenylisobenzofuran Adduct of Tricyclo[3.3.3.0^{2,6}] undec-2(6)-ene (Preparation of a Strained Olefin with *in Situ* Trapping).[21] A solution of tricyclo[3.3.3.0^{2,6}]undecane-2,6-diol (455 mg, 2.5 mmol) in tetrahydrofuran (20 mL, distilled from lithium aluminum hydride) was cooled to 0° under argon. *n*-Butyllithium (2.07 mL of 2.17 *M* hexane solution, 4.49 mmol) was added slowly by syringe with stirring. After 5 minutes thiocarbonyldiimidazole (775 mg, 4.3 mmol) was added in portions with stirring. The reaction mixture was then refluxed for 5 hours, cooled, and poured into water. The water was extracted with methylene chloride (100 mL), and the organic phase was washed twice with water, dried, and evaporated to give a dark semisolid. This material was dissolved in benzene and filtered through a short Florisil column to give 461 mg (2.06 mmol, 82%) of crystalline thionocarbonate. Recrystallization from methylene chloride–heptane gave an analytical sample with mp 186.5–187°; IR (methylene chloride) 2930, 2880, 1475, 1340, 1305, 1290, 1208, 1197, and 1181 cm^{-1}; NMR ($CDCl_3$) δ 1.6–2.8 (m).

A portion of the above thionocarbonate (209 mg, 0.93 mmol) and 1,3-diphenylisobenzofuran (276 mg, 1.02 mmol) were dissolved in triethyl phosphite (4 mL). The solution was deoxygenated under water aspirator pressure and placed under a nitrogen atmosphere. The reaction mixture was refluxed for 36 hours, the volatiles were removed under vacuum, and the residue was dissolved in methylene chloride. A methylene chloride solution of maleic anhydride was added until the fluorescent yellow color of the excess trapping agent was discharged. The mixture was evaporated to dryness, dissolved in benzene, and then filtered through a short column of Florisil. The eluted material was recrystallized from methylene chloride–methanol to give the title compound (310 mg, 0.74 mmol, 80%) as colorless crystals, mp 245–246° dec; IR (methylene chloride) 2940, 2915, 2875, 1603, 1494, 1460, 1343, 1300, 1012, and 993 cm^{-1}; NMR ($CDCl_3$) δ 1.0–2.3 (15 *H*, m), 2.55 (1 *H*, m), 7.10 (4 *H*, m), and 7.3–7.9 (10 *H*, m).

3-*O*-Benzyl-5,6-dideoxy-1,2-*O*-isopropylidene-α-D-xylohex-5-enofuranose (Preparation of an Unsaturated Sugar).[166] 3-*O*-Benzyl-1,2-*O*-isopropylidene-α-D-glucofuranose (33.8 g, 0.109 mol) was placed in a 250-mL, round-bottomed flask, and 40 mL (36 g, 0.240 mol) of ethyl orthoformate and 2 mL of acetic acid were added. The mixture was stirred magnetically and heated for 6 hours at reflux temperature. Excess reagents were then removed by evaporation under diminished pressure at 80–100°. Three 100-mL portions of toluene were evaporated from the product to remove traces of reagents, and the resultant mixture of diastereoisomeric orthoformic esters was used in the next step without further purification.

Triphenylacetic acid (0.5 g) was added to the syrupy mixture of *ortho* esters, and the flask was fitted with a take-off condenser. The mixture was stirred and heated for 6 hours at 170°, during which time ethanol distilled from the mixture. The resultant product was dissolved in 300 mL of ether; solid sodium bicarbonate was added, and the mixture was kept overnight. The mixture was filtered, and the filtrate was washed twice with saturated aqueous sodium bicarbonate, dried with anhydrous magnesium sulfate, and evaporated. The product was distilled at 0.2 torr, and the pure alkene was collected over the range 124–128°; yield 29.1 g (0.105 mol, 96%); bp 111–115°/0.05 torr; $[\alpha]_D^{20}$ − 66° (*c* 3%, ethanol); TLC, R_f 0.40 (silica gel G, 3:1 v/v cyclohexane–diethyl ether).

***trans*-Cyclooctene (Acid-Catalyzed 2-Methoxy-1,3-dioxolane Decomposition).**[96] *trans*-Cyclooctane-1,2-diol (385 mg, 2.7 mmol), methyl orthoformate (0.5 mL, 4.5 mmol), and benzoic acid (20 mg, 0.2 mmol) were heated at 90–100° for 2 hours. The methanol and excess orthoformate were then removed *in vacuo*. Benzoic acid (50 mg, 0.4 mmol) was added, and the mixture was heated at 160–170°. The olefin thus produced was distilled, as were carbon dioxide and methanol. The distillate was placed in dichloromethane, washed with aqueous sodium bicarbonate, and dried (sodium sulfate). Concentration gave crude *trans*-cyclooctene quantitatively. Distillation at 90–100° (bath temperature)/95 mm gave the title compound (155 mg, 1.4 mmol, 52%), which showed a single peak on gas chromatography; IR (neat) 3020, 1650, and 982 cm^{-1}.

(*Z*)-6-Methyl-5-undecene (Acetic-Anhydride-Promoted Decomposition of a 2-Dimethylamino-1,3-dioxolane).[97] A mixture of (5*R*,6*S*)-6-methylundecane-5,6-diol (606 mg, 3.0 mmol) and *N*,*N*-dimethylformamide dimethyl acetal (3.0 mL) was stirred at room temperature for 17 hours. After the volatile material was removed under vacuum, the resulting dioxolane derivative was heated in acetic anhydride (3.0 mL) at 130° for 17 hours. The reaction mixture was allowed to cool to room temperature and was then poured into water. The product was extracted with *n*-pentane. The organic phase was washed with saturated aqueous sodium bicarbonate, saturated brine, and water and then dried (with sodium sulfate). Concentration of the filtrate and column chromatography using *n*-pentane as an eluent gave (*Z*)-6-methyl-5-undecene (43 mg, 2.4 mmol, 80% yield) as a clear oil, bp 130° (bath temperature, 24 mm); TLC, R_f 0.80 (*n*-hexane); IR (neat) 1660–1675 (w), 1385 (m), and 840 cm^{-1} (w); NMR (CCl_4) δ 1.65 (3*H*, s) and 5.03 (1*H*, bt).

***trans*-Cyclooctene (2-Phenyl-1,3-dioxolane Anionic Fragmentation Route).**[115] A mixture of *trans*-cyclooctane-1,2-diol (49.9 g, 0.35 mol), benzaldehyde (73 g, 0.60 mol; freshly redistilled), and concentrated sulfuric acid (0.5 mL) in toluene (200 mL) was heated under reflux (Dean–Stark trap under nitrogen) for 12 hours. After addition of sufficient solid potassium carbonate to neutralize the sulfuric acid, the toluene and benzaldehyde were removed *in vacuo*. The residue was distilled to give 10-phenyl-9,11-*trans*-dioxabicyclo[6.3.0]undecane (72 g), bp 148–150°/0.5 mm, which slowly solidified. It was obtained as needles (60.8 g, 0.26 mol, 75%), mp 44–45° (from 100 mL of light petroleum); NMR δ 1.0–2.4 (12*H*, m), 3.9–4.1 (2*H*, m), 5.88 (1*H*, s) and 7.1–8.7 (5*H*, m).

n-Butyllithium in hexane (37 mL, 0.062 mol) was added by syringe to an ice-cooled solution of the above dioxolane (7.2 g, 0.031 mol) in light petroleum (30 mL) under nitrogen. After 14 hours at 20° the mixture was added to water

(200 mL) and the product was isolated with light petroleum. The extracts were washed with aqueous silver nitrate (3 × 50 mL; 10%), and the combined aqueous extracts were poured onto ammonia (50 mL) and ice. The hydrocarbon liberated was isolated with light petroleum (bp 30–40°). Removal of solvent gave *trans*-cyclooctene (2.48 g, 0.023 mol, 73%), homogeneous by gas chromatographic analysis (polyethylene glycol column); IR (film) 1645, 1195, 990, 935, 850, 825, and 792 cm^{-1}.

***cis*-1,2-Dimethylcyclododec-1-ene (Reductive Elimination of Cyclic Phosphoramidate).**[121] To a stirred solution of 500 mg (2.2 mmol) of *cis*-1,2-dimethylcyclododecane-1,2-diol in 50 mL of tetrahydrofuran was added 2.3 mL of 2.1 *M* ethereal methyllithium (4.8 mmol). After 30 minutes, 15 mL of hexamethylphosphoramide was added. Stirring was continued for 15 minutes, and 0.60 g (3.73 mmol) of *N*,*N*-dimethylamidophosphorodichloridate[168] in 10 mL of tetrahydrofuran was added dropwise. After the mixture was allowed to stir overnight, water was added and the product was extracted with ethyl acetate. Removal of solvent afforded a white solid that was recrystallized from *n*-hexane to give 570 mg (88%) of the cyclic *N*,*N*-dimethylphosphoramidate of *cis*-1,2-dimethylcyclododecane-1,2-diol, mp 131–138°; NMR ($CDCl_3$) δ 2.75 [6*H*, d, $J = 10$ Hz, $N(CH_3)_2$] and 1.47 (6*H*, s, CH_3).

To a stirred solution of the above compound (1.2 g, 0.0037 mol) in 10 mL of tetrahydrofuran and 100 mL of liquid ammonia was added 200 mg (30 mmol) of lithium wire in small pieces. After 1.5 hours, 1 g of ammonium chloride was added and the ammonia was allowed to evaporate. Water was added, and the product was extracted with diethyl ether to give 710 mg (3.66 mmol, 98% yield) of a 90:10 mixture of *cis*- and *trans*-1,2-dimethylcyclododec-1-ene according to gas chromatographic analysis (peak enhancement) and spectral comparison with authentic samples.

***cis*-1,2-Dimethylcyclododec-1-ene (Deoxygenation with Potassium Hexachlorotungstate).**[129,131] Tungsten hexachloride (59.7 g, 0.151 mol) and potassium iodide (49.90 g, 0.300 mol, ground and dried for 4 hours at 120° under vacuum) were combined in a 250-mL round-bottomed flask. The flask was evacuated and placed in a silicon oil bath at 130° for 3.5 days. The solid, now deep red–purple, was transferred to a 200-mL flask, leaving most of the iodine as a cake on the walls and in the gas inlet joint of the 250-mL flask. The 200-mL flask was placed in a silicon oil bath at 230° for about 2 hours while evacuating through a liquid nitrogen trap. The remaining dark red–purple solid, which pulverized easily, weighed 71.28 g (0.150 mol, 100% yield).

cis-1,2-Dimethyl-1,2-cyclododecanediol (2.104 g, 9.22 mmol) was dissolved in 100 mL of tetrahydrofuran in a 250-mL three-necked flask equipped with reflux condenser, magnetic stirrer, and nitrogen atmosphere, and flamed out while under vacuum prior to use. This was titrated (1,10-phenanthroline indicator) with 1.6 *N* methyllithium in diethyl ether (ca. 12 mL), and then dipotassium hexachlorotungstate (8.98 g, 0.0189 mol) was added. The reaction mixture was heated under reflux for 4 hours and added to 75 mL of water, which was

1.5 *M* in sodium tartrate and 2.0 *M* in sodium hydroxide. This was shaken vigorously and extracted four times with a total of 200 mL of *n*-hexane. The combined organic layers were concentrated under vacuum, dried over sodium sulfate, and further evaporated to an oil. This was taken up in *n*-pentane and passed through a silica gel column (4.5 × 15 cm, *n*-pentane) with monitoring of fractions by GC on a UCW 98 column at 160°. Concentration yielded *cis*-1,2-dimethylcyclododec-1-ene (1.04 g, 5.36 mmol, 58%), which contained about 2% of an impurity and 5% of *trans*-1,2-dimethylcyclododec-1-ene (by GC on Carbowax 20M-silver nitrate column at 110°). An analytical sample prepared by preparative GC showed IR (neat) 2970, 2920, 2850, 1460, 1455, 1440, 1373, 1340, 1215, 1190, 1155, 810, and 720 cm^{-1}; NMR ($CDCl_3$) δ 1.37 (16 *H*, br s), 1.61 (6 *H*, s), 2.08 (4 *H*, br t).

Cyclohexylidenecyclohexane (Deoxygenation with Titanium Reagent).[123] A stirred slurry of titanium trichloride (1.103 g, 7.15 mmol) in 40 mL of dry tetrahydrofuran was refluxed for 1.5 hours with potassium (1.3 g, 0.0332 mol) under an inert atmosphere. The black suspension was then cooled slightly, and bicyclohexyl-1,1′-diol (356 mg, 1.8 mmol) in 4 mL of tetrahydrofuran was added. After a further 16 hours at reflux, the reaction mixture was cooled and quenched by the slow addition of methanol. The quenched mixture was diluted with water and extracted with diethyl ether, and the diethyl ether layer was dried and concentrated. Column chromatography of the residue afforded cyclohexylidenecyclohexane (251 mg, 1.5 mmol, 85% yield), mp 52–53°.

Methyl 4,6-*O*-Benzylidene-2,3-dideoxy-α-D-*erythro*-hex-2-enopyranoside (Deoxygenation by the Nucleophilic Displacement Procedure).[166] A vigorously stirred mixture of methyl 4,6-*O*-benzylidene-2,3-di-*O*-*p*-tolylsulfonyl-α-D-glucopyranoside (60 g, 0.102 mol), sodium iodide (750 g, 5 mol), zinc dust (360 g, 5.5 mol), and 1.4 L of *N*,*N*-dimethylformamide was refluxed for 2.5 hours. The mixture was cooled below 100°, and 1 L of water was added. After the mixture was cooled to room temperature, it was mixed with 1 L of dichloromethane and the mixture was filtered. The organic layer was separated; the aqueous layer was extracted twice with 200-mL portions of dichloromethane, and the combined organic extracts were washed twice with 1-L portions of water. The dried (magnesium sulfate) organic phase was evaporated to give the title compound mixed with a small amount of starting material. Diethyl ether (400 mL) was added to dissolve the product, and the mixture was filtered to remove most of the residual starting material. Evaporation of the filtrate and recrystallization of the residue from absolute methanol gave the pure title compound (13 g, 0.052 mol, 52%), mp 117–118°; $[\alpha]_D^{22}$ + 130° (*c* 0.5% in chloroform).

Methyl 2,3,4,6-Tetradeoxy-2,6-diethoxycarbonylamino-α-D-*erythro*-hex-3-enopyranoside (Reductive Elimination of a Vicinal Dimesylate).[153] A mixture of sodium (94 mg, 4.1 mmol) and naphthalene (520 mg, 4.1 mmol) in 8 mL of tetrahydrofuran was stirred at room temperature for 2.5 hours under nitrogen. A portion (4 mL) of this solution was added dropwise to a solution of the 3,4-dimesylate of methyl 2,6-dideoxy-2,6-diethoxycarbonylamino-α-D-glucopyranoside (100 mg, 0.2 mmol) in 3 mL of tetrahydrofuran. The green color of the solution was maintained throughout addition. The mixture was stirred for 15 minutes, treated successively with 0.3 mL of methanol and a small amount of water, and evaporated under vacuum to dryness. The residue was extracted with chloroform several times, and the combined extracts were dried and evaporated to give a mixture of the product and naphthalene that was placed on a column of silica gel with benzene. After the column was washed with benzene, the residue was eluted with chloroform. Evaporation of the solvent and recrystallization of the residue from ethanol gave 40 mg (0.13 mmol, 66%) of the title compound; mp 141–142° (prisms); NMR ($CDCl_3$) δ 3.28 (3 *H*, s, OCH_3), 5.72 (2 *H*, s, H-3,4).

Methyl 2,6-Dibenzyloxycarbonylamino-2,3,4,6-tetradeoxy-3-eno-α-D-glycopyranoside (Decomposition of a Vicinal Dixanthate Ester Induced by Tin Hydride).[155] To a solution of methyl 2,6-dibenzyloxycarbonylamino-2,6-dideoxy-α-D-glucopyranoside (2 g, 4.4 mmol) and carbon disulfide (5 mL) in dimethyl sulfoxide (10 mL) was added dropwise 5 *N* sodium hydroxide (5 mL) at 15° under nitrogen. After the solution had stirred for 20 minutes, methyl iodide (10 mL) was added portionwise to the resulting red solution and stirring was continued for 45 minutes. The solvent was evaporated under vacuum and the residue was extracted with ethyl acetate. The extract was washed with brine, dried, and evaporated, giving 2.92 g of oil. Column chromatography afforded 2.7 g (4.3 mmol, 97%) of methyl 2,6-dibenzyloxycarbonylamino-2,6-dideoxy-3,4-di-*O*-[(methylthio)thiocarbonyl]-α-D-glucopyranoside as an oil; IR (KBr) 3450, 1730, 1520, 1220, and 1060 cm^{-1}; NMR ($CDCl_3$) δ 2.40 and 2.48 (3 *H*, s, each), 3.32 (3 *H*, s), 5.05 (4 *H*, s), 5.20 (2 *H*, s), and 7.28 (10 *H*, s).

To a refluxing toluene (7 mL) solution of the above compound (1.3 g, 2.1 mmol) was added, over a period of 1 hour, a solution of tributylstannane (2.7 g, 9.3 mmol) in 10 mL of toluene. The mixture was refluxed for another hour, and, after cooling, the solvent was evaporated under vacuum. The residue was purified by chromatography, giving 690 mg (1.6 mmol, 77%) of the title compound as an amorphous powder, mp 147° (from methanol). An analytical sample was obtained by further purification by preparative TLC; IR (KBr) 3320, 1690, 1540, 1300, 1270, 1240, 1040, and 1010 cm^{-1}; NMR ($CDCl_3$) δ 3.40 (3 *H*, s), 4.82 (1 *H*, d J = 4 Hz), 5.14 (4 *H*, s), 5.68 (2 *H*, s), and 7.35 (10 *H*, s).

Methyl 2,6-Di-*O*-benzyl-3,4-dideoxy-α-D-*erythro*-hex-3-enopyranoside (Phosphorus Reagent Procedure).[158] A mixture of methyl 2,6-di-*O*-benzyl-α-D-glycopyranoside (0.830 g, 2.2 mmol), triphenylphosphine (2.3 g, 8.8 mmol), imidazole (0.032 g, 4.4 mmol), and triiodoimidazole (1.58 g, 3.5 mmol) was heated under reflux for 3 hours. The mixture was allowed to cool, toluene (50 mL) was added, and the toluene layer was decanted into a stirred solution of saturated aqueous sodium bicarbonate (300 mL). The residue was dissolved in a little acetone, and the resulting solution was also added to the sodium bicarbonate solution. The mixture was stirred for 10 minutes and then transferred to a separatory funnel. The organic phase was washed with aqueous sodium thiosulfate, aqueous sodium bicarbonate, and water, and was then dried with magnesium sulfate, filtered, and concentrated. The residue was dissolved in toluene that was gently warmed to dissolve triphenylphosphine oxide. The solution was chromatographed [Merck aluminum oxide 90 (II–III); toluene] to afford, after concentration, 0.72 g (2.1 mmol, 95%) of the title compound; $[\alpha]_D^{22}$-30° (chloroform); NMR ($CDCl_3$) δ 3.48 (OCH_3), 3.49 (H-6, H-6′), 4.1–4.2 (H-2), 4.25–4.35 (H-5), 4.8 (H-1), 5.76–5.77 (H-3, H-4), $J_{H\text{-}1,H\text{-}2} = 3.7$ Hz, $J_{H\text{-}5,H\text{-}6} = 5.1$ Hz.

In an alternative procedure the triiodoimidazole was replaced with iodoform (2 equivalents per equivalent of starting diol).[159] If the above experiments involving triphenylphosphine are conducted on a larger scale, it is advisable to precipitate the bulk of the triphenylphosphine oxide at the end of the reaction by adding diethyl ether to the toluene solution before proceeding to the chromatographic purification.

TABULAR SURVEY

An attempt has been made to include all of the published examples through December 1982 of the deoxygenation of vicinal diols to olefins involving cyclic species **1**. It is likely that some references have been overlooked because of

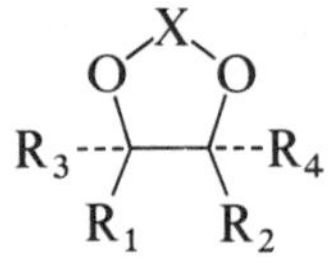

X
O O
R_3 R_4
R_1 R_2
1

difficulties in searching the literature. For comparison purposes, examples of vicinal diol deoxygenations *not* involving cyclic intermediates have also been included. In older methods involving reactions of vicinal disulfonate esters, the tabulation is almost certainly incomplete.

The diols or polyols are listed according to increasing carbon number with diol and olefin stereochemistry indicated when known. The methods used to effect deoxygenation are symbolized by the letters A–L, which are defined as follows, referring to the corresponding sections in the text:

A. Desulfurization of 1,3-dioxolane-2-thiones (Corey–Winter reaction and modifications thereof).

B. Decomposition of 2-ethoxy-, 2-acyloxy-, and 2-dialkylamino-1,3-dioxolanes and related cyclic orthoesters (Eastwood reaction and modifications thereof).

C. Base-induced fragmentation of 2-phenyl-1,3-dioxolanes.
D. Reduction-elimination of cyclic phosphates and phosphoramidates.
E. Transition-metal-based deoxygenations (titanium, tungsten).
F. Pyrolysis of norbornadienone and quadricyclanone ethylene ketals and related routes to 2-carbena-1,3-dioxolanes (discussed under A).
G. Deoxygenation involving cyclic sulfoxylates.
H. Nucleophilic displacement reactions (Tipson–Cohen procedure).
I. Reduction–elimination reactions of vicinal disulfonate esters.
J. Tin-hydride-induced decomposition of vicinal dixanthate esters.
K. Pyrolysis at 200–240° with oxalic acid (discussed under B).
L. Reactions involving phosphorus halides or hydrogen fluoride.

The reaction conditions, when available, have been summarized. Product ratios and/or overall percentage yields are recorded; a dash (—) indicates that no yield was given in the reference. When there is more than one reference, the experimental data are taken from the first one, and the remaining references are listed in numerical order.

The abbreviations used in the table are:

Ac	acetyl
Ar	aryl
Bz	benzyl
COD	1,5-cyclooctadiene
DBPD	1,3-dibenzyl-2-phenyl-1,3,2-diazaphospholidine
Diglyme	bis(2-methoxyethyl) ether
DMAP	4-dimethylaminopyridine
DME	1,2-dimethoxyethane
DMF	*N,N*-dimethylformamide
DMPD	1,3-dimethyl-2-phenyl-1,3,2-diazaphospholidine
DMSO	dimethyl sulfoxide
Ether	diethyl ether
HMPA	hexamethylphosphoramide
IMID	imidazole
Ms	methanesulfonyl
Py	pyridine
TCDI	*N,N*-thiocarbonyldiimidazole
THF	tetrahydrofuran
TMD	triiodoimidazole
Ts	*p*-toluenesulfonyl

TABLE I. OLEFIN SYNTHESIS BY DEOXYGENATION OF VICINAL DIOLS

C_n	Diol	Method	Conditions	Products and Yields (%)	Refs.
C_2	$HOCH_2CH_2OH$	F	1. Quadricyclone, $(C_2H_5O)_3CH$, TsOH, CH_2Cl_2 2. 200°	C_2H_4 (ca. 30)	8
		C	1. C_6H_5CHO 2. C_6H_5Li, ether, 25°	C_2H_4 (ca. 85)	105,106
		C	1. ArCHO 2. RLi, 60°	C_2H_4 (—)	107
		B	1. $(C_2H_5O)_3CH$ 2. CH_3CO_2H 3. Neat, 120–130°	C_2H_4 (29)[a]	87
C_3	$HOCH_2CHOHCH_2Cl$	B	1. $(C_2H_5O)_3CH$ 2. CF_3COCF_3 3. 190–200°, 3 hr	$CH_2{=}CHCH_2Cl$ (—)	102
		B	1. $(C_2H_5O)_3CH$ 2. CH_3CO_2H 3. Neat, 120–140°	$CH_2{=}CHCH_2Cl$ (65)[a]	87
	$HOCH_2CHOHCH_3$	B	1. $(C_2H_5O)_3CH$ 2. CH_3CO_2H 3. Neat, 110–120°	$CH_3CH{=}CH_2$ (42)[a]	87
	$HOCH_2CHOHCH_2OH$	B	1. $(C_2H_5O)_3CH$, 120–140°, 3 hr 2. 200–220°, 4 hr	$CH_2{=}CHCH_2OH$ (72)	86
C_4	(I) [cyclobutane-1,2-diol with R_1, R_2 on C-3]			(II) [cyclobutene with R_1, R_2 on C-3]	
	I, $R_1 = R_2 = H$	A	1. TCDI, $C_6H_5CH_3$, 110° 2. $(CH_3O)_3P$, 110°, 24 hr	II, $R_1 = R_2 = H$ (68)	83
	I, $R_1 = H$, $R_2 = CH_3$	A	1. TCDI, $C_6H_5CH_3$, 110° 2. $(n\text{-}C_4H_9O)_3P$, 110°, 24 hr	II, $R_1 = H$, $R_2 = CH_3$ (57)	83
	I, $R_1 = H$, $R_2 = C_2H_5$	A	"	II, $R_1 = H$, $R_2 = C_2H_5$ (62)	83
	I, $R_1 = R_2 = CH_3$	A	"	II, $R_1 = R_2 = CH_3$ (53)	83
	I, $R_1 = H$, $R_2 = C_3H_7\text{-}i$	A	"	II, $R_1 = H$, $R_2 = C_3H_7\text{-}i$ (60)	83
	I, $R_1 = H$, $R_2 = C_4H_9\text{-}t$	A	"	II, $R_1 = H$, $R_2 = C_4H_9\text{-}t$ $CH_2{=}CHC(C_4H_9\text{-}t){=}CH_2$	83

Substrate	Method	Conditions	Product (% yield)	Refs.
I, $R_1 = R_2 = -CH_2CH_2-$	A	1. TCDI, $C_6H_5CH_3$, 110° 2. $(n\text{-}C_4H_9O)_3P$, 115–120°, 44 hr	II, $R_1 = R_2 = -CH_2CH_2-$ (63)	36
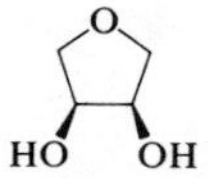 HO OH	K	$(CO_2H)_2$, 200–240°, 44 hr	 (58)	88
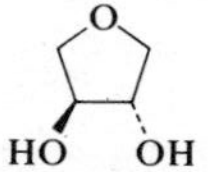 HO OH	K	$(CO_2H)_2$, 200–240°	" (47)	88
meso-$CH_3CHOHCHOHCH_3$	A	1. TCDI, $C_6H_5CH_3$, 110° 2. $(C_2H_5O)_3P$, 150°, 4 hr	*cis*-$CH_3CH{=}CHCH_3$ (59)	10
	I	1. CH_3SO_2Cl, Py. 0° 2. Na, naphthalene, −50°	$CH_3CH{=}CHCH_3$ (62–71) (*cis*:*trans*, 1:3)	152
	K	$(CO_2H)_2$, 200–240°	$CH_3CH{=}CHCH_3$ (70) (*cis*:*trans*, 2:5)	88
d(−)-$CH_3CHOHCHOHCH_3$	A	1. TCDI, $C_6H_5CH_3$, 110° 2. $(C_2H_5O)_3P$, 150°, 6 hr	*trans*-$CH_3CH{=}CHCH_3$ (68)	10
	A	1. TCDI, $C_6H_5CH_3$, 110° 2. *hv* (254 nm), $(n\text{-}C_4H_9O)_3P$, 25°, 4 hr	$CH_3CH{=}CHCH_3$ (67) (*cis*:*trans*, 1:3)	10
d,l-$CH_3CHOHCHOHCH_3$	F	1. Hexachlorocyclopentadiene, KOH, 44 hr 2. $C_6H_5C{\equiv}CH$ 3. 140–150°	*trans*-$CH_3CH{=}CHCH_3$ (—)	9
	I	1. CH_3SO_2Cl, Py, 0° 2. Na, naphthalene, 25°	$CH_3CH{=}CHCH_3$ (46–52) (*cis*:*trans*, 29:71)	152
$CH_3CHOHCHOHCH_3$[a]	B	1. $(C_2H_5O)_3CH$ 2. CH_3CO_2H 3. Neat, 120–125°	$CH_3CH{=}CHCH_3$[b] (65)[a]	87
$HOCH_2CH_2CHOHCH_2OH$	B	1. $(C_2H_5O)_3CH$, 130–150°, 1 hr 2. 200°	$CH_2{=}CHCH_2CH_2OH$ (85)	86
$HOCH_2CHOHCHOHCH_2OH$	H	1. TsCl, Py 2. NaI, acetone, reflux, 3 hr	$CH_2{=}CHCH{=}CH_2$ (—)	135
C_5				
HO OH	K	$(CO_2H)_2$, 200–240°	Cyclopentene (—)	88

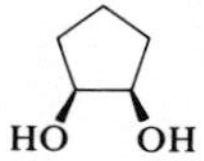

TABLE I. OLEFIN SYNTHESIS BY DEOXYGENATION OF VICINAL DIOLS (*Continued*)

C_n	Diol	Method	Conditions	Products and Yields (%)	Refs.
C_5	HO OH (cyclopentane-1,2-diol)	K	$(CO_2H)_2$, 200–240°	Cyclopentene (—)	88
	HO, OH, OH (tetrahydropyran triol)	B	1. $(C_2H_5O)_3CH$ 2. CH_3CO_2H 3. 200°	HO (dihydropyran) (72)	169
	$HOCH_2$ (lactone) HO OH	B	1. $(C_2H_5O)_3CH$, THF, reflux, 12 hr 2. 220°, 40 torr	$HOCH_2$ (butenolide) (68)	165
	CH_2OH (tetrahydrofuran) HO OH	H	1. TsCl, Py, 0°, 24 hr 2. C_6H_5COCl, Py 3. NaI, Zn, DMF, 170°	$CH_2OCOC_6H_5$ (dihydrofuran) (65)	140
	$HOCH_2CH_2CH_2CHOHCH_2OH$	B	1. $(C_2H_5O)_3CH$, 130–160°, 1 hr	$CH_2=CHCH_2CH_2CH_2OH$ (79)	86
C_6	OH, OH (oxabicyclic diol)	A	1. TCDI 2. $(C_2H_5O)_3P$, 110°, 120 hr	(oxabicyclic diene) (60)[a]	82
	$CH_2=CHCHOHCHOHCH=CH_2$ (*meso*:*d*,*l* 3:2)	A	1. TCDI, C_6H_6, reflux, 30 min 2. $(C_6H_5O)_3P$, 50°, 96 hr	$CH_2=CHCH=CHCH=CH_2$ (I) (63) (*cis*:*trans*, 1:4)	80
		H	1. TsCl, Py 2. NaI, $C_2H_5OCH_2CH_2OCH_2CH_2OH$, 100°, 62 hr	I (68) (*cis*:*trans*, 1:9)	80
	CH_3OCH_2 (lactone) HO OH	B	1. $(C_2H_5O)_3CH$, THF, reflux, 12 hr 2. 176°, 14 hr	CH_3OCH_2 (butenolide) (67)	165

Reactant	Method	Reagents and Conditions	Product(s) (Yield, %)	Refs.
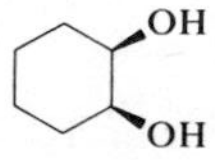 cis-1,2-cyclohexanediol	A	1. TCDI 2. $(C_4H_9O)_3P$, 200°, 24 hr	Cyclohexene (81)	17
	A	1. TCDI 2. $(i\text{-}C_8H_{17}O)_3P$, 150°, 24 hr	Cyclohexene (63)	10
	B	1. $(CH_3)_2NCH(OCH_3)_2$, CH_2Cl_2, 25° 2. CH_3I, 100°	" (90)	90
	B	1. $(C_2H_5O)_3CH$, 110–135°, 1.5 hr 2. 160°	" (75)	86
	K	$(CO_2H)_2$, 200–240°	" (19)	88
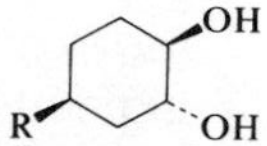 (I)			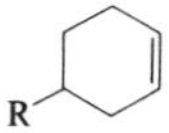 (II)	
I, R = H	A	1. TCDI 2. $Fe(CO)_5$, xylene, reflux, 2 hr	II, R = H (11)[a]	56
I, R = H	B	1. $(C_2H_5O)_3CH$, 160°, 12 hr 2. 200–220°, 3 hr	No reaction	86
I, R = H	K	$(CO_2H)_2$, 200–240°	II, R = H (11)	88
I, R = H	I	1. MsCl, Py, 0° 2. Na, naphthalene, 25°	II, R = H (75)	152
I, R = C_4H_9-*t*	I	1. MsCl, Py, 0° 2. Na, naphthalene, 25°	II, R = C_4H_9-*t* (39–44)	152
I, R = C_4H_9-*t*	I	1. MsCl, Py, 0° 2. Na, anthracene, 25°	II, R = C_4H_9-*t* (64–73)	152
$(CH_3)_2COHCOH(CH_3)_2$	A	1. KOH, CS_2; CH_3I 2. $(C_2H_5O)_3P$, 150°, 46 hr	$(CH_3)_2C{=}C(CH_3)_2$ (29)	10
	A	1. KOH, CS_2; CH_3I 2. *hν* (254 nm), $(C_4H_9O)_3P$, 100°, 2.7 hr	" (32)	10
	A	1. *n*-C_4H_9Li, THF 2. CS_2, heat; CH_3I, heat 3. $Fe(CO)_5$, $C_6H_5CH_3$, 100°, 24 hr	" (3)	26
	B	1. $(C_2H_5O)_3CH$, 125–140° 2. HOAc, 150–160°, 5 hr	" (64–92)	86
	B	1. $(C_2H_5O)_3CH$; HOAc 2. 130–140°	" (93)[a]	87

TABLE I. OLEFIN SYNTHESIS BY DEOXYGENATION OF VICINAL DIOLS (*Continued*)

C_n	Diol	Method	Conditions	Products and Yields (%)	Refs.
C_6	$(CH_3)_2COHCOH(CH_3)_2$	F	1. $Cl_2C{=}N^+(CH_3)_2Cl^-$ 2. $TsNHNH_2$, $(C_2H_5)_3N$, THF, reflux, 2 hr 3. NaH; 290°, vacuum	" (50)	11
	$(CH_3)_2COHCOH(CD_3)_2$	B	1. $(C_2H_5O)_3CH$ 2. HOAc, 175°, 8 hr	$(CH_3)_2C{=}C(CD_3)_2$ (54)	170
	$(CH_3)CD_3COHCOH(CH_3)CD_3$	B	1. $(C_2H_5O)_3CH$ 2. HOAc, 175°, 8 hr	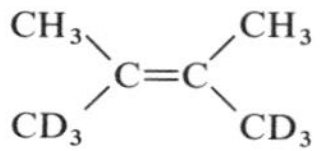 $CH_3(CD_3)C{=}C(CH_3)CD_3$ (56)	170
	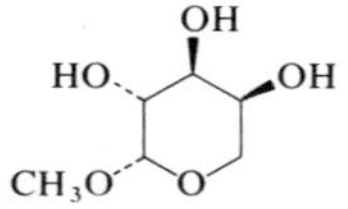 HO, OH, OH, CH_3O, O (pyranoside)	B	1. $(C_2H_5O)_3CH$ 2. HOAc 3. 200°	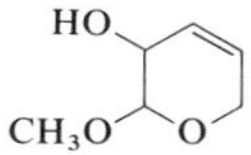 HO, CH_3O, O (dihydropyran) (39)	169
	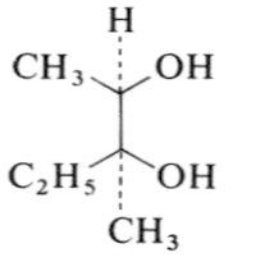 I: H, CH_3, OH / C_2H_5, OH, CH_3 + II: CH_3, H, OH / C_2H_5, OH, CH_3	A	1. n-C_4H_9Li; CS_2; CH_3I 2. $(CH_3O)_3P$, heat	(*Z*)-3-Methyl-2-pentene (III), (*E*)-3-methyl-2-pentene (IV) III + IV (—) (III:IV, 99:5)	161
		B	1. $(C_2H_5O)_3CH$ 2. HOAc, heat	III + IV (—) (III:IV, 99:1)	161
	(I:II, 99:1)	C	1. C_6H_5CHO, H^+ 2. n-C_4H_9Li	III + IV (—) (III:IV, 99:1)	161
	I + II (I:II, 1:99)	A	1. n-C_4H_9Li; CS_2; CH_3I 2. $(CH_3O)_3P$, heat	III + IV (—) (III:IV, 4:96)	161
		B	1. $(C_2H_5O)_3CH$ 2. HOAc, heat	III + IV (—) (III:IV, 2:98)	161
	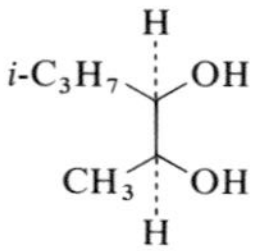 H, i-C_3H_7, OH / CH_3, OH, H	A	1. TCDI 2. $Ni(COD)_2$, DMF, 65°, 46 hr	(*Z*)-4-Methyl-2-pentene (I), (*E*)-4-methyl-2-pentene (II) I (79)[a] + II (<1)[a]	55
		F	1. $Cl_2C{=}\overset{+}{N}(CH_3)_2Cl^-$ 2. $TsNHNH_2$, $(C_2H_5)_3N$, THF 3. NaH; 290°, vacuum	I + II (—) (I:II, 1:1.2)	12

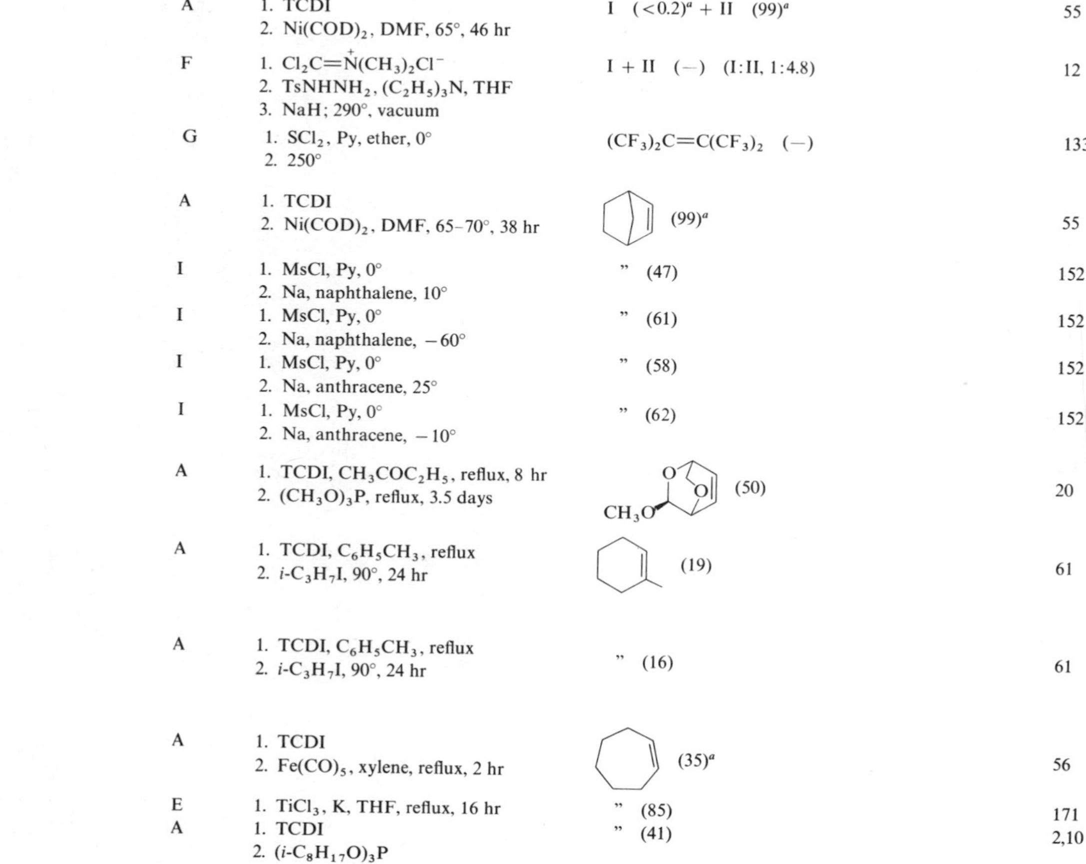

Diol	Method	Conditions	Product(s) and Yield(s) (%)	Refs.
i-C_3H_7CH(OH)CH(OH)CH_3	A	1. TCDI 2. $Ni(COD)_2$, DMF, 65°, 46 hr	I $(<0.2)^a$ + II $(99)^a$	55
	F	1. $Cl_2C{=}\overset{+}{N}(CH_3)_2Cl^-$ 2. $TsNHNH_2$, $(C_2H_5)_3N$, THF 3. NaH; 290°, vacuum	I + II (—) (I:II, 1:4.8)	12
$(CF_3)_2COHCOH(CF_3)_2$	G	1. SCl_2, Py, ether, 0° 2. 250°	$(CF_3)_2C{=}C(CF_3)_2$ (—)	133
C_7	A	1. TCDI 2. $Ni(COD)_2$, DMF, 65–70°, 38 hr	$(99)^a$	55
	I	1. MsCl, Py, 0° 2. Na, naphthalene, 10°	" (47)	152
	I	1. MsCl, Py, 0° 2. Na, naphthalene, −60°	" (61)	152
	I	1. MsCl, Py, 0° 2. Na, anthracene, 25°	" (58)	152
	I	1. MsCl, Py, 0° 2. Na, anthracene, −10°	" (62)	152
	A	1. TCDI, $CH_3COC_2H_5$, reflux, 8 hr 2. $(CH_3O)_3P$, reflux, 3.5 days	(50)	20
	A	1. TCDI, $C_6H_5CH_3$, reflux 2. i-C_3H_7I, 90°, 24 hr	(19)	61
	A	1. TCDI, $C_6H_5CH_3$, reflux 2. i-C_3H_7I, 90°, 24 hr	" (16)	61
	A	1. TCDI 2. $Fe(CO)_5$, xylene, reflux, 2 hr	$(35)^a$	56
	E	1. $TiCl_3$, K, THF, reflux, 16 hr	" (85)	171
	A	1. TCDI 2. $(i$-$C_8H_{17}O)_3P$	" (41)	2,10

TABLE I. Olefin Synthesis by Deoxygenation of Vicinal Diols (*Continued*)

C_n	Diol	Method	Conditions	Products and Yields (%)	Refs.
C_7		A	1. TCDI 2. $(CH_3O)_3P$, 110°, reflux, 24 hr	 [structure][c,d] (71)	2,10,17
		A	1. TCDI 2. $(i\text{-}C_8H_{17}O)_3P$, N_2 sweep	"[c,e] (13)	17
	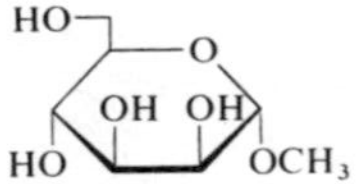 [structure: HO, OH, OH, HO, OCH_3]	C	1. C_6H_5CHO 2. $n\text{-}C_4H_9Li$, 2 eq, THF, −30°	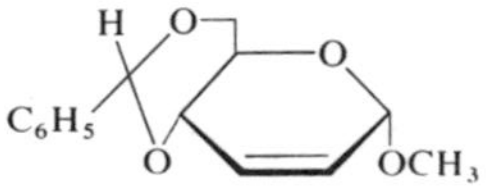 [structure: H, C_6H_5, O, O, O, OCH_3] (1)	116
	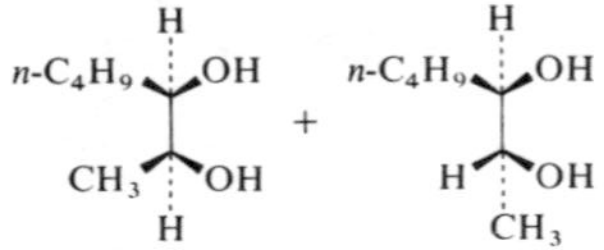 [structures I + II: $n\text{-}C_4H_9$, H, OH, CH_3] (I:II, 99:1)	A	1. $n\text{-}C_4H_9Li$; CS_2; CH_3I 2. $(CH_3O)_3P$, heat	*cis*-2-Heptene (III), *trans*-2-heptene (IV), (—) (III:IV, 96:4)	161
		B	1. $(C_2H_5O)_3CH$ 2. HOAc, heat	III + IV (—) (6:94)	161
	I + II (I:II, 4:96)	A	1. $n\text{-}C_4H_9Li$; CS_2; CH_3I 2. $(CH_3O)_3P$, heat	III + IV (—) (6:94)	161
		C	1. C_6H_5CHO, H^+ 2. C_4H_9Li	III + IV (—) (5:95)	161
		B	1. $(C_2H_5O)_3CH$ 2. HOAc, heat	III + IV (—) (5:95)	161
C_8	$C_6H_5CHOHCH_2OH$	B	1. $(C_2H_5O)_3CH$, $C_6H_5CO_2H$, heat 2. 140–150°, 5 hr	$C_6H_5CH{=}CH_2$ (52)	172
	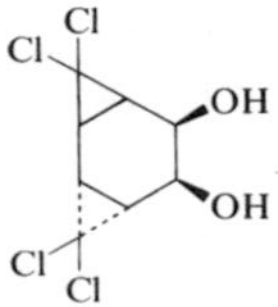 [structure: Cl, Cl, OH, OH, Cl, Cl]	A	1. TCDI, $C_6H_5CH_3$, reflux, 2 hr 2. $(CH_3O)_3P$, reflux, 80 hr	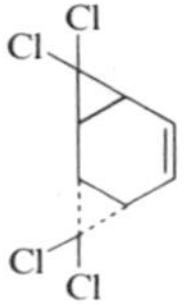 [structure: Cl, Cl, Cl, Cl] (36)	172a

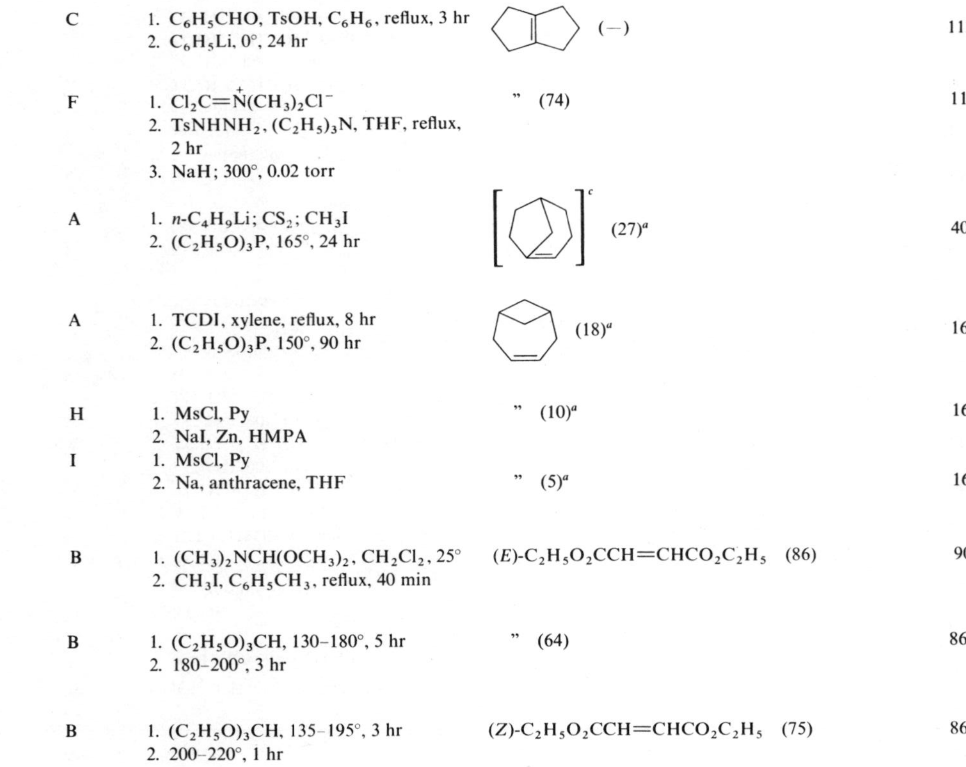

Starting material	Method	Reagents	Product(s) and Yield(s) (%)	Refs.
	C	1. C_6H_5CHO, TsOH, C_6H_6, reflux, 3 hr 2. C_6H_5Li, 0°, 24 hr	(—)	112
	F	1. $Cl_2C{=}\overset{+}{N}(CH_3)_2Cl^-$ 2. $TsNHNH_2$, $(C_2H_5)_3N$, THF, reflux, 2 hr 3. NaH; 300°, 0.02 torr	" (74)	11
	A	1. n-C_4H_9Li; CS_2; CH_3I 2. $(C_2H_5O)_3P$, 165°, 24 hr	[ ][c] (27)[a]	40
	A	1. TCDI, xylene, reflux, 8 hr 2. $(C_2H_5O)_3P$, 150°, 90 hr	(18)[a]	164
	H	1. MsCl, Py 2. NaI, Zn, HMPA	" (10)[a]	164
	I	1. MsCl, Py 2. Na, anthracene, THF	" (5)[a]	164
	B	1. $(CH_3)_2NCH(OCH_3)_2$, CH_2Cl_2, 25° 2. CH_3I, $C_6H_5CH_3$, reflux, 40 min	(E)-$C_2H_5O_2CCH{=}CHCO_2C_2H_5$ (86)	90
	B	1. $(C_2H_5O)_3CH$, 130–180°, 5 hr 2. 180–200°, 3 hr	" (64)	86
	B	1. $(C_2H_5O)_3CH$, 135–195°, 3 hr 2. 200–220°, 1 hr	(Z)-$C_2H_5O_2CCH{=}CHCO_2C_2H_5$ (75)	86

TABLE I. OLEFIN SYNTHESIS BY DEOXYGENATION OF VICINAL DIOLS (*Continued*)

C_n	Diol	Method	Conditions	Products and Yields (%)	Refs.
C_8	OH, OH	F	$CHCl_3$, CH_2Cl_2, NaOH, H_2O, $(C_2H_5)_3\overset{+}{N}CH_2C_6H_5Cl^-$	(25)[f]	13
		A	1. TCDI 2. $Ni(COD)_2$, DMF	" (82)[a]	55
		A	1. TCDI 2. CH_3I, DME, 90°, 8 hr 3. Mg(Hg), THF, 15 min	" (58)	60
		A	1. TCDI, $C_6H_5CH_3$, reflux, 2 hr 2. i-C_3H_7I, 90°, 19 hr	" (26)	60
		A	1. TCDI, $C_6H_5CH_3$, reflux, 2 hr 2. i-C_3H_7I, 90°, 19 hr 3. Mg(Hg), THF, 25°	" (34)	60
		I	1. MsCl, Py 2. Na, naphthalene, 25°	" (48)	60
		I	1. MsCl, Py 2. Na, anthracene, 25°	" (50)	60
	OH, OH	I	1. MsCl, Py 2. Na, anthracene, 25°	" (31)	60
		I	1. MsCl, Py 2. Na, naphthalene, 25°	" (66)	152
		I	1. MsCl, Py 2. Na, anthracene, 25°	" (72)	152
		A	1. TCDI, $C_6H_5CH_3$, reflux, 2 hr 2. i-C_3H_7I, 90°, 19 hr	" (28)	60
		A	1. TCDI, $C_6H_5CH_3$, reflux, 2 hr 2. i-C_3H_7I, 90°, 19 hr 3. Mg(Hg), THF, 25°	" (44)	60

Reactant	Method	Reagents and Conditions	Product(s) (% Yield)	Refs.
	A	1. TCDI 2. $Ni(COD)_2$, DMF, 70°, 45 hr	" (99)[a]	55
	A	1. TCDI 2. $(i\text{-}C_8H_{17}O)P$, argon stream	 (59)	2,10
	A	1. TCDI 2. $(CH_3O)_3P$, 110°, 30 hr	" (62)[c]	2,10
	A	1. TCDI 2. $(i\text{-}C_8H_{17}O)_3P$, nitrogen stream	" (64)[g]	17,42
	B	1. $(CH_3O)_3CH$, $C_6H_5CO_2H$, 100°, 2 hr 2. 170°	" (53)	96
	C	1. C_6H_5CHO, H^+ 2. $n\text{-}C_4H_9Li$, n-hexane, 20°, 14 hr	" (75)[a]	115
	C	1. C_6H_5CHO, TsOH, C_6H_6 reflux, 42 hr 2. $n\text{-}C_4H_9Li$, petroleum ether, 25°, 18 hr	" (20)[h]	109
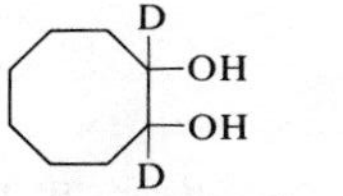 D, OH, OH, D (Mixture of isomers)	C	1. C_6H_5CHO, TsOH, $C_6H_5CH_3$, reflux, 6 hr 2. $n\text{-}C_4H_9Li$, n-hexane-n-pentane, 25°, 43 hr	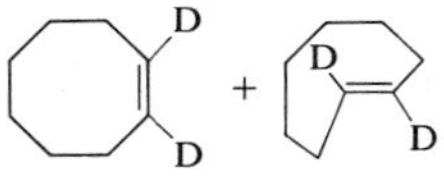 D, D (33) + D, D (14)	173
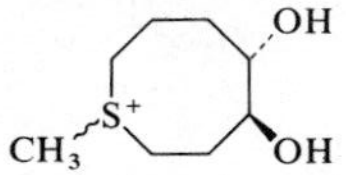 CH_3, S^+, OH, OH	C	1. C_6H_5CHO, $ZnCl_2$, 70° 2. $LiAlH_4$, THF 3. $n\text{-}C_4H_9Li$, THF, −10°	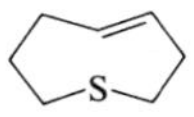 S (36)	111
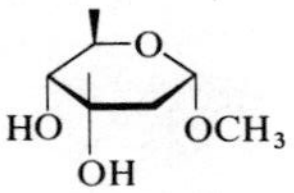 O, HO, OH, OCH_3	A	1. TCDI 2. $(CH_3O)_2POSi(CH_3)_3$, 130–140°, 4 hr	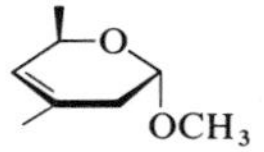 O, OCH_3 (77)	63
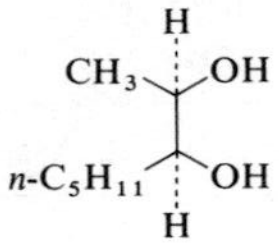 CH_3, H, OH, $n\text{-}C_5H_{11}$, H, OH	I	1. MsCl, Py 2. Na, naphthalene, 30°	*cis*-2-Octene (I), *trans*-2-octene (II) I + II (18), (I:II, 16:84)	152
	I	1. MsCl, Py 2. Na, anthracene, 30°	I + II (22), (I:II, 17:83)	152

TABLE I. Olefin Synthesis by Deoxygenation of Vicinal Diols (*Continued*)

C_n	Diol	Method	Conditions	Products and Yields (%)	Refs.
C_8	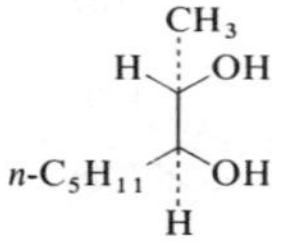 H, OH, CH_3 / n-C_5H_{11}, OH, H (structure)	I	1. MsCl, Py 2. Na, naphthalene, 40°	I + II (19), (I:II, 1:3)	152
		I	1. MsCl, Py 2. Na, anthracene, 0°	I + II (17), (I:II, 15:85)	152
		I	1. MsCl, Py 2. Na, anthracene, 40°	I + II (16), (I:II, 19:81)	152
	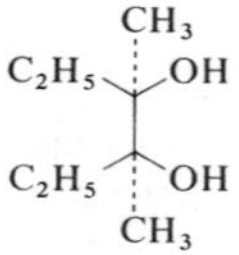 C_2H_5, OH, CH_3 / C_2H_5, OH, CH_3 (structure)	A	1. n-C_4H_9Li; CS_2; CH_3I 2. $(CH_3O)_3P$, heat	(*Z*)-3,4-Dimethyl-3-hexene (I), (*E*)-3,4-dimethyl-3-hexene (II) (—), (I:II, 98:2)	161
		B	1. $(CH_3O)_3CH$ 2. n-C_4H_9Li	I + II (60–90), (I:II, 98:2)	161
		C	1. C_6H_5CHO, H^+ 2. n-C_4H_9Li	I + II (—), (I:II, 98:2)	161
	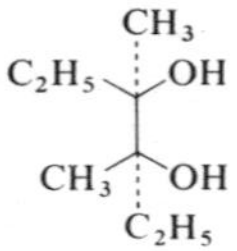 C_2H_5, OH, CH_3 / CH_3, OH, C_2H_5 (structure)	A	1. K, dioxane; CS_2, 70°; CH_3I, 70° 2. $(C_2H_5O)_3P$, reflux, 48 hr	(*E*)-3,4-Dimethyl-3-hexene (—)	174
C_9	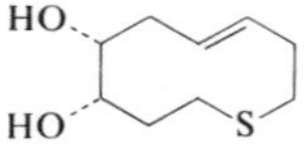 HO, HO, S (cyclic diol structure)	C	1. C_6H_5CHO 2. n-C_4H_9Li, THF, −20°	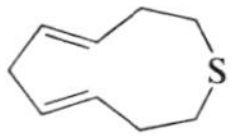 S (structure) (58)[a]	175
	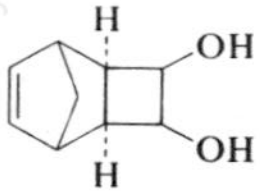 H, OH, OH, H (bicyclic diol structure)	A	1. TCDI, $C_6H_5CH_3$, reflux, 2 hr 2. CH_3I, 90°, 12 hr 3. Zn, C_2H_5OH, H_2O, 85°, 24 hr	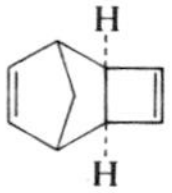 H, H (structure) (77)	61
		A	1. TCDI, $C_6H_5CH_3$, reflux, 2 hr 2. i-C_3H_7I, reflux, 30 hr 3. Zn, C_2H_5OH, H_2O, reflux, 24 hr	" (50)	61

Reactant	Method	Conditions	Product (Yield, %)	Refs.
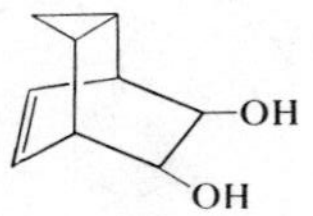 —OH, OH	A	1. TCDI, $C_6H_5CH_3$, heat 2. $Fe(CO)_5$, 100°	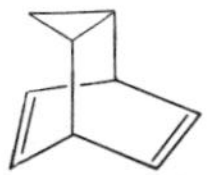 (30)	56,176
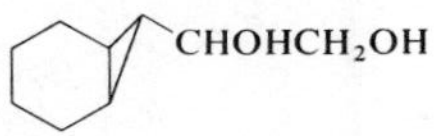 $CHOHCH_2OH$	A	1. TCDI, $C_6H_5CH_3$, 110°, 1 hr 2. $(CH_3O)_3P$, reflux, 30 hr	 (44)	177
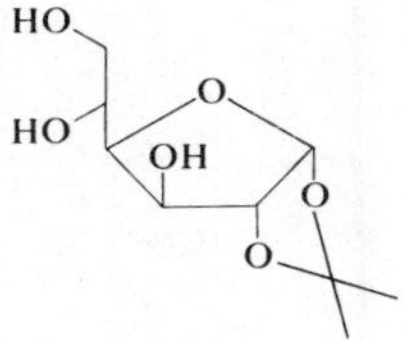 HO, HO, O, OH, O, O	A	1. TCDI, CH_3COCH_3, reflux, 1 hr 2. $(CH_3O)_3P$, reflux, 60 hr	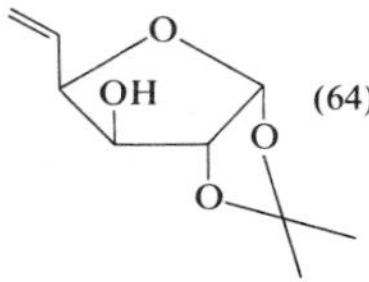 O, OH, O, O (64)	178,179
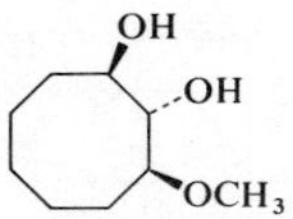 OH, OH, OCH_3	C	1. C_6H_5CHO, TsOH, $C_6H_5CH_3$, heat 2. n-C_4H_9Li, THF	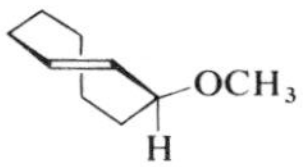 OCH_3, H (7)	113
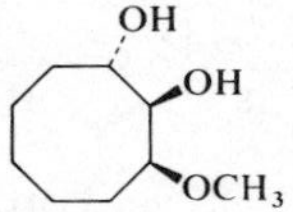 OH, OH, OCH_3	C	1. C_6H_5CHO, TsOH, $C_6H_5CH_3$, heat 2. n-C_4H_9Li, THF	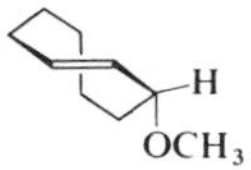 H, OCH_3 (60)	113
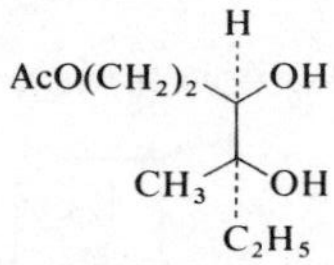 H, $AcO(CH_2)_2$, OH, CH_3, OH, C_2H_5	A	1. TCDI, $C_6H_5CH_3$, 100°, 6 hr 2. $(CH_3O)_3P$, 130°, 50 hr	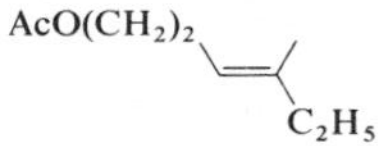 $AcO(CH_2)_2$, C_2H_5 (55)	49
C_{10} H, C_6H_5, OH, CH_3O_2C, OH, H	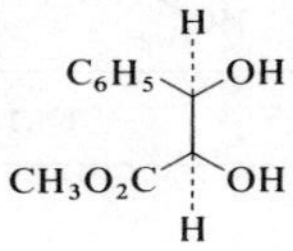 A	1. TCDI, C_6H_5, reflux, 0.5 hr 2. $(CH_3O)_3P$, reflux, 65 hr	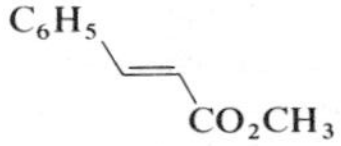 C_6H_5, CO_2CH_3 (46)	81

TABLE I. Olefin Synthesis by Deoxygenation of Vicinal Diols (*Continued*)

C_n	Diol	Method	Conditions	Products and Yields (%)	Refs.
C_{10}	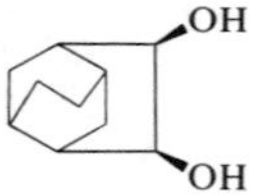 OH, OH	A	1. TCDI, $C_6H_5CH_3$, reflux, 45 min 2. $(CH_3O)_3P$, 135°, 110 hr	 (48)	37
	$C_6H_5CH_2OCH_2CHOHCH_2OH$	B	1. $(C_2H_5O)_3CH$ 2. CF_3COCF_3	$C_6H_5CH_2OCH_2CH{=}CH_2$ (—)	102
	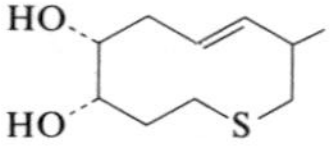 HO, HO, S	C	1. C_6H_5CHO 2. n-C_4H_9Li, THF, −20°	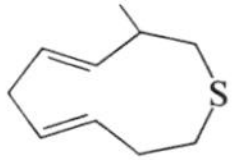 S (78)[a]	175
	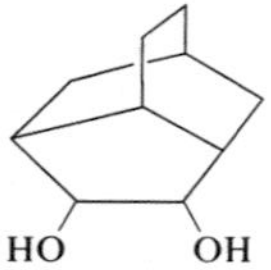 HO, OH	A	1. TCDI, $C_6H_5CH_3$, reflux, 45 min 2. $(CH_3O)_3P$, 135°, 110 hr	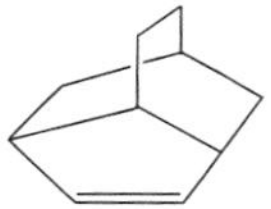 (40)	37
		H	1. p-$BrC_6H_4SO_2Cl$, Py 2. NaI, diglyme, reflux, 30 min	" (55)	37
	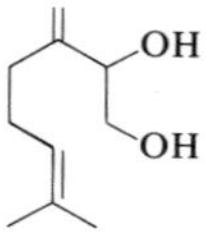 OH, OH	B	1. $(CH_3)_2NCH(OCH_3)_2$ 2. Ac_2O, 110°	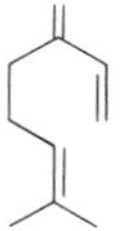 (<20)	162
		L	1. Cu_2Br_2; PBr_3, ether 2. Zn, 0°	" (58)	162,163
	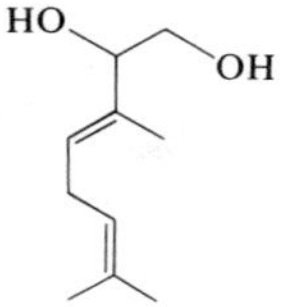 HO, OH	L	1. Cu_2Br_2; PBr_3, ether 2. Zn, 0°	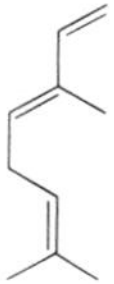 (66)	162,163

Reactant	Method	Conditions	Product(s) and Yield(s) (%)	Refs.
	C	1. C_6H_5CHO, TsOH 2. n-C_4H_9Li, THF, $-80°$	(3)	114
	E	$TiCl_3$, K, THF, reflux, 16 hr	(80)	171
	E	$TiCl_3$, K, THF, reflux	No reaction	171
	B	1. $(C_2H_5O)_3CH$, 2,4,6-$(CH_3)_3C_6H_2CO_2H$, 145°, 20 hr 2. 170°, 72 hr	(76)	93
+	E	$TiCl_3$, K, THF, reflux, 5 hr	" (81)	171
	E	$TiCl_3$, K, THF, reflux, 5 hr	" (60)	171
	A	1. TCDI, THF, reflux, 0.5 hr 2. $(CH_3O)_3P$, reflux, 24 hr	(68)	
	A	1. TCDI, THF, reflux, 0.5 hr 2. CH_3I; Zn, C_2H_5OH, reflux, 12 hr	" (38)	62
	A	1. TCDI, THF, reflux, 0.5 hr 2. CH_3I; $Cr(OAc)_2$, C_2H_5OH, 100°, 12 hr	" (30)	62
	B	1. $(C_2H_5O)_3CH$, HOAc, 115–125°, 2 hr 2. $C_6H_5CO_2H$, 150–170°, 5 hr	" (72)[a]	94
	J	1. NaH, IMID, CS_2; CH_3I, THF 2. $(n$-$C_4H_9)_3SnH$, $C_6H_5CH_3$	" (62)[a]	154

TABLE I. OLEFIN SYNTHESIS BY DEOXYGENATION OF VICINAL DIOLS (*Continued*)

C_n	Diol	Method	Conditions	Products and Yields (%)	Refs.
C_{10}	(cis-cyclodecane-1,2-diol: OH, OH)	A	1. TCDI, $C_6H_5CH_3$, reflux, 30 min 2. $(CH_3O)_3P$, 110°, 72 hr	(cis-cyclodecene) (73)	1,10
		B	1. $(CH_3O)_3CH$, $C_6H_5CO_2H$, 100°, 2 hr 2. 170°, 1 hr	" (67)	96
	(trans-cyclodecane-1,2-diol: OH, OH)	A	1. TCDI, $C_6H_5CH_3$, reflux, 30 min 2. $(CH_3O)_3P$, 110°, 70 hr	(trans-cyclodecene) (60)	10
	n-C_7H_{15}(H)(OH)C–C(OH)(H)CH_3 (*erythro*)	D	1. n-C_4H_9Li, 2 eq 2. $C_2H_5OP(O)Cl_2$ 3. Li–NH_3	n-C_7H_{15}(H)C=C(H)CH_3 I (13) + n-C_7H_{15}(H)C=C(CH_3)H II (10)	121
		D	1. n-C_4H_9Li, 2 eq 2. $(CH_3)_2NP(O)Cl_2$ 3. Li–NH_3	I (26) + II (14)	121
		D	1. n-C_4H_9Li, 2 eq 2. $(CH_3)_2NP(O)Cl_2$ 3. Na–xylene	I (80) + II (20)	121
	n-C_7H_{15}(H)(OH)C–C(HO)(H)CH_3 (*threo*)	D	1. n-C_4H_9Li, 2 eq 2. $C_2H_5OP(O)Cl_2$ 3. Li–NH_3	I (3) + II (22)	121
		D	1. n-C_4H_9Li, 2 eq 2. $(CH_3)_2NP(O)Cl_2$ 3. Li–NH_3	I (5) + II (37)	121
	meso-n-$C_4H_9CHOHCHOHC_4H_9$-n	E	$TiCl_3$, K, THF, reflux, 16 hr	5-Decene (75) (*cis*:*trans*, 2:3)	171
	d,l-n-$C_4H_9CHOHCHOHC_4H_9$-n	E	$TiCl_3$, K, THF, reflux, 16 hr	" (80) (*cis*:*trans*, 9:91)	171
		A	1. TCDI, DMF, heat 2. $(CH_3O)_3P$, heat	No reaction	180

	Reactant	Method	Conditions	Product (Yield, %)	Refs.
	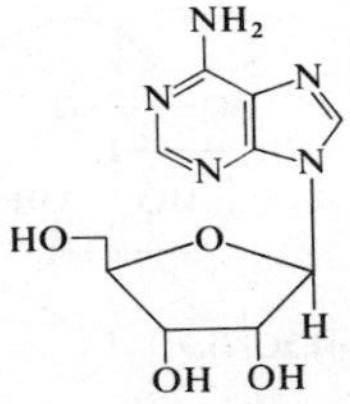	A	1. TCDI, DMF, heat 2. Sponge nickel	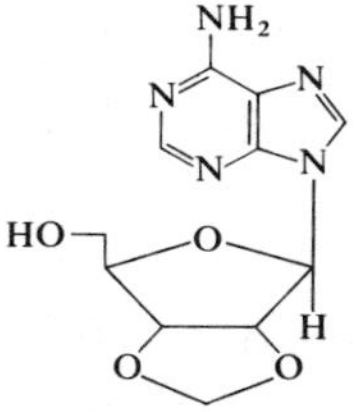 (2)	180
C_{11}	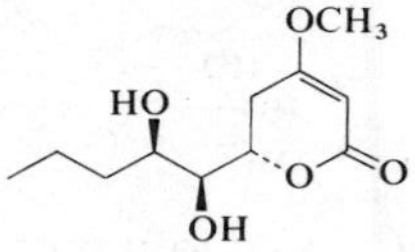	A	1. TCDI, $C_6H_5CH_3$, reflux, 1 hr 2. $(CH_3O)_3P$, reflux, 90 hr	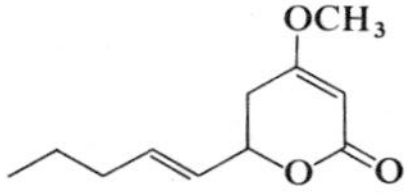 (31)	46
	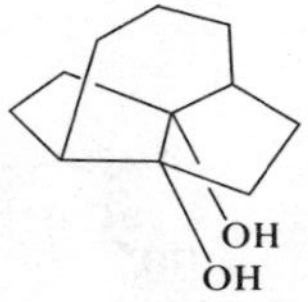	A	1. n-C_4H_9Li, TCDI, THF, reflux 2. $(C_2H_5O)_3P$, reflux, 20 hr	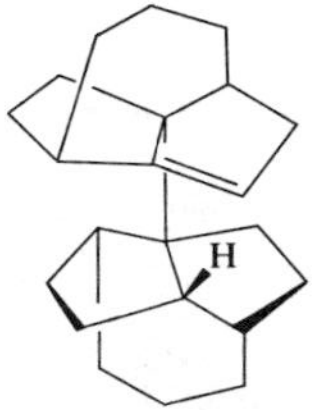 (82)	21
		A	1. n-C_4H_9Li, TCDI, THF, reflux 2. $(C_2H_5O)_3P$, reflux, 20 hr	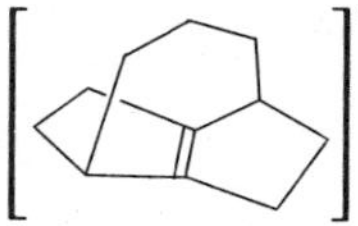 [ ][c] (66)	21
	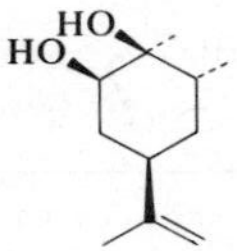	B	1. $(CH_3)_2NCH(OCH_3)_2$, DMF, 150°, 1 hr 2. Ac_2O, heat, 3.5 hr	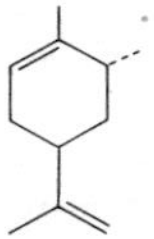 (36)	99

TABLE I. OLEFIN SYNTHESIS BY DEOXYGENATION OF VICINAL DIOLS (*Continued*)

C_n	Diol	Method	Conditions	Products and Yields (%)	Refs.
C_{11}	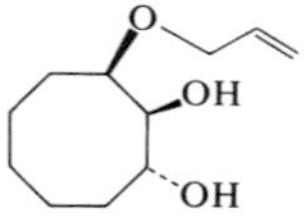 OH, OH	C	1. C_6H_5CHO, TsOH, $C_6H_5CH_3$, reflux, 6 hr 2. DMSO, t-C_4H_9OK 3. HgO–$HgCl_2$, CH_3COCH_3, H_2O 4. n-C_4H_9Li, THF, 20 min	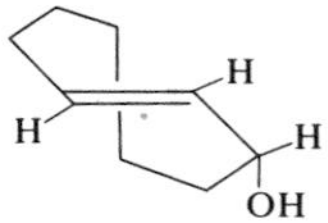 H, H, H, OH (21)	113
	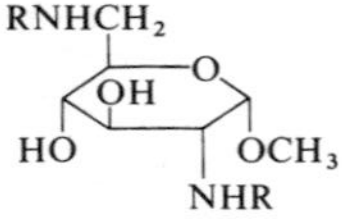 $RNHCH_2$, OH, HO, OCH_3, NHR ($R = CO_2CH_3$)	H	1. MsCl, Py, 25° 2. NaI, Zn, DMF, 98°, 3 hr	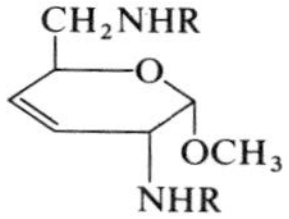 CH_2NHR, OCH_3, NHR (83)	142
	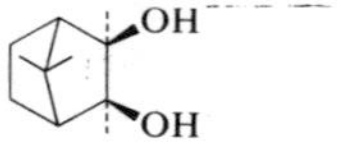 OH, OH	B	1. $(C_2H_5O)_3CH$, AcOH, 130–145°, 18 hr 2. 155–180°, 60 hr	 (69)	93
		A	1. TCDI, $C_6H_5CH_3$, heat 2. $(C_2H_5O)_3P$, heat	" (trace)	93
C_{12}	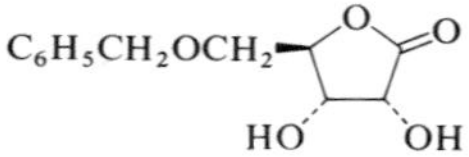 $C_6H_5CH_2OCH_2$, HO, OH	B	1. $(C_2H_5O)_3CH$, THF, reflux, 12 hr 2. AcOH, diglyme, 200°, 7 hr	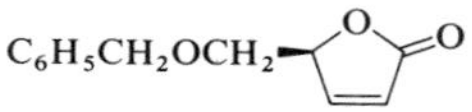 $C_6H_5CH_2OCH_2$ (66)	165
	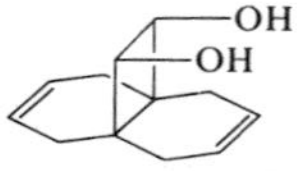 OH, OH	A	1. TCDI, $C_6H_5CH_3$, reflux, 40 min 2. $(CH_3O)_3P$, reflux, 26 hr	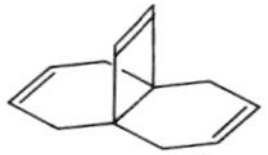 (26)	39
	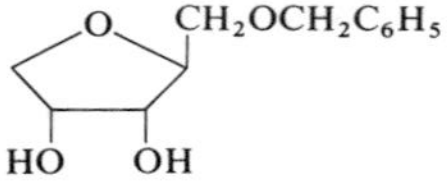 $CH_2OCH_2C_6H_5$, HO, OH	H	1. TsCl, Py, 25° 2. NaI, Zn, DMF, 170°, 2 hr	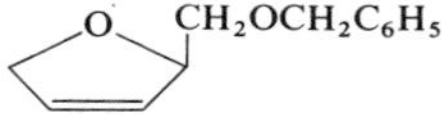 $CH_2OCH_2C_6H_5$ (82)[a]	140
	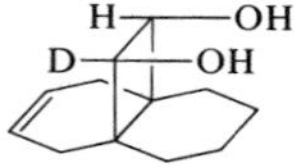 H, OH, D, OH	A	1. TCDI, $C_6H_5CH_3$, reflux 2. $(C_2H_5O)_3P$, reflux, 72 hr	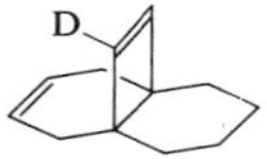 D (12)	38

Reactant	Method	Conditions	Product (Yield, %)	Refs.
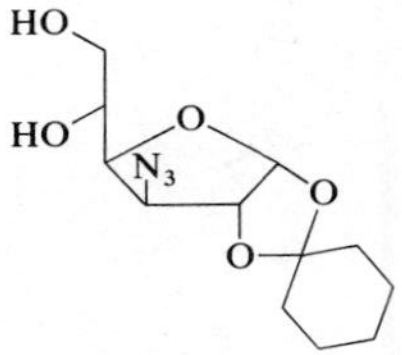 [structure]	H	1. MsCl, Py, 0° 2. NaI, DMF, H_2O, reflux, 1 hr	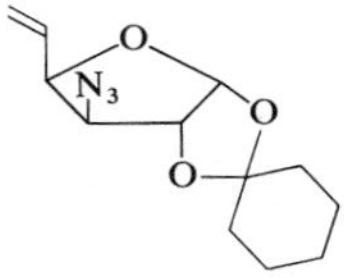 [structure] (72)	149
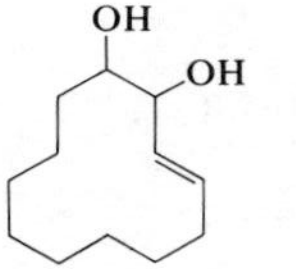 [structure]	L	1. Cu_2Br_2, ether 2. PBr_3, −78° to 0° 3. Zn, 0–25°	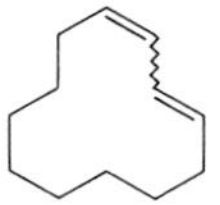 [structure] (70)	162,163
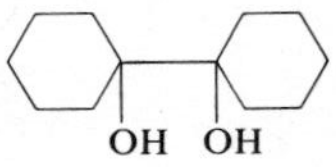 [structure]	E	$TiCl_3$, K, THF, reflux, 16 hr	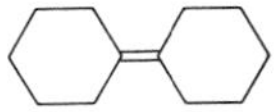 [structure] (85)	171
	A	1. K salt of diol, CS_2, 79° 2. CH_3I, 70° 3. $(CH_3O)_3P$, 110°, 70 hr	" (30)	2,10
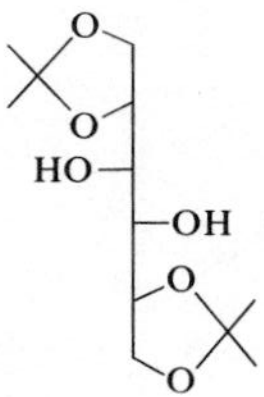 [structure]	A	1. n-C_4H_9Li, THF 2. CS_2, CH_3I 3. $(CH_3O)_3P$, reflux, 70 hr	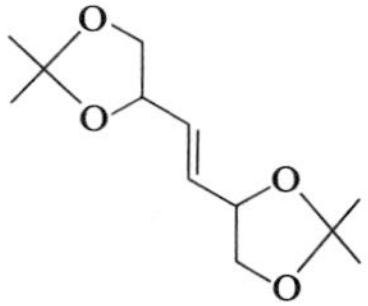 [structure] (22)	27
	B	1. $(CH_3O)_2CHN(CH_3)_2$ 2. Ac_2O, 130°, 3 hr	" (~100)	89
	H	1. MsCl, Py 2. NaI, Zn, DMF, reflux, 5 hr	" (64)[a]	139
	H	1. TsCl, Py 2. KSeCN, DMF, reflux, 4 hr	" (low yield)	28,148
	J	1. IMID, NaH; CS_2; CH_3I, THF 2. $(n\text{-}C_4H_9)_3SnH$, $C_6H_5CH_3$, reflux	" (59)	154

TABLE I. OLEFIN SYNTHESIS BY DEOXYGENATION OF VICINAL DIOLS (*Continued*)

C_n	Diol	Method	Conditions	Products and Yields (%)	Refs.
C_{12}	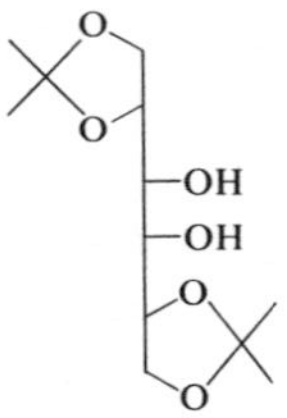	A	1. n-C_4H_9Li, THF 2. CS_2; CH_3I 3. $(CH_3O)_3P$, reflux, 70 hr	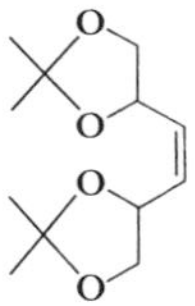 (40)	27
		H	1. TsCl, Py 2. KSeCN, DMF, reflux, 4 hr	" (45)[a]	28,148
		A	1. NaOH, CS_2, 1,4-dioxane 2. I_2, Py 3. $(CH_3O)_3P$, reflux, 70 hr	" (18)	28
		B	1. $(CH_3)_2NCH(OCH_3)_2$, CH_2Cl_2, 25° 2. CH_3I, $C_6H_5CH_3$, reflux, 10 min	" (95)	90
		A	1. $CSCl_2$, CH_2Cl_2, DMAP, 0°, 1 hr 2. DMPD, 40°, 20 hr	" (82)	25
	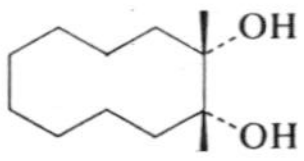 OH OH	D	1. n-C_4H_9Li 2. $C_2H_5OP(O)Cl_2$ 3. Li–NH_3	 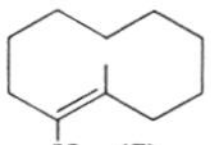 I (50) + II (7)	121
		D	1. n-C_4H_9Li 2. $(CH_3)_2NP(O)Cl_2$ 3. Li–NH_3	I (73) + II (8)	121
	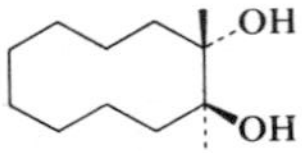 OH OH	D	1. n-C_4H_9Li 2. $(CH_3)_2NP(O)Cl_2$ 3. Li–NH_3	I (10) + II (45)	121
		D	1. n-C_4H_9Li 2. $(CH_3)_2NP(O)Cl_2$ 3. Li–NH_3	I (3) + II (21)	121

Reactant	Method	Conditions	Product(s) and Yield(s) (%)	Refs.
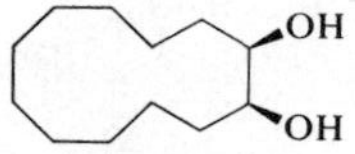 OH, OH	E	1. CH_3Li, 2 eq, THF 2. K_2WCl_6, reflux, 2 days	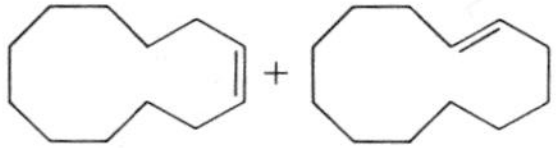 + I (48) II (18)	129
	I	1. MsCl, Py 2. Na–naphthalene, 25°	I (29) + II (44)	152
	I	1. MsCl, Py 2. Na–naphthalene, −15°	I (48) + II (32)	152
	I	1. MsCl, Py 2. Na–anthracene, 25°	I (27) + II (41)	152
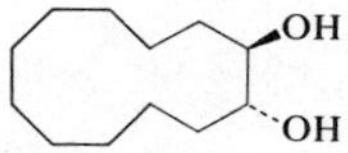 OH, OH	E	1. $2CH_3Li$, THF 2. K_2WCl_6, reflux, 4 days	I (4) + II (46)	129
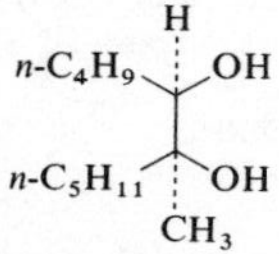 H, n-C_4H_9, OH, n-C_5H_{11}, OH, CH_3	B	1. $(CH_3O)_2CHN(CH_3)_2$, 25°, 12 hr 2. Ac_2O, 130°, 2.5 hr	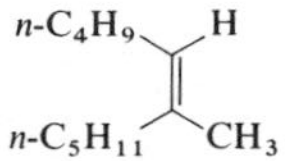 n-C_4H_9, H, n-C_5H_{11}, CH_3 (80)	97
C_{13} 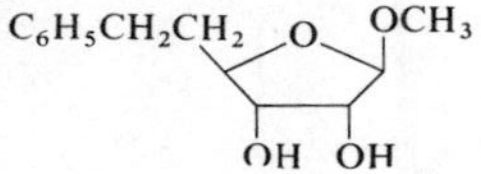 $C_6H_5CH_2CH_2$, O, OCH_3, OH, OH	B	1. $(CH_3O)_2CHN(CH_3)_2$, CH_2Cl_2, 25° 2. CH_3I, $C_6H_5CH_3$, reflux, 30 min	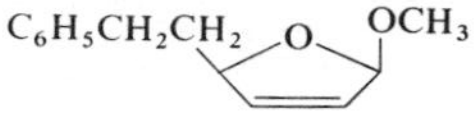 $C_6H_5CH_2CH_2$, O, OCH_3 (89)	90
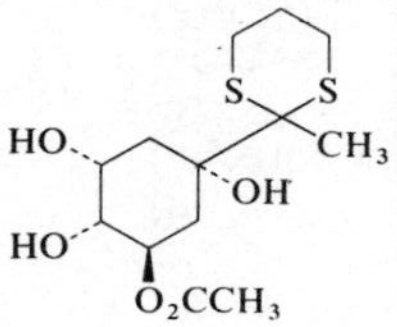 S, S, CH_3, HO, OH, HO, O_2CCH_3	A	1. TCDI, THF, reflux, 1.6 hr 2. $(CH_3O)_3P$, reflux, 9 hr	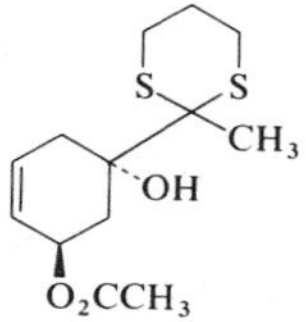 S, S, CH_3, OH, O_2CCH_3 (37)	72
	A	1. TCDI, THF, reflux, 1.6 hr 2. $(CH_3O)_3P$, reflux, 9 hr	No olefin formed	72

TABLE I. OLEFIN SYNTHESIS BY DEOXYGENATION OF VICINAL DIOLS (*Continued*)

C_n	Diol	Method	Conditions	Products and Yields (%)	Refs.
C_{13}	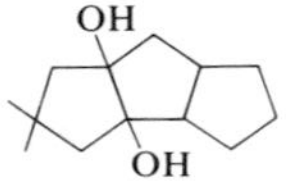 OH, OH	E	$TiCl_3$, K, THF, reflux, 6 days	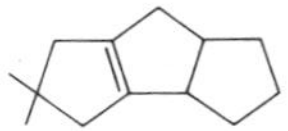 (ca. 100)	181
	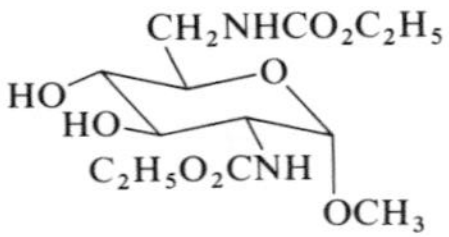 $CH_2NHCO_2C_2H_5$, HO, HO, $C_2H_5O_2CNH$, OCH_3	I	1. MsCl, Py 2. Na, naphthalene, THF	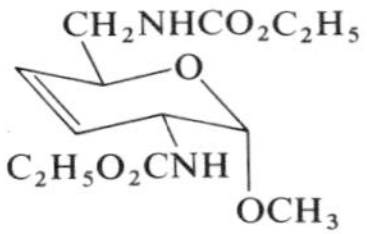 $CH_2NHCO_2C_2H_5$, $C_2H_5O_2CNH$, OCH_3 (65)	153
	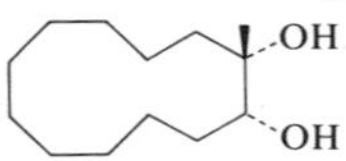 OH, OH	E	1. CH_3Li. 2 eq 2. K_2WCl_6, THF, reflux, 22 hr	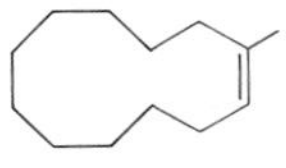 (36) 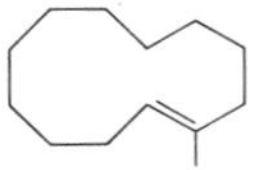 (9)	129, 131
	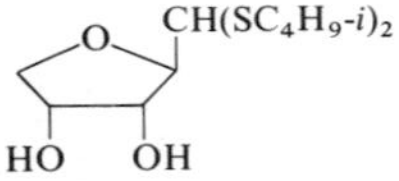 $CH(SC_4H_9\text{-}i)_2$, HO, OH	H	1. TsCl, Py, 25° 2. NaI, Zn, DMF, 170°, 2 hr	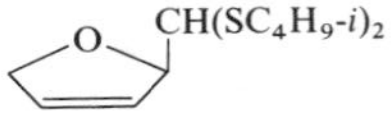 $CH(SC_4H_9\text{-}i)_2$ (73)	140
C_{14}	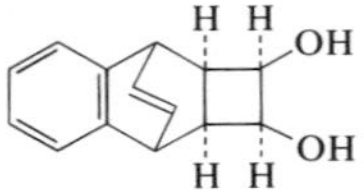 H H OH, H H OH	A	1. TCDI 2. DBMD, 155°, 84 hr	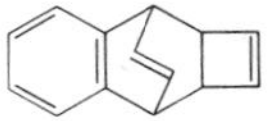 (57)[a]	61
	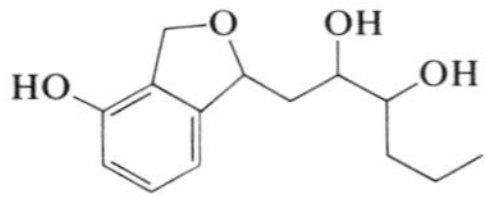 HO, OH, OH	A	1. TCDI, $C_6H_5CH_3$, heat 2. $(C_2H_5O)_3P$, heat	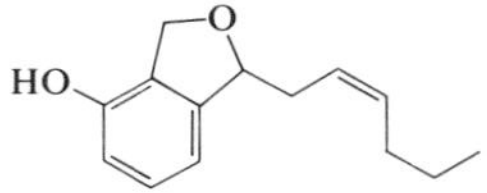 HO (—)	44
	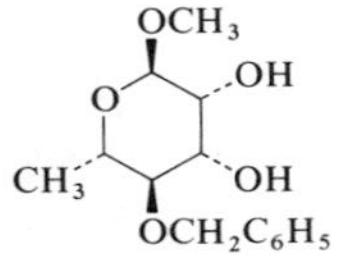 OCH_3, OH, CH_3, OH, $OCH_2C_6H_5$	A	1. $CSCl_2$, CH_2Cl_2, DMAP, 0°, 1 hr 2. DMPD, 40°, 2–24 hr	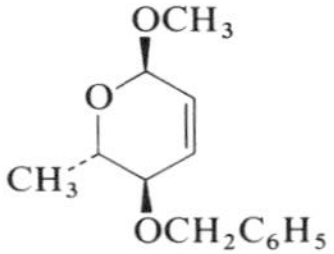 OCH_3, CH_3, $OCH_2C_6H_5$ (68)	25

Reactant	Method	Conditions	Product(s) and Yield(s) (%)	Refs.
$C_6H_5CH_2O$, OCH_3, HO OH (pyranoside diol)	A	1. TCDI, $C_6H_5CH_3$, heat 2. $(CH_3O)_3P$, reflux, 3 days	$C_6H_5CH_2O$, OCH_3 (unsaturated pyranoside) (61)	47
	B	1. $(CH_3O)_2CHN(CH_3)_2$, CH_2Cl_2, 25° 2. CH_3I, $C_6H_5CH_3$, reflux, 30 min	" (95)	90
OCH_3, $C_6H_5CH_2O$, HO OH (pyranoside diol)	A	1. TCDI, $C_6H_5CH_3$, heat 2. $(CH_3O)_3P$, reflux, 3 days	OCH_3, $C_6H_5CH_2O$ (unsaturated pyranoside) (65)	47
RO, RO, RO, OH, OH, OR (cyclitol) (R = CH_3CO—)	A	1. TCDI, CH_3COCH_3, reflux, 2 hr 2. $(CH_3O)_3P$, reflux, 5 hr	RO, RO, OR, OR (cyclohexene) (R = CH_3CO—) (32)	182
	A	1. TCDI 2. *i*-C_3H_7I, reflux, 5 hr; Zn, C_2H_5OH, H_2O	" (60)[a]	61
	A	1. TCDI 2. $(CH_3O)_3P$, 120°, 4 days	No reaction (second step)	61
C_6H_5(H)(OH)C–C(OH)(H)C_6H_5 (*meso*)	A	1. TCDI, $C_6H_5CH_3$, reflux, 30 min 2. $(CH_3O)_3P$, 110°, 72 hr	$C_6H_5CH{=}CHC_6H_5$ (cis) I (88) + $C_6H_5CH{=}CHC_6H_5$ (trans) II (—)	1,10
	A	1. TCDI, $C_6H_5CH_3$, reflux, 30 min 2. Deactivated Raney Ni, dioxane, heat, 100 min	I (65) + II (11)	10
	A	1. TCDI, $C_6H_5CH_3$, reflux, 30 min 2. 1% Na(Hg), dioxane	I (14) + II (9)	10
	A	1. TCDI, $C_6H_5CH_3$, reflux, 30 min 2. Ag(Hg), dioxane, heat, 45 hr	I (23)	10
	A	1. $CSCl_2$, CH_2Cl_2, DMAP, 0°, 1 hr 2. DMPD, 40°, 2–24 hr	I (77)	25
	A	1. TCDI, $C_6H_5CH_3$, reflux, 30 min 2. $Ni(CO)_4$, CH_3COCH_3, 25°, 48 hr	I (14) + II (5)	10

TABLE I. OLEFIN SYNTHESIS BY DEOXYGENATION OF VICINAL DIOLS (*Continued*)

C_n	Diol	Method	Conditions	Products and Yields (%)	Refs.
C_{14}	$C_6H_5CH(OH)CH(OH)C_6H_5$ (stereo structure)	A	1. TCDI, $C_6H_5CH_3$, reflux, 30 min[i] 2. $Fe(CO)_5$, xylene	$C_6H_5CH=CHC_6H_5$ (cis) I + $C_6H_5CH=CHC_6H_5$ (trans) II I (57) + II (17)	56
		A	1. TCDI, $C_6H_5CH_3$, reflux, 2 hr 2. i-C_3H_7I, reflux, 2 hr	II (73)	61
		A	1. TCDI, $C_6H_5CH_3$, reflux, 2 hr 2. I_2, DME, 90°, 16 hr	II (47)	61
		A	1. TCDI, $C_6H_5CH_3$, reflux, 2 hr 2. CH_3I, DME, 90°, 12 hr 3. Mg(Hg), THF, 25°, 1 hr	I (4) + II (80)	61
		B	1. $(C_2H_5O)_3CH$, 95–110°, 2 hr 2. 155–160°, 1 hr	I (84)	86
		B	1. $(C_2H_5O)_3CH$, $C_6H_5CO_2H$, 95–110°, 2 hr 2. 150–170°, 2–3 hr	I (87)	172
		F	$CHCl_3$, CH_2Cl_2, NaOH, H_2O, $(C_2H_5)_3\overset{+}{N}CH_2C_6H_5Cl^-$	I (15) + II (2) + *trans*-stilbene oxide	13
		I	1. NaH; CS_2, IMID, CH_3I, THF 2. $(n$-$C_4H_9)_3SnH$, $C_6H_5CH_3$, reflux	II (72)[a]	154
		E	1. CH_3Li 2. $TiCl_3$[j]	II (31)	125
		C	1. C_6H_5CHO, H^+ 2. n-C_4H_9Li, THF	$C_6H_5CH_2COC_6H_5$ (67)[a] + $C_6H_5CHOHC_4H_9$-n (70)[a]	115
	$C_6H_5CH(OH)CH(OH)C_6H_5$ (stereo structure)	A	1. $CSCl_2$, CH_2Cl_2, DMAP, 0°, 1 hr 2. DMPD, 40°, 2–24 hr	II (89)	25
		A	1. TCDI, $C_6H_5CH_3$, reflux, 20 min 2. $(CH_3O)_3P$, 110°, 80 hr	II (71)	1,10

Reactant	Method	Conditions	Product(s) and Yield(s) (%)	Refs.
	B	1. $(C_2H_5O)_3CH$, $C_6H_5CO_2H$, 95–110°, 2 hr 2. 150–170°, 3 hr	II (76)	172
	B	1. $(CH_3O)_2CHN(CH_3)_2$ 2. Ac_2O, 180°, 2 hr	II (80)	89
	C	1. C_6H_5CHO, H^+ 2. n-C_4H_9Li, THF	II (9)[a] + $C_6H_5CH_2COC_6H_5$ (66)[a] + $C_6H_5CHOHC_4H_9$-n (64)[a]	115
	J	1. NaH; CS_2, IMID; CH_3I, THF 2. $(n$-$C_4H_9)_3SnH$, $C_6H_5CH_3$, reflux	II (77)[a]	154
$(C_6H_5)_2COHCH_2OH$	B	1. $(C_2H_5O)_3CH$, $C_6H_5CO_2H$, heat 2. 150–160°, 3 hr	$(C_6H_5)_2C{=}CH_2$ (50)	172
[structure: OH, OH]	A	1. TCDI, $C_6H_5CH_3$, reflux 2. i-C_3H_7I, reflux, 5 hr 3. Zn, C_2H_5OH, reflux, 14 hr	[structure] (84)[a]	60,61
	A	1. TCDI, $C_6H_5CH_3$, reflux 2. $(CH_3O)_3P$, reflux, 84 hr	No reaction (second step)	60,61
[structure: HO, OH, O, $OCH_2C_6H_5$, O_2CCH_3]	A	1. TCDI, THF 2. $(CH_3O)_3P$, reflux, 4 hr	[structure: O, $OCH_2C_6H_5$, O_2CCH_3] (81)	183
[structure: C_6H_5, O, O, OH, O, H, H, OH, H, OCH_3]	J	1. NaH, IMID; CS_2; CH_3I, THF 2. $(n$-$C_4H_9)_3SnH$, $C_6H_5CH_3$, reflux, 18 hr	[structure: C_6H_5, O, O, O, H, OCH_3] (41)	154
	L	$(C_6H_5)_3P$, CHI_3, IMID, $C_6H_5CH_3$, reflux	" (75)	159
[structure: H, C_6H_5, O, O, H, O, HO, H, OH, H, OCH_3]	L	1. $(C_6H_5)_3P$, IMID, $C_6H_5CH_3$, reflux 2. I_2	[structure: H, C_6H_5, O, O, O, OCH_3, H] (59)	159
	L	$(C_6H_5)_3P$, TMD, $C_6H_5CH_3$, reflux	" (74)	158
	H	1. MsCl or TsCl, Py 2. NaI, Zn, DMF, reflux, 5 min	" (81–85)[a]	144

TABLE I. Olefin Synthesis by Deoxygenation of Vicinal Diols (*Continued*)

C_n	Diol	Method	Conditions	Products and Yields (%)	Refs.
C_{14}	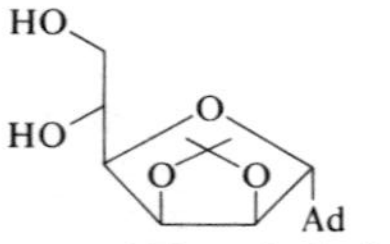 (Ad = adeninyl)	H	1. MsCl, Py 2. NaI, CH_3COCH_3, 100°, 16 hr	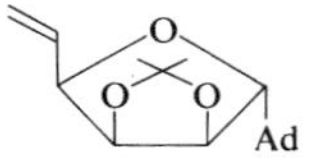 (48)	184
	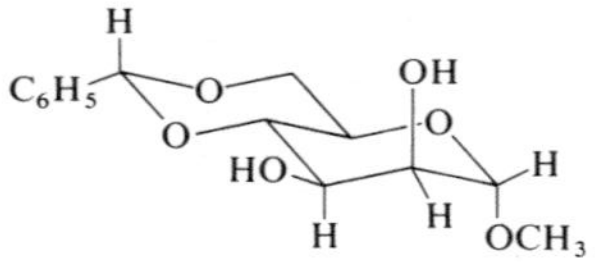	A	1. TCDI, CH_3COCH_3, reflux, 1 hr 2. $(CH_3O)_3P$, reflux, 60 hr	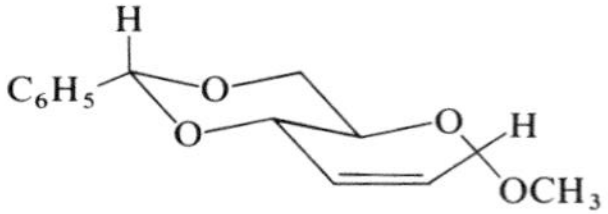 (21)	147
		J	1. NaH, IMID; CS_2, CH_3I, THF 2. $(n\text{-}C_4H_9)_3SnH$, $C_6H_5CH_3$, reflux	" (74)	154
		L	$(C_6H_5)_3P$, CHI_3, IMID, $C_6H_5CH_3$, reflux	" (45)	159
		L	1. $(C_6H_5)_3P$, IMID, $C_6H_5CH_3$, reflux 2. I_2	" (29)	157
		L	$(C_6H_5)_3P$, TMD, $C_6H_5CH_3$, reflux, 2–4 hr	" (45)	158
		L	1. $(C_6H_5)_3P$, IMID, $C_6H_5CH_3$, reflux 2. I_2, Zn	" (56)	157
	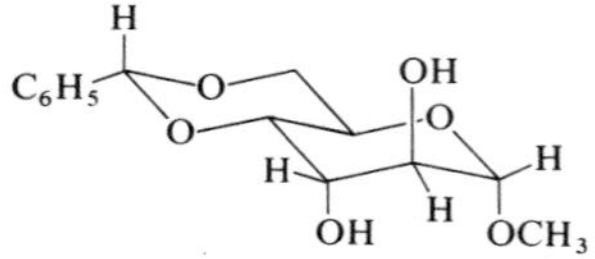	L	$(C_6H_5)_3P$, CHI_3, IMID, $C_6H_5CH_3$, reflux	" (65)	159
	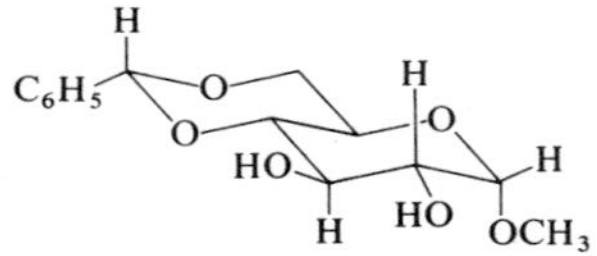	B	1. $(CH_3O)_2CHN(CH_3)_2$, CH_2Cl_2, 25° 2. CH_3I, 120°	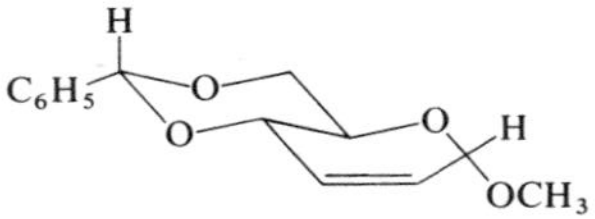 (85)	90

	H	1. MsCl or TsCl, Py 2. NaI, Zn, DMF, reflux	" (45–85)	141,144,147
	H	1. MsCl or TsCl, Py 2. $KSCO_2C_2H_5$, n-C_4H_9OH, reflux, 3 hr	" (35–65)	147,178
	J	1. NaH, IMID; CS_2; CH_3I, THF 2. $(n\text{-}C_4H_9)_3SnH$, $C_6H_5CH_3$, reflux, 18 hr	" (57)	154
	L	$(C_6H_5)_3P$, TMD, IMID, $C_6H_5CH_3$, reflux, 2–4 hr	" (87)	158
	L	1. $(C_6H_5)_3P$, IMID, $C_6H_5CH_3$, reflux 2. I_2	" (76)	157
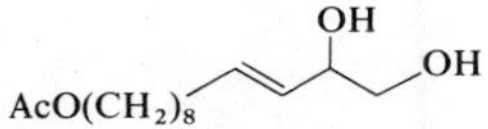	L	1. PBr_3, Cu_2Br_2, ether, −78° to 0°, 1 hr 2. Zn, 0°, 2 hr	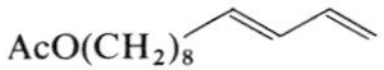 (75)	162,163
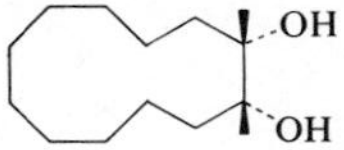	D	1. n-C_4H_9Li 2. $C_2H_5OP(O)Cl_2$ 3. Li–NH_3	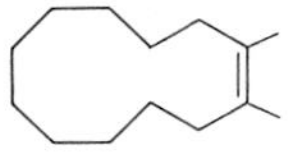 I (71) + 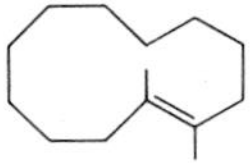 II (5)	121
	D	1. n-C_4H_9Li 2. $(CH_3)_2NP(O)Cl_2$ 3. Li–NH_3	I (77) + II (9)	121
	D	1. n-C_4H_9Li 2. $(CH_3)_2NP(O)Cl_2$ 3. Na-naphthalene	I (61) + II (3)	121
	D	1. n-C_4H_9Li 2. $(CH_3)_2NP(O)Cl_2$ 3. Na–xylene	I (54) + II (<0.5)	121
	D	1. n-C_4H_9Li 2. $(CH_3)_2NP(O)Cl_2$ 3. $TiCl_4$–Mg(Hg)	I (59) + II (9)	121
	D	1. n-C_4H_9Li 2. $(CH_3)_2NP(O)Cl_2$ 3. $TiCl_3$–K	I (71) + II (10)	121
	E	1. CH_3Li, THF 2. K_2WCl_6, reflux, 7 hr	I (69) + II (5)	129

TABLE I. Olefin Synthesis by Deoxygenation of Vicinal Diols (*Continued*)

C_n	Diol	Method	Conditions	Products and Yields (%)	Refs.
C_{14}	OH, OH	E	1. CH_3Li, THF 2. K_2WCl_6, reflux, 9 hr	I (22) + II (50)	129
C_{15}	H, C_6H_5, OH, C_6H_5, OH, CH_3	A	1. $CSCl_2$, CH_2Cl_2, DMAP, 0°, 1 hr 2. DMPD, 40°, 2–24 hr	CH_3, C_6H_5, C_6H_5 (75)	25
	OH, OH	L	1. PBr_3, Cu_2Br_2, ether, −78 to 0°, 1 hr 2. Zn, 0°, 2 hr	(38)	162,163
	OH, OH	L	1. PBr_3, Cu_2Br_2, ether, −78 to 0°, 1 hr 2. Zn, 0°, 2 hr	(49)	162,163
	OAc, O, O, HO, OH, O, O	B	1. $(CH_3O)_2CHN(CH_3)_2$ 2. Ac_2O, heat	OAc, O, O, O, O (—)	98
C_{16}	H, $C_6H_5CH_2$, OH, $C_6H_5CH_2$, OH, H	A	1. TCDI, $C_6H_5CH_3$ 2. $(CH_3O)_3P$, 110°, 50 hr	$C_6H_5CH_2$, H, $C_6H_5CH_2$, H (90)	2,10

Reactant	Method	Conditions	Product(s) and Yield(s) (%)	Refs.
$C_6H_5CH_2CH(OH)CH(OH)CH_2C_6H_5$ (H, OH / H, OH; $CH_2C_6H_5$)	A	1. TCDI, $C_6H_5CH_3$ 2. $(CH_3O)_3P$, 110°, 50 hr	$C_6H_5CH_2CH{=}CHCH_2C_6H_5$ (trans) (96)	2,10
$C_6H_5C(CH_3)(OH)C(OH)(CH_3)C_6H_5$	B	$(C_2H_5O)_3CH$, $C_6H_5CO_2H$, 100°, 2 hr, 190°, 2 hr	$C_6H_5(CH_3)C{=}C(CH_3)C_6H_5$ (I) (100)	96
	G	$(R_2N)_2S$ (R not given)	" (ca. 50)	132
$C_6H_5C(CH_3)(OH)C(OH)(CH_3)C_6H_5$	B	$(C_2H_5O)_3CH$, $C_6H_5CO_2H$, 100°, 2 hr, 190°, 2 hr	$C_6H_5(CH_3)C{=}C(CH_3)C_6H_5$ (II) (83)	96
	G	$(R_2N)_2S$ (R not given)	" (ca. 50)	132
$C_6H_5COH(CH_3)COH(CH_3)C_6H_5$ (mixture of isomers)	E	1. CH_3Li 2. K_2WCl_6, THF	I (44) + II (7)	131
$C_6H_5CH_2O$–furanoside acetonide with $CH(OH)CH_2OH$ side chain	B	1. $(CH_3O)_2CHN(CH_3)_2$, CH_2Cl_2, 24 hr 2. Ac_2O, 130°, 1 hr	$C_6H_5CH_2O$–furanoside acetonide with vinyl group (76)	101
HO, HO (diol side chain)–furanose acetonide, R (R = $C_6H_5CH_2O$)	B	1. $(C_2H_5O)_3CH$, CH_3CO_2H, 115–125°, 3 hr 2. $(C_6H_5)_3CCO_2H$, 150–170°, 5 hr	vinyl furanose acetonide, R (80)	94
$HO(CH_2)_6CH(OH)CH(OH)(CH_2)_7CO_2H$ (HO, H / OH, H)	B	$(C_2H_5O)_3CH$, $C_6H_5CO_2H$, 80–170°, 4 hr	$HO(CH_2)_6CH{=}CH(CH_2)_7CO_2H$ (cis) (95)	92

TABLE I. OLEFIN SYNTHESIS BY DEOXYGENATION OF VICINAL DIOLS (*Continued*)

C_n	Diol	Method	Conditions	Products and Yields (%)	Refs.
C_{16}	HO(CH$_2$)$_6$CH(OH)CH(OH)(CH$_2$)$_7$CO$_2$H	B	$(C_2H_5O)_3CH$, $C_6H_5CO_2H$, 80–170°, 4 hr	HO(CH$_2$)$_6$CH=CH(CH$_2$)$_7$CO$_2$H (94)	92
C_{17}	[structure: CH_3O, H, O, $CH_2CHOHCH_2OH$]	B	$(CH_3O)_2CHN(CH_3)_2$, Ac_2O, 140°	[structure: CH_3O, H, O] (63)	185
	[structure: cyclic thiocarbonate, O, O, S]	A	1. TCDI, $C_6H_5CH_3$, reflux, 1 hr 2. $Fe(CO)_5$, $C_6H_5CH_3$, 100°, 24 hr	[structure] (51)	26,56,58
	n-C_5H_{11}CH=CHCH$_2$CH(OSi(CH$_3$)$_2$C$_4$H$_9$-t)CH(OH)CH$_2$OH	A	1. $CSCl_2$, CH_2Cl_2, DMAP, 0°, 1 hr 2. DMPD, 40°, 2–24 hr	n-C_5H_{11}CH=CHCH$_2$CH(OSi(CH$_3$)$_2$C$_4$H$_9$-t)CH=CH$_2$ (82)	25
C_{18}	$C_6H_5CH{=}CHCHOHCHOHCH{=}CHC_6H_5$	L	P_2I_4, 25°, 30 min, ether	$C_6H_5CH{=}CHCH{=}CHCH{=}CHC_6H_5$ (90)	126
	[structure: OH, OH]	A	1. TCDI, $C_6H_5CH_3$, reflux, 1 hr 2. $Fe(CO)_5$, $C_6H_5CH_3$, 100°, 24 hr	[structure] (3)	26,58
	[structure: OH, OH]	A	1. TCDI, $C_6H_5CH_3$, reflux, 0.5 hr 2. $(n\text{-}C_4H_9)_3P$, $C_6H_5CH_3$, 100°, 96 hr	[structure] (9)	26,56–58
		A	1. TCDI, $C_6H_5CH_3$, reflux, 0.5 hr 2. $Fe(CO)_5$, xylene, reflux, 2 hr	" (23)	26,56–58
		A	1. TCDI, $C_6H_5CH_3$, reflux, 0.5 hr 2. $(C_2H_5O)_3P$, 165°, 36 hr	[structure] (43) + [structure] (32)	26,56–58

	Reactant	Method	Conditions	Product (Yield, %)	Refs.
	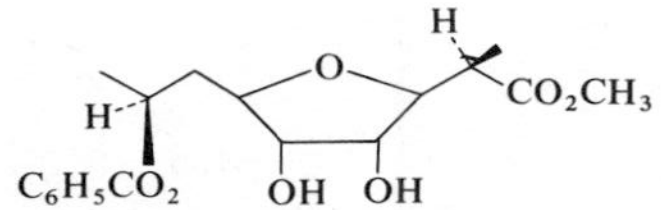 $C_6H_5CO_2$, H, OH, OH, O, H, CO_2CH_3	B	1. $(CH_3O)_2CHN(CH_3)_2$ 2. Ac_2O, heat	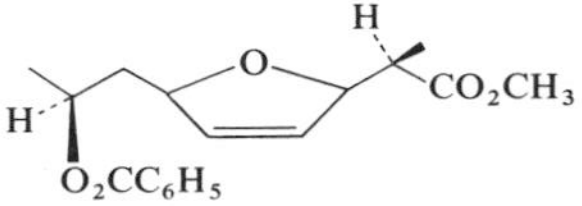 H, $O_2CC_6H_5$, O, H, CO_2CH_3 (78)	100
	n-$C_{16}H_{33}CHOHCH_2OH$	I	1. MsCl, Py 2. Na–naphthalene, 25°	n-$C_{16}H_{33}CH{=}CH_2$ (66)	152
		I	1. MsCl, Py 2. Na–anthracene, 25°	" (62)	152
C_{19}	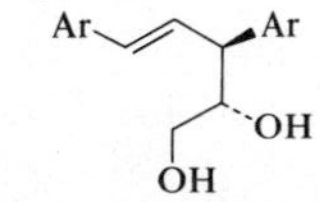 Ar, Ar, OH, OH (Ar = p-$CH_3OC_6H_4$)	B	$(C_2H_5O)_3CH$, heat	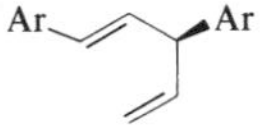 Ar, Ar (—)	186
	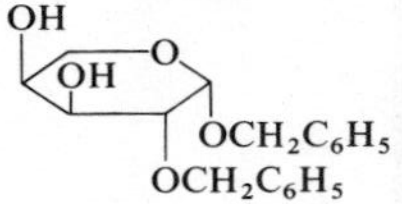 OH, OH, O, $OCH_2C_6H_5$, $OCH_2C_6H_5$	A	1. n-C_4H_9Li, THF, CS_2, CH_3I 2. $(CH_3O)_3P$, reflux, 70 hr	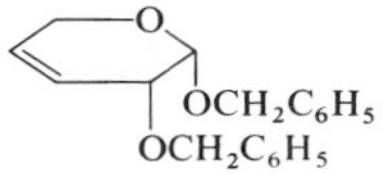 O, $OCH_2C_6H_5$, $OCH_2C_6H_5$ (52)	27
	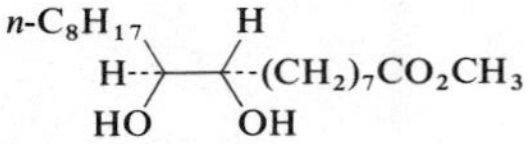 n-C_8H_{17}, H, H, HO, OH, $(CH_2)_7CO_2CH_3$	B	$(C_2H_5O)_3CH$, $C_6H_5CO_2H$, 120–170°	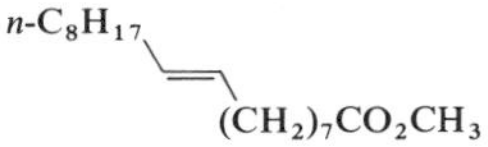 n-C_8H_{17}, $(CH_2)_7CO_2CH_3$ (95)	91
	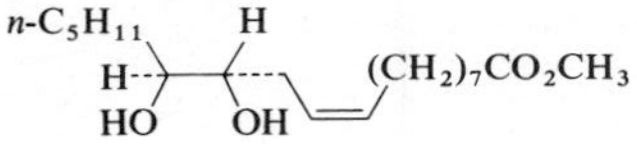 n-C_5H_{11}, H, H, HO, OH, $(CH_2)_7CO_2CH_3$	A	1. TCDI, $C_6H_5CH_3$, reflux, 3 hr 2. $(CH_3O)_3P$, 120°, 2 hr	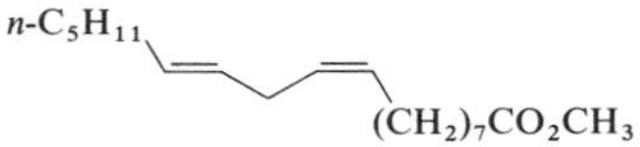 n-C_5H_{11}, $(CH_2)_7CO_2CH_3$ (—)	51
		B	$(C_2H_5O)_3CH$, $C_6H_5CO_2H$, 120–170°	" (82)	91
		H	1. MsCl, Py, 0° 2. NaI, Zn, DMF, reflux, 6 hr	" (—)	51
	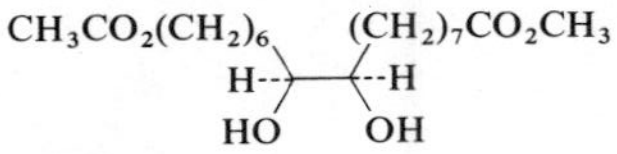 $CH_3CO_2(CH_2)_6$, $(CH_2)_7CO_2CH_3$, H, H, HO, OH	A	1. TCDI, $C_6H_5CH_3$, reflux 2. $(CH_3O)_3P$, 120°, 70 hr	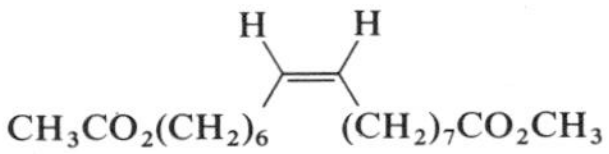 H, H, $CH_3CO_2(CH_2)_6$, $(CH_2)_7CO_2CH_3$ (83)	52
	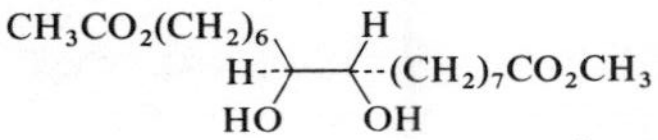 $CH_3CO_2(CH_2)_6$, H, H, $(CH_2)_7CO_2CH_3$, HO, OH	A	1. TCDI, $C_6H_5CH_3$, reflux 2. $(CH_3O)_3P$, 120°, 70 hr	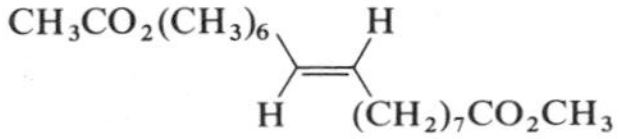 $CH_3CO_2(CH_3)_6$, H, H, $(CH_2)_7CO_2CH_3$ (83)	52

TABLE I. OLEFIN SYNTHESIS BY DEOXYGENATION OF VICINAL DIOLS (*Continued*)

C_n	Diol	Method	Conditions	Products and Yields (%)	Refs.
C_{20}	$(C_6H_5)_2COHCHOHC_6H_5$	B	1. $(C_2H_5O)_3CH$, $C_6H_5CO_2H$, 110–120°, 5 hr 2. 150–160°, 5 hr	$(C_6H_5)_2C{=}CHC_6H_5$ (69)	172
	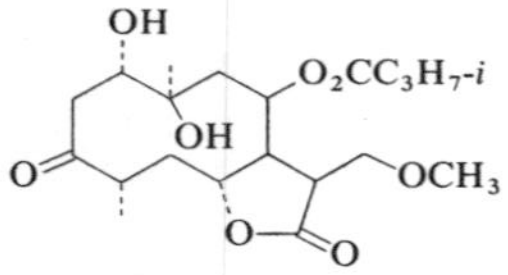 [structure: OH, OH, $O_2CC_3H_7$-*i*, OCH_3, O]	H	NaI, $(CH_3)_3SiCl$, CH_3CN, 25°, 5 min	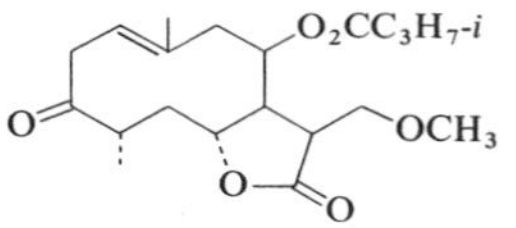 [structure: $O_2CC_3H_7$-*i*, OCH_3, O] (90)	151
	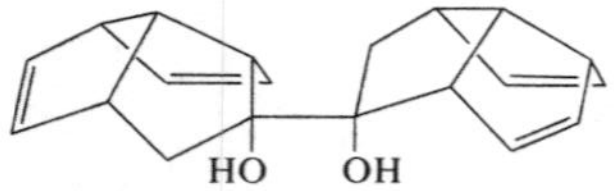 [structure: HO, OH]	A	1. KH; CS_2, 70°; CH_3I, 60° 2. $(C_2H_5O)_3P$, reflux, 2 days	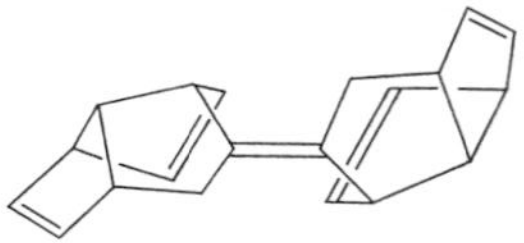 [structure] (36)	22
	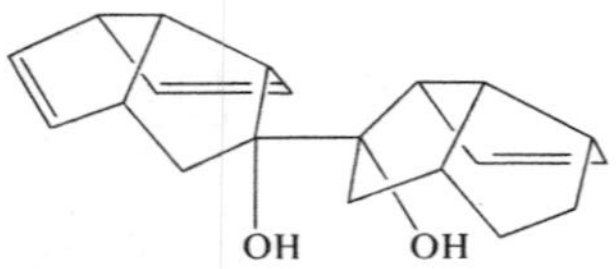 [structure: OH, OH]	A	1. KH; CS_2, 70°; CH_3I, 60° 2. $(C_2H_5O)_3P$, reflux, 2 days	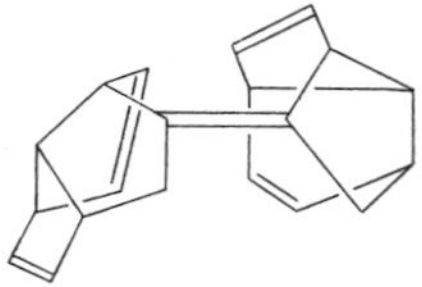 [structure] (48)	22
	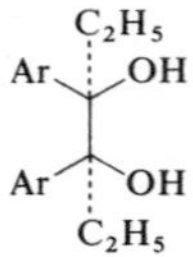 C_2H_5, Ar, OH; Ar, OH, C_2H_5 (Ar = *p*-$CH_3OC_6H_4$)	B	$(C_2H_5O)_3CH$, $C_6H_5CO_2H$, 170°, 1 hr	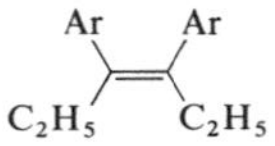 Ar, Ar, C_2H_5, C_2H_5 (86)	96
	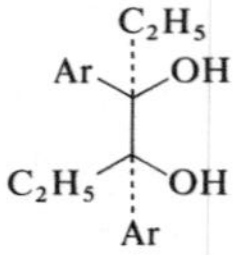 C_2H_5, Ar, OH; C_2H_5, OH, Ar (Ar = *p*-$CH_3OC_6H_4$)	B	$(C_2H_5O)_3CH$, $C_6H_5CO_2H$, 180°, 2 hr	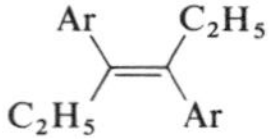 Ar, C_2H_5, C_2H_5, Ar (37)	96
	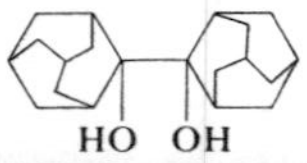 [structure: HO, OH]	A	1. *n*-C_4H_9Li, THF 2. CS_2, heat; CH_3I, heat 3. $Fe(CO)_5$, $C_6H_5CH_3$, 100°, 25 hr	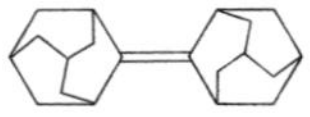 [structure] (23)	26

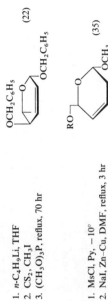

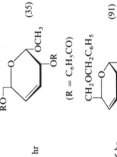

C_{21}

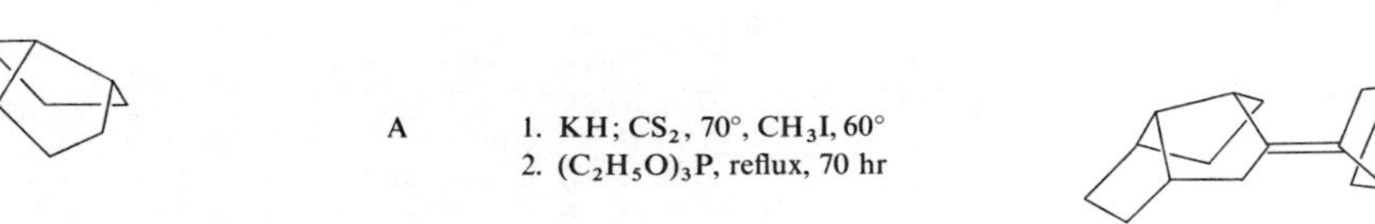

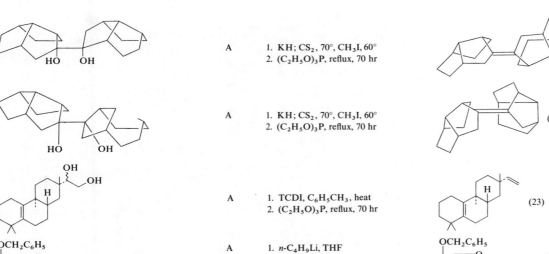

Reactant	Method	Conditions	Product(s) and Yield(s) (%)	Refs.
	A	1. KH; CS_2, 70°, CH_3I, 60° 2. $(C_2H_5O)_3P$, reflux, 70 hr	(54)	22
	A	1. KH; CS_2, 70°, CH_3I, 60° 2. $(C_2H_5O)_3P$, reflux, 70 hr	(73)	22
	A	1. TCDI, $C_6H_5CH_3$, heat 2. $(C_2H_5O)_3P$, reflux, 70 hr	(23)	187
	A	1. n-C_4H_9Li, THF 2. CS_2, CH_3I 3. $(CH_3O)_3P$, reflux, 70 hr	(22)	27
(R = C_6H_5CO)	H	1. MsCl, Py, −10° 2. NaI, Zn–Cu, DMF, reflux, 3 hr	(35) (R = C_6H_5CO)	143
	H	1. MsCl, Py, 25° 2. NaI, Zn, DMF, 130–150°, 1.5 hr	(91)	142

TABLE I. OLEFIN SYNTHESIS BY DEOXYGENATION OF VICINAL DIOLS (*Continued*)

C_n	Diol	Method	Conditions	Products and Yields (%)	Refs.
C_{21}	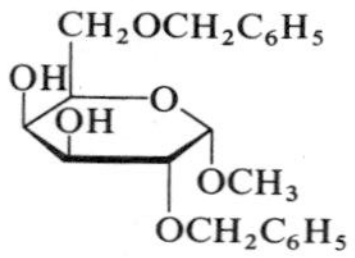 [structure: pyranoside diol with $CH_2OCH_2C_6H_5$, OH, OH, OCH_3, $OCH_2C_6H_5$]	L	$(C_6H_5)_3P$, TMD, IMID, $C_6H_5CH_3$, reflux, 2–4 hr	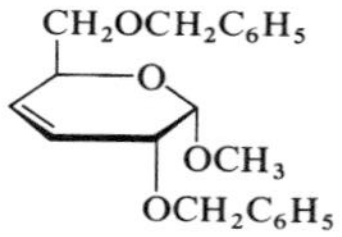 [structure: unsaturated pyranoside with $CH_2OCH_2C_6H_5$, OCH_3, $OCH_2C_6H_5$] (95)	142
	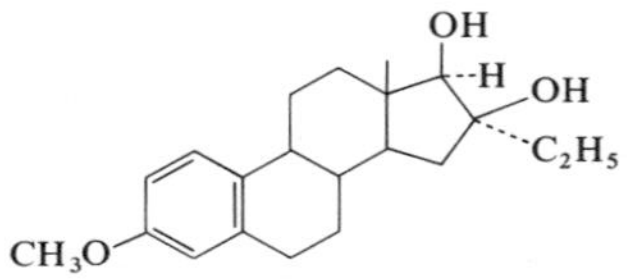 [structure: steroid diol with CH_3O, OH, H, OH, C_2H_5]	B	1. $(C_2H_5O)_3CC_6H_5$, CH_2Cl_2, TsOH, 25°, 2 hr 2. Zn, HOAc, reflux, 6 hr	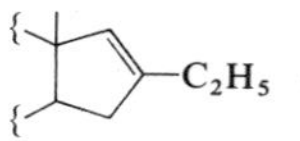 [structure: cyclopentene with C_2H_5] (86)[a]	103
		B	1. $(CH_3O)_3CH$, CH_2Cl_2, TsOH, 25°, 2 hr 2. Zn, HOAc, reflux, 4 hr	" (69)[a]	103
		B	1. $(C_2H_5O)_3CH$, CH_2Cl_2, TsOH, 25°, 2 hr 2. Zn, HOAc, reflux, 4 hr	" (70)[a]	103
		B	1. $(C_2H_5O)_3CCH_3$, CH_2Cl_2, TsOH, 25°, 2 hr 2. Zn, HOAc, reflux, 4 hr	" (92)[a]	103
		B	1. $(C_2H_5O)_3CC_2H_5$, CH_2Cl_2, TsOH, 25°, 2 hr 2. Zn, HOAc, reflux, 4 hr	" (90)[a]	103
		B	1. Ac_2O, HCO_2H 2. Zn, HOAc, reflux, 4 hr	" (67)[a]	103
		B	1. $(C_2H_5CO)_2O$, Py 2. Zn, HOAc, reflux, 4 hr	" (91)[a]	103
		B	1. C_6H_5COCl, Py 2. Zn, HOAc, reflux, 6 hr	" (83)[a]	103
		B	1. Ac_2O, Py 2. Zn, HOAc, reflux, 4 hr	" (92)[a]	103
	n-$C_{18}H_{37}OCH_2CHOHCH_2OH$	A	1. TCDI, C_6H_6, $NaOCH_3$, CH_3OH, reflux, 6 hr 2. $(CH_3O)_3P$, reflux	n-$C_{18}H_{37}OCH_2CH{=}CH_2$ (98)	50

Reactant	Method	Conditions	Product(s) and Yield(s) (%)	Refs.
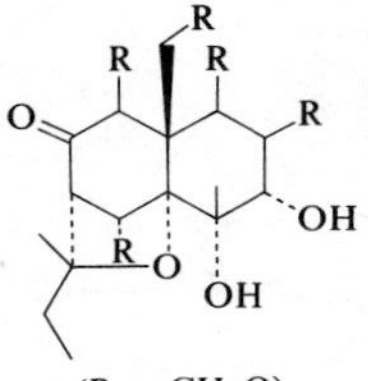 (R = CH_3O)	B	$(C_2H_5O)_3CH$, $C_6H_5CO_2H$, 160°, 13 hr	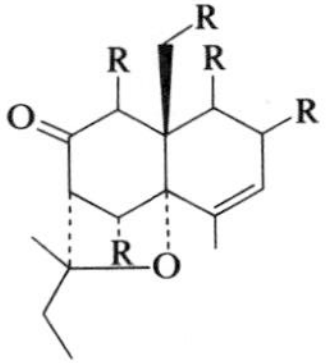 (—)	189
C_{22}	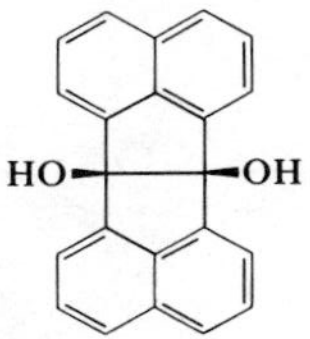 A L	1. $CSCl_2$, IMID, xylene, 100°, 12 hr 2. $[(C_2H_5)_2N]_3P$, reflux, 1 hr	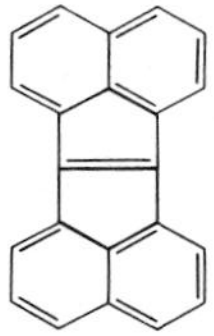 (30)	24
	L L	P_2I_4, ether, 12 hr HF, −70 to 25°	" (16–96) " (68)	24 24
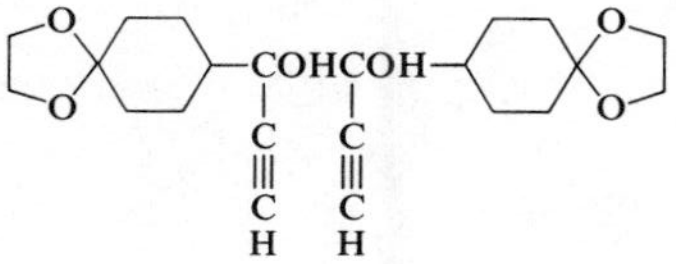	L	P_2I_4, Py, CS_2, 25°, 2 hr	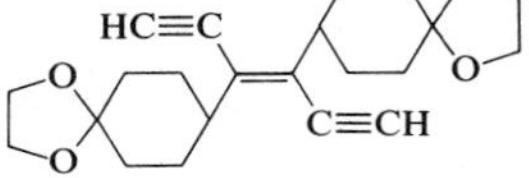 (43)	156
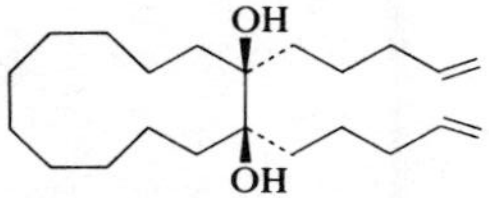	E	1. CH_3Li, 2 eq, THF 2. K_2WCl_6	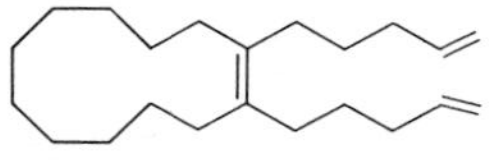 (50)	130
n-$C_{20}H_{41}CHOHCH_2OH$	E	1. CH_3Li, 2 eq, THF 2. K_2WCl_6, reflux, 2 days	n-$C_{20}H_{41}CH{=}CH_2$ (44)	129
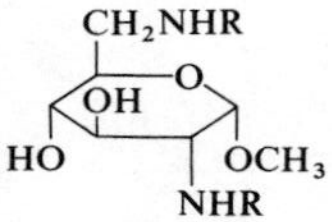 C_{23} (R = $CO_2CH_2C_6H_5$)	J	1. CS_2, DMSO 2. NaOH; CH_3I 3. $(n\text{-}C_4H_9)_3SnH$, $C_6H_5CH_3$, reflux, 2 hr	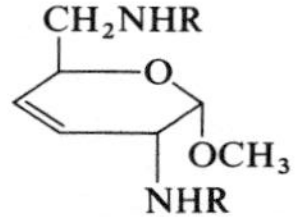 (78)	155

TABLE I. OLEFIN SYNTHESIS BY DEOXYGENATION OF VICINAL DIOLS (*Continued*)

C_n	Diol	Method	Conditions	Products and Yields (%)	Refs.
C_{23}	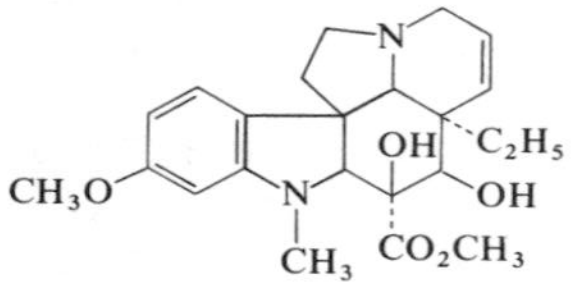 (structure: CH_3O, N, OH, C_2H_5, OH, CH_3, CO_2CH_3)	A	1. TCDI, $CH_3COC_2H_5$, reflux 2. Raney Ni, THF, reflux	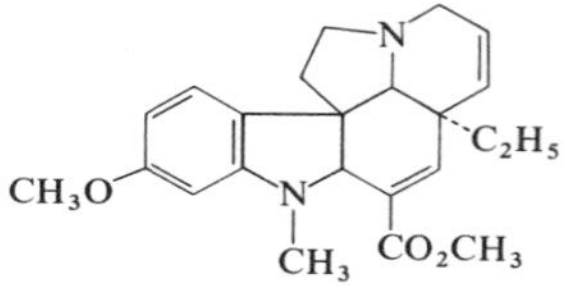 (structure: CH_3O, N, C_2H_5, CH_3, CO_2CH_3) (64)	53
	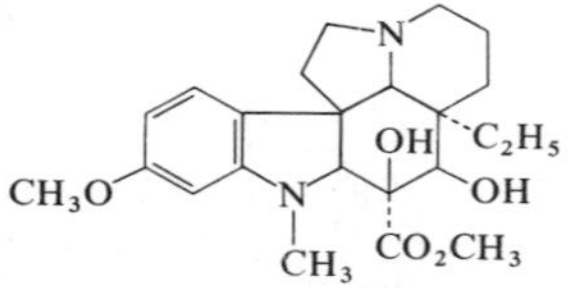 (structure: CH_3O, N, OH, C_2H_5, OH, CH_3, CO_2CH_3)	A	1. TCDI, $CH_3COC_2H_5$, reflux 2. Raney Ni; THF, reflux	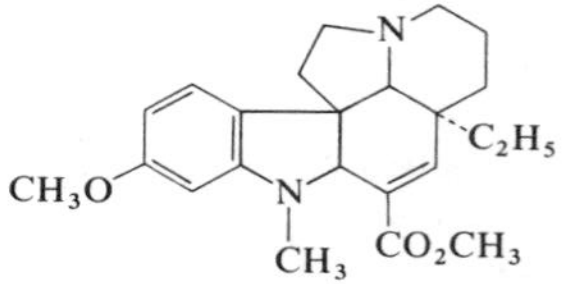 (structure: CH_3O, N, C_2H_5, CH_3, CO_2CH_3) (74)	53,54
C_{24}	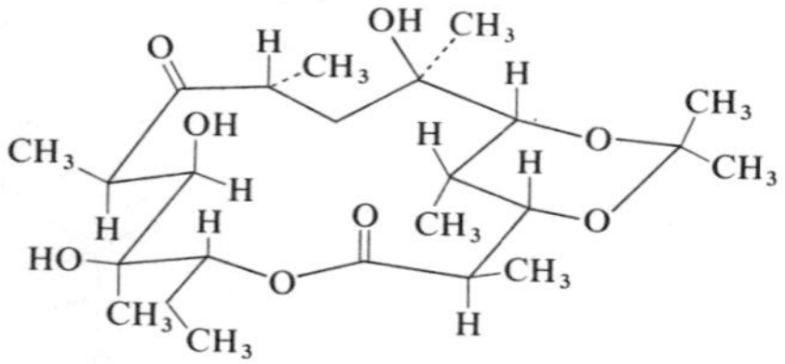 (structure)	A	1. $CSCl_2$, $CHCl_3$, DAP, 25°, 3.5 hr 2. DMPD	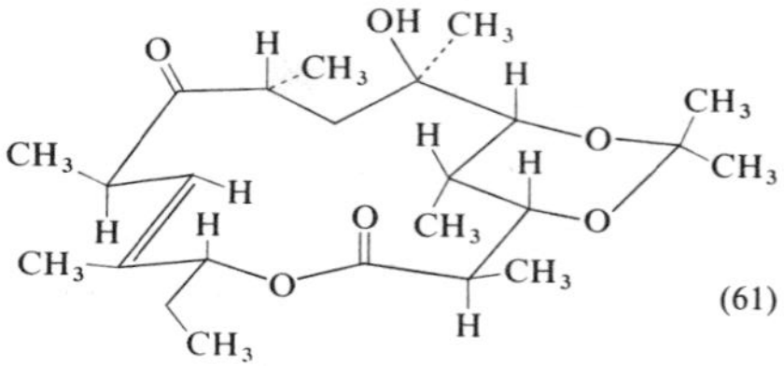 (structure) (61)	25
	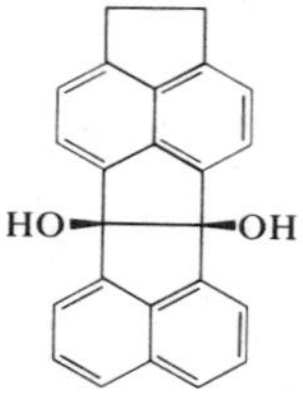 (structure: HO, OH)	A	1. $CSCl_2$, IMID, xylene, 100°, 12 hr 2. $[(C_2H_5)_2N]_3P$, reflux	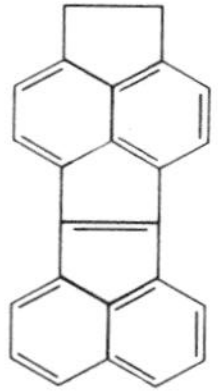 (structure) (30)	24
		L	P_2I_4, 12 hr, ether	" (49)	24
		L	HF, −70 to 25°	" (42)	24

Reactant	Method	Conditions	Product (Yield %)	Refs.
HO OH (bicyclic diol)	E	1. CH_3Li, 2 eq, 2. K_2WCl_6, reflux, 5 hr	(74)	129
	E	$TiCl_3$, $LiAlH_4$	" (—)	124
C_{25} $NHCOC_6H_5$, $C_6H_5CO_2CH_2$, HO OH	A	1. TCDI, CH_3COCH_3, reflux, 4 hr 2. $(CH_3O)_3P$, reflux, 4 hr	$NHCOC_6H_5$, $C_6H_5CO_2CH_2$ (11)	79
	H	1. MsCl, Py, 0° 2. NaI, Zn, DMF, reflux, 8 hr	" (7)	79
$CH_2OC(C_6H_5)_3$, OCH_3, HO OH	A	1. TCDI, $CH_3COC_2H_5$, 110°, 6 hr 2. $(CH_3O)_3P$, 125°, 4 hr	$CH_2OC(C_6H_5)_3$, OCH_3 (62)	75
C_{26} HO OH	A	1. $CSCl_2$, IMID, xylene, 100°, 12 hr 2. $[(C_2H_5)_2N]_3P$, reflux, 1 hr	(30)	24
$(C_6H_5)_2COHCOH(C_6H_5)_2$	L	HF, −70 to 25°	" (32)	24
	E	$TiCl_3$, $LiAlH_4$	$(C_6H_5)_2C{=}C(C_6H_5)_2$ (—)	124

TABLE I. OLEFIN SYNTHESIS BY DEOXYGENATION OF VICINAL DIOLS (*Continued*)

C_n	Diol	Method	Conditions	Products and Yields (%)	Refs.
C_{27}	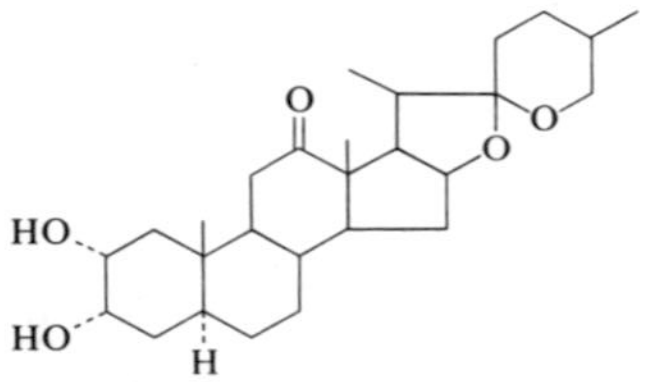 HO, HO, H, O	H	1. MsCl, Py 2. NaI, CH_3COCH_3, 100°, 18 hr	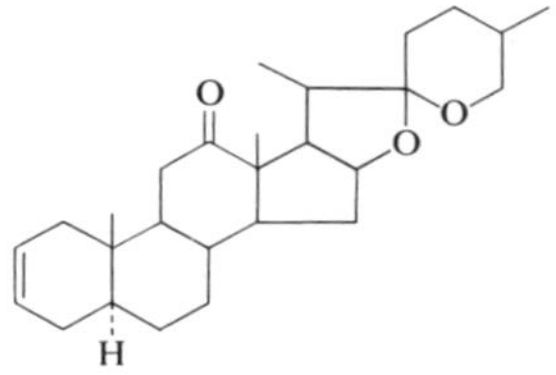 O, H (90)[a]	138
	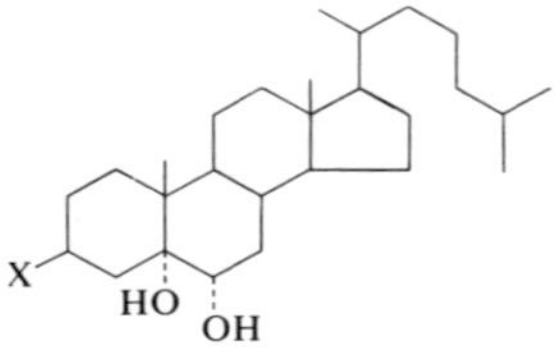 X, HO, OH	H	NaI, $(CH_3)_3SiCl$, CH_3CN, 25°, 20–30 min	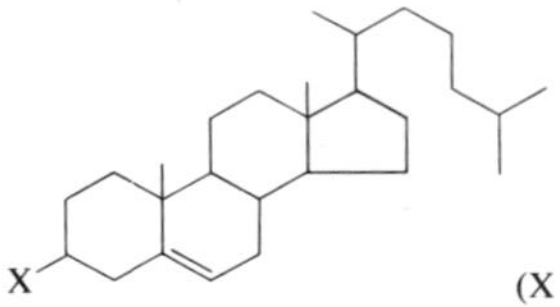 X (X = H, 96) (X = OH, 80)	151
	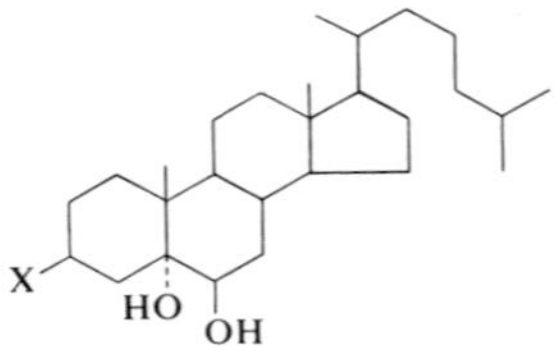 X, HO, OH	H	NaI, $(CH_3)_3SiCl$, CH_3CN, 25°, 20–30 min	" (X = H, 82) (X = OH, 85)	151
C_{28}	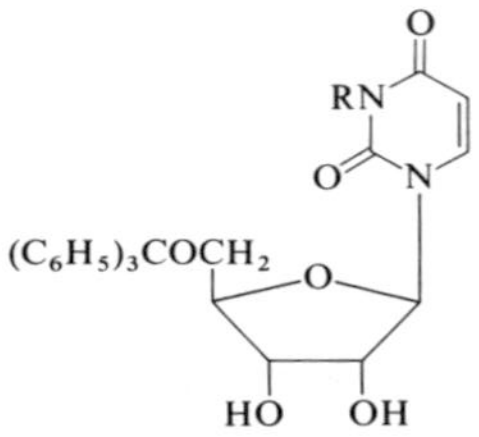 RN, O, N, O, $(C_6H_5)_3COCH_2$, HO, OH	A	1. TCDI, THF, 25°, 5 days 2. Raney Ni, CH_3COCH_3, C_2H_5OH, reflux, 6 hr	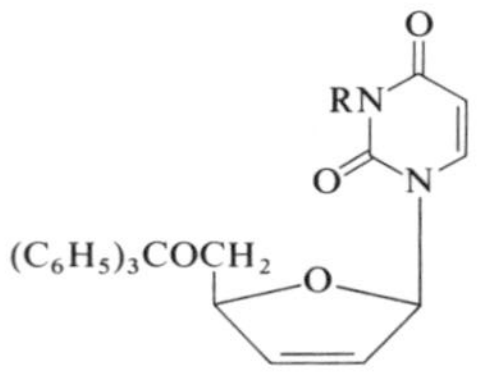 RN, O, N, O, $(C_6H_5)_3COCH_2$ R = H (14)	77

Carbon	Reactant	Method	Conditions	Product(s) and Yield(s) (%)	Refs.
	CH_3O, HO, OH (structure)	A	1. TCDI, THF, 25°, 5 days 2. $(CH_3O)_3P$, reflux, 65 hr	" R = CH_3 (80)	77
		E	1. CH_3Li, 2 eq 2. K_2WCl_6, THF, reflux, 46 hr	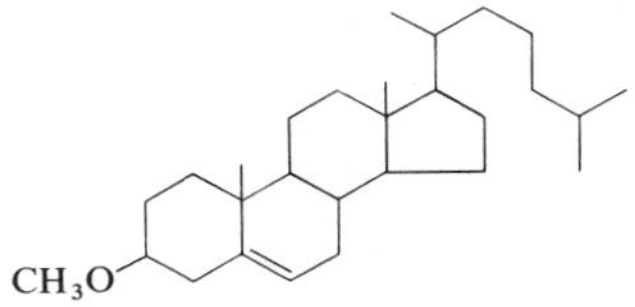 CH_3O (structure) (54)	131
		H	NaI, $(CH_3)_3SiCl$, CH_3CN, 25°, 15 min	" (95)	151
	CH_3O, HO, OH (structure)	H	NaI, $(CH_3)_3SiCl$, CH_3CN, 25°, 10 min	" (98)	151
	HO, HO, H (structure)	E	$TiCl_3$, K, THF, reflux, 16 hr	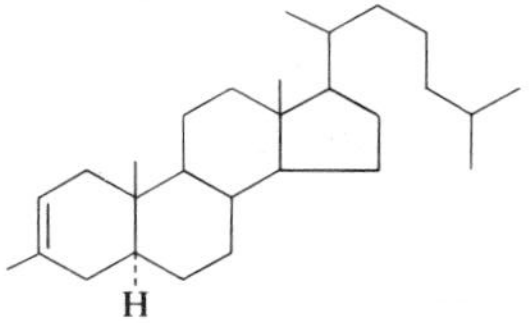 H (structure) (80)	171
C_{29} 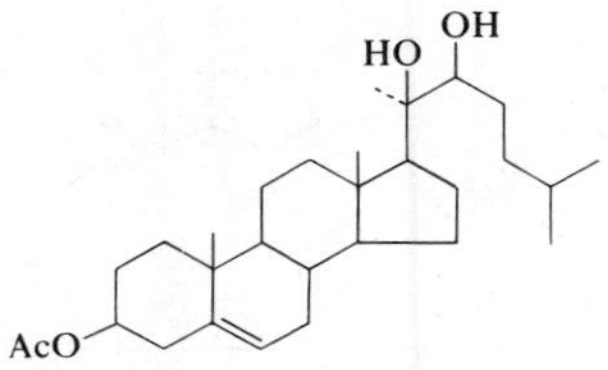	AcO, HO, OH (structure)	B	1. $(C_2H_5O)_3CH$, $C_6H_5CO_2H$, 110°, 3 hr 2. 170–175°, 15 min	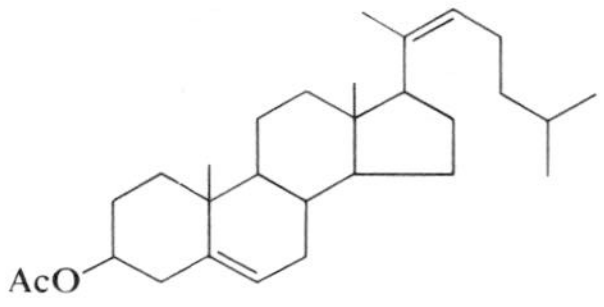 AcO (structure) (48)	95

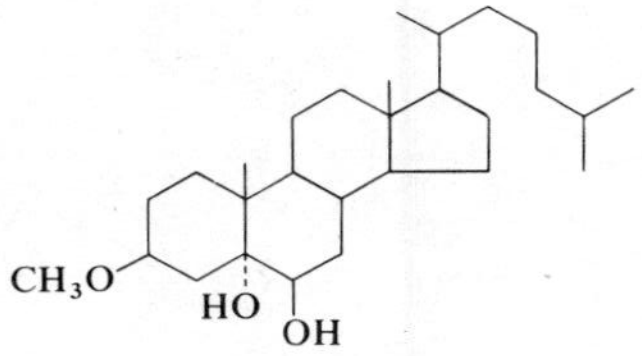

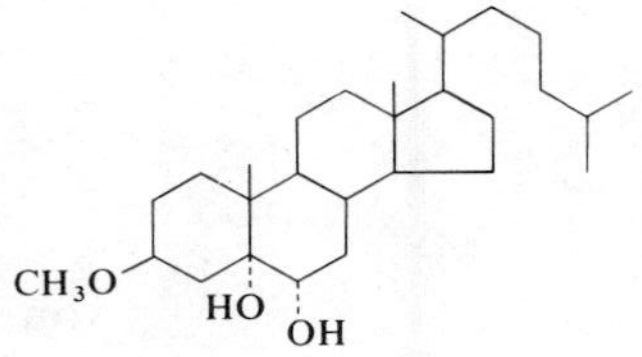

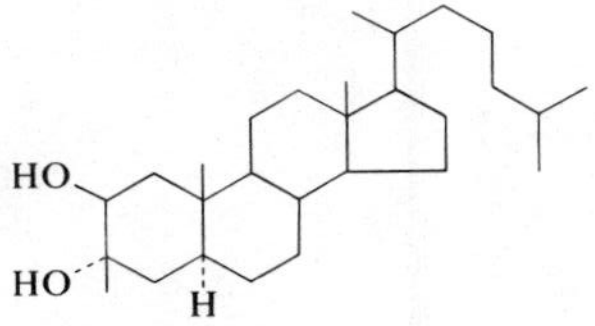

TABLE I. OLEFIN SYNTHESIS BY DEOXYGENATION OF VICINAL DIOLS (*Continued*)

C_n	Diol	Method	Conditions	Products and Yields (%)	Refs.
C_{30}	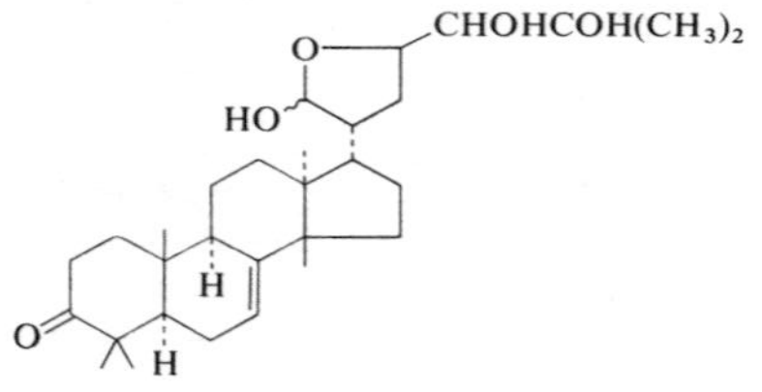 $CHOHCOH(CH_3)_2$; HO; O; H; H; O	A	1. TCDI, $C_6H_5CH_3$, heat 2. $(CH_3O)_3P$, heat 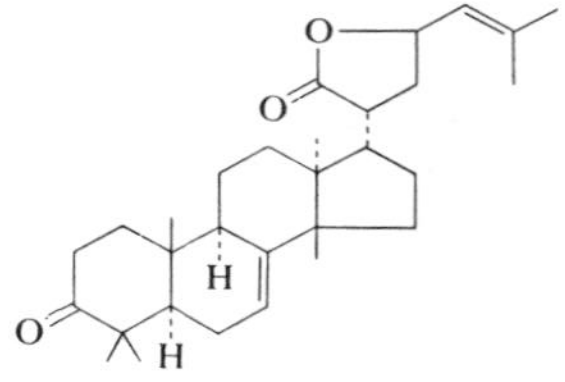	O; O; H; H; O (—)	45
	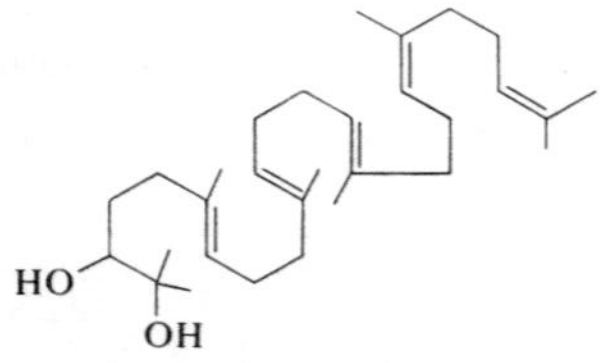 HO; OH	A	1. TCDI, $C_6H_5CH_3$, 110°, 5 hr 2. $(CH_3O)_3P$, 130°, 90 hr	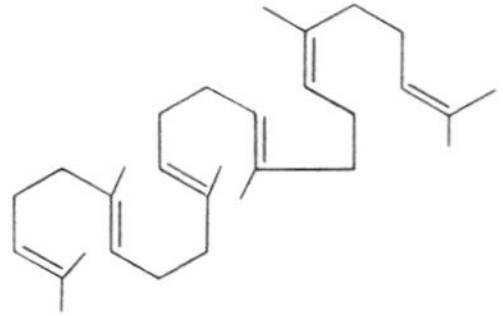 (59)	49
C_{33}	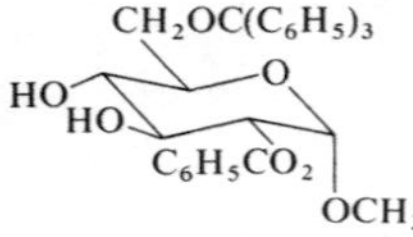 $CH_2OC(C_6H_5)_3$; HO; HO; O; $C_6H_5CO_2$; OCH_3	H	1. MsCl, Py, −10° 2. NaI, Zn–Cu, DMF, reflux, 3 hr 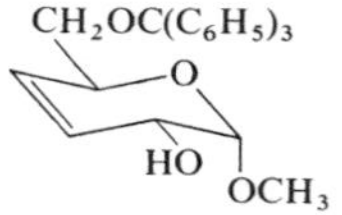	$CH_2OC(C_6H_5)_3$; O; HO; OCH_3 (74)	143
	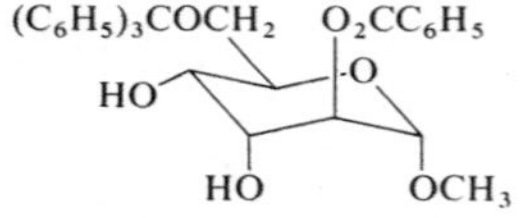 $(C_6H_5)_3COCH_2$; $O_2CC_6H_5$; HO; O; HO; OCH_3	H	1. MsCl, Py, 10° 2. NaI, Zn–Cu, DMF, reflux, 6 hr	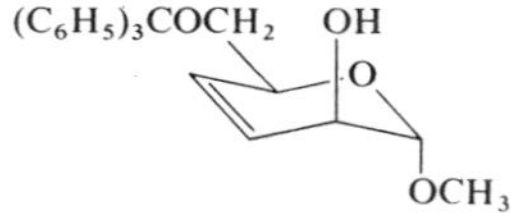 $(C_6H_5)_3COCH_2$; OH; O; OCH_3 (—)	143
C_{34}	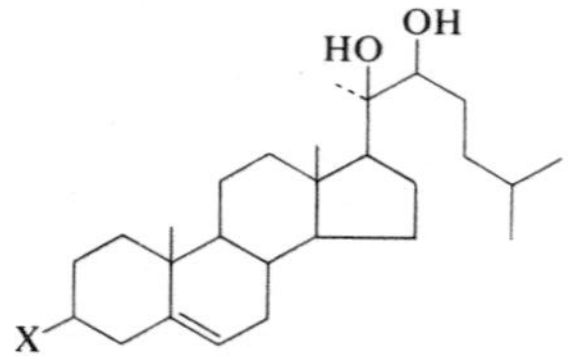 HO; OH; X $(X = C_6H_5CO_2)$	A	1. TCDI, Py, 110°, 12 hr 2. $(C_2H_5O)_3P$, reflux, 12 hr 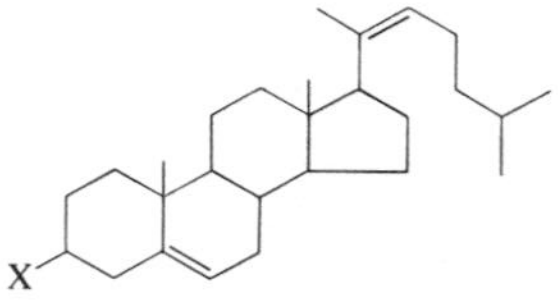	X (82)	48

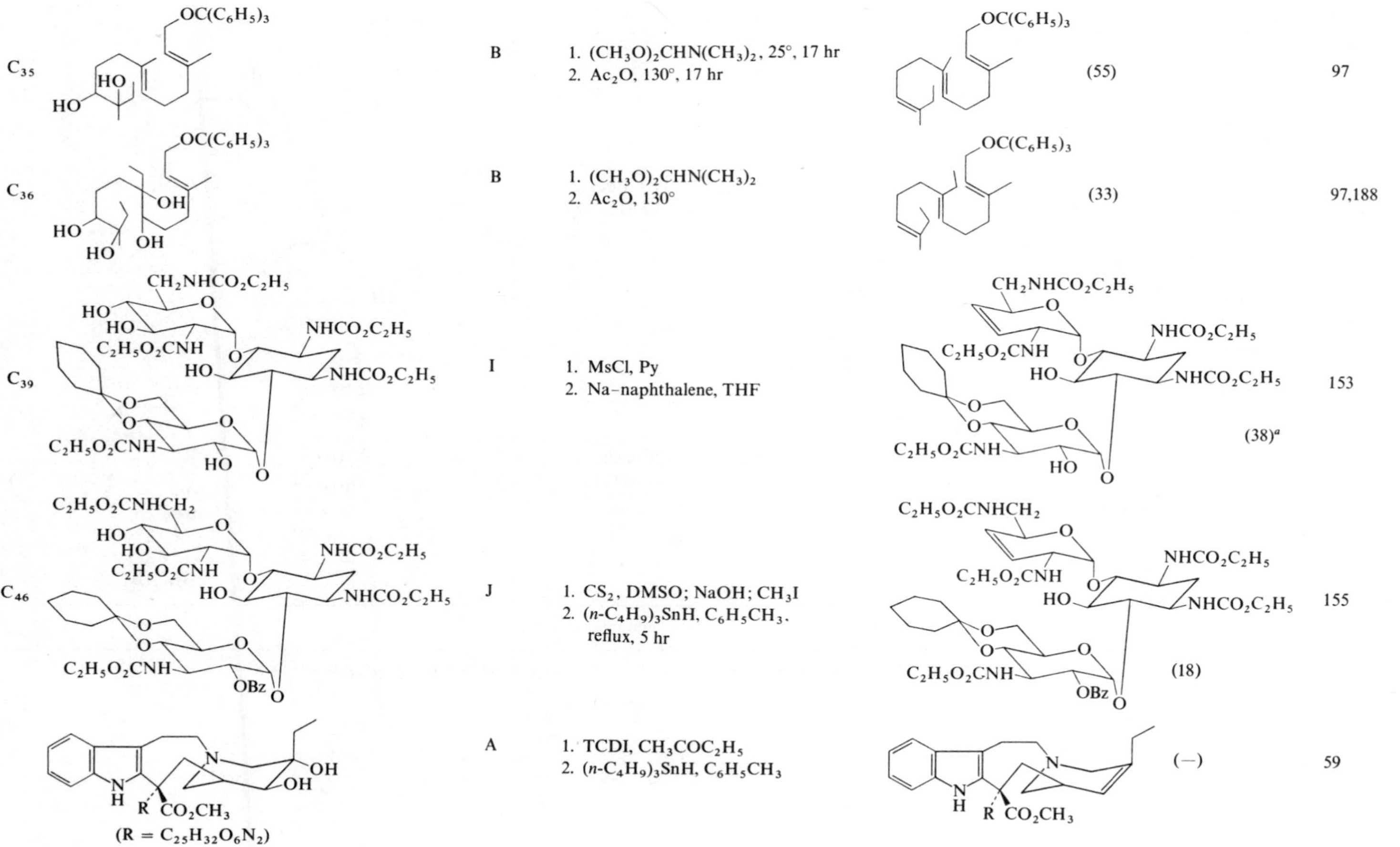

Formula	Method	Conditions	Product (Yield, %)	Refs.
C_{35}	B	1. $(CH_3O)_2CHN(CH_3)_2$, 25°, 17 hr 2. Ac_2O, 130°, 17 hr	(55)	97
C_{36}	B	1. $(CH_3O)_2CHN(CH_3)_2$ 2. Ac_2O, 130°	(33)	97,188
C_{39}	I	1. MsCl, Py 2. Na–naphthalene, THF	(38)[a]	153
C_{46}	J	1. CS_2, DMSO; NaOH; CH_3I 2. $(n\text{-}C_4H_9)_3SnH$, $C_6H_5CH_3$, reflux, 5 hr	(18)	155
($R = C_{25}H_{32}O_6N_2$)	A	1. TCDI, $CH_3COC_2H_5$ 2. $(n\text{-}C_4H_9)_3SnH$, $C_6H_5CH_3$	(—)	59

TABLE I. OLEFIN SYNTHESIS BY DEOXYGENATION OF VICINAL DIOLS (*Continued*)

C_n	Diol	Method	Conditions	Products and Yields (%)	Refs.
C_{53}	[structure: CH_2NHR, HO, HO, NHR, O, NHR, $HOCH_2$, O, OH, OH, OH, NHCO–HOCH–CH_2–$RNHCH_2$]; $R = CO_2Bz$	I	1. MsCl, Py 2. Na–naphthalene, THF	[structure: CH_2NHR, O, NHR, NHR, O, $HOCH_2$, OH, OH, OH, NHCO–HOCH–CH_2–$RNHCH_2$]; $R = CO_2Bz$ (25)	156
C_{59}	[structure: $TsNHCH_2$, O, OH, HO, NHTs, O, NHTs, OH, NHTs, O–CH_2, O, NHTs, O, O, OH]	H	1. $C_6H_5CH_2SO_2Cl$, Py, $-5°$, 18 hr 2. NaI, DMF, 100°, 15 min	[structure: $TsNHCH_2$, O, NHTs, O, NHTs, OH, NHTs, O–CH_2, O, NHTs, O, O, OR] (R = H, 56)	145
		H	1. MsCl, Py, 60°, 18 hr 2. NaI, DMF, 100°, 30 min	" (R = Ms, 44)	145

C_{64}

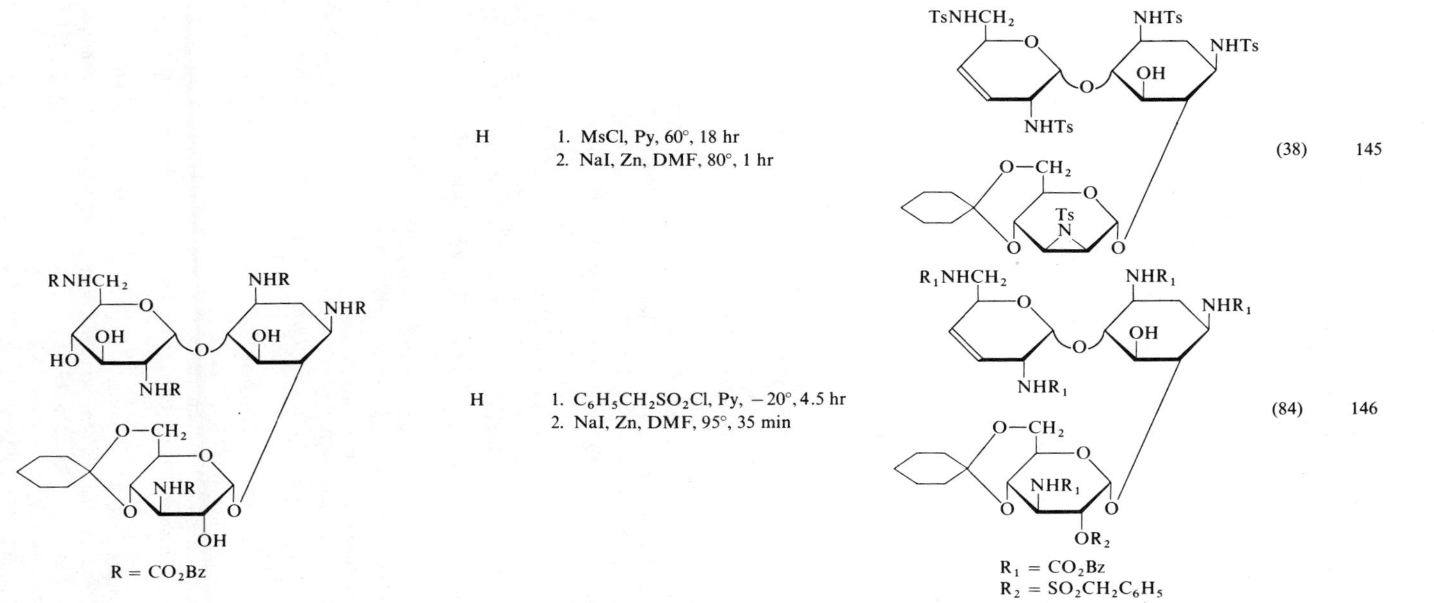

H	1. MsCl, Py, 60°, 18 hr 2. NaI, Zn, DMF, 80°, 1 hr		(38)	145
H	1. $C_6H_5CH_2SO_2Cl$, Py, −20°, 4.5 hr 2. NaI, Zn, DMF, 95°, 35 min		(84)	146

[a] The yield is for the final step only.
[b] The stereochemistry was not indicated.
[c] The product was isolated as the 2,5-diphenyl-3,4-isobenzofuran adduct.
[d] The product was formed by *in situ* trapping.
[e] The product was formed by external trapping.
[f] The product was isolated as the dichlorocarbene adduct.
[g] The optical purity was >99 % from the optically active diol.
[h] The optical purity was 78 % from the optically active diol.
[i] These reaction conditions are assumed.
[j] The reaction conditions were not given.

REFERENCES

[1] E. J. Corey and R. A. E. Winter, *J. Am. Chem. Soc.*, **85**, 2677 (1963).

[2] E. J. Corey, F. A. Carey, and R. A. E. Winter, *J. Am. Chem. Soc.*, **87**, 934 (1965).

[3] R. K. Mackie, in *Organophosphorus Reagents in Organic Synthesis*, J. I. G. Cadogan, Ed., Academic Press, London, 1979, Chapter 8.

[4] E. Block, *Reactions of Organosulfur Compounds*, Academic Press, New York, 1978, p. 229; E. Block, *Aldrichimica Acta*, **11**, 51 (1978).

[5] E. Block, *J. Chem. Ed.*, **48**, 814 (1971).

[6] S. G. Wilkinson, in *Comprehensive Organic Chemistry*, D. H. R. Barton and W. D. Ollis, Eds., Pergamon Press, Oxford, 1979, Vol. I, p. 693.

[7] K. Heyns and J. Feldmann, *Starke*, **32**, 40 (1980).

[8] D. M. Lemal, R. A. Lovald, and R. W. Harrington, *Tetrahedron Lett.*, **1965**, 2779; D. M. Lemal, E. P. Gosselink, and A. Ault, *ibid.*, **1964**, 579.

[9] K. Mackenzie, *J. Chem. Soc.*, **1964**, 5710.

[10] R. A. E. Winter, *Ph.D. Thesis*, Harvard University, 1965 [*Diss. Abstr. B* **27**, 2305 (1967)].

[11] W. T. Borden, P. W. Concannon, and D. I. Phillips, *Tetrahedron Lett.*, **1973**, 3161.

[12] W. T. Borden and L. H. Hoo, *J. Am. Chem. Soc.*, **100**, 6274 (1978).

[13] P. Stromquist, M. Radcliffe, and W. P. Weber, *Tetrahedron Lett.*, **1973**, 4523.

[14] D. Feller, E. R. Davidson, and W. T. Borden, *J. Am. Chem. Soc.*, **103**, 2558 (1981).

[15] G. Bianchi, C. DeMicheli, and R. Gandolfi, *Angew. Chem., Int. Ed. Engl.*, **18**, 721 (1979).

[16] W. C. Davies and W. J. Jones, *J. Chem. Soc.*, **1929**, 33.

[17] J. I. Shulman, *Ph.D. Thesis*, Harvard University, 1970 [*Diss. Abstr. Int. B*, **32**, 3268 (1971)].

[18] E. J. Corey and G. Märkel, *Tetrahedron Lett.*, **1967**, 3201.

[19] G. Scherowsky and J. Weiland, *Chem. Ber.*, **107**, 3155 (1974).

[20] P. Koll and F. S. Tayman, *Chem. Ber.*, **112**, 2296 (1979).

[21] R. Greenhouse, W. T. Borden, T. Ravindranathan, K. Hirotsu, and J. Clardy, *J. Am. Chem. Soc.*, **99**, 6955 (1977); R. Greenhouse, T. Ravindranathan, and W. T. Borden, *ibid.*, **98**, 6738 (1976).

[22] L. A. Paquette, I. Itoh, and W. B. Farnham, *J. Am. Chem. Soc.*, **97**, 7280 (1975).

[23] C. Párkányi and A. Vystrcil, *Roczn. Chem.*, **39**, 931 (1965) [*C. A.*, **64**, 720 (1966)].

[24] R. H. Mitchell, T. Fyles, and L. M. Ralph, *Can. J. Chem.*, **55**, 1480 (1977).

[25] E. J. Corey and P, B. Hopkins, *Tetrahedron Lett.*, **23**, 1979 (1982).

[26] J. Daub, G. Endress, V. Erhardt, K. H. Jogun, J. Kapples, A. Laumer, R. Phiz, and J. J. Stezowski, *Chem. Ber.*, **115**, 1787 (1982).

[27] A. H. Haines, *Carbohydr. Res.*, **1**, 214 (1965).

[28] G. O. Aspinall, N. W. H. Cheetham, J. Furdora, and S. C. Tam, *Carbohydr. Res.*, **36**, 257 (1974).

[29] B. S. Shasha, D. Trimnell, and W. M. Doane, *Carbohydr. Res.*, **32**, 349 (1974).

[30] W. K. Anderson and R. H. Dewey, *J. Am. Chem. Soc.*, **95**, 7161 (1973).

[31] H. M. Fishler and W. Hartmann, *Chem. Ber.*, **105**, 2769 (1972).

[32] T. J. Pullukat and G. Urry, *Tetrahedron Lett.*, **1967**, 1953.

[33] C. Larsen, K. Steliou, and D. N. Harpp, *J. Org. Chem.*, **43**, 337 (1978); C. Larsen and D. N. Harpp, *ibid.*, **45**, 3713 (1980).

[34] O. A. Ossanna, Jr., *Diss. Abstr.*, **24**, 1835 (1963).

[35] A. Schlosser and P. Weiss, *Synthesis*, **1970**, 251.

[36] W. Hartmann, L. Schrader, and D. Wendisch, *Chem. Ber.*, **106**, 1076 (1973).

[37] M. Tichý and J. Sicher, *Tetrahedron Lett.*, **1969**, 4609; *Coll. Czech. Chem. Commun.*, **37**, 3106 (1972).

[38] L. A. Paquette, R. P. Micheli, and J. M. Photis, *J. Am. Chem. Soc.*, **99**, 7911 (1977).

[39] L. A. Paquette and J. C. Philips, *Tetrahedron Lett.*, **1967**, 4645; L. A. Paquette, J. C. Philips, and R. E. Wingard, Jr., *J. Am. Chem. Soc.*, **93**, 4516 (1971).

[40] J. A. Chong and J. R. Wiseman, *J. Amer. Chem. Soc.*, **94**, 8627 (1972).

[41] P. E. Sonnet, *Tetrahedron*, **36**, 557 (1980).

[42] E. J. Corey and J. I. Shulman, *Tetrahedron Lett.*, **1968**, 3655.

[43] L. Hevesi, J. B. Nagy, A. Krief, and E. G. Derouane, *Org. Magn. Res.*, **10**, 14 (1977); A. Krief, L. Hevesi, J. B. Nagy, and E. G. Derouane, *Angew. Chem., Int. Ed. Engl.*, **16**, 100 (1977).

[44] S. Iwasaki, H. Muro, K. Sasaki, S. Nozoe, S. Okuda, and Z. Sato, *Tetrahedron. Lett.*, **1973**, 3537.

[45] C. W. L. Bevan, D. E. U. Ekong, T. G. Halsall, and P. Toft, *Chem. Commun.*, **1965**, 636.

[46] W. J. McGahren, G. A. Ellestad, G. O. Morton, M. P. Kunstmann, and P. Mullen, *J. Org. Chem.*, **38**, 3542 (1973).

[47] A. H. Haines, *Carbohydr. Res.*, **21**, 99 (1972).

[48] M. Koreeda, N. Koizumi, and B. A. Teicher, *J. Chem. Soc., Chem. Commun.*, **1976**, 1035.

[49] A. Krief, *Comp. Rend. Acad. Sci. Paris C*, **275**, 459 (1972).

[50] S. Ramachandran, R. V. Panganamala, and D. G. Cornwell, *J. Lipid Res.*, **10**, 465 (1969).

[51] F. D. Gunstone and F. R. Jacobsborg, *Chem. Phys. Lipids*, **9**, 112 (1972).

[52] L. Hevesi, J. Hontoy, A. Krief, J. Lubochinsky, and B. Lubochinsky, *Bull. Soc. Chim. Belg.*, **84**, 709 (1975).

[53] J. P. Kutney, K. K. Chan, W. B. Evans, Y. Fujise, T. Honda, F. K. Klein, and J. P. de Souza, *Heterocycles*, **6**, 435 (1977).

[54] J. P. Kutney, U. Bunzli-Trepp, K. K. Chan, J. P. de Souza, Y. Fujise, T. Honda, J. Katsube, F. K. Klein, A. Leutwiler, S. Morehead, M. Rohr, and B. R. Worth, *J. Am. Chem. Soc.*, **100**, 4220 (1978).

[55] M. F. Semmelhack and R. D. Stauffer, *Tetrahedron Lett.*, **1973**, 2667.

[56] J. Daub, V. Trautz, and U. Erhardt, *Tetrahedron Lett.*, **1972**, 4435.

[57] J. Daub and U. Erhardt, *Tetrahedron Lett.*, **28**, 181 (1972).

[58] J. Daub, U. Erhardt, J. Kappler, and V. Troutz, *J. Organometal. Chem.*, **69**, 423 (1974).

[59] J. P. Kutney, T. Honda, P. M. Kazmaier, N. J. Lewis, and B. R. Worth, *Helv. Chim. Acta*, **63**, 366 (1980).

[60] E. Vedejs and E. S. C. Wu, *Tetrahedron Lett.*, **1973**, 3793.

[61] E. Vedejs and E. S. C. Wu, *J. Org. Chem.*, **39**, 3641 (1974).

[62] D. H. R. Barton and R. V. Stick, *J. Chem. Soc., Perkin Trans. 1*, **1975**, 1773.

[63] K. Funaki, K. Takeda, and E. Yoshii, *Tetrahedron Lett.*, **23**, 3069 (1982).

[64] D. P. Bauer and R. S. Macomber, *J. Org. Chem.*, **41**, 2640 (1976).

[65] S. I. Shiguro and S. Tejima, *Chem. Pharm. Bull. (Tokyo)*, **16**, 2040 (1968).

[66] A. Schönberg and S. Nickel, *Chem. Ber.*, **64**, 2323 (1931); D. Seebach, *Angew. Chem., Int. Ed. Engl.*, **8**, 639 (1968); N. C. Gonnella, M. V. Lakshmikantham, and M. P. Cava, *Syn. Commun.*, **9**, 17 (1979).

[67] E. Vedejs, M. J. Arnost, J. M. Dolphin, and J. Eustache, *J. Org. Chem.*, **45**, 2601 (1980).

[68] D. M. Lemal and E. H. Banitt, *Tetrahedron Lett.*, **1964**, 245.

[69] D. B. J. Easton, D. Leaver, and T. J. Rawlings, *J. Chem. Soc., Perkin Trans. 1*, **1972**, 41.

[70] K. Hatanaka, S. Tanimoto, T. Oida, and M. Okano, *Tetrahedron Lett.*, **22**, 5195 (1981).

[71] A. Schönberg and L. V. Varhga, *Ber. Dtsch. Chem. Ges.*, **178** (1930).

[72] I. Dyong, R. Hermann, and R. Matter, *Chem. Ber.*, **113**, 1931 (1980).

[73] D. Trimnell, W. M. Doane, C. R. Russell, and C. E. Rist, *Carbohydr. Res.*, **17**, 319 (1971).

[74] F. N. Jones and S. Andreades, *J. Org. Chem.*, **34**, 3011 (1969).

[75] P. Koll and S. Deyhim, *Chem. Ber.*, **111**, 2913 (1978).

[76] D. Horton and C. G. Tindall, Jr., *J. Org. Chem.*, **35**, 3558 (1970).

[77] W. V. Ruyle, T. Y. Shen, and A. A. Patchett, *J. Org. Chem.*, **30**, 4353 (1965).

[78] J. J. Fox and I. Wempen, *Tetrahedron Lett.*, **1965**, 643.

[79] K. Anzai and M. Matsui, *Agr. Biol. Chem.*, **37**, 345 (1973).

[80] H. P. Figeys and M. Gelbcke, *Bull. Soc. Chim. Belg.*, **83**, 381 (1974).

[81] C. Sandris, *Tetrahedron*, **24**, 3589 (1968).

[82] H. Prinzbach and H. Babsch, *Angew. Chem., Int. Ed. Engl.*, **14**, 753 (1975).

[83] W. Hartmann, H.-M. Fischler, and H.-G. Heine, *Tetrahedron Lett.*, **1972**, 853.

[84] E. Lette and D. H. Lucast, *Tetrahedron Lett.*, **1976**, 3401.

[85] M. Z. Haq, *J. Org. Chem.*, **37**, 3015 (1972).

[86] G. Crank and F. W. Eastwood, *Aust. J. Chem.*, **17**, 1392 (1964).

[87] A. P. M. van der Veek and F. H. van Putten, *Tetrahedron Lett.*, **1970**, 3951; J. W. Scheeren, A. P. M. van der Veek, and W. Stevens, *Rec. Trav. Chim.*, **88**, 195 (1969).

[88] H. Wynberg and A. Kraak, *J. Am. Chem. Soc.*, **83**, 3919 (1961).

[89] F. W. Eastwood, K. J. Harrington, J. S. Josan, and J. L. Pura, *Tetrahedron Lett.*, **1970**, 5223.

[90] S. Hanessian, A. Bargiotti, and M. LaRue, *Tetrahedron Lett.*, **1978**, 737.

[91] H. Rakoff and E. A. Emken, *Lipids*, **12**, 760 (1977).

[92] A. N. Singh, V. V. Mhaskar, and S. Dev, *Tetrahedron*, **34**, 595 (1978).

[93] A. W. Burgstahler, D. L. Boger, and N. C. Naik, *Tetrahedron*, **32**, 309 (1976).

[94] J. S. Josan and F. W. Eastwood, *Carbohydr. Res.*, **7**, 161 (1968).

[95] C. Byon and M. Gut, *J. Org. Chem.*, **41**, 3716 (1976).

[96] T. Hiyama and H. Nozaki, *Bull. Chem. Soc. Jpn.*, **46**, 2248 (1973).

[97] A. Yasuda, S. Tanaka, H. Yamanoto, and H. Nozaki, *Bull. Chem. Soc. Jpn.*, **52**, 1701 (1979).

[98] G. Stork, T. Takahashi, I. Kawamoto, and T. Suzuki, *J. Am. Chem. Soc.*, **100**, 8272 (1978).

[99] A. Itoh, S. Ozawa, K. Oshima, S. Sasaki, H. Yamamoto, T. Hiyama, and H. Nozaki, *Bull. Chem. Soc. Jpn.*, **53**, 2357 (1980).

[100] K. M. Sun and B. Fraser-Reid, *Can. J. Chem.*, **58**, 2732 (1980).

[101] R. E. Ireland and J-P. Vevert, *Can. J. Chem.*, **59**, 572 (1981).

[102] R. A. Braun, *J. Org. Chem.*, **31**, 1147 (1966).

[103] G. Goto, *Bull. Chem. Soc. Jpn.*, **50**, 186 (1977).

[104] G. I. Moss, G. Crank, and F. W. Eastwood, *Chem. Commun.*, **1970**, 206.

[105] P. S. Wharton, G. A. Hiegel, and S. Ramaswami, *J. Org. Chem.*, **29**, 2441 (1964).

[106] L. J. Nehmsmann, *Diss. Abstr.*, **23**, 1929 (1962).

[107] K. D. Berlin, B. S. Rathore, and M. Peterson, *J. Org. Chem.*, **30**, 226 (1965).

[108] T. L. V. Ulbricht, *J. Chem. Soc.*, **1965**, 6649.

[109] T. Argtani, Y. Nakanisi, and H. Nozaki, *Tetrahedron*, **26**, 4339 (1970).

[110] J. N. Hines, M. J. Peagram, E. J. Thomas, and G. H. Whitham, *J. Chem. Soc., Perkin Trans. 1*, **1973**, 2332.

[111] V. Ceré, A. Guenzi, S. Pollicino, E. Sandri, and A. Fava, *J. Org. Chem.*, **45**, 261 (1980).

[112] W. T. Borden and T. Ravindranathan, *J. Org. Chem.*, **36**, 4125 (1971).

[113] G. H. Whitham and M. Wright, *J. Chem. Soc. (C)*, **1971**, 886.

[114] A. F. Thomas and W. Pawlak, *Helv. Chim. Acta*, **54**, 1822 (1971).

[115] J. N. Hines, M. J. Peagram, G. H. Whitham, and M. Wright, *Chem. Commun.*, **1968**, 1593.

[116] A. Klemer and G. Rodemeyer, *Chem. Ber.*, **107**, 2612 (1974).

[117] J. Hine, L. G. Mahone, and C. L. Liotta, *J. Am. Chem. Soc.*, **89**, 5911 (1967).

[118] M. Jones, P. Temple, E. J. Thomas, and G. H. Whitham, *J. Chem. Soc., Perkin Trans. 1*, **1974**, 433.

[119] S. Ranganathan, D. Ranganathan, P. V. Ramachandra, M. K. Mahanty, and S. Bamezai, *Tetrahedron*, **37**, 4171 (1981).

[120] J. A. Marshall and M. E. Lewellyn, *Syn. Commun.*, **5**, 293 (1975); *J. Org. Chem.*, **42**, 1311 (1977).

[121] J. A. Marshall and M. Lewellyn, *J. Am. Chem. Soc.*, **99**, 3508 (1977).

[122] S. Nakayama, M. Yashifugi, R. O. Kazaki, and N. Inamoto, *J. Chem. Soc., Perkin Trans. 1*, **1973**, 2069.

[123] J. E. McMurry and M. P. Fleming, *J. Org. Chem.*, **41**, 896 (1976).

[124] J. E. McMurry and M. P. Fleming, *J. Am. Chem. Soc.*, **96**, 4708 (1974).

[125] K. B. Sharpless, R. P. Hanzlik, and E. E. van Tamelen, *J. Am. Chem. Soc.*, **90**, 209 (1968).

[126] R. Kuhn and A. Winterstein, *Helv. Chim. Acta*, **11**, 106 (1928); R. Kuhn and K. Wallenfels, *Chem. Ber.* **71**, 1899 (1938); R. Kuhn and K. L. Scholler, *Chem. Ber.*, **87**, 598 (1954).

[127] H. M. Walborsky and M. P. Murari, *J. Am. Chem. Soc.*, **102**, 426 (1980); A. L. Baumstark, C. J. McCloskey, T. J. Tolson, and G. T. Syriopoulos, *Tetrahedron Lett.*, **1977**, 3003.

[128] H. M. Walborsky and H. H. Wust, *J. Am. Chem. Soc.*, **104**, 5807 (1982).

[129] K. B. Sharpless and T. C. Flood, *J. Chem. Soc., Chem. Commun.*, **1972**, 370; K. B. Sharpless, M. A. Umbreit, M. T. Nieh, and T. C. Flood, *J. Am. Chem. Soc.*, **94**, 6538 (1972).

[130] J. A. Marshall, R. E. Bierenbaum, and K-H. Chung, *Tetrahedron Lett.*, **1979**, 2081.

[131] T. C. Flood, *Ph.D. Thesis*, Massachusetts Institute of Technology, 1972.

[132] D. H. Harpp and K. Steliou, in *Topics in Organic Sulfur Chemistry*, M. Tisler, Ed., University Press, Ljubljana, 1978, p. 105 (8th International Symposium on Organic Sulphur Chemistry).

[133] G. W. Astrologes and J. C. Martin, *J. Am. Chem. Soc.*, **98**, 2895 (1976).

[134] T. Yamazaki, K. Matsuda, H. Sugiyama, S. Seto, and N. Yamaoka, *J. Chem. Soc., Perkin Trans. 1*, **1977**, 1981.

[135] R. S. Tipson and L. H. Cretcher, *J. Org. Chem.*, **8**, 95 (1943).

[136] R. S. Tipson, *Adv. Carbohydr. Chem.*, **8**, 201 (1953).

[137] S. J. Angyal and P. T. Gilham, *J. Chem. Soc.*, **1958**, 375.

[138] H. L. Slates and N. L. Wendler, *J. Am. Chem. Soc.*, **78**, 3749 (1956); N. L. Wendler, H. L. Slates, and M. Tishler, *J. Am. Chem. Soc.*, **74**, 4894 (1952).

[139] R. S. Tipson and A. Cohen, *Carbohydr. Res.*, **1**, 338 (1965); *Adv. Carbohydr. Chem.*, **8**, 107 (1953).

[140] J. Cleophax, J. Hildesheim, and S. D. Gero, *Bull. Soc. Chim. Fr.*, **1967**, 4111; J. Cleophax and S. D. Gero, *Tetrahedron Lett.*, **1966**, 5505.

[141] B. Fraser-Reid and B. Boctor, *Can. J. Chem.*, **47**, 393 (1969).

[142] S. Umezawa, Y. Okazaki, and T. Tsuchiya, *Bull. Chem. Soc. Jpn.*, **45**, 3619 (1972).

[143] N. L. Holder and B. Fraser-Reid, *Can. J. Chem.*, **51**, 3357 (1973).

[144] T. Yamazaki, H. Sugiyama, N. K. Matsuda, and S. Seto, *Carbohydr. Res.*, **50**, 279 (1976).

[145] T. Miyake, T. Tsuchiya, S. Umezawa, and H. Umezawa, *Carbohydr. Res.*, **49**, 141 (1976).

[146] T. Nishimura, T. Tsuchiya, S. Umezawa, and H. Umezawa, *Bull. Chem. Soc. Jpn.*, **50**, 1580 (1977).

[147] E. L. Abano, D. Horton, and T. Tsuchiya, *Carbohydr. Res.*, **2**, 349 (1966).

[148] T. van Es, *Carbohydr. Res.*, **5**, 282 (1967).

[149] H. Ohruti and S. Emoto, *Carbohydr. Res.*, **10**, 221 (1969).

[150] B. K. Radatus and I. S. Clarke, *Synthesis*, **1980**, 47.

[151] N. C. Barua and R. P. Sharma, *Tetrahedron Lett.*, **23**, 1365.

[152] J. C. Carnahan, Jr., and W. D. Closson, *Tetrahedron Lett.*, **1972**, 3447; J. C. Carnahan, Jr., Ph.D. Thesis, SUNY—Albany, 1973; J. C. Carnahan, Jr., *Diss. Abstr.*, **34**, 4282–B (1974).

[153] T. Hayashi, M. Takeda, H. Saeki, and E. Ohki, *Chem. Pharm. Bull.* (*Tokyo*), **25**, 2134 (1977).

[154] A. G. M. Barrett, D. H. R. Barton, and R. Bielski, *J. Chem. Soc., Perkin Trans. 1*, **1979**, 2378; A. G. M. Barrett, D. H. R. Barton, R. Bielski, and S. W. McCombie, *J. Chem. Soc., Chem. Commun.*, **1977**, 866.

[155] T. Hayashi, T. Iwaoka, N. Takeda, and E. Ohki, *Chem. Pharm. Bull.* (*Tokyo*), **26**, 1786 (1978).

[156] H. H. Inhoffen, K. Radscheit, U. Stache, and V. Koppe, *Justus Liebigs Ann. Chem.*, **684**, 24 (1965).

[157] P. J. Garegg and B. Samuelsson, *Synthesis*, **1979**, 469.

[158] P. J. Garegg and B. Samuelsson, *Synthesis*, **1979**, 813.

[159] M. Bessodes, E. Abushanab, and R. P. Panzica, *J. Chem. Soc., Chem. Commun.*, **1981**, 26.

[160] C. E. Slemon and P. Yates, *Can. J. Chem.*, **57**, 304 (1979); J. E. McMurry and W. Choy, *J. Org. Chem.*, **43**, 1800 (1978).

[161] M. Guisnet, I. Plouzennec, and R. Maurel, *Compt. Rend.*, **274C**, 2102 (1972).

[162] A. Yasuda, S. Tanaka, H. Yamamoto, and H. Nozaki, *Bull. Chem. Soc. Jpn.*, **52**, 1752 (1979).

[163] S. Tanaka, A. Yasuda, H. Yamamoto, and H. Nozaki, *J. Am. Chem. Soc.*, **97**, 3252 (1975).

[164] R. Gleiter, P. Bischof, W. E. Volz, and L. A. Paquette, *J. Am. Chem. Soc.*, **99**, 8 (1977).

[165] P. Camps, J. Cardellach, J. Font, R. M. Ortuno, and O. Ponsati, *Tetrahedron*, **38**, 2395 (1982).

[166] D. Horton, J. K. Thompson, and C. G. Tindall, Jr., *Methods Carbohydr. Chem.*, **6**, 297 (1972).

[167] M. K. Das and J. J. Zuckerman, *Inorg. Chem.*, **10**, 1028 (1971).

[168] E. N. Walsh and A. D. F. Toy, *Inorg. Synth.*, **7**, 69 (1963).

[169] G. W. J. Fleet and M. J. Gough, *Tetrahedron Lett.*, **23**, 4509 (1982).

[170] K. R. Kopecky and J. H. van de Sande, *Can. J. Chem.*, **50**, 4034 (1972).

[171] J. E. McMurry, M. P. Fleming, K. L. Kees, and L. R. Krepski, *J. Org. Chem.*, **43**, 3255 (1978).

[172] J. S. Josan and F. W. Eastwood, *Aust. J. Chem.*, **21**, 2013 (1968).

[172a] M. A. Hashem and P. Weyerstahl, *Tetrahedron*, **37**, 2473 (1981).

[173] U. H. Andrews, J. E. Baldwin, and M. W. Grayston, *J. Org. Chem.*, **47**, 287 (1982).

[174] W. Reeve and D. R. Kuroda, *J. Org. Chem.*, **45**, 2305 (1980).

[175] V. Ceré, E. Dalcanale, C. Paolucci, S. Pollicino, E. Sandri, L. Lunazzi, and A. Fava, *J. Org. Chem.*, **47**, 3540 (1982).

[176] J. Daub, J. Kappler, K.-P. Krenkler, S. Schreiner, and V. Trantz, *Justus Liebigs Ann. Chem.*, **1977**, 1730.

[177] S. Bien and J. Segal, *J. Org. Chem.*, **42**, 3983 (1977).

[178] D. Horton and W. N. Turner, *Tetrahedron Lett.*, **1964**, 2531.

[179] D. Horton and W. N. Turner, *Carbohydr. Res.*, **1**, 444 (1966).

[180] G. L. Tong, W. W. Lee, and L. Goodman, *J. Org. Chem.*, **30**, 2854 (1965).

[181] J. E. Pauw and A. C. Weedon, *Tetrahedron Lett.*, **23**, 5485 (1982).

[182] T. L. Nagabhushan, *Can. J. Chem.*, **48**, 383 (1970).

[183] I. Dyong, N. Jersch, and Q. Lam-Chi, *Chem. Ber.*, **112**, 1859 (1979).

[184] L. M. Lerner, *J. Org. Chem.*, **37**, 470 (1972).

[185] S. Takano, C. Kasahara, and K. Ogasawara, *J. Chem. Soc., Chem. Commun.*, **1981**, 637.

[186] C. R. Enzell, Y. Hirose, and B. R. Thomas, *Tetrahedron Lett.*, **1967**, 793.

[187] J. D. Connolly, R. McCrindle, R. D. H. Murray, A. J. Renfrew, K. H. Overton, and A. Melera, *J. Chem. Soc.* (*C*), **1966**, 268.

[188] S. Tanaka, H. Yamamoto, H. Nozaki, K. B. Sharpless, R. C. Michaelson, and J. D. Cutting, *J. Am. Chem. Soc.*, **96**, 5254 (1974).

[189] K. Sugiura, Y. Shizuri, K. Yamada, and Y. Hirata, *Tetrahedron Lett.*, **1975**, 2307.

AUTHOR INDEX, VOLUMES 1–30

CHAPTER AND TOPIC INDEX, VOLUMES 1–30

Many chapters contain brief discussions of reactions and comparisons of alternative synthetic methods which are related to the reaction that is the subject of the chapter. These related reactions and alternative methods are not usually listed in this index. In this index the volume number is in **BOLDFACE**, the chapter number in ordinary type.

SUBJECT INDEX, VOLUME 30

Since the table of contents provides a quite complete index, only those items not readily found from the contents page are listed here. Numbers in BOLDFACE refer to experimental procedures.